AF678874

Breathborne Biomarkers and the Human Volatilome

Breathborne Biomarkers and the Human Volatilome

Second Edition

Edited by

Jonathan Beauchamp

Cristina Davis

Joachim Pleil

ELSEVIER

Elsevier
Radarweg 29, PO Box 211, 1000 AE Amsterdam, Netherlands
The Boulevard, Langford Lane, Kidlington, Oxford OX5 1GB, United Kingdom
50 Hampshire Street, 5th Floor, Cambridge, MA 02139, United States

Copyright © 2020 Elsevier B.V. All rights reserved.
Chapter 39: Copyright © 2020 Published by Elsevier B.V. All rights reserved.

No part of this publication may be reproduced or transmitted in any form or by any means, electronic or mechanical, including photocopying, recording, or any information storage and retrieval system, without permission in writing from the publisher. Details on how to seek permission, further information about the Publisher's permissions policies and our arrangements with organizations such as the Copyright Clearance Center and the Copyright Licensing Agency, can be found at our website: www.elsevier.com/permissions.

This book and the individual contributions contained in it are protected under copyright by the Publisher (other than as may be noted herein).

Notices
Knowledge and best practice in this field are constantly changing. As new research and experience broaden our understanding, changes in research methods, professional practices, or medical treatment may become necessary.

Practitioners and researchers must always rely on their own experience and knowledge in evaluating and using any information, methods, compounds, or experiments described herein. In using such information or methods they should be mindful of their own safety and the safety of others, including parties for whom they have a professional responsibility.

To the fullest extent of the law, neither the Publisher nor the authors, contributors, or editors, assume any liability for any injury and/or damage to persons or property as a matter of products liability, negligence or otherwise, or from any use or operation of any methods, products, instructions, or ideas contained in the material herein.

Library of Congress Cataloging-in-Publication Data
A catalog record for this book is available from the Library of Congress

British Library Cataloguing-in-Publication Data
A catalogue record for this book is available from the British Library

ISBN: 978-0-12-819967-1

For information on all Elsevier publications visit our website at https://www.elsevier.com/books-and-journals

Publisher: Susan Dennis
Acquisitions Editor: Kathryn Eryilmaz
Editorial Project Manager: Chiara Giglio
Production Project Manager: Kumar Anbazhagan
Cover Designer: Alan Studholme

Typeset by TNQ Technologies

Working together to grow libraries in developing countries
www.elsevier.com • www.bookaid.org

Introducing people to innovations, step by step—the legacy of Anton Amann

On January 6, 2015, Professor Anton Amann passed away. He was one of the few polymaths in the true sense of the word. In his late (and busy) years of work, Anton was able to position the city of Innsbruck, Austria, as an international center for breath research. As the son of a well-known industrialist in Austria's western province of Vorarlberg, Anton's chemistry studies should have served as a technical basis for him to take over his father's company, but Anton demonstrated an extraordinary talent in several scientific fields, even at an early stage of his studies, which led him to pursue an academic scientific career. With his business skills—inherited as a member of an old entrepreneurial family—and his widespread academic knowledge, Anton knew how to handle a variety of challenging tasks that he encountered during his lifetime. Professor Amann's rich scientific oeuvre reflects the renaissance heyday of his hometown Hohenems, where Christian and Jewish communities had developed an economically flourishing area.

Anton Amann was born in 1956 during the peak of Vorarlberg's textile industry, which his father's company had been a part of. The "Junker family," as his family was known—in accordance with a custom in western Austria in which the family name is supplemented by a "house" name—lived in the town of Hohenems in a Spanish Villa, a jewel of Art Nouveau designed by the architect Hanns Kornberger in 1904. Oriental ornaments on the walls of this building are reminiscent of the southern origins of the Amann family. In keeping with the bourgeois style of the Amann family, Anton attended the Jesuit high school "Stella Matutina" in Feldkirch, where he graduated in 1974.

When Anton was a young chemistry student at ETH Zurich, Switzerland, he received a phone call from his father that would change his career path: the family textile factory had no future. This news slammed the door to Anton's entrepreneurship, but opened up the world of science to him. After completing his chemistry degree in 1978, Anton was able to implement his extraordinary mathematical talent in theoretical chemistry using the methods of mathematical physics. Anton was a member of a research group headed by Professor Hans Primas, a theoretical chemist and unique human being who was appointed as a professor at ETH Zurich despite having only high school qualifications and no regular university degree. Several Nobel Prizes are the indirect result of Professor Primas' work, including that of his doctoral student Richard Ernst. During this time, Anton's research focused on identifying quantum mechanical observers, the Jauch-Piron states of quantum logics, molecular chirality, and the spectroscopy of single molecules. For both his diploma and doctoral theses, Anton was awarded the highest accolade of the University, the ETH Silver Medal.

After achieving his professorial qualification (habilitation) in 1991 at ETH Zurich, Anton submitted a second habilitation thesis to the University of Innsbruck in 1997, where he switched from theoretical to analytical chemistry. With an appointment at the Experimental Anesthesia Laboratory of the University Clinic for Anesthesia and Intensive Care Medicine, Anton gradually succeeded in combining his biomathematical expertise with avant-garde medical questions. This culminated in the question of how human exhaled air can be used for medical diagnostics. Anton worked closely with Professor Werner Lindinger, the then Director of the Institute of Ion Physics—who bequeathed a rich scientific heritage in the field of mass spectrometry to the University of Innsbruck—exploring the application of breath research for the newly developed real-time mass spectrometer, PTR-MS. Over the course of his academic career, Professor Amann developed a large foundation of basic research by cooperating with many institutions, including diverse departments within the Medical University Innsbruck, several hospitals and clinics in Innsbruck and Vorarlberg, health insurance companies, the working group for preventive and social medicine in Vorarlberg, and the Vorarlberg University of Applied Sciences. In addition to this plethora of collaborators, the knowledge cultivated in Anton's research group in dealing with research equipment was unrivaled at that time. Anton's interdisciplinary approach saw use and expertise in manifold techniques, including PTR-MS and PTR-TOF, SIFT-MS, GC-MS combined with SPME, SPE, and TD or NT, as well as IMS; many research groups commit themselves to only one or few of the techniques mentioned. In order to give breath research an international platform for networking, in 2004, Anton organized and hosted the conference "Breath Gas Analysis for Medical Diagnostics," which took place in Dornbirn, Austria with the support of the province government of Vorarlberg and the University of Applied Sciences in Dornbirn. Anton knew how to network with many colleagues known previously only through publications but whom he had never met in person. Referring to those early years, Anton later asserted "*Most people can only be introduced to innovations in small steps.*"

In 2005, Professor Amann founded the International Association for Breath Research (IABR), with a founding ceremony held at Schönbrunn Palace in Vienna, Austria. IABR brought together a worldwide alliance of over 30 leading research groups in respiratory gas analysis, with Anton heading this organization as its speaker (chairperson) for almost a decade. From 2006 to 2009, Professor Amann coordinated the EU-funded project "BAMOD," which dealt with fundamental questions about the detection of lung cancer in exhaled breath. This was the first project of many leading European experts in the field of respiratory gas analysis for biomarker detection. In 2007, together with Mel Rosenberg, Anton cofounded the *Journal of Breath Research*, published by the Institute of Physics (Bristol, UK), which he managed as editor-in-chief. At this time, the math and science class of the Austrian Academy of Sciences became aware of Anton, recognizing his outstanding knowledge and skills. The Academy founded the Institute of Breath Research located in Dornbirn, Austria, and Professor Amann was appointed as its first scientific director. This institute, which represents Anton's later working life, was taken over by the University of Innsbruck in 2014. In addition, Dr. Bernhard Lang acted as a generous donor for Anton's research activities, and his early support was the foundation of much of the activities reported above.

Anton was a dedicated Roman Catholic and loved baroque church traditions. He shared his cannon of Christian values with us—and his love of life was ingrained in his work rather than leisure time. In 2010, Anton was presented with the Marie Curie Medal of the Polish Chemical Society, conferred by its president Professor Bogusław Buszewski, which he considered to the "knighthood diploma" for his tireless scientific work. In addition to his many publications, international awards, and grants, this medal was a moment of full appreciation of Anton's life achievements by his family and peers. The ceremony was strikingly emotional, with numerous international guests, leaders from universities, politics, and business, and Anton's wife and children present. Upon receiving this accolade, Anton commented of his joy to be under the patronage of Marie Sklodowska Curie *"because of her persistence, and impatience."*

Anton's employees, and also his private environment, benefited to an unusual extent from his virtues as a scientist, friend, and family member. In addition to his academic knowledge, Anton was a talented pianist, historian, and linguist. He had extensive knowledge of ancient Greek and Latin. Anton was fascinated by biographies of numerous historical figures, including those of the secretary of Alexander the Great or Napoleon Bonaparte's chief adviser. As a native of the city of Hohenems, Anton liked works on the Counts of Hohenems; some Roman Catholic cardinals, as well as prince-archbishops of Salzburg, were family members. The family's affinity for science and the arts, as well as their strategic skills, formed the breeding ground for numerous comments at Anton's institute every day. In a figurative sense, Professor Amann implemented the renaissance concept of his hometown in his scientific work: the freedom of science and its teaching was responsible for a large part of Anton's quality of life. Anton Amann was granted the opportunity to pursue his numerous research interests, also thanks to many good employees and his worldwide cooperation network. In September 2015, after his untimely passing, the Anton Amann Medal was created to honor his scientific life achievements and his friendship to us; the award is conferred by IABR to outstanding scientists in the field of breath research who have followed Anton's legacy of rigorous and groundbreaking science.

Dear Anton, thank you for serving as an example to us all!

Marco Freek
Medical University Innsbruck
Innsbruck, January 2020

Anton Amann
* 20/6/1956–6/1/2015 †

Contents

PART E Beyond human breath

Contributors

Sebastian Abegg
Particle Technology Laboratory, Department of Mechanical and Process Engineering, ETH Zurich, Zurich, Switzerland

Waqar Ahmed
School of Biological Sciences, Faculty of Biology, Medicine and Health, University of Manchester, Manchester, United Kingdom

Yaser Alkhalifah
Department of Chemistry, Loughborough University, Loughborough, United Kingdom

Alexander Apolonski
Max Planck Institute of Quantum Optics, Garching, Germany; Ludwig-Maximilians-Universität München, Garching, Germany

Heather D. Bean
School of Life Sciences, Arizona State University, Tempe, AZ, United States

Jonathan D. Beauchamp
Fraunhofer Institute for Process Engineering and Packaging IVV, Freising, Germany

Olof Beck
Department of Clinical Neuroscience, Karolinska Institute, Stockholm, Sweden

Amalia Z. Berna
Department of Pediatrics, Washington University School of Medicine, Saint Louis, MO, United States

Andras Bikov
North West Lung Centre, Manchester University NHS Foundation Trust, Manchester, United Kingdom

Eva Borras
Department of Mechanical and Aerospace Engineering, University of California, Davis, CA, United States

Paul Brinkman
Department of Respiratory Medicine, Amsterdam UMC, University of Amsterdam, Amsterdam, The Netherlands

Emma Brodrick
IMSPEX, Ty Menter, Cardiff, United Kingdom

Massimo Corradi
Department of Medicine and Surgery, University of Parma, Parma, Italy

Simona M. Cristescu
Department of Molecular and Laser Physics, Institute for Molecules and Materials, Radboud University, Nijmegen, The Netherlands

Raquel Cumeras
Metabolomics Platform, Institut d'Investigació Sanitària Pere Virgili (IISPV), CERCA, Universitat Rovira i Virgili (URV), Reus, Spain; Biomedical Research Centre in Diabetes and Associated Metabolic Disorders (CIBERDEM), ISCIII, Madrid, Spain; Genome Center, University of California, Davis, CA, United States

Cristina E. Davis
Department of Mechanical and Aerospace Engineering, University of California, Davis, CA, United States

Michael D. Davis
Wells Center for Pediatric Research, Pulmonology, Allergy, and Sleep Medicine, Riley Hospital for Children at Indiana University, Indianapolis, IN, United States

Ben de Lacy Costello
Centre for Research in Biosciences, University of the West of England, Frenchay, United Kingdom

Corrado Di Natale
Department of Electronic Engineering, University of Rome Tor Vergata, Rome, Italy

Silvano Dragonieri
Department of Respiratory Diseases, University of Bari, Bari, Italy

Raed Dweik
Department of Pulmonary, Allergy and Critical Care Medicine, Respiratory Institute, Cleveland Clinic, Cleveland, OH, United States

Peter P. Egeghy
Office of Research and Development, United States Environmental Protection Agency, Research Triangle Park, NC, United States

Gary A. Eiceman
Department of Chemistry and Biochemistry, New Mexico State University, Las Cruces, NM, United States

Jean-François Focant
Organic and Biological Analytical Chemistry Group, MolSys Research Unit, University of Liège, Liège, Belgium

Stephen Fowler
School of Biological Sciences, Faculty of Biology, Medicine and Health, University of Manchester, Manchester, United Kingdom; Manchester Academic Health Science Centre and NIHR Manchester Biomedical Research Centre, Manchester University Hospitals NHS Foundation Trust, Manchester, United Kingdom

Matthias Frank
Lawrence Livermore National Laboratory, Livermore, CA, United States; Department of Biochemistry and Molecular Medicine, University of California, Davis, School of Medicine, Sacramento, CA, United States

M. Ariel Geer Wallace
National Exposure Research Laboratory, United States Environmental Protection Agency, Research Triangle Park, NC, United States

Ramin Ghorbani
Department of Applied Physics and Electronics, Umeå University, Umeå, Sweden

Peter Gierschner
Department of Anesthesia and Intensive Care, Rostock University Medical Center, Rostock, Germany

Roger Giese
Department of Pharmaceutical Sciences and Barnett Institute, Northeastern University, Boston, MA, United States

Oliver Gould
Centre for Research in Biosciences, University of the West of England, Frenchay, United Kingdom

Andreas T. Güntner
Particle Technology Laboratory, Department of Mechanical and Process Engineering, ETH Zurich, Zurich, Switzerland; Department of Endocrinology, Diabetes, and Clinical Nutrition, University Hospital Zurich, Zurich, Switzerland

Klaus Hackner
Department of Pneumology, University Hospital Krems, Karl Landsteiner University of Health Sciences, Krems, Austria; Department of Internal Medicine II, Division of Cardiology, Medical University of Vienna, Vienna, Austria

Hossam Haick
Department of Chemical Engineering and Russell Berrie Nanotechnology Institute, Technion – Israel Institute of Technology, Haifa, Israel

Peter Hamm
University Hospital for Anaesthesia and Intensive Care, Medical University of Innsbruck, Innsbruck, Austria

George B. Hanna
Department of Surgery and Cancer, Imperial College, St. Mary's Campus, London, United Kingdom

Jens Herbig
IONICON Analytik Gesellschaft m.b.H., Innsbruck, Austria

Jane E. Hill
Chemical and Biological Engineering, University of British Columbia, Vancouver, BC, Canada

Marieann Högman
Department of Medical Sciences, Respiratory, Allergy and Sleep Research, Uppsala University, Uppsala, Sweden

Jens M. Hohlfeld
Fraunhofer Institute for Toxicology and Experimental Medicine ITEM, Hannover, Germany; German Center for Lung Research (DZL—BREATH), Hannover, Germany; Respiratory Medicine, Hannover Medical School, Hannover, Germany

Olaf Holz
Fraunhofer Institute for Toxicology and Experimental Medicine ITEM, Hannover, Germany; German Center for Lung Research (DZL—BREATH), Hannover, Germany

Alan W. Jones
Department of Clinical Pharmacology, University Hospital, University of Linköping, Linköping, Sweden

Julian King
Institute for Breath Research, University of Innsbruck, Dornbirn, Austria

Heike U. Köhler
Institute of Molecular Pathogenesis at 'Friedrich-Loeffler-Institut' (Federal Research Institute for Animal Health), Jena, Germany

Anne Küntzel
Institute of Molecular Pathogenesis at 'Friedrich-Loeffler-Institut' (Federal Research Institute for Animal Health), Jena, Germany

Jiayi Lan
Department of Chemistry and Applied Biosciences, ETH Zurich, Zürich, Switzerland

Zsofia Lazar
Department of Pulmonology, Semmelweis University, Budapest, Hungary

Lauri Lehtimäki
Allergy Centre, Tampere University Hospital, Faculty of Medicine and Health Technology, Tampere University, Tampere, Finland

Michael C. Madden
United States Environmental Protection Agency, Human Studies Facility, Chapel Hill, NC, United States

Andrei Malinovschi
Department of Medical Sciences, Clinical Physiology, Uppsala University, Uppsala, Sweden

Santiago Marco
Department of Electronics and Biomedical Engineering, University of Barcelona, Barcelona, Spain; Signal and Information Processing for Sensing Systems, Institute for Bioengineering of Catalonia, The Barcelona Institute of Science and Technology, Barcelona, Spain

Christopher A. Mayhew
Institute for Breath Research, University of Innsbruck, Dornbirn, Austria; Molecular Physics Group, School of Physics and Astronomy, University of Birmingham, Birmingham, United Kingdom

Mitchell M. McCartney
Department of Mechanical and Aerospace Engineering, University of California, Davis, CA, United States

James P. McCord
National Exposure Research Laboratory, United States Environmental Protection Agency, Research Triangle Park, NC, United States

Markus Metsälä
Department of Chemistry, University of Helsinki, Helsinki, Finland

Alain Michils
Department of Chest Medicine, Erasme University Hospital, Université Libre de Bruxelles, Brussels, Belgium

Wolfram Miekisch
Department of Anesthesia and Intensive Care, Rostock University Medical Center, Rostock, Germany

Justin J. Miller
Department of Pediatrics, Washington University School of Medicine, Saint Louis, MO, United States

Paweł Mochalski
Institute for Breath Research, University of Innsbruck, Dornbirn, Austria; Institute of Chemistry, Jan Kochanowski University, Kielce, Poland

Anil S. Modak
AModak2003@yahoo.com

Morad K. Nakhleh
Owlstone Medical Ltd., Cambridge, United Kingdom

Leena A. Nylander-French
Department of Environmental Sciences and Engineering, Gillings School of Global Public Health, Univeristy of North Carolina, Chapel Hill, NC, United States

Audrey R. Odom John
Department of Pediatrics, Washington University School of Medicine, Saint Louis, MO, United States; Department of Molecular Microbiology, Washington University School of Medicine, Saint Louis, MO, United States; The Children's Hospital of Philadelphia, Philadelphia, PA, United States

Francisco Blanco Parte
Fossil Ion Technology, Madrid, Spain

Joachim D. Pleil
Department of Environmental Sciences and Engineering, Gillings School of Global Public Health, University of North Carolina, Chapel Hill, NC, United States

Silvia Ranzieri
Department of Medicine and Surgery, University of Parma, Parma, Italy

Norman M. Ratcliffe
Centre for Research in Biosciences, University of the West of England, Frenchay, United Kingdom

Petra E. Reinhold
Institute of Molecular Pathogenesis at 'Friedrich-Loeffler-Institut' (Federal Research Institute for Animal Health), Jena, Germany

Terence H. Risby
Department of Environmental Health and Engineering, Bloomberg School of Public Health, Johns Hopkins University, Baltimore, MD, United States

Dorota Ruszkiewicz
Centre for Analytical Science, Department of Chemistry, Loughborough University, Loughborough, United Kingdom

Veronika Ruzsanyi
Institute for Breath Research, University of Innsbruck, Dornbirn, Austria

Stefan W. Ryter
Department of Medicine, Weill Cornell Medical Center, New York, NY, United States

Dahlia Salman
Centre for Analytical Science, Department of Chemistry, Loughborough University, Loughborough, United Kingdom

Michael Schivo
Center for Comparative Respiratory Biology and Medicine, University of California, Davis, School of Medicine, Davis, CA, United States; Department of Internal Medicine, University of California, Davis, School of Medicine, Sacramento, CA, United States; VA Northern California Health Care System, Mather, CA, United States

Florian M. Schmidt
Department of Applied Physics and Electronics, Umeå University, Umeå, Sweden

Jochen K. Schubert
Department of Anesthesia and Intensive Care, Rostock University Medical Center, Rostock, Germany

Katharina Schwarz
Fraunhofer Institute for Toxicology and Experimental Medicine ITEM, Hannover, Germany

David Smith
Trans Spectra Limited, Newcastle-under-Lyme, United Kingdom

Agnieszka Smolinska
Department of Pharmacology and Toxicology, NUTRIM School of Nutrition and Translational Research, Maastricht University, Maastricht, The Netherlands

Jon R. Sobus
National Exposure Research Laboratory, United States Environmental Protection Agency, Research Triangle Park, NC, United States

Steven F. Solga
Department of Medicine, Perelman School of Medicine, University of Pennsylvania, Philadelphia, PA, United States

Lisa A. Spacek
Department of Medicine, Sidney Kimmel Medical College, Thomas Jefferson University, Philadelphia, PA, United States

Patrik Španěl
J. Heyrovský Institute of Physical Chemistry, Academy of Sciences of the Czech Republic, Prague, Czech Republic

Georgios Stavropoulos
Department of Pharmacology and Toxicology, NUTRIM School of Nutrition and Translational Research, Maastricht University, Maastricht, The Netherlands

Pierre-Hugues Stefanuto
Organic and Biological Analytical Chemistry Group, MolSys Research Unit, University of Liège, Liège, Belgium

Matthew A. Stiegel
Occupational and Environmental Safety Office, Duke University Medical Center, Durham, NC, United States

Gerald Teschl
Faculty of Mathematics, University of Vienna, Vienna, Austria

Susanne Teschl
University of Applied Sciences Technikum Wien, Vienna, Austria

C. L. Paul Thomas
Centre for Analytical Science, Department of Chemistry, Loughborough University, Loughborough, United Kingdom

Karl Unterkofler
Institute for Breath Research, University of Innsbruck, Dornbirn, Austria; University of Applied Sciences Vorarlberg, Dornbirn, Austria

Marc P. van der Schee
Owlstone Medical Ltd., Cambridge, United Kingdom

Frederik-Jan van Schooten
Department of Pharmacology and Toxicology, NUTRIM School of Nutrition and Translational Research, Maastricht University, Maastricht, The Netherlands

Guillermo Vidal-de-Miguel
Fossil Ion Technology, Madrid, Spain

Rotem Vishinkin
Department of Chemical Engineering and Russell Berrie Nanotechnology Institute, Technion – Israel Institute of Technology, Haifa, Israel

Helmut Wiesenhofer
Institute for Breath Research, University of Innsbruck, Dornbirn, Austria

Antony J. Williams
National Center for Computational Toxicology, United States Environmental Protection Agency, Research Triangle Park, NC, United States

Laura C. Yeates
Olivet Nazarene University, Bourbonnais, IL, United States

Delphine Zanella
Organic and Biological Analytical Chemistry Group, MolSys Research Unit, University of Liège, Liège, Belgium

Renato Zenobi
Department of Chemistry and Applied Biosciences, ETH Zurich, Zürich, Switzerland

Foreword

The earliest studies of volatile organic compounds (VOCs) present in the body that appear in exhaled breath and urine were carried out as long ago as 1972 by the Nobel Laureate Linus Pauling. The first concerted review of breath analysis and related topics appeared some decades later in 2005 in the book *Breath Analysis for Clinical Diagnosis and Therapeutic Monitoring (World Scientific)*, which described the analytical instrumentation developed and breath sampling methodologies together with several case studies that showed the potential of breath analysis in physiology and medicine. Further developments of more sensitive innovative analytical techniques and an increase in the numbers of scientists and clinicians attracted to the field prompted the publication in 2013 of the follow-up treatise entitled *Volatile Biomarkers: Non-invasive Diagnosis in Physiology and Medicine* (Elsevier). This showed the further rapid advancement of breath analysis research and the increased number of creative and imaginative researchers attracted to the field. Both of these texts comprise a compilation of chapters written by scientists and clinicians researching in the field on VOC analysis. They were both coedited by the late Anton Amann and myself. A tribute to Professor Amann that recognizes his seminal contributions to breath research and his unrivaled promotion of the subject is included in the present treatise.

This volume, "*Breathborne Biomarkers and the Human Volatilome*," is edited by three scientists who have published widely in the field of VOC analyses of exhaled breath and other gaseous media. Collectively, they have knowledge spanning the entire field of VOC analysis research. Hence, they have collaborated to produce a comprehensive text that comprises some 39 chapters written by recognized workers in breath analysis and related topics. The focus of the text is on the latest discoveries and developments in not only the field of VOC analysis of various biogenic media, principally exhaled breath, but also of the headspace of urine and cell cultures. The premise is that blood contains a wide variety of VOC metabolites, which appear in exhaled breath and hold promise as biomarkers of health status and for the diagnosis and management of systemic diseases. A biomarker is defined as "*a characteristic that is reliably and accurately measured and evaluated as an indicator of normal biological processes, pathogenic processes, or pharmacologic responses to an intervention*." Good examples are the concentrations of blood cholesterol (nonvolatile) as an indicator of cardiovascular risk and exhaled nitric oxide (volatile) as an indicator of airway inflammation. So a great deal of research is now being carried out by scientists and clinicians to identify breath VOCs with the expectation that some will eventually be accepted and exploited as biomarkers. But it has to be recognized that in addition to the endogenous VOCs produced by normal or disease-related metabolism, i.e. metabolites, exhaled breath can also include exogenous compounds resulting from the inhalation of ambient air. Furthermore, breath VOCs can be produced by symbiotic or pathogenic bacteria in the gut and the oral cavity and also as the by-products of ingested food, medications, drugs, or toxins. Hence, not all breath VOCs are metabolites and not all metabolites are necessarily biomarkers. The challenge of breath analysis is to realize analytical techniques that can identify and quantify, to an acceptable accuracy and precision, the numerous breath VOCs and to discover their origins and their association, if any, with normal metabolic processes or with pathophysiology/specific diseases.

Initially, it was hoped that particular pathophysiological conditions can be signaled by single biomarker molecules in breath such as NO and CO, the respective concentrations of which are associated with the severity of asthma and chronic obstructive pulmonary disease. Much work has been done on these associations, and chapters that summarize the latest developments are included in this book. However, these single biomarker associations are, as yet, a rare circumstance, and now it is tacitly assumed that complex diseases are much more likely to be recognized by combinations of VOCs in breath, as appears to be so for breath biomarkers of cancer. Nevertheless, a great deal of effort has been given to the quantification of individual metabolites in exhaled breath, such as ammonia (ostensibly associated with anemia), acetone (diabetes), pentane (inflammation) and hydrogen cyanide (*Pseudomonas aeruginosa* infection). Yet the evidence so far obtained for these associations is not sufficiently convincing to project them into clinical use.

The present book covers many aspects of VOC analyses of exhaled breath and other media, including urine and stool, indicating the progress that has been made during the few years since the publication of the previous and first book in this series. Some of the chapters are focused, others are more broadly based. There are chapters on the development of analytical instrumentation, including the mass spectrometer (MS)−based devices SIFT-MS, PTR-MS, IMS, SESI-MS, and GC-MS, and electronic nose sensors and optical spectroscopic methods. The sophisticated MS devices are able to analyze the many VOCs in a breath sample and also target single breath metabolites. It is becoming clear that combinations of these various analytical methods, especially combinations of SIFT-MS or PTR-MS with GC-MS, offer the most profitable approach to VOC research and discovery. This has spawned the new subscience of "volatilomics," which embraces many of the chapters in this book.

Unfortunately, the physical size of the current MS devices means they are unsuitable for the clinical environment and point-of-care application, although work is in train to produce smaller instruments without diminishing sensitivity and versatility. The available sensors and spectroscopic devices are much smaller and easier to use, but they are normally configured to detect single or a small number of breath VOCs. Even so, they have great potential for breath biomarker monitoring in the home and at the point-of-care if their stability and sensitivity can be improved. It is important to appreciate that breath VOC metabolites are present at concentrations that range from a few hundred in parts per billion by volume, ppb_v, to sub-ppb_v, and this is demanding on many instruments, especially considering that precision and accuracy are important in medical diagnostics.

Ultimately, for any VOC or their pattern to be accepted as a *bone fide* biomarker, it must be confirmed by independent laboratories and by clinical studies. This demands universal sampling and standardization methods and a proper appreciation of the physiological modeling. There are chapters in the present text that discuss these important aspects of breath analysis research. It is shown how MS techniques are being productively used for wide-ranging studies of VOCs in breath, in the headspace of urine and stool and skin emanations of humans, and VOCs from the waste products of ruminates and marine mammals. The importance of good study design and data analysis is stressed for meaningful data to be obtained and accepted by other workers. Too often, the temptation is to employ advanced statistics to interpret inconclusive data, which may result in unwarranted conclusions.

Further indications of the comprehensive approach brought to VOC analysis by this text are chapters on the influence of respiratory bacteria and viruses on breath metabolites, the VOCs released by cultures of known pulmonary pathogens, breath VOCs originating from ingested drugs and medication, and breath ethanol in law enforcement. Health and safety practice and occupational medicine issues are reviewed in chapters on the detection of exogenous breath VOCs related to human exposure to toxic gases and vapors and urban search and rescue following natural disasters that exploit the human volatilome.

Some final remarks are in order to indicate the way forward for VOC research and for breath research in particular. Studies involving hospitalized patients have involved either direct analysis of breath collected into polymer bags (mostly Tedlar and Nalophan) using SIFT-MS or PTR-MS or by preconcentration of trace VOCs present in the collected breath sample onto SPME or TD sorbents, followed by offline GC-MS analysis using bulky and remotely situated analytical instruments. This is an undesirable situation given the uncertainties involved in differential adsorption onto bag surfaces. Real-time analyses of single breath exhalations of several compounds can be done simultaneously using SIFT-MS, PTR-MS, and SESI-MS as long as the VOCs are at sufficiently high concentrations (currently >1 ppb_v). Additionally, the end-tidal portion of the exhaled breath can be selected and analyzed. However, it is important to realize that alveolar breath passing through the oral cavity can become contaminated by bacterial and enzymatic activity, which can be a serious consideration when metabolites are at sub-ppb_v levels. Nevertheless, nose-exhaled breath can now be separately analyzed in just a few seconds for some VOCs, which diminishes oral contamination. But a real desire is to achieve direct breath analysis in clinical practice. So a major challenge is to realize analytical instrumentation that is small and readily transported and able to be operated in the clinic, hospital ward, and even intensive therapy units. Direct analyses of exhaled breath of ventilated patients in the operating theater and intensive therapy units are already being explored with promising results. This would surely result in the more accurate quantification that is demanded by clinicians and surgeons.

What seems certain is that only through careful experimental work and innovation will noninvasive breath analysis and the identification of reliable biomarkers be realized and be added to the diagnostic armory of medical practitioners. Breath analysis is still in its infancy, but it has great potential for development into a valuable diagnostic aid. The present excellent compilation of chapters clearly describes the current status of the subject, highlighting those areas where significant and individual contributions can be made to promote this exciting subject. The editors of this text are to be congratulated on recruiting the many expert contributors who have comprehensively described the manifold aspects of this burgeoning subject, which will surely attract more scientists and clinicians to the field.

David Smith, FRS
Newcastle-under-Lyme
Staffordshire, UK
April 2020.

Preface

Breathborne Biomarkers and the Human Volatilome serves as an update to the 2013 book *Volatile Biomarkers: Non-Invasive Diagnosis in Physiology and Medicine* edited by Anton Amann and David Smith. The technologies and concepts for using human exhaled breath as a biological fluid for analysis akin to blood and urine have been evolving rapidly, and the past 7 years since the first edition was published have seen many developments and discoveries, so the time for this new edition is pertinent. Despite the dynamics within the field, central aspects and core concepts remain, as is reflected in the similarity of many of the overarching topics herein with those covered in the first edition. In this second edition, we have completely overhauled the contents and structure of the original book and have realigned all chapters in relation to scope and layout to reflect the state of the art of breath research. Each chapter in this book represents a self-contained treatise on its topic. Broadly, chapters commence with an introduction of concepts, then present relevant applications and discoveries, and wrap up by reporting on recent innovations and predictions of future trends. Chapters do not contain exhaustive references lists but cite key studies for the interested reader to follow up. Chapters from the first edition have been restructured and updated to align with this scope. Both new and revised chapters have been written by the very experts who have performed the latest research and are amongst the leading specialists in their specific disciplines.

In addition to the structural overhaul, this revised edition contains a series of new breath-related topics that simply did not exist—or were only in their infancy—7 years ago, including cellular respiration, sampling standardization initiatives, high-resolution mass spectrometry, expanded concepts for nontargeted analysis, the human breath exposome, clinical phenotyping, and open-sourced data compilations. Two topics are covered in feature-length chapters, namely Chapter 24 on biomarkers in stool and urine, essentially because it covers two topics, and Chapter 34 on canine olfaction, because it offers a comprehensive treatment of this topic in relation to disease biomarkers that is hitherto lacking in the literature. Furthermore, chapters are grouped thematically, covering topics on general concepts and modeling, inorganic and nonvolatile compounds, analytical platforms, established breath tests, applications beyond human breath, field applications, and study design and interpretation of data.

For most academics or physicians not directly working in the field, exhaled breath measurement is largely considered to relate only to gas-phase analysis; after all, breath consists primarily of ambient air (nitrogen, oxygen, argon, and carbon dioxide), two major human metabolites (isoprene and acetone), and traces of perhaps thousands of environmental and metabolic organic compounds. As technology has advanced, however, the aerosol, vapor, and particle constituents have also become important features of breath research beyond gas-phase analyses. This second edition has incorporated the analysis of exhaled breath condensate and aerosols (EBC and EBA, respectively), and exhaled particles, into the different research chapters, especially for new topics, such as in respiratory infections or discovery analysis using secondary electrospray ionization. Until recently—even within the field—breathborne compounds of exogenous (environmental) origin were considered a nuisance and a source of confounding variables. While this largely holds true when seeking to find endogenous disease biomarkers, exogenous compounds have value themselves by providing information of previous exposures (such as ethanol or drug use) or with causal links to systemic imbalances through adverse outcome pathways. Nevertheless, the search for endogenous (metabolic) compounds as biomarkers of current health state—ideally indicating preclinical effects that lead to early disease intervention—remains a central pillar of breath research.

Another theme woven into the chapters of this revised edition is the concept of the *breath volatilome* for gas-phase constituents, and the expanded *breath exposome*, which includes chemicals beyond the gas-phase. These "omics" concepts treat chemical expression more as a pattern, rather than a collection of individual constituents. In fact, a number of the chapters propose the use of pattern perturbation as a probative diagnostic to complement the "one compound—one effect" thought process that has previously dominated preclinical disease research. Further, processing these expanding and extensive complex volatilome datasets requires advanced and sophisticated mathematical modeling and mining tools, which are covered in detail in this edition.

On the whole, this new edition covers all of the basic topics from the original book, adds newly developed concepts and topics, and completely updates the major methods and technologies currently in use for understanding human breath as an important biological fluid. The renaming of the book title reflects these changes and the expanded scope: *Breathborne* replaces *Volatile* with respect to biomarkers to indicate relevance beyond volatile organic compounds; reference to the *Human Volatilome* illustrates the expansion beyond breath; the allusion to diagnostics in medicine was omitted as this represents only part of this edition.

We, the editors, are immensely grateful to our friends and colleagues in the breath research community, the editorial board of *Journal of Breath Research* (JBR), and members of the International Association of Breath Research (IABR) for their input, advice, and overall support for this complex and important undertaking. The authors of the chapters in the ensuing pages are authoritative experts in many aspects of breath analysis and related disciplines, and we are indebted to them for their time and dedication in writing the overviews and guidance presented in this book; we acknowledge and assert here that an edited book is only as good as the quality of the individual contributions. Finally, we would like to thank David Smith, editor of the first edition and a formidable presence in the field of breath research over the past two decades, for handing over the editorial baton to us for this second edition. This revised edition is dedicated to the memory of the late Anton Amann, whose passion for breath analysis and dedication in bringing likeminded researchers together led to the establishment of IABR, the creation of JBR, and the first dedicated compendium of breath research in the form of the first edition of this book; this revised edition builds on his legacy. In this context, we would additionally like to thank Marco Freek for contributing a befitting and affectionate tribute to Anton in the preceding pages.

As of 2020, this book represents the most comprehensive single resource and compendium for our field of work. We hope that it serves as a useful and informative reference source—and enjoyable reading material—not only for all those who are interested in breath research, from the enthusiastic graduate student to the experienced academic researcher, but also to the casual reader with a general interest in the concepts of breath analysis.

Jonathan D Beauchamp, Freising, Germany
Cristina E Davis, Davis, CA
Joachim D Pleil, Durham, NC
March 2020

PART

Concepts and modeling

A

CHAPTER

Breath biomarkers and the exposome

1

Joachim D. Pleil[1], Matthew A. Stiegel[2], Jonathan D. Beauchamp[3]

[1]*Department of Environmental Sciences and Engineering, Gillings School of Global Public Health, University of North Carolina, Chapel Hill, NC, United States;* [2]*Occupational and Environmental Safety Office, Duke University Medical Center, Durham, NC, United States;* [3]*Fraunhofer Institute for Process Engineering and Packaging IVV, Freising, Germany*

Chapter outline

Breathborne Biomarkers and the Human Volatilome. https://doi.org/10.1016/B978-0-12-819967-1.00001-3
Copyright © 2020 Elsevier B.V. All rights reserved.

1.1 Overview

Human exhaled breath is a biological medium that is rich in content. It comprises a complex mixture of gases, aerosols, and particles that come from both exogenous (environmental) and endogenous (metabolic) sources. The most abundant components of exhaled breath are the fixed gases from inhalation, namely nitrogen, oxygen, and argon, as well as water vapor and carbon dioxide. Besides these main constituents, breath contains a mixture of inorganic and organic compounds at trace concentration levels. Some of these breathborne compounds have the potential to offer insights into the present physiology and state of health of an individual. Such biological markers, or *biomarkers*, are of great interest because of the relative practicalities of breath analysis: providing a sample is simple, painless (noninvasive), and can be done on demand, breath is inexhaustible allowing for repeated sampling, and analysis can be potentially performed immediately at point-of-care (PoC). Furthermore, sampling does not require privacy or medical personnel, and typically no potentially infectious waste is generated. There has been moderate success in discoveries and applications of breathborne biomarkers, but despite its apparent simplicity, many challenges must be overcome before a useable test is developed.

Returning to its composition, exhaled breath predominantly comprises the two primary components of our atmosphere, namely molecular nitrogen (N_2) and oxygen (O_2), which directly reflect the inspired air and collectively make up 88.2% of breath. While inhaled N_2 is exhaled relatively unperturbed (78.1% in the atmosphere vs. 74.5% in breath), O_2 is depleted due to its uptake in venous blood (20.1% in the atmosphere vs. 13.7% in breath).[1] Carbon dioxide (CO_2) is a by-product of human metabolism and is present in exhaled breath at 5.3%, which is much higher than the current atmospheric levels of about 410 parts per million (ppm)[a], or 0.04%. Water vapor (H_2O) makes up the final major constituent of breath, at ~6.2%.[1] The remaining fraction of breath comprises several inorganic species and several hundred volatile organic compounds (VOCs), whereby the exact makeup depends on physiology, state of health, diet, drug intake, and (recent) environmental exposure.

These trace components are of particular interest—and thereby at primary focus—for biomarker discovery.[b] In addition to these gas-phase compounds, exhaled breath contains aerosols (water droplets and particles). These hold potential clues on physiology and illness through the presence of water soluble organics, low volatility compounds, and macromolecules that can be analyzed in liquid phase when condensed as exhaled breath condensate (EBC) or after collection of dried surface impactions as exhaled breath aerosol (EBA) (see Chapter 7). Further, physical

[a] https://www.esrl.noaa.gov/gmd/ccgg/trends/.

[b] Exhaled CO_2 is not a trace constituent but is widely used in clinical practice throughout the world to monitor breathing in patients, such as in the ICU. As such, it might be considered the most commonly used exhaled breath biomarker.

properties of exhaled particles (dimensions, number of particles, etc.) can provide indications of respiratory performance and dysfunction (see Chapter 8).

Focusing on the volatile fraction of breath, over 870 VOCs of diverse nature have been reported in breath.[2] Together with volatiles originating from other bodily media, such as urine and stool (see Chapter 24), sweat (see Chapter 25), saliva, and blood—which number at least 1840 compounds[2]—these collectively form a volatile signature that is referred to as the *human volatilome*.

This chapter introduces the concept of the human volatilome and breath analysis, briefly reviews current tests using breathborne biomarkers, and discusses the concept and practical aspects of the human exposome, which plays an intrinsic role in defining the human volatilome and in the manifestation of volatile biomarkers in exhaled breath.

1.2 The human volatilome and breath analysis

The human volatilome is a complex and dynamic system in which the origins of each VOC can be varied and manifold. The distribution of different classes of compounds within the human volatilome for specific bodily fluids is illustrated in Chapter 31 (Fig. 31.1). Overall, exhaled VOCs can be categorized as *endogenous*, i.e., compounds originating within the body, or *exogenous*, i.e., foreign compounds from the environment; many compounds can have endogenous and exogenous contributions. Commensal and pathogenic microorganisms residing in the body emit volatiles that contribute to the volatilome (see Chapter 26); there is currently no consensus as to whether the microbiome should be attributed to exogenous or endogenous sources due to the foreign nature of some pathogens yet the symbiosis of the commensal microorganisms within the human body.

Endogenous compounds derive from normal metabolic processes but can also arise from an imbalance in the body in relation to disease or (organ) dysfunction; these are the compounds typically sought as biomarkers. In contrast, exogenous compounds can enter the body via inhalation (ambient trace constituents and pollutants), through ingestion (food, drugs), or by dermal uptake. Collectively, the latter contribute to what is referred to as the *human exposome*, a concept that is considered later in this chapter. Clearly, the origin of any compound present in breath—or any other bodily medium, for that matter—is rarely singular, with many contributing factors. This is illustrated conceptually in Fig. 1.1, which depicts the varied sources of individual VOCs. Some volatiles might have a dominant source, with only minor contributions from other routes, whereas others might derive from many minor sources. It is analytically challenging—if at all possible—to tease out the different origins of potential markers. Exogenous compounds might also be metabolized by the human body and subsequently measured in their altered form.

Public perception of exhaled breath analysis has long been focused on police evidentiary measurement of alcohol intoxication, as described in Chapter 29. More recently, breath gas analyses have been used to assess occupational or

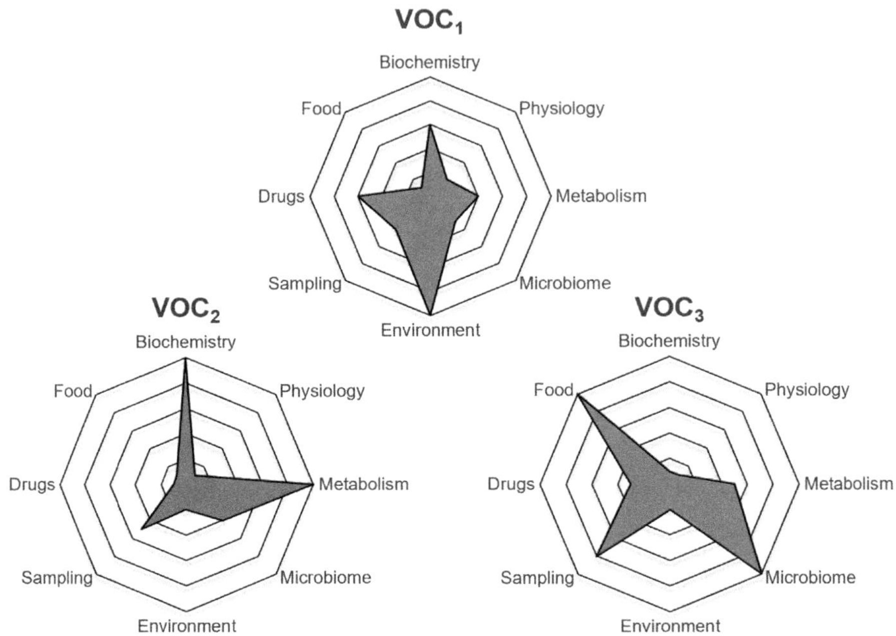

FIGURE 1.1

Radar plot conceptually illustrating the different potential sources that contribute to individual volatile organic compounds (VOCs) in exhaled breath, whereby areas represent concentrations.

environmental exposures to airborne contaminants from industry, cars, water disinfection, and commercial products via detection of VOCs (see Chapters 32 and 33). The differences between inhaled and exhaled VOCs have also been used to assess potential dose. The advent of methods for measuring constituents in EBC and EBA has provided access to semivolatile chemicals from the environment, including polycyclic aromatic hydrocarbons (PAHs), pesticides, plasticizers, pharmaceuticals, and personal consumer products, among others.

In terms of endogenous compounds, exhaled acetone and isoprene measurements were the first attempts at understanding exhaled markers of metabolism and physiology (see Chapter 3). High levels of exhaled acetone were considered a marker for uncontrolled diabetes, and isoprene changes from baseline were found to be indicative of exercise, sleep cycle, and cholesterol metabolism. Other compounds were also explored, such as exhaled ethane and pentane, for which changes were linked to oxidative stress. The most prominent single biomarker in current use for purposes of disease diagnostics is fractional exhaled nitric oxide (F_ENO), which has become a major factor in assessing respiratory diseases in humans, as described in Chapters 4 and 5. Patterns of trace-level gas-phase aldehydes, ketones, and alcohols have been used in various physiology and preclinical comparisons. Condensed-phase breath samples have yielded probative information of inflammation and pulmonary damage via cytokine, chemokine, and fatty acids measurements.

The hypothesis and justification for exploring breathborne biomarkers is that a disease state would induce a perturbation of a general pattern from "baseline" conditions. This could be described as observing a *remarkable* event in an otherwise *unremarkable* profile. The "holy grail" in disease screening is, of course, to find that one very special and unique compound that announces "lung cancer!" or "kidney disease!" etc., well before any standard clinical test. Such an unambiguous flag could save lives through early medical intervention. Regrettably, finding a single unambiguous biomarker for a specific disease has largely eluded the breath research community so far, and chasing this goal has led to frustration and skepticism that breath analysis has value in clinical practice. There are notable exceptions, such as the progress made with the analysis of changes in F_ENO as an indicator of preclinical asthma and other health states, but in general, common biomarkers are present in all samples, albeit often at different concentrations. Furthermore, it is worthy of note that variance among people is often large, which renders any individual (single) measurements ambiguous, as is discussed in Chapter 37.

Yet, all is not lost. Among breath researchers, it is now widely thought that a subtle pattern shift in a group of biomarkers, or perhaps even the whole exposome, is more likely to signal an overall perturbation of the human biochemistry that can be exploited for diagnostic purposes. This approach has been implemented in numerous studies using case–control epidemiology and longitudinal (time-trend) analyses, for instance, in the Heartsbreath test for screening heart transplant rejection[3] (see Section 1.3.7). Other pattern-based methods have been developed for cancer screening with varying degrees of success. The current state of the art is exploratory, but detecting changes in the breath exposome for cancer screening is considered a promising area of research.

1.3 Breathborne biomarkers

Human exhaled breath has been of interest in medical research for almost two and a half centuries since the discoveries of exhaled oxygen and carbon dioxide in 1784 by Lavoisier and Laplace. More recently, research on the existence of breathborne biomarkers of disease has intensified, especially in the past two decades or so. Nevertheless, only a handful of tests have demonstrated suitable biomarker robustness for transition into clinical practice or as diagnostic aids in other applications, with accompanying approval by the Food and Drug Administration (FDA) of the United States or the European Medicines Association (EMA) of the European Union. These validated and/or approved tests, as well as others in clinical practice use, are summarized in Table 1.1.

The ensuing subsections of this chapter will provide a brief overview of these tests, with further details of some of the tests covered in other chapters of this book, as cross-referenced where appropriate. Other breath tests that have not reached approval stage or are currently not in use in clinical practice are not covered in this chapter, but many promising tests or notable applications are introduced or discussed in other chapters of this book.

Table 1.1 Overview of current breath-based tests that have received approval or are in clinical use as diagnostic aids.

Test	Marker	Target/Disease	Description
Capnography	CO_2	Ventilation/breathing	CO_2, as a human metabolic by-product, has a distinct profile during breathing, and its measurement is used to monitor breathing in infirm patients, e.g., in the intensive care unit. Perturbations in this profile can indicate blood circulation problems or impairment of ventilation equipment.
F_ENO	NO	Asthma	Fraction of exhaled nitric oxide is a marker of type 2 inflammation in asthma that can aid in diagnosis.
Breathalyzer	Ethanol	Alcohol intake (law enforcement applications to determine driving under the influence of alcohol)	Breath ethanol is linearly correlated to blood ethanol and can therefore be used as an indirect measure of blood ethanol concentration. This test is routinely used in law enforcement to ascertain driving under the influence of alcohol.
Hydrogen breath test (HBT)	H_2	Lactase deficiency	The HBT requires ingestion of lactose (typically 25−100 g) with subsequent monitoring of exhaled hydrogen over a period of 2 h. A significant increase in exhaled H_2 indicates lactose intolerance.
Carbon monoxide	CO	Neonatal jaundice	An impaired conjugation of bilirubin in the liver of neonates leads to its accumulation in the blood, and accordingly, to symptoms of jaundice. A concomitant increase in exhaled carbon monoxide concentrations provides indications of jaundice in newborns.
Urea breath tests (UBTs)	$^{13}CO_2$	*Helicobacter pylori* infection	The UBT involves ingestion of ^{13}C-labeled urea, which is metabolized by urease

Table 1.1 Overview of current breath-based tests that have received approval or are in clinical use as diagnostic aids.—*cont'd*

Test	Marker	Target/Disease	Description
			activity by *H. pylori* when present in the gut to produce $^{13}CO_2$ that can be subsequently detected in exhaled breath, thereby indicating the presence of infection.
Gastric emptying breath test (GEBT)	$^{13}CO_2$	Gastroparesis	The GEBT utilizes ^{13}C-labeled spirulina to determine delayed gastric emptying by monitoring exhalation of $^{13}CO_2$ over time.
Maximum liver function capacity (LiMAx)	$^{13}CO_2$	Liver function	LiMAx targets activity of cytochrome P450 1A2 (CYP1A2) enzyme in the liver by monitoring the metabolizing capacity of ^{13}C-methacetin via exhaled $^{13}CO_2$.
Heartsbreath	Alkanes	Heart transplant rejection	Oxidative stress in heart transplant recipients can be measured by alkanes in exhaled breath and an ensuing model can predict grade 3 organ rejection. The model is based on what is termed "breath methylated alkane contour" (BMAC), which comprises C4–C20 alkanes and monomethylated alkanes.

1.3.1 Capnography

Capnography is the measurement of carbon dioxide (CO_2) during breathing. More specifically, capnography determines the partial pressure of CO_2 throughout a breathing cycle, which has a characteristic profile, in order to monitor respiration, as depicted in Chapter 2 (Fig. 2.1). Deviations in this profile can indicate complications in relation to blood circulation or breathing. In the intensive care unit (ICU), for instance, capnography is a critical parameter for observing ventilated patients and can indicate instances of ventilation failure, such as through malposition of intubation tubes. As stated earlier, CO_2 can be considered the most commonly used breath biomarker.

1.3.2 Exhaled nitric oxide in asthma

Nitric oxide was discovered in exhaled breath in 1991 and soon thereafter was observed to be elevated in airway inflammation and asthma. A diagnostic breath test for asthma based on fraction of exhaled nitric oxide (F_ENO) emerged only after comprehensive studies to characterize this inorganic compound, including the discovery of flow dependency of its concentration. The latter is evident in the use of the term F_ENO_{50}, whereby the subscript 50 specifies its measurement at a flow of 50 mL/min. Great effort was made in standardizing the detection of exhaled NO, including guidelines that were published in 1999 by the American Thoracic Society (ATS) and later revised in 2005 in a joint publication with the European Respiratory Society (ERS). Approval of F_ENO as a clinical test for airway inflammation in asthma was granted by the FDA in 2003. The use of F_ENO in asthma and other airway illnesses is covered extensively in Chapters 4 and 5.

1.3.3 Ethanol breath test for law enforcement

To the general public, the term "breath test" will invariably invoke an image of a police officer offering a handheld device to a driver of a halted vehicle with a demand to provide a breath sample. The ethanol breath tests used in law enforcement to ascertain drunk driving is in widespread and routine use. Breath ethanol after consumption of alcohol directly correlates to blood ethanol concentrations and can therefore be used as an indirect measure of when limits have been exceeded. This topic is covered in detail in Chapter 29.

1.3.4 Hydrogen breath test for lactase deficiency

Hypolactasia, or lactose intolerance, is a condition in which the sugar lactose cannot be readily digested due to insufficient lactase enzymes in the gut. Fermentation of accumulated lactose by bacteria in the cecum leads to the production of hydrogen, which is excreted from the body via exhalation.[4] The hydrogen breath test (HBT) exploits this metabolism by detecting exhaled hydrogen over a 2-h period after ingestion of typically 25, 50, or 100 g lactose, whereby concentrations are significantly elevated in patients with hypolactasia. The HBT is exempt from FDA approval because the substrate (lactose) is considered a common ingredient of food that is typically consumed in large quantities on a daily basis. The test is in routine use in clinical practice to offer evidence of lactose intolerance or malabsorption.

1.3.5 Carbon monoxide in neonatal jaundice

Neonatal hyperbilirubinemia, or jaundice, is a condition in newborn infants that arises when heme oxygenase is increased, resulting in an accumulation of bilirubin in the blood. Endogenous carbon monoxide (CO) in humans occurs—as a major source—as a by-product of heme oxygenase dependent degradation of heme to biliverdin, which is subsequently converted to bilirubin by biliverdin reductase. When the liver of a newborn is unable to readily conjugate and thereby excrete bilirubin from heme catabolism, the resulting accumulation of bilirubin in the bloodstream is accompanied by elevated CO concentrations in exhaled breath. As such, exhaled CO ($_{E}CO$) can be used as a proxy to determine bilirubin accumulation and, by association, to indicate hyperbilirubinemia.[5] Analysis of CO in the exhaled breath of infants is a common clinical diagnostic procedure to determine neonatal jaundice. The test received an FDA waiver in 2018. Further details of this breath test are reported in Chapter 6.

1.3.6 Urea breath test for *Helicobacter pylori* infection

The urea breath test (UBT) involves ingestion of an isotopically labeled substrate, i.e., a probative marker, to examine metabolic activity associated with infection. Specifically, ^{13}C-labeled urea is ingested, which is metabolized through urease activity by *Helicobacter pylori* present in the gut. This metabolism results in the production of labeled carbon dioxide, i.e., $^{13}CO_2$, which can be detected in exhaled breath and is indicative of *H. pylori* infection. The UBT is a well-established breath test that appeared on the market over 30 years ago. It received both FDA and EMA approvals in 1997. Further details of the test are provided in Chapter 17.

1.3.7 Gastric emptying breath test for gastroparesis

As with the UBT, the gastric emptying breath test (GEBT) utilizes labeled substrates, in this case ^{13}C-spirulina. Rapid absorption of these compounds and their subsequent metabolism to $^{13}CO_2$ in the liver allows gastric emptying of food intake to be determined via measurement of exhaled $^{13}CO_2$. When these analyses are made over time, the ensuing profile provides indications of delayed gastric emptying, or gastroparesis. The GEBT received FDA approval in 2015. Further details of the test are given in Chapter 17.

1.3.8 Maximum liver function capacity breath test

Another ^{13}C-labeled substrate-based test exists in the form of the maximum liver function capacity (LiMAx) breath test. In this test, liver function can be ascertained by monitoring the metabolizing capacity of the liver for ^{13}C-methacetin that targets cytochrome P410 1A2 (CYP1A2) enzymes. The LiMAx test was approved in Europe by the EMA in 2017. Further information about the LiMAx test is given in Chapter 17.

1.3.9 Heart transplant rejection breath test

The oxidative stress in heart transplant recipients can be determined by measuring C_4–C_{20} alkanes and monomethylated alkanes in exhaled breath and creating what has been named a "breath methylated alkane contour." When placed into a 3D model, this test allows for the detection of grade 3 organ rejection.[6] Termed "Heartsbreath," this breath test received FDA approval in 2004.

1.4 Targeted versus untargeted breath analysis

The "new" field of breath research primarily explores the concept of nontargeted analysis (NTA) wherein samples are analyzed using broad-ranging methods with no preconception of potential markers. In short, samples are analyzed to gather as much qualitative information as possible on each sample. Broadly speaking, the idea is to document "features" in the analytical data, such as peaks in chromatograms from gas or liquid chromatography–mass spectrometry (GC-MS or LC-MS, respectively), without necessarily knowing their exact identity or quantity. This approach, also referred to as *top-down*, is in stark contrast to the traditional methods for assessing specific chemicals from discrete sources as in targeted or regulatory analyses, sometimes termed *bottom-up*. The concept of targeted *bottom-up* versus nontargeted *top-down* strategies for assaying biomarkers is illustrated in Fig. 1.2: from the left, specific compounds from the environment are targeted and then sought in the population to assess exposures; from the right, biological media are analyzed without preconception and then interpreted for all manner of health states or exposures.

The increasing widespread implementation of NTA approaches, especially for breath media, is a direct result of the growing development of state-of-the-art instrumentation, especially comprehensive GC-MS and high-resolution mass spectrometry (see Chapters 15 and 16, respectively). Depending on the exact method and instrument, modern NTA of breath can provide chemical information for hundreds, sometimes thousands, of chemicals from a single sample. It is important to note that NTA does not necessarily identify the exact chemical eliciting a chromatographic feature but rather gathers sufficient information (mass spectrum, retention time or index, etc.) to document the same feature across samples[8]. As such, this approach is the main tool contributing to the documentation of breathborne compounds related to health state and the human exposome, either volatile compounds in the gas-phase or less volatile or highly polar chemicals in condensed phase samples. The Environmental Protection Agency (EPA) in the United States has established an NTA *discovery database* containing over 875,000 exogenous and endogenous chemicals (at the time of writing) that includes a special subset list specific to the volatile breath exposome containing ~1075 triaged volatile compounds that have been documented in gas-phase human breath. This database—the *US EPA CompTox Chemicals Dashboard*—is regularly updated with new entries and serves as a clearinghouse for information on every chemical listed, including elemental composition, structure, references, measurement methods, mass spectra, toxicity, metabolism, source, and so on. Ultimately, the dashboard is expected to help define the total exhaled breath exposome, especially as the breath community continues to provide

FIGURE 1.2

Visual description of (left) targeted *bottom-up* versus (right) nontargeted *top-down* analyses for assessing health state or human exposures.

Adapted from [7].

new information. Furthermore, continuous expansion of the dashboard to provide new subset lists that are curated for accuracy, for instance, containing volatiles associated with bacteria and pathogens, will be of great value to breath researchers in identifying potential sources of breathborne volatiles. The US EPA CompTox Chemicals Dashboard is the dedicated subject matter of the final chapter of this book.

1.5 The human exposome

Examining the breath tests reported in the previous section, it becomes apparent that half of those tests make use of exogenous compounds, either by their direct analysis (breath ethanol) or by measurement of their metabolic by-products ($^{13}CO_2$ from probative ^{13}C-labeled substrates). Use or examination of unlabeled exogenous VOC probes[9] or compounds can be similarly made to explore metabolism or enzymatic processes, or organ impairment, as recently observed in exhaled exogenous limonene in association with liver cirrhosis[10] (see Chapter 10, Section 10.6). Such studies demonstrate that exogenous compounds do not always represent confounders that should be filtered from datasets, but may instead be of utility to explore metabolism. Equally, examining such compounds or their metabolic by-products in exhaled breath can be of use in studies on drug intake (see Chapter 30), occupational exposures (see Chapter 32) or via intentional exposure (chamber) studies (see Chapter 33). The latter is discussed in the following within the context of exposure science.

Research on exposure has focused on the cumulative effects of the total environment on human metabolism and health state. The fundamental discipline, *exposure science*, is becoming an academic specialty that is distinct from toxicology, epidemiology, and biology.[11] Exposure science has been defined as the study of "*... human contact with chemical, physical, or biological agents occurring in their environments* [that] *advances knowledge of the mechanisms and dynamics of events either causing or preventing adverse health outcomes*".[12] This discipline encompasses the study of the contribution of the external environment to the human system, as well as the chemicals already present within human biological media; these are collectively referred to as the *human exposome*.[13]

1.5.1 A brief history of human exposome research

The concept of the human exposome was first proposed by Christopher Wild in 2005 when he was director of the International Agency of Research on Cancer (IARC).[13] The broad implementation of the human exposome concept began in earnest in 2008 when the National Institute of Environmental Health Science (NIEHS) in the United States established the Center for Exposure Biology at University of California in Berkeley, CA.[c] By 2011, Balshaw, Lioy, Rappaport, and others were publishing regularly about human exposome topics, Pleil and colleagues were exploring data visualization strategies, Wishart was looking at toxicity issues, and in 2016, Stiegel et al. linked inflammatory markers to exposures and cardiopulmonary parameters.[14–17] In the midst of this flurry of exposome research activity, in 2013, the European Helix Project[d] was initiated to focus on the "early life exposome," and the Hercules Exposome Research Center[e] was launched jointly at Emory University and Georgia Tech University in Atlanta, GA, to investigate environmental exposures and community wellbeing.

1.5.2 Defining the human exposome concept

Within the exposure and biological health communities, the human exposome was first considered mostly as an exposure assessment tool to look for targeted environmental chemicals. Early exposome research examined blood and urine as the biological media to measure environmental metabolites, but this soon expanded to wide-ranging NTA studies wherein essentially all life-based molecules that did not fall into the genome category were broadly included in the exposome.

The general concept of the exposome considers all of the chemicals in the body that represent environmental exposures over the lifetime of an individual, as well as chemical metabolites and reaction products, by-products of the microbiome, and

[c] https://circle.berkeley.edu/research/exposome/.
[d] http://www.projecthelix.eu/.
[e] https://emoryhercules.com/.

systemic response molecules. In short, the exposome represents everything that is "not the genome" in human systems biology.[18,19] In contrast to the genome, however, the exposome fluctuates throughout the human lifetime due to constant influence by changes in the environment, human activities, and the individual's current metabolism and health state at the molecular, cellular, and organ levels of organization, representing the overall concept of human systems biology.[7]

So, the obvious question raised is: why study such a convoluted moving target? The exposome concept is important because current research suggests that incidence of cancers, as well as autoimmune, cardiovascular, pulmonary, and other chronic diseases, is not just bad luck but is largely attributable—estimated at 70%–90%—to environmental factors (i.e., the human exposome) through the genome-environment interaction (G×E) and the epigenome.[20] This observation appeared surprising because the sequencing of the human genome, announced in 2000, and the subsequent associations of function with specific genes were thought to be the final explanation for human disease incidence. Instead, the genome results and expected unambiguous associations to disease have been elusive,[21] leading to efforts to describe the chemical adverse outcome pathways (AOPs) preceding and manifesting disease.[22]

To understand the underlying issue, consider familial relatives, such as a father and son pair: on average, they have a probability of sharing 50% of the same genetic traits, but they have very little likelihood of sharing the same exposome because the father has a generation head-start of lifetime exposure. Furthermore, only when the father and the son are collocated do they share the same short-term exposures, and even then, they are likely to absorb exogenous chemicals at different rates and process them with different efficiencies as a result of individual physiologies. In short, there is little expectation that their exposomes are as similar as their genomes over time, and so their long-term health outcomes—and the associated AOPs—are much more likely to be different, too. In contrast, consider contemporary occupants of a locality or community that have grown up and lived for a long time in the local environment. If long-term health outcome is indeed driven by the exposome, as suggested by the research, then the extensive periods commonly spent in the local environment should be much more important than whether or not the people are genetically related.

The overall human exposome has proven extremely useful in screening for exposures that are unknown a priori and subsequently tracking these over time. Certainly targeted (bottom-up) analyses are definitive; one cannot breathe out an exogenous compound without having been previously exposed to it. The exposome becomes particularly important in nontargeted (top-down) analyses, wherein exogenous compounds appear concomitantly with metabolites and other biochemical, which can lead to the discovery of new environmental chemicals not previously noticed in targeted analyses. This exposure screening approach has value in identifying compounds and developing lists for targeted approaches, as well as in understanding how such exposures may affect normal metabolic pathways.

1.5.3 The exhaled breath exposome

The exploration of breath as a biological fluid for cataloging the human exposome is relatively new. Breath research had historically been applied to assess uptake and pharmacokinetics of targeted VOCs from the environment to help assess emissions (automobiles, industry, manufacturing) and their ambient exposures, either via defined exposure chamber assessments (see Chapter 33) or through occupational exposure (see Chapter 32). Because breath sampling is noninvasive, does not require medical personnel, and does not typically generate potentially infectious wastes, breath-based methods have become preferred over blood and urine collection, especially for public health applications. In so doing, breath-based sampling has been exploited for access to VOCs and associated kinetics measurements in relation to pharmacokinetics and toxicokinetics. Strategies are now being implemented to describe absorption, distribution, metabolism, and excretion (ADME) of VOCs more precisely, and new methods utilizing EBC and EBA are being used to explore the breath exposome for proteins, cytokines, chemokines, fatty acids, and other semi- and non-volatile compounds.

A major challenge for defining the breath exposome is that different fractions of breath (gas, condensate, aerosol) require different sampling and analytical methods, which then focus on different subsets of the exposome. Furthermore, as technology improves, and different laboratories implement new instrumentation, the list of identified compounds diverges. Finally, the pattern of compounds in breath may vary depending on the respective sensitivity and specificity of compound analyses. As such, the "breath exposome" is not a fixed list of chemicals, but rather a chemical and/or statistical concept that can be exploited to potentially tease out information from group or longitudinal comparisons to screen for exposures or preclinical disease states, as discussed in Chapter 37. Ultimately, the goal in breath analysis is to use a baseline of the *unremarkable* breath exposome to search for perturbations in specific samples—i.e., *remarkable* breath—that might indicate an AOP or other preclinical signature.

One of the classic exhaled breath applications for exposures screening was the measurement of methyl tertiary butyl ether (MTBE), a gasoline additive introduced in the United States in the mid-1990s with the intention to reduce lead additives, automobile CO emissions, and secondary ozone formations. The low odor threshold of MTBE prompted public concerns about exposures, and a concerted research effort was implemented to assess the levels of environmental exposure. Serendipitously, MTBE and its primary human metabolite, tertiary butyl alcohol (TBA), are both sufficiently volatile to be excreted in the gas-phase of human breath, which allowed relatively trivial confirmation and quantitation of exposures. Further, breath analyses of these compounds enabled the development of one of the early classical pharmacokinetic ADME models for environmental chemicals.[23,24]

1.6 Data treatment tools

Complex breath compound datasets relating to the human volatilome or exposome require suitable tools to extract useable information to their fullest extent. Sophisticated data processing steps are the subject matter of other chapters of this book (see Chapters 37 and 38), but a few additional tools with particular relevance to exposome datasets are covered in the section below.

1.6.1 Summary statistics

Probably the most useful aspect of cataloging the breath exposome is to develop summary statistics of prevalence and patterns of exhaled chemicals. As mentioned above, restrictions to some specific subset of the exposome that is dependent on the sampling and analytical strategies are always in place, but various statistical approaches can be developed within that regime to address the question of "what is normal or expected?," i.e., to characterize *unremarkable* data and identify it as such.

Briefly, the task is to develop some quantitative measures of prominent features in the exposome, such as concentrations series of known compounds, relative responses of unknown features, or compound ratios, etc. The next step is to test for distribution (normal, lognormal), missing values (left-censored, unbalanced data), and metafactors (age, ethnicity, sex) that might influence the outcomes (see Chapter 37). At this point, the expected levels of individual exposome features can be tabulated by central tendency and confidence levels. Any future measurements can then be interpreted with respect to their position within this "expected" range, that is, are they outliers and do they reach a specified level of concern, or are they still within an acceptable range of the central tendency?

1.6.2 Data visualization

Another valuable tool in dealing with complex data is pattern visualization. Related techniques were originally developed to explore relationships among measured quantities and their respective groups. The most common visualization tool is the *heatmap*, which is a matrix plot to visualize quantitative changes in a dataset, either over time or with respect to different cohorts (or individuals therein). As an example, an exhaled breath matrix dataset is plotted in a square array whereby the x-axis identifies the sample and the y-axis identifies the chemical; each square of this array is color coding according to its concentration.[25]

A heatmap visualization of a dataset is presented in Fig. 1.3, which displays a series of alcohols and aldehydes in EBC of three male and four female individuals, grouped by sampling time before and after exposure to clean air and city levels of diesel exhaust. From the heatmap, it becomes immediately apparent that females overall have lower levels of these compounds in EBC, as indicated by the trend

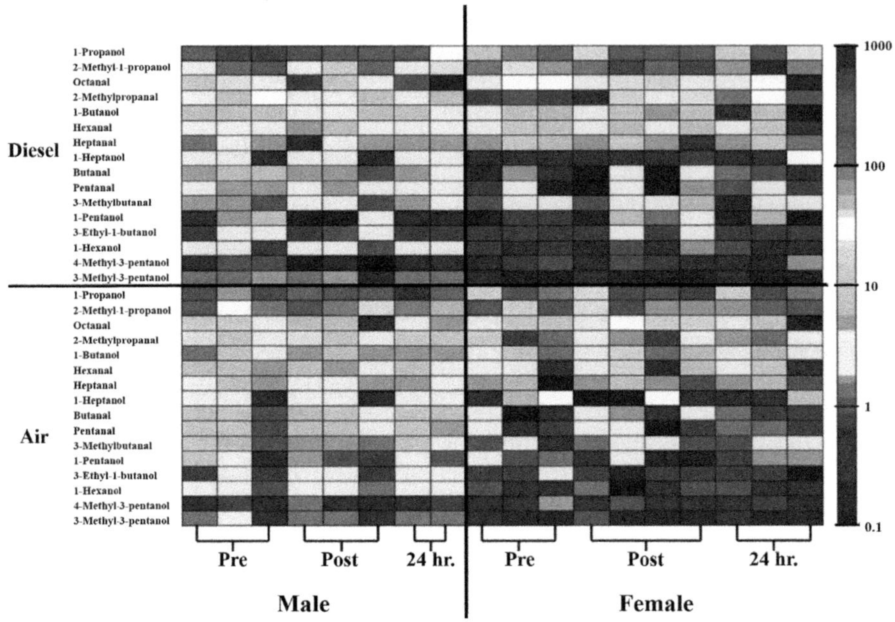

FIGURE 1.3

A representative heatmap for data visualization purposes. The visualized dataset is of biomarkers in exhaled breath condensate (EBC) of seven individuals, split between males (n = 3) and females (n = 4), whereby the x-axis is partitioned by sex of individuals and subdivided into pre-, (immediate) post-, and 24 h postexposures to clean air or dilute diesel exhaust. The y-axis represents lists of exhaled alcohols and aldehydes, grouped by exposure scenario. The quantitation color scheme is plotted on the right vertical column with blue representing low concentrations and red high concentrations (arbitrary units).

Reproduced from [25].

of darker blue compared to the male datasets columns. Further, different compounds (comparing rows) evidently have greater differences from each other, in that propanol and 2-methyl-1-propanol trend toward red compared to other compounds. Finally, the exposure scenario does not exhibit any obvious effect. Certainly, these data could be expressed as a table of summary statistics, leading to similar conclusions, but the heatmap is a good start to get an overview of the data before embarking on complex mathematical analyses.

Accordingly, the value of the heatmap is to explore and explain data relationships without the constraints of statistical assumptions. The heatmap can be arranged in any fashion to demonstrate relationships among the data, for example, by arranging the x-axis (samples) in blocks of male versus female individuals, or in order of youngest to oldest individuals, or other configurations. The y-axis can be arranged in blocks of compounds with some commonality, for example, the first five rows could be aldehydes, the next five ketones, the next five exogenous aromatics, and so forth. Once arranged and drawn, the color coding often clearly flags cluster relationships.

1.7 Summary

The complex and dynamic nature of the human volatilome presents a considerable challenge in breath biomarker discovery. More than 870 VOCs have been discovered in breath, with diverse origins, largely the environment, human physiology, the microbiome, or other sources. Nontargeted breath analysis studies are confronted with the task of identifying "remarkable" samples among a backdrop of "unremarkable" samples, whereby the former might exhibit only minor differences to the latter in the form of quantitative and/or qualitative perturbations. To complicate matters, individual biomarkers that are unique to a specific disease seemingly do not exist. Nevertheless, a handful of breath tests have reached sufficient maturity to be of clinical value, such as fraction of exhaled nitric oxide (F_ENO) as a diagnostic aid in asthma. Other breath tests utilize isotopically labeled substrates and exploit in vivo metabolism to ascertain impairment or infection, such as the urea breath test to identify *H. pylori* infection. Tests involving probative molecules demonstrate the value of using exogenous compounds to explore metabolic pathways. Environmental exposure science is one field in which breath analysis holds particular promise. The realization that disease manifestation is largely driven by environmental exposures has led to focused research on AOPs in relation to the exposome to decipher causal links of exposure with disease.

The human breath exposome is becoming a valuable resource for assessing existing and new environmental exposures, as well as deducing if such exposures are affecting an individual's systems biology and ultimately their health state. The basic premise of the human exposome is the existence of a chemical baseline upon which environmental insults play out, both in the targeted exogenous profile and the resulting metabolic changes. The applications of exposome perturbations analysis are still in their infancy as the variability between and within humans is often greater than the effects of an AOP. Nevertheless, as researchers expand the spectrum of known constituents of the breath exposome, the longitudinal perturbations and the case–control effects pattern shifts become more distinct. The development of comprehensive reference libraries, such as the US EPA CompTox Chemicals Dashboard, and inclusion of observed breathborne compounds, represent valuable resources that will aid in identifying biomarkers. Exploration of the human volatilome in the context of the exposome will further our understanding of the environmental or physiological origins of biomarker compounds and ultimately lead to knowledge on the AOPs and biochemical mechanisms that lead to disease.

Acknowledgments

The authors are thankful for the background materials and a long history of excellent advice from Ariel Wallace, Jon Sobus, Peter Egeghy, Michael Madden, and Antony Williams from US EPA, Terence Risby from Johns Hopkins University, Stephen Rappaport from University of California, Berkeley, and Anil Modak. We acknowledge Wolfram Miekisch from University Medicine Rostock, Germany, for permission to use his conceptual idea of representing potential sources contributing to individual VOCs in the volatilome as a radar plot, as depicted in Fig. 1.1.

References

1. Comroe JH. *Physiology of respiration: an introductory text*. 2nd ed. Chicago, IL: Year Book Medical Publishers; 1974.
2. de Lacy Costello B, Amann A, Al-Kateb H, Flynn C, Filipiak W, Khalid T, et al. A review of the volatiles from the healthy human body. *J Breath Res* 2014;**8**(1):014001.
3. Phillips M, Boehmer JP, Cataneo RN, Cheema T, Eisen HJ, Fallon JT, et al. Heart allograft rejection: detection with breath alkanes in low levels (the HARDBALL study). *J Heart Lung Transplant* 2004;**23**(6):701–8.
4. Metz G, Peters T, Jenkins DA, Newman A, Blendis L. Breath hydrogen as a diagnostic method for hypolactasia. *Lancet* 1975;**305**(7917):1155–7.
5. Cohen RS, Ostrander CR, Cowan BE, Stevens GB, Hopper AO, Stevenson DK. Pulmonary excretion rates of carbon monoxide using a modified technique: differences between premature and full-term infants. *Neonatology* 1982;**41**(5–6):289–93.
6. Phillips M, Boehmer JP, Cataneo RN, Cheema T, Eisen HJ, Fallon JT, et al. Prediction of heart transplant rejection with a breath test for markers of oxidative stress. *Am J Cardiol* 2004;**94**(12):1593–4.
7. Pleil JD, Stiegel MA. Evolution of environmental exposure science: using breath-borne biomarkers for "discovery" of the human exposome. *Anal Chem* 2013;**85**(21):9984–90.
8. Sobus JR, Wambaugh JF, Isaacs KK, Williams AJ, McEachran AD, Richard AM, et al. Integrating tools for non-targeted analysis research and chemical safety evaluations at the US EPA. *J Expo Sci Environ Epidemiol* 2018;**28**(5):411–26.
9. Gaude E, Nakhleh MK, Patassini S, Boschmans J, Allsworth M, Boyle B, et al. Targeted breath analysis: exogenous volatile organic compounds (EVOC) as metabolic pathway-specific probes. *J Breath Res* 2019;**13**(3):032001.
10. Fernández del Río R, O'Hara ME, Holt A, Pemberton P, Shah T, Whitehouse T, et al. Volatile biomarkers in breath associated with liver cirrhosis—comparisons of pre- and post-liver transplant breath samples. *EBioMedicine* 2015;**2**(9):1243–50.
11. Pleil JD, Blount BC, Waidyanatha S, Harper M. Establishing exposure science as a distinct scientific discipline. *J Expo Sci Environ Epidemiol* 2012;**22**(4):317–9.

12. Barr DB. Human exposure science: a field of growing importance. *J Expo Sci Environ Epidemiol* 2006;**16**(6):473.
13. Wild CP. Complementing the genome with an "exposome": the outstanding challenge of environmental exposure measurement in molecular epidemiology. *Cancer Epidem Biomar* 2005;**14**(8):1847–50.
14. Lioy PJ, Rappaport SM. Exposure science and the exposome: an opportunity for coherence in the environmental health sciences. *Environ Health Perspect* 2011;**119**(11): a466–7.
15. Stiegel MA, Pleil JD, Sobus JR, Madden MC. Inflammatory cytokines and white blood cell counts response to environmental levels of diesel exhaust and ozone inhalation exposures. *PLoS One* 2016;**11**(4):e0152458.
16. Stiegel MA, Pleil JD, Sobus JR, Stevens T, Madden MC. Linking physiological parameters to perturbations in the human exposome: environmental exposures modify blood pressure and lung function via inflammatory cytokine pathway. *J Toxicol Environ Health A* 2017;**80**(9):485–501.
17. Wishart D, Arndt D, Pon A, Sajed T, Guo AC, Djoumbou Y, et al. T3DB: the toxic exposome database. *Nucleic Acids Res* 2014;**43**(D1):D928–34.
18. Miller GW, Jones DP. The nature of nurture: refining the definition of the exposome. *Toxicol Sci* 2014;**137**(1):1–2.
19. Rappaport SM. Implications of the exposome for exposure science. *J Expo Sci Environ Epidemiol* 2011;**21**(1):5–9.
20. Rappaport SM, Smith MT. Environment and disease risks. *Science* 2010;**330**(6003): 460–1.
21. Bromberg Y. Building a genome analysis pipeline to predict disease risk and prevent disease. *J Mol Biol* 2013;**425**(21):3993–4005.
22. Pleil JD, Beauchamp JD, Miekisch W, Funk WE. Adapting biomarker technologies to adverse outcome pathways (AOPs) research: current thoughts on using in vivo discovery for developing in vitro target methods. *J Breath Res* 2015;**9**(3):039001.
23. Lindstrom AB, Pleil JD. Alveolar breath sampling and analysis to assess exposures to methyl tertiary butyl ether (MTBE) during motor vehicle refueling. *J Air Waste Manag Assoc* 1996;**46**(7):676–82.
24. Pleil JD, Kim D, Prah JD, Rappaport SM. Exposure reconstruction for reducing uncertainty in risk assessment: example using MTBE biomarkers and a simple pharmacokinetic model. *Biomarkers* 2007;**12**(4):331–48.
25. Pleil JD, Stiegel MA, Sobus JR, Liu Q, Madden MC. Observing the human exposome as reflected in breath biomarkers: heat map data interpretation for environmental and intelligence research. *J Breath Res* 2011;**5**(3):037104.

CHAPTER

Breath sampling and standardization

2

Jonathan D. Beauchamp[1], Wolfram Miekisch[2]

[1]*Fraunhofer Institute for Process Engineering and Packaging IVV, Freising, Germany;* [2]*Department of Anesthesia and Intensive Care, Rostock University Medical Center, Rostock, Germany*

Chapter outline

2.1 Overview

As a concept, breath analysis continues to give the impression of being a new field of research, or at the very least, a novel diagnostic approach. This is despite the fact that contemporary breath research has steadily increased since the early analytical investigations by Linus Pauling in the 1970s. Half a century later, the field is still very much in its infancy, and it is without a doubt that the lack of standardized pracp: tices, and the resulting disparity of research findings, has played a primary role in holding back progress and is one of the main reasons for the principal absence of breath tests in the clinical practice. Sampling and standardization are intrinsically linked; even standardized approaches in recruiting suitable patient cohorts, in analyzing samples, or in data treatment and interpretation are of little use if the sampling procedure is flawed.

Breathborne Biomarkers and the Human Volatilome. https://doi.org/10.1016/B978-0-12-819967-1.00002-5
Copyright © 2020 Elsevier B.V. All rights reserved.

Until relatively recently, sampling of breath was more or less considered a necessary step as a means to an end in order to supply the analyzer with a sample, with little (or no) consideration given to how the sampling procedure itself might affect the sample. Lately, however, increasing awareness of how critical the sampling step is in obtaining robust data has spread throughout the breath research community, and many researchers working in breath analysis have established their own optimized sampling approaches for quality assurance. Breath sampling has many facets, and each parameter requires careful consideration to ensure sample validity and utility. Best practice in sampling depends on the nature of the sample being collected, for instance, breath gas, exhaled breath condensate or aerosols (EBC and EBA, respectively; see Chapter 7), or exhaled particles (see Chapter 8). The targeted biomarker—if known a priori—might equally dictate the sampling procedure, for example, as is the case for analyzing fraction of exhaled nitric oxide (F_ENO) as a diagnostic aid for asthma, whereby the exhalation flow is critical for its meaningful detection (see Chapters 4 and 5). The analytical platform for detecting breathborne compounds will also prescribe how samples are collected. While some instrumentation is capable of directly analyzing samples during exhalation, such as in soft chemical ionization mass spectrometry (SCIMS) approaches (see Chapters 9 and 10), other platforms require sample workup prior to analysis, such as in gas chromatography−mass spectrometry (GC-MS; see Chapter 15), thereby dictating the need for the collection and transfer of a sample to the instrumentation. If the latter is required, storage becomes an issue, whereby the storage system (adsorptive medium, bag, or canister) dictates the storage conditions and duration. Because of the manifold considerations that must be accounted for in breath sampling, introducing standardized approaches will be of great benefit to align outcomes in breath research.

Standardization in breath research is not a trivial task, owing the wide scope of the field. The multifaceted aspects of sampling have been alluded to above—and will be discussed below—but when factors such as disease, physiological effects, targeted biomarker compound, analytical platform, and data mining strategy are thrown into the mix, standardization attempts become a colossal and seemingly almost futile undertaking. Clearly, a one-size-fits-all approach for standardizing breath analysis is neither feasible nor conducive to innovation and developments in the field, which have been forthcoming at considerable pace in recent years. A more appropriate and realistic way forward is to compartmentalize tasks along the breath analysis workflow and to establish standardized practices for each of these steps. This chapter commences with a comprehensive review of various breath sampling approaches, and will highlight considerations on standardization and how this might be established in the field of breath research.

2.2 Sampling

Sampling approaches for exhaled breath can take on many forms. In general, these can be categorized as *offline* or *online*: in the former, samples are collected for transfer and subsequent measurement by the analytical platform, whereas in the latter, breath is sampled directly by the analyzer and measured immediately, obviating the need for the use of a sample storage medium. Accordingly, the mode of sampling is primarily dictated by the analytical platform used for biomarker detection, although other logistical factors might also be of influence, such as a preference for offline sampling despite the real-time measurement capabilities of the analyzer. Furthermore, consideration must be made on the phase of exhalation that is specifically of interest, i.e., either gas from the oral cavity, mixed expiratory breath, or end-tidal volume. Additionally, beyond the materials and protocols used in sampling, other factors, such as physiology or time of day, can influence breath profiles, all of which need to be taken into account in order to achieve reproducibility in collected samples.

In general, the dilution and contamination of breath samples and the loss of analytes during sampling and storage procedures have resulted in huge variations of substance concentrations in different studies reported in the literature. As such, well-defined and reproducible breath samples are a critical prerequisite for breath analysis and clinical studies. These considerations are presented and discussed.

2.2.1 Exhalation phases

In humans, a single exhalation does not proceed as a uniform flow and volume of a homogenous mixture, but rather consists of several subphases, as depicted in Fig. 2.1 in a representative profile of exhaled carbon dioxide (CO_2). These transitional phases of a breathing cycle are

- *Phase I*: end of inhalation and deadspace[a] gas
- *Phase II*: mixed expiratory gas, comprising deadspace gas and gas from the lungs
- *Phase III*: end-tidal volume, most closely mirroring alveolar gas

In breath analysis, the phase sampled depends on the application: if the production of volatile organic compounds (VOCs) by bacteria in the oral cavity is of interest in relation to halitosis or oral hygiene, then phase I should be sampled. Phase II will be most relevant when exploring compounds that are generated in the upper airways, for instance, nitric oxide (NO) or carbon monoxide (CO). Phase III is representative of gas in the lungs and is thereby of most interest when targeting bloodborne volatiles, either endogenous or exogenous. It should be noted that end-tidal breath sampled at the mouth (or nose) during exhalation cannot be considered to be pure alveolar gas because the gas necessarily passes through and interacts with the trachea and upper airway physiology, where it can undergo

[a] The term "deadspace" can refer to the *anatomical deadspace*, which is gas in the oral cavity, upper airway, and trachea (where no gas exchange takes place), or the *analytical deadspace*, which is the gas that stagnates in the sampling lines between exhalations; deadspace in breath analysis always comprises both aspects.

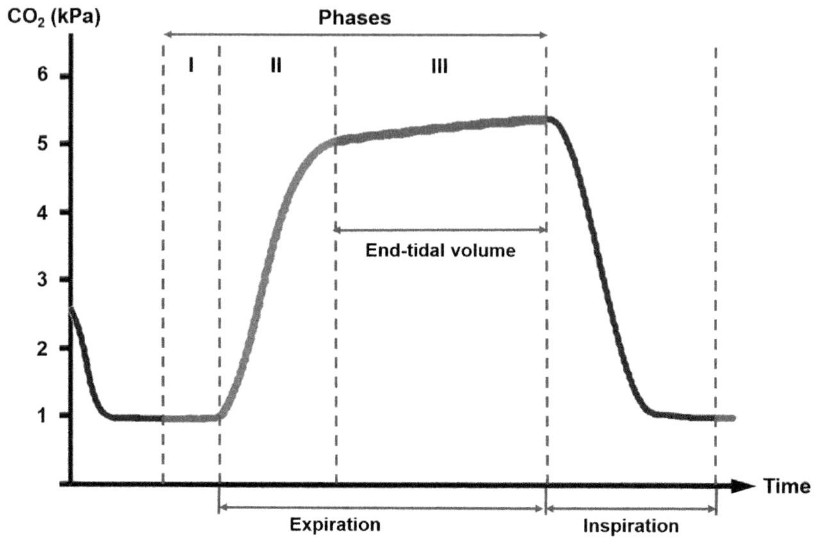

FIGURE 2.1

The transitional phases of a breathing cycle, as illustrated in relation to the profile of carbon dioxide (CO_2), comprising phase I, end of inspiration and deadspace gas; phase II, mixed expiratory gas; and phase III, end-tidal volume.

changes. Nevertheless, for practical purposes, the end-tidal volume sampled during phase III is the best representation of alveolar gas that can be collected noninvasively.

Putting an interest in oral volatiles aside, breath research primarily focusses on seeking biomarkers of disease or exposure that are present in the systemic circulation and thereby exhaled after blood—gas exchange in the alveolar of the lungs (see Chapter 1). In practice, the collection of breath is achieved either as mixed expiratory gas, which comprises all three phases of exhalation, including deadspace gas, or as end-tidal gas, i.e., phase III only. Essentially, the latter would always be preferable, as this phase most closely represents the alveolar gas of interest. Bloodborne volatiles have been observed to be a factor two to three higher in abundance in end-tidal gas compared to mixed expiratory samples, and end-tidal breath contains the lowest abundances of contaminants from deadspace.[1,2]

Although end-tidal sampling is desirable in breath analysis, invariably mixed expiratory gas sampling is performed, most often due to its simplicity or an absence of suitable control mechanisms. Sampling without controlled identification of the respiratory phases bears the risk of dilution by deadspace gas during mixed expiratory gas sampling. As the concentrations of volatiles present in exhaled breath are in the ppb to ppt range[b], analytical results are easily affected by impurities from inspired air or by dilution with deadspace gas. Accordingly, in order to achieve exclusive phase III sampling, some form of control is needed. Controlled sampling is a key requirement for reliable analysis of breath biomarkers by ensuring focus on the phase of interest and thereby minimizing confounders, and may help unveil the origin of breathborne volatiles by comparison of VOC profiles during the expiratory phases.[3]

2.2.2 Controlled sampling

The simplest controlled approach to sampling is to time an exhalation, but samples collected using this procedure suffer from large variations in their analytical outcomes.[3] A better strategy is to base sampling on physiological parameters, such as partial pressure of exhaled CO_2 (or oxygen, O_2), temperature, or flow. Typically, any parameter used to identify and separate phases of respiration must be recordable at minimum intervals of 200 ms. Monitoring exhaled CO_2—termed capnography—is particularly conducive for controlled breath sampling; CO_2 represents a direct marker of gas exchange in the alveoli and thereby exhibits a clear respiratory phase profile that allows the discrimination between alveolar (end-tidal) and nonalveolar (deadspace, airways) phases, as depicted in Fig. 2.1. The detection of CO_2 can be achieved with high frequency using mainstream commercial capnometer devices, as is routinely performed in clinical practice; indeed CO_2 can be considered the most widely utilized breathborne biomarker, as discussed in Chapter 1. Controlled end-tidal sampling by means of exhaled CO_2 concentrations has been reported as best suited when targeting bloodborne VOC biomarkers.[2]

2.2.3 Analytical platforms

The most common and suitable analytical platforms currently in use for the analysis of breathborne volatiles are introduced and discussed in other chapters of this book (see Chapters 9–16), thus will not be covered at length in this chapter. Technologies include (high-resolution) mass spectrometry (MS) approaches, including SCIMS or GC-MS, spectroscopic techniques, and sensors. Broadly speaking, breath analyzers[c] fall into two categories, namely *offline* systems, which require sample workup, such as preconcentration prior to analysis, and *online* instrumentation, which can analyze

[b] Units of concentration—strictly speaking, gas-phase volume mixing ratios—used in this chapter (and largely throughout this book) are ppt, parts per trillion; ppb, parts per billion; and ppm, parts per million, by volume, whereby the suffix v (as in ppb_v) to refer to volume has been omitted for conciseness.

[c] Although the term "breath analyzer" is used here, most laboratory-based systems are suitable for a wide range of applications, such as atmospheric chemistry, food science, or industrial monitoring, to name but a few; essentially, the specific origin of a VOC, whether from human, plant, or machine, is of little relevance to the analyzer.

samples directly without manipulation. The latter typically exhibit measurement frequencies in the range of Hertz or sub-Hertz, making them suitable for direct analysis of breath during exhalation. The fastest among these devices can achieve breath-by-breath resolution, thereby making it possible to discriminate between the different phases of exhalation. Some detection systems are compact and portable, allowing their use at point-of-care (PoC), such as sensor systems or spectroscopic devices, whereas others are large benchtop systems more suitable for the laboratory applications, such as MS-based analyzers.

The specifications and performance of any analytical platform in terms of sensitivity, accuracy, reproducibility, and reliability are essential for achieving valid and robust breath volatiles data. An exhaustive introduction and discussion of instrumental performance is beyond the scope of this chapter, although this aspect is covered in brief in relation to standardization, below. Regardless of specifications, however, the nature of breath sampling and the sample transfer to the analyzer dictates the quality of all subsequent steps, from the sample analysis to the subsequent data processing and interpretation.

2.2.4 Sampling for offline analysis

Many analytical platforms do not offer sufficiently fast detection to achieve breath-resolved analysis or are too bulky for bedside or PoC applications. Thus, there is a need to collect discrete breath samples and transfer these individually to the analyzer. This requires additional sampling steps in order to ensure the collection of well-defined and reproducible breath samples. The most commonly used breath sampling techniques for offline analysis employ either (inert) polymer bags[2,4,5] (and less often, canisters[6]) to collect and store gas-phase samples or sorbent materials[7] that bind volatile molecules at ambient conditions and retain these until heated to high temperatures in a process termed thermal desorption (TD). Since many breathborne volatiles are susceptible to changes during (prolonged) storage, particularly in the gas phase, and especially in bags (see below), often the two approaches are used complementarily, whereby gas-phase breath samples collected in bags or canisters are then transferred onto sorbent traps. In this scenario, sampling with immediate trapping onto sorbents is preferable to gas-phase storage in a bag with transfer onto sorbents some time later, due to the greater likelihood of compromising bag-stored samples, as discussed below.

Focusing on sorbent-based sampling techniques, these can be broadly split into two approaches, namely adsorption (or TD) tubes, whereby sample gas is flushed through a tube that is packed with sorbent materials that extract volatiles from the gas phase, or distribution-based methods, in which sorbent-coated fibers are passively exposed to gas-phase samples and trap volatiles as they interact with the fiber. In both of these techniques, volatiles are preconcentrated (or enriched) onto the sorbent materials, which offers an additional advantage in sampling. Specifically, as breathborne volatiles are mainly present at only trace concentrations, this presents an analytical challenge as concentrations approach the detection limits of instrumentation. As such, many analytical platforms, especially laboratory-based systems, require that volatile substances from breath samples are preconcentrated before analysis. Therefore, sorbent-based sampling approaches simultaneously solve two issues, namely the transfer of a breath sample to the analyzer and an enrichment of the sample prior to its analysis.

A further important point to note is that in general, whether sampling into a receptacle or onto a sorbent material, or both, the high humidity of exhaled breath must be taken into account, as water vapor can affect sample stability in bags or the adsorption capacity of sorption-based systems, as well as saturating cryo-focusing systems during TD analysis.

Bags

Breath sampling into (inert) bags is a relatively simple undertaking and as such has been a very popular sampling approach in breath analysis over the past decades, and is still in use today; blowing into a bag is easy for the patient, it is fast, and the sample can be easily transported to the laboratory for analysis. This simplicity can be deceptive, however, as there are several shortcomings in sampling with bags. In relation to sampling itself, an important consideration is the potential impact on breathborne VOC profiles through the high resistance encountered when blowing into the small orifice of the bag sampling port, which is discussed below in the context of physiological considerations. Moreover, stability of bag-stored breath samples is a critical issue, as is the material of the bag, which may interact with breath compounds.

Most bag sampling approaches utilize polymer bags, predominantly Teflon, Tedlar, or Nalophan.[8] Despite their simplicity and ease of use, samples stored in bags are readily compromised through losses and artifacts, which limits their suitability for breath sampling. Specifically, several studies have shown that many VOCs are unstable when stored in bags, with significant losses of compounds due to diffusion through the bags, even on timescales of several hours, and certainly within 24 h, although the degree of losses depends on the type of bag.[8–10] The stability has been observed to be compound dependent, with some substances, such as hydrocarbons, being relatively stable, whereas other volatiles, such as oxygenated compounds, exhibiting pronounced decreases over time.[11] Clearly, this phenomenon can have huge repercussions on the outcome of breath research trials; if care is not taken to align storage duration and conditions of bags containing breath samples, then any qualitative or quantitative differences observed between samples might merely relate to sample losses, rather than being of physiological or clinical relevance.

Conversely, diffusion of ambient VOCs into the bags can lead to artifacts. This was clearly demonstrated in a study that measured acetone and hexanal in a Tedlar bag filled with clean air (zero air) during storage in a contaminated environment.[9] Concentrations of acetone and hexanal inside the bag were continuously monitored while open vials of these compounds as reference standards were placed next to the bag housed in an enclosed environment (see Fig. 2.2). A marked increase in both compounds inside the bag was observed within minutes of contamination, indicating a rapid diffusion into the bag sample. In practical terms, this highlights the potential

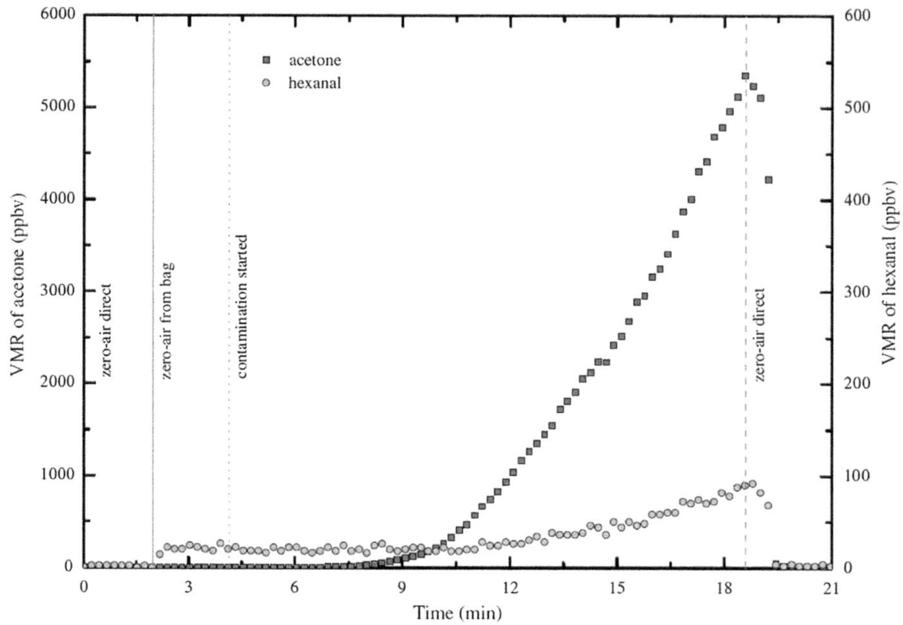

FIGURE 2.2

Measurement of acetone (left scale) and hexanal (right scale) in a Tedlar bag filled with clean air (zero air) and stored in an environment with high gas-phase concentrations of these two compounds. Ambient contamination was achieved by placing open vials of reference standards of acetone and hexanal within an enclosed space containing the bag. Analyses were performed using proton transfer reaction—mass spectrometry (PTR-MS), with sampling switching between zero air direct and then zero air from within the bag before and during ambient contamination. A marked increase in the two compounds inside the bag, especially acetone, is observed within several minutes of placing the vials next to the bag.

Reproduced from [9].

for bag samples to become compromised through the presence of gas-phase environmental constituents, thus their storage in contaminated areas, such as in the clinical setting or wet chemistry laboratories, will almost certainly introduce artifacts in such samples. Furthermore, the polymer material of the bag might emit compounds, such as phenol, that could confound the analyses of breath samples.[9]

Based on the observations of these phenomena, it is generally accepted that the use of polymer bags for breath sampling is acceptable only when storage periods are short, ideally within minutes, but certainly less than a few hours. In general, bag sampling with prompt transfer of the sample to an analyzer for immediate analysis, or onto a sorbent material for preconcentration and subsequent analysis, will incur the least distortions of the sample and is considered acceptable.

Sorbent traps

The type of sorbent material suited for sampling is dictated by the physicochemical properties of the VOCs of interest, whereby a suitable sorbent will ensure efficient

Table 2.1 Comparative overview of the performances of sorbent-based extraction techniques for sampling breath volatiles.

Extraction method	Tubes	SPE	SPME	NTME
Extraction range	low ppb	ppt–ppb	low ppb	ppt–ppb
Sample volume (mL)	200–1000	200–5000	10–20	20–100
Sampling time (min)	5–10	10–30	1–5	1–5
Stability	weeks–months	hours–days	min–hours	hours–weeks
Humidity effects	moderate	high	low	moderate–low

Tubes, *thermal desorption tubes;* SPE, *solid-phase extraction;* SPME, *solid-phase microextraction;* NTME, *needle trap microextraction.*
Adapted from [12].

capture of volatiles and reduce or avoid breakthrough or memory effects. Carbon molecular sieves (e.g., Carboxen 1021 or 1000) or graphitized carbon (e.g., Carbopack X) are best suited for adsorption of highly volatile compounds, but can suffer from irreversible adsorption and may be affected by the high water content of exhaled breath. Organic polymers (e.g., Tenax TA or GR), on the other hand, have shown relatively low breakthrough volumes for highly volatile compounds and are least affected by high water vapor content. Often, combinations of sorbent materials—referred to as multibed traps—offer an excellent compromise by trapping a wide range of compounds and exhibiting a high tolerance to water vapor, which is achieved by combining strong sorbents, such as graphitized carbon or molecular sieves, with a protective layer of a weaker organic polymer adsorbent, such as Tenax.

Sorbent trap-based sampling for breath analysis applications come in two varieties, namely extraction and microextraction techniques. The regular extraction techniques encompass TD tubes and solid-phase extraction (SPE), whereas microextractions include solid-phase microextraction (SPME) and needle trap microextraction (NTME). The most common approach in breath analysis is the use of TD tubes, often loaded not only from bag samples but also from direct exhalations using specialized devices. The technique of NTME is relatively new and particularly suitable to breath sampling, combining advantages of SPE and SPME. A comparison of performances of the different techniques is given in Table 2.1.

Regular sorbent extraction

A major disadvantage of regular sorbent trap extraction techniques is the relatively high sampling volume needed for reliable capture of volatiles and subsequent sensitive analysis, whereby typical volumes range from 200 mL to 5 L (see Table 2.1), although this parameter is also coupled to the analytical detection method used and its detection limits. At the latter end of this range, the associated lengthy sampling times are not only an unnecessary burden to patients, but can elicit changes in physiological parameters that result in a shift in the breath volatiles profile. Furthermore, sampling larger volumes can lead to issues relating to the high water vapor content. Although sensitivity typically increases with sample volume, the long sampling times required for SPE and the subsequent complex and time-consuming desorption steps make it a comparably slow method; as such, SPE is infrequently used in breath analysis.

As stated above, TD tubes are widely used in breath research due to several advantages they offer over other devices. In breath applications, TD tubes typically contain either Tenax TA or a combination of Tenax TA/Carbograph as multibed traps, offering good trapping efficiency for a wide range of volatiles. Further advantages are their robustness, with physical breakages unlikely and associated easy transportation, and their ability to achieve relatively good stability of trapped compounds when handled and stored appropriately. The general consensus for storing adsorbent tubes is to keep them cool and minimize storage time. Extensive trials on sample stability in stored adsorbent tubes have demonstrated reliable storage for 14 days at 4°C for Tenax TA adsorption tubes loaded with 76 VOCs[13] and acceptable stability for at least 1.5 months storage at −80°C for Tenax TA/Carbograph tubes containing breath samples.[14]

In relation to transferring breath onto adsorbent tubes, commercial devices are available, which simplify sampling by incorporating capnography to enable selective sampling of the appropriate phase of exhalation, as discussed above, with some systems drawing samples simultaneously onto several tubes to provide sample replicates. Such devices have been subjected to systematic assessments and are generally considered to provide reliable samples, although instrumental operating and sampling parameters have been shown to be critical in this endeavor.[15,16]

Microextraction

Microextraction techniques, such as SPME or NTME, combine the advantages of small sample volumes and fast and controlled sample preparation with in situ sampling and preconcentration in one step. SPME and NTME techniques can be differentiated by their modes of extraction,[17] namely distribution versus exhaustive techniques.

Nonexhaustive techniques, such as SPME, rely on a distribution-based mechanism, whereby the extracted amount of a compound will solely depend on its concentration in the sample, provided the sample volume is much greater than that of the sorbent. The degree of sorption is characterized by a distribution constant, which is defined through the physiochemical properties of both the analyte and the sorbent. When these conditions are met, relatively small volumes of breath (typically 10–20 mL) can be extracted without any loss in sensitivity compared to larger volumes. In SPME, once equilibrium is reached between the volatile and the sorbent material, a subsequent change (increase) in the sample volume will not affect (increase) the extracted amount. The typical concentration range of extraction by distribution-based techniques is low ppb, with SPME capable of extracting some 100 VOCs in this range. One shortcoming of these distribution-based approaches is that the number of available sorbents is limited.

As an exhaustive technique, NTME represents a promising new approach for robust and reproducible breath sample preparation that combines the advantages of SPE and SPME.[17,18] Like SPME, NTME only requires small sample volumes (typically 10–100 mL), but with the advantage of improving sensitivity by increasing the sample volume (similar to SPE). The analytical performance of NTME crucially depends on the intrinsic properties of the sorbent material, desorption temperatures, and sampling conditions,[19] thus these parameters must be carefully chosen for optimum extraction of volatiles, which can be achieved in the ppt to ppb range.[20] Furthermore, exhaustive techniques can provide preconcentration for a broad range of VOCs. Low sample volumes, straightforward sampling and desorption, and good sample stability facilitate the application of NTME in breath research. Another important feature of NTME that makes it particularly suited to breath sampling and analysis is the compatibility with PoC requirements in the clinical environment. Well-controlled samples can be collected via NTME within minutes—even in an automated fashion—and the needle traps can easily be transported and stored.[20] Similar to TD tubes, commercial devices for simplified sampling that incorporate capnography are available.[21]

2.2.5 Sampling for online analysis

Although offline analysis of breath is currently the most widely implemented approach in breath research, online analysis has key advantages and is growing traction, especially with the rapid developments in the field of sensor systems (see Chapter 13) and optical spectroscopy (see Chapter 14) that can be manufactured as compact and portable devices and thereby enable PoC analysis. Benchtop or laboratory-based online systems, such as ion mobility spectrometry (IMS; see Chapter 11), or selected ion flow tube–mass spectrometry (SIFT-MS) and proton transfer reaction–mass spectrometry (PTR-MS)[22] (see Chapters 9 and 10, respectively), have been the driving technologies in characterizing rapid changes in volatilome profiles, e.g., during exercise (see Chapter 3), from therapeutic interventions, or in response to drug intake. Indeed, the key advantages of online analytical systems over offline platforms are not only their ability for breath-by-breath resolution analyses, but also their capability for continuous analysis, i.e., repeated measurements in short succession. Taken together, these features offer the possibility to record highly resolved breath profiles over an extended duration. One application for which this combination is particularly advantageous is pharmacokinetics. This has been demonstrated in studies using PTR-MS and IMS, whereby a volatile constituent of a pharmaceutical preparation was monitored in breath following its intake, allowing for the kinetics of its excretion from the body to be characterized, as is further reported in Chapters 10 and 11, respectively.

In terms of sampling, most online analytical platforms have tailored systems that comprise an orifice or tubing to which a mouthpiece is attached for breath sampling, with the downstream end directly connected to the sampling inlet of the analyzer. The exact configuration and sampling flow are instrument-specific, but common features are a heated inlet system and the use of inert sampling lines, such as Teflon PFA (perfluoroalkoxy) or PTFE (polyfluorotetraethylene) tubing, polyether ether ketone

(PEEK) capillaries, or passivated/deactivated stainless steel tubing. Both of these features (heating and use of inert materials) are important to ensure efficient and quick transfer of highly volatile and labile compounds, both polar and nonpolar, to the detection system by minimizing bleeding, carry-over, and losses to wetted surfaces. A secondary advantage of using inert materials is that they typically do not emit compounds that could contaminate the sample gas and confound the analytical data.

A further beneficial feature of many online systems is the possibility to perform side-stream sampling, whereby sample gas can be extracted continuously and in relatively small amounts from a primary flow, i.e., exhaled breath, with only minimal or no perturbation. This means that exhaled breath can be sampled without flow restrictions that could otherwise cause potential changes in volatile profiles associated with physiology (see Section 2.2.6). Additionally, the capability of many online analytical platforms for breath-resolved detection obviates the need for controlled sampling to separate phases of exhalation. While some devices incorporate a capnometer into their breath sampling interfaces, such as in secondary electrospray ionization (see Chapter 12), others exploit signals from endogenous compounds, for instance, acetone (or CO_2, if detectable) or water vapor, such as in PTR-MS and SIFT-MS (see Chapters 9 and 10, respectively) to identify exhalation phases; these signals are then utilized in postprocessing to select representative data. One device for online sampling that is worthy of mention here is the buffered end-tidal (BET) sampler, which was developed for PTR-MS but is compatible with other analyzers.[23] This sampling interface extends the time for analysis of the end-tidal phase of exhalation (as depicted in Fig. 10.2) and improves signal stability; this device is further discussed in Chapter 10.

2.2.6 Influence of physiological parameters

Physiological changes during sampling, for example, associated with hemodynamic effects, breathing patterns, or respiratory flow, may cause significant changes in the profiles of exhaled VOCs. It is therefore imperative that the potential influence of physiological parameters is taken into account during breath analysis to avoid these factors overriding the biochemical, metabolic, or pathological effects of interest in exhaled volatile profiles. Pulmonary ventilation and perfusion influence alveolar gas exchange, thus also affect VOC concentrations in exhaled breath. Indeed, simple changes in physiology can have sudden and pronounced effects on breath VOC profiles. Recent studies have demonstrated that hemodynamic changes induced by different breathing patterns or body postures alter breath profiles.[24,25] Specifically, isoprene concentrations have been obversed to change immediately after altering body positions, whereas acetone was not affected.[24]

Similarly, changes in ventilation patterns, such as hyper- or hypoventilation, or ventilation maneuvers, such as deep inhalations or forced exhalations, initiate an immediate change in breath VOC patterns. Breath holding, for example, can induce a two- to threefold increase in the concentrations of certain volatiles VOCs.[25] Breathing against resistance also leads to altered VOC patterns in breath, so forced breathing or breathing against resistance, e.g., blowing through tubing of narrow diameter, or filling bags via a small orifice, must be avoided during sampling.[26] Side-stream sampling can overcome this limitation. Further, implementing control mechanisms, such as paced breathing from visual or audible clues, can help to reduce interbreath variability.

Other important aspects of sampling are oral versus nasal sampling, which has been studied extensively, primarily in relation to ammonia from the oral cavity.[27] In addition to exploring its oral origins, studies have highlighted the case of ammonia to raise concern about the potential for compounds that primarily originate in the oral cavity to influence (and confound) mouth-exhaled breath samples that are of systemic interest (see Chapter 9). Finally, a recent study explored whether selected breathborne compounds exhibit a circadian rhythm in their exhaled concentrations and this was found to be the case for several compounds, including isoprene and F_ENO in asthma sufferers.[28]

2.2.7 Extraneous confounders

Breath analyses to date have sought primarily to discover endogenous biomarkers of disease. In this context, exogenous compounds represent confounding features in a breath profile. More recently, the interest in exploring exogenous compounds has grown, such as in human exposome research (Chapter 1), drugs studies (Chapter 30), occupational exposures (Chapter 32), and exposure studies on environmental stressors (Chapter 33). Nevertheless, external influences on breath samples must be taken into consideration when endogenous disease biomarkers are targeted, especially in highly contaminated environments, such as in the clinical settings where gas-phase compounds associated with disinfectants, detergents, and emissions from plastics and one-way materials are typically present at high concentrations.[29]

In addition to errors introduced during sampling or profile shifts from physiological changes, as discussed above, there are many extraneous factors that can affect the volatile composition of exhaled breath. These have been reviewed in the literature,[30] but in general, xenobiotic compounds can derive from ambient air, diet, drugs and medication, and other environmental stressors. Taking smoking as an example, the compound acetonitrile is a marker for biomass burning and its presence in breath is highly correlated with tobacco smoking activity, akin to being a breathborne marker to identify cigarette smokers. High levels of ethanol in breath is indicative of alcohol intake, which can often be detected without instrumentation when smelled on the breath (see Chapter 29). These are two relatively unambiguous and distinct examples, but a large number of the many hundreds of VOCs in exhaled breath certainly are of exogenous origin.

The challenge lies in teasing out which compounds are systemic and which are xenobiotic. There is no simple solution to achieve this, but requesting and recording information on recent meals, medication, and other activities might offer clues as to the presence of some compounds in a breath profile. Another strategy that is increasingly used in clinical breath studies is to recruit patients' household partners or family members as controls, whereby the intention here is to attain samples that are matched in terms of inhalation exposure when cohabiting, and dietary exposure through shared meals, among other factors. Further, there is a growing focus on longitudinal studies in which an individual has his/her own control over time in view of monitoring disease progression or the efficacy of therapeutic interventions. This presents a novel situation in which transient exposures that result in a temporary perturbation in the exhaled breath profile through the presence of exogenous compounds might no longer pose a problem because this will be easily identifiable as an outlier event. Compare this situation with the case of a resident hospital patient who supplies just one sample of breath for a dataset on a particular disease; any extraneous factors that have influenced the breath profile on that particular day would not be so easily identifiable as a red herring. At the very least, a thorough control of the acutely inspired fraction is of critical importance for robust sampling and data analysis in (clinical) breath analysis in order to isolate environmental confounders.

2.3 Standardization

Breath analysis is multifaceted, so standardization efforts must cater for this. As mentioned in the overview of this chapter, the diversity of breath research in terms of targeted disease and compound(s), and analytical technique with the associated sampling procedure, makes it impossible to establish a single set of parameters that is applicable to all breath analysis endeavors.[31] Standardization in breath research currently does not exist, although guidelines on best practices have been formulated for specific applications, such as F_ENO and EBC (see Chapters 4, 5, and 7, respectively). Recommendations on other areas and aspects of breath research have been made in this book, for example, in relation to canine olfaction (Chapter 34) or clinical study design (Chapter 35).

In general, standardization efforts in breath research require a compartmentalized—or stepwise—approach, as has been put forward previously.[32] In this proposed framework, each aspect of a breath study, from the design through to the data evaluation, is treated as a discrete entity, as depicted in Fig. 2.3. By establishing best practices within each step, interstudy comparisons of equivalent aspects can be made more easily.

In this proposed strategy, *standard parameters*, *specifications*, and *practices* are generated within the community by researchers, engineers, and physicians working on similar tasks. Within this framework, the *standard parameters* are measurable quantities, such as limit of detection (LOD) or sensitivity of the analyzer. The *standard specifications* are quantities that characterize a standard parameter, which is either widely applicable or compound-specific; in the case of LOD, the specification might be, say, 5 ppt for acetone. The *standard practices* represent best practices that fulfill all specifications, such as the procedure of capnography for controlled end-tidal sampling.

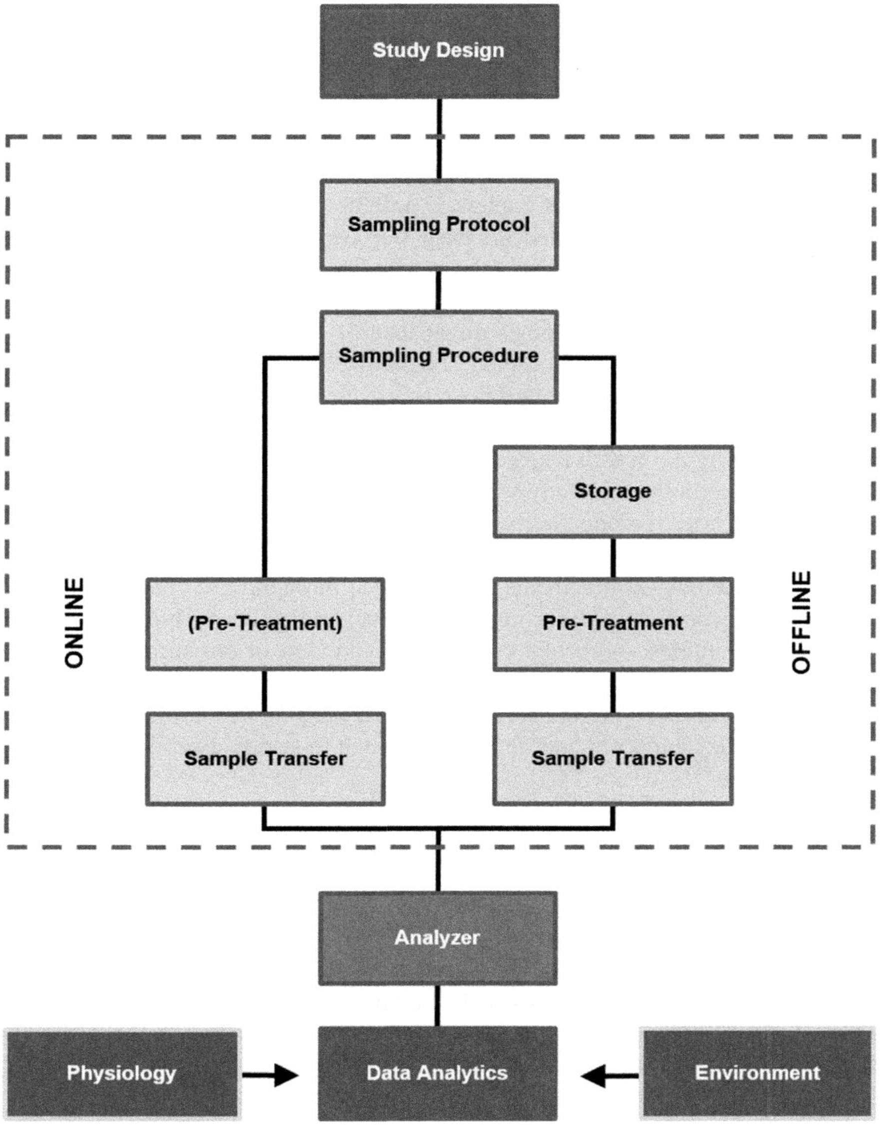

FIGURE 2.3

Flow diagram of compartmentalized steps within a breath study workflow, whereby standardized practices are established within each step to offer better interstudy reproducibility and data comparability.

Modified from [32].

Returning to the flow diagram of Fig. 2.3, each step (box) is treated independently, with related parameters, specifications, and practices defined and subsequently adhered to in practice. Clearly, each of these steps can be divided further into subsections, for instance, in relation to different aspects of study design (target population, cohorts, constraints, etc.; see Chapter 35), sampling procedure (exhalation phase, controlled sampling, etc.), analyzer (operating parameters, sensitivity, LOD, etc.), data analytics (see Chapters 37 and 38), and so forth. When the respective standards have been defined for each step (or substep), these can be implemented by independent research laboratories to reduce variability and ensure comparability for equivalent steps in breath research studies. Any discrepancies that arise in the resulting outcomes might then be traced back to individual steps that were not aligned with the accepted, standardized procedures. The whole process is iterative, whereby newly suggested standards that demonstrate superiority over existing standards replace them. This strategy is geared toward self-definition and self-regulation by the breath research community. Over time, individual aspects of processes along the breath study workflow will align, leading to lower variability and therefore better reproducibility in outcomes. Further, by establishing a framework with dynamic content, rather than a fixed set of rules, this allows for innovation in the field, whereby developments are not stifled through rigid constraints.

This framework concept for introducing standardization in breath research is slowly gaining interest. A current ongoing study in view of this strategy is working toward establishing a benchmark for sampling techniques and analytical platforms, as partly reported in the literature,[33] as well as in Chapter 11. Further details of this compartmentalized concept and iterative approach to standardization can be found in the literature.[32]

2.4 Summary

The heterogeneity of exhaled breath underlines the importance of well-documented and controlled sampling procedures, with the necessity for specific sampling strategies adapted for targeted diseases and/or compounds. Exhalation is a mixture of different phases, which is mirrored by different VOC patterns in each phase. For bloodborne substances, sampling the end-tidal volume, which most closely reflects gas at the alveoli, is of primary interest, and it is recommended that sampling be undertaken with appropriate control, such as capnography, or with other parameters that identify this phase in the resulting breath profile data. For substances generated mainly in the upper airways, modified settings may be reasonable.

On top of the complex mixture of breath, and the variety of diseases or environment stressors that are targeted, there exist manifold procedures to sample breath. Sampling can be done with direct, immediate measurement, as is the case in online analysis, or by collection and subsequent transfer to the analyzer, potentially including storage and/or preconcentration, as is conducted in offline analysis. Various methods are available to address the latter, with storage of volatiles either in the gas-phase or on sorbent materials—also of diverse nature and differing configurations—that can affect outcomes. Adapted sampling strategies should take into account effects of physiology and potential (exogenous) confounders, such as acute or previous environmental loads or medications.

The analyzer itself can equally have an impact on outcomes of breath studies. Limits of detection, compound-dependent sensitivity, target specificity, and system drift, among other factors, can all affect data quality to varying degrees. In contrast to laboratory-based analytical methods, volatile breath profile data from fast online methods operating at sub-Hertz frequency detection can be segmented postprocessing to select relevant phases, such as end-tidal volume or inspiratory phases.

A proposed strategy to cope with the multifarious aspects of breath research in view of introducing standardization is to establish a framework that encompasses different procedural steps in a breath study workflow. This approach is geared toward optimizing self-contained procedures in an iterative process. By compartmentalizing individual aspects and treating these specifically, it enables each step or substep to be applied appropriately, according to the study target. This stepwise approach also allows additional steps to be introduced (or removed) in relation to the needs of a particular endeavor and permits further innovation in the field, in terms of sampling approaches, instrumental improvements and developments, and data analytics strategies.

The successful outcome of any breath research pursuit primarily rests on the quality of the samples collected; a set of compromised samples will lead to misinterpretation of the ensuing data. Great technological improvements have been made in the field in terms of sampling technologies, analytical instrumentation, and data processing algorithms in recent years, as outlined in the different chapters of this book. The breath analysis field is slowly aligning procedures through transparency and exchange; as consensus on sampling strategies become established, more discoveries on breathborne biomarkers are inevitable.

References

1. Schubert JK, Esteban-Loos I, Geiger K, Guttmann J. In vivo evaluation of a new method for chemical analysis of volatile components in the respiratory gas of mechanically ventilated patients. *Technol Health Care* 1999;**7**(1):29–37.
2. Schubert JK, Spittler K-H, Braun G, Geiger K, Guttmann J. CO_2-controlled sampling of alveolar gas in mechanically ventilated patients. *J Appl Physiol* 2001;**90**(2):486–92.
3. Miekisch W, Kischkel S, Sawacki A, Liebau T, Mieth M, Schubert JK. Impact of sampling procedures on the results of breath analysis. *J Breath Res* 2008;**2**:026007.
4. Larstad MA, Toren K, Bake B, Olin AC. Determination of ethane, pentane and isoprene in exhaled air—effects of breath-holding, flow rate and purified air. *Acta Physiol* 2007; **189**(1):87–98.
5. Risby TH, Sehnert SS. Clinical application of breath biomarkers of oxidative stress status. *Free Radic Biol Med* 1999;**27**(11–12):1182–92.

6. Pleil JD, Lindstrom AB. Collection of a single alveolar exhaled breath for volatile organic-compounds analysis. *Am J Ind Med* 1995;**28**(1):109–21.
7. Sanchez JM, Sacks RD. Development of a multibed sorption trap, comprehensive two-dimensional gas chromatography, and time-of-flight mass spectrometry system for the analysis of volatile organic compounds in human breath. *Anal Chem* 2006;**78**(9):3046–54.
8. Mochalski P, Wzorek B, Sliwka I, Amann A. Suitability of different polymer bags for storage of volatile sulphur compounds relevant to breath analysis. *J Chromatogr B Analyt Technol Biomed Life Sci* 2009;**877**(3):189–96.
9. Beauchamp J, Herbig J, Gutmann R, Hansel A. On the use of Tedlar bags for breath-gas sampling and analysis. *J Breath Res* 2008;**2**(4):046001.
10. Steeghs MML, Cristescu SM, Harren FJM. The suitability of Tedlar bags for breath sampling in medical diagnostic research. *Physiol Meas* 2007;**28**(1):73.
11. Miekisch W, Herbig J, Schubert JK. Data interpretation in breath biomarker research: pitfalls and directions. *J Breath Res* 2012;**6**(3):036007.
12. Miekisch W, Trefz P, Bergmann A, Schubert JK. Microextraction techniques in breath biomarker analysis. *Bioanalysis* 2014;**6**(9):1275–91.
13. Harshman SW, Mani N, Geier BA, Kwak J, Shepard P, Fan M, et al. Storage stability of exhaled breath on Tenax TA. *J Breath Res* 2016;**10**(4):046008.
14. Kang S, Thomas CLP. How long may a breath sample be stored for at −80°C? A study of the stability of volatile organic compounds trapped onto a mixed Tenax:Carbograph trap adsorbent bed from exhaled breath. *J Breath Res* 2016;**10**(2):026011.
15. Doran SLF, Romano A, Hanna GB. Optimisation of sampling parameters for standardised exhaled breath sampling. *J Breath Res* 2017;**12**(1):016007.
16. Harshman SW, Pitsch RL, Davidson CN, Scott AM, Hill EM, Smith ZK, et al. Characterization of standardized breath sampling for off-line field use. *J Breath Res* 2019;**14**(1):016009.
17. Lord HL, Zhan W, Pawliszyn J. Fundamentals and applications of needle trap devices: a critical review. *Anal Chim Acta* 2010;**677**(1):3–18.
18. Mieth M, Kischkel S, Schubert JK, Hein D, Miekisch W. Multibed needle trap devices for on site sampling and preconcentration of volatile breath biomarkers. *Anal Chem* 2009;**81**(14):5851–7.
19. Trefz P, Kischkel S, Hein D, James ES, Schubert JK, Miekisch W. Needle trap micro-extraction for VOC analysis: effects of packing materials and desorption parameters. *J Chromatogr A* 2012;**1219**:29–38.
20. Mieth M, Schubert JK, Groger T, Sabel B, Kischkel S, Fuchs P, et al. Automated needle trap heart-cut GC/MS and needle trap comprehensive two-dimensional GC/TOF-MS for breath gas analysis in the clinical environment. *Anal Chem* 2010;**82**(6):2541–51.
21. Trefz P, Rösner L, Hein D, Schubert JK, Miekisch W. Evaluation of needle trap micro-extraction and automatic alveolar sampling for point-of-care breath analysis. *Anal Bioanal Chem* 2013;**405**(10):3105–15.
22. Smith D, Španěl P, Herbig J, Beauchamp J. Mass spectrometry for real-time quantitative breath analysis. *J Breath Res* 2014;**8**(2):027101.
23. Herbig J, Titzmann T, Beauchamp J, Kohl I, Hansel A. Buffered end-tidal (BET) sampling-a novel method for real-time breath-gas analysis. *J Breath Res* 2008;**2**(3):037008.
24. Sukul P, Trefz P, Kamysek S, Schubert JK, Miekisch W. Instant effects of changing body positions on compositions of exhaled breath. *J Breath Res* 2015;**9**(4):047105.

25. Sukul P, Trefz P, Schubert JK, Miekisch W. Immediate effects of breath holding maneuvers onto composition of exhaled breath. *J Breath Res* 2014;**8**(3):037102.
26. Sukul P, Schubert JK, Kamysek S, Trefz P, Miekisch W. Applied upper-airway resistance instantly affects breath components: a unique insight into pulmonary medicine. *J Breath Res* 2017;**11**(4):047108.
27. Chen W, Metsälä M, Vaittinen O, Halonen L. The origin of mouth-exhaled ammonia. *J Breath Res* 2014;**8**(3):036003.
28. Wilkinson M, Maidstone R, Loudon A, Blaikley J, White IR, Singh D, et al. Circadian rhythm of exhaled biomarkers in health and asthma. *Eur Respir J* 2019;**54**(4):1901068.
29. Kischkel S, Miekisch W, Sawacki A, Straker EM, Trefz P, Amann A, et al. Breath biomarkers for lung cancer detection and assessment of smoking related effects — confounding variables, influence of normalization and statistical algorithms. *Clin Chim Acta* 2010;**411**(21):1637–44.
30. Beauchamp J. Inhaled today, not gone tomorrow: pharmacokinetics and environmental exposure of volatiles in exhaled breath. *J Breath Res* 2011;**5**(3):037103.
31. Beauchamp JD, Pleil JD. Simply breath-taking? Developing a strategy for consistent breath sampling. *J Breath Res* 2013;**7**(4):042001.
32. Herbig J, Beauchamp J. Towards standardization in the analysis of breath gas volatiles. *J Breath Res* 2014;**8**(3):037101.
33. Malásková M, Henderson B, Chellayah PD, Ruzsanyi V, Mochalski P, Cristescu SM, et al. Proton transfer reaction time-of-flight mass spectrometric measurements of volatile compounds contained in peppermint oil capsules of relevance to real-time pharmacokinetic breath studies. *J Breath Res* 2019;**13**(4):046009.

CHAPTER

Physiological modeling of exhaled compounds

3

Julian King[1], Paweł Mochalski[1,2], Gerald Teschl[3], Susanne Teschl[4], Christopher A. Mayhew[1,5], Ramin Ghorbani[6], Florian M. Schmidt[6], Karl Unterkofler[1,7]

[1]*Institute for Breath Research, University of Innsbruck, Dornbirn, Austria;* [2]*Institute of Chemistry, Jan Kochanowski University, Kielce, Poland;* [3]*Faculty of Mathematics, University of Vienna, Vienna, Austria;* [4]*University of Applied Sciences Technikum Wien, Vienna, Austria;* [5]*Molecular Physics Group, School of Physics and Astronomy, University of Birmingham, Birmingham, United Kingdom;* [6]*Department of Applied Physics and Electronics, Umeå University, Umeå, Sweden;* [7]*University of Applied Sciences Vorarlberg, Dornbirn, Austria*

Chapter outline

Breathborne Biomarkers and the Human Volatilome. https://doi.org/10.1016/B978-0-12-819967-1.00003-7
Copyright © 2020 Elsevier B.V. All rights reserved.

3.1 Overview

There is nothing more practical than a good theory. (David Hilbert[a]).

Apart from diagnostics and therapeutic monitoring, the detection of volatile organic compounds (VOCs) as biochemical probes in exhaled breath has potential use in assessing dynamics of physiological function (e.g., via cycle ergometer stress tests, in an intraoperative setting, or in a sleep laboratory), pharmacodynamics (drug testing), or quantifying body burden in response to environmental exposure (e.g., in occupational health). Furthermore, as has been demonstrated in the Nobel Prize–winning work of Furchgott, Ignarro, and Murad on the small inorganic molecule nitric oxide (NO), trace gases can actively participate in the regulation of physiological events.[1] This renders breath VOCs as intriguing measures for examining fundamental endogenous processes.

While sufficiently accurate and fast instrumental techniques are vital to exploit this huge potential, several methodological issues must be addressed before exhaled breath analysis can supplement or replace blood tests. Specifically, a crucial yet underestimated aspect of exhaled breath analysis concerns the blood–gas kinetics of the VOCs under scrutiny, as well as their systemic distribution and physiological flow within the human body. Much can be learnt in this context from the cumbersome development of the test for fraction of exhaled NO (F_ENO), which nowadays ranks among the most successful breath tests based on an endogenous compound (see the next two chapters). This paradigmatic example shows that a purely explorative search for VOC biomarkers is not sufficient: in order to make breath tests operational, a thorough quantitative understanding of the underlying gas-exchange mechanisms and their sensitivity with respect to physiological factors is required. In particular, such knowledge can then contribute to the implementation of unified sampling protocols, similar to the guidelines for determining exhaled lower respiratory tract nitric oxide and nasal nitric oxide (see next chapter).

Due to the current lack of standardized sampling recommendations in breath analysis (see Chapter 2), end-tidal breath collection is usually adopted as best sampling practice. Accordingly, the first part of an exhalation—corresponding to the gas volume of the anatomical deadspace and hence not participating in gas exchange according to classical pulmonary inert gas elimination theory—is discarded and only the last exhalation segment (considered to best reflect alveolar air) is collected for analysis (see Fig. 2.1). In sophisticated sampling systems, end-tidal breath extraction is triggered by virtue of simultaneously measured carbon dioxide (CO_2) concentrations and/or respiratory flow (see Chapter 2). While such configurations are an important contribution to establishing reproducible breath sampling, they generally give little insight into major experimental questions, such as:

[a] Also attributed to Immanuel Kant.

- What is the quantitative relationship between concentrations of VOCs in exhaled breath and their underlying endogenous (e.g., blood) concentrations?
- Does the end-tidal concentration of a VOC reflect its alveolar concentration?
- How does the variability of exhaled VOC concentrations relate to varying physiological conditions?

In this context, mathematical modeling and simulation can be employed to establish mechanistic descriptions of the gas-exchange processes governing the VOC under scrutiny. Once such a model has been developed, it becomes possible to understand the key relationships underlying the physiological behavior of the compound and to identify its defining parameters. Such a quantitative treatment provides a sound conceptual framework for experimental design and decision-making, and can aid substantially in preventing misinterpretations of empirical results.

This chapter will address the questions above by presenting models of the exhalation patterns of selected compounds, with a focus on the two most abundant, endogenous trace gases found in human breath (besides CO_2), namely:

- isoprene, which has a low solubility in blood (as reflected by a small blood:gas partition coefficient of $\lambda_{b:air} \approx 0.95$), is lipophilic and presumed to be correlated with cholesterol biosynthesis, and
- acetone, which is relatively much more soluble in blood ($\lambda_{b:air} \approx 340$) and has been linked to fat catabolism.

Isoprene and acetone can be viewed as prototypical examples of two major classes of VOCs, despite not covering the whole spectrum of physicochemical properties. Furthermore, these two compounds rank among the most notable VOCs studied in the context of breath gas analysis and thus represent natural choices for a more detailed assessment of observable trace gas behavior in exhaled breath. In addition to isoprene and acetone, models are presented and discussed for exhaled carbon monoxide (CO), methane, and the anesthetic agent sevoflurane.

3.2 Experimental basis from real-time analyses

The range of measurement techniques employed for breath gas analytical investigations is extremely diverse and each method comes with its specific strengths and weaknesses. The modeling work covered in this chapter is based predominantly on breath analyses performed via direct mass spectrometry, specifically proton transfer reaction—mass spectrometry (PTR-MS), or via optical spectroscopy in the case of CO and methane. Importantly, when the *humid* breath matrix is analyzed by PTR-MS, the abundance of water cluster reagent ions in the instrument reaction chamber is significantly elevated, leading to proton transfer or ligand switching reactions from the water cluster to the analyte molecules. As such, it is essential to accommodate for these reactions in data normalization processes to ensure an accurate model of the ensuing data; this phenomenon is discussed in Chapter 10.

The real-time capability of PTR-MS, allowing for concentration measurements of VOCs with breath-by-breath resolution (i.e., on a timescale of <0.5 s), makes this technique ideal for delivering data for modeling purposes. The possibility of generating high sample rate data can be viewed as an essential requirement for relating short-term changes in breath VOC concentrations to quick physiological variations (e.g., in blood or ventilatory flow). This can be examined experimentally, for instance, by combining real-time trace gas measurements with data streams reflecting hemodynamic and respiratory parameters.[2]

Using real-time analytical tools, end-tidal VOC concentration profiles can be examined in relation to several decisive physiological factors during rest, exercise challenges on a cycle ergometer, or in a sleep laboratory setting.[3,4] The ensuing synchronized data streams offer an indispensable phenomenological basis pertinent to any quantitative modeling approach.

3.3 Modeling theory

In classical pulmonary inert gas elimination theory, the relationship between the blood concentrations of VOCs and their respective concentrations in the gaseous phase is captured by the well-known Farhi equation.[5] This equation predicts a direct proportionality between the alveolar concentration C_A of a VOC (presumed to be accessible from end-tidal exhalation segments) and its concentrations C_a and $C_{\bar{v}}$ in arterial and mixed venous blood, respectively. Here, the associated proportionality factor depends on the substance-specific blood:gas partition coefficient $\lambda_{b:air}$ (describing the diffusion equilibrium at the alveolar–capillary interface according to Henry's law), alveolar ventilation $\dot{V}_A$ (governing the transport of the compound through the respiratory tree), and cardiac output $\dot{Q}_c$ (controlling the rate at which the VOC is delivered to the lungs), viz.,

$$C_{measured} = C_A = \lambda_{b:air} C_a = \frac{C_{\bar{v}}}{\lambda_{b:air} + \frac{\dot{V}_A}{\dot{Q}_C}} \tag{3.1}$$

An elementary but particularly useful tool for deriving such formulae is to employ the fundamental physical concept of mass balance coupled with a compartmental model design. Within a physiological framework, the compartmental approach to mass transfer traditionally consists of dividing the body into an ensemble of localized anatomic structures (like lung, liver, kidney, etc.) or roughly homogenous tissue volumes (lumping together tissue groups of a given type, such as fat, muscle, viscera, bone, etc.) in which the investigated VOC can reasonably be postulated to behave uniformly. These compartments are subsequently interconnected via the arterial and venous network, yielding a simplified representation of the human organism that reflects known or hypothesized physiology and biochemistry.[6,7] A typical example of such a model is shown in Fig. 3.1.

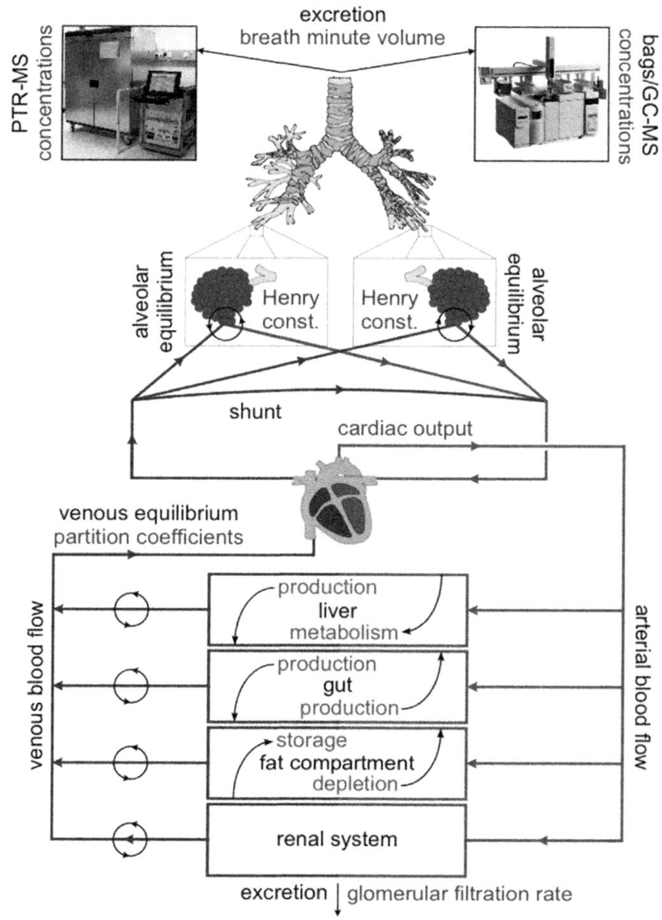

FIGURE 3.1

Physiologically-based compartment model for the distribution of an endogenous compound within the human body. The spectrum of model parameters can roughly be divided into fixed, measured, and unknown variables. The first class typically comprises quantities that have been determined a priori and are assumed to remain constant during the course of experimentation, such as approximate tissue volumes or partition coefficients. Information of this type has been compiled by several investigators.[28,33] Measured parameters include quantities that are accessible by virtue of the given experimental configuration (blood flow, ventilation), while unknown parameters (e.g., production, metabolism, storage, and depletion rates) must be extracted by calibrating the model using available experimental data (e.g., compound concentrations derived from proton transfer reaction—mass spectrometry (PTR-MS) or gas chromatography—mass spectrometry (GC-MS) analysis).

Since their introduction in the 1930s,[8] compartment models have received widespread attention in the biomedical sciences. The reason for this popularity is that the model structures bear a direct correspondence with the underlying physiology, thereby maintaining a clear interpretation of the involved parameters. Moreover, by writing down the associated mass balance relationships,[b] the underlying model structure can be expressed directly as a set of ordinary differential equations (ODEs) of the form

$$\dot{\mathbf{x}}(t) = \mathbf{g}(\mathbf{x}(t), \mathbf{u}(t), \boldsymbol{\vartheta}), \quad \mathbf{x}(t_0) = \mathbf{x}_0.$$

These equations describe the evolution of a state vector $\mathbf{x}(t)$, belonging to an open set $X \subseteq \mathbb{R}^n$, within an observation interval $[t_0, t_{\max}]$. The right side $\mathbf{g}$ usually is a continuously differentiable function with respect to its arguments and depends on a set of measured, possibly time-varying variables $\mathbf{u} \in U \subseteq \mathbb{R}^q$ that are postulated to be uniformly bounded over $[t_0, t_{\max}]$, as well as on a vector $\boldsymbol{\vartheta} \in \Theta \subseteq \mathbb{R}^p$, lumping together a set of constant, unknown model parameters. The latter must be estimated from available process data (possibly together with some components of the initial conditions $\mathbf{x}_0$). In the present setting, $\mathbf{x}$ typically reflects a vector of molecular concentrations of the investigated trace gas within the tissue compartments introduced;$\mathbf{u}$ stands for external inputs (blood/respiratory flows, temperature, etc.) that can be modified by experimentation; $\boldsymbol{\vartheta}$ includes kinetic constants, such as endogenous production or metabolic rates. Furthermore, information on the model dynamics can be collected via the (scalar) measurement equation

$$y(t) = h(\mathbf{x}(t), \boldsymbol{\vartheta}),$$

where h is some function defining the observable model output (i.e., the measurable breath concentration). Specific examples for the framework introduced above are presented in the following sections.

It is instructive to note that the Farhi Eq. (3.1) represents a simple form of such a compartment model, involving only one gas-exchange unit (the lungs) and assuming steady-state conditions throughout the measurement (i.e., the rate of gas build-up/ depletion in the lungs is 0). It is therefore not surprising that this formulation might be inadequate for explaining non–steady-state behavior, e.g., in response to relatively quick physiological transitions (for instance, during exercise). Indeed, the profiles of isoprene and acetone have been shown to considerably depart from the trend predicted by Eq. (3.1).[2] More specifically, at the onset of exercise, the ventilation–perfusion ratio $\dot{V}_A/\dot{Q}_c$ increases drastically and hence, other factors being equal, a more or less pronounced drop in the respective breath concentrations could be expected[c].

---

[b] Stating that the rate at which the compartmental amount of a VOC changes per unit time equals the rate of mass transfer into the compartment less the rate of removal from the compartment.

[c] Such a drop can in fact be observed for some endogenous breath VOCs, e.g., butane (cf. Fig. 3.2 in Ref. 12).

As will be illustrated in the following, in the case of isoprene this discrepancy can be explained by reference to a peripheral (extrahepatic) production of the compound, which is realized by complementing the basic Farhi model with two systemic compartments affecting the time evolution of the mixed venous concentration $C_{\bar{v}}$ according to their fractional perfusion. For acetone, a mechanistic description of exhaled profiles can be achieved by taking into account pre- and postalveolar gas exchange,[2] which amounts to dividing the pulmonary compartment into two parts, the alveolar tract and the conducting airways.

3.4 Isoprene

Due to its volatility and low affinity for water and blood, exchange of endogenous isoprene occurs mainly in the alveolar region. Physical activity causes marked changes in exhaled breath isoprene concentrations of humans.[2,9,10] Extensive real-time measurements during moderate workload ergometer challenges have demonstrated an exercise-induced increase of breath isoprene concentrations by a factor of 2–3 in mixed exhaled breath,[10] and by a factor of 3–4 in end-tidal breath.[2] While this behavior has been explained in reference to a delay in the rise of ventilation with respect to the rise in perfusion at the start of exercise (i.e., by assuming a temporary drop in the average pulmonary ventilation–perfusion ratio), recent evidence points toward a substance-specific washout from an isoprene buffer tissue within the human body.

3.4.1 Experiments to examine endogenous isoprene

Exercise bouts at constant workload interrupted by breaks of variable duration lead to markedly different heights of the characteristic breath isoprene concentration peak at the onset of pedaling, despite an almost identical behavior of cardiac output and alveolar ventilation throughout all workload segments.[2] Additional information can be distilled from such profiles by performing one-legged ergometer experiments at moderate workloads,[11] which appear to favor the hypothesis of a peripheral tissue source affecting breath isoprene output. More specifically, a switch of the working leg after a short break following an ergometer exercise segment at 50 W results in an immediate recovery of the initial peak height, whereas continuing the exercise with the same leg leads to a washout effect similar to the two-legged case.

Experiments carried out by stimulating muscle groups with specific weights provide indications of the tissues involved in this behavior.[13] Breath isoprene concentration profiles during repeated arm-crank exercises with intermediate pauses qualitatively resemble the trends observed during ergometer challenges. Compared to two-legged ergometer exercise, however, arm-crank exercise induces much lower peak isoprene concentrations, despite similar time courses of the ventilation–perfusion ratio within both exercise regimes. From these experiments it might hence be concluded that a major part of isoprene variability during exercise phases can be ascribed to an increased fractional perfusion of the working locomotor muscles, eventually leading to higher isoprene concentrations in mixed venous blood at the onset of physical activity. This rationale is in accordance with the predominant physiological role of working muscles during exercise. Collectively, the skeletal muscles constitute up to 40%–45% of body weight, which is more than any other single organ. At rest, about 10%–15% of cardiac output is distributed to skeletal muscle, while during strenuous exercise the latter may receive more than 80% of the total blood flow. This underlines the importance of skeletal muscle as one of the predominant factors in systemic hemodynamics.[14]

3.4.2 A physiological model for breath isoprene dynamics

In light of the experiments outlined above, the mathematical model of isoprene distribution presented here adopts the assumption of a peripheral source of isoprene in the body. Following the methodological steps described at the beginning of Section 3.3, the lung is modeled by a single alveolar unit and the systemic part consists of two homogenous compartments: a richly perfused tissue compartment (including intestines, liver, brain, connective muscles, and skin) and a peripheral tissue compartment. The model structure is sketched in Fig. 3.2. For the associated compartmental mass balance equations, and model validation and estimation, the reader is referred to the original publication.[11] In order to correlate the model output to experimental outcomes, it is postulated that the measurable breath isoprene concentration corresponds to the alveolar level C_A.

While production and metabolic elimination of isoprene occurs in both systemic compartments, peripheral tissue takes the role of an isoprene buffer and is postulated to contain the working muscle groups, which receive a disproportionally high fraction of the systemic blood flow during exercise. According to the simulation in Fig. 3.3, at rest the peripheral compartment is characterized by high isoprene concentrations resulting from extrahepatic production at a constant rate. Due to the relatively small fractional blood flow for these tissues, however, the mixed venous concentrations are mainly governed by the lower isoprene content of the venous blood returning from the richly perfused tissue group. At the start of exercise, the fractional perfusion q_{per} in the periphery increases and consequently the mixed venous concentrations become dominated by peripheral venous return. The isoprene concentration peak visible in breath can thus be interpreted as a direct consequence of the associated increase in mixed venous blood concentrations.

3.4.3 Practical implications of isoprene model

Dynamic isoprene breath profiles corroborate the view that the breath output of certain VOCs can vary quickly with the physiological status of an individual. While a high sensitivity with respect to hemodynamics, respiration, food intake, etc., imposes strict regulations on the breath sampling protocol for such compounds, it can also provide further insights into their metabolic and physiological fate within

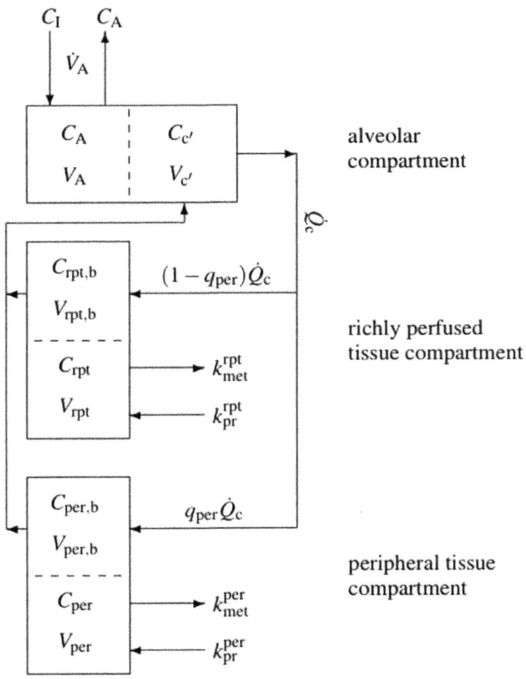

FIGURE 3.2

Schematic representation of the model structure used to describe the dynamics of isoprene concentrations (C) in various parts of the human body. The latter is divided into three distinct functional units (with volumes V): alveolar/end-capillary compartment (gas exchange), richly perfused tissue (metabolism and production), and peripheral tissue (storage, metabolism, and production). Fractional blood flow to the periphery is denoted by q_{per}, while k_{pr} and k_{met} represent constant production and (linear) metabolic rates, respectively. Dashed boundaries indicate a diffusion equilibrium. Subscripts denote the following: A, alveolar; b, blood; c', end-capillary; I, inhaled; per, peripheral tissue; rpt, richly perfused tissue.

Reproduced from. [11]

the human body. In particular, for VOCs stemming from a specific endogenous (tissue) source, a change in the respective breath output should be observable if that source is activated, stimulated, or depressed. Using an analogous line of argument, exercise scenarios on an ergometer provide data that support the idea that some peripheral (muscle) tissue may play a role in isoprene synthesis. Recently, this viewpoint has been further strengthened by observing drastically reduced breath and blood isoprene levels in late-stage Duchenne muscle dystrophy patients.[15] Additionally, recent data show that primary human muscle cells possess the same repertoire for cholesterol biosynthesis as liver cells, suggesting that skeletal muscle can contribute markedly to endogenous isoprene production.[36] While these findings

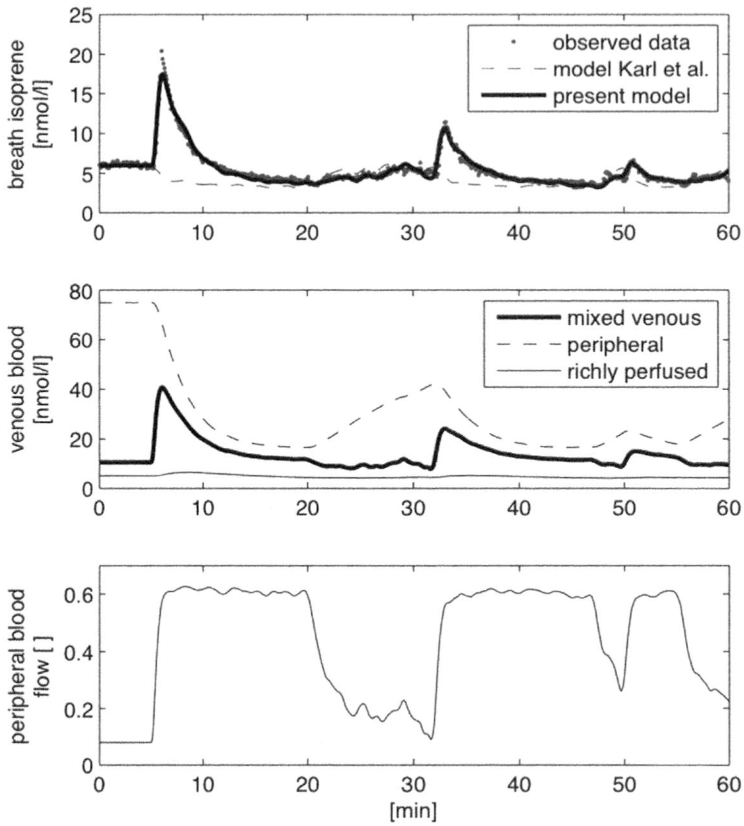

FIGURE 3.3

Top panel: model response fitted to the isoprene breath data in Ref. 11, Protocol A. Center panel: predicted concentrations in mixed venous blood and venous blood returning from the peripheral and richly perfused tissue groups. Bottom panel: predicted profile of fractional blood flow q_{per} to peripheral tissue.

Reproduced from. [11]

are in contrast to previous work postulating a purely hepatic origin of isoprene,[16] they provide interesting new perspectives on the age dependence of isoprene output,[17] as well as its reduction under statin therapy.[9] Furthermore, the modeling study presented here opens up the possibility for a mechanistic interpretation of future experimental findings and challenges new hypotheses.

3.4.4 Extending the isoprene model with two compartments

The classical Farhi equation is derived from a single compartment model for the lung, which is assumed to be in equilibrium with an abstract body compartment lacking any explicit production or metabolism of VOCs. This can be extended using a simple dynamic two-compartment model comprising body and lung that additionally takes the inhaled VOC concentrations into account.[18] This model connects the exhaled breath concentration of systemic VOCs with small blood:air partition coefficients with physiological parameters, such as endogenous production rates and metabolic rates. The validity of this two-compartment model has been tested with data obtained for isoprene and inhaled deuterated isoprene-D5.

The two-compartment model can be extended further using a three-compartment model, which retains the body compartment but splits the lung into the upper and lower airways.[19] This model takes into account the influence of the upper airways on exhaled concentrations of VOCs with higher blood:air partition coefficients, such as acetone or ethanol. For such VOCs the classical Farhi equation is not valid due to airway gas-exchange mechanisms and the fact that exhaled breath gas concentrations do not directly reflect alveolar concentrations.

3.5 Acetone

In contrast to isoprene, acetone is both highly blood and water soluble. Hydrophilic compounds like acetone have been demonstrated to interact with the water-like mucus membrane lining the conductive airways, an effect which has become known as wash-in/wash-out behavior.[20] Briefly, the term wash-in/wash-out refers to the fact that while "fresh" air is being inhaled, it becomes enriched with hydrophilic VOCs stored in the airway surface walls of the peripheral bronchial tract, thus leading to a decrease of the respective pressure/tension gradient between the gas phase and capillary blood in the alveolar space. This causes an effective reduction of the driving force for gas exchange in the alveoli and minimizes the unloading of the capillary VOC level. Correspondingly, during exhalation the aforementioned diffusion process is reversed, with a certain amount of hydrophilic VOCs being stripped from the gas stream and redepositing onto the previously depleted mucus layer. As a phenomenological consequence, exhaled breath concentrations of such VOCs tend to be diminished on their way from the deeper respiratory tract to the airway opening. The resulting discrepancies between the "true" alveolar and measured breath concentrations can be substantial (even if breath samples are drawn in a strictly standardized manner, e.g., by employing CO_2-controlled sampling) and will depend on a variety of factors, such as airway temperature profiles, airway perfusion, and volumetric flow.

3.5.1 Experiments to examine endogenous acetone

As in the case of isoprene, end-tidal acetone breath profiles during moderate ergometer challenges contradict the qualitative behavior anticipated from classical pulmonary inert gas elimination theory. End-tidal acetone levels in response to exercise generally resemble the profile of ventilatory flow, showing abrupt increases and decreases in the range of 10%–40% at the onsets and ends of the individual workload periods, respectively.[2] Similarly, during hyperventilation breath acetone levels tend to increase, whereas a (slight) decrease might be expected from Eq. (3.1) (due to a drastic increase in the ventilation–perfusion ratio). Further evidence for the inadequacy of classical pulmonary inert gas elimination theory with respect to describing the blood–breath relationships of hydrophilic compounds can be derived from isothermal rebreathing experiments.[21] The heuristic intention leading to isothermal rebreathing is to create an experimental situation diminishing the influence of the wash-in/wash-out mechanisms discussed above. This can be accomplished by "closing the respiratory loop," i.e., by continuous reinspiration and reexhalation of a fixed mass of air from a rebreathing receptacle (e.g., a sampling bag), causing the air stream to equilibrate with the mucosa linings over the entire respiratory cycle.[22] Additionally, warming the rebreathing volume to body temperature will ensure a similar solubility of these VOCs in the regions of the pulmonary tract, alveoli, and airways. End-tidal breath profiles of highly water soluble VOCs, such as acetone and methanol, may increase by up to 50% during isothermal rebreathing,[21] until they approach a plateau value that corresponds to the underlying alveolar concentration. Again, this behavior sharply contrasts with the formalism underlying Eq. (3.1).

3.5.2 A physiological model for breath acetone dynamics

The experimental findings described above strongly suggest that airway gas exchange should be taken into account as a major determinant affecting the breath levels of hydrophilic VOCs. From a modeling viewpoint, this can be achieved by dividing the pulmonary tract into two separate compartments: an airway (bronchial) space, and the alveolar space. The former comprises both a gas phase and a mucus membrane, which is assumed to have the physical properties of water and acts as a reservoir for acetone.[23,24] Part of the acetone dissolved in this layer is transferred to the bronchial circulation, whereby a major fraction of the associated venous drainage is postulated to join the pulmonary veins via the postcapillary anastomoses.[25] The alveolar region itself is again represented by a single homogeneous alveolar unit. Similar to Section 3.4, the systemic part of the model consists of two functional units: a liver compartment, where acetone is endogenously produced and metabolized, and a passive tissue compartment, representing an effective storage volume. The latter basically lumps together tissue groups with similar blood:tissue partition coefficients $\lambda_{b:tissue} \approx 1.38$, such as richly perfused tissue, muscles, and skin.[24,26] A schematic representation of the model structure is presented in Fig. 3.4. For the associated compartmental mass balance equations, the reader is referred to the original publication.[27] In order to correlate the model output to experimental outcomes, it is postulated that the measurable breath acetone concentration corresponds to the bronchial level C_{bro}.

A direct implication of the wash-in/wash-out mechanisms discussed at the beginning of this section is that during free tidal breathing at rest, the amount of hydrophilic VOCs reaching the alveolar region with each inhalation is approximately equal to the amount extracted with each exhalation. In other words, under these

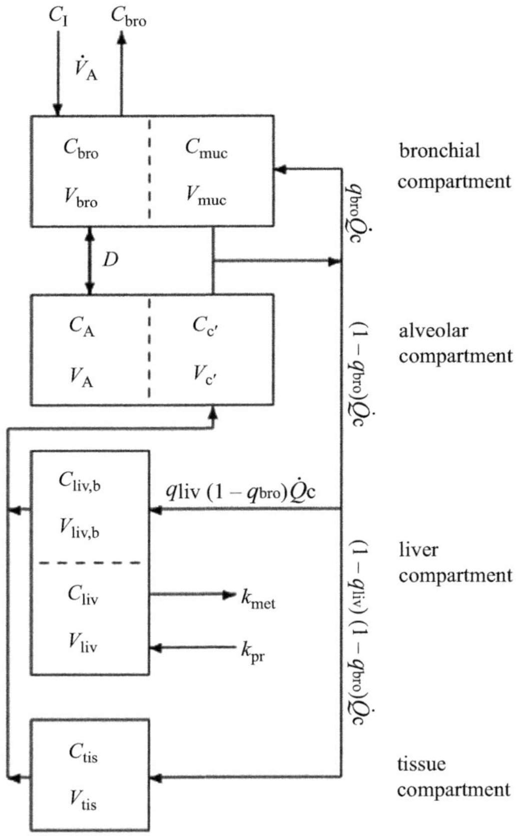

FIGURE 3.4

Schematic representation of the model structure used for describing the dynamics of acetone concentrations (C) in various parts of the human body. The latter is divided into four distinct functional units (with volumes V): bronchial/mucosal compartment (gas exchange), alveolar/end-capillary compartment (gas exchange), liver (metabolism and production), and tissue (storage). Fractional blood flows are denoted by q, while k_{pr} and k_{met} represent constant production and (linear) metabolic rates, respectively. Dashed boundaries indicate a diffusion equilibrium. Subscripts denote as follows: A, alveolar; b, blood; bro, bronchial; c', end-capillary; I, inhaled; liv, liver; muc, mucosal; tis, tissue.

Reproduced from. [27]

circumstances the bronchioalveolar net exchange of such compounds over one breathing cycle will be close to zero and a major part of the overall pulmonary gas exchange (e.g., up to 96% as in the case of acetone[26]) can be ascribed to the conducting airways. As a consequence, a concentration gradient develops along the respiratory tree, with higher levels of the respective VOC found in the proximal (alveolar) parts. Increasing ventilatory flow rates and volume minimizes equilibration times with the more distal mucus layers and hence counteracts the saturation and depletion of the gas stream during inhalation and exhalation, respectively. Consequently, the location of gas exchange is continuously shifted toward the alveolar region and the contribution of the latter becomes more pronounced. In particular, the end-tidal (measured) breath concentration approaches values closer to the alveolar level.

This rationale can be incorporated into the model by including a diffusion term of the form $D(C_{\text{bro}}-C_{\text{A}})$ into the underlying mass balance equations. This expression reflects the net gas exchange between the alveolar and the bronchial tract over a complete breathing cycle (i.e., the conductance parameter D is close to zero during free tidal breathing at rest and increases with ventilatory flow).

The relationships outlined here are summarized in Fig. 3.5, which shows a calibrated version of the model with the observable model output (i.e., the bronchial concentration) fitted to the experimental data of Ref. 27, Protocol B. It is interesting to note that the alveolar acetone concentration remains relatively constant throughout the entire experiment, i.e., the short-term variability observable in breath acetone concentrations can almost exclusively be attributed to airway gas exchange. Similarly, only minute changes in the underlying blood and tissue concentrations are expected to occur.

For the sake of completeness, it should be mentioned that another important factor affecting the gas exchange of hydrophilic VOCs is airway temperature. In particular, temperature changes will affect the solubility of acetone and similar compounds in the mucus surface of the respiratory tree.[20] For example, the local solubility of acetone in the mucus layer increases from $\lambda_{\text{muc:air}}(37°\text{C}) \approx 392$ in the lower respiratory tract to $\lambda_{\text{muc:air}}(32°\text{C}) \approx 498$ at the mouth, thereby accentuating the *washout* of gas stream acetone during the exhalation phase, cf. Section 3.5.1. Warming the airways to body core temperature can minimize this effect by ensuring a uniform solubility distribution throughout the airways. This is consistent with an increase of the measured breath concentrations during isothermal rebreathing experiments.[21,22] For further details on how to incorporate explicit temperature dependences into the mechanistic model presented above, the reader is referred to Ref. 27, which also further reports on the validation of the model, e.g., during exposure scenarios.

3.5.3 Practical implications of acetone model

The major conclusion that can be drawn from quantitatively modeling exhalation dynamics of hydrophilic VOCs is that for this class of compounds the measured end-tidal breath concentrations and the underlying alveolar levels must not be taken as identical. For perspective, the experimental evidence outlined above indicates that for acetone during free tidal breathing at rest the associated values can differ by a factor of more than 1.5. In particular, this implies that multiplying end-tidal breath concentrations with the substance-specific in vitro blood:gas partition coefficient $\lambda_{\text{b:air}}$ will generally result in an underestimation of the true arterial blood

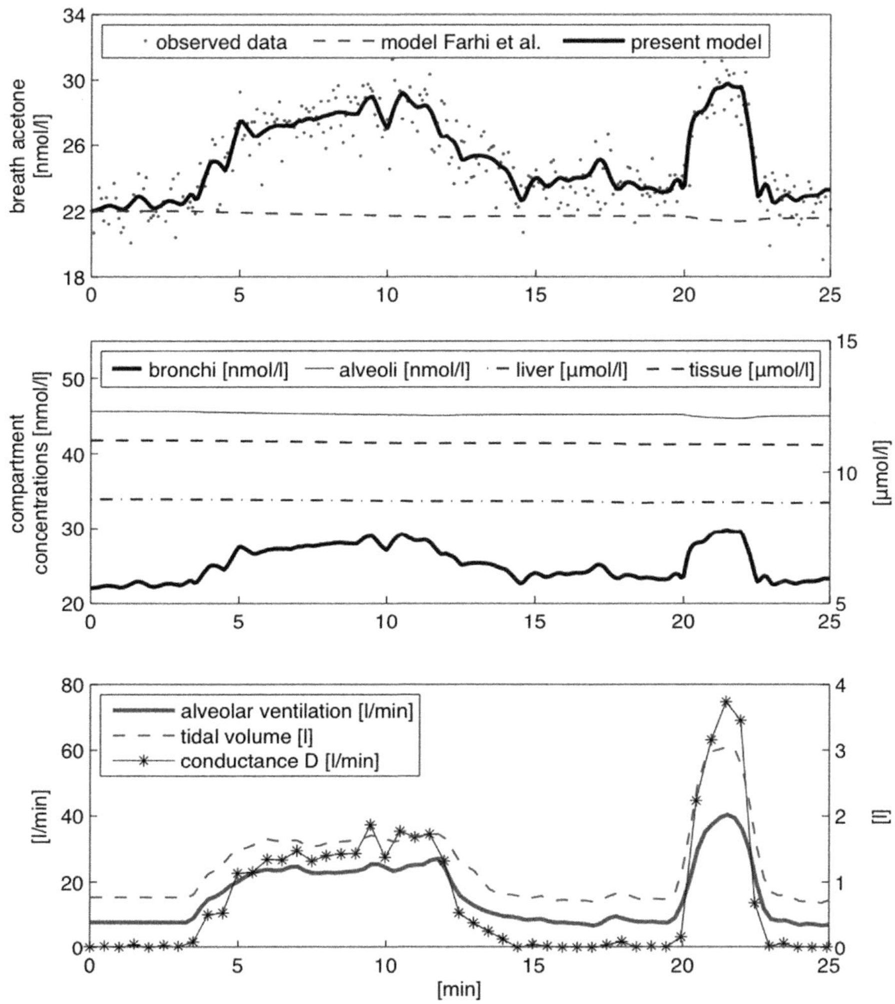

FIGURE 3.5

Top panel: model response fitted to the acetone breath data in Ref. 27, figure 3.2, Protocol B. Center panel: corresponding concentrations in the bronchial, alveolar, liver, and tissue compartments. Bottom panel: measured respiratory parameters and the estimated time course of the conductance parameter *D*, reflecting the net gas exchange between the alveolar and the bronchial tract, as described in the text. Note that *D* is close to zero during free tidal breathing at rest (i.e., the measureable breath acetone levels are almost exclusively determined by the respective levels in the conducting airways) and varies in parallel with ventilatory flow. In particular, higher values of *D* correspond to an increased contribution of the alveoli to total pulmonary gas exchange.

Reproduced from. [27]

concentration of hydrophilic VOCs. In vivo, the (arterial) blood–breath ratio strongly depends on a variety of additional factors that can vary considerably among individuals, including airway temperature and bronchial blood flow (both exerting an inversely monotonous influence on the blood-breath ratio). Modeling provides a good mechanistic account of these relationships. In particular, it can be used to predict the impact of conditions/pathologies that are known to alter airway gas exchange (e.g., in asthma or bronchiectasis) and can provide hints for designing and evaluating sampling protocols aimed at improving the reliability and information content of breath measurements involving hydrophilic VOCs.

3.6 Carbon monoxide

For many breath biomarkers, exhaled concentrations do not necessarily represent the alveolar concentration in equilibrium with blood, but also depend on the lung diffusion properties and the gas exchange in airways and oral/nasal cavities. This is the case for nitric oxide (NO), as outlined in Chapter 4. Advanced analytical techniques today allow accurate measurements of individual breath cycles[30,31] and have demonstrated that the shape of single-exhalation profiles carries information about the location and nature of the gas exchange. Physiological parameters can be retrieved by fitting a model of the gas-exchange dynamics in the respiratory tract to the experimental data.[32–34]

Exhaled carbon monoxide ($_{E}CO$), which is a potential biomarker for noninvasive assessment of (systemic and local) oxidative stress, chronic respiratory diseases, and exposure to air pollution (see Chapter 6), has been subject of rapid analyses to characterize its dynamics in breath. A laser-based CO sensor with controlled on-line breath sampling was employed to acquire $_{E}CO$ exhalation profiles with high time-resolution and precision.[30,31] The ensuing data allowed a trumpet model of the lung with axial diffusion (TMAD) to be developed to simulate the CO gas exchange during inhalation, breath-holding, and exhalation, and characterize corresponding $_{E}CO$ profiles.[32,33] The morphological model considers the bifurcating structure of the lung spreading over 24 generations, from trachea to alveoli, and incorporates comprehensive gas mixing mechanisms. The inhaled ambient CO concentration, average inhalation and exhalation flow rates, and average inhaled and exhaled air volumes were experimentally measured during breath sampling and served as model input parameters.

From data fitting to the $_{E}CO$ profiles, the maximum CO fluxes, average lung diffusing capacities, and expected equilibrium concentrations of CO during systemic elimination could be extracted for both airway and alveolar region. It was found that changes in alveolar flux and diffusing capacity differently affect the expirogram shape, and that a small increase in airway flux is resolvable. Clinical studies with healthy nonsmokers showed that end-tidal $_{E}CO$ is exhalation flow rate dependent due to the diffusion-limited CO gas exchange, that the spread in obtained TMAD parameter values is small, and that there is no contribution from the nasal cavity. After a 2 h exposure to 10 ppm CO, end-tidal $_{E}CO$ increased and was eliminated according to the Coburn–Forster–Kane model, while the lung diffusion capacity remaining unchanged.[35]

3.7 Methane

Methane is a volatile compound with a very low blood:air partition coefficient. As such, breath concentration levels would be expected to behave according to the Farhi equation when blood and breath flow are altered, yet experimental data indicate that this is not true. Taking the example of exercising with a workload of 75 W, the blood flow typically doubles and the breath flow increases fourfold. Consequently, breath methane concentrations should decrease by a factor of two compared to the steady-state level during rest; experiments show a decrease by a factor of three to four. This discrepancy can be explained by the fact that the Farhi equation only takes into account changes in blood flow and breath flow and assumes that the concentration of a volatile compound is identical in every part of the body, thus the relative changes in blood flows in different body regions has no effect. In the case of methane, however, it originates from metabolism of intestinal bacteria in the gut. When exercising, the relative blood flow through the intestine decreases, resulting in an additional decrease in exhaled methane. A simple three-compartment model based on mass balance equations has been developed to describe the dynamics of exhaled methane profiles during exercise on an ergometer and takes into account its production in the gut.[28] In particular, using this model it is possible to estimate the endogenous production rate of methane in the large intestine by measuring exhaled concentrations of methane.

3.8 Sevoflurane

Physiological modeling of exhaled volatiles can equally be applied to compounds that are solely of exogenous origin. King et al. extended the three-compartment model to include a brain compartment to explore the potential of breath analysis in anesthesia.[29] The model describes a monitoring scheme for estimating target site concentrations and cardiac output via real-time breath profiles of the anesthetic agent sevoflurane ($\lambda_{b:air} \approx 0.6$). Simulations of sevoflurane profiles during administration of approximately 3% sevoflurane over 25 min show good agreement with observed data from the literature. This proof-of-concept demonstrates that exhaled breath measurements can potentially contribute to facilitating real-time assessments of the anesthetic state of patients.

3.9 Summary

Breath gas analysis would greatly benefit from a more causal understanding of the kinetics of individual compounds in response to distinct physiological states. Unfortunately, the development of quantitative formulations relating breath concentrations of trace gases to their underlying systemic levels clearly lags behind the enormous analytical progress in the field. This situation calls for the establishment of a proper modeling framework that strives for validated mechanistic descriptions of VOC dynamics in different parts of the organism.

The knowledge of blood:gas partition coefficients $\lambda_{b:air}$ of VOCs (i.e., low or high water solubility) is essential in order to know if measured end-tidal concentrations will be equal to alveolar concentrations (e.g., isoprene), bronchial concentrations (e.g., acetone, ethanol), or a mixture of both (e.g., limonene). In the first case, a two-compartment model is sufficient; in the other cases, at least a two-compartment lung is needed for the quantitative description of exhaled breath concentrations. If, in addition, the relative blood flows in different body regions with different blood concentrations are altered (e.g., through exercise), then further compartments are needed to describe the effects on exhaled breath concentrations.

As an outlook, from an operational point of view, such modeling studies can

- help to identify potential confounding factors affecting the interpretation of breath test results and propose corrective measures.
- guide the ongoing standardization of sampling procedures and provide formal means for evaluating the information content of newly proposed experimental settings.
- enhance the fundamental understanding of the physiological role of VOCs, e.g., by relating phenomenological features of observable breath concentration profiles to specific distribution, transport, biotransformation, and excretion processes of these substances.
- stimulate biochemical investigations exploring the endogenous (e.g., cellular) sources of VOCs.

In particular, the last two points are directed toward further consolidating the potential role of breath gas analysis in general biomonitoring applications.

Acknowledgments

Christopher A. Mayhew and K. Unterkofler greatly appreciate the generous support from the government of Vorarlberg, Austria.

References

1. Barnes PJ. Nitric oxide and airway disease. *Ann Med* 1995;**27**(3):389–93.

2. King J, Kupferthaler A, Unterkofler K, Koc H, Teschl S, Teschl G, Miekisch W, Schubert J, Hinterhuber H, Amann A. Isoprene and acetone concentration profiles during exercise on an ergometer. *J Breath Res* 2009;**3**(2):027006.
3. King J, Kupferthaler A, Frauscher B, Hackner H, Unterkofler K, Teschl G, et al. Measurement of endogenous acetone and isoprene in exhaled breath during sleep. *Physiol Meas* 2012;**33**:413–28.
4. Amann A, Telser S, Hofer L, Schmid A, Hinterhuber H. Exhaled breath gas as a biochemical probe during sleep. In: Amann A, Smith D, editors. *Breath analysis for clinical diagnosis and therapeutic monitoring*. Singapore: World Scientific; 2005. p. 305–16.
5. Farhi LE. Elimination of inert gas by the lung. *Respir Physiol* 1967;**3**(1):1–11.
6. Andersen ME. Physiological modelling of organic compounds. *Ann Occup Hyg* 1991; **35**(3):309–21.
7. Wagner PD. In: Farhi LE, Marsh Tenney S, editors. *Handbook of physiology: section 3: the respiratory system volume IV: gas exchange*. New York: Oxford University Press; 1988. p. 257–81.
8. Teorell T. Kinetics of distribution of substances administered to the body I The extravascular modes of administration. *Arch Int Pharmacodyn Ther* 1937;**57**:205–25.
9. Karl T, Prazeller P, Mayr D, Jordan A, Rieder J, Fall R, Lindinger W. Human breath isoprene and its relation to blood cholesterol levels: new measurements and modeling. *J Appl Physiol* 2001;**91**(2):762–70.
10. Turner C, Spanel P, Smith D. A longitudinal study of breath isoprene in healthy volunteers using selected ion flow tube mass spectrometry (SIFT-MS). *Physiol Meas* 2006; **27**(1):13–22.
11. King J, Koc H, Unterkofler K, Mochalski P, Kupferthaler A, Teschl G, Teschl S, Hinterhuber H, Amann A. Physiological modeling of isoprene dynamics in exhaled breath. *J Theor Biol* 2010;**267**(4):626–37.
12. King J, Mochalski P, Kupferthaler A, Unterkofler K, Koc H, Filipiak W, Teschl S, Hinterhuber H, Amann A. Dynamic profiles of volatile organic compounds in exhaled breath as determined by a coupled PTR-MS/GC-MS study. *Physiol Meas* 2010;**31**(9): 1169–84.
13. Koc H, King J, Teschl G, Unterkofler K, Teschl S, Mochalski P, Hinterhuber H, Amann A. The role of mathematical modeling in VOC analysis using isoprene as a prototypic example. *J Breath Res* 5, 2011. 037102.
14. Mohrman DE, Heller LJ. *Cardiovascular physiology*. 6th ed. New York: Lange Medical Books/McGraw-Hill; 2006.
15. King J, Mochalski P, Unterkofler K, Teschl G, Klieber M, Stein M, et al. Breath isoprene: muscle dystrophy patients support the concept of a pool of isoprene in the periphery of the human body. *Biochem Biophys Res Commun* 2012;**423**:526–30.
16. Stone BG, Besse TJ, Duane WC, Evans CD, DeMaster EG. Effect of regulating cholesterol biosynthesis on breath isoprene excretion in men. *Lipids* 1993;**28**(8):705–8.
17. Kushch I, Arendacka B, Stolc S, Mochalski P, Filipiak W, Schwarz K, Schwentner L, Schmid A, Dzien A, Lechleitner M, Witkovsky V, Miekisch W, Schubert J, Unterkofler K, Amann A. Breath isoprene–aspects of normal physiology related to age, gender and cholesterol profile as determined in a proton transfer reaction mass spectrometry study. *Clin Chem Lab Med* 2008;**46**(7):1011–8.
18. Unterkofler K, et al. Modeling-based determination of physiological parameters of systemic VOCs by breath gas analysis: a pilot study. *J Breath Res* 2015;**9**:036002.

19. Ager C, et al. Modeling-based determination of physiological parameters of systemic VOCs by breath gas analysis, part 2. *J Breath Res* 2018;**12**:036011.
20. Anderson JC, Babb AL, Hlastala MP. Modeling soluble gas exchange in the airways and alveoli. *Ann Biomed Eng* 2003;**31**(11):1402–22.
21. King J, Unterkofler K, Kupferthaler A, Teschl G, Teschl S, Koc H, Hinterhuber H, Amann A. A modeling based evaluation of isothermal rebreathing for breath gas analysis of highly soluble volatile organic compounds. *J Breath Res* 2012;**6**:016005.
22. O'Hara ME, O'Hehir S, Green S, Mayhew CA. Development of a protocol to measure volatile organic compounds in human breath: a comparison of rebreathing and on-line single exhalations using proton transfer reaction mass spectrometry. *Physiol Meas* 2008;**29**(3):309–30.
23. Kumagai S, Matsunaga I. A lung model describing uptake of organic solvents and roles of mucosal blood flow and metabolism in the bronchioles. *Inhal Toxicol* 2000;**12**(6): 491–510.
24. Mörk AK, Johanson G. A human physiological model describing acetone kinetics in blood and breath during various levels of physical exercise. *Toxicol Lett* 2006;**164**(1): 6–15.
25. Lumb A. *Nunn's applied respiratory physiology.* Oxford: Butterworth-Heinemann; 2005.
26. Anderson JC, Lamm WJ, Hlastala MP. Measuring airway exchange of endogenous acetone using a single-exhalation breathing maneuver. *J Appl Physiol* 2006;**100**(3): 880–9.
27. King J, Unterkofler K, Teschl G, Teschl S, Koc H, Hinterhuber H, Amann A. A mathematical model for breath gas analysis of volatile organic compounds with special emphasis on acetone. *J Math Biol* 2011;**63**(5):959–99.
28. Szabó A, et al. Modeling of breath methane concentration profiles during exercise on an ergometer. *J Breath Res* 2016;**10**:017105.
29. King J, Unterkofler K, Amann A, Teschl S, Teschl G. Breath gas analysis for estimating physiological processes using anesthetic monitoring as a prototypic example. *Conf. Proc. IEEE Eng. Med. Biol. Soc.* 2011:1001–4.
30. Ghorbani R, Schmidt FM. Real-time breath gas analysis of CO and CO_2 using an EC-QCL. *Appl Phys B* 2017;**123**(5):144.
31. Ghorbani R, Schmidt FM. ICL-based TDLAS sensor for real-time breath gas analysis of carbon monoxide isotopes. *Optic Express* 2017;**25**(11):12743–52.
32. Ghorbani R, Blomberg A, Schmidt FM. Modeling pulmonary gas exchange and single-exhalation profiles of carbon monoxide. *Front Physiol* 2018;**9**:927.
33. Ghorbani R, Schmidt FM. Fitting of single-exhalation profiles using a pulmonary gas exchange model—application to carbon monoxide. *J Breath Res* 2019;**13**(2):026001.
34. Mountain JE, Santer P, O'Neill DP, Smith NMJ, Ciaffoni L, Couper JH, et al. Potential for noninvasive assessment of lung inhomogeneity using highly precise, highly time-resolved measurements of gas exchange. *J Appl Physiol* 2018;**124**(3):615–31.
35. Ghorbani R. *Real-time breath gas analysis of carbon monoxide : laser-based detection and pulmonary gas exchange modeling [Doctoral thesis].* Umeå: Umeå University; 2018.
36. Grunwald SA, Popp O, Haafke S, Jedraszczak N, Grieben U, Saar K, et al. Statin-induced myopathic changes in primary human muscle cells and reversal by a prostaglandin F2 alpha analogue. *Scientific Reports* 2020;**10**(1):2158.

CHAPTER

Exhaled nitric oxide physiology and modeling

4

Marieann Högman[1], Lauri Lehtimäki[2]

[1]*Department of Medical Sciences, Respiratory, Allergy and Sleep Research, Uppsala University, Uppsala, Sweden;* [2]*Allergy Centre, Tampere University Hospital, Faculty of Medicine and Health Technology, Tampere University, Tampere, Finland*

Chapter outline

4.1 Overview

Nitric oxide (NO) is involved in many processes in the respiratory system and is important for the homeostasis of the epithelium, blood vessels, and neuronal control in the healthy lung. NO is synthesized by NO synthases that convert L-arginine into L-citrulline and NO in the presence of oxygen, nicotinamide adenine dinucleotide phosphate, flavin adenine dinucleotide, flavin mononucleotide, and tetrahydrobiopterin. NO synthases are both constitutive (cNOS) and inducible (iNOS). cNOS is Ca^{2+}- and calmodulin-dependent and at stimulation, such as during airway provocation, they readily release NO that can be detected in exhaled breath. There are two isoforms of cNOS known as neuronal NOS (nNOS) and endothelial/epithelial NOS (eNOS), and these are widely distributed

Breathborne Biomarkers and the Human Volatilome. https://doi.org/10.1016/B978-0-12-819967-1.00004-9
Copyright © 2020 Elsevier B.V. All rights reserved.

in the human tissue. cNOS produces low concentrations of NO (nano to picomolar range), while the iNOS produces high concentrations (micromolar to nanomolar range). iNOS is also found in various human tissues and inflammatory cells, such as macrophages and neutrophils, and is induced by proinflammatory cytokines.

Low concentration of NO regulates several biological processes in the respiratory system. It acts as a neurotransmitter in the iNANC system, induces bronchodilation, and facilitates surfactant production in the alveoli to keep them open and thereby improve gas exchange. Further, it increases mucous secretion which, together with enhanced ciliary motility, facilitates mucous transportation to clear the airways, it regulates pulmonary vascular tone to improve gas exchange, has an antiproliferative effect on airway smooth muscle, is bacteriostatic, and it inhibits virus replication. In contrast to the low concentration of NO from cNOS, iNOS-driven NO production can be cytotoxic at high concentrations. When stimulated, the iNOS-driven production can continue for days and contribute to bronchial hyperreactivity, vasodilation, free radical production, mucous hypersecretion, and ciliary motility inhibition, and can have a proinflammatory effect. This high NO production is also seen in exhaled breath and has been shown to correlate to eosinophilic inflammation. iNOS-driven NO production is known to be sensitive to inhaled corticosteroids (ICSs), as in asthma treatment. As such, exhaled NO is termed an "inflammometer" and can be used to follow response to treatment.

When synthesized, NO can diffuse freely over cell membranes and thereby be detected in exhaled breath. NO production in the respiratory system follows a gradient, with lowest concentrations in the alveolar region of the lung, higher levels in the proximal airway, and highest production in the upper airway and nasal cavity. As such, exhaled NO collected over a single exhalation stems from different sources. When the flow dependency of exhaled NO was discovered,[1,2] it offered the opportunity to model NO production in the lung.

Nitric oxide concentrations in breath vary with the exhalation flow (see Fig. 4.1, center panel). In contrast to carbon dioxide (CO_2), where the concentration peaks at the end phase of exhalation, for NO the maximum concentration occurs at the beginning of the exhalation (see Fig. 4.1, left panel); this indicates that less NO is present in the alveoli. This NO concentration gradient allows the application of Fick's first law of diffusion, proposed in 1856, which can be simplified as follows: NO at low concentration from the alveoli enters and travels along a tube, the airways. During its transition, NO molecules diffuse into the airways from the walls. The degree of diffusion depends on the transfer rate across the airway walls and the concentration surrounding them. Hence, during inhalation air is drawn down into the lung periphery by a pressure difference (convection), the alveoli expand, and gas exchange driven by a concentration gradient (diffusion) takes place. With the movement of air and the addition of NO from conducting airways into the alveolar region, equilibrium of the different gases between alveolar gas and capillary blood is soon reached, before the exhalation phase begins.

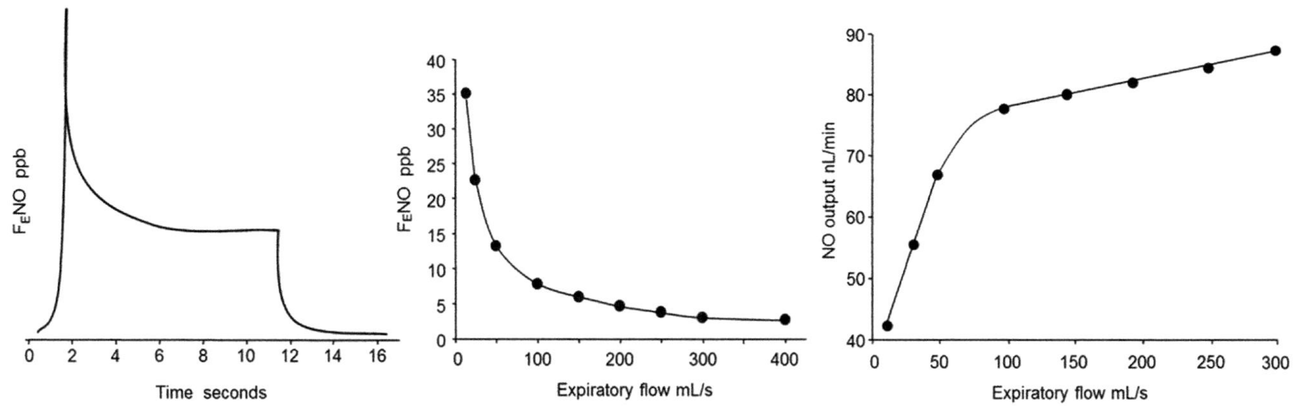

FIGURE 4.1

The left panel shows a typical NO concentration curve during an exhalation flow rate of 50 mL/s. The initial peak in exhaled NO represents the high concentration of NO inhaled from upper airways during nasal inhalation; the central plateau of the curve represents NO reached at a steady exhalation flow; the drop of the curve after the plateau is when exhalation stops and NO rapidly decreases toward environment NO concentrations. The center panel shows the flow dependency of exhaled NO concentration. The plateau value of exhaled NO at different exhalation flow is plotted against flow. The right panel shows the output of NO for each flow; this is the plot that is used for both the linear and the nonlinear methods to solve different NO parameters.

In healthy individuals, the concentration of NO in the alveolar gas phase (C_ANO) is most likely on the same order as that of the alveolar tissue, i.e., close to zero, since blood is an NO sink. A bolus of gas with near-zero NO[a] starts the transport toward the upper airways to be exhaled. During the passage in the airway, the bolus picks up NO from the small and large airways; the increase in NO in the bolus during this phase depends on the duration of contact between exhaled gas and bronchial mucosa. From a system theory perspective, the conducting airways can be modeled as a black box with the addition of NO to the gas flow from the lungs. The experimentally obtained mutual relationships between NO output and NO input for a set of exhalation flows can be used to construct a transfer function to predict the system NO output for any other flow. Instead of a purely mathematical approach, however, it is important to be able to link all the information that defines the system to physical/physiological properties of the human airways parameters. In Fig. 4.1 (right panel), the NO output from the lung is illustrated as a function of exhalation flow, and this information is used for both linear and nonlinear NO models of the lung.

4.2 Two-compartment model

A simplification of the lung can be made based on a two-compartment model. The model consists of one rigid compartment representing the airways (the tube) and one expansile compartment representing the gas exchange areas[3] (see Fig. 4.2). Fick's first law of diffusion can then be applied to the mechanism controlling the transfer of NO to the airway lumen. Measurements of exhaled NO at three or more flows can be exploited to determine three different parameters that define NO production. The abbreviations in current use were agreed upon by international experts of the European Respiratory Society (ERS) and American Thoracic Society (ATS)[4,5] and are listed and defined in Table 4.1.

The plot in Fig. 4.1 (right panel) depicts the NO output as a function of flow. The curve shows two distinct phases, with the initial steeper curve representing the airways portion and the latter gentle slope depicting the alveolar portion of exhaled NO. These two approximately linear curves intercept at a flow of around 50–60 mL/s, just at the recommended flow rate (ERS/ATS), irrespective of a high or low F_ENO_{50} value.[6] In healthy individuals, C_ANO does not correlate with F_ENO_{50}, but F_ENO_{50} correlates well with $J_{aw}NO$, which strengthens the two-compartment model. Other findings that support the two-compartment model are that a position change from sitting to prone or supine position only changes the C_ANO value and not the NO produced in the airways.[7] It is well known that the alveolar volume changes with body position and that the airways are not affected by this change.

[a] In disease state, the NO concentration may be elevated.

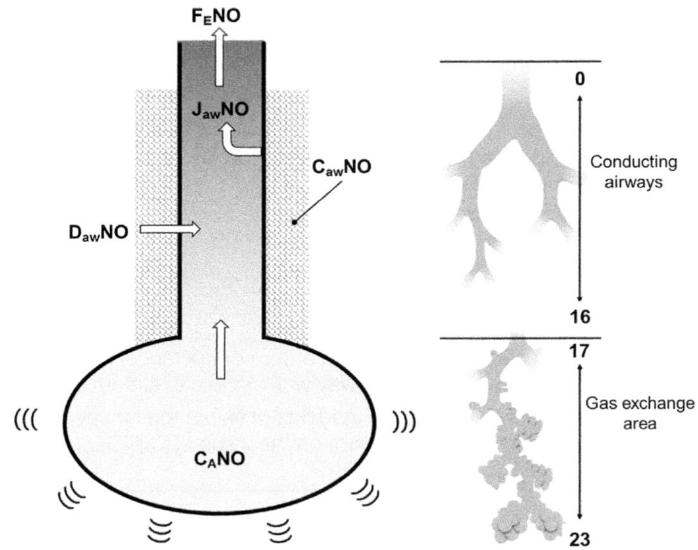

FIGURE 4.2

The two-compartment model of pulmonary NO dynamics is generated from Weibel's model of airways; it divides the lung into a cylindrical tube that represents the conducting airways (airway generations 0–16) and an expansile compartment that represents the gas exchange area of the lung (generations 17–23). Both compartments are surrounded by tissue capable of producing and consuming NO, and the tissue layers are surrounded by blood circulation that serves as an infinite sink of NO. Alveolar gas contains a low concentration of NO (C_ANO) representing the balance between NO produced locally, NO inhaled or diffused from conducting airways, and NO diffused into pulmonary circulation. During exhalation, alveolar gas travels through the conducting airways and is conditioned by NO that diffuses from bronchial mucosa ($J_{aw}NO$) to luminal air, the amount of which is dependent on NO concentration in bronchial mucosa ($C_{aw}NO$) and diffusivity of NO between bronchial mucosa and luminal gas ($D_{aw}NO$). Fractional exhaled NO (F_ENO) concentration is thereby dependent on C_ANO, $C_{aw}NO$, $D_{aw}NO$, and exhalation flow rate (see text for details).

4.3 Different mathematical methods to calculate nitric oxide parameters

Two different methods to calculate the NO parameters have been proposed in the recent ERS technical standard,[5] and both are based on the two-compartment model. In this standard, a calculation spreadsheet is available online for modeling NO. Reference values of pulmonary NO dynamics in healthy subjects[8] are presented in Table 4.2.

The expansile compartment of the lung holds the gas-phase fraction of NO in the alveolar region (C_ANO; in ppb). The rigid conducting airway system

Table 4.1 Abbreviations and definitions of pulmonary NO parameters.

Abbreviation	Definition
F_ENO	*Fractional exhaled NO concentration in the orally exhaled gas*: The exhalation flow is determined in mL/s and is given as a numerical value in the subscript; at the recommended flow of 50 mL/s, the abbreviation becomes F_ENO_{50} (ppb)
C_ANO	*Alveolar NO concentration*: Calculated fractional concentration of NO in the gas phase of the alveolar region (ppb)
$C_{aw}NO$	*Bronchial wall NO concentration*: Calculated tissue concentration of NO of the airway wall (ppb)
$D_{aw}NO$	*Bronchial wall diffusing capacity of NO*: Calculated airway compartment diffusing capacity, transfer factor, or conductance for radial mass transfer of NO from the airway wall to the gas stream (mL/s)
$J_{aw}NO$	*Bronchial NO flux*: Calculated flux of NO in the airway compartment (nL/s). This flux is dependent on the value of C_ANO:$J_{aw}NO = D_{aw}NO \cdot (C_{aw}NO\text{-}C_ANO)$

Table 4.2 F_ENO_{50} and NO parameters in different age groups according to sex (female $n = 165$, male $n = 268$).

Age group	<20 years	20–49 years	≥50 years
F_ENO_{50} (ppb)			
Female	11 (6, 26)	13 (7, 26)	17 (8, 35)
Male	10 (5, 26)	17 (9, 44)	19 (8, 39)
$J_{aw}NO$ (nL/s)			
Female	0.4 (0.2, 1.1)	0.6 (0.3, 1.7)	0.8 (0.2, 1.8)
Male	0.4 (0.1, 1.5)	0.8 (0.4, 2.5)	0.9 (0.3, 2.1)
C_ANO (ppb)			
Female	2.1 (0.7, 3.6)	2.0 (0.3, 3.8)	2.4 (0.9, 3.9)
Male	2.0 (0.2, 3.6)	1.5 (0.3, 3.7)	2.2 (0.4, 3.6)
$C_{aw}NO$ (ppb)			
Female	70 (23, 415)	77 (34, 155)	110 (51, 509)
Male	48 (19, 193)	139 (35, 417)	161 (49, 491)
$D_{aw}NO$ (mL/s)			
Female	7.1 (1.2, 25)	8.3 (4.4, 19)	9.1 (0.9, 21)
Male	11 (1.1, 26)	8.0 (1.8, 20)	5.7 (1.3, 16)

Data are median (5th, 95th percentile) values.
Reproduced from [8].

represented in the model as a single tube is characterized by the airway tissue concentration of NO (wall concentration, $C_{aw}NO$; in ppb) and the total airway compartment diffusing capacity or transfer factor of NO from the airway wall to the gas stream ($D_{aw}NO$; in mL/s). The flux of NO from the wall to the lumen depends on the distance z along the airway and has a total value $J_{aw}NO = D_{aw}NO \cdot (C_{aw}NO\text{-}C_{A}NO)$. At sufficiently high flows, when the transit time of alveolar gas through the airway is low, luminal NO concentration remains close to C_ANO throughout the airway. Physiologically, a calculated C_ANO should be positive. A negative C_ANO is an indication that the method of NO production of the respiratory system is inadequate.

4.3.1 Linear method

The Tsoukias and George linear method[3] uses a linear approximation of the governing exponential equation of the original two-compartment model to calculate $J_{aw}NO$ and C_ANO. In this method, NO concentrations at multiple constant exhalation flow rates of between 100 and 350–400 mL/s are obtained and the output of NO is plotted versus the flows (see Fig. 4.3, left panel). The slope and the intercept of the resulting linear regression line provide estimates of C_ANO and $J_{aw}NO$, and the r-value of the line gives the quality of the fit, which should be >0.95. Note that F_ENO_{50}[b] is not one of the flows that is included in the estimations of the NO parameters.

4.3.2 Nonlinear model

The Högman Merilänen algorithm (HMA)[9] is based on the same two-compartment model used in the linear method, but in this nonlinear method NO values from at least three different flows are required, namely one low flow (≤20 mL/s), one medium flow (100 mL/s), and one high flow (350 mL/s or even 400 mL/s).[5] Note that, again, F_ENO_{50} is not one of the flows that is included in the estimations of the NO parameters, but with HMA this value can be verified from the curve generated from the measured NO values (see Fig. 4.3, right panel). A comparison can be made between estimated and measured F_ENO_{50} as a control of the estimations, whereby the values should be within ±5 ppb of each other.[5,10] Estimations of C_ANO, $D_{aw}NO$, $C_{aw}NO$, and $J_{aw}NO$ can be made using this nonlinear method. An added feature of this approach is to use an algorithm to test if the measured set of data points is mathematically consistent with the model.[5,9] If this is not the case, it indicates measurement errors in the dataset that relates to flow or NO concentration, or both. Additionally, the value of C_ANO must naturally be positive.

[b] F_ENO at 50 mL/s is the recommended flow for single F_ENO measurements.

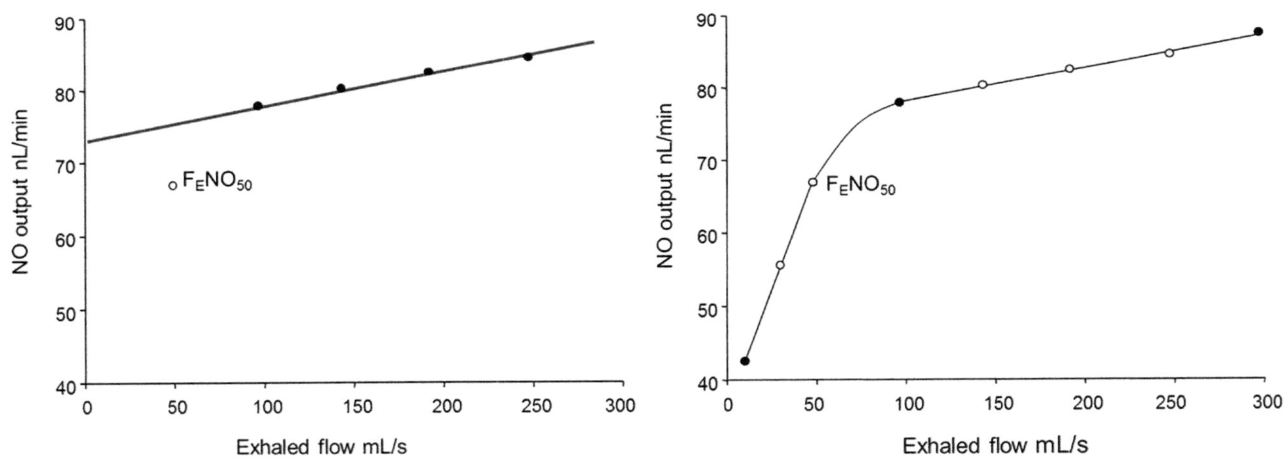

FIGURE 4.3

Tsoukias and George model (left panel). When F_ENO is measured at several exhalation flows of at least 100 mL/s, a linear approximation of the governing equation of the two-compartment model can be used to calculate C_ANO and $J_{aw}NO$. NO output is plotted against exhalation flow, and the slope and the intercept of the regression line between NO output and exhalation flow are approximates of C_ANO and $J_{aw}NO$, respectively. Note that F_ENO_{50} is not one of the flows that is included in the estimations. Högman Meriläinen Algorithm (right panel). To create the nonlinear flow–volume curve, at least three exhalation flows are needed and these flows that span the curve; one at both ends and one in the middle of the curve. Note that F_ENO_{50} is not one of the flows that is included in the estimations.

4.4 Mechanisms altering central and peripheral pulmonary nitric oxide parameters

Extended NO analysis provides an easy tool to assess NO dynamics in both central and peripheral airways. As many different disease processes and pathophysiological phenomena may alter NO dynamics, these parameters will likely not be useful in making any diagnosis on their own, but they add to the clinical decision-making when interpreted in the context of other findings.

Bronchial wall NO concentration, $C_{aw}NO$, is elevated during mucosal inflammation,[11] especially type 2 inflammation in eosinophilic asthma. This is likely the reason why F_ENO_{50} is a good predictor of responsiveness to treatment with ICSs in subjects with asthma. $C_{aw}NO$ also decreases after treatment with ICS,[12] likely due to glucocorticoids decreasing the expression of iNOS.

In contrast to $C_{aw}NO$, bronchial wall diffusing capacity of NO, $D_{aw}NO$, is elevated not only in asthma and allergic rhinitis, but also in some subjects with chronic obstructive pulmonary disease (COPD).[9,11] The exact mechanisms increasing $D_{aw}NO$ are not known, but increased $D_{aw}NO$ may reflect an increase in NO production arising from an increased mucosal surface area or from increased physical diffusivity of NO across mucosa due to bronchial remodeling. Interestingly, ICS treatment does not seem to affect $D_{aw}NO$,[12] but airway obstruction decreases $D_{aw}NO$, probably because mucosal surface area in contact with luminal air decreases while the diameter of bronchi decreases. This is observable as decreased $D_{aw}NO$ after methacholine challenge and increased $D_{aw}NO$ after inhaled β_2 agonists.[13]

Alveolar NO concentration, C_ANO, is also affected by many different mechanisms. Inflammation in lung parenchyma or small airways may increase C_ANO by increasing local NO production. On the other hand, decreased diffusion of NO from alveolar gas to capillaries due to parenchymal fibrosis of defects in pulmonary circulation may also increase C_ANO. Changes in small airway diameter are thought to affect C_ANO by decreasing or increasing back diffusion of NO from the conducting airways to the alveolar compartment.[14]

Considering the wide range of mechanisms affecting central and peripheral NO parameters, it is easy to understand that exhaled NO concentrations can hardly be used on their own to make a diagnosis of any pulmonary disease, but they may be used to support the information provided by other measures and clinical history.

4.5 Nitric oxide parameters in airway diseases

4.5.1 Asthma

The measurement of F_ENO_{50} is very useful in asthma in finding type 2 inflammatory patterns and in predicting response to ICS, as well as in assessing adherence to ICS treatment. Since $J_{aw}NO$ and F_ENO_{50} are highly correlated, $J_{aw}NO$ probably does not provide added clinical value over F_ENO_{50}. However, calculating $C_{aw}NO$ and $D_{aw}NO$ allows a more precise assessment of whether increased $J_{aw}NO$ or F_ENO_{50} are explained by inflammatory activity ($C_{aw}NO$) or physical changes in bronchial mucosa ($D_{aw}NO$). As $D_{aw}NO$ is increased in asthma and is not affected by steroid treatment,[12] $J_{aw}NO$ and F_ENO_{50} may be elevated even in the absence of current inflammatory activity in well-treated patients. Therefore, $C_{aw}NO$ probably provides a better basis to make conclusions on type 2 inflammatory activity and ICS treatment.[15]

Asthma leads to elevated C_ANO,[11] which is likely due to small airway inflammation. Increased C_ANO is associated with more severe asthma and it has been shown to be associated with small airway dysfunction in studies using impulse oscillometry, nitrogen washout techniques, or body plethysmography.[16–18] Extra fine particle ICS deposit more effectively to the peripheral small airways and such drugs have been shown to decrease C_ANO in asthma more effectively than ordinary ICS preparations;[19] C_ANO in asthma therefore seems to be a reliable marker of small airway involvement. Further, the systemic inflammatory component in asthma might be reflected as increased C_ANO, since systemic glucocorticoids reduce C_ANO, even in cases where increasing ICS dose is not effective.[20]

4.5.2 Chronic obstructive pulmonary disease

Nitric oxide parameters are much less studied in COPD than in asthma. Variable levels of F_ENO_{50} have been reported in different groups of patients with COPD. This may be related to the fact that both ICS treatment and active smoking seem to decrease $C_{aw}NO$.[21,22] Interpretation of central NO dynamics is thereby often complicated by smoking among subjects with COPD. Higher levels of $J_{aw}NO$ in untreated subjects with COPD are related to better treatment response to ICS in terms of both symptoms and lung function,[21] but the evidence on the use of NO parameters in predicting treatment responses in COPD is very limited.

Elevated C_ANO has been observed in COPD, but it seems not to be affected by either smoking or glucocorticoids. The mechanisms that increase C_ANO in COPD may be small airway and parenchymal inflammation, as well as a reduced uptake of NO to circulation due to emphysema. Interestingly, C_ANO has been shown to correlate with both small airway dysfunction and pulmonary diffusing capacity.[23] This has also practical consequences, since high levels of C_ANO in subjects with COPD are related to poorer exercise capacity and more prominent exercise-induced hypoxemia.[24,25] The subject of breathborne biomakers for COPD, including F_ENO, is treated at length in Chapter 19.

4.5.3 Parenchymal lung diseases

Interstitial or parenchymal lung diseases are a heterogeneous group of diseases that cause variable degrees of inflammation and fibrosis in pulmonary parenchyma.

In theory, parenchymal lung diseases might increase C_ANO due to an increased NO production in the alveolar region or due to impaired diffusion of NO from alveoli to pulmonary circulation.

In terms of NO parameters, the most widely studied interstitial lung disease (ILD) is that associated with systemic sclerosis (SSc). C_ANO is higher among patients with SSc, and it is further increased in those who have developed ILD, thereby making it a reliable marker of ILD in subjects with SSc.[26] In SSc-ILD, C_ANO correlates with alveolar inflammation activity, degree of pulmonary fibrosis, and degree of impairment in lung function.[26,27] High C_ANO is also related to better treatment response to cyclophosphamide among these patients.

In subjects with nonspecific interstitial pneumonia or idiopathic pulmonary fibrosis (IPF), C_ANO is elevated and correlates with impaired lung function, poorer exercise capacity, and more progressive disease.[28,29] C_ANO is also higher in subjects with pulmonary transplant and further increases in association with the common complication, bronchiolitis obliterans syndrome (BOS).[30] C_ANO has been demonstrated to identify subjects with BOS among those with pulmonary transplant with relatively good specificity (area under the curve [AUC] of the receiver operating characteristic [ROC] curve of 0.82). C_ANO measurement may therefore be a useful marker in different forms of ILD to aid in diagnosis, severity grading, and prediction of disease progression and treatment responses.

Heavy exposure to asbestos fibers or silica dust may cause inflammation and fibrosis of lung parenchyma that leads to pneumoconiosis, known as asbestosis and silicosis, respectively. C_ANO is higher in subjects with asbestosis or silica exposure and correlates with markers of pulmonary inflammation and oxidative stress (see Chapter 32).[31,32] Further, short-term exposure to wood smoke has been shown to increase C_ANO.[33] C_ANO therefore seems to be a nonspecific marker of any inflammatory/irritative reaction due to exposure to several different small particles affecting the lung periphery.

4.6 Nitric oxide parameters in systemic diseases

Systemic diseases similarly affect C_ANO, probably by changes in pulmonary circulation. Crohn's disease, for example, is often associated with pulmonary manifestations in both conducting airways and lung parenchyma on top of the typical intestinal inflammation. C_ANO is elevated in subjects with Crohn's disease and correlates with disease activity.[34]

In liver cirrhosis, pulmonary circulation is often altered, which may lead to impaired gas diffusion or hepatopulmonary syndrome. In subjects with liver cirrhosis, both C_ANO and the production rate of NO in the alveolar compartment are increased, and these correlate with the degree of changes in systemic and pulmonary circulation and impairment of gas diffusion.[35] Based on these findings, increased C_ANO in liver cirrhosis and associated hepatopulmonary syndrome is caused not only by changes in pulmonary circulation, but also by inflammatory activity and increased production of NO locally in pulmonary parenchyma.

It has been speculated that rheumatoid arthritis (RA) might have its immunological origin in the lung. A recent study reports altered NO parameter levels in RA, whereby C_ANO and $C_{aw}NO$ were decreased, while $D_{aw}NO$ was increased in subjects with RA.[36]

4.7 Current clinical use and further research on nitric oxide parameters

To date there have been no formal trials that have compared clinical decision-making (presence or absence of NO) with NO parameters from extended NO analysis in any pulmonary disease. Nevertheless, an abundance of clinical studies shows how NO dynamics are altered in several different pulmonary diseases, and how these alterations relate to pathophysiology, progress, and treatment responses in these diseases. A comprehensive review of clinical applications of F_ENO_{50} is offered in the next chapter.

In relation to asthma, F_ENO_{50} is now recommended by several guidelines for use in its diagnosis and phenotyping. C_ANO seems to be useful in asthma in assessing the degree of small airway involvement that is left undetected by F_ENO_{50}. In many subjects with asthma, F_ENO_{50} may remain elevated regardless of appropriate treatment and adherence to ICS. Measuring $C_{aw}NO$ would allow an estimation of whether an increased F_ENO_{50} relates to inflammatory activity and elevated $C_{aw}NO$, or due to increased $D_{aw}NO$ and nonmodifiable bronchial remodeling. This would help in titrating ICS treatment among these subjects. At present, it seems that the NO parameters reflecting the central airways may not be very useful in COPD, but measuring C_ANO may turn out to be valuable in assessing dysfunction of pulmonary parenchyma and circulation. Further studies are needed to test the clinical value of these concepts.

In SSc-ILD, a measurement of C_ANO seems to be clinically useful in detecting ILD, in assessing the degree of inflammatory activity, and in predicting treatment responses. Nevertheless, there is presently a lack of information on the clinical value of C_ANO in other interstitial lung diseases. Since IPF is the most common of these diseases, further studies are required in order to find out if C_ANO could be used to assess the degree of inflammatory activity and treatment responses in IPF.

4.8 Summary

Nitric oxide is involved in many processes in the respiratory system and is important for the homeostasis of epithelium, blood vessels, and neuronal control in the healthy lung. Many different disease processes and pathophysiological phenomena may alter NO dynamics. Modeling exhaled NO provides an interesting option to gain a deeper understanding of pulmonary NO dynamics, and it may offer further clinical applications where the current F_ENO measurement at a single flow may not be applicable. The model gives alveolar NO concentration, airway wall concentration, airway compartment diffusing capacity, and NO flux in the airway compartment. These parameters will likely not be useful in making any diagnosis on their own, but they add to the clinical decision-making when interpreted in the context of other findings. Modeling exhaled NO is close to clinical use in some diseases but needs further research in other areas before the possible clinical value can be assessed and assured. It took many years since the invention and standardization of F_ENO_{50} for it to be recommended in international practice guidelines in the diagnosis and assessment of asthma. Two of the many methods to calculate NO parameters are now recommended in a recent technical standard, and further clinical trials using these methods are now needed to assess their potential in clinical use.

References

1. Högman M, Strömberg S, Schedin U, Frostell C, Hedenstierna G, Gustafsson LE. Nitric oxide from the human respiratory tract efficiently quantified by standardized single breath measurements. *Acta Physiol Scand* 1997;**159**(4):345–6.
2. Silkoff PE, McClean PA, Slutsky AS, Furlott HG, Hoffstein E, Wakita S, et al. Marked flow-dependence of exhaled nitric oxide using a new technique to exclude nasal nitric oxide. *Am J Respir Crit Care Med* 1997;**155**(1):260–7.
3. Tsoukias NM, George SC. A two-compartment model of pulmonary nitric oxide exchange dynamics. *J Appl Physiol* 1998;**85**(2):653–66.
4. George SC, Högman M, Permutt S, Silkoff PE. Modeling pulmonary nitric oxide exchange. *J Appl Physiol* 2004;**96**:831–9.
5. Horvath I, Barnes PJ, Loukides S, Sterk PJ, Högman M, Olin AC, et al. A European Respiratory Society technical standard: exhaled biomarkers in lung disease. *Eur Respir J* 2017;**49**(4).
6. Högman M, Thornadtsson A, Hedenstierna G, Meriläinen P. A practical approach to the theoretical models to calculate NO parameters of the respiratory system. *J Breath Res* 2014;**8**(1):016002.
7. Verbanck S, Kerckx Y, Schuermans D, de Bisschop C, Guenard H, Naeije R, et al. The effect of posture-induced changes in peripheral nitric oxide uptake on exhaled nitric oxide. *J Appl Physiol* 2009;**106**(5):1494–8.
8. Högman M, Thornadtsson A, Liv P, Hua-Huy T, Dinh-Xuan AT, Tufvesson E, et al. Effects of growth and aging on the reference values of pulmonary nitric oxide dynamics in healthy subjects. *J Breath Res* 2017;**11**:047103.
9. Högman M, Holmkvist T, Wegener T, Emtner M, Andersson M, Hedenström H, et al. Extended NO analysis applied to patients with COPD, allergic asthma and allergic rhinitis. *Respir Med* 2002;**96**:24–30.
10. Lassmann-Klee PG, Lehtimäki L, Lindholm T, Malmberg LP, Sovijärvi ARA, Piirila PL. Converting F_ENO by different flows to standard flow F_ENO. *Clin Physiol Funct Imag* 2019;**39**(5):315–21.

11. Karvonen T, Lehtimäki L. Flow-independent nitric oxide parameters in asthma: a systematic review and meta-analysis. *J Breath Res* 2019;**13**(4):044001.
12. Silkoff PE, Sylvester JT, Zamel N, Permutt S. Airway nitric oxide diffusion in asthma. Role in pulmonary function and bronchial responsiveness. *Am J Respir Crit Care Med* 2000;**161**:1218–28.
13. Cattoni I, Guarnieri G, Tosetto A, Mason P, Scarpa MC, Saetta M, et al. Mechanisms of decrease in fractional exhaled nitric oxide during acute bronchoconstriction. *Chest* 2013; **143**(5):1269–76.
14. Shin HW, George SC. Impact of axial diffusion on nitric oxide exchange in the lungs. *J Appl Physiol* 2002;**93**(6):2070–80.
15. Högman M, Meriläinen P. Guidance for a personal target value of F_ENO in allergic asthma: case report and theoretical example. *Ups J Med Sci* 2013;**118**:59–61.
16. Keen C, Olin AC, Wennergren G, Gustafsson P. Small airway function, exhaled NO and airway hyper-responsiveness in paediatric asthma. *Respir Med* 2011;**105**(10):1476–84.
17. Matsumoto H, Niimi A, Jinnai M, Nakaji H, Takeda T, Oguma T, et al. Association of alveolar nitric oxide levels with pulmonary function and its reversibility in stable asthma. *Respiration* 2011;**81**(4):311–7.
18. van Veen I, Sterk PJ, Schot R, Gauw SA, Rabe KF, Bel EH. Alveolar nitric oxide versus measures of peripheral airway dysfunction in severe asthma. *Eur Respir J* 2006;**27**(5): 951–6.
19. Nicolini G, Chetta A, Simonazzi A, Tzani P, Aiello M, Olivieri D. Both bronchial and alveolar exhaled nitric oxide are reduced with extrafine beclomethasone dipropionate in asthma. *Allergy Asthma Proc* 2010;**31**(5):85–90.
20. Berry M, Hargadon B, Morgan A, Shelley M, Richter J, Shaw D, et al. Alveolar nitric oxide in adults with asthma: evidence of distal lung inflammation in refractory astma. *Eur Respir J* 2005;**25**:989–91.
21. Lehtimäki L, Kankaanranta H, Saarelainen S, Annila I, Aine T, Nieminen R, et al. Bronchial nitric oxide is related to symptom relief during fluticasone treatment in COPD. *Eur Respir J* 2010;**35**(1):72–8.
22. Bazeghi N, Gerds TA, Budtz-Jorgensen E, Hove J, Vestbo J. Exhaled nitric oxide measure using multiple flows in clinically relevant subgroups of COPD. *Respir Med* 2011; **105**(9):1338–44.
23. Hirano T, Matsunaga K, Sugiura H, Minakata Y, Koarai A, Akamatsu K, et al. Relationship between alveolar nitric oxide concentration in exhaled air and small airway function in COPD. *J Breath Res* 2013;**7**(4):046002.
24. McCurdy MR, Sharafkhaneh A, Abdel-Monem H, Rojo J, Tittel FK. Exhaled nitric oxide parameters and functional capacity in chronic obstructive pulmonary disease. *J Breath Res* 2011;**5**(1):016003.
25. Högman M, Thornadtsson A, Bröms K, Janson C, Lisspers K, Ställberg B, et al. Higher alveolar nitric oxide in COPD is related to poorer physical capacity and lower oxygen saturation after physical testing. *Eur Respir J* 2019;**54**(2).
26. Tiev KP, Coste J, Ziani M, Aubourg F, Cabane J, Dinh-Xuan AT. Diagnostic value of exhaled nitric oxide to detect interstitial lung disease in systemic sclerosis. *Sarcoidosis Vasc Diffuse Lung Dis* 2009;**26**(1):32–8.
27. Tiev KP, Hua-Huy T, Riviere S, Le-Dong NN, Febvre M, Cabane J, et al. High alveolar concentration of nitric oxide is associated with alveolitis in scleroderma. *Nitric Oxide* 2013;**28**:65–70.

28. Cameli P, Bargagli E, Refini RM, Pieroni MG, Bennett D, Rottoli P. Exhaled nitric oxide in interstitial lung diseases. *Respir Physiol Neurobiol* 2014;**197**:46–52.
29. Kotecha J, Shulgina L, Sexton DW, Atkins CP, Wilson AM. Plasma vascular endothelial growth factor concentration and alveolar nitric oxide as potential predictors of disease progression and mortality in idiopathic pulmonary fibrosis. *J Clin Med* 2016;**5**(9).
30. Cameli P, Bargagli E, Fossi A, Bennett D, Voltolini L, Refini RM, et al. Exhaled nitric oxide and carbon monoxide in lung transplanted patients. *Respir Med* 2015;**109**(9): 1224–9.
31. Lehtonen H, Oksa P, Lehtimäki L, Sepponen A, Nieminen R, Kankaanranta H, et al. Increased alveolar nitric oxide concentration and high levels of leukotriene B(4) and 8-isoprostane in exhaled breath condensate in patients with asbestosis. *Thorax* 2007; **62**(7):602–7.
32. Sauni R, Oksa P, Lehtimäki L, Toivio P, Palmroos P, Nieminen R, et al. Increased alveolar nitric oxide and systemic inflammation markers in silica-exposed workers. *Occup Environ Med* 2012;**69**(4):256–60.
33. Barregard L, Sallsten G, Andersson L, Almstrand AC, Gustafson P, Andersson M, et al. Experimental exposure to wood smoke: effects on airway inflammation and oxidative stress. *Occup Environ Med* 2008;**65**(5):319–24.
34. Malerba M, Ragnoli B, Buffoli L, Radaeli A, Ricci C, Lanzarotto F, et al. Exhaled nitric oxide as a marker of lung involvement in Crohn's disease. *Int J Immunopathol Pharmacol* 2011;**24**(4):1119–24.
35. Delclaux C, Mahut B, Zerah-Lancner F, Delacourt C, Laoud S, Cherqui D, et al. Increased nitric oxide output from alveolar origin during liver cirrhosis versus bronchial source during asthma. *Am J Respir Crit Care Med* 2002;**165**:332–7.
36. Thornadtsson A, Lind A, Weitoft T, Högman M. Altered levels of exhaled nitric oxide in rheumatoid arthritis. *Nitric Oxide* 2018;**76**:1–5.

PART

Inorganics and non-volatiles

B

CHAPTER

Exhaled nitric oxide in clinical practice

5

Andrei Malinovschi[1], Alain Michils[2], Marieann Högman[3]

[1]*Department of Medical Sciences, Clinical Physiology, Uppsala University, Uppsala, Sweden;* [2]*Department of Chest Medicine, Erasme University Hospital, Université Libre de Bruxelles, Brussels, Belgium;* [3]*Department of Medical Sciences, Respiratory, Allergy and Sleep Research, Uppsala University, Uppsala, Sweden*

Chapter outline

5.1 Overview

The detection of nitric oxide (NO) in exhaled breath in 1991[1] opened a new and broad research field. This placed initial demand on suitable technical solutions for its detection, and the chemiluminescence analyzers in use in environmental applications were soon adopted and adapted for exhaled breath analysis. Detection of NO using these systems was possible in the parts per billion (ppb) range. With the discovery that the fraction of exhaled NO (F_ENO)[a] is elevated in asthma,[2] F_ENO

[a] 50 mL/s is the recommended flow for single F_ENO measurements, referred to as F_ENO_{50}.

Breathborne Biomarkers and the Human Volatilome. https://doi.org/10.1016/B978-0-12-819967-1.00005-0
Copyright © 2020 Elsevier B.V. All rights reserved.

became the most promising inflammatory marker of asthma. Supporting this view were observations of the effect of inhaled corticosteroids (ICS) on F_ENO,[3] whereby F_ENO decreased promptly after starting treatment, then returned to baseline levels when treatment was discontinued; this offered the rationale for F_ENO-guided treatment in asthma and the use of F_ENO to assess adherence to medication. Subsequent research on production mechanisms of exhaled NO led to the development of models to characterize its generation and release under different conditions, as outlined in the previous chapter (Chapter 4). NO is one of the few exhaled biomarkers now in clinical routine use (see Chapter 1).

Several guidelines to standardize F_ENO measurements have been established over the years, and the most recent recommendations were issued jointly by the American Thoracic Society (ATS) and European Respiratory Society (ERS) in 2005.[4] The document summarizes factors that can affect endogenous NO production and recommends suitable procedures for its measurement. An update by the ERS in the form of a technical standard on exhaled biomarkers in lung disease[5] proposes the use of a questionnaire to assess use of tobacco, upper respiratory tract infection, respiratory diseases, chronic diseases, excercise, and intake of tea or coffee, or foods rich in nitrates and nitrites.

A practical guideline for the clinical interpretation of F_ENO and its cutoff values was issued in 2011 by the ATS.[6] In this document, F_ENO was recommended as a diagnostic aid for eosinophilic airway inflammation, and the current understanding is that it is a marker of type 2, interleukin (IL)-4/13–driven airway inflammation, and for determining the effects of steroid treatment responsiveness. According to the recommendations, a F_ENO value >50 ppb (in children >35 ppb) can be indicative of eosinophilic inflammation in the presence of symptoms, therefore responsiveness to ICS is likely. By comparison, F_ENO values in the range 25–50 ppb (in children 20–35 ppb) should be interpreted cautiously and a decision for starting ICS treatment or changing the dose of ICS should additionally take into account symptom burden. Today, F_ENO is finding footing in the general clinical guidelines for asthma diagnosis; the National Institute for Health and Care Excellence (NICE) (United Kingdom) and recent Scottish guidelines for asthma diagnosis and management, for instance, recommend F_ENO measurements for this purpose.[7] Further, the Global Initiative Network for Asthma (GINA) 2019 guidelines for management of difficult-to-treat and severe asthma place F_ENO as a marker to define type 2 inflammatory patterns or to follow or predict response to biological therapies.

5.2 Establishing a standardized approach for exhaled nitric oxide

Reference values of F_ENO from healthy individuals are used as a basis for accurately interpreting values recorded in clinical practice. Many individuals with respiratory symptoms might present F_ENO values within the intermediate range according to clinical cutoffs recommended in the ATS guidelines.[6] There have been numerous publications reporting reference values in children and adults, and these provide strong evidence that men have higher F_ENO levels than women. Age is another important factor to be accounted for when measuring F_ENO, whereby the current opinion is that three phases exist in the evolution of F_ENO.[8] F_ENO appears to increase linearly with somatic growth, in girls between 6 and 14 years and in boys between 6 and 16 years of age. Values then plateau until approximately age 45 years in females and 59 years in males. Thereafter, again F_ENO increases linearly up to the age of 80 years. The Global Lung Function Initiative supported by the ERS has recently started a taskforce to develop reference equations for F_ENO with the aim of generating individualized reference values to aid clinical decisions.

5.3 Lung function and airway caliber

The impact of airway caliber on F_ENO was demonstrated two decades ago when F_ENO was observed to decline with decreasing forced expiratory volume (FEV_1) induced by airway challenge.[9] It was proposed that this was likely due to a decrease in available epithelial surface, which reduces the total amount of NO that diffuses from the airway epithelium into the airway lumen.[10] It was also demonstrated that when the airway challenge procedures involved peripheral airways, the reduction had twice the magnitude compared with methacholine challenge.[11] Investigations on early and late phase reactions following an allergen challenge performed in mild asthmatics further showed that F_ENO may fully lose its ability to reflect airway inflammation if airway caliber is concomitantly reduced.[12] This may explain—at least to some extent—the relatively weak relationship between F_ENO and asthma control that is readily observed in patients suffering from severe asthma with deteriorated lung function.[13] Nevertheless, the correlation between F_ENO and asthma treatment shows significant improvement when concomitant changes in airway caliber and F_ENO are accounted for.[14] The direction of FEV_1 and F_ENO changes might reflect which site of the airways is affected: a decrease in FEV_1 and increase in F_ENO, for example, might reflect a more proximal involvement, while a decrease in both FEV_1 and F_ENO signifies a more distal process in the airways.[14] In individuals who exhibit opposing tendencies in FEV_1 and F_ENO changes over time, F_ENO appears to correlate better with asthma control.[14] These phenomena may be relevant with regard to changes of airway caliber in asthma during exacerbations, yet no such studies have investigated this, to date.

The behavior of F_ENO after bronchodilation can be utilized as a pulmonary function test to identify site of obstruction. In asthma[15] and COPD,[16] individuals with relief of intraacinar airway constriction, for example, exhibit decreased F_ENO that is thought to reflect amplification of NO back diffusion,[17] while subjects with relief of obstruction up to preacinar airways respond with increased F_ENO levels. Further studies are needed to establish the clinical value of measurements of F_ENO before and after bronchodilation in asthma.

5.4 Exhaled nitric oxide in asthma

5.4.1 Asthma diagnosis

Exhaled nitric oxide is elevated in the exhaled breath of patients with asthma.[2] This observation justifies the interest in using F_ENO in asthma diagnosis, yet there is a large variation in F_ENO values in healthy individuals, which has been associated with age, height, sex, allergic sensitization, and smoking. This variability limits the diagnostic value of F_ENO, which is considered moderate. Accordingly, the implementation of F_ENO in guidelines has been deferred. Currently, F_ENO is recommended in asthma diagnosis by NICE in the UK and in the Scottish guidelines for asthma diagnosis and management,[7] but not GINA. The proposed cutoff values for asthma diagnosis by NICE are 40 ppb in adults and 35 ppb in children.

Recent studies have focused on occupational asthma (see Chapter 32) and the potential utility of F_ENO to assess work-related increased inflammation in real-life settings[18] or to assess response to specific inhalation challenges.

5.4.2 Asthma endotypes

The current understanding of asthma is that it is a syndrome with different inflammatory pathways activated in different subtypes. Usually, the inflammation is divided into type 2 and non-type 2 mechanisms. The most studied mechanism is that of type 2 inflammation that is driven through T-helper type 2 cells and innate lymphoid cells of group 2 activation. A simplified description of the immunological mechanisms leading to increased F_ENO levels is given in Fig. 5.1. Several biological

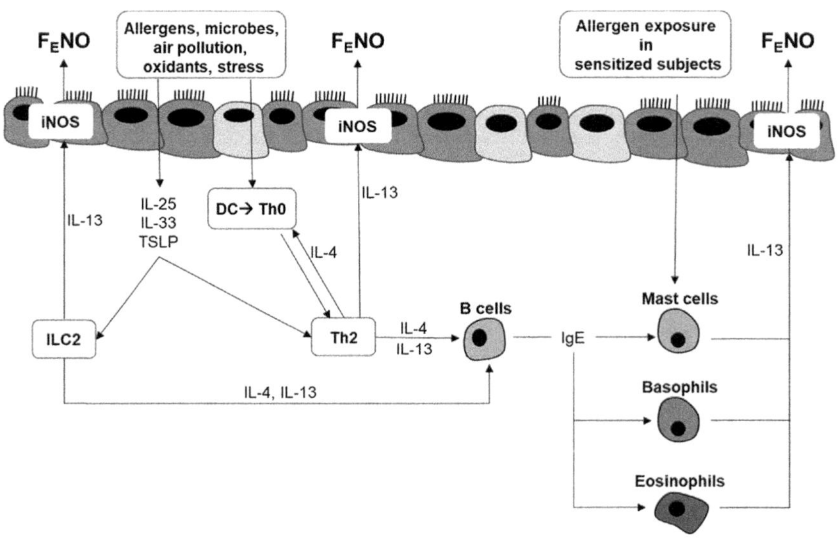

FIGURE 5.1

Production of F_ENO in asthma. *DC*, dendritic cell; *F_ENO*, fraction of exhaled nitric oxide; *IgE*, immunoglobulin E; *IL*, interleukin; *ILC*, innate lymphoid cell; *iNOS*, inducible nitric oxide synthase; *Th*, T-helper cell; *TSLP*, thymic stromal lymphopoietin.

treatments have emerged with the advances in mapping immunological processes that lead to airway inflammation in asthma, especially type 2 inflammation. F_ENO has been recognized by GINA and is recommended in its 2019 guidelines for management of difficult-to-treat and severe asthma as one of the biomarkers for defining type 2 asthma.

5.4.3 Monitoring and guiding therapy

Elevated F_ENO and blood eosinophils (B-Eos) levels may reflect activation of different inflammatory pathways (IL-4/13 and IL-5, respectively) and appear to be related to a higher burden of asthma and different asthma events—including asthma attacks or asthma-related hospital visits—in the general population.[19] These effects have been observed at relatively low levels of $F_ENO \geq 25$ ppb and B-Eos $\geq$300 cells/mm^3 but are more pronounced at higher levels of $F_ENO \geq$50 ppb and B-Eos $\geq$500 cells/mm^3. In severe asthma, a higher likelihood of severe exacerbations was found at levels of $F_ENO \geq$25 ppb and B-Eos $\geq$300 cells/mm^3.[20] The largest decrease in exacerbation rate was found in patients with severe asthma who were treated with dupilumab when exhibiting $F_ENO \geq$25 ppb and B-Eos $\geq$300 cells/mm^3. Similar cutoffs, i.e., $F_ENO \geq$30 ppb and B-Eos $\geq$300 cells/mm^3, have been reported in relation to type 2 inflammation, as assessed by airway epithelial cell transcriptomics.[21]

Inhaled corticosteroids

Inhaled corticosteroids are the primary treatment in asthma to control airway inflammation. Besides the discovery of elevated F_ENO in asthma, another seminal finding in this field was that F_ENO levels may be lowered by the use of ICS and modulated by changing the dose of ICS.[3] This observation led to great interest in F_ENO as a potential marker of airways inflammation control and a monitoring tool to guide asthma therapy. The first study that reported on this application was published in 2005 and demonstrated that F_ENO-guided treatment achieved similar control as symptom-guided treatment, with lower doses of ICS required.[22] Since then, many studies assessing F_ENO-guided treatment have been performed, and it is now generally accepted that F_ENO-guided treatment is superior to standard care in reducing exacerbations, as concluded by a recent metaanalysis,[23] although no significant effect on asthma control or lung function could be found. Some groups of patients, such as aeroallergen-sensitized subjects, may benefit more from F_ENO-guided treatment in reducing exacerbations.[24] Similarly, a pediatric study reported that atopy and baseline F_ENO >35 ppb were useful predictors in F_ENO-guided treatment.[25] Interestingly, a recent study analyzed the utility of F_ENO in identifying patients with nonspecific respiratory symptoms who would benefit most from ICS and found that individuals who exhibited higher F_ENO levels responded better to ICS for asthma control, severity of cough, and FEV_1.[26]

Inhaled corticosteroids titration during pregnancy has been associated with both better disease control, e.g., fewer exacerbations, and lower ICS doses. Moreover, the offspring of these asthma patients treated according to F_ENO-guided asthma management during pregnancy had reduced doctor-diagnosed asthma in follow-ups up to age six.[27]

A large proportion of subjects with asthma do not adhere to their ICS medication. As such, F_ENO is an interesting noninvasive measurement for assessing medication use, before other therapeutical alternatives, such as biological treatment, are considered. F_ENO has been reported to also be useful in assessing compliance of ICS treatment in patients with difficult asthma who display F_ENO >45 ppb at inclusion.[28] Direct observation of treatment demands large resources, but remote assessment of treatment nonadherence via F_ENO measurements has been demonstrated in difficult-to-control asthma; this finding provides clinically important future benefits.[29]

Biological treatment

Anti-Immunoglobulin E (IgE) treatment was the first biological treatment available for asthma, and like with ICS, the use of F_ENO for guiding and/or monitoring therapy has been demonstrated for this treatment. In a large multicenter longitudinal study that explored the potential of type 2 biomarkers, the largest reduction of asthma exacerbations was observed in individuals with F_ENO ≥19 ppb, with an approximate threefold reduction.[30]

Anti-IL-5 treatment targets eosinophils, therefore no effect on F_ENO reduction has been observed using this treatment[31]; instead, B-Eos is used as a biomarker for this treatment, whereby severe asthma patients with B-Eos ≥300 cells identified as suitable candidates for therapy.[32] Nonetheless, baseline F_ENO ≥50 ppb has been associated with a positive response to the treatment.[31]

Two anti-IL-13 drugs have been tested in asthma and proven to reduce F_ENO levels, whereby increased F_ENO appeared to predict response to treatment with lebrikizumab,[33] although the phase 3 studies were negative, and these treatments were therefore discontinued.

Dupilumab is emerging as a more efficient therapy than anti-IL-13 drugs as it binds to the alpha subunit of the IL-4 receptor and consequently blocks the signal transduction pathways that are activated by both IL-4 and IL-13. F_ENO appears to be a useful marker to predict the effect of dupilumab on asthma exacerbations in moderate-to-severe uncontrolled asthma.[34] This effect was observed at relatively low values of F_ENO ≥25 ppb, and to a greater extent at ≥50 ppb, compared to subjects with F_ENO <25 ppb. Furthermore, the combination of both elevated F_ENO (≥25 ppb) and B-Eos (≥150 cells/mm^3) related to the largest exacerbation reduction.[34]

New asthma treatments are currently being tested, such as antithymic stromal lymphopoietin (anti-TSLP). The treatment has been tested in several phase 2 studies and proven efficient with regard to exacerbations and reducing F_ENO, although baseline F_ENO could not predict the effect of anti-TSLP.[35]

5.5 Exhaled nitric oxide in other pulmonary diseases

5.5.1 Chronic obstructive pulmonary disease

Exhaled nitric oxide is much less studied in chronic obstructive pulmonary disease (COPD) than in asthma (see Chapter 19). The interpretation of F_ENO is more problematic in COPD, since the majority of the patients are smokers or ex-smokers who generally have lower F_ENO values than non-smoking individuals (although values increase after smoking cessation). F_ENO values seem to be decreased by ICS in COPD, but this effect appears to be confined to ex-smokers, since a treatment effect of ICS on F_ENO was not observed in current smokers in stable disease.[36] F_ENO does not appear to relate to the stages of disease according to the Global Initiative for Chronic Obstructive Lung Disease (GOLD) but does appear to increase in relation to exacerbations. Individuals with COPD exacerbations, for example, had a mean F_ENO value of 49 ppb before receiving treatment with peroral corticosteroids, compared to 29 ppb in individuals that had already commenced treatment.[37] A good correlation between B-Eos and F_ENO was found at admission but not after a few days of treatment. Similarly, a study performed in patients with stable condition suggests that persistently elevated F_ENO (>20 ppb) relates to a higher risk of exacerbations.[38] Therefore, F_ENO measurement would be beneficial in individual COPD patients, since an early increase might predict an exacerbation regardless of the severity of the disease. More studies are needed on F_ENO in relation to COPD phenotype. Recent metaanalysis suggests that F_ENO might be increased in asthma–COPD overlap (ACO),[39] although the latest GOLD guidelines from 2020 state that asthma and COPD are different disorders and deprecate the term ACO, thus the clinical importance of F_ENO in ACO is questionable.

5.5.2 Chronic cough

Values of F_ENO are increased in cough variant asthma compared to noncough variant asthma and can therefore be used to differentiate between them. In a study with children, the cutoff F_ENO value was 25 ppb to identify cough variant asthma.[40] Recent guidelines on managing chronic cough suggest that noninvasive measures of airway inflammation could help predict response to corticosteroids in patients with cough.[41] A few studies have hypothesized that F_ENO could guide the response to corticosteroids, and although the results range from "not useful at all" to "very useful," this broad spread was most likely due to the different methodologies employed. In terms of response to corticosteroid therapy, in one study 86% of adult patients with F_ENO >25 ppb responded to treatment while only 46% did when F_ENO was <25 ppb.[42] In another study, 88% of individuals had a significant improvement on ICS for F_ENO >35 ppb, compared to only 9% when F_ENO was in reference range.[43] Randomized control studies of F_ENO-guided therapy in chronic cough are needed.

5.5.3 Primary ciliary dyskinesia

Primary ciliary dyskinesia (PCD) is a condition in which the ciliary function is totally or partially impaired. F_ENO levels in PCD patients have been shown to both overlap and be lower than values in healthy subjects.[44] Specifically, nasal NO (F_nNO) is decreased and is therefore useful in assessing the diagnosis of PCD.[45] F_nNO is included in a proposed algorithm in the official ATS clinical practice guidelines for setting the diagnosis of PCD.[45] In this document, it is argued that F_nNO measurements should only be performed with chemiluminescence analyzers using standardized protocols at centers with specific expertise in F_nNO measurements.[46] Different technical methods to measure F_nNO have been proposed, and simultaneous measurement of F_ENO and F_nNO could be useful.[5]

5.5.4 Cystic fibrosis

Cystic fibrosis (CF) is a genetic disorder that is characterized by chronic airway inflammation, increased mucus production, recurrent infections, and the frequent colonization of airways with pathogenic bacteria, such as *Pseudomonas aeruginosa*. Inducible nitric oxide synthase (iNOS) expression has been found to be low in CF patients, and infants with CF have low values of F_ENO, suggesting that F_ENO reduction is seen early in life.[47] Despite the chronic airway inflammation, F_ENO levels in CF sufferers remain generally low. Therefore, low F_ENO could be prognostic of disease progression during *P. aeruginosa* colonization.[48] A recent study suggests that F_ENO is increased in children with CF with allergic bronchopulmonary aspergillosis (ABPA) and that F_ENO can be of value in its diagnosis.[49] NO donors activate the chloride transport in CF airway epithelia in vitro, and treatment with ivacaftor, potentiator of the chloride channel, has been shown to increase F_ENO levels as well as lung function acutely,[50] although F_ENO levels returned to pretreatment values during a 12-month follow-up.[48] Future studies must demonstrate whether F_ENO measurements are useful in the acute phase to observe treatment response.

5.6 Summary

Exhaled nitric oxide, F_ENO, is an established biomarker of asthma that is recommended in guidelines on asthma diagnosis and therapeutic monitoring. Measuring F_ENO offers decision support for asthma diagnosis. In established asthma, F_ENO has a role in assessing adherence to medication and monitoring effect of antiinflammatory medication, it is one of the markers for type 2 inflammation, and it offers support for selection of patients for biological treatment, especially anti-IL-4/13. The current ongoing work in creating reference values for F_ENO, together with further understanding on the intricate relationship between airway caliber and F_ENO, will lead to increased use of exhaled nitric oxide as a reliable biomarker for diagnostic and treatment efficacy assessments. For a better understanding of which F_ENO should be targeted, the concept of personal best F_ENO value is interesting and appears possible to estimate by modeling of NO production in the lung, as described in Chapter 4. With technological advances and the development of new measurement techniques for NO detection, devices are becoming smaller, cheaper, and more feasible for home monitoring. Another area of interest in the use of F_ENO, with promising results, is in chronic cough, but this needs to be further tested in randomized controlled trials that compare F_ENO-guided treatment with best clinical practice. The current clinical value of F_ENO in COPD and CF is limited.

Nasal nitric oxide, F_nNO, measurements are included in guidelines for PCD diagnosis, although current recommendations to use chemiluminescence methods of detection limit its use, thus further studies are needed to explore the use of alternative measurement technologies in PCD diagnosis.

References

1. Gustafsson LE, Leone A, Persson M, Wiklund N, Moncada S. Endogenous nitric oxide is present in the exhaled air of rabbits, Guinea pigs and humans. *Biochem Biophys Res Commun* 1991;**181**:852–7.
2. Alving K, Weitzberg E, Lundberg JM. Increased amount of nitric oxide in exhaled air of asthmatics. *Eur Respir J* 1993;**6**:1368–70.
3. Kharitonov SA, Yates DH, Chung KF, Barnes PJ. Changes in the dose of inhaled steroid affect exhaled nitric oxide levels in asthmatic patients. *Eur Respir J* 1996;**9**:96–201.
4. ATS/ERS recommendations for standardized procedures for the online and offline measurement of exhaled lower respiratory nitric oxide and nasal nitric oxide. *Am J Respir Crit Care Med* 2005;**171**:912–30.
5. Horvath I, Barnes PJ, Loukides S, Sterk PJ, Högman M, Olin AC, et al. A European Respiratory Society technical standard: exhaled biomarkers in lung disease. *Eur Respir J* 2017;**49**(4).
6. Dweik RA, Boggs PB, Erzurum SC, Irvin CG, Leigh MW, Lundberg JO, et al. An official ATS clinical practice guideline: interpretation of exhaled nitric oxide levels (F_ENO) for clinical applications. *Am J Respir Crit Care Med* 2011;**184**:602–15.
7. Kuo CR, Spears M, Haughney J, Smith A, Miller J, Bradshaw T, et al. Scottish consensus statement on the role of F_ENO in adult asthma. *Respir Med* 2019;**155**:54–7.
8. Jacinto T, Malinovschi A, Janson C, Fonseca J, Alving K. Evolution of exhaled nitric oxide levels throughout development and aging of healthy humans. *J Breath Res* 2015;**9**: 036005.
9. de Gouw HWFM, Hendriks J, Woltman AM, Twiss IM, Sterk PJ. Exhaled nitric oxide (NO) is reduced shortly after bronchoconstriction to direct and indirect stimuli in asthma. *Am J Respir Crit Care Med* 1998;**158**:315–9.
10. Verbanck S, Kerckx Y, Schuermans D, Vincken W, Paiva M, van Muylem A. Effect of airways constriction on exhaled nitric oxide. *J Appl Physiol* 2008;**104**:925–30.
11. Van Muylem A, Kerckx Y, Michils A. Axial distribution of nitric oxide airway production in asthma patients. *Respir Physiol Neurobiol* 2013;**185**:313–8.
12. Haccuria A, Michils A, Michiels S, Van Muylem A. Exhaled nitric oxide: a biomarker integrating both lung function and airway inflammation changes. *J Allergy Clin Immunol* 2014;**134**:554–9.
13. Michils A, Baldassarre S, Van Muylem A. Exhaled nitric oxide and asthma control: a longitudinal study in unselected patients. *Eur Respir J* 2008;**31**:539–46.
14. Michils A, Haccuria A, Michiels S, Van Muylem A. Airway calibre variation is a major determinant of exhaled nitric oxide's ability to capture asthma control. *Eur Respir J* 2017;**50**(2).
15. Michils A, Malinovschi A, Haccuria A, Michiels S, Van Muylem A. Different patterns of exhaled nitric oxide response to beta2-agonists in asthmatic patients according to the site of bronchodilation. *J Allergy Clin Immunol* 2016;**137**:806–12.

16. Perez-Bogerd S, Michils A, Malinovschi A, Van Muylem A. COPD patients with peripheral airway obstruction reversibility identified by exhaled nitric oxide. *J Breath Res* 2019;**13**:036002.
17. Kerckx Y, Michils A, Van Muylem A. Airway contribution to alveolar nitric oxide in healthy subjects and stable asthma patients. *J Appl Physiol* 2008;**104**:918–24.
18. van Kampen V, Bruning T, Merget R. Serial fractional exhaled nitric oxide measurements off and at work in the diagnosis of occupational asthma. *Am J Ind Med* 2019; **62**:663–71.
19. Malinovschi A, Fonseca JA, Jacinto T, Alving K, Janson C. Exhaled nitric oxide levels and blood eosinophil counts independently associate with wheeze and asthma events in National Health and Nutrition Examination Survey subjects. *J Allergy Clin Immunol* 2013;**132**:821–7.
20. Soma T, Iemura H, Naito E, Miyauchi S, Uchida Y, Nakagome K, et al. Implication of fraction of exhaled nitric oxide and blood eosinophil count in severe asthma. *Allergol Int* 2018;**67S**:S3–11.
21. Pavlidis S, Takahashi K, Ng Kee Kwong F, Xie J, Hoda U, Sun K, et al. "T2-high" in severe asthma related to blood eosinophil, exhaled nitric oxide and serum periostin. *Eur Respir J* 2019;**53**(1).
22. Smith AD, Cowan JO, Brassett KP, Herbison GP, Taylor DR. Use of exhaled nitric oxide measurements to guide treatment in chronic asthma. *N Engl J Med* 2005;**352**:2163–73.
23. Petsky HL, Cates CJ, Kew KM, Chang AB. Tailoring asthma treatment on eosinophilic markers (exhaled nitric oxide or sputum eosinophils): a systematic review and meta-analysis. *Thorax* 2018;**73**:1110–9.
24. Zeiger RS, Schatz M, Yang SJ, Chen W. Fractional exhaled nitric oxide-assisted management of uncontrolled persistent asthma: a real-world prospective observational study. *Perm J* 2019;**23**.
25. Morphew T, Shin HW, Marchese S, Pires-Barracosa N, Galant SP. Phenotypes favoring fractional exhaled nitric oxide discordance vs guideline-based uncontrolled asthma. *Ann Allergy Asthma Immunol* 2019;**123**:193–200.
26. Price DB, Buhl R, Chan A, Freeman D, Gardener E, Godley C, et al. Fractional exhaled nitric oxide as a predictor of response to inhaled corticosteroids in patients with non-specific respiratory symptoms and insignificant bronchodilator reversibility: a randomised controlled trial. *Lancet Respir Med* 2018;**6**:29–39.
27. Morten M, Collison A, Murphy VE, Barker D, Oldmeadow C, Attia J, et al. Managing Asthma in Pregnancy (MAP) trial: F_ENO levels and childhood asthma. *J Allergy Clin Immunol* 2018;**142**:1765–72.
28. McNicholl DM, Stevenson M, McGarvey LP, Heaney LG. The utility of fractional exhaled nitric oxide suppression in the identification of nonadherence in difficult asthma. *Am J Respir Crit Care Med* 2012;**186**:1102–8.
29. Heaney LG, Busby J, Bradding P, Chaudhuri R, Mansur AH, Niven R, et al. Remotely monitored therapy and nitric oxide suppression identifies nonadherence in severe asthma. *Am J Respir Crit Care Med* 2019;**199**:454–64.
30. Hanania NA, Wenzel S, Rosen K, Hsieh HJ, Mosesova S, Choy DF, et al. Exploring the effects of omalizumab in allergic asthma: an analysis of biomarkers in the EXTRA study. *Am J Respir Crit Care Med* 2013;**187**:804–11.
31. Pavord ID, Korn S, Howarth P, Bleecker ER, Buhl R, Keene ON, et al. Mepolizumab for severe eosinophilic asthma (DREAM): a multicentre, double-blind, placebo-controlled trial. *Lancet* 2012;**380**:651–9.

32. Bleecker ER, Wechsler ME, FitzGerald JM, Menzies-Gow A, Wu Y, Hirsch I, et al. Baseline patient factors impact on the clinical efficacy of benralizumab for severe asthma. *Eur Respir J* 2018;**52**(4).
33. Corren J, Lemanske RF, Hanania NA, Korenblat PE, Parsey MV, Arron JR, et al. Lebrikizumab treatment in adults with asthma. *N Engl J Med* 2011;**365**:1088–98.
34. Castro M, Corren J, Pavord ID, Maspero J, Wenzel S, Rabe KF, et al. Dupilumab efficacy and safety in moderate-to-severe uncontrolled asthma. *N Engl J Med* 2018;**378**: 2486–96.
35. Gauvreau GM, O'Byrne PM, Boulet LP, Wang Y, Cockcroft D, Bigler J, et al. Effects of an anti-TSLP antibody on allergen-induced asthmatic responses. *N Engl J Med* 2014; **370**:2102–10.
36. Högman M, Thornadtsson A, Bröms K, Janson C, Lisspers K, Ställberg B, et al. Different relationships between F_ENO and COPD characteristics in smokers and ex-Smokers. *COPD* 2019;**16**:227–33.
37. Rio Ramirez MT, Juretschke Moragues MA, Fernandez Gonzalez R, Alvarez Rodriguez V, Aznar Andres E, Zabaleta Camino JP, et al. Value of exhaled nitric oxide (F_ENO) and eosinophilia during the exacerbations of chronic obstructive pulmonary disease requiring hospital admission. *COPD* 2018;**15**:369–76.
38. Alcazar-Navarrete B, Ruiz Rodriguez O, Conde Baena P, Romero Palacios PJ, Agusti A. Persistently elevated exhaled nitric oxide fraction is associated with increased risk of exacerbation in COPD. *Eur Respir J* 2018;**51**(1).
39. Hosseini M, Almasi-Hashiani A, Sepidarkish M, Maroufizadeh S. Global prevalence of asthma-COPD overlap (ACO) in the general population: a systematic review and meta-analysis. *Respir Res* 2019;**20**:229.
40. Zhou J, Zhao X, Zhang X, Yu X, Wang Y, Jiang W, et al. Values of fractional exhaled nitric oxide for cough-variant asthma in children with chronic cough. *J Thorac Dis* 2018;**10**:6616–23.
41. Cote A, Russell RJ, Boulet LP, Gibson PG, Lai K, Irwin RS, et al. *Managing chronic cough due to asthma and NAEB in adults and adolescents: CHEST guideline and expert panel report. Chest*. 2020.
42. Lamon T, Didier A, Brouquieres D, Escamilla R, Dupuis M, Guibert N, et al. Exhaled nitric oxide in chronic cough: a good tool in a multi-step approach. *Respir Med Res* 2019;**76**:4–9.
43. Hahn PY, Morgenthaler TY, Lim KG. Use of exhaled nitric oxide in predicting response to inhaled corticosteroids for chronic cough. *Mayo Clin Proc* 2007;**82**:1350–5.
44. Karadag B, James AJ, Gultekin E, Wilson NM, Bush A. Nasal and lower airway level of nitric oxide in children with primary ciliary dyskinesia. *Eur Respir J* 1999;**13**:1402–5.
45. Shapiro AJ, Davis SD, Polineni D, Manion M, Rosenfeld M, Dell SD, et al. Diagnosis of primary ciliary dyskinesia. An official American thoracic society clinical practice guideline. *Am J Respir Crit Care Med* 2018;**197**:e24–39.
46. Shapiro AJ, Dell SD, Gaston B, O'Connor M, Marozkina N, Manion M, et al. Nasal nitric oxide measurement in primary ciliary dyskinesia. A Technical Paper on Standardized Testing Protocols. *Ann Am Thorac Soc* 2020;**17**:e1–12.
47. Korten I, Liechti M, Singer F, Hafen G, Rochat I, Anagnostopoulou P, et al. Lower exhaled nitric oxide in infants with Cystic Fibrosis compared to healthy controls. *J Cyst Fibros* 2018;**17**:105–8.

48. Vincken S, Verbanck S, De Wachter E, Vanderhelst E. Exhaled nitric oxide in stable adult cystic fibrosis patients, during exacerbation and following CFTR-modifying treatment. *Eur Respir J* 2019;**53**(5).
49. Keown K, Abbott S, Kuzeljevic B, Rayment JH, Chilvers MA, Yang CL. An investigation into biomarkers for the diagnosis of ABPA and aspergillus disease in cystic fibrosis. *Pediatr Pulmonol* 2019;**54**:1787–93.
50. Grasemann H, Gonska T, Avolio J, Klingel M, Tullis E, Ratjen F. Effect of ivacaftor therapy on exhaled nitric oxide in patients with cystic fibrosis. *J Cyst Fibros* 2015;**14**:727–32.

CHAPTER

Exhaled carbon monoxide

6

Stefan W. Ryter

Department of Medicine, Weill Cornell Medical Center, New York, NY, United States

Chapter outline

6.1 Overview

The analysis of the chemical composition of human exhaled breath for clinical diagnostics and therapeutic monitoring remains an important area of biomedicine in continuing stages of development. The profiles of exhaled metabolites may provide information on the presence and severity of disease, in particular pulmonary and airways diseases such as asthma, inflammatory lung disease, chronic obstructive pulmonary disease (COPD), idiopathic pulmonary fibrosis (IPF), as well as other human diseases, including metabolic diseases, diseases of the liver and kidney, and cancer. Exhaled breath testing may also provide a means to determine the effectiveness of therapeutic intervention, such as antiinflammatory therapies.

Breathborne Biomarkers and the Human Volatilome. https://doi.org/10.1016/B978-0-12-819967-1.00006-2
Copyright © 2020 Elsevier B.V. All rights reserved.

There are many possibilities for exhaled breath monitoring, as detailed in other chapters in this book. Much research has focused on low molecular weight diatomic gases present in breath, namely nitric oxide (NO) and carbon monoxide (CO). Fraction of exhaled nitric oxide (F_ENO) has been widely studied in the context of human respiratory diseases, including asthma, COPD, and IPF, and has been suggested as a marker of airway inflammation (see Chapters 4 and 5). Similarly, exhaled carbon monoxide ($_ECO$) levels have been monitored in asthma/airways diseases and COPD.[1,2] Furthermore, $_ECO$ has been employed to monitor cigarette smoking in cessation programs.[3] Additional focus areas for exhaled gas research involving either $_ECO$ or F_ENO measurements include cystic fibrosis (CF), diabetes, surgery and critical illness, and others.[1]

This chapter will focus primarily on the potential applications of $_ECO$ measurement in lung disease. Sources of environmental and endogenous CO will be discussed from the perspective of exhaled breath monitoring, with an emphasis on airways inflammation, asthma, smoking, and COPD. The aim of this chapter is to summarize past research in this area in the hope of stimulating interest for further development.

6.2 Carbon monoxide: environmental and biological aspects

6.2.1 Chemical and biochemical properties of carbon monoxide

Carbon monoxide, an odorless, colorless, and tasteless gas, poses a dangerous hazard in the context of environmental and occupational exposures.[4] CO is similar to another endogenously produced gas nitric oxide (NO), with respect to molecular structure, size, and molecular weight (CO, $M_R = 28.01$ g/mol, NO, $M_R = 30.006$ g/mol). Both CO and NO exist in gaseous form at ambient temperature and pressure (boiling points, CO: 191.5°C, NO: 151.8°C; melting points, CO: 205.1°C, NO: 163.6°C) and are water soluble (CO: 0.0015 mol/L, NO: 0.0033 mol/L, at 1.013 bar and 0°C). CO is a stable molecule with little physiological reactivity with respect to non-iron compounds.[4] In contrast, NO, a free radical, contains one unpaired electron, which permits participation in a number of redox reactions under physiological conditions, including reaction with mitochondria-generated superoxide anion to produce the reactive species peroxynitrite.[1]

The compounds CO and NO are similar in their capacity to act as heme ligands, and both gases can bind with heme-containing proteins at the heme iron. CO typically binds only to reduced (ferrous) iron centers, whereas NO may bind to both ferrous and ferric hemes. CO binds avidly to the heme groups of circulating hemoglobin under physiological conditions to form a stable but reversible adduct carboxyhemoglobin (HbCO). This interaction occurs with ~250 times the affinity of molecular oxygen (O_2).[4] Binding of CO to hemoglobin impairs the release of the remaining bound O_2, further impairing O_2 delivery to tissues. The resulting hypoxemia can lead to death in cases of prolonged or excessive exposure to CO.[4]

6.2.2 Environmental carbon monoxide

Carbon monoxide occurs as a common contaminant of indoor and outdoor air. Environmental sources can account for a significant fraction of the CO found in human exhaled breath. Automobile exhaust, burning organic material (including wood and tobacco smoke), and the inefficient combustion of fossil fuels represent the major sources of CO in the ambient air.[4] Indoor levels of CO range from 0.5 to 5 parts per million (ppm) but may reach higher values (up to 30 ppm) at point sources of combustion. CO levels in densely populated urban areas may reach 10–50 ppm.[4] The US National Ambient Air Quality Standard (NAAQS) for CO is <9 ppm as an 8 h nonoverlapping average, and 35 ppm for a 1 h exposure.[4] Cigarette smoke, which contains ~4% CO by volume, represents a significant source of ambient CO. Individual cigarettes may yield an average of 67 mg CO per cigarette. Values of HbCO in heavy smokers may reach as high as 15%–17%.[4]

6.2.3 Endogenous sources of carbon monoxide: the heme oxygenase enzyme system

Endogenously produced CO arises primarily from the enzymatic oxidation of heme catalyzed by the heme oxygenase (HO, E.C. 1:14:99:3) enzymes[5] (Fig. 6.1). While the systemic turnover of hemoglobin likely represents the major source of heme for CO production, heme derived from other hemoprotein sources can contribute to CO generation at a cellular level, whereas a minor fraction of CO production may originate from nonheme sources, such as xenobiotic metabolism.[5]

Heme oxygenase activity provides the rate-limiting step in the heme degradative pathway.[5] HO oxidizes heme at the α-methene bridge carbon in a reaction requiring molecular oxygen and electrons derived from NADPH cytochrome P450 reductase. The HO reaction produces equimolar amounts of CO, iron, and bilirubin-IXα, the latter which is reduced to biliverdin-IXα by NAD(P)H: biliverdin reductase. HO exists as two major isoforms (HO-1, HO-2). The inducible form of heme oxygenase (HO-1), a major cellular stress protein, responds to transcriptional regulation by multiple forms of chemical and physical stress. These include oxidants, electrophilic antioxidant compounds and phytochemicals, heavy metals, ultraviolet-A radiation, thiol-reactive substances, and fluctuations in ambient oxygen tension. HO-1 represents a general marker of oxidative or proinflammatory stress, and its elevated expression has been observed consistently in various lung injury and sepsis models. In contrast, HO-2 is a constitutive isoform expressed highly in testis, liver, brain, and vascular endothelium.[5]

Based on the enzymology of HO, the endogenous production of CO is predicted to have both basal and stress-inducible components, both of which depend on the bioavailability of heme as substrate. Heme serves as an inducer of HO-1

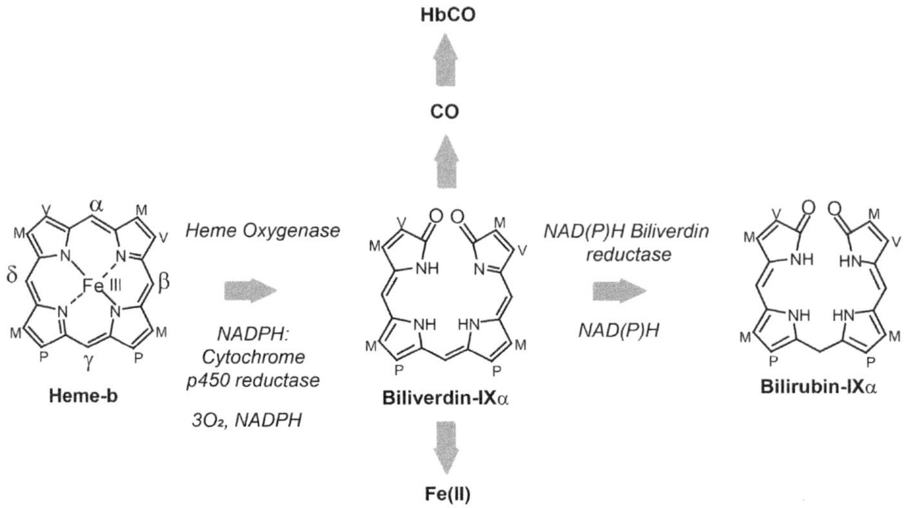

FIGURE 6.1

Heme oxygenase reaction as a major source of endogenous CO. Heme oxygenase (HO; E.C. 1:14:99:3) catalyzes the rate-limiting step in the oxidative conversion of heme-b to biliverdin-IXα. The reaction requires molecular oxygen (O_2) and electrons from NADPH: cytochrome P450 reductase. The HO reaction liberates the α-methene bridge carbon of heme as CO and releases the central heme iron as ferrous iron (Fe II). Biliverdin formed in the HO reaction is converted to bilirubin-IXα by NAD(P)H biliverdin reductase. HO-1 is the inducible form of HO that can be upregulated by inflammation and oxidative stress. The CO released from the HO reaction is available for reaction with hemoprotein targets, mainly represented by complexation with circulating hemoglobin to form carboxyhemoglobin (HbCO).

transcriptional regulation and expression, as well as the major enzymatic substrate. CO generation attributable to endogenous HO enzymatic activity may increase in response to systemic stress, such as inflammation.[5] CO may theoretically arise as the product of inducible HO-1 activity in any cell type, including alveolar macrophages, endothelial cells, airway epithelial cells in the lung, as well as in most cells of other organ tissues, such as liver and kidney, as a consequence of local inflammation or oxidative stress[5] (Fig. 6.2). Exhaled CO likely originates mainly from the alveolus by diffusion from the HbCO pool in the pulmonary circulation. Elevations in HbCO correlate to increases in environmental CO exposure and are also responsive to cigarette smoke exposure. In the absence of environmental CO inhalation exposure, increased HbCO may reflect increased HO activity in peripheral tissues and increased systemic hemoglobin turnover.[5,6]

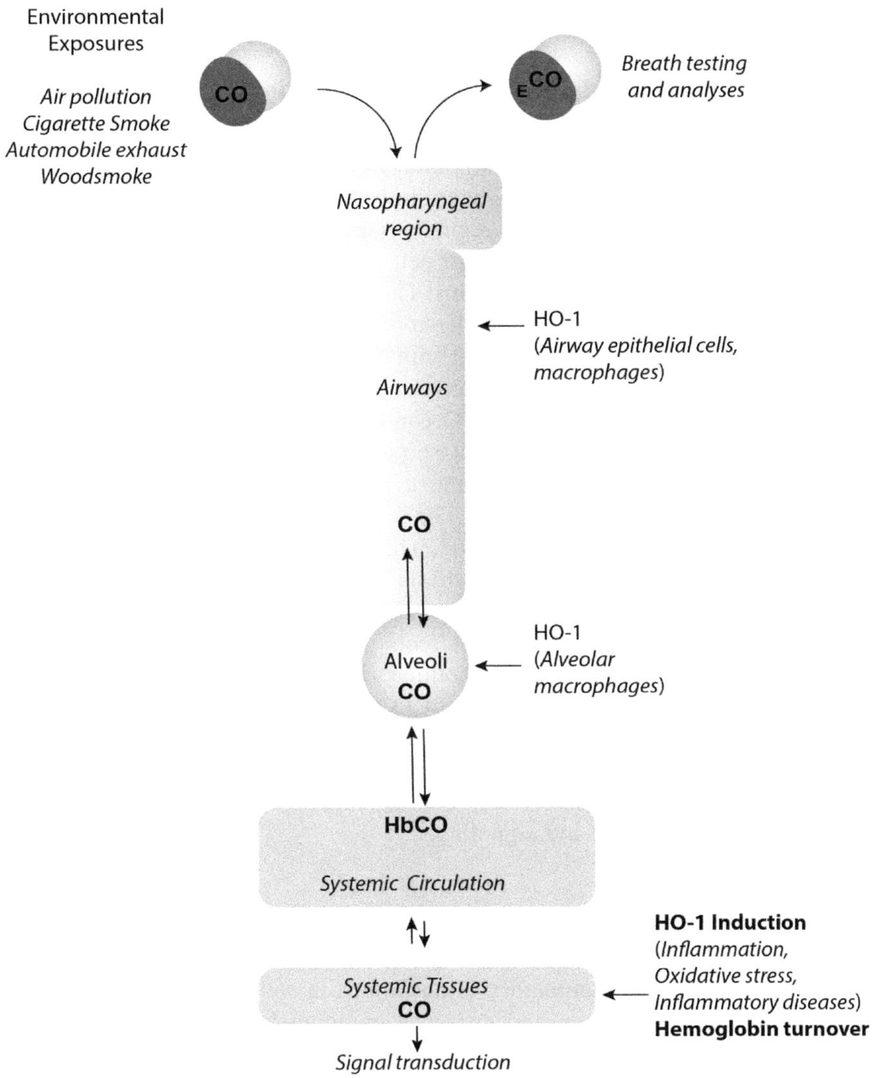

FIGURE 6.2

Biodistribution of CO. Exhaled CO ($_{E}CO$) can be measured in the exhaled breath of humans and is a possible indicator of inflammatory disease. CO can be inspired as a consequence of environmental or occupational exposure to cigarette smoke, wood smoke, automobile exhaust, and air pollution. CO also evolves endogenously in the body as the product of heme oxidation catalyzed by heme oxygenase (HO) enzymes. The inducible form of this enzyme, HO-1, responds to regulation by proinflammatory stress. Thus, increased HO-1 activity in the airways and airway or alveolar macrophages in response to airway inflammation may account in part for changes in $_{E}CO$. CO may also evolve from systemic hemoglobin turnover and the degradation of other hemoproteins. In the systemic circulation, CO exists mainly as carboxyhemoglobin (HbCO) and diffuses passively from the lung through the alveoli. CO may be produced in systemic tissues as a result of increased heme turnover. CO in the systemic tissues may impact cellular signaling and homeostatic processes through interactions with cellular hemoproteins.

6.2.4 Carbon monoxide as an intracellular signaling molecule

Carbon monoxide has been proposed to act as a cellular and physiological mediator on the basis of extensive investigation in cellular and animal studies, as is reviewed elsewhere.[5] These studies have suggested that CO can modulate cellular signal transduction pathways that regulate various cellular programs, including apoptosis and other forms of regulated cell death, autophagy, proliferation, and innate and adaptive immune mechanisms. The mechanisms by which CO can elicit these effects remain incompletely understood. Mitochondria have been proposed as a key proximal target for CO action, with modulation of mitochondrial respiratory chain activity and reactive oxygen species (ROS) generation as key effectors of downstream signaling pathways. CO can bind to numerous cellular hemoproteins, including soluble guanylate cyclase (sGC), inducible nitric oxide synthase (iNOS), NADPH oxidases, cytochrome *c* oxidase, and other targets, such as heme-containing transcription factors. The functional consequences of these interactions remain incompletely understood. The binding of CO to soluble guanylate cyclase results in the increased synthesis of guanosine 3′,-5′-monophosphate (cGMP), albeit with much reduced potency relative to NO. cGMP has been implicated in vaso- and neuromodulatory effects of CO. Furthermore, CO has been implicated in the modulation of cellular ion channel activities, including calcium-dependent potassium channels. CO exposure has been reported to indirectly modulate the activation of mitogen-activated protein kinases (MAPK), which are critical for the transduction or proinflammatory and/or stress signaling. Other potential mechanisms for CO action include the regulation of Toll-like receptor (TLR) trafficking and signaling.[5]

6.2.5 Carbon monoxide as a therapeutic agent in preclinical and clinical applications

Preclinical studies of inflammatory organ injury in rodents and higher animals, such as nonhuman primates and swine, accumulated over two decades, have uncovered beneficial and therapeutic effects of CO when applied endogenously at relatively low ambient concentrations (i.e., 100–250 ppm). These concentrations result in elevations of circulating HbCO. The salutary effects of CO have been demonstrated in numerous injury models, including -but not limited to- organ transplantation/graft rejection, ischemia/reperfusion injury, acute lung and kidney injury, vascular injury, mechanical ventilation–induced lung injury, experimental pulmonary fibrosis, polymicrobial sepsis, nonalcoholic steatohepatitis, preeclampsia, and others. These cyto- and tissue-protective properties of CO have been attributed to potent antiinflammatory effects, as well as other mechanisms, including effects on the microvasculature, the modulation of regulated cell death pathways, the regulation of cellular proliferation and platelet aggregation, as well as pronounced effects on innate and adaptive immunity.[5] Recent studies have attempted to translate these preclinical findings to clinical application. The safety of inhaled CO, for example, has recently been tested in human clinical studies of acute respiratory distress syndrome[a]. Research on the experimental therapeutic and clinical therapeutic aspects of CO has been reviewed elsewhere.[5]

6.3 Carbon monoxide as an exhaled biomarker of pulmonary disease

A number of studies have reported associations between $_{E}CO$ levels and various disease states. Modulations in $_{E}CO$ levels have been observed in neonatal jaundice, critically ill or postsurgical patients, and patients with pulmonary diseases associated with inflammation, including COPD, asthma, CF, and bacterial infections.[1,2,6] The potential application of $_{E}CO$ as a predictive tool for assessing pulmonary inflammation in these disease contexts is discussed below.

6.3.1 Exhaled carbon monoxide

Early studies relying on measurements of $_{E}CO$ were compromised by the relatively insensitive detection methods employed. The $_{E}CO$ levels in exhaled breath were most commonly measured with electrochemical (chemiluminescence) technology.[1,2,6] The values thus obtained correlate with parallel gas chromatographic (GC) analysis, and these sensors are sensitive in the 1−500 ppm range.[1,2] CO may also be detected using laser spectroscopic methods (see Chapter 14), which offer advantages for increased sensitivity (relative to electrochemical detection methods) and the potential to facilitate real-time, online monitoring with breath-to-breath resolution. Infrared laser spectroscopy methods that have been described for CO measurement include cavity leak-out spectroscopy (CALOS) and integrated cavity output spectroscopy (ICOS). More recent advances describe tunable diode laser absorption spectroscopy (TDLAS) methods for CO detection, which can be enhanced by multi-pass cell (MPS) or wavelength modulation spectroscopic (WMS) methods.

6.3.2 Asthma

Asthma is a chronic inflammatory disease of the airways resulting in airflow limitation, which is partially reversible either spontaneously or by therapy. The persistent inflammation associated with this disease causes airway hypersensitivity to allergens and irritants, such as smoke, pollens, or dust. The pathological airway changes observed in asthma include eosinophil granulocyte predominant inflammation, edema, bronchoconstriction, and increased thick mucus secretion. Measured $_{E}CO$ levels were significantly increased in 37 nonsteroid-treated asthmatic patients compared with 37 healthy subjects, whereas $_{E}CO$ levels were not elevated in patients

[a] www.clinicaltrials.gov; NCT03799874; NCT02425579.

who received corticosteroid therapy.[6] Acute exacerbations of asthma caused increases in measured $_{E}CO$ associated with a reduction in peak expiratory flow rate (PEFR). Both increases in $_{E}CO$ and PEFR decline were reversible by oral glucocorticoid treatment.[7] The authors noted a significant correlation between the maximal $_{E}CO$ values and recovery time of PEFR after treatment of acute asthma exacerbations with oral glucocorticoids.[7] Another study similarly reported increased $_{E}CO$ values in untreated asthmatics relative to normal subjects, with the highest values recorded in healthy smokers.[8] $_{E}CO$ levels in asthma patients on β-agonists were decreased in response to a 4 week treatment with inhaled corticosteroids. The $_{E}CO$ values in response to steroids correlated with decreases in sputum eosinophil numbers and improved airway obstruction.[8]

In nonsmoking, steroid naive asthmatics, both $_{E}CO$ and $F_{E}NO$ increased in response to allergen challenge. In dual responders, the peak $_{E}CO$ values occurred at 6 h after challenge, and preceded the peak decline in lung function, as assessed by FEV_1, whereas the maximal $F_{E}NO$ response occurred later (10 h).[9] The $_{E}CO$ values in these patients were not sensitive to stimuli that reduced FEV_1, such as histamine, or to subsequent recovery by β-agonists, suggesting that $_{E}CO$ values were independent of airway function.[9]

In patients with bronchial asthma, $_{E}CO$ levels correlated with arterial HbCO levels.[10] The HbCO levels increased in these patients during exacerbations and returned to control levels in response to oral glucocorticoids.[10] The changes in HbCO concentrations also significantly correlated with FEV_1. In the same study, it was also reported that HbCO levels increased in other pulmonary conditions, such as pneumonia and IPF, and correlated to circulating levels of C-reactive protein.[10]

Inducible HO expression has been examined in the airway macrophages of human asthmatics. Increased HO-1 expression was reported in airway macrophages of patients with asthma relative to healthy controls or patients undergoing steroid treatment.[6] Increased HO-1 expression was associated with increased bilirubin levels in the induced sputum and correlated with increases in $_{E}CO$ values in nonsteroid-treated asthmatics.[6] In contrast, another study reported no change in HO-1 or HO-2 expression or distribution in airway submucosal macrophages and epithelium of mild asthmatics relative to healthy controls. Extended inhaled steroid treatment, which reduced airway eosinophils and bronchial responsiveness to methacholine, reduced $F_{E}NO$ values but not $_{E}CO$ values, and did not alter the expression of either HO-1 or HO-2.[11] These studies suggest that airways HO-1/HO-2 were not a significant source of CO during asthma or in response to treatments.[11]

Increased $_{E}CO$ and $F_{E}NO$ have been recorded in children with persistent asthma, relative to those with episodic asthma or healthy controls. In contrast, $F_{E}NO$ levels were comparably increased in patients with both persistent and infrequent episodic asthma.[12] $_{E}CO$ increased during acute exacerbations in children, which was reversible by combination therapies.[13] In childhood asthma patients, $_{E}CO$ values were independent of expiratory flow rate.[14] Increased $_{E}CO$ levels were reported in patients with a history of allergic seasonal rhinitis and were attributed to increased CO originating from upper airways inflammation.[15]

Differential effects of methacholine and allergen challenges on the levels of $_{E}CO$ and $F_{E}NO$ were examined in a cohort of adult subjects with atopic asthma. Bronchospasm negatively modulated $_{E}CO$ and $F_{E}NO$ values, whereas the inflammatory stimulus of allergen exposure increased $F_{E}NO$.[16]

Despite reports supporting $_{E}CO$ measurement as a biomarker of asthma exacerbation and response to steroid therapy, several studies have reported a lack of $_{E}CO$ changes in asthmatics.[11,17] No change in $_{E}CO$ was recorded in asthmatics, for example, following 30 days of inhalation corticosteroid therapy, despite decreases in airway eosinophil content and bronchial responsiveness to methacholine.[11] A metaanalysis of the literature concluded that $_{E}CO$ levels generally increase in asthmatics relative to normal patients, but only partially reflect disease severity.[18] Furthermore, the authors concluded that $_{E}CO$ levels do not distinguish between degrees of disease control or between steroid-treated and steroid-free patients in cross-sectional studies. Nevertheless, $_{E}CO$ levels were concluded to be generally responsive to corticosteroid treatment. $_{E}CO$ may continue to represent a potentially useful biomarker of airway inflammation and oxidative stress in asthma in nonsmokers.[18]

6.3.3 Chronic obstructive pulmonary disease

Chronic obstructive pulmonary disease is defined by consensus as "a preventable and treatable disease state characterized by airflow limitation that is not fully reversible, usually progressive, and associated with an abnormal inflammatory response of the lungs to noxious particles or gases, primarily caused by cigarette smoking."[19] The Global Initiative for Chronic Obstructive Pulmonary Disease (GOLD) has established a four-point scale for classifying disease severity based on airflow limitation, with grade 1 indicating smokers with mild loss of lung function, and stages 3–4 indicating severe to very severe airflow limitation (see Chapter 19). In the Western nations, cigarette smoking remains a primary cause of COPD, which develops in ~15%–20% of smokers. COPD is a complex disease with unclear pathogenesis, though a number of mechanisms have been proposed. Environmental and occupational exposures to particulate air pollution, chronic microbiological infections, and asthma, may all contribute to COPD progression.[19] Aberrant and persistent airway inflammation is a major component of COPD, characterized by the influx of macrophages and neutrophils. Alveolar macrophages and epithelial cells represent primary targets of cigarette smoke exposure, which react to the oxidant burden by producing proinflammatory cytokines, chemokines, and growth factors. Increased oxidative stress associated with smoking may promote an imbalance between tissue protease and antiprotease activities, which in turn contributes to tissue degeneration characteristic of emphysema. Regulated cell death mechanisms (such as apoptosis, necroptosis) may also contribute to disease pathogenesis.[19]

Cigarette smoke is a complex mixture consisting of gas and tar phases that contain ~4500 chemical constituents, including free radicals, oxidants, aldehydes, and aromatic hydrocarbons, as well as the diatomic gases NO and CO, which are generated during combustion. A number of studies have examined the potential for $_{E}CO$ as a breath marker of smoking status and/or COPD. In smokers without lung function impairment, measured $_{E}CO$ levels increased dramatically relative to nonsmokers. Chronic smokers may display $_{E}CO$ values of up to 20–25 ppm.[20] $_{E}CO$ correlates with the rate of cigarette smoking, and also varies with smoking habit or type of cigarette (i.e., filtered or nonfiltered).[20] $_{E}CO$ is therefore a useful diagnostic parameter to indicate smoking or to monitor smoking cessation.[3,20] In a population survey of smokers from the Copenhagen Heart Study, $_{E}CO$ values did not significantly correlate to lung function or lung function decline in smokers.[21]

Studies of outpatient smokers suggested optimal cutoff values of approximately 6.5 ppm; values above this would indicate recurrent smoking with 90%–96% sensitivity and 80%–93% specificity.[21] Among nonsmokers, $_{E}CO$ values are consistently high in densely populated areas, suggesting that permanent exposure to urban air pollution significantly impacts $_{E}CO$ status and may complicate assessment of smoking status.[22] Furthermore, in asthma and COPD, airway inflammation contributes to $_{E}CO$ values, which makes it difficult to define optimal cutoff values. CO monitoring in smoking remains controversial, with some investigators suggesting that cutoff values should be adjusted among subpopulations, in particular for those subjects who also concurrently have asthma or COPD, or live in dense urban areas.[23]

Studies of patients with COPD reported generally higher $_{E}CO$ values than nonsmokers without COPD. Among those patients with COPD, those who were currently smoking at the time of study had higher $_{E}CO$ values than those who were former smokers.[2] $_{E}CO$ values did not correlate with lung function measurements assessed by FEV_1, or with F_ENO values. It should be noted that the measured F_ENO values were also higher in COPD patients than controls; however, F_ENO levels were reported higher among COPD nonsmokers than COPD smokers. F_ENO values were found to negatively correlate with lung function as assessed by FEV_1.[2]

Arterial HbCO levels have been measured in COPD patients in comparison to patients without airways disease. Ex-smoking COPD patients with stable disease displayed higher arterial HbCO levels than healthy ex-smoking patients. The increases in HbCO correlated with COPD disease severity; with greater HbCO levels recorded in severe COPD, relative to moderately severe disease. Furthermore, HbCO levels further increased in patients with COPD exacerbations relative to stable disease.[24] Increased HbCO correlated with increased $_{E}CO$ at moderate stages of COPD severity (GOLD 2–3) but not in severe COPD (GOLD 4). The reason for this discrepancy is unclear, but may reflect declined lung/airways production of CO or decreased alveolar function and CO diffusion capacity in severe COPD, in combination with increased systemic CO production.[24]

Inducible HO, the enzyme primarily responsible for endogenous CO production, has been implicated in the pathogenesis of COPD. HO-1 is associated with cellular defense mechanisms against oxidative stress,[5,6] and in this regard potentially plays a role in cellular protection against cigarette smoke–induced lung cell injury. Elevated levels of HO-1 were reported in the alveolar spaces of chronic smokers with and without COPD relative to nonsmokers.[25] Increased HO-1 positive inflammatory cells were found during exacerbations of severe COPD relative to stable disease.[26] In contrast, in a study that controlled for smoking status, ex-smoking COPD patients were reported to have reduced HO-1 expression in alveolar macrophages relative to healthy ex-smokers.[27] These observations suggest that elevations of $_{E}CO$ observed in COPD are not necessarily consistent with HO-1 activation, as similarly reported for asthmatics.[6] Further studies are needed to define the relationship between $_{E}CO$ and HO-1 in COPD.

6.3.4 Cystic fibrosis

Cystic fibrosis (CF), is a life-threatening lung disease. It is a genetic disorder caused by a mutation in the gene encoding the CF transmembrane conductance regulator (CFTR). CF is characterized by abnormal mucous buildup in the airways, which can cause pulmonary damage associated with secondary infections. Measured $_{E}CO$ levels were found to be higher in untreated versus oral steroid-treated CF patients.[28] Furthermore, $_{E}CO$ increased in patients during exacerbations of CF, correlating to deterioration of FEV_1, with normalization of the $_{E}CO$ levels after treatment.[29] CO levels correlate with exhaled ethane, a product of lipid peroxidation that serves as an indirect marker of oxidative stress. Both $_{E}CO$ and exhaled ethane were elevated in nonsteroid-treated CF patients, when compared to healthy controls, and these values were decreased by steroid treatment.[30] Increased $_{E}CO$ was detected in children with CF, relative to control patients. In addition to the inflammatory and oxidative stress responses to infection in these patients, $_{E}CO$ may respond to hypoxia. Exercise testing further increased $_{E}CO$ in CF children and correlated with the degree of oxyhemoglobin desaturation, suggestive of an increased HO-1 expression in CF patients during hypoxic states induced by exercise.[31] It should be noted that several studies have reported no significant fluctuation of $_{E}CO$ in CF patients relative to healthy controls.[14,17]

6.3.5 Infectious pulmonary disease

In patients with pneumonia, higher HbCO levels can be measured at the onset of illness, with decreasing values to control levels after antibiotic treatment.[10] Elevated $_{E}CO$ levels were reported in lower respiratory tract infections, and these were decreased by antibiotic therapy.[32] Median $_{E}CO$ levels were higher among patients with bronchiectasis, a condition related to increased neutrophil inflammation, regardless of steroid treatment, although $_{E}CO$ did not correlate with HbCO.[33] Furthermore, $_{E}CO$ levels in upper respiratory tract infections were higher when compared to healthy controls.[14,15]

6.3.6 Sepsis and acute lung injury

The potential application of $_{E}CO$ for the noninvasive monitoring of critically ill patients has also been explored. Elevated $_{E}CO$ values have been measured in mechanically ventilated critical care subjects. In a clinical study of 29 patients on mechanical ventilation with eight healthy nonsmoking controls, average $_{E}CO$ values were higher in mechanically ventilated patients relative to healthy subjects, and correlated with arterial HbCO levels, as well as with serum total and indirect bilirubin levels.[34] HbCO also correlated with total bilirubin measurements. The patient group had variable respiratory function and disease severity. The authors concluded that heme metabolic activity was elevated among ventilated critical care patients versus healthy controls. $_{E}CO$ did not generally correlate with or predict inflammation or disease severity but a trend of higher $_{E}CO$ values was associated with survival.[34] Elevated $_{E}CO$ values were also recorded in mechanically ventilated critically ill patients with sepsis, relative to ventilated but nonseptic controls. Elevated $_{E}CO$ values on the first day of treatment correlated with the probability of patient survival[35].

6.3.7 Neonatal jaundice

Exhaled CO is one of the few breathborne biomarkers that has achieved sufficient robustness to transition to clinical practice (see Chapter 1). Specifically, $_{E}CO$ is in clinical use as a diagnostic parameter for neonatal jaundice, a pediatric disease associated with the excess accumulation of serum bilirubin. Although circulating bilirubin has been identified as a potent antioxidant, its accumulation in neonates poses a potential risk for neurological injury, which warrants its therapeutic removal, commonly by application of phototherapy.[36]

Since CO and biliverdin are produced in equimolar amounts in the heme oxygenase reaction, and biliverdin is quantitatively reduced to bilirubin, the measurement of $_{E}CO$ in breath has been used experimentally as an indexof heme degradation and bilirubin production.[36]

Increased $_{E}CO$ level in the early neonatal period is associated with subsequent hyperbilirubinemia, even in infants without hemolytic disease.[36],[37] End-tidal $_{E}CO$ corrected for ambient CO concentration (ETCOc) has been shown to be useful as a screening test for predicting hyperbilirubinemia (>257 μmol/L) even in the absence of hemolytic diseases.[37] Further studies have also identified ETCOc as a useful index of pediatric hemolytic hyperbilirubinema (hemolytic jaundice).[38] A device, approved by the US Food and Drug Administration (FDA), has been developed and evaluated for assessment of ETCOc in newborns.[39] ETCOclevel ≥2.5 ppm was shown to be a more accurate method to determine the presence of significant hemolysis in newborns, and consequently, to determine the need for phototherapy, than the currently available direct antiglobulin test (DAT).[40]

Overall, measurement of $_{E}CO$ levels offers a rapid noninvasive method for predicting hyperbilirubinemia and hemolysis in newborns.

6.4 Summary

Carbon monoxide is one of the many metabolites that can be measured in the exhaled breath of humans. The monitoring of CO and NO gas and other volatiles in the breath has the advantage over other biomarker-based clinical tests in that these methods are noninvasive and practical for bedside measurements. CO is a common contaminant of outdoor and indoor air, and therefore measurements of CO must be carefully controlled and account for ambient levels. While CO is typically present in breath at ppm levels, a range that can be detected with electrochemical apparatus, instrumentations with greater sensitivity based on laser spectroscopy are currently under development and show promise to bring further advances to the field.

Exhaled CO has been proposed as a general marker of inflammation and may also selectively reflect increased oxidative stress in the lung. Measured $_{E}CO$ levels have been reported to fluctuate in several inflammatory lung diseases and respond generally to antiinflammatory therapies. Nevertheless, the utility and specificity of $_{E}CO$ as a marker of disease remains controversial. Of note, $_{E}CO$ is highly elevated in the exhaled breath of smokers; therefore, $_{E}CO$ is not likely to be a reliable marker of inflammation or of any disease state in patients who smoke. Despite improvements in the standardization and sensitivity of methods to detect $_{E}CO$, the predictive value of $_{E}CO$ measurements as a diagnostic tool in human diseases remains unclear.

Although there has been extensive investigation in $_{E}CO$ in human disease over the past two decades, progress in this area has waned in the past five years. The reason for the apparent decline in progress may be due in part to limitations in the sensitivity of monitoring systems widely available to investigators and the limited availability or cost associated with more sensitive experimental detection systems. Further progress in developing $_{E}CO$ as an index of disease may depend on the continued development of sensitive analytical systems, their increased availability to researchers, and improved practicality for clinical testing and application.

References

1. Kharitonov SA, Barnes PJ. Exhaled markers of pulmonary disease. *Am J Respir Crit Care Med* 2001;**163**(7):1693–722.
2. Montuschi P, Kharitonov SA, Barnes PJ. Exhaled carbon monoxide and nitric oxide in COPD. *Chest* 2001;**120**(2):496–501.
3. Goldstein AO, Gans SP, Ripley-Moffitt C, Kotsen C, Bars M. Use of expired air carbon monoxide testing in clinical tobacco treatment settings. *Chest* 2018;**153**(2):554–62.
4. Von Burg R. Carbon monoxide. *J Appl Toxicol* 1999;**19**(5):379–86.
5. Ryter SW, Choi AM. Targeting heme oxygenase-1 and carbon monoxide for therapeutic modulation of inflammation. *Transl Res* 2016;**167**(1):7–34.
6. Horvath I, Donnelly LE, Kiss A, Paredi P, Kharitonov SA, Barnes PJ. Raised levels of exhaled carbon monoxide are associated with an increased expression of heme oxygenase-1 in airway macrophages in asthma: a new marker of oxidative stress. *Thorax* 1998;**53**(8):668–72.

7. Yamaya M, Sekizawa K, Ishizuka S, Monma M, Sasaki H. Exhaled carbon monoxide levels during treatment of acute asthma. *Eur Respir J* 1999;**13**(4):757–60.
8. Zayasu K, Sekizawa K, Okinaga S, Yamaya M, Ohrui T, Sasaki H. Increased carbon monoxide in exhaled air of asthmatic patients. *Am J Respir Crit Care Med* 1997;**156**(4 Pt 1):1140–3.
9. Paredi P, Leckie MJ, Horvath I, Allegra L, Kharitonov SA, Barnes PJ. Changes in exhaled carbon monoxide and nitric oxide levels following allergen challenge in patients with asthma. *Eur Respir J* 1999;**13**(1):48–52.
10. Yasuda H, Yamaya M, Yanai M, Ohrui T, Sasaki H. Increased blood carboxyhaemoglobin concentrations in inflammatory pulmonary diseases. *Thorax* 2002;**57**(9):779–83.
11. Lim S, Groneberg D, Fischer A, Oates T, Caramori G, Mattos W, Adcock I, Barnes PJ, Chung KF. Expression of heme oxygenase isoenzymes 1 and 2 in normal and asthmatic airways: effect of inhaled corticosteroids. *Am J Respir Crit Care Med* 2000;**162**(5):1912–8.
12. Uasuf CG, Jatakanon A, James A, Kharitonov SA, Wilson NM, Barnes PJ. Exhaled carbon monoxide in childhood asthma. *J Pediatr* 1999;**135**(5):569–74.
13. Ohara Y, Ohrui T, Morikawa T, He M, Yasuda H, Yamaya M, et al. Exhaled carbon monoxide levels in school-age children with episodic asthma. *Pediatr Pulmonol* 2006;**41**(5):470–4.
14. Beck-Ripp J, Latzin P, Griese M. Exhaled carbon monoxide is not flow dependent in children with cystic fibrosis and asthma. *Eur J Med Res* 2004;**9**(11):518–22.
15. Andersson JA, Uddman R, Cardell LO. Increased carbon monoxide levels in the nasal airways of subjects with a history of seasonal allergic rhinitis and in patients with upper respiratory tract infection. *Clin Exp Allergy* 2002;**32**(2):224–7.
16. Sethi JM, White AM, Patel SA, Dinella JV, Calhoun WJ, Choi AM. Bronchoprovocation testing in asthma: effect on exhaled monoxides. *J Breath Res* 2010;**4**(4):047104.
17. Zetterquist W, Marteus H, Johannesson M, Nordval SL, Ihre E, Lundberg JO, et al. Exhaled carbon monoxide is not elevated in patients with asthma or cystic fibrosis. *Eur Respir J* 2002;**20**(1):92–9.
18. Zhang J, Yao X, Yu R, Bai J, Sun Y, Huang M, et al. Exhaled carbon monoxide in asthmatics: a meta-analysis. *Respir Res* 2010;**11**:50.
19. Barnes PJ, Shapiro SD, Pauwels RA. Chronic obstructive pulmonary disease: molecular and cellular mechanisms. *Eur Respir J* 2003;**22**(4):672–88.
20. Fabricius P, Scharling H, Lokke A, Vestbo J, Lange P. Exhaled CO, a predictor of lung function? *Respir Med* 2007;**101**(3):581–6.
21. Deveci SE, Deveci F, Acik Y, Ozan AT. The measurement of exhaled carbon monoxide in healthy smokers and non-smokers. *Respir Med* 2004;**98**(6):551–6.
22. Maga M, Janik MK, Wachsmann A, Chrząstek-Janik O, Koziej M, Bajkowski M, et al. Influence of air pollution on exhaled carbon monoxide levels in smokers and non-smokers. A prospective cross-sectional study. *Environ Res* 2017;**152**:496–502.
23. Sato S, Nishimura K, Koyama H, Tsukino M, Oga T, Hajiro T, et al. Optimal cutoff level of breath carbon monoxide for assessing smoking status in patients with asthma and COPD. *Chest* 2003;**124**(5):1749–54.
24. Yasuda H, Yamaya M, Nakayama K, Ebihara S, Sasaki T, Okinaga S, et al. Increased arterial carboxyhemoglobin concentrations in chronic obstructive pulmonary disease. *Am J Respir Crit Care Med* 2005;**171**(11):1246–51.

25. Maestrelli P, El Messlemani AH, De Fina O, Nowicki Y, Saetta M, Mapp C, Fabbri LM. Increased expression of heme oxygenase (HO)-1 in alveolar spaces and HO-2 in alveolar walls of smokers. *Am J Respir Crit Care Med* 2001;**164**(8 Pt 1):1508–13.
26. Tsoumakidou M, Tzanakis N, Chrysofakis G, Siafakas NM. Nitrosative stress, heme oxygenase-1 expression and airway inflammation during severe exacerbations of COPD. *Chest* 2005;**127**(6):1911–8.
27. Slebos DJ, Kerstjens HA, Rutgers SR, Kauffman HF, Choi AM, Postma DS. Haem oxygenase-1 expression is diminished in alveolar macrophages of patients with COPD. *Eur Respir J* 2004;**23**(4):652–3. Author reply 653.
28. Paredi P, Shah PL, Montuschi P, Sullivan P, Hodson ME, Kharitonov SA, et al. Increased carbon monoxide in exhaled air of patients with cystic fibrosis. *Thorax* 1999;**54**(10): 917–20.
29. Antuni JD, Kharitonov SA, Hughes D, Hodson ME, Barnes PJ. Increase in exhaled carbon monoxide during exacerbations of cystic fibrosis. *Thorax* 2000;**55**(2):138–42.
30. Paredi P, Kharitonov SA, Leak D, Shah PL, Cramer D, Hodson ME, Barnes PJ. Exhaled ethane is elevated in cystic fibrosis and correlates with carbon monoxide levels and airway obstruction. *Am J Respir Crit Care Med* 2000;**161**(4 Pt 1):1247–51.
31. Horvath I, Borka P, Apor P, Kollai M. Exhaled carbon monoxide concentration increases after exercise in children with cystic fibrosis. *Acta Physiol Hung* 1999;**86**(3–4):237–44.
32. Biernacki WA, Kharitonov SA, Barnes PJ. Exhaled carbon monoxide in patients with lower respiratory tract infection. *Respir Med* 2001;**95**(12):1003–5.
33. Horvath I, Loukides S, Wodehouse T, Kharitonov SA, Cole PJ, Barnes PJ. Increased levels of exhaled carbon monoxide in bronchiectasis: a new marker of oxidative stress. *Thorax* 1998;**53**(10):867–70.
34. Morimatsu H, Takahashi T, Maeshima K, Inoue K, Kawakami T, Shimizu H, et al. Increased heme catabolism in critically ill patients: correlation among exhaled carbon monoxide, arterial carboxyhemoglobin, and serum bilirubin IXalpha concentrations. *Am J Physiol Lung Cell Mol Physiol* 2006;**290**(1):L114–9.
35. Zegdi R, Perrin D, Burdin M, Boiteau R, Tenaillon A. Increased endogenous carbon monoxide production in severe sepsis. *Intensive Care Med* 2002;**28**(6):793–6.
36. Stevenson DK, Vreman HJ. Carbon monoxide and bilirubin production in neonates. *Pediatrics* 1997;**100**(2 Pt 1):252–254.R.
37. Okuyama H, Yonetani M, Uetani Y, Nakamura H. End-tidal carbon monoxide ispredictive for neonatal non-hemolytic hyperbilirubinemia. *Pediatr Int* 2001;**43**(4):329–33.
38. Christensen RD, Malleske DT, Lambert DK, Baer VL, Prchal JT, Denson L/E, Gerday E, Weaver Lewis KA, Shepherd JG. Measuring end-tidal carbon monoxide of jaundicedneonates in the birth hospital to identify those with hemolysis. *Neonatology* 2016;**109**(1):1–5.
39. Bhutani VK, Maisels MJ, Schutzman DL, Castillo Cuadrado ME, Aby JL, Bogen DL, Christensen RD, Watchko JF, Wong RJ, Stevenson DK. Identification of risk forneonatal haemolysis. *Acta Paediatr* 2018;**107**(8):1350–6.
40. Elsaie AL, Taleb M, Nicosia A, Zangaladze A, Pease ME, Newton K, Schutzman DL. Comparison of end-tidal carbon monoxide measurements with direct antiglobulintests in the management of neonatal hyperbilirubinemia. *J Perinatol* 2020.

CHAPTER

Exhaled breath condensate and aerosol

7

Michael D. Davis

Wells Center for Pediatric Research, Pulmonology, Allergy, and Sleep Medicine, Riley Hospital for Children at Indiana University, Indianapolis, IN, United States

Chapter outline

7.1 Overview

During inhalation and exhalation, the airways warm and humidify breath to body temperature and pressure, saturated (BTPS), which is approximately 37°C and 100% humidified. BTPS exhaled breath is composed mainly of water vapor that is carried by the humidified breath (>99.99%) and aerosolized particle droplets that evolve from changes in alveolar surface tension and liberated airway surface liquid (ASL) (<0.01%), as described in Chapter 8 (see Fig. 8.1). When exhaled breath contacts a surface that is cooler than its temperature, exhaled breath condensate (EBC) forms. This fluid contains biomarkers dissolved in the water vapor fraction of breath, as well as those trapped in aerosolized droplets. The humidified and aerosolized fractions of EBC can be separated by collection of the aerosol particles in filters prior to contact with a condensing surface. When collected separately, this aerosolized particle fraction of EBC is referred to as exhaled breath aerosol (EBA).[1–3] As a point of note, the terminology in this field can lead to confusion. Specifically, an aerosol is generally defined as a suspension of water droplets and particulate matter, thus technically exhaled breath contains a gas phase and an

Breathborne Biomarkers and the Human Volatilome. https://doi.org/10.1016/B978-0-12-819967-1.00007-4
Copyright © 2020 Elsevier B.V. All rights reserved.

aerosol phase. Aerosol in the context of this chapter, i.e., EBA, is the dry fraction of exhaled aerosols that has been collected; this can either be separate from the water droplet fraction (collected as EBC) or include any suspensions within the vapor fraction that remain once the droplets dissolve or evaporate. When considering the physical properties of exhaled aerosols, these are commonly referred to as exhaled particles, as is the subject matter of the next chapter.

For context, consider the Venn diagram in Fig. 7.1, wherein the outside circle represents all exhaled breath (gas phase, water vapor, and aerosolized particles), the larger inner circle represents that fraction of the condensed phase liquids, vapors, semivolatile and heavy organic compounds, aerosols, and particles, and the smallest subset encompasses only that fraction that could stick to a surface and not evaporate. This diagram is not to scale; it has been estimated that the condensate circle represents about 5%–6% of all breath, and that the aerosols represents about 0.01%–0.05% of all breath. Certainly, these numbers are estimates that depend upon the study, the variability of subjects, and other measurement factors. These estimates vary depending on the calculations of mass fraction versus concentrations by volume.

Interest in EBC as a source of potential biomarkers has increased significantly over the past two decades due to its safe and ease of collection.[1,3,4] Although several collection techniques have been described and standardized for nearly all age groups and disease states, the composition of EBC varies between collection method and

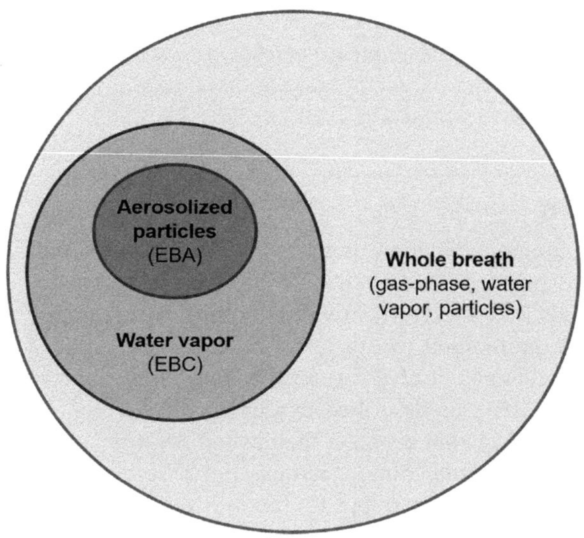

FIGURE 7.1

Venn diagram of fractions of whole breath (not to scale). The *outer circle* represents all (100%) of gas-phase breath, water vapor, and aerosolized particles, and the *inner circles* represent ~5.3% exhaled breath condensate (EBC) and 0.01%–0.05% exhaled breath aerosol (EBA), respectively.

patient population. Therefore, a working knowledge of variables associated with EBC collection is crucial in order to maintain sample integrity.

Exhaled breath condensate is extremely dilute and often contains biomarkers at the lower end of instrumental limits of detection (LOD) with current technology; moreover, EBA is diluted at least 1000-fold within EBC. This makes analysis challenging and samples particularly susceptible to contamination. Despite this, hundreds (if not thousands) of chemical compounds have been discovered in EBC. Many of these compounds have been validated for evaluating disease states, and some have been approved for clinical use.[4] As with biomarkers in the gas phase of exhaled breath, however, the majority of EBC biomarkers are currently used in research rather than clinical practice. One distinction between EBC and EBA is the capability for capturing the volatile organics fraction of breath in EBC. This is of particular interest because volatile organic compounds (VOCs), such as alcohols, ketones, and aldehydes, are of great interest as markers of biological processes but can be very difficult to assess in standard gas-phase breath samples.[5]

From a practical perspective, it is more difficult to collect whole EBC in the field than just the aerosol subset, i.e., just EBA. Field EBC collection requires an infrastructure including some form of coolant (typically wet or dry ice), laboratory space to defrost and transfer liquid samples to particular vials, and then the necessary logistics to ship the sample to the analytical laboratories under refrigerated conditions. Aerosols, on the other hand, are typically collected from masks or from hard surfaces. Samples can comprise either the entire mask (e.g., standard hospital mask) or a wipe from a smooth surface from the interior of a respirator. As such, there is no need for any special infrastructure or refrigerated transport. Conversely, EBC analyses are easier in the laboratory as the samples are bigger and are already in liquid form. By comparison, EBA samples from masks require extraction with use of specialized solvents, with subsequent preconcentration for analysis.[5–7]

This chapter introduces the concept of EBC and EBA, outlines sampling procedures, and provides an overview of EBC/EBA composition and validated as well as prospective biomarkers.

7.2 Composition of exhaled breath condensate

Similar to serum, urine, or saliva, EBC is a biofluid in which biomarkers can be identified. EBC contains many compounds in its water vapor and aerosolized particle fractions. These include (exogenous) compounds inhaled into the airways, released from airway epithelia, and diffused across the alveolar–capillary membrane. Several systems for classifying fractions of EBC have been previously proposed based on the types of potential biomarkers found in each fraction,[8] including

- Volatile and nonvolatile
- Inorganic and organic
- pH- or redox-relevant and nonreactive
- Proteinaceous and nonproteinaceous

The most substantially different fractions of EBC are the *aerosolized particle* and *water vapor* fractions. EBC represents both fractions, but when its aerosolized fraction is collected separately from the water vapor fraction, it is referred to as EBA. These two fractions contain significantly different compounds and can be collected separately from one another.[9] The water vapor fraction contains water-soluble compounds capable of being dissolved and carried in the water of BTPS gas. EBA contains larger molecules that are trapped in aerosolized droplets.

7.3 Collection and fractionation of exhaled breath condensate

7.3.1 General collection principles

In theory, collection of EBC is a relatively simple process—in short, EBC forms whenever we breathe on a cold surface (such as a mirror or window). Three variables control this process, namely (1) the volume of exhaled breath available (minute ventilation during collection; V), (2) the surface area of the condenser (SA), and (3) the temperature gradient between the exhaled breath and the condenser surface (T). Increasing any one of these variables will increase the yield of EBC collected. It is worthy of note that a change in the condenser temperature will alter the composition of collected EBC.[10] Based on these principles, homemade devices for EBC collection have been established and used for several decades.[1,4,11,12] Because of this, it is important to standardize EBC collection methods to ensure comparable results between collections. Several commercially available devices allow a more standardized approach to be used between operators; an overview of these devices is given in Table 7.1.

Temperature is one of the most important variables to consider during EBC collection. When collected at temperatures above 0°C, EBC is collected in a liquid state; below 0°C, it is collected as "wet ice"; below −40°C it is collected as dry crystals or "snow." The largest volume of EBC collected will occur at the lowest condenser temperatures, as these maximize the temperature gradient between the condenser surface and exhaled breath. The optimal collection temperature therefore

Table 7.1 Current commercial EBC collection systems.

EBC collection system	Manufacturer	Description
Turbo DECCS	Medivac	Reusable, portable device with single-use collection interface with a thermoelectric cooling system. Requires mains power.
RTube and RTubeVent	Respiratory Research, Inc.	Lightweight, single-use, portable condenser system with coolant sleeves that must be refrigerated prior to use. The RTubeVent is designed for use during mechanical ventilation.

depends on the compounds of interest and the volume of EBC needed, generalized as follows:

1. VOCs in EBC are more readily absorbed into liquid-phase water than ice and therefore are best collected at temperatures slightly above freezing.[13]
2. larger molecules present in the scant aerosol fraction of EBC are better trapped by freezing condensate rapidly and therefore are best collected at the coldest temperatures possible.
3. when both volatile and aerosolized compounds are desired, collection of two samples at different temperatures is likely the optimal approach.

Ideal collection of EBC occurs during regular, tidal breathing. Regular breathing patterns help to standardize aerosol droplet size and volume during collection and are less strenuous for subjects.[3,7,14] Of note, patients may hyperventilate during coached breathing and care must be taken to avoid this. Technical standards for the collection of exhaled biomarkers have been proposed by the American Thoracic Society (ATS) and the European Respiratory Society (ERS) and an update of these standards is ongoing.[15,16] These standards should be considered alongside the recommendations in this chapter when collecting EBC and EBA.

7.3.2 Collection during mechanical ventilation

Exhaled breath condensate can be collected from subjects of all ages and disease states during mechanical ventilation.[1,17,18] For subjects who are unable to follow commands and voluntarily breathe into a collection device, it may be easier to collect EBC during mechanical ventilation. Collection during mechanical ventilation follows the same principles as described above, although there are several additional concerns that must be considered in this setting. These concerns include patient/staff safety and sample integrity.

Mechanical ventilation is life-support equipment and care must always be taken when interfacing with such equipment. Collecting EBC in this setting requires placement of the condenser in the expiratory limb of the ventilator. Any obstruction of the expiratory limb by the condenser will result in cessation of mechanical ventilation and potential barotrauma to the patient. Opening the ventilator circuit to place and remove the condenser results in a temporary cessation of mechanical ventilation, which some subjects may not tolerate. When the ventilator circuit is open, pathogens can be introduced to the patient or pathogens may be transmitted from the patient to the collection staff. To address these concerns, the following considerations must be made:

1. Prior to use with any model of ventilator, condensers should be tested to ensure they do not obstruct ventilation during collection.
2. Clinical staff should always be consulted prior to collection of EBC from mechanically ventilated patients.
3. Patients who are deemed too unstable for temporary interruption of mechanical ventilation should not have condensers placed or removed.

4. Patients should be preoxygenated prior to interruption of mechanical ventilation to place or remove condensers.
5. All condensers placed within the ventilator circuit should be sterilized prior to each use.
6. Collection staff should use aseptic technique during placement and removal of condensers.
7. Collection staff should wear personal protective equipment appropriate to prevent transmission of droplet-based pathogens.

It is worth noting that EBC can be collected from the exhaust port of some ventilators, which is outside of the ventilator circuit. The aerosolized fraction is lost during this method of collection, as it will be trapped in expiratory filters present in mechanical ventilators.

7.3.3 Fractional collection

The water vapor and aerosolized particle fractions of EBC can be separated by filtration.[2,3,9,13] Commercially available bacterial filters trap exhaled aerosols and prevent them from contaminating respiratory medical equipment, such as pulmonary function machines. These filters are readily available in most hospitals and clinical settings. Even simpler, reproducible levels of exhaled aerosol have been collected on simple surgical masks worn by subjects during tidal breathing.[3] When collecting EBA by these methods, the temperature gradient is not relevant as the aerosols are mechanically trapped by the filter or mask. These methods can also be used in tandem with a condenser (Fig. 7.2) to allow for separate and simultaneous EBC and EBA collection.

7.4 Biomarkers in exhaled breath condensate

7.4.1 Biomarkers in the water vapor fraction of exhaled breath condensate

The majority of biomarkers that have been discovered in EBC are present in its water vapor fraction. This is because it composes the majority of the fluid volume of whole EBC and therefore is capable of carrying a higher volume of compounds than EBA.[1] The water vapor fraction of EBC also contains systemic compounds from within the body that are capable of diffusing across the alveolar–capillary membrane in the lungs, such as VOCs, which have varying origins throughout the body.

Several of the most validated biomarkers in EBC are present in its water vapor fraction. EBC pH, for example, has been standardized and approved by the US Food and Drug Administration (FDA) for clinical use in the management of patients with gastric reflux. Its determinants are ionic components, including lactic, acetic, and formic acid, all of which are water soluble.[19] Similarly, EBC hydrogen peroxide (H_2O_2) has been standardized and now can be measured by a commercially available, point-of-care (PoC) research test.[20] A brief list of clinically approved well-characterized biomarkers in this fraction of EBC is presented in Table 7.2.

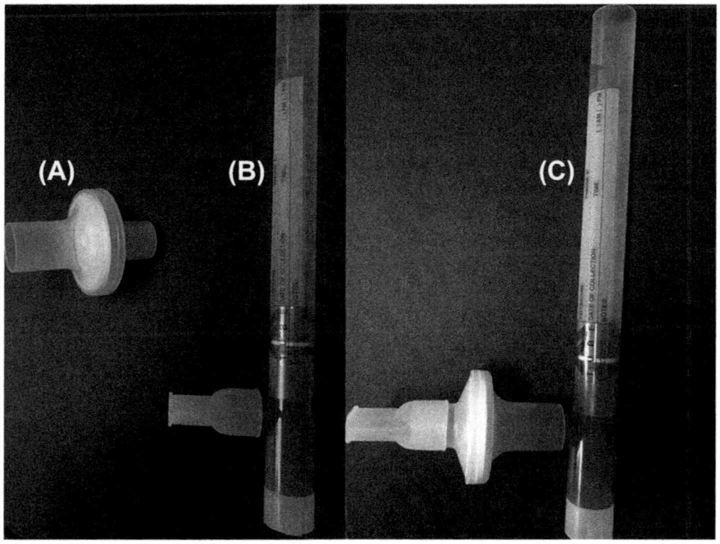

FIGURE 7.2

Simple EBC and EBA collection systems, separate (left) and in tandem (right). (A) A common spirometry filter designed to trap exhaled aerosols; (B) a commercially available EBC collection device; (C) these devices combine to allow for simultaneous fractional collection of EBC and EBA.

Table 7.2 Current commercial EBC tests.

Target	Use	Test name	Manufacturer
pH	Diagnosis of reflux-induced cough[2,21]	EBC pH	Respiratory Research, Inc.
Tuberculosis bacilli	Detection of TB by biochemical reaction	TB breathalyzer	Rapid Biosensor Systems
Hydrogen peroxide	Measurement of airway inflammation and oxidant stress[20]	Inflammacheck	Exhalation Technology

Other biomarkers in EBC have been validated that are not yet clinically approved. Most of these biomarkers have been used to study airway diseases, particularly inflammatory airway diseases.[1,4,22] Glutathione in EBC is associated with airway inflammation and oxidative stress. When evaluating glutathione, its reduced and oxidized states can be studied as a ratio that eliminates dilution effects between samples of EBC.[1,2,8,22] Higher oxides of nitrogen are increased in allergic disease and inflammatory airway diseases.[4,23,24] 8-isoprostane is increased in inflammatory and interstitial lung diseases; further, it is clearly associated with oxidative stress and cigarette smoke exposure.[1,4,22,25,26] Metabolomic patterns in EBC differ between healthy, inflammatory, and infectious airways.[1,4,11,22,27,28] Additionally, EBC has been explored for its potential biomarker consituents in relation to occupational or environmental exposures, as is discussed in Chapters 32 and 33, respectively.

Although initial reports regarding the measurement of EBC cytokines led to some controversy over conflicting results, modern analytical techniques have allowed for validation of a variety of these peptides in inflammatory, fibrotic, and infectious airway diseases.[1,4,22,29] Several cytokines in EBC have also been associated with systemic diseases, including sepsis and inflammatory bowel disease.[1,30] Similarly, glutathione has been observed to be increased in alcoholism.[4] Although not comprehensive, a list of these compounds is presented in Table 7.3. Several of these biomarkers have also been isolated in EBA and are described in Section 7.4.2.

7.4.2 Biomarkers in exhaled breath aerosol

Within the past decade, interest in EBA has grown.[3,7,9] The exact source of EBA is unknown, but its presence as aerosolized particle droplets has been verified in vivo by particle counters and it likely originates from ASL (see Chapter 8). This theory of EBA origination is supported by the similarity in compounds found in both EBA and bronchoalveolar lavage fluid (BALF).[14] Results of EBC particle counts studies and measurements of the dilution of EBC suggest that EBA represents less than 0.1% of liquid EBC.[31] The scant nature of this fraction of EBC has proven challenging to analysis, and reproducible results of biomarkers in EBA depend on meticulous collection and incredibly sensitive assays. Nevertheless, several cytokines have been well characterized in EBA.[14] The analysis of other compounds in EBC and

Table 7.3 Validated EBC biomarkers.

Target	Disease/Disorder
Nitrite/nitrate/higher oxides of nitrogen	Increased in asthma, allergic disease, and inflammatory airway diseases.[4,23,24]
8-Isoprostane	Increased in inflammatory airway diseases, interstitial lung diseases, and oxidative stress. Increased by cigarette smoke exposure.[1,4,22,25,26]
Cytokines	Leukotrienes—Increased in asthma and exercise-induced bronchoconstriction. Also increased in pediatric inflammatory bowel diseases.[4,22,30] Interleukins/Chemokines—Increased in inflammatory airway diseases, pneumonia, and pulmonary fibrosis. Also increased in systemic inflammatory syndromes, such as sepsis.[1,4,22,29,30] Tumor necrosis factors—Increased in lung cancer, pulmonary fibrosis, and inflammatory airway diseases.[4,22,29]
Glutathione	Increased during airway inflammation, airway oxidative stress, and alcoholism.[4]
Metabolomic patterns	Significantly different patterns noted between healthy controls, patients with asthma, and patients with cystic fibrosis.[1,4,11,22,27,28]

EBA has progressed in the past few years to extend to nontargeted analysis using state-of-the-art instrumentation including liquid chromatography coupled to high-resolution mass spectrometry (LC-HRMS). Herein, researchers have found a variety of heretofore unknown compounds in exhaled breath, including fatty acids, some exogenous compounds, and a series of larger organic molecules ascribed to life processes in the molecular weight range from 330 to 1733 Da.[32]

The larger molecules contained in EBA, especially cytokines and chemokines, are of great interest to clinicians and researchers. Many of these compounds have been validated in biofluids that require invasive techniques to acquire and are biohazardous, such as blood and BALF.[14] The ability to measure them noninvasively and with minimal biohazard in exhaled breath would be advantageous. Of note, pathogens that are transmitted by airborne droplets are present in EBA. As with any biofluid, universal precautions should be observed while handling EBA; however, EBA presents a lower risk than other biofluids, such as blood of BALF, since it is diluted significantly by the nonbiohazardous water vapor fraction of EBC.

7.5 Summary

Exhaled breath condensate and aerosol are promising, minimally invasive, low-risk sources of potential biomarkers. Research on both of these matrices will continue to expand knowledge on health and disease by facilitating evaluations in patients and disease states in which conventional biomarker matrices are suboptimal. Although this chapter provides an overview of EBC and a list of validated biomarkers, it is by no means comprehensive; reports of newly discovered compounds in EBC are published monthly and will likely only increase as analytical technology improves.

Standardization of the collection and analysis of EBC and its biomarkers is necessary to improve their usefulness to researchers, clinicians, and, most importantly, patients. Some of this work is already ongoing by researchers who recognize that breathing technique (volume, flow rate, cadence) needs to be standardized to make EBC and EBA samples generalizable across different studies and laboratories.[15,16] Sampling recommendations have been made by the ATS/ERS. The International Association of Breath Research (IABR) now has a focus group that is assigned the task of coordinating and standardizing sampling and analysis for different fractions of breath. Once appropriately standardized, breath biomarkers should be translatable between groups and facilities.

Overall, research in EBC and EBA is progressing at a rapid rate. This is accelerated by the evolution of sampling and analytical methodology that now allows expansion of routine measurements of targeted VOCs into the realm of exhaled proteins, inflammatory markers, semi- and nonvolatile organics, viruses, bacteria, and a variety of chemicals derived from drugs and other exogenous sources. Further, the low-risk biohazardous nature of EBC and EBA compared to many other body fluid matrices continues to increase interest in their application for noninvasive biomarker exploration.

References

1. Davis MD, Montpetit AJ. Exhaled breath condensate: an update. *Immunol Allergy Clin* 2018;**38**(4):667–78.
2. Hunt J. Exhaled breath condensate: an evolving tool for noninvasive evaluation of lung disease. *J Allergy Clin Immunol* 2002;**110**(1):28–34.
3. Wallace MAG, Pleil JD. Evolution of clinical and environmental health applications of exhaled breath research: review of methods and instrumentation for gas-phase, condensate, and aerosols. *Anal Chim Acta* 2018;**1024**:18–38.
4. Kuban P, Foret F. Exhaled breath condensate: determination of non-volatile compounds and their potential for clinical diagnosis and monitoring. *Anal Chim Acta.* 2013;**805**: 1–18.
5. Pleil JD, Hubbard HF, Sobus JR, Sawyer K, Madden MC. Volatile polar metabolites in exhaled breath condensate (EBC): collection and analysis. *J Breath Res* 2008;**2**(2): 026001.
6. Pleil JD, Wallace MAG, Madden MC. Exhaled breath aerosol (EBA): the simplest non-invasive medium for public health and occupational exposure biomonitoring. *J Breath Res* 2018;**12**(2):027110.
7. Winters BR, Pleil JD, Angrish MM, Stiegel MA, Risby TH, Madden MC. Standardization of the collection of exhaled breath condensate and exhaled breath aerosol using a feedback regulated sampling device. *J Breath Res* 2017;**11**(4):047107.
8. Horvath I, Hunt J, Barnes PJ, Alving K, Antczak A, Baraldi E, et al. Exhaled breath condensate: methodological recommendations and unresolved questions. *Eur Respir J* 2005;**26**(3):523–48.
9. Pleil JD, Wallace MAG. New breath related topics: sample collection for exhaled breath condensate and aerosol, development of real-time medical alerts, measurement of artificial atmospheres, and analysis of legalized cannabis product. *J Breath Res* 2018;**12**(3): 039001.
10. Zamuruyev KO, Borras E, Pettit DR, Aksenov AA, Simmons JD, Weimer BC, et al. Effect of temperature control on the metabolite content in exhaled breath condensate. *Anal Chim Acta* 2018;**1006**:49–60.
11. Zamuruyev KO, Aksenov AA, Pasamontes A, Brown JF, Pettit DR, Foutouhi S, et al. Human breath metabolomics using an optimized non-invasive exhaled breath condensate sampler. *J Breath Res* 2016;**11**(1):016001.
12. Zamuruyev KO, Schmidt AJ, Borras E, McCartney MM, Schivo M, Kenyon NJ, et al. Power-efficient self-cleaning hydrophilic condenser surface for portable exhaled breath condensate (EBC) metabolomic sampling. *J Breath Res* 2018;**12**(3):036020.
13. Vaughan J, Ngamtrakulpanit L, Pajewski TN, Turner R, Nguyen TA, Smith A, et al. Exhaled breath condensate pH is a robust and reproducible assay of airway acidity. *Eur Respir J* 2003;**22**(6):889–94.
14. Stiegel MA, Pleil JD, Sobus JR, Morgan MK, Madden MC. Analysis of inflammatory cytokines in human blood, breath condensate, and urine using a multiplex immunoassay platform. *Biomarkers* 2015;**20**(1):35–46.
15. American Thoracic S, European Respiratory S. ATS/ERS recommendations for standardized procedures for the online and offline measurement of exhaled lower respiratory nitric oxide and nasal nitric oxide, 2005. *Am J Respir Crit Care Med* 2005;**171**(8): 912–30.
16. Horvath I, Barnes PJ, Loukides S, Sterk PJ, Hogman M, Olin AC, et al. A European Respiratory Society technical standard: exhaled biomarkers in lung disease. *Eur Respir J* 2017;**49**(4).

17. Walsh BK, Davis MD, Hunt JF, Kheir JN, Smallwood CD, Arnold JH. The effects of lung recruitment maneuvers on exhaled breath condensate pH. *J Breath Res* 2015;**9**(3): 036009.
18. Walsh BK, Mackey DJ, Pajewski T, Yu Y, Gaston BM, Hunt JF. Exhaled-breath condensate pH can be safely and continuously monitored in mechanically ventilated patients. *Respir Care* 2006;**51**(10):1125–31.
19. Greenwald R, Ferdinands JM, Teague WG. Ionic determinants of exhaled breath condensate pH before and after exercise in adolescent athletes. *Pediatr Pulmonol* 2009;**44**(8): 768–77.
20. Neville DM, Fogg C, Brown TP, Jones TL, Lanning E, Bassett P, et al. Using the inflammacheck device to measure the level of exhaled breath condensate hydrogen peroxide in patients with asthma and chronic obstructive pulmonary disease (the EXHALE pilot study): protocol for a cross-sectional feasibility study. *JMIR Res Protoc* 2018;**7**(1):e25.
21. Hunt J, Yu Y, Burns J, Gaston B, Ngamtrakulpanit L, Bunyan D, et al. Identification of acid reflux cough using serial assays of exhaled breath condensate pH. *Cough* 2006;**2**:3.
22. Konstantinidi EM, Lappas AS, Tzortzi AS, Behrakis PK. Exhaled breath condensate: technical and diagnostic aspects. *Sci World J* 2015;**2015**:435160.
23. Aldakheel FM, Bourke JE, Thomas PS, Matheson MC, Abramson MJ, Hamilton GS, et al. NOx in exhaled breath condensate is related to allergic sensitization in young and middle-aged adults. *Clin Exp Allergy* 2019;**49**(2):171–9.
24. Ratnawati, Morton J, Henry RL, Thomas PS. Exhaled breath condensate nitrite/nitrate and pH in relation to pediatric asthma control and exhaled nitric oxide. *Pediatr Pulmonol* 2006;**41**(10):929–36.
25. Aldakheel FM, Thomas PS, Bourke JE, Matheson MC, Dharmage SC, Lowe AJ. Relationships between adult asthma and oxidative stress markers and pH in exhaled breath condensate: a systematic review. *Allergy* 2016;**71**(6):741–57.
26. Montuschi P, Ciabattoni G, Paredi P, Pantelidis P, du Bois RM, Kharitonov SA, et al. 8-Isoprostane as a biomarker of oxidative stress in interstitial lung diseases. *Am J Respir Crit Care Med* 1998;**158**(5 Pt 1):1524–7.
27. Aksenov AA, Zamuruyev KO, Pasamontes A, Brown JF, Schivo M, Foutouhi S, et al. Analytical methodologies for broad metabolite coverage of exhaled breath condensate. *J Chromatogr B Analyt Technol Biomed Life Sci* 2017;**1061–1062**:17–25.
28. Schmidt AJ, Borras E, Nguyen AP, Yeap D, Kenyon NJ, Davis CE. Portable exhaled breath condensate metabolomics for daily monitoring of adolescent asthma. *J Breath Res* 2020;**14**(2):026001.
29. Majewski S, Tworek D, Szewczyk K, Kurmanowska Z, Antczak A, Gorski P, et al. Epithelial alarmin levels in exhaled breath condensate in patients with idiopathic pulmonary fibrosis: a pilot study. *Clin Res J* 2019;**13**(10):652–6.
30. Krenke K, Peradzynska J, Lange J, Banaszkiewicz A, Lazowska-Przeorek I, Grzela K, et al. Inflammatory cytokines in exhaled breath condensate in children with inflammatory bowel diseases. *Pediatr Pulmonol* 2014;**49**(12):1190–5.
31. Papineni RS, Rosenthal FS. The size distribution of droplets in the exhaled breath of healthy human subjects. *J Aerosol Med* 1997;**10**(2):105–16.
32. Pleil JD, Wallace MAG, McCord J, Madden MC, Sobus J, Ferguson G. How do cancer-sniffing dogs sort biological samples? Exploring case-control samples with non-targeted LC-Orbitrap, GC-MS, and immunochemistry methods. *J Breath Res* 2019;**14**(1):016006.

CHAPTER

Exhaled particles

8

Katharina Schwarz[1], Olaf Holz[1,2], Jens M. Hohlfeld[1,2,3]

[1]*Fraunhofer Institute for Toxicology and Experimental Medicine ITEM, Hannover, Germany;* [2]*German Center for Lung Research (DZL—BREATH), Hannover, Germany;* [3]*Respiratory Medicine, Hannover Medical School, Hannover, Germany*

Chapter outline

8.1 Overview

The human lung is a very complex organ as it contains a huge variety of about 40 different cell types. The endothelial side of the lung is perfused by the entirety of blood, and regulation of perfusion is very diligent in order to accomplish adaptation and synchronization with ventilation. The epithelial side of the lung is highly complex and dictated by securing its main function, which is gas exchange. In this respect, alveolar type I cells create a huge inner surface, while alveolar type II cells predominantly produce pulmonary surfactant that lines the inner surface of the lung

Breathborne Biomarkers and the Human Volatilome. https://doi.org/10.1016/B978-0-12-819967-1.00008-6
Copyright © 2020 Elsevier B.V. All rights reserved.

and comprises the main constituent of the epithelial lining layer. The majority of epithelial lining liquid originates from the alveolar compartment. As the cellular composition of the epithelium in the upper airways is more complex, however, the epithelial lining layer is supplied with additional components during its passage up the bronchial tree. Here, cells like club cells, airway basal cells, ciliated and non-ciliated bronchial epithelial cells, and gland cells contribute constituents to the epithelial lining layer.

The major constituent of the epithelial lining layer in the alveoli is pulmonary surfactant, which is a mixture of phospholipids and surfactant-associated proteins.[1] Pulmonary surfactant secures the openness of alveoli and airways by reducing surface tension at the air—liquid interface. While this reduction of surface tension is dynamic in the alveoli, reaching near zero values for surface tension at end-expiration, surface tension in the airways is higher but surfactant is still a prerequisite for the openness of the airways. Despite the function of pulmonary surfactant, there is a tendency for airway collapse when airways become smaller in the lung periphery and whenever surfactant dysfunction occurs, especially during exhalation. Even without disease-related surfactant dysfunction, airways become narrow at end-expiration, leading to airway wall ("compliant") and fluid collapse[1] and then upon inhalation, these areas reexpand and can "burst" open creating tiny aerosols. This phenomenon, referred to as "surface film disruption" or "fluid film rupture," is now thought to be the primary mechanism for aerosol formation, although upper airway turbulence during forceful exhalations may contribute.[2,3] This collapse and reexpansion of the epithelial lining liquid in the small (i.e., lower) airways and alveoli is described and discussed in detail in this chapter.

Breath analysis is possibly one of the oldest forms of diagnosis, because already the ancient Greeks used breath odors, such as the sweet smell of diabetic ketoacidosis, the fishy smell of breath associated with liver illness, or the urine-like odor of kidney disease; as such, the early use of breath analysis was based on volatile organic compounds (VOCs).[4] When exhaled breath condensate (EBC) was introduced as a novel form of breath analysis, by allowing the measurement of nonvolatile components, a joint EBC taskforce of the European Respiratory Society (ERS) and American Thoracic Society (ATS) recommended intensifying research into the mechanisms of emissions of nonvolatile components from the lung.[5] This stimulated several independent research groups to elaborate on the mechanism of emissions of nonvolatile components by exhaled particles, which is reviewed in detail below.

Exhaled particles originating from the epithelial lining layer are not only a useful specimen to fingerprint the biochemical composition of the epithelial lining fluid (ELF), but also carry intriguing physical information on the origin and site of airway closure. In addition, comprehensive analysis of breath-resolved particle emissions might allow airway structure to be determined, although this physical detail is beyond the scope of this chapter.[6] Thus, this chapter introduces the reader to the basic principles of exhaled particles, such as their generation, origin, transport, and redeposition, as a basis for a clear understanding of the resulting profile of particles in exhaled breath. Further, the principles of how to sample exhaled particles, elucidate their principal components, and potentially use this information for the diagnosis of lung diseases are discussed.

8.2 Aerosol generation and transport in the lung

The following overview on the origin and formation focuses on particle generation and transport in the lung periphery during normal/quiet oral breathing. This activity is usually relevant for the sampling protocols for breath aerosol or condensate sampling for subsequent biochemical analysis. Droplet generation mechanisms in the upper airways related to other activities, such as speaking or coughing, are not considered here, but are reviewed in the literature, e.g., by Bake et al.[7]

In principle, possible mechanisms of aerosol generation in the lung during normal breathing are (i) droplet formation at the air–liquid interface due to shear forces at high velocities in the upper airways and (ii) a reopening of terminal airway structures associated with a break-up of lung lining fluid bridges,[8] as depicted in Fig. 8.1.

In recent years, several independent research groups have systematically investigated concentrations and sizes of breath particles with the aim of gaining better insights into the mechanism of droplet formation. According to the studies of Johnson and Morawska, Almstrand et al., and Schwarz et al.,[2,9–11] particles emitted during normal breathing are mainly in the submicron range, with a maximum size of 0.3–0.4 μm and only a very small fraction >1 μm. The tidal or end-expiratory volume was identified to be the main factor determining the number of exhaled particles in the individual subject regardless of the health status of the lungs. Breath holding at the end of the exhalation maneuver further increases the amount of droplets generated.[12]

These results have consistently led to the conclusion that the reopening of collapsed airways is the dominant mechanism for droplet generation during normal breathing. The surface tension between the air lumen and the lung lining liquid is considered to be the dominant driver for terminal airway collapse. Specifically, a decreasing transmural pressure at the end of exhalation leads to surface tension–driven instabilities that can result in the structural collapse of the very small airways or the formation of fluid bridges in the larger airways.[1] With the subsequent inhalation maneuver, i.e., with reexpansion, collapsed airways are reopened with the associated generation of small droplets (Fig. 8.1). These droplets are emitted in the subsequent exhalation.[2] These findings are consistent with the observation that airway closure is confined to very small airways with dimensions below about 0.5 mm, and the resulting derivation of the so-called closing volume in the human lung more than 50 years ago by Dollfuss et al. and Macklem et al. (reviewed in Bake et al.[7]).

More detailed investigations into the mechanism of droplet formation were conducted using computational fluid dynamic simulations.[13,14] In these studies, the fluid

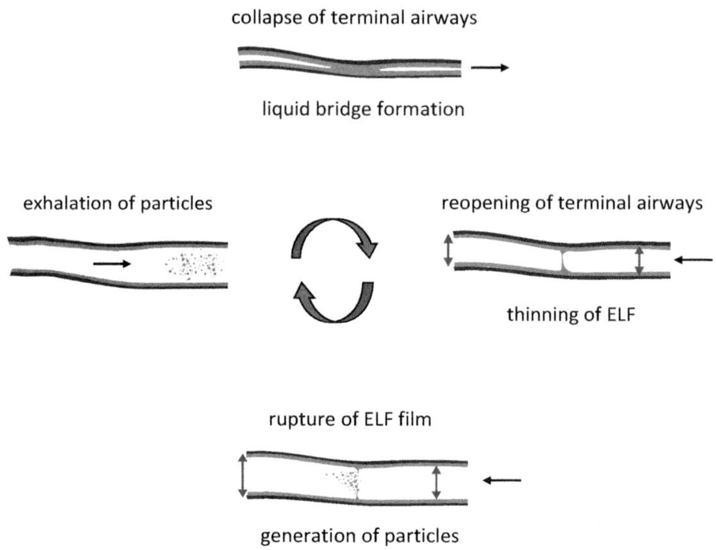

FIGURE 8.1

Assumed cyclic mechanism of breath aerosol generation during normal breathing. Reopening of collapsed peripheral airways is associated with a break-up of the epithelial lining fluid (ELF) film that covers the lung. The small arrows indicate the respective airflow direction.

bridge break-up was described as a rupture of a thin fluid film following instabilities, thereby resulting in the generation of small, mainly submicron droplets. The theoretically calculated count median diameter was determined to be about 0.4 μm; this value very nicely fits experimental data and is largely independent of the variation of the density and the rheological properties of the fluid.

Considering the corresponding particle residence times in the lung periphery of several seconds, and the relevant dimensions of the peripheral lung structure of about 200–500 μm, size-dependent redeposition processes might influence the number and size spectrum of exhaled aerosols. Work conducted by the authors[14] has been undertaken to investigate the particle transport and redeposition process of endogenously generated droplets in the lung. The time-resolved analysis of the droplet size spectrum during an individual exhalation maneuver at larger tidal volumes showed a shift toward smaller droplet sizes with efficient removal of particles in the supermicron range with increasing exhalation duration. Theoretical investigations into the penetration of particles generated in the peripheral lung up to the mouth using the ICRP[a] model[15] support the results. In addition, these findings have been confirmed experimentally by investigations under conditions of

[a] International Commission on Radiological Protection.

microgravity during parabolic flights.[16] Particles around 1 μm in size and larger are efficiently redeposited in the alveolar region due to gravitational settling, droplets smaller than about 0.1 μm are deposited in the lung periphery due to diffusion processes. These findings were in very good agreement with the results for the particle size spectrum experimentally observed, especially when considering that droplet size measurement usually takes place at a relative humidity below 100%, so that a shrinkage by a factor of about two is assumed.[13,17,18]

8.3 Physical characteristics of breath particles

Several clinical studies have been conducted with healthy individuals[19] and different patient populations with the aim of gaining a better understanding of the parameters that determine breath aerosol properties and for assessing the diagnostic potential of exhaled breath particles.

In general, particle number was found to be variable but very low, with concentrations about three orders of magnitude below typical concentrations in ambient air. In contrast, particle size distribution during normal at-rest breathing has been observed to be similar for all individuals investigated, independent of health status (Fig. 8.2), with diameters primarily in the submicron range.[9,17]

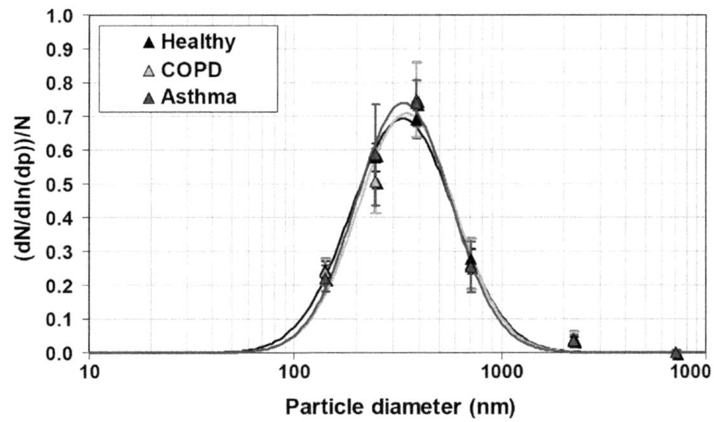

FIGURE 8.2

Exhaled particle size distribution averaged over a whole exhalation for healthy controls (n = 16, *black line and symbols*), chronic obstructive pulmonary disease (COPD) patients (n = 25, *blue/light gray line and symbols*), and subjects with allergic asthma (n = 11, red/ *dark gray line and symbols*). N = particle number, p = particle diameter, ln = natural logarithm.

Reproduced from [9].

The exhaled particle numbers and size distributions, and their correlations to lung function parameters for healthy volunteers in comparison to subjects with chronic obstructive pulmonary disease (COPD) and asthma, have been investigated in detail.[9,19] An exponential rise in the particle emission was observed with increasing tidal volume for all subjects, while breathing flow rates were found to be only of minor importance. For the individual subject, the emission level and the corresponding relationship with tidal volume was highly reproducible, yet there were considerable interindividual differences in the number emission level by about two orders of magnitude, which were independent of the health status of the lungs.[9,14,20] For rhinovirus-infected subjects in comparison to healthy subjects, similar general findings regarding variability of emission level and its dependency on health status were observed.[21]

A pharmacologically induced obstruction of the central airways did not lead to a change in the particle emission pattern. This is in agreement with the concept of peripheral airway collapse and generation of very small droplets not being affected by the larger airway structures. The obstruction of the peripheral airways in COPD also does not lead to a significant change, neither in the emitted number of particles nor in the size distribution averaged per breath compared to the healthy subjects.[9]

Extensive correlation analyses do not show any clear correlation between the number emission level and lung function parameters, such as total lung capacity (TLC), functional residual capacity (FRC), residual volume (RV), or demographic/biomedical parameters (e.g., age). Weak trends were only observed in healthy subjects, toward increasing emission level with increasing age[2,14,19] and decreasing ratio of FRC to TLC,[9] again supporting the hypothesis of peripheral airway closure as a relevant mechanism of aerosol generation during normal breathing. The lack of any relationship between emission level and obstructive airway diseases (COPD and asthma) is presumably related to the very complex interplay of morphology, ventilation, and lung lining liquid properties that occur in the context of these diseases.

Considering the low particle number emission and the small droplet sizes generated, even by maximizing the aerosol output by performance of deep or forced exhalation maneuvers over a longer collection time, e.g., 10 min, the aerosol output will only be between about a few ng up to several hundred ng.[9,22]

8.4 Particle sampling

A number of procedures have been developed to collect breath aerosols from the small airways. Aerosols from this location contain nonvolatile compounds and water, despite being generally referred to as particles.[7] The nature and heterogeneity of these microdroplets with respect to shape, size, and volume has been recently demonstrated using atomic force microscopy (AFM).[23]

8.4.1 Exhaled breath condensate

Exhaled breath condensate (EBC) was the first broadly used medium to explore nonvolatile markers from the peripheral lung. EBC sampling uses the process of freezing to collect aerosol droplets either in cooled pipes or specific cooled condensers (see previous chapter). As reported earlier, an ERS taskforce has focused on exhaled biomarkers in EBC, and a recent report lists three points that contribute to the limited reproducibility of the method, namely[24] (i) lack of standardized collection methods, (ii) EBC biomarker concentrations near detection limits, and (iii) the absence of a valid dilution factors. In addition, most available EBC collection methods do not count the number of particles or enable a differentiation of their size, which are factors with known large interindividual differences that are not controlled in EBC sampling, thereby adding to the variability of this method.[9]

8.4.2 Filter-based collection

Several small and very simple devices are available that collect breath aerosols with specifically designed polymer filters with low flow resistance (see Fig. 30.3, Chapter 30).[25] These filters retain larger particles and their simple use appears to be especially suited for drug monitoring, e.g., in traffic controls.

8.4.3 Impactor-based collection

To increase the concentration of nonvolatile compounds in breath samples and to avoid the large dilution encountered with EBC collection, devices using impactors were developed (Fig. 8.3).[26]

With impactor-based systems, the number of collected particles is maximized by concentrating them onto a solid substrate. These substrates can be directly used for mass spectrometric analysis or can be dissolved in a small amount of liquid to be analyzed with other techniques. The impactor from the particles in exhaled air (PExA) system is currently the only commercially available device. As outlined in the schematic of Fig. 8.3, the device measures the sampled air volume and includes a particle counter with eight size interval channels (range 0.41−4.55 μm). Based on the latter, it is possible to estimate the mass of the sampled particles. PExA recommends a specific breathing pattern, called the airway reopening maneuver,[28] to collect as many particles per breath as possible. The air volume and the accumulated mass of each breath are displayed online to allow the control and optimal training of the subject or volunteer performing breath sampling.

From the direct comparison of samples collected by the SenseAbuse device and the PExA instrument from patients undergoing methadone supplementation, it can be concluded that the filter collection technology is more likely to sample larger particles as compared to the collection by impaction.[27] This is confirmed by a recent publication about an improved filter device (BreathExplor[25]).

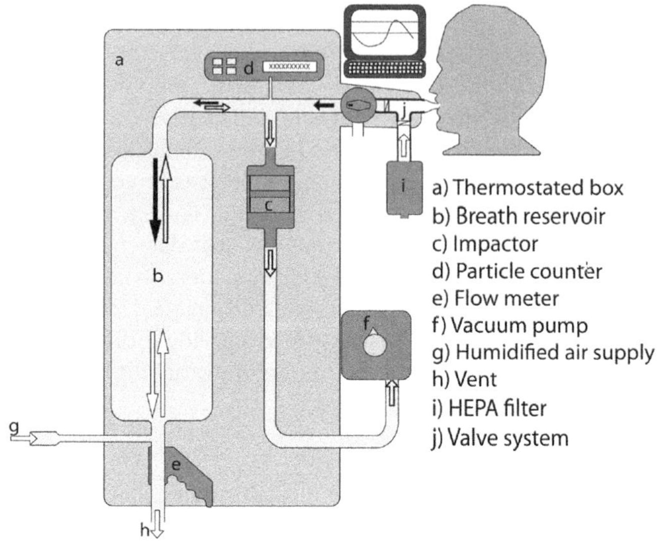

FIGURE 8.3

A schematic presentation of the particles in exhaled air (PExA) instrument.

Reproduced from [27].

8.4.4 Electrostatic collector

A recent publication introduces a novel electrostatic collector to allow for a more detailed morphological breath particle analysis. To achieve this, breath particles are pumped into a collector through a tube and around a tungsten needle to charge the particles with ions. The charged particles are then collected on a conducting substrate that is connected to a negative pole of a power supply. The "dry residue" particles on this substrate can then be directly characterized using AFM.[23]

8.5 Particle composition

Assuming a diameter of average mass for exhaled droplets of around 0.8 μm and particle numbers to be as variable as 150–35,000 per breath with maneuvers that enforce particle generation, it is calculated that every single breath contains around 40 pg –10 ng particles mass. Correspondingly, it is estimated that the amount of collected particles from the peripheral airways is in the range pg/L exhaled air.[7] Therefore, the analysis of this specimen requires very sensitive methods.

8.5.1 Lipid composition

Using mass spectrometry it was demonstrated that the phospholipid composition of exhaled particles, such as phosphatidylcholine, phosphatidylglycerol, and phosphatidylinositol, is very similar to ELF.[26] Others have confirmed that exhaled particles contain phospholipids and proteins similar to surfactant.[7] Lower concentrations of dipalmitoylphosphatidylcholine (DPPC), the major component of surfactant, in particles generated by cough compared to by the reopening maneuver[28] are in line with the hypothesis that the particles are derived mainly from the small peripheral airways.

8.5.2 Proteins

In a first proteomic analysis of pooled breath samples from 12 healthy volunteers, corresponding to >3000 L, it was possible to identify 124 different proteins.[29] The total protein content was estimated to be 0.33 ng particles/L exhaled air. The identified protein spectrum showed a high degree of overlap with proteins found in bronchoalveolar lavage fluid (BAL). The most abundant proteins were albumin with 26%, surfactant protein SP-A with 12%, and immunoglobulins with 14%. Using the very sensitive SomaLogic analysis platform, it was recently shown that up to 207 proteins can be robustly found in exhaled breath particles.[30]

8.5.3 Drugs

Testing drug abuse in nonvolatile components of exhaled breath has mainly been performed using EBC (see Chapter 30) or with methods using specific filters to collect breath particles.[31] The easy sampling procedure offered by devices like SenseAbues or BreathExplor appears to be especially suited for roadside sample collection when blood sampling is not possible.

8.5.4 Workplace exposure and pollutants

Studies that have used exhaled particles or EBC to explore or monitor occupational exposure are covered in Chapter 32. An example of such research is a study of nine subjects that reported increased levels of manganese in four out of nine subjects following experimental welding fume exposure. The particles were analyzed by inductively coupled plasma mass spectrometry (ICP-MS) to detect metals of interest.[32]

8.6 Application in disease

As breath particles are derived from the peripheral lung, their number, physical characteristics, and composition are of interest to investigate diseases with suspected small airways contribution. Currently, published studies have focused on particle numbers, particle mass, and on the analysis of abundant proteins, such as albumin and SP-A, which are generally detectable in breath particles with standard immunological assays.

8.6.1 Asthma

Schwarz et al. studied ten patients with asthma before and after methacholine challenge and found no difference in particle emissions despite the significant change in airway obstruction.[9] Recently, Carpaij and coworkers showed that asthmatics and healthy individuals differ with respect to exhaled PExA mass, and that particle mass appears to be related to the degree of asthma remission.[33] Patients with complete remission emitted similar levels to healthy controls, while the levels of patients with clinical remission and those with persistent asthma showed lower levels. In line with this, the level of exhaled breath particles was found to be reduced in asthma patients following exposure to birch pollen.[34] Neither this study nor the large study from Soares et al.[35] observed differences in SP-A or albumin between healthy and asthmatics due to either asthma severity or seasonal variations.

8.6.2 Chronic obstructive pulmonary disease

Ambiguous results have been reported with respect to exhaled particle numbers from COPD patients. Schwarz et al. showed no difference in particle numbers per breath between healthy subjects and patients with COPD. In their study all subjects used identical standardized breathing maneuvers with a tidal volume to vital capacity ratio of 0.2.[9] Lärstad et al. reported marked lower particle numbers/L exhaled breath using the "airway closure" maneuver in which all subjects exhaled to their individual level of residual volume.[36] Potentially, the reopening maneuver is less effective in COPD patients due to increasing hyperinflation in more severe COPD patients. Very little data on the biochemical composition of breath particles in COPD currently exist, although first evidence suggests that the level of SP-A is slightly lower in patients with COPD.[36]

8.6.3 Lung transplantation and mechanical ventilation

A recently published study shows that connection of the PExA device to a mechanical ventilator is feasible.[37] In this study, the six patients who developed a primary graft dysfunction (PGD) showed increases in particle numbers during mechanical ventilation before clinical symptoms were detectable. The patients who did not develop a PGD did not show this increase. There is also first evidence that a lower level of SP-A is associated with a higher risk of developing bronchiolitis obliterans syndrome following lung transplantation.[38]

8.7 Summary

It is very well established that submicron particles are generated within the lung when films that contain epithelial lining liquid rupture during reopening of the airways. This offers an intriguing diagnostic opportunity by utilizing both the physical and biochemical information that is emitted from the lung with every breath. Importantly, intraindividual reproducibility is excellent, offering repeat measurements for longitudinal assessment of lung health status. By comparison, interindividual particle output—due to low particle numbers—is variable, with a span of two orders of magnitude between subjects, thus quantification of biomarkers in absolute values is hampered. Also, it is important to acknowledge that particle emission is dependent on the breathing pattern. While tidal breathing maneuvers favor particle emission from the lung periphery, forced expiration can lead to upper airway contamination with considerably larger particles and biased biomarker concentrations.

Undoubtedly, highly sensitive analytical methods are needed for reliable assessments of low abundant biomarkers in the ELF, such as cytokines and chemokines. In contrast, biochemical analysis of surfactant phospholipids and larger proteins, such as immunoglobulins, surfactant proteins, or albumin, has given promising results. Analysis of drug levels with highly sensitive methods might also open a novel way to assess pulmonary pharmacokinetics. Current endeavors to improve sampling methods, either through filter or impaction-based methods, or novel electrostatic capture, as well as the whole EBC fraction, will lead to new insights in particle emissions in the exhaled breath in both healthy and diseased populations.

Given the fact that exhaled particles are generated in the lung periphery, there is an emerging diagnostic potential for small airway disease. While applications in asthma and chronic obstructive lung disease are obvious, future diagnostic use might intensify inclusion of bronchiolitis in transplantation, airway remodeling in fibrotic lung disease, and small airways disease for environmental or occupational hazards.

Finally, future research needs to unravel the unresolved mystery of significant interindividual differences in particle emission. In addition, improvements in sensitivity of biomarker analysis will contribute to more widespread applications of exhaled particles in lung research and treatment of lung diseases. Despite these remaining challenges, exhaled particles are an intriguing modality with huge potential for noninvasive diagnosis of small airway diseases.

References

1. Hohlfeld J, Fabel H, Hamm H. The role of pulmonary surfactant in obstructive airways disease. *Eur Respir J* 1997;**10**(2):482–91.
2. Johnson GR, Morawska L. The mechanism of breath aerosol formation. *J Aerosol Med Pulm Drug Deliv* 2009;**22**(3):229–37.
3. Pleil JD, Wallace MAG, Madden MC. Exhaled breath aerosol (EBA): the simplest non-invasive medium for public health and occupational exposure biomonitoring. *J Breath Res* 2018;**12**(2):27110.
4. Paschke KM, Mashir A, Dweik RA. Clinical applications of breath testing. *Med Rep* 2010;**2**(56).
5. Horváth I, Hunt J, Barnes PJ, Alving K, Antczak A, Baraldi E, et al. Exhaled breath condensate: methodological recommendations and unresolved questions. *Eur Respir J* 2005;**26**(3):523–48.

6. Schwarz K, Koch W, Biller H, Hohlfeld JM. Single breath particle characterization to detect small airway alterations: American Thoracic Society International Conference A5748. *Am J Respir Crit Care Med* 2012;**185**.
7. Bake B, Larsson P, Ljungkvist G, Ljungström E, Olin A-C. Exhaled particles and small airways. *Respir Res* 2019;**20**(1):8.
8. Gebhart J, Anselm J, Heyder J, Stahlhofen W. The human lung as aerosol generator. *J Aerosol Med* 1988;**1**:196–7.
9. Schwarz K, Biller H, Windt H, Koch W, Hohlfeld JM. Characterization of exhaled particles from the human lungs in airway obstruction. *J Aerosol Med Pulm Drug Deliv* 2015; **28**(1):52–8.
10. Almstrand A-C, Bake B, Ljungström E, Larsson P, Bredberg A, Mirgorodskaya E, et al. Effect of airway opening on production of exhaled particles. *J Appl Physiol* 2010;**108**(3): 584–8.
11. Morawska L, Johnson GR, Ristovski ZD, Hargreaves M, Mengersen K, Corbett S, Chao CYH, Li Y, Katoshevski D. Size distribution and sites of origin of droplets expelled from the human respiratory tract during expiratory activities. *J Aerosol Sci* 2009;**40**(3): 256–69.
12. Holmgren H, Gerth E, Ljungström E, Larsson P, Almstrand A-C, Bake B, et al. Effects of breath holding at low and high lung volumes on amount of exhaled particles. *Respir Physiol Neurobiol* 2013;**185**(2):228–34.
13. Haslbeck K, Schwarz K, Hohlfeld JM, Seume JR, Koch W. Submicron droplet formation in the human lung. *J Aerosol Sci* 2010;**41**:429–38.
14. Schwarz K, Biller H, Windt H, Koch W, Hohlfeld JM. Characterization of exhaled particles from the healthy human lung–a systematic analysis in relation to pulmonary function variables. *J Aerosol Med Pulm Drug Deliv* 2010;**23**(6):371–9.
15. International commission on radiological protection (ICRP). In: *ICRP publication 66 - human respiratory tract model for radiological protection: areport of a task group of He international commission on radiological protection*; 1994.
16. Schwarz K, Dunkhorst W, Lödding H, Windt H, Koch W. Investigations into the mechanism and site of droplet generation in the human lung. In: *Abstract book of the International Aerosol Conference 2010, Helsinki*; 2010 (8).
17. Holmgren H, Ljungström E, Almstrand A-C, Bake B, Olin A-C. Size distribution of exhaled particles in the range from 0.01 to 2.0 μm. *J Aerosol Sci* 2010;**41**(5):439–46.
18. Holmgren H, Bake B, Olin A-C, Ljungström E. Relation between humidity and size of exhaled particles. *J Aerosol Med Pulm Drug Deliv* 2011;**24**(5):253–60.
19. Bake B, Ljungström E, Claesson A, Carlsen HK, Holm M, Olin A-C. Exhaled particles after a standardized breathing maneuver. *J Aerosol Med Pulm Drug Deliv* 2017;**30**(4): 267–73.
20. Schwarz K. *Zur Entstehung, Charakterisierung und diagnotischen Nutzbarkeit in der menschlichen Lunge endogen generierter exhalierter Aerosole*. Technical University of Clausthal, Clausthal-Zellerfeld; 2013 [Ph.D. dissertation].
21. Fabian P, Brain J, Houseman EA, Gern J, Milton DK. Origin of exhaled breath particles from healthy and human rhinovirus-infected subjects. *J Aerosol Med Pulm Drug Deliv* 2011;**24**(3):137–47.
22. Larsson P, Mirgorodskaya E, Samuelsson L, Bake B, Almstrand A-C, Bredberg A, et al. Surfactant protein A and albumin in particles in exhaled air. *Respir Med* 2012;**106**(2): 197–204.

23. Morozov VN, Mikheev AY. A collection system for dry solid residues from exhaled breath for analysis via atomic force microscopy. *J Breath Res* 2017;**11**(1):16006.
24. Horváth I, Barnes PJ, Loukides S, Sterk PJ, Högman M, Olin A-C, et al. A European Respiratory Society technical standard: exhaled biomarkers in lung disease. *Eur Respir J* 2017;**49**(4).
25. Seferaj S, Ullah S, Tinglev Å, Carlsson S, Winberg J, Stambeck P, et al. Evaluation of a new simple collection device for sampling of microparticles in exhaled breath. *J Breath Res* 2018;**12**(3):36005.
26. Almstrand A-C, Ljungström E, Lausmaa J, Bake B, Sjövall P, Olin A-C. Airway monitoring by collection and mass spectrometric analysis of exhaled particles. *Anal Chem* 2009;**81**(2):662–8.
27. Ljungkvist G, Ullah S, Tinglev Å, Stein K, Bake B, Larsson P, et al. Two techniques to sample non-volatiles in breath-exemplified by methadone. *J Breath Res* 2017;**12**(1): 16011.
28. Larsson P, Bake B, Wallin A, Hammar O, Almstrand A-C, Lärstad M, et al. The effect of exhalation flow on endogenous particle emission and phospholipid composition. *Respir Physiol Neurobiol* 2017;**243**:39–46.
29. Bredberg A, Gobom J, Almstrand A-C, Larsson P, Blennow K, Olin A-C, et al. Exhaled endogenous particles contain lung proteins. *Clin Chem* 2012;**58**(2):431–40.
30. Östling J, Van Geest M, Viklund E, Gustafsson P, Mirgorodskaya E, Olin A-C. Exploring particles from exhaled air as a new source for protein biomarkers from the airways. *Eur Respir J* 2017;**50**.
31. Beck O, Ullah S, Kronstrand R. First evaluation of the possibility of testing for drugged driving using exhaled breath sampling. *Traffic Inj Prev* 2019;**20**(3):238–43.
32. Bredberg A, Ljungkvist G, Taube F, Ljungström E, Larsson P, Mirgorodskaya E, et al. Analysis of manganese and iron in exhaled endogenous particles. *J. Anal. At. Spectrom.* 2014;**29**(4):730–5.
33. Carpaij OA, Muiser S, Bell AJ, Kerstjens HAM, Galban CJ, Fortuna AB, et al. Assessing small airways dysfunction in asthma, asthma remission and healthy controls using particles in exhaled air. *ERJ Open Res* 2019;**5**(4).
34. Larsson P, Lärstad M, Bake B, Hammar O, Bredberg A, Almstrand A-C, et al. Exhaled particles as markers of small airway inflammation in subjects with asthma. *Clin Physiol Funct Imag* 2017;**37**(5):489–97.
35. Soares M, Mirgorodskaya E, Koca H, Viklund E, Richardson M, Gustafsson P, et al. Particles in exhaled air (PExA): non-invasive phenotyping of small airways disease in adult asthma. *J Breath Res* 2018;**12**(4):46012.
36. Lärstad M, Almstrand A-C, Larsson P, Bake B, Larsson S, Ljungström E, et al. Surfactant protein A in exhaled endogenous particles is decreased in chronic obstructive pulmonary disease (COPD) patients: a pilot study. *PLoS One* 2015;**10**(12):e0144463.
37. Broberg E, Hyllén S, Algotsson L, Wagner DE, Lindstedt S. Particle flow profiles from the airways measured by PExAdiffer in lung transplant recipients who develop primary graft dysfunction. *Exp Clin Transplant* 2019;**17**(6):803–12.
38. Ericson PA, Mirgorodskaya E, Hammar OS, Viklund EA, Almstrand A-CR, Larsson PJ-W, et al. Low levels of exhaled surfactant protein A associated with BOS after lung transplantation. *Transplant Direct* 2016;**2**(9):e103.

PART C

Analytics

CHAPTER

Selected ion flow tube mass spectrometry

9

David Smith[1], Patrik Španěl[2], George B. Hanna[3], Raed Dweik[4]

[1]*Trans Spectra Limited, Newcastle-under-Lyme, United Kingdom;* [2]*J. Heyrovský Institute of Physical Chemistry, Academy of Sciences of the Czech Republic, Prague, Czech Republic;* [3]*Department of Surgery and Cancer, Imperial College, St. Mary's Campus, London, United Kingdom;* [4]*Department of Pulmonary, Allergy and Critical Care Medicine, Respiratory Institute, Cleveland Clinic, Cleveland, OH, United States*

Chapter outline

9.1 Overview

Selected ion flow tube mass spectrometry (SIFT-MS) was first realized as a viable technique for the analysis of trace gases in ambient air in 1996 and its potential for direct online, real-time analysis became immediately obvious. Instrumental development has gone through several phases from a large laboratory-based device to the current mobile commercial analytical instruments. SIFT-MS can be exploited for real-time analysis of volatile trace gases of biological origin, especially in breath,[1] but also in the headspace of biological fluids, particularly urine,[2] and trace compounds emitted by mammalian cell[3] and bacterial cultures[4] of clinical significance. The earliest research using SIFT-MS for breath analysis and related topics

Copyright © 2020 Elsevier B.V. All rights reserved.

has been reviewed elsewhere.[5] This chapter briefly introduces the technique and then reviews the latest discoveries and developments in SIFT-MS research of clinical relevance.

Selected ion flow tube-mass spectrometry relies on the chemical ionization by selected reagent ions, H_3O^+, NO^+, and $O_2^{+\bullet}$, of trace analyte compounds in air/breath samples introduced at a controlled flow rate into helium carrier gas passing rapidly through a flow tube.[1] Reactions occur between the selected reagent ion and the composite analyte molecules that produce characteristic product (analyte) ions and their currents (count rates) are measured by a downstream analytical mass spectrometer/detection system. The identified analyte ions are related to specific neutral trace compounds in the air/breath from an in-depth knowledge of the ion-molecule chemistry. Their count rates, coupled with physical parameters, such as carrier (helium) gas and sample gas flow rates, are converted to concentrations of the neutral trace analyte molecules in the sample. To achieve accurate trace gas analysis, a proper appreciation of the complex ion chemistry occurring in SIFT-MS analytical reactors is needed, especially when humid samples are being analyzed. Thus, an ion kinetics library has been built based on extensive studies of the reactions of the reagent ions with a wide range of volatile organic compounds (VOCs). SIFT-MS allows online, real-time analysis of the trace gases in humid media at atmospheric pressure down to concentrations in parts-per-billion by volume, ppb or lower.[6] Thus, directly exhaled breath can be analyzed, which avoids sample collection into bags or onto surface or cryogenic traps that can compromise the sample through losses, artifacts, or incomplete recovery (see Chapter 2).

The raw data from a SIFT-MS analysis of a breath/air sample is usually in the form of a mass spectrum acquired by the downstream analytical mass spectrometer over a chosen range of mass-to-charge ratio, *m/z*. The peaks on the spectrum are those of the reagent ions and the analyte ions. The kinetics library relates the analyte ions to the VOCs in the sample and treatment of the ion counts and the known kinetics of the reagent ion/VOC reactions allows the quantification of the VOCs. This is called the full scan (FS) mode of analysis. For rapid and precise targeted analyses, the count rates of selected analyte ions are monitored. This is called the multiple ion monitoring (MIM) mode of analysis and is chosen when analyzing single-breath exhalations for several chosen metabolites. An additional valuable feature of online analysis of single exhalations is that on the inhalation cycle through the mouth, the ambient air adjacent to the sampling port can be analyzed, which is very valuable for breath analysis. SIFT-MS is especially suited to direct exhaled breath analysis, principally because of its practical simplicity, as is required for a patient-friendly tool in the clinical environment. Single sequential breath exhalations can be followed over periods of many hours allowing biochemical kinetics to be studied with a large data flow and good temporal definition.[7]

Notwithstanding these unique online analysis features of SIFT-MS, sample collection followed by offline analysis can be achieved with care, noting the well-reported pitfalls in this procedure and the likely errors and uncertainties that can result (see Chapter 2). Indeed, much data on breath analysis have been obtained by breath collection into polymer bags (predominantly Tedlar or Nalophan) followed by SIFT-MS analysis, because signal integration times can be much longer than are possible for single-breath exhalations. In this manner, both FS and MIM spectra can be accumulated, providing commensurately greater analytical sensitivity. Also, preconcentration of the trace metabolites in collected breath samples using solid-phase microextraction (SPME; widely used together with gas chromatography−mass spectrometry [GC-MS]), can be used in combination with SIFT-MS as the analytical method. Similarly, thermal desorption (TD) has been used together with SIFT-MS.[8] A shortcoming of SIFT-MS is that it can be difficult to identify some neutral analyte molecules in mixtures, especially structural isomers, from only their analyte ions. Then, coordinated SPME-GC-MS or TD-GC-MS with SIFT-MS experiments can be exploited, the former providing more positive identification of neutral analytes and the latter providing more accurate quantification as long as the analytical ion chemistry is understood.

9.2 The search for and quantification of single-breath metabolites related to specific disease and infection

The earlier work was summarized in the SIFT-MS chapter in the previous book in this series[5] and needs only to be mentioned in this chapter. Briefly, it involved the detailed construction of reference ranges for normal breath metabolites. Much emphasis was placed on ammonia and acetone and their variations with age, gender, and diet. The seminal work on the detection and quantification of hydrogen cyanide (HCN) emitted by the bacterium *Pseudomonas aeruginosa* (PA) and its detection in the exhaled breath of children with cystic fibrosis (CF) was reported and its medical significance highlighted. Further work has been carried out to seek and accurately quantify single VOC breath metabolites, the importance of which is emphasized in a recent paper.[9]

9.2.1 Ammonia

For decades, interest in breath ammonia has been developing with the anticipation that its analysis will be useful in clinical practice, especially in nephrology. A detailed study of mouth-exhaled ammonia and nose-exhaled ammonia shows that much of the breath ammonia is generated in the oral cavity by the action of enzymes on salivary urea.[10] It must be concluded that mouth-exhaled ammonia is always elevated above a concentration that would be equilibrated with blood ammonia. Nose-exhaled ammonia largely originates at the alveolar interface and so its concentration more closely relates to the expected alveolar blood ammonia concentration. Ingestion of proteins results in increased blood/saliva urea, as does the presence of *Helicobacter pylori* infection, which ultimately results in increased mouth-exhaled ammonia. So elevated mouth-exhaled ammonia may be due to abnormally high blood urea, a high pH of the saliva/mouth/airways mucosa, poor oral hygiene, or a combination of these.[10] A recent review article considers the potential value of breath ammonia measurements in patients with chronic kidney disease as a potential estimator of the severity of uremia.[11] However, the clinical value of mouth-exhaled ammonia is doubtful and nose-breath sampling is more sensible. Finally, a further point is that laboratory air often contains appreciable concentrations of ammonia, which is inhaled during the normal breathing cycle and partially exhaled, apparently contributing to "endogenous" ammonia. A careful study of the partial retention coefficients of several inhaled (normally exogenous) compounds in exhaled breath has been carried out using online SIFT-MS analyses, including ammonia and other common exhaled breath VOCs.[12]

9.2.2 Acetone

Breath acetone has attracted much attention, principally because of its proposed association with blood glucose and diabetes, yet there is no convincing experimental evidence that this association is valid.[13] A serious compounding factor is diet. A SIFT-MS study has shown that taking a ketogenic diet can result in greatly increased breath acetone.[14] The further aspects of breath acetone vis-à-vis diabetes are discussed in more detail by Turner et al.[15] A very interesting recent SIFT-MS study on the restriction of diet by bariatric surgery[16] has shown that exhaled acetone increased significantly whereas some other breath ketones actually decreased after dieting and surgery. It was concluded that breath acetone can be potentially clinically useful as a noninvasive nutritional assessment in obese patients.

Breath acetone as measured by SIFT-MS and single mixed metal oxide (MOx) sensors in a group of type 1 diabetes patients during hypoglycemic glucose clamps was found to correlate linearly with blood glucose, although concentrations differed greatly between patients with the same blood glucose. It is therefore unlikely that breath analysis can entirely replace blood glucose testing.[17]

9.2.3 Hydrogen cyanide

Since the first discovery of HCN above in vitro cultures of PA using SIFT-MS direct analysis, much work has been carried out to detect and quantify this compound in exhaled breath of CF patients. The early seminal work was reported in the SIFT-MS chapter in the previous edition of this book.[5] Since then, a concerted effort to solidify the early results has been carried out involving analysis of the breath of cohorts of patients with and without CF. Additionally, much in vitro work has been carried out on genotypically different strains of PA, which showed that HCN and other volatile compounds, including 2-aminoacetophenone, are produced by most strains but in different degrees. A separate detailed study using coordinated SIFT-MS and SPME-GC-MS analyses showed that methyl thiocyanate was emitted by 36 strains of PA and was also present in the exhaled breath of 28 CF patients. Further to these studies, HCN and 2-aminoacetophenone emitted by PA cultured under planktonic and biofilm conditions have been studied using SIFT-MS analysis. This major effort on both in vitro PA cultures and on HCN in exhaled breath amounted to more than a decade of work, which is summarized in a short review[18] and well demonstrated by the timeline presented in Fig. 9.1.

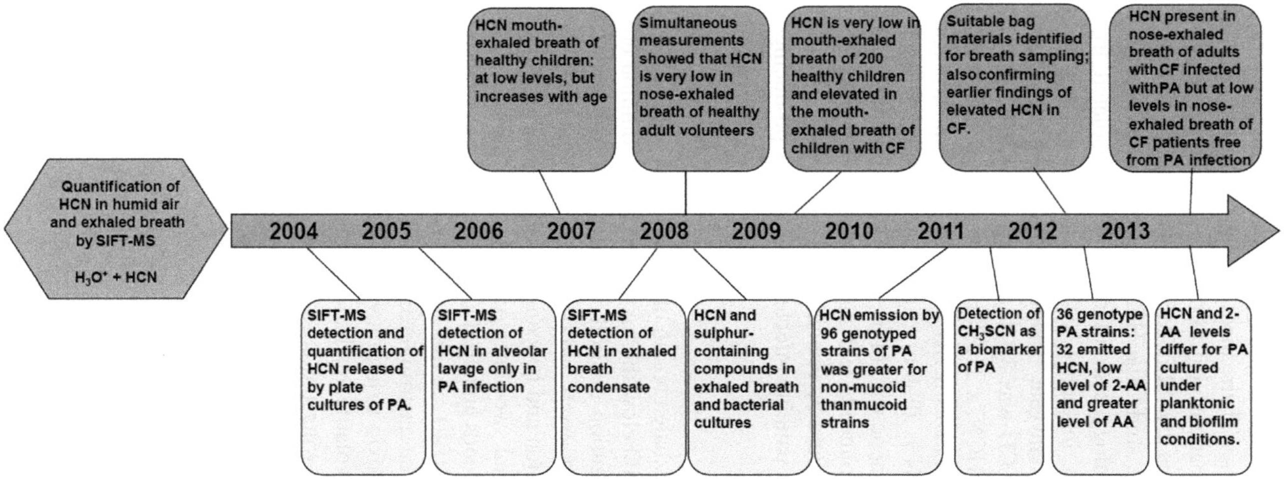

FIGURE 9.1

The timeline representing the milestones of the research program directed toward the study of gaseous hydrogen cyanide (HCN) as a biomarker of *Pseudomonas aeruginosa* (PA) infection in cystic fibrosis (CF). The upper part of the scheme traces the in vivo studies with human subjects and the lower part traces the related parallel in vitro studies.

Reproduced from [18].

In order to investigate the real clinical value of HCN as a biomarker of PA infection in CF, a multicenter study involving eight hospitals in central England was carried out.[19] Thus, 233 children with CF who were initially free from PA infection were followed for 2 years. Their median (interquartile) age range was 8.0 (5.0–12.2) years. At each patient visit to the clinic, an exhaled breath sample was collected for HCN analysis by SIFT-MS. In total, 2055 breath samples were analyzed. At the end of the study, the HCN concentrations were compared to the results of routine microbiology surveillance. PA was isolated from 71 children during the study with an incidence (95% confidence interval [CI]) of 0.19 (0.15–0.23) cases per patient-year. Using a random-effects logistic model, the estimated odds ratio (95% CI) was 3.1 (2.6–3.6), which showed that for a 1 ppb increase in exhaled breath HCN, a 212% increase in the odds of PA infection was expected. The sensitivity and specificity were estimated at 33% and 99%, respectively. Exhaled breath HCN is thus a specific biomarker of new PA infection in children with CF, but the genetic variability of PA strains means that in some patients the elevation is less pronounced, leading to lower diagnostic sensitivity (false negatives). Hence, HCN should be complemented by other biomarker compounds to avoid false negatives for patients infected with strains producing less HCN.

9.2.4 Methanol

Methanol is present in the exhaled breath of all people at concentrations of typically 100–400 ppb, which is well below the toxic limit. This common breath metabolite has received much less attention than other breath VOCs. It is contained in some foods and drinks, such as ripe fruits and fruit juices, which when ingested increases the methanol in the circulation and hence in the exhaled breath. It is also formed in the hydrolysis of the artificial sweetener Aspartame, methyl-L-α-aspartyl-L-phenylalaninate, which is used in foods and drinks and is considered to be safe at an acceptable daily intake (ADI) of 40 mg per kg of body weight. This compound is completely hydrolyzed in the gastrointestinal tract to aspartic acid, phenylalanine, and methanol, each being toxic at high levels. Thus, a study was carried out using SIFT-MS to quantify the volatile methanol component in the exhaled breath of ten healthy volunteers following the ingestion of a single ADI dose of aspartame scaled according to their body weight.[20] Breath methanol concentrations were observed to increase in all volunteers by 1082 ± 205 ppb from their preingestion values, which ranged from 193 to 436 ppb, to peak values ranging from 981 to 1622 ppb, which slowly decreased with time. These observations agree quantitatively with a predicted increase of 1030 ppb estimated using a one-compartment model of uniform dilution of the methanol generated from a known amount of aspartame throughout the total body water (including blood). In summary, an ADI dose of aspartame leads to a three to sixfold increase of blood methanol concentration above the individual baseline values.

9.2.5 Pentane

The hydrocarbon n-pentane has for some time been considered as a valuable biomarker of oxidative stress, lipid peroxidation, and inflammation in the body,

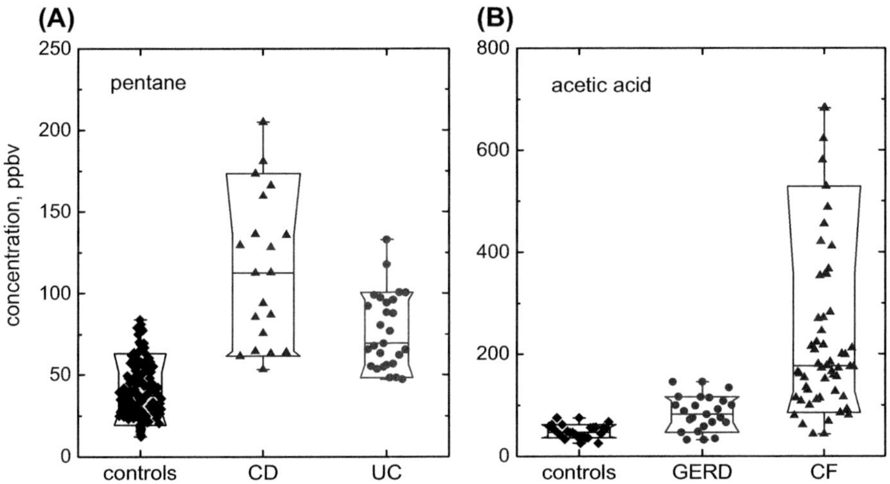

FIGURE 9.2

Box-and-whisker plots including the individual data points for breath concentrations of metabolites obtained using SIFT-MS. (A) Breath pentane for controls, Crohn's disease (CD), and ulcerative colitis (UC) cohorts. (B) Acetic acid in controls, patients with gastroesophageal reflux disease (GERD), and patients with cystic fibrosis (CF).

Reproduced from [21,22].

including inflammatory bowel disease (IBD). SIFT-MS was employed in a pilot study to explore pentane in the exhaled breath of a small cohort of patients with Crohn's disease (CD) and ulcerative colitis (UC), and a few healthy volunteers. Pentane was found to be significantly elevated in the breath of both the CD (mean 114 ppb) and the UC (mean 84 ppb) patients relative to the healthy controls (mean 40 ppb). Recently, a more detailed study has been carried out involving 136 CD patients and 51 UC patients and 14 controls,[21] which confirmed that the median concentration of pentane is elevated in the breath of the IBD patients compared to those of the healthy controls, as can be seen in Fig. 9.2A. This method of pentane analysis, which avoids sample storage, can ultimately form the basis of noninvasive screening of inflammatory processes, including IBD.

9.2.6 Acetic acid

Gastroesophageal reflux disease (GERD) is one of the most common causes of chronic cough. A simple in vitro model based on pork tissue samples exposed to artificial gastric fluid showed, via SPME-GC-MS analysis, that the only VOC significantly increased above the exposed tissue sample was acetic acid.[23] A study of the mouth-exhaled breath of 22 GERD patients using direct SIFT-MS analyses revealed that end-expiratory concentration of acetic acid (median 85 ppb) was significantly higher than that in the breath of a control group (median 48 ppb).[23] It was concluded that breath acetic acid may be useful for noninvasive diagnostics of GERD, which results from a lowering of the pH of the lining of the airways by reflux acid.

Subsequently, a study of the VOCs present in the exhaled breath of 58 CF patients was carried out.[22] The study revealed that the acetic acid vapor concentration in the samples was significantly elevated in the exhaled breath of the CF patients, independent of the PA infection status; see Fig. 9.2B. The cause for this may again be a decreased pH of the mucus lining the CF airways like that for the GERD patients, but with the acetic acid concentration reaching much higher values. Thus, it was proposed that noninvasive measurement of breath acetic acid concentration could serve as an indicator of the acidity of the CF airways mucosa.

9.3 Full scan spectra identifying breathprints of disease or disorders

The driving force for the research to find patterns or combinations of several VOCs in exhaled breath that allow identification of disease or disorders is the recognition that single-breath biomarkers are unlikely to achieve this. Yet this might be challenged given the success of some single-breath biomarkers as noninvasive indicators of adverse physiological conditions, as discussed above. Nevertheless, much research work on complex conditions such as cancer and some metabolic disorders has not yet revealed single-breath biomarkers that are specific to certain diseases. Much work has been directed toward the development of sensor arrays (electronic noses) that can be trained to recognize "*smellprints*" in exhaled breath, which allow identification of certain diseases or disorders (see Chapter 13). However, this technology is not well-suited to identify the specific compounds that contribute to a recognized pattern. The mass spectrometry approach, on the other hand, allows the identification of specific individual compounds in the breath.

Both approaches have been used to analyze breath and have revealed the strengths and weaknesses of both. Thus, while the electronic nose is easy to use, it does not lend itself to identify specific molecules, whereas mass spectrometry approaches are good at identifying specific compounds but are not intuitive or easy to use. Nevertheless, it has been found that SIFT-MS is particularly suited for breath analysis but requires a new approach for data analysis to make it potentially clinically useful. By approaching each compound, or analyte ion on the SIFT-MS FS mass spectrum, as its own sensor, patterns or "*breathprints*" can be recognized in a way similar to how the sensor arrays recognize smellprints. But unlike pattern recognition by sensor arrays, the major strength of this approach is that the individual components that contribute to each pattern can be recognized. Thus, unique breathprints have been identified in SIFT-MS FS spectra for the analysis of breath from patients with liver disease (fetor hepaticus) or with heart and lung disease. The advantage of SIFT-MS (over the electronic nose) is that it is possible to analyze these patterns to identify single molecules in the breath of these patients and link them to the underlying pathobiology of the disease, as illustrated by the breathprint pattern (heatmap) in Fig. 9.3.

A good example of this approach is single exhaled breath metabolomic analysis by which a unique breathprint in patients with acute decompensated heart failure

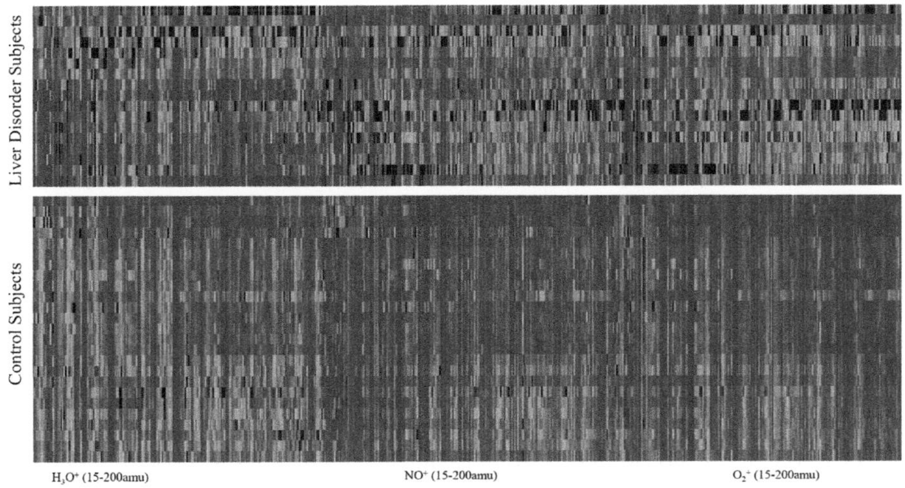

FIGURE 9.3

Heatmap of SIFT-MS full scan mass spectral breath data obtained for liver disease and control subjects using the three reagent ions (red [black in print version] = high analyte ion signal, blue [white in print version] = low analyte ion signal). Analyte ion *m/z* corresponds to individual columns and each subject corresponds to one row. The liver disorder subjects have much higher concentrations of exhaled metabolites compared to controls.

Data from [24].

(ADHF) has been identified.[25] The objective of that study was to establish a breathprint that distinguishes between patients with ADHF and non–heart failure (HF) cardiovascular diagnoses. In this study, the VOC composition of exhaled gas in patients with HF compared with non-HF patients admitted to the hospital was evaluated. Single-breath exhalations from all study participants were analyzed by SIFT-MS, which included quantification of six preselected VOCs (their analyte ions) and by obtaining FS spectra from *m/z* 14 to 200 using H_3O^+, NO^+, and O_2^+ reagent ions. The findings from a training set were tested in a validation set of 36 patients admitted with ADHF. It was seen that acetone and pentane were significantly elevated in HF patients versus controls. Discriminant analysis utilizing preselected VOCs and selected FS ion peaks correctly classified 100% of subjects as HF versus control. The results were validated by a 36-patient validation cohort. Thus, these SIFT-MS breath analyses demonstrated a signature in HF patients distinct from control patients, which may have clinical utility in the diagnosis and monitoring of these patients.

Continuing this theme, breathprints in patients with liver disease have been studied and novel breath biomarkers in alcoholic hepatitis (AH) were identified.[24] To investigate whether breath VOCs correlate with the clinical diagnosis and with the severity of liver disease in patients with AH, patients with liver disease were prospectively recruited. The study population was divided into two groups: liver cirrhosis with AH (n = 40) and liver cirrhosis with acute decompensation (AD) from etiologies other than alcohol (n = 40). A healthy control group without liver disease was also identified (n = 43). Using SIFT-MS, precise identification of VOCs in the breath, measured in the ppb range, was achieved for all subjects. Of the 14 preselected breath VOCs, six were elevated in patients with liver disease compared to healthy control. Those compounds included 2-propanol, acetaldehyde, acetone, ethanol, pentane, and trimethylamine (TMA). The mean concentrations of TMA, acetone, and pentane, in particular, in the exhaled breath, were remarkably higher in patients with AH compared to those with AD and in healthy volunteers (all $P<.001$). With the use of receiver operating characteristic (ROC) curves, a model was developed for the diagnosis of AH that included the breath levels of TMA, acetone, and pentane (designated as the TAP model). Thus, TAP provided excellent prediction accuracy for the diagnosis of AH (with an area under the curve [AUC] of the ROC curve of 0.93) with 97% sensitivity and 72% specificity for a TAP score of 28. The levels of exhaled breath TMA moderately correlated with the severity of AH, as presented by the Model for End-stage Liver Disease (MELD) score. Furthermore, it was apparent that isoprene and ethanol in the breath were associated with transplant-free survival in AH.

Several other related studies were able to identify recognizable breathprints for patient with hypertension,[26] inflammatory bowel disease,[27] renal failure,[28] obesity, and esophageal cancer, among others. Thus, this breathprint approach provides a noninvasive method for the diagnosis of several disease states and maybe helpful in the screening for such conditions.

Finally in this subsection, SIFT-MS has been used to search for volatile breath biomarkers in CF patients' breath whose lungs and airways are infected by PA.[29] The exhaled breath of 20 PA-infected patients and 38 PA-negative patients was analyzed and special attention was given to the positive identification and accurate quantification of 16 VOCs. The diagnostic sensitivity and specificity of the concentrations of any of the 16 compounds taken individually were found to be low. When a linear combination of the concentrations of all 16 VOCs was used to construct an optimized ROC curve using a linear logistic model, the diagnostic separation of PA-infected patients relative to the PA-negative patients was apparently good in terms of the derived sensitivity (89%) and specificity (86%), with an AUC of 0.91. Four compounds were revealed by the linear logistic model as significant, viz. malondialdehyde, isoprene, phenol, and acetoin. While such a metabolomics approach to optimize the ROC curve is widely used in breath analysis, it can engender misleading indications. Therefore, it was concluded that the results of the linear logistic model analyses are of limited immediate clinical value. The identified compounds should rather be considered as a stimulus for further independent studies involving larger patient cohorts.

9.4 Breath biomarkers of cancer

The earliest SIFT-MS research focusing on cancer involved the study of the VOCs released from in vitro cultures of mammalian lung cancer cells in the hope it would stimulate the search for breath biomarkers. This revealed that lung cancer cell lines release acetaldehyde molecules in amounts proportional to the number of cells in the growth medium.[3] This was followed up by a study of acetaldehyde release in a 3D model system, which clearly showed that this compound was released at much greater amounts than in the usual 2D model for the same number of cells in the culture.[30] Further studies have explored the influence of peroxidative stress (reactive oxygen species [ROS]) on cancer cells by quantifying aldehydes in the headspace of cultures using SIFT-MS.[31] Propanal, butanal, pentanal, hexanal, heptanal, and malondialdehyde were identified in the culture headspace, each reaching peak concentrations during the time of exposure to ROS and eventually reducing as the reactants were depleted in the culture. Pentanal and hexanal were the most abundant in the culture headspace, reaching concentrations of a few hundred ppb. The results of these experiments demonstrated that peroxidation of cancer cells in vitro can be monitored and evaluated by direct real-time analysis of the volatile aldehydes by SIFT-MS. The analysis of volatiles in the headspace of cell cultures is examined in detail in Chapter 26.

The most thorough and comprehensive study of VOCs emitted in breath and from bodily fluids using SIFT-MS has focused on gastroesophageal cancer. The earliest research involved a SIFT-MS study of the headspace of gastric content fluid, which may be a contributory factor in the development of upper gastrointestinal diseases.[32] Seven VOCs, viz. acetone, formaldehyde, acetaldehyde, hexanoic acid, hydrogen sulfide, hydrogen cyanide, and methyl phenol, were found to be significantly different between cancer and healthy groups by the Mann—Whitney U test. A binary logistic regression optimization was conducted to describe a combination of selected VOCs (acetaldehyde, formaldehyde, hydrogen sulfide, and methyl phenol) that best discriminates cancer patients from healthy controls, achieving the optimized ROC with AUC = 0.9. This result suggested that a VOC profile rather than a single biomarker may be preferable in the molecular-orientated diagnosis of gastroesophageal cancer.

Continuing this theme, SIFT-MS analysis was exploited for the quantification of VOCs in the headspace of urine samples, which were retrieved from three groups of patients with gastroesophageal cancer, noncancer diseases of the upper gastrointestinal tract, and a healthy cohort.[33] The concentrations of acetaldehyde, acetone, acetic acid, hexanoic acid, hydrogen sulfide, and methanol were significantly increased in the cancer cohort compared with healthy controls, while the concentration of phenol was significantly decreased. Again, a binary logistic regression ROC optimization was applied for a combination of six VOCs (acetaldehyde, acetone, acetic acid, hexanoic acid, hydrogen sulfide, and methanol) to discriminate the cancer patients from noncancer controls, achieving an AUC of 0.904. This result indicates that VOC profiling may be suitable in identifying those at high risk of gastroesophageal cancer.

These studies of VOCs from urine have recently been complemented by investigating the effect of hydration status of healthy individuals using the concentration of phenol in their urine.[34] It was seen that hydration status of the donor volunteers (assessed using urinary osmolality and creatinine concentration) affects urinary headspace phenol concentrations as measured via SIFT-MS. A strong correlation between headspace phenol concentration and both the urinary osmolality and urinary creatinine concentration has been demonstrated. This study suggests that for the proper measurement of VOC concentrations in urine by headspace measurements, it is essential to account for the osmolality (or creatinine concentration) of the urine. Volatiles in urine (and feces) are further discussed in Chapter 24.

The next logical development was to study the VOCs in exhaled breath. Thus, as before, SIFT-MS was used to quantify the VOCs in the exhaled breath from three groups of patients, viz., those with esophageal—gastric cancer, noncancer diseases of the upper gastrointestinal tract, and a healthy upper gastrointestinal tract cohort.[35] A total of 17 VOCs was investigated in this study. The concentrations of four VOCs, hexanoic acid, phenol, methyl phenol, and ethyl phenol, were found to be significantly different between esophageal—gastric cancer and positive control groups. A binary logistic regression optimization achieved an AUC of 0.91 for the ROC curve. These breath analysis results highlight the potential of VOC profiling as a noninvasive test for esophageal—gastric cancer. Further studies on gastroesophageal cancer have examined the concentration of selected metabolites in exhaled breath in five patients before and after surgery.[36] Eight metabolites in exhaled breath were specifically investigated. No significant change was observed for propanol and acetic acid, whereas acetone and hexanoic acid mean breath concentrations were found to increase after surgery, and the concentrations of ammonia, ethanol, HCN, and isoprene were found to decrease in the same patients. Concentrations of hexanoic acid were found to be significantly different before and after recovery from surgery.

9.5 In vitro studies of volatile compound emissions from pulmonary pathogens

Respiratory tract infection by bacteria or fungi is a major cause of morbidity and mortality across the world, particularly in children. Diagnosis mostly relies on cough swabs, which are unreliable, or bronchoalveolar lavage (BAL) samples taken during bronchoscopy, which is an invasive procedure and requires a general anesthetic. Due to these issues, a child-friendly method of detecting lower airway pathogens is needed. Breath analysis for VOCs that are specific to certain lung and airway pathogens is an attractive noninvasive diagnosis if it can be realized (see Chapter 21). This is the driving force for in vitro studies of VOCs emitted by pulmonary pathogens. Therefore, following the thorough studies of the VOCs emitted by cultures of PA bacteria, as reported above, SIFT-MS has been used for real-time analyses of the headspace above cultures of a range of well-known respiratory pathogens. The initial study involved clinical isolates of *Staphylococcus aureus* (SA), *Streptococcus pneumoniae* (SP), and *Haemophilus influenza* (HI).[37] Six volatile compounds, mainly alcohols, ketones, and aldehydes, were found to be elevated in the headspace of SA cultures and eight were elevated in the SP cultures. Only indole and ethanol were somewhat elevated in the headspace of some of the HI cultures. Principal component analyses (PCA) indicated that the SA cultures clearly separated into two distinct groups in terms of their volatile compound emissions. This may relate to genetic or phenotypic differences. It was postulated that the very efficient production of ethanol and acetaldehyde by SA and SP may provide indicators of airways infection by these bacteria if the abundances of these compounds are elevated in exhaled breath above those levels expected due to their normal endogenously produced concentrations. This study was extended to the investigation of the VOCs emitted by in vitro cultures of the common respiratory fungus *Aspergillus fumigatus* (AF).[38] It was seen that ammonia and the organosulfur compounds methanethiol, dimethyl sulfide, and dimethyl disulfide are produced by AF cultures. These may be sufficient to allow for noninvasive detection of the AF in the airways of infected patients by breath analysis. AF also efficiently absorbs and metabolizes acetaldehyde, butanal, and pentanal from the supportive medium (brain–heart infusion broth in these studies). Preliminary studies of the volatile compounds emitted by cocultures of AF with PA, SA, and SP revealed that the biomarker HCN (for PA) is not compromised by the presence of AF, and the organosulfur compounds (for AF) are not compromised by the presence of SA or SP.

Subsequently, studies were carried out to characterize the volatile metabolites produced by genotypically diverse strains of the *Stenotrophomonas* genus.[39] VOCs emitted from 15 clinical and five environmental strains belonging to different genogroups of *Stenotrophomonas maltophilia* (n = 18) and *Stenotrophomonas rhizophila* (n = 2) cultured in Mueller–Hinton Broth (MHB) liquid media were investigated. The headspace cultures were analyzed by both SPME-GC-MS and SIFT-MS. Several VOCs were detected at high concentration, including ammonia, propanol, dimethyl sulfide, and dimethyl disulfide. The GC-MS measurements showed that all 15 clinical strains produced similar headspace VOCs, and SIFT-MS quantification showed that the rates of production of the VOCs by the genotypically distinct strains were very similar. All in vitro cultures of both the *Stenotrophomonas* species were characterized by efficient production of two isomers of methyl butanol, which can be described by known biochemical pathways and which is absent in other pathogens, including PA. These in vitro data indicate that methyl butanol isomers may be exhaled breath biomarkers of *S. maltophilia* lung infection in patients with CF.

A recent SIFT-MS study continued the search for breath biomarkers of lung and airways infection in patients with CF.[40] In this comprehensive study, the VOCs released by PA cultures and others belonging to the *Burkholderia cepacia* complex, *Staphylococcus aureus*, and *Stenotrophomonas maltophilia* were analyzed. These opportunistic pathogens are generally harmless to healthy people but they may cause serious infections in patients with severe underlying disease or impaired immunity, such as CF patients. In order to minimize the effect of differences in the culture media composition, all strains were cultured in three different liquid media. Many VOCs were detected and quantified, including series of alcohols, aldehydes, ketones, carboxylic acids, and some organosulfur compounds, as detailed in the published paper. Multivariate statistical analysis reveals that the four taxa can be well discriminated by the differences in the headspace VOC concentration profiles. The compounds that should be targeted in breath as potential biomarkers of airway infection were identified for each of these taxa of CF pathogens.

As previously mentioned, these detailed studies involving the very powerful combination of SIFT-MS and GC-MS analyses are intended to identify volatile compounds in exhaled breath that can be used as noninvasive biomarkers of the particular pathogens. This basic in vitro work has yet to be followed up with breath analysis from patients with clinically recognized respiratory diseases, but the potential is there for a valuable contribution to respiratory medicine.

9.6 Summary

Selected ion flow tube-mass spectrometry is recognized as a valuable method for the analysis of air and breath. Its special feature is the real-time, quantitative analysis of single-breath exhalations, which allows continuous monitoring of the VOCs on time scales of seconds, through minutes and hours, as long as the VOC concentrations exceed a few ppb_V. Thus, the common breath metabolites can be monitored in large cohorts of patients and volunteers in short time. The perceived inadequacy of the current SIFT-MS instruments are their sensitivity, typically 1 ppb_V/s, but research is in train that will greatly improve the sensitivity to sub-ppb_V/second levels in the next generation of smaller instruments. Much effort is also being given to the development of the TD/SIFT-MS combination as a method for high-throughput screening of breath samples. The analysis of structural isomers is also a problem, but the powerful combination of GC-MS and SIFT-MS deals with this, the former more positively identifying some compounds in a mixture, while the latter can accurately quantify the identified compounds as long as the analytical ion chemistry is understood.

Thus, with the anticipated more sensitive and smaller SIFT-MS instruments, the identification and accurate quantification of volatile biomarkers of disease that are likely to be present in exhaled breath at sub-ppb_V levels will be possible online avoiding the problems associated with sample collection. Then the analysis of specific metabolites could be done at the bedside, in the ITU and even in the primary care surgery. In this way, breath analysis can become a very valuable support to already established clinical diagnostics.

References

1. Španěl P, Smith D. Progress in SIFT-MS; breath analysis and other applications. *Mass Spectrom Rev* 2011;**30**:236–67.
2. Smith D, Španěl P, Holland TA, Al Singari W, Elder JB. Selected ion flow tube mass spectrometry of urine headspace. *Rapid Commun Mass Spectrom* 1999;**13**:724–9.
3. Smith D, Wang TS, Sule-Suso J, Španěl P, El Haj A. Quantification of acetaldehyde released by lung cancer cells in vitro using selected ion flow tube mass spectrometry. *Rapid Commun Mass Spectrom* 2003;**17**:845–50.

4. Thorn RMS, Reynolds DM, Greenman J. Multivariate analysis of bacterial volatile compound profiles for discrimination between selected species and strains in vitro. *J Microbiol Methods* 2011;**84**:258–64.
5. Španěl P, Smith D. Recent SIFT-MS studies of volatile compounds in physiology, medicine and cell biology. In: Amann A, Smith D, editors. *Volatile biomarkers: non-invasive diagnosis in physiology and medicine*. Boston: Elsevier; 2013. p. 49–76.
6. Ross B. Sub-parts per billion detection of trace volatile chemicals in human breath using selected ion flow tube mass spectrometry. *BMC Res Notes* 2008;**1**:41.
7. Smith D, Spanel P. Status of selected ion flow tube MS: accomplishments and challenges in breath analysis and other areas. *Bioanalysis* 2016;**8**:1183–201.
8. Hryniuk A, Ross BM. Detection of acetone and isoprene in human breath using a combination of thermal desorption and selected ion flow tube mass spectrometry. *Int J Mass Spectrom* 2009;**285**:26–30.
9. Smith D, Spanel P. On the importance of accurate quantification of individual volatile metabolites in exhaled breath. *J Breath Res* 2017;**11**:047106.
10. Spanel P, Smith D. What is the real utility of breath ammonia concentration measurements in medicine and physiology? *J Breath Res* 2018;**12**:027102.
11. Davies SJ, Španěl P, Smith D. Breath analysis of ammonia, volatile organic compounds and deuterated water vapor in chronic kidney disease and during dialysis. *Bioanalysis* 2014;**6**:843–57.
12. Španěl P, Dryahina K, Smith D. A quantitative study of the influence of inhaled compounds on their concentrations in exhaled breath. *J Breath Res* 2013;**7**:017106.
13. Smith D, Španěl P, Fryer AA, Hanna F, Ferns GAA. Can volatile compounds in exhaled breath be used to monitor control in diabetes mellitus? *J Breath Res* 2011;**5**:022001.
14. Španěl P, Dryahina K, Rejskova A, Chippendale TWE, Smith D. Breath acetone concentration; biological variability and the influence of diet. *Physiol Meas* 2011;**32**:N23–31.
15. Turner C. Potential of breath and skin analysis for monitoring blood glucose concentration in diabetes. *Expert Rev Mol Diagn* 2011;**11**:497–503.
16. Boshier PR, Fehervari M, Markar SR, et al. Variation in exhaled acetone and other ketones in patients undergoing bariatric surgery: a prospective cross-sectional study. *Obes Surg* 2018;**28**:2439–46.
17. Walton C, Patel M, Pitts D, et al. The use of a portable breath analysis device in monitoring type 1 diabetes patients in a hypoglycaemic clamp: validation with SIFT-MS data. *J Breath Res* 2014;**8**:037108.
18. Smith D, Španěl P, Gilchrist FJ, Lenney W. Hydrogen cyanide, a volatile biomarker of *Pseudomonas aeruginosa* infection. *J Breath Res* 2013;**7**:044001.
19. Gilchrist FJ, Belcher J, Jones AM, et al. Exhaled breath hydrogen cyanide as a marker of early *Pseudomonas aeruginosa* infection in children with cystic fibrosis. *ERJ Open Res* 2015;**1**. 00044-2015.
20. Spanel P, Dryahina K, Vicherková P, Smith D. Increase of methanol in exhaled breath quantified by SIFT-MS following aspartame ingestion. *J Breath Res* 2015;**9**:047104.
21. Dryahina K, Smith D, Bortlík M, Machková N, Lukáš M, Španěl P. Pentane and other volatile organic compounds, including carboxylic acids, in the exhaled breath of patients with Crohn's disease and ulcerative colitis. *J Breath Res* 2017;**12**:016002.
22. Smith D, Sovová K, Dryahina K, Doušová T, Dřevínek P, Španěl P. Breath concentration of acetic acid vapour is elevated in patients with cystic fibrosis. *J Breath Res* 2016;**10**:021002.

23. Dryahina K, Pospisilova V, Sovova K, et al. Exhaled breath concentrations of acetic acid vapour in gastro-esophageal reflux disease. *J Breath Res* 2014;**8**:037109.
24. Hanouneh IA, Zein NN, Cikach F, et al. The breathprints in patients with liver disease identify novel breath biomarkers in alcoholic hepatitis. *Clin Gastroenterol Hepatol* 2014;**12**:516–23.
25. Samara MA, Tang WHW, Cikach F, et al. Single exhaled breath metabolomic analysis identifies unique breathprint in patients with acute decompensated heart failure. *J Am Coll Cardiol* 2013;**61**:1463–4.
26. Cikach FS, Tonelli AR, Barnes J, et al. Breath analysis in pulmonary arterial hypertension. *Chest* 2014;**145**:551–8.
27. Rieder F, Kurada S, Grove D, et al. A distinct colon-derived breath metabolome is associated with inflammatory bowel disease, but not its complications. *Clin Transl Gastroenterol* 2016;**7**:e201.
28. Demirjian S, Paschke KM, Wang XF, Grove D, Heyka RJ, Dweik RA. Molecular breath analysis identifies the breathprint of renal failure. *J Breath Res* 2017;**11**:026009.
29. Spanel P, Sovova K, Dryahina K, Dousova T, Drevinek P, Smith D. Do linear logistic model analyses of volatile biomarkers in exhaled breath of cystic fibrosis patients reliably indicate *Pseudomonas aeruginosa* infection? *J Breath Res* 2016;**10**:036013.
30. Rutter AV, Chippendale TWE, Yang Y, Španěl P, Smith D, Sule-Suso J. Quantification by SIFT-MS of acetaldehyde released by lung cells in a 3D model. *Analyst* 2013;**138**:91–5.
31. Shestivska V, Rutter AV, Sule-Suso J, Smith D, Spanel P. Evaluation of peroxidative stress of cancer cells in vitro by real-time quantification of volatile aldehydes in culture headspace. *Rapid Commun Mass Spectrom* 2017;**31**:1344–52.
32. Kumar S, Huang JZ, Cushnir JR, Španěl P, Smith D, Hanna GB. Selected ion flow tube-MS analysis of headspace vapor from gastric content for the diagnosis of gastro-esophageal cancer. *Anal Chem* 2012;**84**:9550–7.
33. Huang JZ, Kumar S, Abbassi-Ghadi N, Španěl P, Smith D, Hanna GB. Selected ion flow tube mass spectrometry analysis of volatile metabolites in urine headspace for the profiling of gastro-esophageal cancer. *Anal Chem* 2013;**85**:3409–16.
34. Sovova K, Wiggins T, Markar SR, Hanna GB. Quantification of phenol in urine headspace using SIFT-MS and investigation of variability with respect to urinary concentration. *Anal Methods* 2015;**7**:5134–41.
35. Kumar S, Huang JZ, Abbassi-Ghadi N, Španěl P, Smith D, Hanna GB. Selected ion flow tube mass spectrometry analysis of exhaled breath for volatile organic compound profiling of esophago-gastric cancer. *Anal Chem* 2013;**85**:6121–8.
36. Huang JZ, Kumar S, Boshier PR, Wakefield S, Cushnir JR, Hanna GB. Breath analysis using SIFT-MS to assess metabolic status in patients after gastro-oesophageal cancer surgery- a pilot study. *Curr Anal Chem* 2013;**9**:584–92.
37. Chippendale TWE, Gilchrist FJ, Španěl P, Alcock A, Lenney W, Smith D. Quantification by SIFT-MS of volatile compounds emitted by in vitro cultures of *S. aureus*, *S. pneumoniae* and *H. influenzae* isolated from patients with respiratory diseases. *Anal Methods* 2014;**6**:2460–72.
38. Chippendale TWE, Gilchrist FJ, Španěl P, Alcock A, Lenney W, Smith D. Quantification by SIFT-MS of volatile compounds emitted by *Aspergillus fumigatus* cultures and in co-culture with *Pseudomonas aeruginosa*, *Staphylococcus aureus* and *Streptococcus pneumoniae*. *Anal Methods* 2014;**6**:8154–64.

39. Shestivska V, Dryahina K, Nunvar J, et al. Quantitative analysis of volatile metabolites released in vitro by bacteria of the genus Stenotrophomonas for identification of breath biomarkers of respiratory infection in cystic fibrosis. *J Breath Res* 2015;**9**:027104.
40. Dryahina K, Sovova K, Nemec A, Spanel P. Differentiation of pulmonary bacterial pathogens in cystic fibrosis by volatile metabolites emitted by their in vitro cultures: *Pseudomonas aeruginosa*, *Staphylococcus aureus*, *Stenotrophomonas maltophilia* and the *Burkholderia cepacia* complex. *J Breath Res* 2016;**10**:037102.

CHAPTER

Proton transfer reaction—mass spectrometry

10

Christopher A. Mayhew[1,2], Jens Herbig[3], Jonathan D. Beauchamp[4]

[1]*Institute for Breath Research, University of Innsbruck, Dornbirn, Austria;* [2]*Molecular Physics Group, School of Physics and Astronomy, University of Birmingham, Edgbaston, Birmingham, United Kingdom;* [3]*IONICON Analytik Gesellschaft m.b.H., Innsbruck, Austria;* [4]*Fraunhofer Institute for Process Engineering and Packaging IVV, Freising, Germany*

Chapter outline

10.1 Overview

Proton transfer reaction—mass spectrometry (PTR-MS) belongs to the class of soft chemical ionization mass spectrometry (SCIMS) techniques that include selected ion flow tube—mass spectrometry (SIFT-MS) and atmospheric chemical ionization—mass spectrometry (APCI-MS), among others, and is characterized by its ability to detect trace volatile organic compounds (VOCs) directly and quantitatively in real time. The technique evolved after several key advances in analytical chemistry, commencing with the mid-1960s introduction of chemical ionization by Munson and Field[1], and the development of the flowing afterglow and flow drift tube techniques by Fehsenfeld, Ferguson, Schmeltekopf and colleagues at the National Oceanic and Atmosphere Administration (NOAA) in Boulder, Colorado.[1a,2] One key colleague was Werner Lindinger, who would become the inventor of PTR-MS. The flowing afterglow systems spawned two

[1a] E.E. Ferguson, F.C. Fehsenfeld, and A.L. Schmeltekopf, Flowing afterglow measurements of ion-neutral reactions, In "Advances in Atomic and Molecular Physics" D.R. Bates and I. Estermann, eds., Vol. 5, pp. 1—56, Academic Press, New York (1969).

Breathborne Biomarkers and the Human Volatilome. https://doi.org/10.1016/B978-0-12-819967-1.00010-4
Copyright © 2020 Elsevier B.V. All rights reserved.

similar techniques: SIFT-MS and a variant thereof that imposed an electric (drift) field on the flow tube, termed flow drift tube (SIFDT-MS). The drift tube systems were originally geared toward studying the kinetic energy effects on ion—molecule reactions. Then, in the 1990s, it became apparent that knowledge of specific ionization reaction rates for individual molecules could conversely be utilized to determine the concentrations of trace molecules in a gas sample: this realization was the birth of online SCIMS as an analytical detection tool. While SIFT-MS has been studied extensively for its detection of VOCs, primarily in breath (see Chapter 9), SIFDT-MS underwent a transformation that replaced the front-end mass filter—used for reagent ion preselection—with an ion source that relied on ion—molecule chemistry to generate an abundance of hydronium (H_3O^+), and consequently a focus on proton transfer reaction ionization alone.[3] PTR-MS, as it became known, was implemented in many fields of application that were of relevance to VOC detection, including breath research.

A PTR-MS principally comprises three elements: an ion source; a reaction chamber (the drift tube); and a detection system (mass spectrometer coupled to a detector). A schematic of the original instrument identifying these three regions is shown in Fig. 10.1. In this original design, a quadrupole mass spectrometry was used to separate the reagent and product ions according to their ratio of mass-to-charge, *m/z*.

The operating principle of PTR-MS—as the name implies—is based on proton transfer reactions, whereby a proton from the reagent hydronium (protonated water) is transferred upon collision to a neutral target molecule present in the reaction chamber. Critically for breath analysis, for which the sample gas is of high humidity,

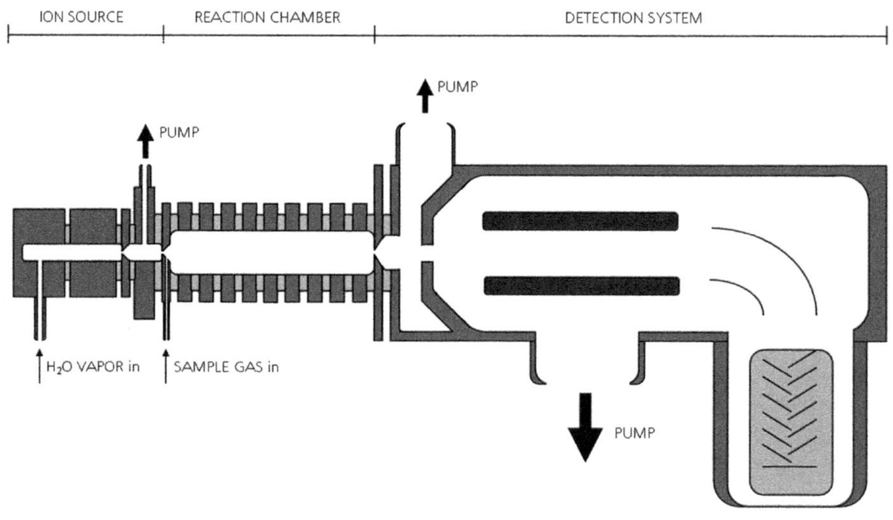

FIGURE 10.1

Schematic depiction of the construction of a PTR-MS instrument, comprising three distinct regions, namely an ion source, a reaction chamber, and a detection system.

Fraunhofer IVV.

analyte molecules can additionally be ionized via proton transfer or ligand switching reactions from water cluster ions equally present in the reaction chamber. Specifically, the proton transfer reaction can be described as follows:

$$H_3O^+ \cdot (H_2O)_n + M \xrightarrow{k} MH^+ + (n+1)H_2O \quad (n=0 \text{ and } 1),$$

where M is the neutral target molecule and k represents the bimolecular reaction rate coefficient.

It should be noted that the PTR-MS signatures of volatiles may differ in terms of sensitivity and type of product ions between those found in breath to those contained in air samples. This results from the higher humidity and CO_2 content of breath, and possibly a higher oxygen content, as occurs in clinical settings or when using breathing apparatus (e.g., in scuba diving).[4,5] Trefz et al. investigated more than 20 VOCs, including aldehydes, ketones, and aromatic compounds, with different sample humidity and CO_2 and low O_2 (0%–20%) concentrations.[4] A large difference in the measured VOC intensities between the normal and humid operating conditions was observed, whereas CO_2 and O_2 concentrations had little influence. In a subsequent study, for which the oxygen level was further elevated, and in which in vivo experiments involving healthy volunteers and mechanically ventilated animals were undertaken, almost all the PTR-MS signatures for a set of investigated compounds were affected. Ion signal intensity reductions of more than 50% were observed when O_2 reached close to 90%.[5] These studies highlight that the effect of the breath matrix needs to be taken into account for reliable quantitation in breath analysis.

The especially powerful features of PTR-MS are (i) proton transfer reactions from H_3O^+ occur at the collisional rate to all VOCs whose proton affinities are greater than that of water; (ii) the main constituents of air (N_2, O_2, CO_2, and Ar) do not elicit proton transfer from hydronium, allowing the use of air as a sample carrier gas (humid samples also do not present a problem); and (iii) the instrument has a linear detection range over seven orders of magnitude, meaning that compounds at ultratrace concentrations ($pptv_v$; the limit of detection [LOD] of the system) can be detected simultaneously with those present at upper-trace concentrations (low ppmv; the upper linearity range of detection) within the same sample. This latter aspect has great benefit for breath volatiles, which are present over a broad range of concentrations between species and as they dynamically change (see Chapter 2).

A detailed treatise on the ion chemistry, collisional energies, compound-dependent reaction rate coefficients, and quantitation in PTR-MS, regardless of application, is given in the literature[6] and will not be discussed in detail in this chapter. Instead, the objective of this chapter is to highlight selected key PTR-MS studies that have had a significant impact on the field of breath analysis. This chapter commences with an overview of the historical developments in the field, then focusses on particularly innovative applications that have delivered new insights to breath research, presents and discusses technical developments in PTR-MS and peripheral tools, and concludes with an outlook of prospective future developments. It should be noted that PTR-MS has been implemented in diverse areas that are relevant to biomarker detection, from the analyses of volatiles in urine, stool, and skin (see Chapters 24 and 25, respectively) to cell headspace (or "cell breath," see Chapter 26), or in critical care medicine (see Chapter 18), but these applications are not included in this chapter, as they are covered elsewhere in this book.

10.2 From quadrupole to time-of-flight and beyond

Before presenting and discussing key research highlights of PTR-MS for breath analysis, it is important to first appreciate the instrumentation developments and improvements since its inception in the mid-1990s. When it first emerged, the detection system of the instrument was equipped with a quadrupole mass spectrometer (subsequently referred to specifically for PTR-MS as QMS). In QMS, ions are transferred to the detector—in the case of PTR-QMS typically a secondary electron multiplier (SEM)—discretely according to their *m/z*. This selection is achieved via specific electric field settings of the quadrupole mass filter rods. As such, in order to channel (analyte) ions of different *m/z* to the detector, the settings need to be altered accordingly and in succession. The QMS can be operated in one of two modes, namely multiple ion detection (MID) or scan mode. In the former, individual *m/z* can be selected for sequential analysis, thus the measurement of only a few individual analytes within a sample can be performed more rapidly than a large number of analytes; in the latter, the QMS scans over an *m/z* range in integer steps, allowing for continuous transition of the *m/z* to the detector, with measurement (cycle) times dependent on the range and the analysis time per *m/z* interval (also referred to as "dwell time" or "integration time"). Cycle times for MID mode for a handful of compounds can be seconds, whereas in scan mode tens of seconds are typically needed. An important point to note is that the LOD of the system depends on the dwell time, i.e., shorter dwell times will elicit noisier signals, and consequently higher LODs.

A development from 2004 was to replace the QMS of the PTR-MS with a time-of-flight (TOF) mass spectrometer (subsequently referred to as PTR-TOF-MS).[7] In TOF-MS, ions that are pulsed into a field-free flight chamber are separated owing to differences in their velocities, which depend on *m/z*, and reach the detector (typically a microchannel plate; MCP) at different times, whereupon they elicit a signal; with knowledge of the flight time of the ions and the dimensions of the flight chamber (distance traveled by the ions), the *m/z* of the incident reagent and product ions can be determined with a high mass resolving power. As such, isobaric protonated compounds, i.e., those with different elemental composition but the same nominal mass, can be distinguished when their full-width at half maximum (FWHM) peak separation is on the order of 0.010—0.015 (*m/z* dependent, with mass resolving power of commercial instruments typically of 5000—6000)[8,9]. An additional factor for accurate peak position determination is the intensity of the target peak and its adjacent peaks. Further details about high-resolution MS (HR-MS) and associated peak deconvolution and identification is provided in Chapter 16. Given that the time-of-flight of the ions in the flight chamber is very short, on the order of tens of μs, rapid detection is possible. Thus, PTR-TOF-MS is of particular benefit in breath analysis applications, as it enables rapid full-mass spectral detection (at several Hertz measurement frequency). This is beneficial for achieving breath-resolution measurements, with high mass spectral resolution providing a high confidence in the assignment of the observed product ions, which aids compound identification. Use of QMS is still employed in other fields (as well as in SIFT-MS; see Chapter 9), but current PTR-MS—related breath applications almost exclusively utilize TOF-MS due to the substantial benefits discussed above.

More recent technological developments have included the implementation of diverse focusing systems to increase ion yield, via more efficient ion extraction from the reaction chamber, or by focusing the extracted ions into a dense beam after exiting the drift tube, or both. This is achieved by incorporating radiofrequency-driven funnels, or placing an additional mass spectrometer (quadrupole, hexapole) immediately upstream of the TOF-MS, yielding order of magnitude detection sensitivity improvements of up to several thousand counts-per-second (cps) signal intensity per ppb_V and LODs in the ppq_V range.[10–12] A recent development utilizes a newly configured gas inlet and ion source together with a novel drift field geometry with rapid sampling rates, which greatly extends reaction times to provide higher sensitivity.[13] Furthermore, this reconfigured PTR-MS system is capable of detecting semivolatiles and radicals, thereby expanding the analytical range for novel real-time breath monitoring.

10.3 Early breath—related applications

Returning to the beginning of the PTR-MS research timeline, the first paper published a quarter of a century ago had breath analysis as its focus, with a comparison of the exhaled breath of smokers and nonsmokers.[14] This research demonstrated that the exogenous compounds benzene and acetonitrile are elevated in the breath of smokers, with the latter compound remaining elevated for a considerable time (up to a week) after cessation of smoking. The importance of this seminal paper should be stressed: it not only demonstrated the applicability of PTR-MS for breath analysis in general but equally represented the first pharmacokinetics study of exogenous compounds via direct analysis of breath, albeit via offline analysis of bag samples rather than online breath-resolved measurements. A major strength of PTR-MS, of course, is its capacity for online quantitative analysis without the need for prior sample workup, offering the opportunity for breath-resolved analytics. A demonstration of this real-time capability for analyzing exhaled breath is given in a paper from 1997, which presents data on the metabolism of isopropanol into acetone, as determined in exhaled breath, among other applications.[15] This was the first time that kinetics studies of this nature could be made by direct exhalation into an analytical instrument, and demonstrated the benefits of online VOC analysis.

Further early breath studies undertaken by the same group—the inventors of PTR-MS at the University of Innsbruck, Austria—focused on the exhalation of methanol after consumption of alcohol and fruits,[16] endogenous isoprene,[17,18] and direct and passive smoking.[19,20] Other early research covered diverse applications, from atmospheric to food chemistry, which is not dealt with in this chapter, but is reviewed at length in the literature.[6]

10.4 Real-time breath analysis

Although the collection of breath samples into bags or syringes for PTR-MS analysis has been extensively used for untargeted discovery purposes, it is those breath studies for which online breath analysis is required that PTR-MS really comes into its own. Such studies allow rapid changes in breath volatile concentrations to be monitored without the need for sample preparation and collection, as is required in analyses by other techniques, such as gas chromatography−mass spectrometry (GC-MS; see Chapter 15). This online approach has applications to not only real-time noninvasive diagnosis but also for use in personalized medicine, rapidly diagnosing bacterial infection or drug overdoses, pharmacokinetics, and investigations of the effects of movement and exercise on volatile constituents of breath, to name but a few areas where sudden changes in breath VOC concentrations can occur. Breath studies over regular time intervals can be used to determine the efficacy of treatment, to monitor changes in an individual metabolism, and to determine drug or medication concentrations and their effects.

For the early studies of real-time breath analysis reported above, the volunteer simply breathed into the inlet of the PTR-MS. Although simple, this presents problems with reproducibility and restricts analysis time, which is particularly problematic for PTR-QMS, which intrinsically requires longer analysis time than PTR-TOF-MS. The introduction of a buffered end-tidal (BET) sampling system for direct connection to the PTR-MS inlet was of great benefit to breath analysis.[21] The system allows individual breath and breath-by-breath analyses by buffering the end-tidal fraction of the breath, thereby extending the time for sampling and improving signal quality. The first use of a BET sampling system in conjunction with PTR-TOF-MS demonstrated the possibility to measure all detectable VOCs up to *m/z* 320 in just a single exhalation and with a limit of detection in the sub-ppb$_v$ range.[22] Fig. 10.2 illustrates the quality of the data that is obtained with such a system by comparing several inhalation/exhalation cycles without the BET and one exhalation using the BET sampling device.

A noteworthy additional advantage of direct breath analysis in PTR-MS is that during the inhalation cycle—regardless of whether the person inhales through or away from the sampling mouthpiece—ambient air is automatically analyzed, i.e., in the between-breath intervals. This is clearly observable in Fig. 10.2 in the

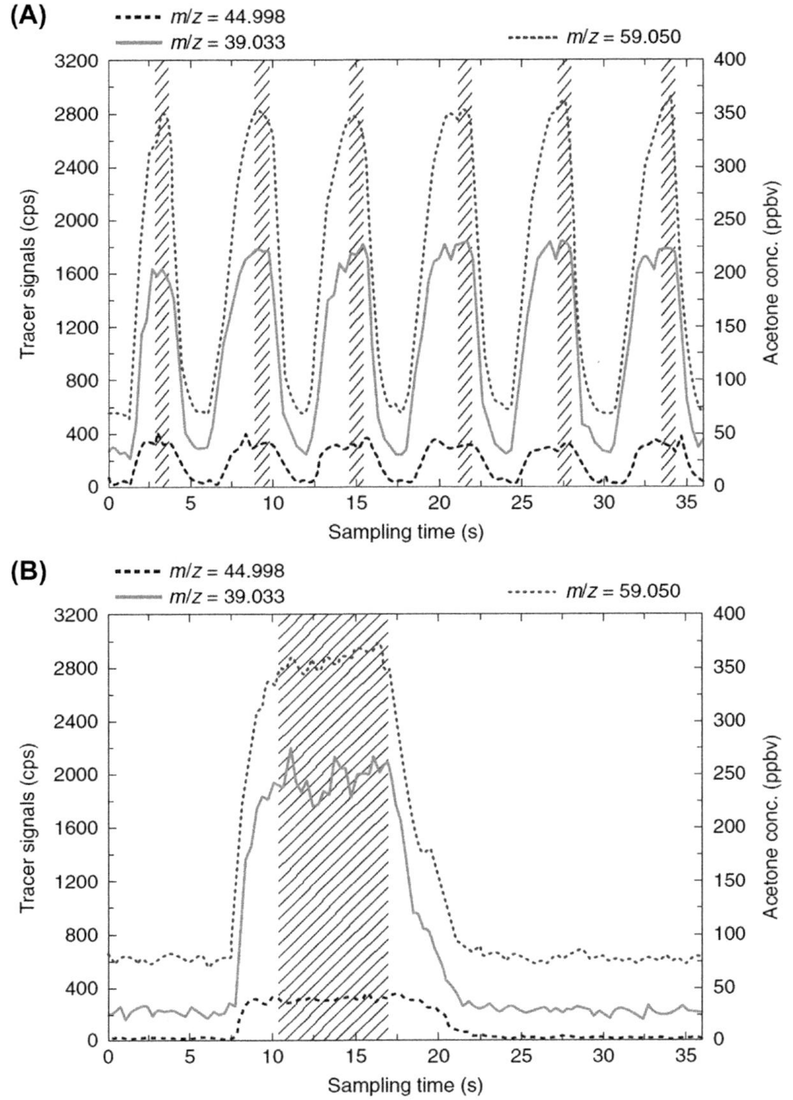

FIGURE 10.2

Comparison of signal levels for three ions measured using a PTR-TOF-MS for (A) a series of inhalations/exhalations using direct online sampling and (B) a buffered end-tidal (BET) measurement from a single direct online sampling exhalation. For (A) the test person exhaled and inhaled through a heated sampling tube. The shaded areas mark the sampling time for the end-tidal fraction of the exhaled breath. The three species highlighted are $H_3O^+(H_2O)$ (measured via its isotopologue at *m/z* 39.033 owing to detection saturation of the ion peak at *m/z* 37.029), CO_2H^+ (*m/z* 44.998),[a] and protonated acetone (*m/z* 59.050). For direct online sampling typically less than 1 s of end-tidal values can be used, whereas a single exhalation using BET yields ~ 7 s of end-tidal measurements. Although observed with very low signal intensities, the temporal change in the CO_2H^+ intensity can be used to accurately determine the transitionary (changeover) phase from deadspace to end-tidal phases of exhalation. Thus, basically, PTR-MS users have an in-built fast response capnometer.

Reproduced from [22].

nonshaded areas of the graphs where the compound concentrations are at their lowest. Given the capacity for ambient air constituents to confound exhaled breath profiles, especially when concentrations of the former are high (see Chapter 2), this beneficial concurrent capture of potential environmental confounders offers an intrinsic quality control check on the exhaled breath data.

In a clinical setting, the real-time advantages of PTR-MS for the breath analysis of exogenous compounds were first demonstrated in an early study by Harrison et al. in 2003.[23] In the study, the researchers established that propofol (2,6-diisopropyl phenol), a routinely used intravenous anesthetic, could be detected in real-time during surgery by monitoring a patient's exhaled breath, thereby providing a "proof of concept of reality to the conjecture of pulmonary propofol elimination and its measurement."[24] Importantly, Harrison and coworkers demonstrated the feasibility of real-time measurements of propofol and its metabolites in expired breath and suggested that alveolar propofol concentrations could be used to determine blood concentrations, with the potential of providing important information on the depth of unconsciousness to the anesthetist. This discovery that propofol can be detected in exhaled breath has inspired many other studies investigating correlations between breath and plasma concentrations, so that real-time monitoring of the anesthetic may become viable in an analogous manner to that for volatile anesthetic agents.

Proton transfer reaction-mass spectrometry has also been used to investigate inhalation anesthetics on breath postsurgery. Fernández del Río et al. reported a PTR-MS study of the elimination characteristics of postoperative isoflurane ($CF_3CHClOCHF_2$) using end-tidal breath samples from patients who had undergone major surgery.[25] Of interest, isoflurane was detected in the breath samples of these patients even weeks after the surgical procedures. This raises a question as to the impact of the length of time these anesthetics remain in a human body on a patient's health—and associated potential cognitive and physical impairment—and on the duration and type of surgery.

10.5 Exogenous versus endogenous breath volatiles

The ingestion, injection, absorption, or inhalation of an exogenous compound and the subsequent monitoring of its temporal washout profile and the production of any associated metabolites using PTR-MS holds considerable potential for following metabolic processes with applications to the detection of disease or for monitoring the effectiveness of therapeutic drugs in the body. The first detailed pharmacokinetics PTR-MS measurements to determine the washout characteristics of an ingested compound is reported in a paper by Beauchamp et al.[26] They provided

[a] CO_2 does not undergo proton transfer reaction with hydronium in the drift tube, but when present at high concentrations—as is the case in exhaled breath—it becomes protonated through high energy collisions at the transfer lens from the drift tube to the mass spectrometer.

a real-time gas analysis of breath after the ingestion of a eucalyptol-containing capsule. Their work nicely illustrated the online capabilities of PTR-MS to record fast temporal changes in breath concentrations of a swallowed exogenous compound.

Temporal changes in the breath concentrations of ingested compounds have also been investigated using isotopically labeled compounds, which permit the emerging volatile metabolites to be easily distinguished from their naturally occurring, highly abundant counterparts in human breath.[27] The first reported PTR-MS study was by Winkler et al., who used medical gelatin capsules in their PTR-MS pharmacokinetic study of ingested isotope-labeled ethanol. They found that d_3-ethanol ($CD_3{-}CH_2{-}OH$) was in part metabolized to deuterated acetone and isoprene. Another key study was reported by Ruzsanyi et al., who monitored the conversion of d_3-2-propanol-1,1,1 (dissolved in water and then drank) to d_3-acetone by alcohol dehydrogenase.[28] The use of isotope labeling allowed the d_3-acetone (*m/z* 62.08) to be easily detected beside the normal and extremely high acetone breath level. It should be noted, however, that the use of deuterium labeled compounds has limitations owing to D/H exchange processes in the drift tube of the PTR-MS instrument,[29] which could be circumvented by using (the typically more expensive) ^{13}C-labeled compounds.

The usefulness of exogenous compounds (isotopically labeled or not) for targeted breath analysis lies in the fact that they often lead to volatiles that are not normally present in breath, thereby significantly reducing the effects of confounding factors and "voodoo" statistics, all too present when dealing with endogenous volatile compounds (see Chapter 18). Given the complexity with endogenous compounds, it is not surprising that after several decades of untargeted exhaled breath research, no clinical diagnostic techniques using endogenous VOCs have been established in the clinical setting. It is still not known whether there are any endogenous volatiles or patterns of endogenous volatiles which will be specific enough for disease diagnosis. The origins and function of many endogenous VOCs are not known. Without such knowledge, their use as biomarkers for disease diagnosis is limited. Another complication is that volatile metabolites can often be attributed to several metabolic processes, making them less specific.

The use of PTR-MS in untargeted discovery studies is particularly challenging due to limitations in the unequivocal identification of individual compounds in complex samples. Unlike GC-MS, which utilizes chromatographic retention times and exploits characteristic fragmentation patterns elicited by ionization at high energy (70 eV) against reference standards or libraries (see Chapter 15), the soft ionization of PTR-MS leads to only one or several product ions per compound, thereby limiting confidence in identifying potential biomarkers. In contrast, PTR-MS really excels in targeted approaches, where compound identities are known a priori. Several PTR-MS studies have been instrumental in revealing hitherto unknown behaviors of the two most abundant VOCs found in breath, namely isoprene and acetone. Specifically, in a series of pioneering studies, breath-resolved analyses of these compounds by PTR-MS provided data of sufficient resolution to allow the dynamics of isoprene and acetone in breath to be modeled mathematically (reported in detail in Chapter 3). One of the key findings of this series of studies is that endogenous isoprene appears to be stored in muscle tissue, which is released upon activation of that muscle, with stores being depleted during exercise and replenished after a period of rest. In general, PTR-MS is a powerful analytical tool for characterizing perturbations of induced changes in a time-resolved manner.

Another area where the detection of endogenous VOCs has a major application is in urban search and rescue (see Chapter 31). PTR-MS has played a crucial role in determining key trace endogenous volatiles contained in breath, which combined with the detection of CO_2 provides a unique chemical signature of life. This represents a very powerful analytical technique to find entrapped or hidden people.

Finally, the versatility of PTR-MS in breath-related research can be demonstrated by its implementation in yet another wholly different field—a literal field in the form of a football pitch. Moving on from breath analyses of individuals, PTR-MS has been employed to analyze what has been termed "crowd breath,"[30] whereby sampling of ambient air in crowded environments might offer insights into group responses to stimuli. This approach has been investigated, as mentioned, to explore VOCs in the ambient air of a stadium during a football (soccer) game,[31] and subsequently in a cinema during repeat showings of movies of different genres, where correlations between concentrations of specific *m/z* signals and certain movie scenes including injury, mystery, and suspense were observed.[32] These findings indicate the phenomenon of physiological group responses to emotional state induced by visual and audio stimuli.

10.6 Disease biomarkers

Successful PTR-MS breath research investigating diseases to find unique biomarkers are limited, but there are a few notable studies that have been published. One deals with kidney disease by Kohl et al. and the others report work on liver disease by Fernández del Río et al.

In the kidney disease study, the researchers used PTR-QMS in a clinical trial to measure selected *m/z* of patients before and after renal transplantation.[33] The data revealed a product ion at a nominal *m/z* 115 that correlated with creatinine, the marker used to assess kidney function. Subsequent analyses of selected breath samples by PTR-TOF-MS ascertained an exact *m/z* 115.1113; although the volatile eliciting the signal at this *m/z* was not specifically identified, its elemental composition was determined to be associated with either a C_7 ketone or a branched C_7 aldehyde.

Liver disease is a stealthy killer: by the time symptoms appear, the patient is already extremely ill. A simple test that can detect early-stage liver disease is needed. Changes in lifestyle and medication can then reverse early-stage damage. PTR-MS breath studies of liver disease have been successful in identifying volatiles that are biomarkers of chronic liver disease, namely limonene, methanol, and 2-pentanone, with limonene showing the best diagnostic capability.[34] The importance of this groundbreaking study cannot be stressed enough. While most biomarker studies seek emergence or perturbations of (unique) endogenous metabolic products or signatures, the latter study identified an exogenous compound that correlated with disease state. Specifically, an impaired metabolism of this xenobiotic substance by the damaged liver means that it is detected in breath in much higher concentrations than in healthy individuals. This establishes a new direction for breath research—already exploited in labeled substrate studies (see Chapter 17)—by utilizing exogenous VOC (or EVOC) probes to explore metabolism and potential disease-related imbalances.[35]

Another highlight of the aforementioned liver disease study is that the unambiguous identification of the three volatiles associated with liver disease resulted from using patients as their own controls and then monitoring their breath pre- and post-liver transplant. Importantly, these PTR-MS studies have identified limonene as a reliable volatile biomarker in breath, which could form the basis for the establishment of a breath-based test to provide a rapid and noninvasive diagnosis, screening or therapeutic monitoring of early-stage liver disease. Furthermore, the researchers demonstrated that by monitoring the washout of limonene following transplant, graft liver function can be noninvasively assessed. Future development and exploitation of this work hold enormous promise for the improved healthcare of patients suffering from impaired liver function, the implications of which are far-reaching both for patients and for healthcare providers.

10.7 Complementary technologies and emerging applications

Although PTR-MS has many advantages as an analytical device, its major disadvantage is that without compound preseparation the *m/z* of the product ions in a mass spectrum cannot be unambiguously assigned to a volatile. This has been partly overcome with the advent of high-resolution PTR-TOF-MS instruments, which has enabled peak positions of the product ions to be determined with great accuracy, thereby limiting the chemical make-up of a given product ion. To improve its chemical specificity, GC and fast-GC (minutes) have been developed for PTR-MS,[36,37] but this addition does minimize the real-time analysis advantage, which is needed when concentrations of breath volatiles change rapidly. Furthermore, as with GC-MS, considered to be the gold analytical standard for breath analysis (see Chapter 15), the reliance of chromatographic columns makes it difficult to handle a large range of VOC functionality (oxygenates, sulfur compounds, etc.). Nevertheless, the recent development of fast-GC coupled to PTR-MS provides a compromise between real-time measurements, ensuring reasonably fast analysis (within approximately 90 s), while still taking advantage of limited preseparation of compounds to improve the analytical specificity of PTR-MS. Further, the mode of analysis can be switched between direct and GC sampling, thereby allowing for real-time analysis with intermittent chromatographic separation to ascertain the presence of isomeric compounds.

In breath analysis, the overriding strength of PTR-MS is in its capability for breath-by-breath resolution analysis, as discussed earlier in this chapter. A secondary advantage of this technology is direct sampling and analysis, without sample workup. In the broader context of breath research, this finds direct use in in vitro model analyses, such as determining cellular VOC emissions for signature analytes or analyte profiles (see Chapter 26). In such applications, the time constraints are dominated by sample handling, rather than the analyses themselves. As such, the recent coupling of a robotic autosampler to PTR-MS, to enable automated static headspace sampling of sample vials, will be beneficial for rapid throughput analysis in this field.[38]

As with any analytical technology, calibrating the system is imperative for data quality assurance. With the broad and numerous types of VOCs present in exhaled breath (see Chapter 1), accurate quantitation is challenging due to limited availability of stable and reliable gas standards. PTR-MS is often pitched as a pseudoabsolute analytical method, meaning that concentrations of analytes in a sample can be determined based on first-order reaction kinetics calculations.[39] Although ascertaining compound concentrations through calculation is better than no quantitation at all, accurate detection after instrumental calibration is preferred. The development of a calibration system that allows the generation of defined gas-phase concentrations for target compounds in bespoke liquid-phase mixtures has greatly broadened the range of calibration compounds for PTR-MS[40]; this is an important development in breath research with PTR-MS. Notably, unlike some detection tools (e.g., sensor-based systems), PTR-MS achieves robust performance stability over extended periods, thereby obviating the need for daily calibrations.

An emerging field in breath analysis is the implementation of novel chemical sensors, which are a particularly promising possibility for medical or health applications because they are cheap to produce, small, and have good but not high sensitivities (see Chapter 13). Due to their small size, chemical sensors can be integrated into tabletop or handheld devices for point-of-care (PoC) use or placed in smartphones for personalized at-home applications. A major current limitation for chemical sensors is that only a limited number of breath volatiles can be detected by some sensor types at one time. A second limitation of some sensor types is their low chemical specificity and long-term stability. For technological advancements in chemical sensors or, for that matter, ion mobility spectrometry (IMS) systems, PTR-MS can be expected to play a key role in their characterization and validation. This has been recently illustrated in a proof-of-principle study presenting the possibilities of detecting entrapped humans with chemical sensor arrays.[41] In that study, PTR-MS was used to compare the sensitivity and selectivity of an inexpensive sensor array, designed to detect three key volatiles in human breath, namely acetone, ammonia, and isoprene, by tailoring nanostructured, chemoresistive gas sensors with distinct selectivities. The responses of the sensors in the array were converted to concentrations using the results obtained from the PTR-MS measurements. This demonstrated the successful use of PTR-MS for validation purposes in breath-related technological developments.

A final advancement in PTR-MS analyses that is worthy of mention is the recent development of an interface that allows for the chemical composition of aerosols to be analyzed in real time.[42] This chemical analysis of aerosols online (CHARON) system has been demonstrated in applications in the field of atmospheric chemistry, but given the high abundance of particles exhaled with each breath (see Chapters 7 and 8), there is a potential wealth of chemical information waiting to be discovered by PTR-MS equipped with a CHARON interface; time will tell what discoveries, if any, will emerge.

10.8 Summary

The British scientist Sir Humphrey Davy (1778–829) remarked that "*nothing tends so much to the advancement of knowledge as the application of a new instrument.*" This could not be truer than for the application of PTR-MS to breath analysis. New areas of medical sciences have been discovered by taking advantage of its real-time capabilities, such as the insightful observation of the dramatic rise in isoprene levels immediately at the start of exercise, for which an explanation is still wanting, or the breath-by-breath monitoring of VOC concentrations so that the end-tidal phase can be easily observed and hence analyzed.

Owing to its high sensitivity and real-time capabilities, the trapping and concentration of large volumes of breath (which ill patients may be unable to provide), and as required for other MS analytical devices, such as GC-MS, are not required with PTR-MS. This eliminates or reduces issues of surface adsorption losses, contamination and permeability, thermal decomposition, and catalytic reactions on surfaces, which can make the analysis of breath samples unreliable. The real-time capability of PTR-MS means that transportation of breath samples to specialized laboratories, for which there is an inevitable risk for the introduction of errors, is not required. Simply put, PTR-MS is an ideal laboratory-based analytical platform for discovery in breath research.

This chapter has highlighted that PTR-MS is a powerful technique for assessing and monitoring volatiles in breath, and particularly for real-time analysis, with high sensitivity and good selectivity, for either targeted or untargeted research. However, owing to its size, complexity, power requirements, and cost, it is unlikely that it will be directly used in clinical environments for patient care. For this, other analytical tools that are user-friendly, portable, compact, and inexpensive are required. In this regard, instrumentation such as IMS and chemical sensors hold great promise, but these will need further development if they are to be used for real-time breath analysis and for them to be routinely adopted in clinical or PoC settings.

References

1. Munson MSB, Field FH. Chemical ionization mass spectrometry. I. General introduction. *J Am Chem Soc* 1966;**88**(12):2621−30.
2. Lindinger W, Albritton DL, Fehsenfeld FC, Schmeltekopf AL, Ferguson EE. Flow−drift tube measurements of kinetic energy dependences of some exothermic proton transfer rate constants. *J Chem Phys* 1975;**62**(9):3549−53.
3. Lagg A, Taucher J, Hansel A, Lindinger W. Applications of proton transfer reactions to gas analysis. *Int J Mass Spectrom Ion Process* 1994;**134**(1):55−66.
4. Trefz P, Schubert JK, Miekisch W. Effects of humidity, CO_2 and O_2 on real-time quantitation of breath biomarkers by means of PTR-ToF-MS. *J Breath Res* 2018;**12**(2): 026016.
5. Trefz P, Pugliese G, Brock B, Schubert JK, Miekisch W. Effects of elevated oxygen levels on VOC analysis by means of PTR-ToF-MS. *J Breath Res* 2019;**13**(4):046004.
6. Ellis AM, Mayhew CA. *Proton transfer reaction mass spectrometry: principles and applications*. 1 ed. Chichester, UK: John Wiley & Sons Ltd.; January 2014. p. 366.
7. Blake RS, Whyte C, Hughes CO, Ellis AM, Monks PS. Demonstration of proton-transfer reaction time-of-flight mass spectrometry for real-time analysis of trace volatile organic compounds. *Anal Chem* 2004;**76**(13):3841−5.
8. Zardin E, Tyapkova O, Buettner A, Beauchamp J. Performance assessment of proton-transfer-reaction time-of-flight mass spectrometry (PTR-TOF-MS) for analysis of isobaric compounds in food-flavour applications. *LWT Food Sci Technol* 2014;**56**: 153−60. 2013.
9. Jordan A, Haidacher S, Hanel G, Hartungen E, Märk L, Seehauser H, et al. A high resolution and high sensitivity proton-transfer-reaction time-of-flight mass spectrometer (PTR-TOF-MS). *Int J Mass Spectrom* 2009;**286**(2−3):122−8.
10. Brown PA, Cristescu SM, Mullock SJ, Reich DF, Lamont-Smith CS, Harren FJM. Implementation and characterization of an RF ion funnel ion guide as a proton transfer reaction chamber. *Int J Mass Spectrom* 2017;**414**:31−8.
11. Krechmer J, Lopez-Hilfiker F, Koss A, Hutterli M, Stoermer C, Deming B, et al. Evaluation of a new reagent-ion source and focusing ion−molecule reactor for use in proton-transfer-reaction mass spectrometry. *Anal Chem* 2018;**90**(20):12011−8.
12. Sulzer P, Hartungen E, Hanel G, Feil S, Winkler K, Mutschlechner P, et al. A proton transfer reaction-quadrupole interface time-of-flight mass spectrometer (PTR-QiTOF): high speed due to extreme sensitivity. *Int J Mass Spectrom* 2014;**368**:1−5.
13. Breitenlechner M, Fischer L, Hainer M, Heinritzi M, Curtius J, Hansel A. PTR3: an instrument for studying the lifecycle of reactive organic carbon in the atmosphere. *Anal Chem* 2017;**89**(11):5824−31.
14. Jordan A, Hansel A, Holzinger R, Lindinger W. Acetonitrile and benzene in the breath of smokers and non-smokers investigated by proton transfer reaction mass spectrometry (PTR-MS). *Int J Mass Spectrom Ion Process* 1995;**148**(1−2):L1−3.

15. Jordan A, Hansel A, Warnecke C, Holzinger R, Prazeller P, Vogel W, et al. "On-line" Spurengasanalyse im ppt-Bereich und ihre Anwendungen auf Gebieten der Medizin, Lebensmittelforschung und Luftqualität (in German). *Ber Nat-Med Verein Innsbruck* 1997;**84**:7–17.
16. Taucher J, Lagg A, Hansel A, Vogel W, Lindinger W. Methanol in human breath. *Alcohol Clin Exp Res* 1995;**19**(5):1147–50.
17. Taucher J, Hansel A, Jordan A, Fall R, Futrell JH, Lindinger W. Detection of isoprene in expired air from human subjects using proton-transfer-reaction mass spectrometry. *Rapid Commun Mass Spectrom* 1997;**11**(11):1230–4.
18. Karl T, Prazeller P, Mayr D, Jordan A, Rieder J, Fall R, et al. Human breath isoprene and its relation to blood cholesterol levels: new measurements and modeling. *J Appl Physiol* 2001;**91**(2):762–70.
19. Karl T, Jordan A, Hansel A, Holzinger R, Lindinger W. Benzene and acetonitrile in smokers and nonsmokers. *Ber Nat-Med Verein Innsbruck* 1998;**85**:7–15.
20. Prazeller P, Karl T, Jordan A, Holzinger R, Hansel A, Lindinger W. Quantification of passive smoking using proton-transfer-reaction mass spectrometry. *Int J Mass Spectrom* 1998;**178**(3):L1–4.
21. Herbig J, Titzmann T, Beauchamp J, Kohl I, Hansel A. Buffered end-tidal (BET) sampling-a novel method for real-time breath-gas analysis. *J Breath Res* 2008;**2**(3): 037008.
22. Herbig J, Muller M, Schallhart S, Titzmann T, Graus M, Hansel A. On-line breath analysis with PTR-TOF. *J Breath Res* 2009;**3**(2):027004.
23. Harrison GR, Critchley ADJ, Mayhew CA, Thompson JM. Real-time breath monitoring of propofol and its volatile metabolites during surgery using a novel mass spectrometric technique: a feasibility study. *Br J Anaesth* 2003;**91**(6):797–9.
24. Kharasch Evan DMDPD. Every breath you take, we'll be watching you. *Anesthesiology* 2007;**106**(4):652–4.
25. Fernández del Río R, O'Hara ME, Pemberton P, Whitehouse T, Mayhew CA. Elimination characteristics of post-operative isoflurane levels in alveolar exhaled breath via PTR-MS analysis. *J Breath Res* 2016;**10**(4):046006.
26. Beauchamp J, Kirsch F, Buettner A. Real-time breath gas analysis for pharmacokinetics: monitoring exhaled breath by on-line proton-transfer-reaction mass spectrometry after ingestion of eucalyptol-containing capsules. *J Breath Res* 2010;**4**(2):026006.
27. Winkler K, Herbig J, Kohl I. Real-time metabolic monitoring with proton transfer reaction mass spectrometry. *J Breath Res* 2013;**7**(3):036006.
28. Ruzsanyi V, Lederer W, Seger C, Calenic B, Liedl KR, Amann A. Non-13CO2targeted breath tests: a feasibility study. *J Breath Res* 2014;**8**(4):046005.
29. Mochalski P, Mirmigkou S, Unterkofler K, Sulzer P, Mayhew CA, Märk TD. PTR-MS studies of the reactions of H_3O^+ with a number of deuterated volatile organic compounds and the subsequent sequential reactions of the primary product ions with water under normal and humid drift tube conditions: implications for use of deuterated compounds for breath analysis. *Int J Mass Spectrom* 2019;**436**:65–70.
30. Williams J, Pleil J. Crowd-based breath analysis: assessing behavior, activity, exposures, and emotional response of people in groups. *J Breath Res* 2016;**10**(3):032001.
31. Veres PR, Faber P, Drewnick F, Lelieveld J, Williams J. Anthropogenic sources of VOC in a football stadium: assessing human emissions in the atmosphere. *Atmos Environ* 2013;**77**:1052–9.

32. Williams J, Stönner C, Wicker J, Krauter N, Derstroff B, Bourtsoukidis E, et al. Cinema audiences reproducibly vary the chemical composition of air during films, by broadcasting scene specific emissions on breath. *Sci Rep* 2016;**6**:25464.
33. Kohl I, Beauchamp J, Cakar-Beck F, Herbig J, Dunkl J, Tietje O, et al. First observation of a potential non-invasive breath gas biomarker for kidney function. *J Breath Res* 2013; **7**(1):017110.
34. Fernández del Río R, O'Hara ME, Holt A, Pemberton P, Shah T, Whitehouse T, et al. Volatile biomarkers in breath associated with liver cirrhosis —comparisons of pre- and post-liver transplant breath samples. *EBioMedicine* 2015;**2**(9):1243—50.
35. Gaude E, Nakhleh MK, Patassini S, Boschmans J, Allsworth M, Boyle B, et al. Targeted breath analysis: exogenous volatile organic compounds (EVOC) as metabolic pathway-specific probes. *J Breath Res* 2019;**13**(3):032001.
36. Romano A, Fischer L, Herbig J, Campbell-Sills H, Coulon J, Lucas P, et al. Wine analysis by FastGC proton-transfer reaction-time-of-flight-mass spectrometry. *Int J Mass Spectrom* 2014;**369**:81—6.
37. Ruzsanyi V, Fischer L, Herbig J, Ager C, Amann A. Multi-capillary-column proton-transfer-reaction time-of-flight mass spectrometry. *J Chromatogr A* 2013;**1316**(0): 112—8.
38. Capozzi V, Yener S, Khomenko I, Farneti B, Cappellin L, Gasperi F, et al. PTR-ToF-MS coupled with an automated sampling system and tailored data analysis for food studies: bioprocess monitoring, screening and nose-space analysis. *JoVE* 2017;(123):e54075.
39. Beauchamp J, Herbig J, Dunkl J, Singer W, Hansel A. On the performance of proton-transfer-reaction mass spectrometry for breath-relevant gas matrices. *Meas Sci Technol* 2013;**24**(12):125003.
40. Beauchamp J, Herbig J. Proton-transfer-reaction time-of-flight mass spectrometry (PTR-TOFMS) for aroma compound detection in real-time: technology, developments, and applications. In: Guthrie B, Beauchamp J, Buettner A, Lavine BK, editors. *The chemical sensory informatics of food: measurement, analysis, integration. ACS symposium series. 1191*. Washington, DC: American Chemical Society; 2015. p. 235—51.
41. Güntner AT, Pineau NJ, Mochalski P, Wiesenhofer H, Agapiou A, Mayhew CA, et al. Sniffing entrapped humans with sensor arrays. *Anal Chem* 2018;**90**(8):4940—5.
42. Eichler P, Müller M, D'Anna B, Wisthaler A. A novel inlet system for online chemical analysis of semi-volatile submicron particulate matter. *Atmos Meas Tech* 2015;**8**(3): 1353—60.

CHAPTER

Ion mobility spectrometry 11

Dahlia Salman[1], Gary A. Eiceman[2], Dorota Ruszkiewicz[1], Veronika Ruzsanyi[3], Emma Brodrick[4], C. L. Paul Thomas[1]

[1]*Centre for Analytical Science, Department of Chemistry, Loughborough University, Loughborough, United Kingdom;* [2]*Department of Chemistry and Biochemistry, New Mexico State University, Las Cruces, NM, United States;* [3]*Institute for Breath Research, University of Innsbruck, Dornbirn, Austria;* [4]*IMSPEX, Ty Menter, Cardiff, United Kingdom*

Chapter outline

11.1 Overview

The origins of ion mobility spectrometry (IMS) can be traced to investigations on the structure of matter around 1900 and later studies from 1910 to 1950 in gaseous electronics and on ionization phenomena and atmospheric ions.[1] Investigations into long-range, ion–neutral interactions, established five decades earlier by Langevin, were refined in the 1950s after using the mobility of ions in gases at 1–10 Torr.[2] After a burst of development of ionization detectors for gas chromatography (GC) in the early 1960s,[3] methods derived from these detectors were joined in 1970 with drift tubes developed for the characterization of ions by mobility.[4] This introduced a new method for chemical measurements of substances with high speed, superb detection limits, and convenient operation without vacuum systems. Owing to

Breathborne Biomarkers and the Human Volatilome. https://doi.org/10.1016/B978-0-12-819967-1.00011-6
Copyright © 2020 Elsevier B.V. All rights reserved.

favorable chemistry of ionization of chemical warfare agents and explosives, handheld analyzers for military establishments and benchtop analyzers for security checkpoints in airports were developed and distributed in large numbers (more than 50,000 and 20,000 analyzers, respectively) and are still in widespread use today.[5] IMS might have become a niche technique for these applications if it were not for innovations after 1980 in drift tube technology,[6] with high-speed response and reproducible mobility spectra. Along with improvements in understanding ionization chemistry at ambient pressure, and models to interpret qualitative and quantitative responses, refinements in technology provided a foundation for subsequent metabolomics measurements by IMS. The additional use of a GC frontend interface offered a unique analytical platform; it was simple and small-sized with low power consumption, making it ideal for real-time monitoring applications, such as detection of metabolites in exhaled breath, among other applications. The benefits of IMS for breath monitoring in a range of applications, from metabolite detection to monitoring therapeutic endpoints in lung infections, were recognized in the late 1990s by Baumbach et al., who pioneered methods and instrumentation based on a combination of GC with IMS drift tubes (GC-IMS).[7] A legacy of these studies was the commercialization of GC-IMS and a steady record of refinements and impacts in metabolomics[8] and breath analysis.[9]

A new method of ion mobility that employed strong electric fields with high frequency waveforms was demonstrated in 1993[10] and resulted in three technical progeny, including cylindrical analyzers called field asymmetric IMS (FAIMS),[11] small planar embodiments called differential mobility spectrometry (DMS),[12] and ultra-small structures termed ultra-FAIMS.[13] A common principle of these mobility analyzers was the motion of ion swarms in asymmetric waveforms at kHz to MHz frequencies and amplitudes producing fields of 30,000 V/cm or more; a direct current bias voltage was used to selectively control the transmission of ions in a flow of gas through a narrow gap. While cylindrical units were attached to mass spectrometers for biomolecule studies, the planar DMS analyzers were developed for measurements with small molecules, both as GC-DMS and DMS-mass spectrometry (MS). Overall, there is currently a limited choice of IMS instruments for measurements of small molecules in metabolomics applications, and some have a record of use for breath analyses. The principles and practices of IMS in general is reported at length in the literature.[1,14,15] The discussion in this chapter is confined to the principles and practices as found in IMS analyzers with low electric field ($<$400 V/cm) cylindrical drift tubes for their specific application in breath and related research.

11.2 Principles of ion mobility spectrometry

Ion mobility spectrometry has been used in human metabolomics studies[16–18] for the rapid characterization of volatile substances in breath, principally using sample prefractionation by GC.[7,19] Substances delivered to an ion mobility analyzer after GC elution undergo two consecutive processes, namely the formation of a gas ion(s) derived from a volatile substance, followed by the mobility determination of these ions in an electric field. Pressures inside IMS analyzers are typically

4–19 Torr (~5–25 mbar).[20] The practice of IMS as relevant to metabolomics occurs commonly in a supporting atmosphere of purified air at ambient pressure, as emphasized in discussions below on the principles and technology for small or portable instruments.

The first step in response (i.e., ionization of analytes) is the formation of ions from an analyte (M) in the vapor phase, commonly with chemical ionization methods, where M undergoes collisions with a reagent or reactant ion (often a hydrated proton, $H^+(H_2O)_n$). This interaction can form a product ion, e.g., a protonated monomer ($MH^+(H_2O)_{n-1}$), through displacement of H_2O in the hydrated proton by M. The hydrated proton can originate from a beta source or one of several nonradioactive alternatives, including electrical discharges[21] or X-ray sources.[22] Photoionization is another method of ionization for GC-IMS,[23–25] with benefits particularly for measuring substances with small dipoles and delocalized electrons, including alkanes, alkenes, and aromatic hydrocarbons, which form product ions through electron detachment to $M^{+}\cdot$. Once formed, the second step of response in ion mobility characterization of a substance is mobility analysis, where product ions pass through an electric field as ion swarms with characteristic drift velocities (v_d) over a fixed distance.

Ions move as swarms due to high collision frequencies of ions with neutrals of the supporting atmosphere maintaining low axial or longitudinal diffusion compared to the speed of swarm motion through the electric field. Upon reaching a detector, ion swarms produce a signal by collision on a detector, i.e., the ion mobility spectrum. Swarm drift velocity (v_d) is established by the time of arrival, or drift time (t_d), for a swarm in the drift region at constant pressure, electric field strength (E), and temperature. Drift velocity, when normalized to electric field strength (E), provides the mobility coefficient, i.e., $K = v_d/E$, and K, when corrected for temperature and pressure, is expressed as the reduced mobility coefficient (K_o). Mobility coefficients, or K_o values, can provide refined descriptions of collisional cross sections (Ω) of simple ions measured at reduced pressures in nonclustering atmospheres. Recently, the concept of ion size (mass) and shape (as Ω) has been employed with large biological ions at low pressures to study structure–function relationships and the dynamics of ion transformations.[26] In contrast, the relationships between K_o and Ω for small ions at ambient pressure with clustering atmospheres, such as air, are complex, and calculations of collisional cross sections have been applied rarely for breath metabolomics.[27] This can be attributed to complications from the formation of adducts between an ion of interest, e.g., the $MH^+(H_2O)_n$ of an alcohol, with impurities or polar constituents in the supporting gas atmosphere. Consequently, K_o is governed by the composition of the drift gas (with potential vulnerabilities to gas impurities inside an IMS analyzer) and thus presents significant complications in developing reference tables for ion cross sections. In practice, drift times or $1/K_o$ values obtained from mobility spectra under specific experimental conditions are used for the interpretation of IMS responses to metabolites. Other possibilities of extracting analytical information (more than drift time or K_o) from mobility spectra have been demonstrated using computational methods, including neutral network analyses, though such methods are underdeveloped in metabolomics studies (see

Chapter 38). A broad assessment of ion mobility measurements today is that drift times or K_o values from mobility spectra are useful for chemical analyses, particularly when combined with chromatographic retention where processes of ionization are controlled.

When coupled to IMS, GC offers the additional benefit of separating the individual constituents of complex samples through chromatographic retention, thereby aiding compound identification. A still larger importance of GC prefractionation of complex samples may be ascribed to the simplicity of ionization chemistry and mobility analysis when constituents of a sample are delivered individually to the ion source compared to the simultaneous introduction of multiple components in a sample mixture. This makes the formation of product ions reliable without cross reactivity with mixtures and quantitatively predictable without vapor concentration overload, as is expected when all constituents of a sample are delivered at once to the ion source. In practice, the dispersion of chemical information of a mixture in datasets of ion intensity, retention time, and drift time may enable the detection of target metabolites and computational processing of signals to extract such information if found at low abundances. The patterns generated by such datasets also enable untargeted analysis of metabolomics, as demonstrated by many published studies, such as for lung cancer diagnostics.[28]

11.3 Metabolomics and clinical practice

Measurements in urgent care or routine clinical practice require small, portable, and easy-to-use devices with a quick and simple answer, often just “yes” or “no” (red or green light indicator), and IMS seems to be a promising candidate for online diagnostic testing and/or treatment monitoring.

11.3.1 Clinical trials

Ion mobility spectrometry has been used in many previous studies for the detection of volatile organic compounds (VOCs) in the breath of patients with lung or heart diseases[28–30]. In 2005, IMS coupled to a multicapillary column frontend interface (MCC-IMS) was used in a pilot study to detect lung cancer from 36 patients and 54 healthy controls, whereby 25 VOCs were reported to enable the significant separation of the two groups with a classification accuracy of 98.7%. A subsequent study investigated the ability of IMS to separate different lung infections and respiratory exacerbations, such as chronic obstructive pulmonary disease (COPD), bronchiectasis, and pneumonia.[28] MCC-IMS was then reported with the use of cyclohexanone as a biomarker to differentiate COPD (with or without lung cancer) from healthy individuals, with a sensitivity and specificity of 60% and 91%, respectively.[30] In a pilot study of nine patients, IMS was used for the detection of sarcoidosis using only 10 mL breath sampling volume; the analysis showed promising data, with different distribution of metabolites in the breath of patients with confirmed sarcoidosis to those with unspecific mediastinal lymph node enlargement.[29]

11.3.2 Pilot and exploratory studies

Commercial GC-IMS technology exists that is tailored to breath analysis and allows patients' breath to be analyzed at the bedside or in a community setting due to high portability and an integrated gas recirculation system that obviates the need of an external gas supply. This commercial GC-IMS breath analysis system, marketed as Breathspec, allows direct breath sampling from a single breath provided through a disposable mouthpiece. The system draws the breath into the sample loop of the analyzer, where 1 mL of breath is analyzed within 10 min. This device was recently used to develop a noninvasive breath test to diagnose respiratory tract infection (RTI) with the aim of combating antimicrobial resistance (AMR) (unpublished work). The increasing threat of AMR to global public health will require concerted action across all government sectors and society, with the risk of compromising the success of major surgery or chemotherapy without effective antibiotics. As a result, new portable technologies such as GC-IMS are needed to gather real-time patient data and thereby help clinicians to make a choice on when to administer or preserve antibiotics in their treatment. The GC-IMS breath analysis system was trialed in the United Kingdom within eight National Health Service (NHS) clinical environments: two community general practices (GPs) and six central hospitals. Thirteen analyzers were integrated into standard medical care pathways, with the instruments deployed in accident and emergency, medical assessment units, and GP surgeries. The trial was a multicenter prospective, observational, longitudinal cohort study in adult patients with suspected upper or lower respiratory tract infection (URTI/LRTI) to identify a spectrum of biomarkers in exhaled breath to discriminate participants with bacterial RTI from participants without. The GC-IMS systems were tested in both the community and hospital settings, where 1102 patients with suspected RTIs provided two breath samples for analysis in both positive and negative analysis modes. Of these subjects, 451 males and 661 females were recruited (of which 652 were smokers), with an age range of 18–100 years (mean 50 ± 23 years). Initial data analysis indicates that the GC-IMS data can separate diagnostic groups well, thus this observational study confirmes the utility of exhaled breath analysis to distinguish between patients with bacterial RTI and those without. Further, the analysis shows that the features associated with upper and low RTI are different, and demographic parameters, such as sex and smoking, are factors that affect diagnostic performance. This may indicate that the disease–host interaction is linked to the nature of the disease and lifestyle of the host. When these factors are taken into account, the performance of the system is higher, which highlights the need to consider these in future diagnostic platform applications. The study demonstrates that GC-IMS has the potential to provide the prescribing clinician with diagnostic results at the time of appointment, with prospective benefits for patients in terms of health and quality of life.

Several other clinical studies have employed GC-IMS, including a recent study that aimed to evaluate the use of breath biomarkers in breathlessness patients in order to differentiate between common but different cardiorespiratory diseases, such as asthma (in both adults and children cohorts), acute heart failure, community-acquired pneumonia, COPD, and aged-matched healthy participants. The project was carried out across two adult and one children acute admission units, and used five different offline and online analytical platforms, including MCC-GC-IMS and GC-MS. Breath samples were collected for both GC-MS and GC-IMS analysis from patients with self-reported acute exacerbation within 24 h of admission, and another set of breath samples at the follow-up appointment (up to 6 months postdischarge). The project recruited 455 patients and is currently in the data analysis phase.

11.4 Drug metabolism and pharmacokinetics

One promising use of IMS is the detection of VOCs for isotopically labeled compounds, permitting fast, noninvasive monitoring of substances or metabolites of ingested drugs or other therapies. The detection of *Helicobacter pylori* infection, for instance, is now feasible through the ^{13}C-urea breath test (see Chapter 17). Alternatively, VOCs appearing in human breath following the administration of a pharmaceutical preparation could be specific to the administered formulation and/or as a result of a related metabolic pathway of the drug constituents, thus breath analysis using IMS offers the potential opportunity to monitor the kinetics of such compounds without the specific need for isotopic labeling.

Despite the promising approaches in monitoring drug metabolism and pharmacokinetics, it is important to note that most of the current pharmacokinetics applications focus on metabolite analysis in blood and plasma, with IMS technology mainly applied in pharmacokinetics as a preseparation technique with liquid chromatography (LC) multiple ion monitoring.[31]

One recent example of the use of IMS in a clinical setting is the monitoring of intravenous anesthetic propofol during surgery.[32] The standard method for the determination of propofol concentrations in plasma is liquid chromatography—mass spectrometry (LC-MS), requiring more than 30 min for single sample analysis, with extra time to collect and prepare the sample.[33] A recent study explored the use of IMS inside an operating room for determining intraoperative blood propofol levels within 2 min using negative polarity and with a photoionization source containing acetone reagent gas.[34] This study demonstrated with nearly 100 surgical patients that IMS was an excellent platform to characterize the pharmacokinetics of propofol. Efforts using IMS to detect propofol in breath have also been reported,[35] and Perl et al. described IMS measurements of propofol concentrations in respired air with predictions of serum levels. Research into the measurement of breath propofol resulted recently in the exciting commercial development of an IMS-based medical device for this application.

The utility of MCC with GC-IMS for pharmacokinetic breath experiments was explored in a 2013 study on the kinetics of release of eucalyptol in breath after ingestion of a capsule containing this compound.[36] The washout characteristics of repeated experiments for a single volunteer were reproducible, but interindividual differences were observed among the candidates, despite the use of the same experimental protocol and instrumentation.

Research into breath pharmacokinetics experiments has been extended and strengthened through the *Peppermint initiative* that currently involves 16 partners, seven European countries, eight breath measurement techniques, and six breath sampling systems. The aim of this ongoing study is to produce comparable datasets between laboratories to determine a set of benchmark values through a standardized breath experiment. The experiment was designed to impose well-defined changes in the breath profile of a participant after ingestion of a 200 mg peppermint oil capsule. The central concept of the *Peppermint initiative* is that the capsule dissolves in the stomach of the participant and releases different constituents of peppermint oil into the body, which enter the bloodstream and become detectable in exhaled breath (Fig. 11.1). Concentrations of these specific compounds decrease in breath over several hours as they are eliminated from the body, which discloses details on the metabolism of the constituent compounds of the peppermint oil capsule through the release of associated VOCs.

The *Peppermint initiative* most importantly has given insights into the reproducibility of breath measurements, with a proposed mathematical logarithmic washout model (Fig. 11.2, top), as well as providing benchmarking values for the expected duration of the measure change in breath, reflecting the sensitivity of the methods employed.

Plots of cumulative frequencies, supported by box−whisker plots of response magnitudes and data distribution, provide information about the characteristics of the pharmacokinetics data acquired and enable an evaluation of the overall attributes of the method being used. In Fig. 11.3, an example is shown for the comparison of the cumulative frequency distribution of all of the features obtained from GC-IMS analyses of peppermint washout from two independent laboratories, which offers an indication of data fidelity. These findings are based on breath samples collected 60 min postingestion of peppermint oil and represent the medium for each in the cohort. Although both centers yielded similar endpoints, it is clear that one dataset

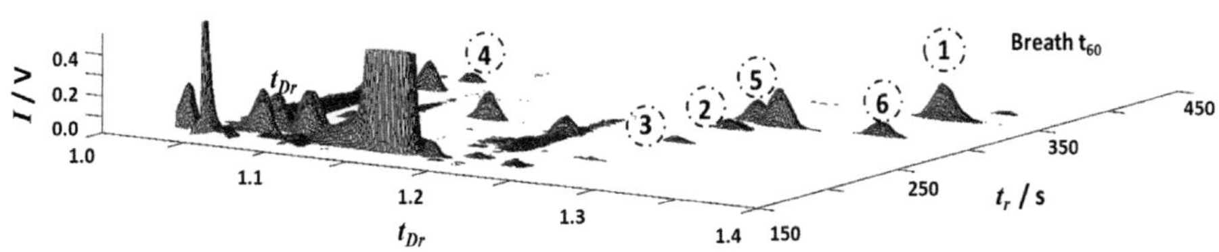

FIGURE 11.1

Example of cutout section GC-IMS 3D surface plot of a breath sample collected 60 min postingestion of a peppermint oil capsule, highlighting features related to the capsule constituents: 1—eucalyptol, 2—β-pinene, 3—α pinene, 4—unidentified metabolite, 5 and 6—limonene monomer and dimer. Note: t_r—retention time [s], t_{Dr}—relative drift time, I—intensity [V].

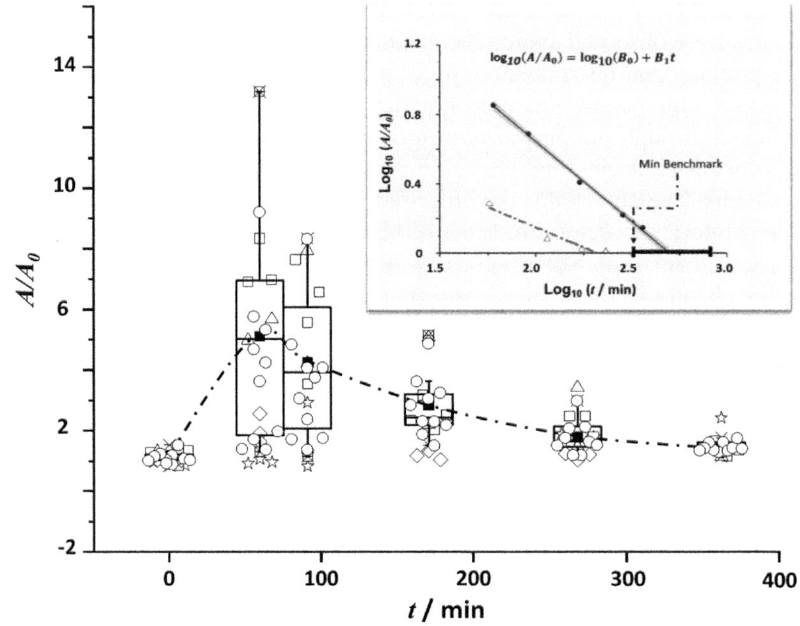

FIGURE 11.2

Box–whisker plot of the eucalyptol breath washout, showing the change in peak area versus time based on GC-IMS results obtained from 23 participants in the peppermint experiment. Note: different symbols indicate data collected in different laboratories. Inset: mean logarithmic model of the washout (*black curve*), reflecting reproducible sampling and established current benchmark range for limit of detection; any measurements below the minimum *red curve (dashed black in print version)* indicates that improvements are needed to fit within the current standard.

contains almost twice as many features from the same type of breath sample. Further, these features were predominantly of low abundance. This corresponded to the smaller number of detected peppermint-related features on average (highlighted in black in Fig. 11.3). Understanding the performance of each IMS instrument and statistically monitoring and controlling it throughout the study is therefore essential. This is particularly important if a study involves a biomarker discovery phase in a clinical setting where the environment is not controlled and the phenotypic variabilities are unknown.

11.5 Recent developments in ion mobility spectrometry technology

While commercial instruments for IMS measurements in metabolomics applications exist, research and development worldwide is broadly advancing the practice and technology of IMS. In addition, understandings of the principles of ion mobility,

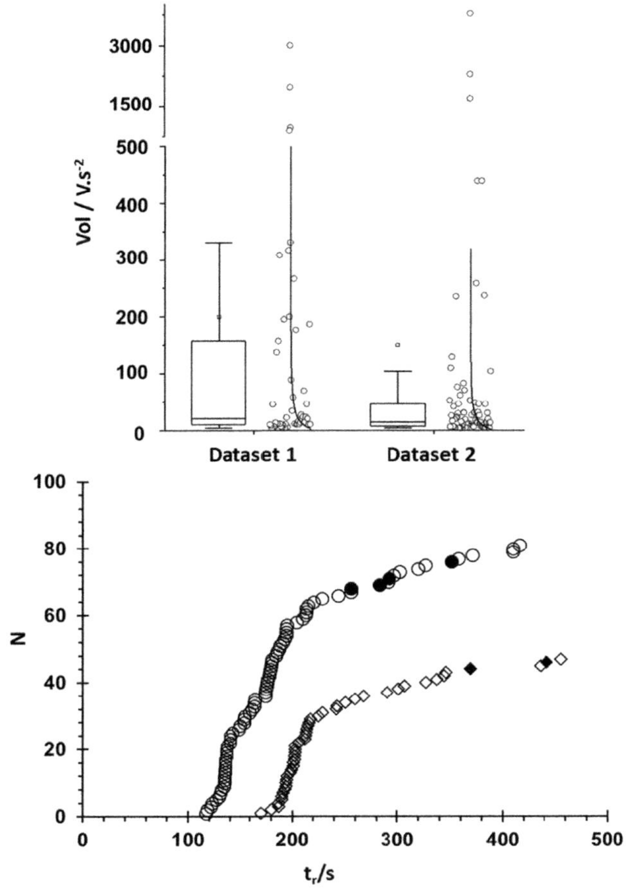

FIGURE 11.3

Example of data fidelity. Top: box-whisker plot of features intensities and their distribution. Bottom: cumulative frequency of features across the analytical space. Note: V—volume in V/s^2; tr—retention time; N—number of features.

refinements in instrumentation, and robustness in clinical applications have improved significantly during the past decade. Some examples of these are described below, though the selection is selective and the discussion is not extensively documented.

11.5.1 Nonradioactive ion sources

Chemical ionization using β-emission into air or nitrogen has been the standard method of ionization in IMS for decades and provides stable and favorable responses to molecules of strong interest in breath metabolomics. The continued or long-term future use of β-sources, however, is now limited by governmental regulations and institutional compliance for radioactivity. Some investigators or manufacturers have extended the functional use of radioactive sources by using amounts of radioactivity below regulated levels; nonetheless, a major trend in IMS technology is the replacement of ^{63}Ni, ^{3}H, and ^{241}Am with nonradioactive alternatives. Most commercial explosive trace detectors based on IMS today are provided with nonradioactive ion sources, including corona discharges,[21] X-rays,[22] photoionization,[23–25] or dielectric barrier discharges.[37] These or other nonradioactive ion sources will likely be migrated or adapted for IMS analyzers in clinical or metabolomics applications.

11.5.2 Advances in resolving power

The quality and value of mobility spectra in metabolomics measurements are directly associated with the resolving power of the drift tube, established principally by spectral peak widths that are governed by time of ion injection into the drift region. Until recently, drift tubes exhibited resolving powers of 20–40, but these have been extended to as much as 100 through improvements in ion injectors. Developments are ongoing worldwide on computational modeling and experimental innovation with ion shutters. Increases in resolving power widely throughout the breath metabolomics community should improve the quality and content of datasets over those obtained with analyzer technology of past and even recent years.

11.5.3 Tandem mobility spectrometers

The combination of multiple quadrupole mass analyzers in series transformed mass spectrometry, with the development of tandem MS, has enabled previously unavailable specificity of response and limits of detection. In recent years, a similar development in IMS has occurred with tandem ion mobility spectrometers operating in air at ambient pressure.[38] As with tandem MS, these tandem IMS analyzers have reactive stages to fragment (thermally or electrically) the precursor ions, which are mobility isolated in a first stage. The fragment ions are mobility characterized in a second mobility stage, producing field-induced fragmentation (FIF) spectra or thermally-induced dissociation spectra. Current findings suggest that FIF spectra exhibit fragmentation patterns that are characteristic of a chemical family portending a transition in IMS from a selective detector to a molecular identifier. Although reactive stage tandem IMS is at a laboratory stage of development, such new capabilities in mobility measurements could introduce another dimension of analytical quality and value into datasets for metabolomics and in clinical applications.

11.6 Summary

Technological developments in IMS over the years have allowed its diversified implementation from its early (and continual) use in military and security venues to an online metabolite detection tool for human breath metabolomics studies.

GC-IMS is capable of measuring hundreds of volatile metabolites in human breath and can aid the study of human biology and uncover metabolic pathways relevant to phenotypic characteristics of healthy or disease cohorts. Refined IMS measurements became possible as the research community developed a better understanding of ambient pressure ionization and more specialized data processing and interpretation tools. Furthermore, interfaces between GC and IMS enable prefractionation of sample constituents before the mobility stage, which allows chemical characterization of individual compounds within complex mixtures of VOCs in exhaled human breath. Chromatographic retention time combined with K_0 mobility analysis and ion intensity is the current state-of-the-art in ion mobility measurements. Commercial GC-IMS allows patients' breath to be analyzed at the bedside or in the community settings. Several recent clinical studies and projects have used IMS either as a biomarker/pattern discovery tool or as a fast detection technology for targeted biomarker analysis.

References

1. Eiceman G, Karpas Z, Hill Jr H. *Ion Mobility Spectrometry.* Boca Raton: CRC Press; 2014.
2. Mason EA, Schamp HW. Mobility of gaseous lons in weak electric fields. *Ann Phys* 1958;**4**:233–70.
3. Poole CF. Ionization-based detectors for gas chromatography. *J Chromatogr A* 2015; **1421**:137–53.
4. Cohen MJ, Karasek FW. Plasma chromatography —A new dimension for gas chromatography and mass spectrometry. *J Chromatogr Sci* 1970;**8**:330–7.
5. Eiceman G, Stone J. Ion mobility spectrometry in homeland security. *Anal Chem* 2004; **76**:390A–7A.
6. Baim MA, Hill HH. Tunable selective detection for capillary gas chromatography by ion mobility monitoring. *Anal Chem* 1982;**54**:38–43.
7. Ruzsanyi V, Baumbach JI, Sielemann S, et al. Detection of human metabolites using multi-capillary columns coupled to ion mobility spectrometers. *J Chromatogr A* 2005; **1084**:145–51.
8. Vautz W, Slodzynski R, Hariharan C, et al. Detection of metabolites of trapped humans using ion mobility spectrometry coupled with gas chromatography. *Anal Chem* 2013;**85**: 2135–42.
9. Horsch S, Baumbach JI, Rahnenführer J. Statistical analysis of MCC-IMS data for two group comparisons-an exemplary study on two devices. *J Breath Res* 2019;**13**:36011.
10. Buryakov IA, Krylov EV, Nazarov EG, et al. A new method of separation of multiatomic ions by mobility at atmospheric pressure using a high-frequency amplitude-asymmetric strong electric field. *Int J Mass Spectrom Ion Process* 1993;**128**:143–8.
11. Barnett DA, Ells B, Guevremont R, et al. Application of ESI-FAIMS-MS to the analysis of tryptic peptides. *J Am Soc Mass Spectrom* 2002;**13**:1282–91.
12. Miller R, Eiceman G, Nazarov E, et al. A novel micromachined high-field asymmetric waveform-ion mobility spectrometer. *Sensor Actuator B Chem* 2000;**67**:300–6.
13. Shvartsburg AA, Smith RD, Wilks A, et al. Ultrafast differential ion mobility spectrometry at extreme electric fields in multichannel microchips. *Anal Chem* 2009;**81**:6489–95.

14. Shvartsburg AA. *Differential ion mobility spectrometry*. CRC Press; 2008 [Epub ahead of print December 2008].
15. Nazarov E, Postlethwaite T, Markoski K, et al. Planar differential mobility spectrometry for clinical breath diagnostics. In: Piraino F, Selimović S, editors. *Diagnostic devices with microfluidics*. CRC Press; 2017.
16. Handa H, Usuba A, Maddula S, et al. Exhaled breath analysis for lung cancer detection using ion mobility spectrometry. *PLoS One* 2014;**9**:e114555.
17. Basanta M, Jarvis RM, Xu Y, et al. Non-invasive metabolomic analysis of breath using differential mobility spectrometry in patients with chronic obstructive pulmonary disease and healthy smokers. *Analyst* 2010;**135**:315.
18. Covington JA, van der Schee MP, Edge ASL, et al. The application of FAIMS gas analysis in medical diagnostics. *Analyst* 2015;**140**:6775–81.
19. Steinbach J, Goedicke-Fritz S, Tutdibi E, et al. Bedside measurement of volatile organic compounds in the atmosphere of neonatal incubators using ion mobility spectrometry. *Front Pediatr* 2019;**7**:248.
20. Langejuergen J, Allers M, Oermann J, et al. High kinetic energy ion mobility spectrometer: quantitative analysis of gas mixtures with ion mobility spectrometry. *Anal Chem* 2014;**86**:7023–32.
21. Tabrizchi M, Khayamian T, Taj N. Design and optimization of a corona discharge ionization source for ion mobility spectrometry. *Rev Sci Instrum* 2000;**71**:2321–8.
22. Cochems P, Kirk AT, Bunert E, et al. Fast pulsed operation of a small non-radioactive electron source with continuous emission current control. *Rev Sci Instrum* 2015;**86**:065102.
23. Walendzik G, Baumbach JI, Klockow D. Coupling of SPME with MCC/UV–IMS as a tool for rapid on-site detection of groundwater and surface water contamination. *Anal Bioanal Chem* 2005;**382**:1842–7.
24. Baim MA, Eatherton RL, Hill HH. Ion mobility detector for gas chromatography with a direct photoionization source. *Anal Chem* 1983;**55**:1761–6.
25. Chen C, Jiang D, Li H. UV photoionization ion mobility spectrometry: fundamentals and applications. *Anal Chim Acta* 2019;**1077**:1–13.
26. Zhang X, Quinn K, Cruickshank-Quinn C, et al. The application of ion mobility mass spectrometry to metabolomics. *Curr Opin Chem Biol* 2018;**42**:60–6.
27. Han H, Wang H, Jiang H, et al. Corona discharge ion mobility spectrometry of ten alcohols. *Chin J Chem Phys* 2009;**22**:605–10.
28. Westhoff M, Litterst P, Freitag L, et al. Ion mobility spectrometry: a new method for the detection of lung cancer and airway infection in exhaled air? First results of a pilot study. *Chest* 2005;**128**:155S.
29. Westhoff M, Litterst P, Freitag L, et al. Ion mobility spectrometry in the diagnosis of sarcoidosis: results of a feasibility study. *J Physiol Pharmacol* 2007;**58**(Suppl. 5):739–51.
30. Westhoff M, Litterst P, Maddula S, et al. Differentiation of chronic obstructive pulmonary disease (COPD) including lung cancer from healthy control group by breath analysis using ion mobility spectrometry. *Int J Ion Mobil Spectrom* 2010;**13**:131–9.
31. Ren T, Li R, Meng X, et al. Differential mobility spectrometry followed by tandem mass spectrometry with multiple ion monitoring for bioanalysis of eptifibatide in rat plasma. *J Pharmaceut Biomed Anal* 2018;**151**:260–5.

32. Wang X, Zhou Q, Jiang D, et al. Ion mobility spectrometry as a simple and rapid method to measure the plasma propofol concentrations for intravenous anaesthesia monitoring. *Sci Rep* 2016;**6**:37525.
33. Sørensen LK, Hasselstrøm JB. Simultaneous determination of propofol and its glucuronide in whole blood by liquid chromatography—electrospray tandem mass spectrometry and the influence of sample storage conditions on the reliability of the test results. *J Pharmaceut Biomed Anal* 2015;**109**:158—63.
34. Xiao Y, Wang X, Li E, et al. Rapid determination of intraoperative blood propofol concentration in operating theatre by dopant-enhanced neutral release and negative photoionization ion mobility spectrometry. *Anal Chim Acta* 2020;**1098**:47—55.
35. Perl T, Carstens E, Hirn A, et al. Determination of serum propofol concentrations by breath analysis using ion mobility spectrometry. *Br J Anaesth* 2009;**103**:822—7.
36. Ruzsanyi V. Ion mobility spectrometry for pharmacokinetic studies—exemplary application. *J Breath Res* 2013;**7**:046008.
37. Hu J, Li W, Zheng C, et al. Dielectric barrier discharge in analytical spectrometry. *Appl Spectrosc Rev* 2011;**46**:368—87.
38. Shokri H, Vuki M, Gardner BD, et al. Reactive tandem ion mobility spectrometry with electric field fragmentation of alcohols at ambient pressure. *Anal Chem* 2019;**91**:6281—7.

CHAPTER

Secondary electrospray ionization

12

Jiayi Lan[1], Francisco Blanco Parte[2], Guillermo Vidal-de-Miguel[2], Renato Zenobi[1]

[1]*Department of Chemistry and Applied Biosciences, ETH Zurich, Zürich, Switzerland;* [2]*Fossil Ion Technology, Madrid, Spain*

Chapter outline

12.1 Overview

As a rule of thumb, the biological significance of metabolites increases with molecular weight. Thus, detecting and quantifying compounds with higher mass in breath is of great interest, yet these tend to be present at trace concentrations due to their low vapor pressures, and by association, low volatility. Ideally, instruments targeting such species should be suited to detect molecules with concentrations in the ppt range; however, the concentrations of metabolites normally reported in the literature are several orders of magnitude higher. This mismatch shows that there is room for improvement.

Secondary electrospray ionization (SESI) coupled with mass spectrometry (MS) has shown much promise for detecting low volatility species. Although this method emerged only within the last 12 years, its sensitivity, relatively broad detection range, and modularity (possibility to be coupled with various types of commercial MS instruments) enable SESI to be used in many kinds of applications. In the early days of electrospray ionization (ESI), Fenn and coworkers found that trace levels of

Breathborne Biomarkers and the Human Volatilome. https://doi.org/10.1016/B978-0-12-819967-1.00012-8
Copyright © 2020 Elsevier B.V. All rights reserved.

contaminants from the gaseous environment were very efficiently ionized, and this phenomenon was subsequently exploited in several pioneering studies. Gamez et al., for instance, showed that the pharmacokinetics of valproic acid, an antiepileptic drug requiring therapeutic monitoring, could be detected in human breath in real time,[1] and Sinues et al. found entire families of fatty acids in breath.[2]

12.2 Technical aspects

12.2.1 Ionization mechanism

The study of the ionization mechanism involved in SESI was motivated by the promising results obtained with the first ionization sources. Early debates revolved around the nature of the charging agent, with two mechanisms proposed, namely (i) vapor—droplet interaction, where analytes are dissolved in charged droplets and reemitted as ions as the droplet evaporates, and (ii) gas-phase ion molecule reactions, in which ionized water clusters collide with the analyte molecules and transfer their charge in an almost instantaneous process, analogous to atmospheric pressure chemical ionization (APCI).[3] Today, high-performance SESI sources operate very close to the boiling point of the electrospray liquid, resulting in such rapid evaporation of the droplets that the ionization process is dominated by gas-phase proton transfer in a very similar manner to that of APCI.

In this context, the reaction rate is proportional to the concentrations of analyte molecules and protonated clusters, both of which are proportional to the partial pressure. As a result, the reaction rate scales with the square of the pressure. Because of this, SESI exhibits very high ionization efficiency, i.e., the ratio of ions generated in the ionizer to the number of neutral molecules entering the ion source. In addition, because ionization occurs in the atmospheric pressure region, located outside of the interface to the vacuum chamber of the mass spectrometer, losses of low volatility species to the walls of the interface do not occur.

In regular APCI, the cascade of reactions initiated by the high-energy ions originating in the discharge generates protonated water clusters and leads to the fragmentation of analyte molecules. In contrast, in SESI water clusters originate from droplets with no involvement of high-energy ions, resulting in only very low fragmentation of the analytes, even when the spray is directed toward the inlet of the instrument to enhance the ionization efficiency.

Current mechanistic studies aim at finding optimum charging solutions.[3] One of the most commonly used electrospray charging solutions is formic acid (0.1% v/v) in water, where the acid is needed to render the liquid conductive. This pair forms an azeotropic mixture, for which both components evaporate at a similar rate, resulting in a particularly stable composition at high temperatures. Experimental evidence shows that humidity plays a key role in the ionization mechanism and greatly improves the ionization efficiency.[4] Serendipitously, this is particularly useful for breath analysis because breath is a very humid matrix and its water partial pressure is well defined.

12.2.2 Secondary electrospray ionization development

Initial efforts to optimize SESI focused on the ionization efficiency, which has a direct impact on the signals produced for a given analyte concentration. The coupled kinetics, electrohydrodynamic, and diffusion equations govern the dynamics of the charging of the analyte ions in the expanding electrospray plume. The complex equations describing these phenomena can be solved in the proximity of the electrospray tip, and the solution shows that the concentration ratio of analyte ions over analyte neutral molecules is uniform and species dependent.[5] This suggests that SESI is a quantitative technique, although this remains to be demonstrated. The ionization and ion sampling efficiency can be improved by ensuring that ions are directed toward the mass spectrometer inlet or by reducing the dilution of neutral molecules in the sample gas.

The first configuration to implement these concepts improved the ionization efficiency, and hence the sensitivity, by a factor of 50–100.[6] This was achieved by introducing a set of plate electrodes at high voltage between the spray and the MS inlet, which serves two purposes: (i) they separate the gas entering the ion source and the gas sampled by the MS and (ii) they produce electric fields that accelerate the ions toward the MS. However, introducing these electrodes came at the cost of increasing the complexity of the geometry, which became vulnerable to contaminants that accumulated on the surfaces exposed to the analyte vapors, including the electrodes and insulators. The resulting ionizer performed very well when freshly cleaned, and demonstrated the validity of the approach—namely that a reduction in mixing and dilution, and an optimization of the ion transmission, improves sensitivity—but background levels rose rapidly during aging, resulting in only a moderate improvement of the limit of detection.

Based on these preliminary prototypes, the later commercial development, which was named Super SESI and produced by Fossiliontech (Madrid, Spain), aimed at reducing background by using only high purity, inert materials and minimizing the area exposed to the vapors. This was achieved by eliminating the internal electrodes and insulators, which greatly reduced background signals. Maintaining separate flows of the analyte carrier gas and MS injection flow to minimize dilution and maximize ion transmission to the MS without the use of electrodes required a carefully designed internal geometry. In this geometry, laminar flows prevent turbulent mixing and channel the neutral molecules to the ionization region in which the ions are subsequently guided with the aid of electric fields to the MS inlet. A simulation tool that captures all of the relevant processes was developed to understand the configuration in detail, and to optimize the geometry. This model revealed that the friction between the ions and the neutral gas induces a toroidal vortex swirling about the electrospray, which was responsible for turbulent mixing in previous designs. The electrode-free configuration stabilizes this vortex and uses it to enhance ionization efficiency by (i) maintaining laminar flow to prevent turbulent mixing, thus no dilution, and (ii) the vortex acting as a conveyor belt that carries the molecules to the ionization region and the ensuing ions to the MS (Fig. 12.1). In addition,

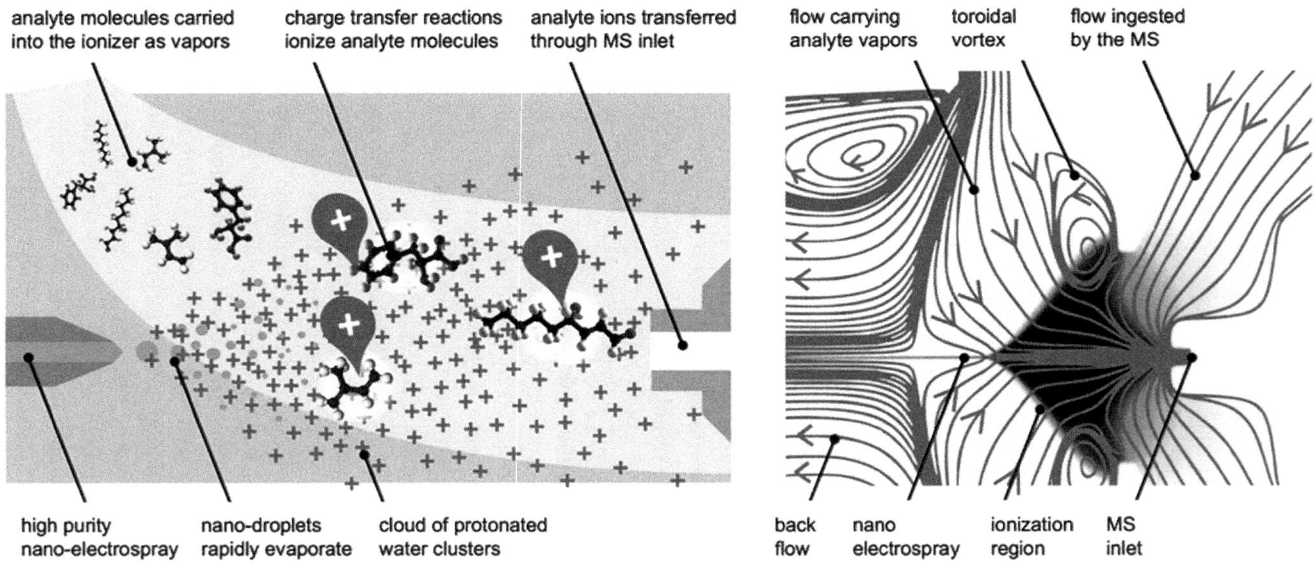

FIGURE 12.1

(Left) Principle of operation of SESI. Analytes enter the ionizer in the vapor phase and react with the charging agents produced by the electrospray to produce analyte ions that are transferred to the MS. (Right) Electrode-free SESI configuration, showing the streamlines (*gray lines*), the analyte ion concentration (dark brown to white), and the different regions of the optimized configuration. Full color available in online version.

a backflow stream is used to sweep contaminants desorbing from the walls away from the ionization region to lower the instrumental background noise.

Other features that improve the background include a design that allows easy cleaning with solvents, an automated flushing of the system with high purity nitrogen during operation, and its operation at high temperature. The combined effect of these features allows very low volatility species to be detected. Today, commercial SESI sources coupled with high-sensitivity mass spectrometers routinely detect species with vapor pressures as low as 2×10^{-7} bar, including fatty acids, fatty alkenals, amino acids,[7] and many more species that remain unidentified. With a dynamic range of 10^5, between 4,000 and 10,000 peaks can be detected in one single exhalation, which highlights the importance of high-resolution MS.

12.2.3 Standardization procedures and data interpretation

While having the possibility to detect thousands of different species is very useful for biomarker discovery, this increases the risk of having spurious correlations that are statistically significant ("false positives"). It is thus of paramount importance to be very vigilant about confounding factors in the design of experimental procedures and studies. In this regard, a key advantage of SESI and other real-time approaches is that no sample manipulation is required, which eliminates some potential sources of noise in the data. The risk of confounding variables being present in the data can be further reduced by restricting the total number of features (analytical signals) being fed to the data-mining algorithms. This can be achieved by defining higher signal thresholds and by considering only the common peaks that appear in a majority of subjects.

Establishing and implementing standardized procedures has been identified as a key need for broadening the use of SESI for breath analysis, for enabling multicenter studies, and for checking the reproducibility of results. In SESI, one approach for standardized sampling is to employ an interface that assists and guides the exhalation maneuver and aligns the ensuing datasets. One such device on the market, the Exhalion (Fossiliontech, Madrid, Spain), is equipped with CO_2 and gas flow sensors, which provides capnography and spirometry data for SESI-MS breath analysis. By synchronizing the CO_2 and SESI-MS data, the end-tidal fraction (see Chapter 2) can be differentiated by integrating MS signals only after the CO_2 concentration rises above a specified threshold, e.g., 3%. A study by Singh et al. that investigated the different sources of data variability in SESI-MS found that the breathing pattern imposed by exhaling through this interface triggers a transient evolution, where consecutive exhalations produce increasing or decreasing signals (species dependent) until they reach a steady state after about six exhalations.[8] For the species that have been identified in the past, this trend strongly correlated with the Ostwald blood–air partition coefficient, $\lambda_{b:a}$, suggesting that this is due to the metabolites originating from within the respiratory tract. This study also showed that once the steady state is reached, the intersubject variability was more important than the intrasubject variability or the time drift of the instrument when the number of features was restricted to 2255 by imposing high intensity thresholds.

12.3 Breath as an alternative biofluid for metabolomics

Metabolomics is increasingly employed to determine the metabolic state of a living organism, and plasma and urine are the most widely used samples for this purpose. Exhaled breath contains a wealth of biochemical information, which has the potential to be used as an alternative to urine and plasma for untargeted metabolomic studies and diagnostics. The validity of using breath strongly depends on the analytical methods. In order to be comparable with standard metabolomics methods, three aspects of SESI breath analysis are pivotal: the metabolome coverage, the quantification accuracy, and the ability of compound identification.

Metabolome coverage can limit the power of a metabolomics study. Unlike fluid-based metabolomics, breath mostly consists of volatile and semivolatile compounds. With SESI, molecular weights up to 1000 Da can be detected in breath, resulting in a relatively deep coverage of metabolites and enabling individual breath phenotyping in real time. In an early study, Sinues et al. used SESI to profile breath from 11 subjects at 36 time points, resulting in a matrix of 2928 metabolic features from these individuals.[9] A principal component analysis/correspondence analysis (PCA/CA) algorithm was then used to reduce the dimensionality and a k-nearest neighbor classifier was trained to classify spectra from different individuals. A recognition score of 76% was achieved in their study, indicating the existence of breathprints that persist over time for individuals. Interestingly, "daily noise" was also observed for several compounds. These fluctuations were attributed to the biological clock, indicating the potential of using SESI for studying circadian rhythms. Although SESI shows great ability to obtain deep metabolite coverage for individual phenotyping, it is worth noting the limitation of SESI in detecting some types of metabolites, for example, nonpolar or large molecules.

The quantification accuracy of SESI breath analysis has been benchmarked against standard metabolomics profiling in plasma.[7] Since most of the molecules that occur in breath are volatile or semivolatile plasma components, low volatility compounds are only present when they have a relatively high concentration in plasma. Consider alanine as an example: the limit of detection by SESI was found to be 0.6 ppt, and with a blood–air partitioning coefficient (k_H) of 6×10^7 mol/L/atm, the equivalent concentration in blood, and thereby minimum concentration at which it can be detected, is 36 ppm.

To characterize low volatility compounds, the method must exhibit high sensitivity. In combination with high-resolution MS (HRMS; see Chapter 16), SESI has the ability to quantify compounds in breath vapor down to the ppt concentration range. Garcia-Gomez et al. demonstrated the detection of eight low volatility amino acids in breath vapor with a coefficient of variation of $<10\%$.[7] The amino acid concentrations in breath correlated well with their concentrations in plasma (with k_H

taken into account), enabling SESI breath monitoring to be used as a noninvasive alternative to determine plasma amino acid concentrations. The study monitored longitudinal amino acid profiles over 12 h and observed changes in concentration over time. Notably, some of the amino acids exhibited higher variations in breath than in plasma, presumably due to their low plasma concentrations, which reach the limit of detection.

Another crucial component of untargeted metabolomics is to determine the chemical identity of the compounds in order to derive biological knowledge from the studies. Since data acquisition is usually made without prior knowledge of compounds, a particular strength of SESI is that thousands of features with distinct *m/z* values can be generated in untargeted metabolomics. When SESI is coupled to HRMS instrumentation, the accurate mass of the individual molecules detected can be measured, thus the elemental composition may be determined. Additionally, fragmentation and tandem MS analysis of the product ions offers elucidation of putative structures.

A study on exhaled breath condensate (EBC) using SESI with ultrahigh-pressure liquid chromatography (UHPLC) MS/MS analysis determined the presence of a series of putative products from the ω-oxidation pathway, as confirmed using synthetic reference standards.[10] Despite the complexity of SESI-MS/MS spectra (see Fig. 12.2), compound identities could be determined by comparing the ionization fragments with those of the standards. If the accurate mass, the MS/MS spectra acquired online by SESI, and the chromatographic retention times all match with data from standards, this gives high confidence in compound identities.

The above-mentioned studies form a solid basis for using SESI in human breath profiling. Breath profiling with SESI can provide detailed and accurate quantification that is comparable to plasma-based metabolomics, and the molecules present in breath can be further identified through comparison with offline methods. These properties of SESI enable its implementation as an online method to perform metabolic state monitoring, drug metabolism monitoring, and biomarker discovery.

12.4 Applications

Over the past decade, SESI has matured from an experimental technology into a tool for biological/clinical research. In the following sections, the state of the art of exhalomics with SESI-MS is highlighted for several important areas of application, including pathway monitoring, drug pharmacokinetic profiling, and biomarker discovery.

12.4.1 Real-time monitoring of metabolic pathways and circadian rhythms

Metabolites are often the first components of an organism that are influenced by external (e.g., environmental) factors. The ability to monitor metabolic changes in

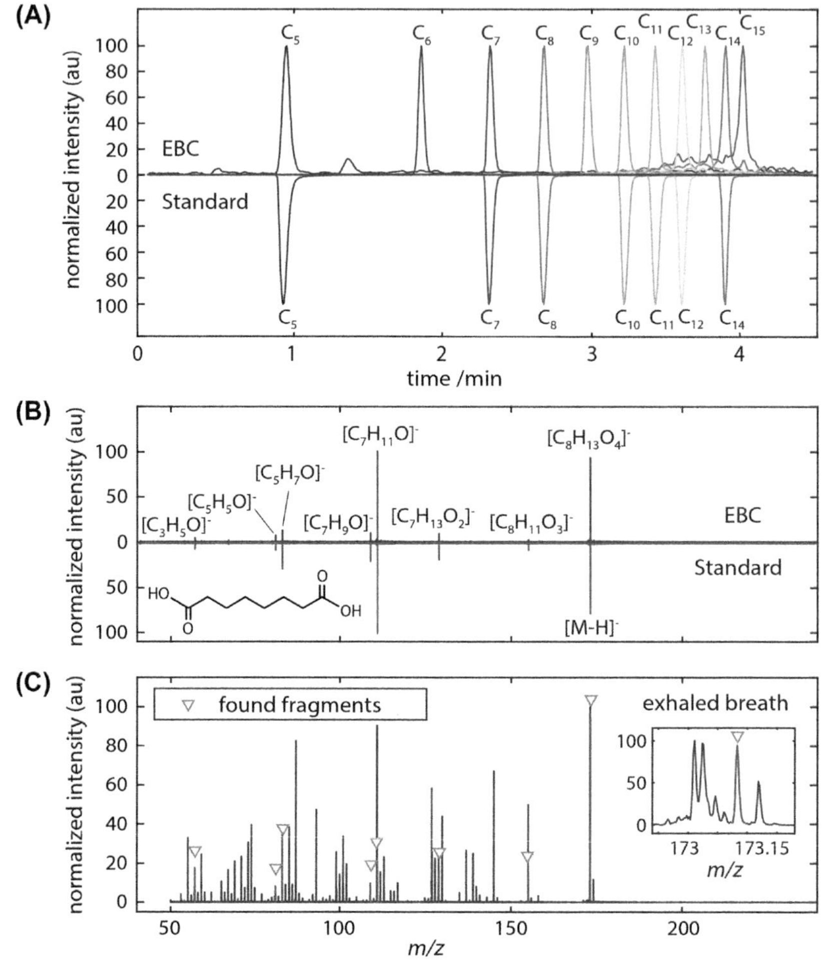

FIGURE 12.2

(A) Chromatograms of the [M-H]$^-$ ions of linear, saturated dicarboxylic acids with chain lengths of C5–C15 in a UHPLC-MS run of exhaled breath condensate (EBC) (top) and chemical standards (bottom). The compounds were confirmed by matching retention time and MS/MS fragments with the standards. (B) MS/MS spectrum of octanedioic acid in EBC (top) and standard (bottom). (C) Online MS/MS spectrum of exhaled breath (precursor: 173.10 ± 0.35 Da). The inset shows a zoom of the precursor mass. As no chromatographic separation was performed, multiple parent *m/z* are selected, explaining the abundance of other fragments. Nonetheless, the high-resolution time-of-flight analyzer makes it possible to clearly identify all fragments of octanedioic acid, supporting the presence of this compound in breath. Full color available in online version.

Reproduced from [10].

real time is key to understanding the dynamics of a biological system. Methods have been established for real-time metabolome profiling in the liquid phase, such as a real-time sampler coupled with LC-MS/MS to monitor bacterial growth. Due to the technical difficulty and the requirement for preseparation, however, these methods are not widely used. SESI-MS is well suited to monitoring metabolic pathways in real time because changes in the body's metabolic state cause fluctuations in metabolite concentrations in breath. Breath samples can be taken in a real-time, noninvasive manner, and then introduced to the MS through SESI without pretreatment. In the aforementioned study by Gaugg et al.,[10] products of the ω-oxidation pathway were identified and characterized, including families of ω-hydroxyalkanoic acids, ω-oxoalkanoic acids, and alkanedioic acids. The signal levels of the identified compounds correlated significantly between or within families, indicating that they are connected by a common transformation cascade. The study provides a basis for using SESI for online metabolic monitoring in which the metabolites of a selected pathway are detectable and identifiable, and their signal intensities are significant and biologically valid.

An in vivo pathway monitoring study using SESI to detect isotopically labeled metabolites from the tricarboxylic acid cycle (TCA) in human exhaled breath was performed by Rioseras et al.[11] Compounds detected included fumaric, succinic, malic, keto glutaric, oxaloacetic, and aconitic acids, and these were found to have high pairwise correlations, indicating that they are regulated by the same metabolic cycle. The measured compounds were also highly variable among individuals, while abundances of oxaloacetic acid, malic acid, and cis-aconitic acid additionally exhibited diurnal variations.

Besides monitoring single metabolic pathways, SESI has been employed to monitor fluctuations in breath relating to circadian rhythms. As previously discussed, human breathprints are stable over extended periods, but compound abundances may exhibit a time-of-day dependency. In a study by Sinues et al., the breath of three individuals was continuously monitored over 24 h[12] using SESI and 111 *m/z* features were selected for circadian analysis. In one particular case, the signal at *m/z* 118.0675 (protonated indole) exhibited not only a circadian variation but also distinct circadian patterns among the three individuals. A partial least squares (PLS) model was trained to predict the breathing time points of individual data points, delivering a correlation coefficient of 0.85 between the predicted and the experimental breathing times. The study provided evidence for the existence of individual-specified circadian rhythm and identified the variation within human breathprints that are introduced by the inner body clock, as well as offered insights into chronobiological processes.

12.4.2 Determining drug pharmacokinetics in breath

Drug pharmacokinetics has long been used as a key criterion in drug approval to determine the clearance of a drug by living organisms. Current methods to determine pharmacokinetics do not offer high time resolution, with only intermittent sampling of biofluids (urine or plasma), or a need to sacrifice animals at each time point for sampling their plasma and/or organs. The time resolution of current methods is therefore limited by the sampling frequency. SESI-MS is a suitable tool for monitoring drug pharmacokinetics because of its high sampling rate and noninvasive nature. Li et al. reported a method to determine drug pharmacokinetics of ketamine with SESI using a mouse model.[13] The signal intensities of ketamine and its metabolites were continuously monitored in the exhaled breath of mice using an accumulation window of 10 s. The half-life measured by SESI in breath agreed well with the values obtained from plasma. Further, the time when ketamine reached its highest abundance in breath was delayed compared to plasma, indicating that the pharmacokinetic pattern obtained from breath reflects the metabolites in the lungs rather than in plasma. The metabolic patterns of primary and secondary metabolites of ketamine can be clearly differentiated in breath by their abundance. This pharmacokinetics study using SESI shows that real-time breath analysis is a promising strategy to monitor drug pharmacokinetics, with advantages in terms of high time resolution, noninvasive sampling, and eliminating interindividual variabilities.

Combining the two approaches of pharmacokinetics analyses with circadian rhythm studies, Sinues et al. used SESI-MS to determine ketamine pharmacokinetics in mice.[14] By comparing the pharmacokinetic profiles of the metabolites of ketamine after drug administration, it was observed that an evening injection of ketamine led to a faster metabolism than a morning injection. This observation was further confirmed with pharmacokinetic profiles of Bmal1 (a mammalian circadian clock master gene) knockout mice, where the wild-type mice showed distinct metabolic profiles with morning/evening injections and the knockout mice did not. Besides ketamine and its metabolites, daily variations of 24 compounds correlated with ketamine. The study suggests that the time of dosing is an important factor to be considered in pharmacokinetics in future clinical development.

In a study reported by Gaugg et al., metabolic changes after inhaling a bronchodilator medication were monitored in human subjects.[15] One group of patients ($n = 50$) inhaled salbutamol and another group ($n = 48$) inhaled a placebo spray as control. Although salbutamol itself could not be accurately quantified with SESI due to issues with mass resolving power, other quantified metabolites were used as indications of the metabolic state after drug administration. A total of 131 m/z features (metabolites) were observed to change following inhalation of salbutamol, while none of these features was significant in the placebo spray control group. In addition to these intergroup differences, a clear intragroup variability was identified, reflecting the diverse drug responses of individuals. Significant metabolites were clustered into several groups according to their metabolic patterns, whereby the compounds in each group were chemically similar. By using UPLC-MS/MS, compounds within one of the groups were structurally assigned to fatty acids, indicating lipolysis stimulation after inhalation of salbutamol, which is in agreement with previous data obtained in plasma.[16] Although the pharmacokinetic profile of the drug was not obtained in the study, it demonstrates the ability of breath analysis by SESI to track metabolic processes triggered by drug intake, and to identify individual drug metabolism patterns. The study further highlights the need for a sufficiently high mass resolution when employing MS in order to characterize a targeted drug unambiguously in human breath.

12.4.3 Identifying biomarkers and diagnosis of diseases

The initial focus of metabolomics studies was on biomarker discovery, which aims to identify metabolites that correlate with certain diseases or environmental exposures. In this context, it is attractive to perform in situ diagnostics in a noninvasive manner, which perfectly fits the scope of SESI-MS. As in metabolomics in general, biomarker discovery in breath suffers from two shortcomings, namely poor biomarker identification and limited validation. Several aspects of breath analysis make biomarker discover a particular challenge: (1) the composition of breath is highly affected by medication, lifestyle, and environmental factors, (2) specific biomarkers identified in other biofluids might be below the limit of detection in breath analysis, and (3) there is a difference between in vitro discovery and in vivo diagnostics. Attention must be paid to study designs and statistics. To illustrate this, five examples of biomarker discovery for several diseases using SESI-MS are discussed, namely: bacterial lung infections, cystic fibrosis, breast cancer, chronic obstructive pulmonary disease (COPD), and lung cancer.

An early study to detect bacterial lung infections was reported by Zhu et al.[17] The authors first obtained breathprints of three strains of bacteria in vitro, and then used these breathprints to identify the infecting strains in vivo. Three strains were used in this study, *P. aeruginosa* PAO1, *P. aeruginosa* FRD1, and *S. aureus* RN450. The volatilome of different strains of bacteria were profiled using cultures, and breath was measured in mice infected with these strains and in uninfected controls. Significant *m/z* features were found for all bacterial cultures and mice models, but surprisingly only ~30% of those features overlapped between the bacteria and the corresponding mice infection models. By comparing the infected mice and uninfected controls, 21 features were identified as breathprint markers of bacterial infection, regardless of the infection strains. Nevertheless, 18 *m/z* features were shown to be unique in mice infected with *S. aureus*, 19 *m/z* peaks were common between *P. aeruginosa* PAO1 and *P. aeruginosa* FRD1, and 10 *m/z* peaks showed distinct behavior between the two species. Though restricted to bacteria cultures and mice models, the study provided a method to identify infecting pathogens at the strain level, as well as illustrating the gap between biomarker discovery using in vitro and in vivo models. The value of in vitro cell-based studies is further discussed in Chapter 26.

Breath analysis using SESI has been employed to investigate lung infection in patients with cystic fibrosis (CF), as reported by Gaisl et al.[18] A total of 3273 *m/z* features were detected in the study, of which 49 features exhibited significant differences between CF patients and healthy controls. UPLC-MS/MS and in silico fragmentation was used to identify the 49 potential biomarkers and ascertain the crucial physiological changes driving the observed differences. At the molecular level, fatty acids and analogs differed in the breath of CF patients, which is in line with previous findings in tissue.[19] To identify the infecting bacteria, breath signals from patients were also correlated with six bacterial strains, and 11 *m/z* features were found to correlate significantly between patients and a *Stenotrophomonas maltophilia* culture. A support vector machine model was subsequently generated using the breath signals to predict CF, with a prediction accuracy of 77.1% obtained. This study generated molecular level information as a valuable resource for disease diagnosis, monitoring, and therapy guidance.

Besides study designs and technical considerations, the validity and robustness of identified biomarkers highly depends on the selected statistical approach used to evaluate SESI-MS data. Statistical approaches must be carefully adapted for biomarker discovery with SESI-MS. A study reported by Sinues et al. provided a bioinformatics workflow to identify biomarkers in exhaled breath of breast cancer patients ($n = 14$) versus healthy controls ($n = 11$).[20] Breath signals from SESI-MS analysis were analyzed using an informatics workflow. The breathprint data initially underwent quality assessment, feature filtering, batch correction, and missing-data imputation in order to eliminate technical variations while preserving the biological variation to the maximum extent. Of the 296 *m/z* features extracted from breath signals and retained after preprocessing, the dataset was reduced to 35 significant *m/z* features between the groups after a Wilcoxon class comparison test was applied. A linear support vector machine model was trained with a bootstrap feature selection in order to predict breast cancer, and the classification accuracy was evaluated by a leave-one-out (LOO) cross validation. The final classifier achieved a sensitivity of 0.93 and a specificity of 0.91, and included eight *m/z* features that could be used to further functional interpretation. The bioinformatics approach developed in this study provided a general guideline for breath analysis acquired by SESI-MS and an indication for choosing suitable statistical tools to discover robust biomarkers among breath metabolites. Further discussion on data-mining approaches is covered in Chapter 38.

In addition to applications for biomarker discovery, SESI-MS can be utilized to monitor disease-induced perturbations of an organism at the molecular level. Gaugg et al., for example, identified biomarkers for COPD with or without exacerbations.[21] Using a similar computational workflow approach to that outlined above, 545 *m/z* features were found to be different between the group with frequent exacerbations and the nonexacerbation group. Of these, eight *m/z* features were selected as predictors of exacerbations using a relief feature–selection algorithm. Most of these features were fatty acids that belong to the ω-oxidation pathway, which were found to be downregulated in the exacerbation group. This fact connects the increasing leukotriene B4 and prostaglandins in inflammatory processes and the disturbance of ω-oxidation as their major degradation pathway. Another class of compounds that was significantly altered between the two groups was the nitro-aromatic compounds. Compounds of this class are all upregulated in the exacerbation group, indicating a promoted bacterial cell killing and increased airway inflammation. In this study, biomarkers for exacerbations were identified in a population of COPD patients, while metabolic changes were also deduced from breath data. Breath biomarkers for COPD is further discussed in Chapter 19.

Beyond breath, SESI-MS can be utilized to characterize small molecules in other biofluids for biomarker discovery. Gao et al.,[22] for example, used SESI-MS to pinpoint the presence of non—small cell lung cancer in human sputum. Compounds, including glutamine, phenylalanine, tyrosine, and glucose were characterized, and arginine, were quantified in sputum samples with external calibration. By using PCA on the breath data, sputum signals from lung cancer patients were distinguishable from sputum signals of healthy controls. Phosphatidylcholine and dipalmitoyl phosphatidylcholine were identified to be the main components responsible for differentiation. Although the study does not contain strict statistics for biomarker discovery or validation, it demonstrated the possibility of achieving absolute quantification with SESI-MS, which is particularly valuable for developing further diagnostic methods based on biomarkers identified by SESI.

12.5 Summary

The ability to monitor metabolic changes in real time is key to understanding the dynamics of perturbations in biological systems, which could be a manifestation of disease. Exhaled breath is an excellent medium for this purpose: it is readily available, can be sampled in a noninvasive manner that is painless for the patient, and presents a less complicated matrix compared to blood or urine. Metabolites from the blood partitioning via the alveoli into the airways depends on their volatility, polarity, solubility in fatty tissue, and Henry's partition constant. Many compounds, however, are only present in exhaled breath at very low concentrations, in the ppb to ppt range. SESI, in particular in combination with state-of-the art, high-resolution atmospheric pressure inlet mass spectrometers, has shown promise for the online analysis of exhaled breath due to its high sensitivity and its ability to detect compounds with high molecular weights. The range of compounds that are ionized well in SESI are partly complementary to those that are detected with other real-time chemical ionization MS techniques, such as selected ion flow tube MS (SIFT-MS; see Chapter 9) or proton transfer reaction MS (PTR-MS; see Chapter10). SESI-MS is increasingly used in the clinical setting and beyond. The first step in most clinical studies is an exploratory study, for example, a cross-sectional or longitudinal study. During the data acquisition, continuous quality monitoring is necessary. Data processing converts, filters, and normalizes the data, which is then analyzed by statistical methods to extract candidate biomarkers. The molecular identity of the structure of the extracted "features" is then elucidated, with the goal of achieving unambiguous compound identification. Once the molecules are identified, a metabolomics analysis of the identified panel of molecules is often a logical next step. Such exploratory studies designed to identify a panel of putative biomarkers for a certain condition (e.g., the presence of a disease) should then be followed by a validation study to generate clinically meaningful biomarkers. Once this has been accomplished, a quantitative assay can be developed for implementation into daily clinical practice. At the moment, SESI-MS is not at this stage, however, a growing role for SESI-MS in clinical applications is anticipated, especially once the correlation with blood-based metabolomics has been established, biomarkers have been validated, and standardization and quantitation issues have been resolved.

Conflict of interest

Francisco Blanco and Guillermo Vidal work for the company Fossiliontech, which commercializes the Super SESI source and the Exhalion interface.

References

1. Gamez G, Zhu L, Disko A, Chen H, Azov V, Chingin K, et al. Real-time, in vivo monitoring and pharmacokinetics of valproic acid via a novel biomarker in exhaled breath. *Chem Commun* 2011;**47**(17):4884–6.
2. Sinues PM, Fernández de la Mora J. Direct analysis of fatty acid vapors in breath by electrospray ionization and atmospheric pressure ionization-mass spectrometry. *Anal Chem* 2008;**80**(21):8210–5.
3. Sinues PM, Criado E, Vidal G. Mechanistic study on the ionization of trace gases by an electrospray plume. *Int J Mass Spectrom* 2012;**313**:21–9.
4. Berchtold C, Meier L, Steinhoff R, Zenobi R. A new strategy based on real-time secondary electrospray ionization and high-resolution mass spectrometry to discriminate endogenous and exogenous compounds in exhaled breath. *Metabolomics* 2014;**10**(2):291–301.
5. Vidal-de-Miguel G, Herrero A. Secondary electrospray ionization of complex vapor mixtures. Theoretical and experimental approach. *J Am Soc Mass Spectrom* 2012;**23**(6):1085–96.
6. Vidal-de-Miguel G, Macia M, Pinacho P, Blanco J. Low-sample flow secondary electrospray ionization: improving vapor ionization efficiency. *Anal Chem* 2012;**84**(20):8475–9.
7. Garcia-Gomez D, Gaisl T, Bregy L, Cremonesi A, Sinues PM, Kohler M, et al. Real-time quantification of amino acids in the exhalome by secondary electrospray ionization-mass spectrometry: a proof-of-principle study. *Clin Chem* 2016;**62**(9):1230–7.
8. Singh KD, Tancev G, Decrue F, Usemann J, Appenzeller R, Barreiro P, et al. Standardization procedures for real-time breath analysis by secondary electrospray ionization high-resolution mass spectrometry. *Anal Bioanal Chem* April 15, 2019:1–6.
9. Sinues PM, Kohler M, Zenobi R. Human breath analysis may support the existence of individual metabolic phenotypes. *PLoS One* April 3, 2013;**8**(4):e59909.
10. Gaugg MT, Bruderer T, Nowak N, Eiffert L, Martinez-Lozano Sinues P, Kohler M, et al. Mass spectrometric detection of omega-oxidation products of aliphatic fatty acids in exhaled breath. *Anal Chem* September 14, 2017;**89**(19):10329–34.

11. TejeroRioseras A, Singh KD, Nowak N, Gaugg MT, Bruderer T, Zenobi R, et al. Real-time monitoring of tricarboxylic acid metabolites in exhaled breath. *Anal Chem* May 16, 2018;**90**(11):6453–60.
12. Sinues PM, Tarokh L, Li X, Kohler M, Brown SA, Zenobi R, et al. Circadian variation of the human metabolome captured by real-time breath analysis. *PLoS One* December 29, 2014;**9**(12):e114422.
13. Li X, Sinues PM, Dallmann R, Bregy L, Hollmén M, Proulx S, et al. Drug pharmacokinetics determined by real-time analysis of mouse breath. *Angew Chem Int Ed* June 26, 2015;**54**(27):7815–8.
14. Sinues PM, Kohler M, Brown SA, Zenobi R, Dallmann R. Gauging circadian variation in ketamine metabolism by real-time breath analysis. *Chem Commun* February 14, 2017; **53**(14):2264–7.
15. Gaugg MT, Engler A, Nussbaumer-Ochsner Y, Bregy L, Stöberl AS, Gaisl T, et al. Metabolic effects of inhaled salbutamol determined by exhaled breath analysis. *J Breath Res* September 13, 2017;**11**(4):046004.
16. Oomen JM, van Rossum CT, Hoebee B, Saris WH, Van Baak MA. β2-Adrenergic receptor polymorphisms and salbutamol-stimulated energy expenditure. *J Clin Endocrinol Metabol* April 1, 2005;**90**(4):2301–7.
17. Zhu J, Bean HD, Wargo MJ, Leclair LW, Hill JE. Detecting bacterial lung infections: in vivo evaluation of in vitro volatile fingerprints. *J Breath Res* January 10, 2013;**7**(1): 016003.
18. Gaisl T, Bregy L, Stebler N, Gaugg MT, Bruderer T, García-Gómez D, et al. Real-time exhaled breath analysis in patients with cystic fibrosis and controls. *J Breath Res* April 30, 2018;**12**(3):036013.
19. Njoroge SW, Laposata M, Katrangi W, Seegmiller AC. DHA and EPA reverse cystic fibrosis-related FA abnormalities by suppressing FA desaturase expression and activity. *J Lipid Res* February 1, 2012;**53**(2):257–65.
20. Sinues PM, Landoni E, Miceli R, Dibari VF, Dugo M, Agresti R, et al. Secondary electrospray ionization-mass spectrometry and a novel statistical bioinformatic approach identifies a cancer-related profile in exhaled breath of breast cancer patients: a pilot study. *J Breath Res* September 21, 2015;**9**(3):031001.
21. Gaugg MT, Nussbaumer-Ochsner Y, Bregy L, Engler A, Stebler N, Gaisl T, et al. Real-time breath analysis reveals specific metabolic signatures of COPD exacerbations. *Chest* 2019;**156**(2):269–76.
22. Gao XF, Xiao Y, Dai Y. Direct analysis of human sputum for differentiating non-small cell lung cancer by neutral desorption extractive electrospray ionization mass spectrometry. *Anal Sci* September 10, 2018;**34**(9):1067–71.

CHAPTER

Sensor systems

13

Hossam Haick[1], Rotem Vishinkin[1], Corrado Di Natale[2], Santiago Marco[3,4]

[1]*Department of Chemical Engineering and Russell Berrie Nanotechnology Institute, Technion – Israel Institute of Technology, Haifa, Israel;* [2]*Department of Electronic Engineering, University of Rome Tor Vergata, Rome, Italy;* [3]*Department of Electronics and Biomedical Engineering, University of Barcelona, Barcelona, Spain;* [4]*Signal and Information Processing for Sensing Systems, Institute for Bioengineering of Catalonia, The Barcelona Institute of Science and Technology, Barcelona, Spain*

Chapter outline

13.1 Overview

Optical spectroscopy and mass spectrometry are effective methods for identification and quantification of the human volatilome. Despite their exceptional performance, however, these techniques have significant limitations that prevent their extended usage in the medical practice. Expensive and bulky apparatuses are the primary reason.

Breathborne Biomarkers and the Human Volatilome. https://doi.org/10.1016/B978-0-12-819967-1.00013-X
Copyright © 2020 Elsevier B.V. All rights reserved.

In addition, the operation of these instruments requires a high level of expertise, as well as preconcentration techniques and time-consuming analysis in some cases. Furthermore, due to variations in the available analytical tools and different sampling methods, it is challenging to determine unique patterns of volatile organic compounds (VOCs) for individual diseases. Utilization of the obtained results promotes affordable and field-orientated methods, including development of nanomaterial-based sensor arrays for volatilome detection. Nanomaterial-based sensor arrays are excellent candidates for disease diagnosis and screening as point-of-care (PoC) tools and treatment monitoring in real-world conditions. Such systems have exceptional advantages, which include (1) low-cost fabrication and operation; (2) easy, low-powered and portable operation; (3) increased specificity and sensitivity by an informed choice of nanomaterials and exploitation of synergistic sensing effects; and (4) large surface-to-volume ratios of the nanomaterial signals, providing higher sensitivity and faster response and recovery times.

In order to promote the development of sensor arrays toward real applications, the following threshold requirements must be fulfilled: (1) well-known stability and repetitive responses; (2) ultrasensitive VOC detection in humidity-saturated environments (e.g., exhaled breath and sweat); (3) elimination of effects of confounding factor on the results (e.g., due to sex, viral infection, etc.). This chapter introduces the topic of nanoarrays for the detection and classification of diseases using volatilomic biomarkers, reviews recent advances, and discusses the development of validated predictive models toward disease diagnosis.

13.2 Translating analytical volatile organic compound data for sensor development

Harnessing the VOC data obtained from variations of chromatography and/or spectrometry methods allows each compound to be studied separately and used to estimate the likelihood that it can serve as an individual biomarker of disease or as a part of a disease-associated pattern. Broza et al.[1] have summarized the available literature to map disease-related VOCs and their origin. Such maps aid in sensor development, choice of sensing layers for the relevant health condition, and development of suitable sampling methods in relation to VOC sources that fit real-world clinical use. As an example, halitosis is diagnosed only from exhaled breath, and most of the compounds are organosulfur, which focuses the developers toward selective or semiselective sensing layers for volatile sulfur compounds (VSCs).

In order to achieve detection of VOCs by sensors, numerous modified nanomaterials have been reported as transduction elements, including nanoparticles (NPs), silicon nanowires (SiNWs), carbon nanotubes (CNTs), metal oxides (MOxs), polymers, graphene, and many others.[1,2] Recognition of the VOCs can be achieved by two main sensing strategies (see Fig. 13.1): (1) selective detection of (pre-)identified VOCs (lock-and-key like) or (2) semiselective/cross-reactive sensors in conjunction with pattern recognition algorithms, which are sensitive to a wide spectrum of VOCs.

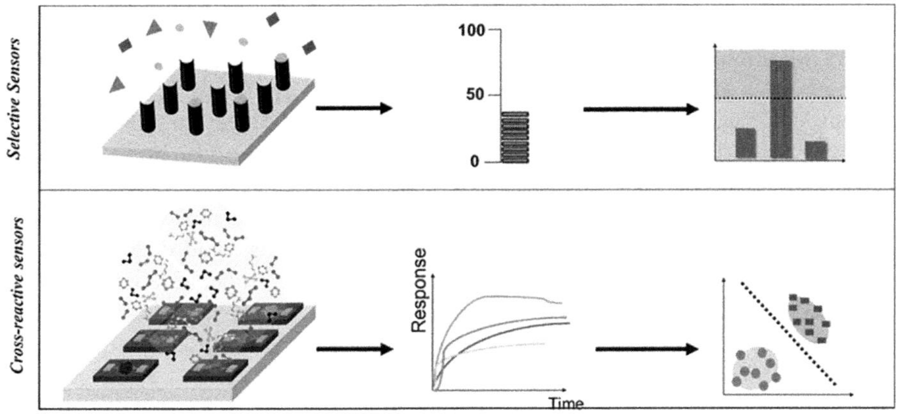

FIGURE 13.1

Schematics illustrating nanomaterial-based sensors for detecting disease-related volatile organic compounds: (top) selective nanoscale sensing approach and (bottom) cross-reactive sensing approach.

13.2.1 Selective nanomaterial-based sensors

The approach of using nanomaterial-based sensors focuses on the detection of specific and well-defined targeted VOCs in the presence of background gaseous species in human samples (Fig. 13.1, top). In order to target a specific preidentified VOC, there is a need for fabrication of highly selective nanomaterials. Selectivity can be obtained with respect to a restricted number of gases or for selected sensitive materials; for instance, palladium (Pd) layers for hydrogen (H_2) detection is one of those cases.

To design selective sensors, it is highly conducive to observe how nature develops selective receptors. Generally, these are sequences of peptides where nonspecific interactions, such as hydrogen bonds, are spatially arranged in order to meet the steric configuration of the molecule of interest. Molecular imprinting is a somewhat simple method to implement the same approach toward an increase in sensitivity of gas sensors.[3] In nature, the development of selective receptors is performed for molecules of fundamental importance, such as pheromones for insects or molecular transmitters among cells. In these cases, the type and number of target molecules is very well defined and has remained unchanged for most of the evolution of the species. In the literature, there are several examples for detection of specific VOCs by selective sensors, including nitric oxide (NO), ammonia (NH_3), acetone, and hydrogen sulfide (H_2S). Other examples of selective detection of VOCs can be found elsewhere.[1] Nevertheless, currently there are no VOCs that can be used to uniquely diagnose a single disease, which limits the use of this approach for a small number of potential VOCs to be detected in breath. An additional limitation is driven from the need to synthesize separately selective nanomaterials for each VOC, mainly for nonpolar VOCs.[4] On top of this, the majority of nanomaterial-based sensors require high operation temperatures or low humidity conditions, which confines their usability as a PoC apparatus for field practice.

13.2.2 Sensor arrays for volatilomic applications

Arrays of semiselective/cross-reactive sensors for volatilomics are an emerging strategy. In conjunction with predictive models, this approach is able to detect various VOCs with a single sensor array (Fig. 13.1, bottom).[5,6] Each sensor responds differently to individual VOCs in the sample, which aids in the evaluation of the volatilome in a qualitative and semiquantitative manner, while selectivity is achieved by predictive methods. Despite the lower sensitivity of such systems to a specific analyte compared to a selective sensor, this approach allows a rapid diagnostic when detecting multicomponents under variable humidity and temperature conditions or in the presence of other confounding parameters. While this approach lacks the ability to identify each VOC separately, VOCs can be detected and used for future classification of similar patterns. In cases of multiple VOC changes, as in systemic diseases, this is the most suitable of the two approaches.

Characterizing such arrays begins with exposing the array to "artificial mixtures" that simulate the key components in real samples along with exposure to single VOCs. This step provides a comparison of the characteristics of the sensors in terms of selectivity, sensitivity, and specificity in response to single VOCs versus the same analyte mixed with other compounds. Furthermore, this allows iterative feedback to be generated during the array optimization without the intervention of disruptive parameters, such as humidity and temperature. After the first step, the array is exposed to real samples, allowing further improvements of the sensor array and the predictive model in order to achieve a successful identification of compounds in the sample of interest.

13.3 Sensors for volatile organic compound detection

Modern gas sensors are equivalent to electronic devices whose parameters are variable—rather than fixed—in relation to the chemical composition of the atmosphere at which the sensors are exposed.[7] Sensor development benefits from progress made in microelectronics, thus the development of the latter is considered of utmost importance in order to expand the analytical capabilities of sensors toward low-cost and ubiquitous systems. Such advances are considered of great importance in the medical field to accelerate diagnoses and to implement the principles of personalized medicine. For these reasons, medical applications are driving large parts of the current development of gas sensors.[8] Independent of their nature, gas sensors can be represented as the combination of a sensitive material with a basic sensor, as shown in Fig. 13.2.

The sensitive material is a solid-state layer, either organic or inorganic, that interacts with compounds in the target gas. These interactions induce a change in certain physical properties of the sensitive material. A basic sensor is then applied to detect one of these quantities. Although there is a countless number of sensitive materials, in practice, the physical properties for which a basic sensor is available are

FIGURE 13.2

Schematic structure of a generic gas sensor: The target gas interacts with a sensitive material and as a consequence some physical properties of the material change. This could be the mass (Δm), the surface potential ($\Delta \Phi$), the conductivity ($\Delta \sigma$), the dielectric constant ($\Delta \varepsilon$), or the optical absorbance ($\Delta \alpha$). A basic sensor is then applied to transduce one of these quantities into the parameter of an electronic device. These can be resistance (ΔR), capacitance (ΔC), or transconductance (Δg_m). Finally, the basic sensor is connected in an electronic circuit to establish signals (e.g. a voltage [V]) that are a function of the concentration of the target gas constituents.

limited to electrical impedance, surface potential, mass, and optical spectra (absorbance and/or emission). Since basic sensors are few in number and type, their most significant and essential part is the sensitive material. An important requirement of this material is that it must be both sensitive and selective in relation to the target species and technologically compatible with the basic sensor. The principles of gas sensors are numerous, but they can be classified into two main families:

(1) *Adsorption-based systems*: The adsorption of molecules onto a sensing layer can be driven by a number of different interactions. These are traditionally divided into physisorption and chemisorption according to the strength of the forces involved. Physisorption is ruled by weak electrostatic dipolar interactions, including dipole–dipole, dipole–induced dipole, or induced dipole–induced dipole. A particular kind of physisorption is the hydrogen bond between molecules containing hydrogen atoms bonded to electronegative atoms. Chemisorption may occur as a consequence of adsorption; it consists of a chemical bond engaged between the sensing layer and the adsorbed airborne molecule.

(2) *Catalysis-based systems*: The other family of gas sensors includes those based on heterogeneous catalysis in which the target gas interacts with a molecule (usually oxygen) previously bonded onto the sensitive material surface. Adsorption processes are typical in organic-sensitive materials (molecular or polymer films), while heterogeneous catalytic processes are more pertinent to metals and MOxs. It is noteworthy to state that catalysis is temperature activated.

Interactions with the sensing layers are detected through the measurement of a physical quantity whose magnitude is affected by the interaction. For each quantity, there is a number of basic sensors that can be coupled with the sensitive layer and connected to an appropriate measurement circuit. In most situations, the physical quantity is proportional to the number of absorbed molecules. As such, the isotherm describing the adsorption onto the sensing layer is readily translated into the magnitude of the physical quantity. However, the physical quantity introduces an additional variable. In mass sensors, the sensor response is the product of the number of adsorbed molecules and their respective molecular mass. In conductive sensors, variation in conductivity can be affected, among other causes, by the number of electrons exchanged in a chemisorption process, the electric dipole of adsorbed molecules, or even the change of volume of the sensitive material. Often a single target molecule can elicit several effects.

Sensitivity and selectivity are the most important features of gas sensors. Sensitivity is defined as the derivative of the sensor response with respect to the concentration of a target compound. Selectivity is achieved when the sensitivity with respect to one of the species dominates over any other. Considering the described principles of interactions and the multitude of chemical species, the development of selective gas sensors is complicated. In breath analysis, with the exception of some restricted cases (e.g., acetone or ammonia), the number and type of molecules involved with a particular condition or disease is very large and the clinical significance of very similar molecules may be different. For instance, the correlation of the abundance of methylated alkanes with a disease can change with the position of the methyl group.[9] Consequently, the identification of a disease, such as lung cancer, is not related to well-specified molecules, but rather to the molecular pattern.[10] This picture is similar to the working principle of the sense of olfaction. Olfactory receptors did not evolve into specific systems, rather they preserved a certain cross-selectivity, which makes each of them sensitive to multiple molecules such that the same molecule can be sensed by many receptors. Odors are univocally identified by response patterns elicited by more than one kind of receptor in a process that is known as combinatorial selectivity. The power of combinatorial selectivity is well described in humans, where a reservoir of little more than 300 different receptors can distinguish over a trillion odors of different quality and quantity.

The application of the combinatorial selectivity paradigm to gas sensors began with the seminal work of Persaud and Dodds[11] and resulted in the current concept of electronic noses. The main process in the electronic nose is the compression of the pattern of volatile compounds into the pattern of sensors signals (see Fig. 13.3). Of course, the compression reduces the information content, therefore

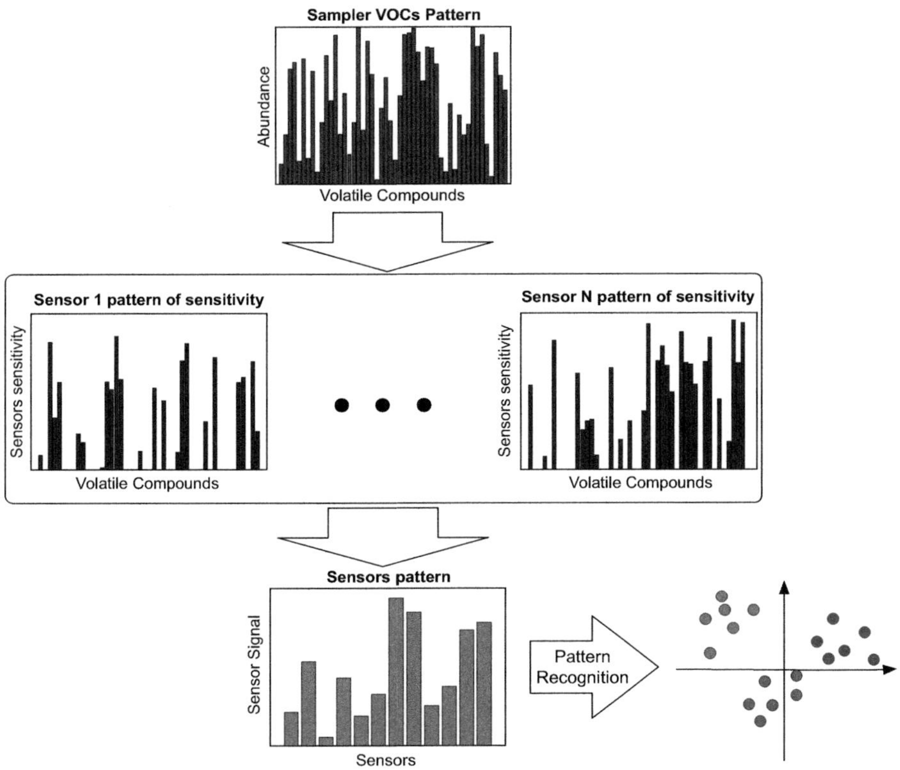

FIGURE 13.3

Principle of combinatorial selectivity: The pattern of VOCs in the environment meets the sensitivity pattern of each sensor of the array. As a consequence, each sensor expresses a response. The collection of responses forms a pattern of signals that encodes the totality of VOCs into a small dimensional vector. If the relevant information is retained in the process, the application of a pattern recognition algorithm can sort the patterns into classes, with each class representing a meaningful category of samples. In breath analysis, the samples are the breath of individuals and the classes should correspond to the diagnosis.

it is meaningful only if the relevant information in the volatile compounds pattern is retained by the sensor array. As such, sensors are required to be sensitive to the more important compounds and their selectivity toward these compounds must be sufficiently different for each sensor. The latter requirement necessitates sensitive materials where the sensitivity can be easily adjusted with minimal changes of the synthesis process. Organic chemistry offers infinite possibilities to configure molecules to achieve this requirement. A rational process can be provided by some molecular families, such as porphyrinoids. These are synthetic variations of porphyrins, a molecule used by nature for fundamental tasks, such as electron transfer, oxygen carrier, and catalysis. The sensitivity of porphyrins can be tuned to change their basic constituents, such as the aromatic ring, the metal atom complexed at the center of the macrocycle, and the lateral substituents.[12]

13.3.1 Conductive sensors

In some sensing materials, interaction with gases or volatile compounds changes their conductivity. Conductive sensors share a common structure of an insulator substrate on which conductive electrodes are deposited. The sensitive material is then deposited on top of the substrate and electrodes. The overall resistance depends not only on the conductivity of the sensitive material but also on the geometry of the electrodes pair. Often, in order to adjust the overall resistance, electrodes are shaped as an interdigitated pair. Depending on the requirements for sensor features, the electrodes may have different size and spacing, ranging from the nanometer to the micrometer scale.

Metal oxide sensors

Changes in conductivity are more easily measurable in materials characterized by a limited number of mobile charge carriers, such as semiconductors. In these materials, even modest variations of the number of charge carriers may cause large changes in conductivity. The standard materials used for gas sensing are transition MOxs. The majority charges define the main character of the gas sensors: *n-type* (such as zinc oxide, tin dioxide, titanium dioxide, or iron (III) oxide) are mainly sensitive to reducing gases, i.e., low oxidation potential, and *p-type* (such as nickel oxide, cobalt oxide) respond most to oxidizing gases. Among those, the most studied and widely applied are n-type semiconductors, in particular tin dioxide (SnO_2).

Metal oxide sensors work on the principle of surface distribution of oxygen atoms. At sufficiently high temperature (typically $>200°C$), dissociative adsorption sites for molecular oxygen (O_2) are activated on the oxide surface. Oxygen is ionosorbed and forms a surface concentration of negative charges provided by the conduction band of SnO_2. This results in a surface depletion of electrons, with a consequent surface barrier analog to the Schottky barrier observed in a metal–semiconductor junction. The adsorption of O_2 elicits a decrease in the conductivity of the semiconductor. In the presence of a reducing gas, the interaction of the molecules with the surface oxygen causes a decrease in the concentration of absorbed O_2 molecules. The electrons previously transferred to the adsorbed oxygen are then injected back to the conduction band, and the conductivity of the semiconductor increases. These factors make the electric resistance of the material dependent on the gas concentration. The sensitivity can be modified by adding tiny layers of catalytic metal atoms on the surface. The sensors are kept at the necessary temperature by a heater located on the back of the substrate. The necessity to heat the sensor is reflected in the larger power consumption compared to other sensors.

Due to their widespread availability, SnO_2-based sensors have been used in a number of different applications, including breath analysis. For instance, an array of these sensors was demonstrated to be selective with respect to various respiratory diseases, such as chronic obstructive pulmonary disease (COPD) and lung cancer.[13] A nanoarray composed of four different metal-doped SnO_2 films was used for detecting formaldehyde in a gaseous mixture of different VOCs, such as acetone, ethanol, and NH_3, with 90% relative humidity, mimicking human breath. The detection limit of the array was 3 ppb for formaldehyde at 400°C.[14] In another study, loaded Pd−SnO_2 yolk−shell sensing film coated with a thin catalytic Co_3O_4 layer presented high selectivity toward benzene, toluene, xylene, and ethanol.[15]

Tin oxide is only one of many materials with interesting sensing properties. For instance, ε-WO_3, a crystalline phase of tungsten oxide, was found to have enough sensitivity and selectivity to enable detection of acetone in breath.[16] Villi-like WO_3 nanostructure-based sensor presented high and selective responses toward low concentration of NO that is associated with asthmatic disorders.[17] Arrays of MOx have also been used in clinical studies for detecting different diseases from both in vitro and breath samples, including breast cancer, prostate cancer, squamous cell carcinoma of the head, bacteria detection, heart failure, Alzheimer's disease, diabetes mellitus, and others.[1]

Metal oxides can also perform as a transistor sensor. Most of the properties of metal−oxide field-effect transistors (MOSFETs) depend on the difference between the work function of the metal gate and the semiconductor. This difference can be modulated by the presence of a layer of electric dipoles at the metal−oxide interface. The adsorption of electric dipoles at this interface results in a voltage drop that is additive with respect to the gate voltage. The overall effect in the MOSFET is a change in the drain-source current, even if the applied gate voltage remains constant. The first chemically sensitive MOSFET sensors were hydrogen (H_2)-sensitive devices with a palladium gate.[18] Palladium (Pd), like other noble metals, has catalytic properties that allow the H_2 molecules to dissociate into atomic hydrogen at the gate surface. In palladium, however, hydrogen atoms can diffuse through the metal to reach the oxide surface, where they form an ordered layer of electric dipoles. Anomalous concentrations of H_2 in breath may signal several diseases. PdFET is the device of choice for sensitive and selective detection of H_2. The original design evolved from replacing silicon with different materials, such as graphene. The introduction of nanostructured semiconductors, such as SiNWs, largely extended the appeal of FET structures for chemical sensors development.

Nanomaterial-based conductive sensors

The advantages of nanostructured materials in chemoresistors is evident in MOx semiconductors. In these sensors, the active region is the surface depletion layer formed by ionosorption processes. Nanostructured sensors in the form of nanocrystals, or more recently, nanowires, or nanoribbons, completely eliminate the bulk of the semiconductor, so all of the material contributes to the gas sensing.[19] Nanostructures can be prepared with different materials and in different shapes. Hybrid materials formed by inorganic NPs (MOx semiconductors or metals) coated by an organic layer are particularly interesting.

In contrast, monolayer-capped metallic nanoparticles (MCNPs) are fully exploited for volatilome studies. In MCNP, the metal core provides the electronic conductivity, while the nonconductive organic matrix provides selective binding sites for organic molecules. MCNPs with gold (Au) and platinum (Pt) cores are the preferred choice for application in chemiresistors, even though a wider variety of metal core or semiconducting material types can be considered.[1,20] Other parameters affecting the sensing include the particle shape (i.e., sphere, cube, star, rod), size (1−100 nm), and fabrication method. The sensing layer composition (i.e., alkylthiols, dialkyl disulfides, DNA, proteins, alkanethiols, and others) is critical, and numerous reports have been published.[1] Film morphology, which affects sensing properties, is another critical factor. MCNPs can be deposited on various substrates, both solid and flexible.[1] Development of flexible MCNP sensors is currently at the peak of growth, as is the integration of such sensors in various systems. Furthermore, MCMNPs can be integrated into single variable sensors, such as chemiresistors, as described here, or into multivariable sensors, such as resonant or nonresonant multivariable electrical sensors and material- or structure-based multivariable photonic sensors.[21]

The application of MCNPs to breath volatilome studies is broad and well-studied. Arrays of MCNPs linked with various sensing layers were tested for detection of a wide range of diseases, including lung cancer, hepatocellular carcinoma, breast cancer, ovarian cancer, gastric cancer, colorectal cancer, head and neck cancer, pulmonary arterial hypertension, intestinal diseases, renal diseases, Parkinson's and Alzheimer's diseases, multiple sclerosis, tuberculosis, and many others.[1]

Other important materials used in nanomaterial-based conductive sensors are carbon nanostructures, such as CNTs. Single-wall carbon nanotubes (SWNTs) deposited across interdigitated electrodes can detect very low concentrations of VOCs. The detection capability is attributed to a direct charge transfer to individual semiconducting SWNT conductivity, with additional electron hopping effects on intertube conductivity through physically adsorbed molecules between SWNTs.[22] Functionalization of CNTs with various materials may enhance the sensing abilities and overcome limitations, such as poor sensitivity due to low adsorption energy or low affinity of the tested gas, lack of selectivity, or long recovery time. Nevertheless, low detection of nonpolar VOCs and increased sensitivity to water are problematic, affecting the limit of detection of gases in breath samples.[23] Sensors based on CNTs were investigated as CNTFET devices for gas-sensing applications in laboratory conditions, mostly focusing on two gases, NO_2 and NH_3. Integration of CNTs into flexible sensors for gas-sensing applications is well studied, and many reports have been published, mostly on CNT-based chemiresistors. Many studies include functionalized MCNPs with CNTs in the same array in order to increase the clinical performance.

Resistive sensors based on purely organic materials are rare, as only few organic materials are conductive. A method to transduce the absorption into a layer of nonconductive organic material, such as a polymer, into a change of electric resistance was introduced some years ago. These sensors are made by dispersing conducting particles, such as carbon black grains, in an insulating polymeric matrix. The conductivity of the mixture relates to the charge transfer between the grains suspended inside the polymer. The absorption of VOCs elicits a swelling of the polymer, resulting in an increase in the mean distance between carbon black grains, thereby increasing the total resistance. This technology was the basis for a commercial device, known as Cyranose (previously distributed by Smiths Detection, USA, and now commercialized by Sensigent, USA), which has been used in several breath analysis applications, such as lung cancer diagnosis.[24] Another interesting approach is the combination of MCNPs with polymers to create flexible healing devices. Such arrays include different MCNPs linked with thiol ligands.[25] The flexible sensor was built with three layers, namely a self-healable polymer substrate from self-healing disulfide cross-linked polyurethane, a self-healable silver–polymer composite, and the sensing layer based on MCNPs. The VOCs tested in this system have been reported as being emitted from breath and skin, thus indicating a potential application in healthcare and diagnostics.[25] The sensors were demonstrated to have excellent healing properties within 3 h from being scratched. Furthermore, these self-healing sensors gave ultradetection and classification of VOCs when used in combination with pattern recognition methods, both before and after scratching.

Silicon nanowires are also excellent candidates for gas-sensing applications as chemically gated FET devices, being highly compatible with very–large-scale integration (VLSI) processes and complementary metal–oxide–semiconductor (CMOS) technologies. Although the sensing capabilities of bare SiNWs to polar VOCs are excellent, detection of nonpolar VOCs remains problematic. In order to overcome this, functionalization of the SiNWs with different coating layers is the easiest solution. Nevertheless, there are only few reports on the detection of VOCs in the gas phase. A number of systematic studies on the fundamentals and applications of functionalized SiNW-based FETs for VOC detection has been published.[1] Furthermore, arrays of modified SiNW FETs have been shown to successfully discriminate VOCs linked to gastric cancer and lung cancer from environmental confounding factors, in both real and simulated scenarios.[26]

13.3.2 Mass sensors

A change in mass is an intrinsic consequence of absorption. Nevertheless, to enable chemical sensing, the change in mass must be measured with great sensitivity. Piezoelectricity offers a basis for the fabrication of mass detectors with sufficient resolution, such as the quartz crystal microbalance (QMB). This type of sensor uses a surface that is coated with a sensitive material that undergoes a change in mass upon absorption of molecules onto the sorbent layer, allowing for an estimation of the number of molecules adsorbed.

Quartz crystal microbalances are thin slabs of quartz coated with metal electrodes (e.g., gold) on the opposite faces. The crystallographic cut activates the thickness shear resonance mode, and due to the piezoelectric effect, the mechanical resonator is equivalent to an electric resonating circuit. The fundamental frequency is inversely proportional to the thickness of the quartz plate. This value falls in the range of a few tens of megahertz. Any mass loading onto the quartz surfaces increases the energy dissipation and reduces the resonance frequency. In the low-perturbation regime, the relationship between the change in frequency (Δf) and the change in mass (Δm) is ruled by the Sauerbrey equation:

$$\Delta f = -\frac{2f_0^2}{A\sqrt{\mu_q \rho_q}}\Delta m \tag{13.1}$$

where f_0 is the unperturbed resonance frequency, A is the active area, and μ_q and ρ_q are the quartz shear module and density, respectively. When connected to a circuit, the QMB forces the frequency of an electric signal to follow its resonant frequency. As a result, the change in mass can be appreciated by measuring the change in frequency of the electric signal. It is relevant to note that among the electric quantities, frequency is measured with the best accuracy and resolution in the range 0.1 Hz–20 MHz. As an example, a 20 MHz QMB with an active diameter of 7 mm achieves a sensitivity of −0.58 Hz/ng; the limit of detection that corresponds to a frequency accuracy of 0.1 Hz is about 0.2 ng.[27]

Coating the surfaces of the QMB with a sensitive layer turns it into a gas sensor. In this regard, it should be noted that the Sauerbrey equation is strictly valid only for rigid coatings that do not store elastic energy. The performance of a QMB system can be controlled by different coating layers selected from a wide range of MOx, polymers, and other nanomaterials.

Arrays of metalloporphyrins-coated QMBs were among the first sensors to be used for breath analysis. These sensors provided the first evidence of lung cancer detection with an array of gas sensors.[28] Similar sensors have been applied to various volatilome studies both in vivo and in vitro. Among in vivo applications, it is worth highlighting the identification of therapy progress in tuberculosis.[29] In vitro applications have focused on differentiation of stem cells[30] or the identification of transmissible malaria infection stages,[31] among other applications.

13.3.3 Optical sensors

Optical detection is an attractive approach to chemical sensor design. A particular benefit of optical sensors is that the sensing material does not need to be in physical contact with the transducer; for this reason, such sensors may be developed independently of transducer technology. Optical sensors can be applied in sensitive materials where the interactions with the analyte elicit variations of either the absorption or the emission spectra. Optical sensors development takes advantage of the progresses of optoelectronic devices, in particular, the detectors made in silicon technology. The continuous growth in performance and cost reduction of individual detectors and image sensors has greatly expanded the possibility to implement optical sensors with commercially available devices, for instance, exploiting the cameras embedded in widely used appliances, such as smartphones. In spite of that, the use of silicon devices restricts the range of wavelengths to the visible and near infrared spectral regions. This greatly limits the sensitive materials available for chemical sensing and the range of interactions that are expected to modify the molecular optical properties. In particular, interactions driven by van der Waals forces should modify the optical features. This assumption can be overcome by the indicator displacement assay approach, where sensitive materials are made by the combination of a nonoptical active, but sensitive, molecule with an optically active molecule. Such optical sensors were also assessed in field tests both in vitro and in vivo. Carey et al.[32] reported on discrimination between both different bacterial species and between the different strains within each species. All strains and their antibiotic resistant forms were successfully classified with 98.8% accuracy within 10 h. Optical sensors array were also used for detection of lung cancer.[33]

13.4 Predictive models of sensor system data

When applied to breath analysis or human volatilomics, sensor systems provide high dimensional outputs. Data analysis utilizes machine learning models trained with the available examples. The high costs associated with collecting human samples often result in limited datasets for predictive model development and validation. In these conditions, models have difficulties to ensure the translation to new samples. It is of key importance to be aware of the main factors that limit the validity of the models and to use proper techniques for model assessment.[34] This section will review the hidden difficulties in the development of machine learning models for volatilome applications of sensors and the recommended best practices for model validation and assessment. Further aspects of data analysis are covered in Chapters 37 and 38.

13.4.1 Signal and data processing architectures: optimization and assessment

Signal and data processing are of increasing importance in omics applications. While in some metabolomics applications data analysis pipelines are tailored for knowledge discovery, such as biomarker discovery and metabolite annotation, in the specific case of chemical sensor systems, signal and data processing architectures are often designed for on-board implementation in smart instruments that eventually can be used in PoC scenarios.[35] Due to the unique characteristics of chemical sensor arrays, proper signal and data processing is the only way to obtain system outputs that are amenable to human interpretation.

From a computational perspective, the raw sensor signals are first preprocessed, features are extracted, and a predictive model is built using machine learning techniques based on available examples. Obviously, all the steps of this basic data processing pipeline should be optimized to minimize prediction errors. It is important to differentiate the phase of model development from that of model operation. During model development, the engineer or data scientist optimizes the data processing architecture, and model parameters are tuned for optimum performance in internal validation. For this purpose, the calibration dataset is split into a training set and an internal validation set (see also Chapter 38). The final model is selected to achieve optimum performance in the internal validation set. In the machine learning terminology, internal validation is used to select the optimum complexity of the predictor. The performance of this final model is assessed over a set of blind samples or an external validation set (as depicted in Fig. 13.4). A large variety of data mining techniques is available to achieve the most efficient use of the samples for internal validation purposes, including hold-out, k-fold, leave-one-out (LOO), random subsampling (RS), repeated k-fold, Latin bootstrapped partitions, and bootstrapping. For external validation, hold-out with blind samples obtained after model development is the preferred option. In some cases, and when samples are scarce, double cross-validation can be performed; under this scheme, all samples in the calibration set may be used for external validation. During model operation, the signal and data processing architecture and model parameters are frozen. The model is ready to evaluate individual samples for diagnosis or screening purposes.

The development of signal and data processing architectures and predictive models based on examples is generally termed sensor system calibration. System calibration is an expensive part of sensor system development for biomedical applications. Predictive models built with machine learning techniques only learn from the examples experienced. For this reason, it is desirable to process many calibration samples. When such samples are of human origin, the cost increases and in many studies less than a few hundred samples are used for model building.[36] The selection of project participants is key to eliminate or reduce confounding factors.[37]

Due to sensor tolerances, a calibration model developed for a sensor array will have a degraded accuracy in a second sensor array, even if this is otherwise identical. For this reason, calibration transfer strategies should be used to minimize the required number of samples for recalibration.[38,39] In this paradigm, only the master instrument is calibrated with a large sample dataset and the slave instruments are

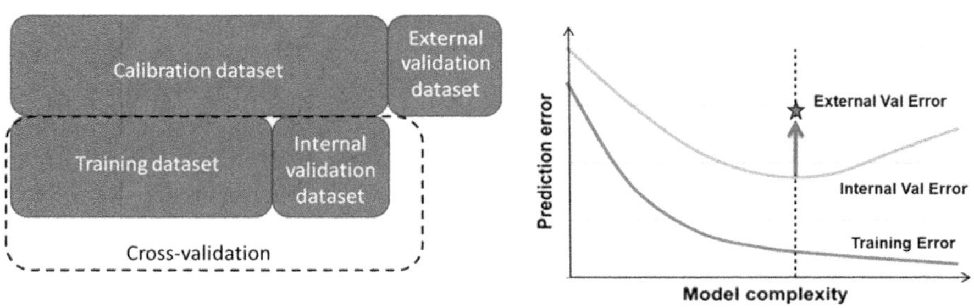

FIGURE 13.4

Cross-validation takes places within the calibration dataset. Minimizing the prediction errors over the internal validation set is the usual approach for model complexity control. Performance assessment in internal validation is mostly overoptimistic.

calibrated only with a reduced sample count. Lately, proposals to obtain general calibration models that reject intersensory variability have been proposed; however, the use of these techniques in biomedical scenarios has not been demonstrated, to date.

13.4.2 The role of signal and data processing

Signal and data processing aims to compensate the shortcomings of low-cost chemical sensors. Multivariate data processing, either for regression or classification problems, is a powerful tool to counteract insufficient selectivity. Drift correction methods are employed to compensate poor system stability, and self-validating architectures with autodiagnostics may serve to detect sensor faults and even provide a corrected operation until maintenance and recalibration is performed.[40] If required, inverse dynamic filters may be used to improve the slow time response of some sensor technologies.[41]

13.4.3 Dangers in predictive model development

Although signal processing and machine learning may improve the metrological behavior of chemical sensor array-based instruments, in general, such systems show poor analytical robustness. In addition to the aforementioned issues with stability-related shifts in data over time, gas sensor arrays typically exhibit hypersensitivity to instrument conditions and operation protocols, including power operation modes. This is well known for MOx sensors. Computationally, the greatest dangers arise from the curse of dimensionality, the problem of overfitting in conjunction with weak validation methods, and the statistical biases (see Chapter 38). According to the World Health Organization, "*Robustness (or ruggedness) is the ability of the procedure to provide analytical results of acceptable accuracy and precision under a variety of conditions. The results from separate samples are influenced by changes in operational or environmental conditions.*"[42]

In the case of chemical sensor systems, it must be considered that the analytical results appear at the output of the instrument only after data processing. Often, the most accurate models in the internal validation carry intrinsic overfitting to the operational conditions of the instrument during the analysis of the calibration samples. Departure from those conditions degrades the performance of the predictive model. As such, it is recommended to weight model robustness over accuracy. In practical PoC scenarios, the operational conditions are often different from those used in system calibration. For this reason, prioritizing robustness with respect to predictive accuracy during model development is desirable. When evaluating system robustness, the following aspects must be considered:

(i) *Sampling*: Accepted standards for breath sampling do not exist (see Chapter 2). Differences in sampling will alter the predictive power of the model. Additionally, breath may contain VOCs that have been inhaled before sample collection (exogenous confounders). This memory effect may span from minutes to hours, depending on the considered analyte (see Chapter 1).

(ii) *Sensitivity to sensor tolerances*: Individual sensors of the same type show response variabilities that may have an impact on model performance.
(iii) *Instrumental time shifts*: Few studies have demonstrated the stability of model performance months after the calibration phase.
(iv) *Sensitivity to geographical location*: Mature studies should be able to show reproducibility in other medical centers with different operators.

A final consideration on data processing is what has been termed "*the most important threat to validity.*"[43] This is a statistical issue that transcends chemical sensor studies but is very relevant for observational studies in metabolomics. According to Ranshoff,[43] "*all observational research should be presumed guilty of bias, until proven innocent.*" Confounding factors are the main cause of bias. Most experienced researchers in metabolomics try to block the usual suspects, i.e., sex, age, medication, and smoking status; however, many others may appear, and in some cases, they are not even known to the researcher. When this occurs, bias is unintentional and unconscious. Goodacre has reported intrinsic and extrinsic factors that may affect the metabolome, where the former includes body composition, tissue turnover, metabolic rate at rest, human genotype, health status, reproductive status, microbiome, and diurnal cycles, and the latter includes diet, drugs, physical activity, environmental exposure, and stress levels.[44] When those factors are not balanced among the control and condition groups, observed differences may result from those conditions instead of the target health condition. This is particularly important in chemical sensor arrays, since metabolite identification is not possible. As a consequence, no additional insight may be gained about the relevance of discriminant features or biomarkers.

13.4.4 Limits of validity of predictive models for volatilomics

The dangers that plague the development of calibration models for chemical sensor arrays in volatilomics have been discussed above. External validation and permutation tests are the recommended practice for model validation, but caution must be taken to avoid errors associated with double cross-validation. While double cross-validation technically provides blind samples for model validation, those blind samples share potential sources of biases present in the calibration dataset; blind samples originate from the same set of individuals and likely have been collected with the exact same protocol and instrument, which limits the validity of the ensuing model.

In general, model validity cannot be claimed beyond the domain of the validation samples. This is extreme in the case of double cross-validation, since the blind samples belong to the same dataset. In order to claim additional extension to the validity of specific findings, it is necessary to carry out stricter validation. To ensure that results are stable in time and robust against instrumental time shifts, or even seasonal changes in metabolism, an extended temporal validation is recommended. The model should be applied to new samples collected at least six months after the study and address the question of whether the instrument and calibration model remain discriminant. Any claims of certain geographical extension for specific findings require multicenter studies involving medical centers and patients from different parts of the world. Although constraints, such as budget and time, can limit or restrict such extended validation, researchers should be aware of the real limits of any model and these should be reported with the results accordingly.

13.5 Summary

Nanomaterial-based sensors are excellent candidates for PoC diagnostics based on detecting disease-related VOCs excreted from human body fluids, especially breath. This chapter summarizes the vast variety of nanomaterials and the predictive models that can tailor-up the diagnostic performances according to the desired application. A particularly promising approach is a design that incorporates hybrid arrays of combined multitransducers, similar to the lab-on-chip approach, that lower the number of the correlations between the responses of the different sensor types and thereby lead to power increase of the data analysis.[45] In order to interface with the Information and Communications Technology (ICT), there is a need for investing in array miniaturization. Such miniaturization can be achieved by microfluidics elements and chip technologies to reduce the sample volume and allow integration with other devices in the current ICT platform. Two additional steps include parallel analyses and data networking by ICT that ensure fast analysis and data transmission to the medical staff in real time.

For embedded instruments, signal and data processing must run in real time and provide human interpretable information to the operator. For this to be useful in PoC scenarios, the instruments should provide maximum robustness in terms of the analytical evaluation of the sample. Tailored signal and data processing combined with strict validation methodologies are essential to improve the performance of nanosensor systems. Additionally, calibration transfer methodologies should be ensured since it is absolutely unrealistic to calibrate each single instrument with biological samples. This should be done only with a master instrument, and there should be reliable methods to transfer this information to a larger set of instruments with minimal calibration effort. Finally, due to the hypersensitivity of sensors and the lack of selectivity, maximum attention should be given to the potential biases introduced by hidden confounding factors.

Acknowledgments

S.M. would like to acknowledge the Department d'Universitats, Recerca i Societat de la Informació de la Generalitat de Catalunya (expedient 2017 SGR 1721); the Comissionat per a Universitats i Recerca del DIUE de la Generalitat de Catalunya; and the European Social Fund (ESF). Additional financial support has been provided by the Institut de Bioenginyeria de Catalunya (IBEC). IBEC is a member of the CERCA Programme/Generalitat de Catalunya.

References

1. Broza YY, Vishinkin R, Barash O, Nakhleh MK, Haick H. Synergy between nanomaterials and volatile organic compounds for non-invasive medical evaluation. *Chem Soc Rev* 2018;**47**.
2. Vishinkin R, Haick H. Artificially intelligent nanoarrays for disease detection via volatolomics. In: *Technical digest – international electron devices meeting*. IEDM; 2015.
3. Dickert FL, Lieberzeit P, Miarecka SG, Mann KJ, Hayden O, Palfinger C. Synthetic receptors for chemical sensors – subnano- and micrometre patterning by imprinting techniques. In: *Biosensors and bioelectronics*; 2004. 1040–4.
4. Goepel W. Chemical sensing, molecular electronics and nanotechnology: interface technologies down to the molecular scale. *Sens Actuators B* 1991;**B4**:7–21.
5. Konvalina G, Haick H. Sensors for breath testing: from nanomaterials to comprehensive disease detection. *Acc Chem Res* 2013;**47**:66–76.
6. Peng G, Hakim M, Broza YY, et al. Detection of lung, breast, colorectal and prostate cancers from exhaled breath using a single array of nanosensors. *Br J Cancer* 2010;**103**: 542–51.
7. Vishinkin R, Haick H. Nanoscale sensor technologies for disease detection via volatolomics. *Small* 2015;**11**.
8. Di Natale C, Paolesse R, Martinelli E, Capuano R. Solid-state gas sensors for breath analysis: a review. *Anal Chim Acta* 2014;**824**:1–17.
9. Peng G, Tisch U, Adams O, et al. Diagnosing lung cancer in exhaled breath using gold nanoparticles. *Nat Nanotechnol* 2009;**4**:669–73.
10. Pennazza G, Santonico M, Martinelli E, D'Amico A, Natale C Di. Interpretation of exhaled volatile organic compounds. In: *Eur respir monogr 2010*. European Respiratory Society; 2010. p. 115–29.
11. Persaud K, Dodd G. Analysis of discrimination mechanisms in the mammalian olfactory system using a model nose. *Nature* 1982;**299**:352–5.
12. Paolesse R, Nardis S, Monti D, Stefanelli M, Di Natale C. Porphyrinoids for chemical sensor applications. *Chem Rev* 2017;**117**:2517–83.
13. De Vries R, Brinkman P, Van Der Schee MP, et al. Integration of electronic nose technology with spirometry: validation of a new approach for exhaled breath analysis. *J Breath Res* 2015;**9**.
14. Güntner AT, Koren V, Chikkadi K, Righettoni M, Pratsinis SE. E-nose sensing of low-ppb formaldehyde in gas mixtures at high relative humidity for breath screening of lung cancer? *ACS Sens* 2016;**1**:528–35.
15. Jeong SY, Yoon JW, Kim TH, et al. Ultra-selective detection of sub-ppm-level benzene using Pd-SnO_2 yolk-shell micro-reactors with a catalytic Co_3O_4 overlayer for monitoring air quality. *J Mater Chem* 2017;**5**:1446–54.
16. Righettoni M, Tricoli A, Gass S, Schmid A, Amann A, Pratsinis SE. Breath acetone monitoring by portable Si:WO_3 gas sensors. *Anal Chim Acta* 2012;**738**:69–75.
17. Moon HG, Choi YR, Shim Y-S, et al. Extremely sensitive and selective NO probe based on villi-like WO_3 nanostructures for application to exhaled breath analyzers. *ACS Appl Mater Interfaces* 2013;**5**:10591–6.
18. Lundstrom I. Hydrogen sensitive mos-structures. *Sens Actuators* 1981;**1**:403–26.

19. Comini E, Baratto C, Faglia G, Ferroni M, Vomiero A, Sberveglieri G. Quasi-one dimensional metal oxide semiconductors: preparation, characterization and application as chemical sensors. *Prog Mater Sci* 2009;**54**:1−67.
20. Tisch U, Haick H. Arrays of chemisensitive monolayer-capped metallic nanoparticles for diagnostic breath testing. *Rev Chem Eng* 2010;**26**:171−9.
21. Potyrailo RA. Toward high value sensing: monolayer-protected metal nanoparticles in multivariable gas and vapor sensors. *Chem Soc Rev* 2017;**46**:5311−46.
22. Li J, Ng HT, Cassell A, et al. Carbon nanotube nanoelectrode array for ultrasensitive DNA detection. *Nano Lett* 2003;**3**:597−602.
23. Peng G, Tisch U, Haick H. Detection of nonpolar molecules by means of carrier scattering in random networks of carbon nanotubes: toward diagnosis of diseases via breath samples. *Nano Lett* 2009;**9**:1362−8.
24. Thriumani R, Zakaria A, Hashim YZHY, et al. A study on volatile organic compounds emitted by in-vitro lung cancer cultured cells using gas sensor array and SPME-GCMS. *BMC Cancer* 2018;**18**.
25. Huynh T-P, Haick H. Self-healing, fully functional, and multiparametric flexible sensing platform. *Adv Mater* 2016;**28**:138−43.
26. Shehada N, Cancilla JC, Torrecilla JS, et al. Silicon nanowire sensors enable diagnosis of patients via exhaled breath. *ACS Nano* 2016;**10**:7047−57.
27. D'Amico A, Di Natale C. A contribution on some basic definitions of sensors properties. *IEEE Sens J* 2001;**1**:183−90.
28. Di Natale C, Macagnano A, Martinelli E, et al. Lung cancer identification by the analysis of breath by means of an array of non-selective gas sensors. *Biosens Bioelectron* 2003; **18**:1209−18.
29. Zetola NM, Modongo C, Matsiri O, et al. Diagnosis of pulmonary tuberculosis and assessment of treatment response through analyses of volatile compound patterns in exhaled breath samples. *J Infect* 2017;**74**:367−76.
30. Capuano R, Spitalieri P, Talarico RV, et al. A preliminary analysis of volatile metabolites of human induced pluripotent stem cells along the in vitro differentiation. *Sci Rep* 2017;**7**.
31. Capuano R, Khomenko I, Grasso F, et al. Simultaneous Proton Transfer Reaction-Mass Spectrometry and electronic nose study of the volatile compounds released by Plasmodium falciparum infected red blood cells in vitro. *Sci Rep* 2019;**9**.
32. Carey JR, Susick KS, Hulkower KI, et al. Rapid identification of bacteria with a disposable colorimetric sensing array. *J Am Chem Soc* 2011;**133**:7571−6.
33. Mazzone PJ. Exhaled breath volatile organic compound biomarkers in lung cancer. *J Breath Res* 2012;**6**:27106.
34. Marco S. The need for external validation in machine olfaction: emphasis on health-related applications Chemosensors and Chemoreception. *Anal Bioanal Chem* 2014; **406**:3941−56.
35. Marco S, Gutierrez-Galvez A. Signal and data processing for machine olfaction and chemical sensing: a review. *IEEE Sens J* 2012;**12**:3189−214.
36. Vinaixa M, Samino S, Saez I, Duran J, Guinovart JJ, Yanes O. A guideline to univariate statistical analysis for LC/MS-based untargeted metabolomics-derived data. *Metabolites* 2012;**2**:775−95.
37. Hauschild A-C, Frisch T, Baumbach J, Baumbach J. Carotta: revealing hidden confounder markers in metabolic breath profiles. *Metabolites* 2015;**5**:344−63.

38. Fonollosa J, Fernández L, Gutiérrez-Gálvez A, Huerta R, Marco S. Calibration transfer and drift counteraction in chemical sensor arrays using Direct Standardization. *Sens Actuators B Chem* 2016;**236**:1044–53.
39. Fernandez L, Guney S, Gutierrez-Galvez A, Marco S. Calibration transfer in temperature modulated gas sensor arrays. *Sens Actuators B Chem* 2016;**231**:276–84.
40. Padilla M, Perera A, Montoliu I, Chaudry A, Persaud K, Marco S. Fault detection, identification, and reconstruction of faulty chemical gas sensors under drift conditions, using Principal Component Analysis and Multiscale-PCA. *Proc Int Jt Conf Neural Netw* 2010: 18–23.
41. Martinez D, Burgués J, Marco S. Fast measurements with MOX sensors: a least-squares approach to blind deconvolution. *Sensors* 2019;**19**:4029.
42. World Health Organization. *Quality assurance of pharmaceuticals: a compendium of guidelines and related materials*, vol. I; 1997.
43. Ransohoff DF. Bias as a threat to the validity of cancer molecular-marker research. *Nat Rev Cancer* 2005;**5**:142–9.
44. Goodacre R. Metabolomics of a superorganism. *J Nutr* 2007;**137**:259S–66S.
45. Pardo M, Kwong LG, Sberveglieri G, et al. Data analysis for a hybrid sensor array. *Sens Actuators B Chem* 2005;**106**:136–43.

CHAPTER

Optical spectroscopy

14

Florian M. Schmidt[1], Markus Metsälä[2], Alexander Apolonski[3,4], Simona M. Cristescu[5]

[1]*Department of Applied Physics and Electronics, Umeå University, Umeå, Sweden;* [2]*Department of Chemistry, University of Helsinki, Helsinki, Finland;* [3]*Max Planck Institute of Quantum Optics, Garching, Germany;* [4]*Ludwig-Maximilians-Universität München, Garching, Germany;* [5]*Department of Molecular and Laser Physics, Institute for Molecules and Materials, Radboud University, Nijmegen, The Netherlands*

Chapter outline

14.1 Overview

State-of-the-art breath gas analysis calls for online sampling and selective real-time biomarker detection with high sensitivity, precision, and accuracy. Real-time breath gas analysis exploits the inherent directness and immediacy of breath gas sampling. The short analysis time leads to improved sample throughput and statistics, which facilitates larger clinical and epidemiological studies. Continuous real-time measurements over longer periods can give accurate information about the uptake and

Breathborne Biomarkers and the Human Volatilome. https://doi.org/10.1016/B978-0-12-819967-1.00014-1
Copyright © 2020 Elsevier B.V. All rights reserved.

elimination of biomarkers in the body and its response to interventions. Moreover, spatiotemporal information on the respiratory tract and lung can be obtained from exhalation profiles recorded during the breathing cycle. Finally, analysis of breath gas immediately after exhalation without preconcentration and storage simplifies sample handling and reduces the risk of sample contamination. Ideally, several biomarkers are measured simultaneously to reliably assess a health condition, but fast and accurate quantitative multispecies detection in exhaled breath is very challenging with any of the current techniques.

The main analytical techniques associated with breath gas analysis today are variants of mass spectrometry (MS) (cf. Chapters 9, 10, and 15). MS is very useful in the exploratory phase of research and for the detection of larger volatile organic compounds (VOCs). However, accurate quantification is difficult, and on-site clinical application is problematic, since MS-based instruments are often large and expensive, and, depending on the type of system, the analysis can be time consuming. Soft ionization MS techniques allow for online, multispecies analysis, but detection of smaller molecules can be challenging.

Thus, it is not surprising that of the few clinically available breath tests, most are based on optical devices. Among the tests approved by the US Food and Drug Administration (FDA) are capnography to detect exhaled carbon dioxide (CO_2), the $^{13}CO_2/^{12}CO_2$ urea breath test for detection of *Helicobacter pylori* bacterial infection (both nondispersive infrared spectroscopy [NDIR]; see Chapter 17), and the fraction of exhaled nitric oxide (F_ENO) test (chemiluminescence; "semioptical"; see Chapters 4 and 5). In general, many of the proven breath biomarkers, in particular nitric oxide (NO), carbon monoxide (CO), and carbon dioxide (CO_2) isotopes, but also ammonia (NH_3) and methane (CH_4), are organic or inorganic molecules of low molecular weight and as such are perfectly suited for optical analysis.

The main advantages of optical spectroscopy include reasonable cost and size, online and real-time capability, reliable quantification, and high sensitivity and selectivity for small molecules and for isotope-resolved measurements. While optical spectroscopy is less suitable for biomarker identification and fingerprinting, it complements MS and helps in driving breath gas analysis into the clinical practice. Optical techniques are also useful in the exploratory research phase, when it comes to detailed physiological studies of low molecular weight biomarkers. Emerging application include real-time detection coupled to physiological modeling, lung function testing, and in vitro studies.

14.2 Basic principles

Depending on the wavelength region used, optical spectroscopy is sensitive to either the electronic structure (ultraviolet and visible, UV-VIS) or the intramolecular bond vibrations (infrared, IR) of an analyte. An optical spectrometer probes transitions between energy levels associated with these electronic or vibrational degrees of freedom.[1] Together with the excitation of molecular vibrations, IR photons also induce changes in the rotational quantum number. Transitions involving changes in both vibrational and rotational states are called rovibrational. Vibrations involve a displacement of the atoms from their equilibrium position with respect to each other. Rotations are connected to the movement of the molecule relative to a set of mutually orthogonal axes of fixed orientation in space, centered on the center of mass of the molecule. The combination of vibrational and rotational excitation leads to a myriad of possible transitions over the IR region for every volatile organic and inorganic compound (Fig. 14.1A). Even a relatively small molecule like water (H_2O) exhibits millions of IR transitions.

Excitation of valence shell electrons or resonant vibrations and rotations occurs via absorption of photons. The highest energy differences are between electronic states, followed by vibrational and then rotational states. The intensity of the IR transitions depends mostly on the number of vibrational quanta that change due to the excitation. The strongest IR transitions are the so-called fundamental transitions in the mid-IR region (Fig. 14.1A), where the vibrational quantum number changes by one, from the ground state to the first excited state. Most organic molecules have strong signatures around 3.5 μm (C—H bond) and/or 6 μm (C=O bond). The overtones (or harmonics) of the fundamental transitions (vibrational quantum number changes by more than one) fall into the near-IR and visible range and become weaker as the vibrational energy increases. Thus, most optical trace gas analyzers probe rovibrational transitions in the mid-IR (3—20 μm) or near-IR

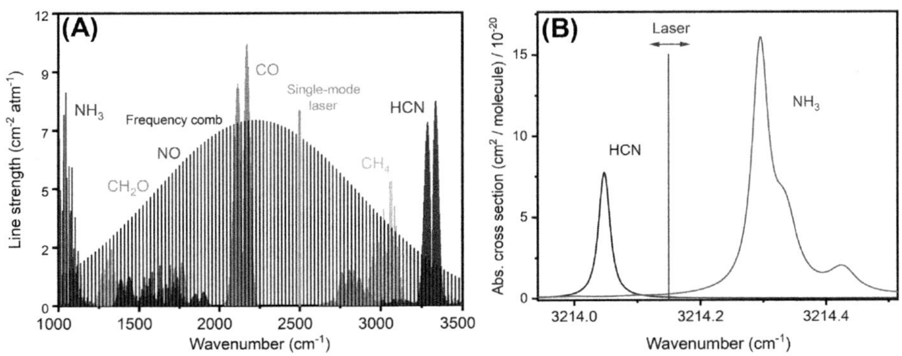

FIGURE 14.1

(A) Absorption spectra of some breath biomarkers. The spectra of an octave-spanning optical frequency comb (black lines) and a single-mode laser (red/dark gray line) are shown to illustrate the difference in spectral coverage. The frequency comb and single-mode laser are included in the graph for illustrative purposes; the vertical axis of the graph relates only to the molecular absorption. The separation between the individual comb lines is exaggerated for clarity. (B) Simulation of a narrowband scan with a tunable mid-IR semiconductor laser across individual rovibrational transitions of hydrogen cyanide (HCN) and ammonia (NH_3). The molecular line data are from the HITRAN database.[2] Wavenumber is defined as the reciprocal of the wavelength of the light in centimeters.

(0.7–3 μm) range. On the other hand, the UV-VIS (0.2–0.7 μm) region has the advantage of being free of strong H_2O bands. Moreover, in both UV-VIS and near-IR, powerful (broadband) lasers and sensitive light detectors are available, while such devices are only just emerging for the mid-IR range (Fig. 14.1A).

The rovibrational molecular spectra described above can be probed optically using light absorption, emission, and scattering. The concentration of a molecule in a gaseous matrix is most conveniently determined with absorption spectroscopy, where the ratio between incident and transmitted light intensity at or in the vicinity of an absorbing molecular transition is measured. The transmitted light intensity, $I(\nu)$, depends on the frequency, ν, of the incident light, I_0, and decreases with absorption path length, L, according to Beer–Lambert's law:

$$I(\nu) = I_0 e^{-\alpha(\nu)L}, \tag{14.1}$$

where $\alpha(\nu)$ is the absorption coefficient and the exponent $\alpha(\nu)L$ is the absorbance. In spectroscopy, the frequency equals the speed of light divided by the wavelength, and the wavenumber is defined as the reciprocal of the wavelength of the light in centimeters. The absorption coefficient can be written as the product of the transition line strength S ($cm^{-2} \cdot atm^{-1}$), the sample pressure p (atm), the analyte concentration (relative to volume) c, and the frequency-dependent absorption line shape function $\chi(\nu)$ (cm) according to

$$\alpha(\nu) = Spc\chi(\nu) . \tag{14.2}$$

For small molecules, the line shape profile $\chi(\nu)$ is a well-defined analytical function with a typical (pressure-dependent) wavenumber width of less than 0.1 cm^{-1} (Fig. 14.1B). As the size of the molecule increases, the number of rotational and vibrational energy levels increases accordingly and the absorption lines start to overlap and create complex broadband absorption features.

Trace species in exhaled breath are usually present at concentrations below parts per million (ppm), while the inorganic molecules H_2O, CO_2, and N_2 are present at percent levels. Thus, high selectivity is needed to detect the trace species, and problems arise in spectral regions where both target molecules and abundant species or VOCs of unknown identity absorb. For narrowband and broadband detection of small molecules (Fig. 14.1), identification and quantification is usually performed by analyzing the individual, unique absorption line shapes and using the spectral line parameters available through databases, such as the High Resolution Transmission (HITRAN) database.[2] Spectral interference with CO_2 and H_2O can be avoided by careful line selection. For broadband detection of larger VOCs, identification and quantification is more challenging, but can also lean on databases, such as the National Institute of Standards and Technology (NIST) library.[3] Strategies to improve the selectivity comprise removal of H_2O[4] and CO_2 or calibration using artificial breath samples.

14.3 Techniques

By far, the most common optical method used in breath gas analysis is absorption spectroscopy. This is primarily because absorption spectroscopy allows the selective and highly sensitive determination of absolute species concentrations in real time (<1 s) without sample preconcentration and frequent calibration procedures (Fig. 14.2). Notable exceptions are exhaled NO detection by chemiluminescence, where light emission due to oxidation of NO to nitrogen dioxide (NO_2) by ozone (O_3) is measured, and isotope-resolved analysis by Raman scattering. Several detailed reviews have been published on laser-based absorption techniques for breath analysis, including extensive compilations of references.[5–7] In the following, the most established and promising techniques are briefly described. A performance comparison is presented in Table 14.1, and the main components of typical experimental setups are shown schematically in Fig. 14.2A.

The detection limit of a technique is determined by the sensitivity of the absorption measurement, which can be enhanced by either increasing the signal-to-noise ratio or by increasing the absorption path length. The former is usually achieved by modulation techniques to avoid laser intensity noise or by choosing a stronger absorbing molecular transition. The effective absorption path length can be increased using multipass cells (MPCs, $L_{eff} < 100$ m) or resonant optical cavities (L_{eff} up to 10 km) (Fig. 14.2A). Although the sample pressure is usually reduced to 50–100 mbar, the large volume of typical MPCs (>0.3 L) may limit the gas exchange and acquisition time, while optical cavities can be made very compact.

Typical narrowband (single-mode) semiconductor light sources include distributed feedback diode (DFB) lasers (0.7–3 μm), interband cascade lasers (ICLs) (3–6 μm) and quantum cascade lasers (QCLs) (4–14 μm). These lasers allow tuning in a narrow wavenumber range (1–5 cm^{-1}) and usually enable simultaneous targeted detection of one to three species with low molecular weight (i.e., well-defined absorption line profiles). To address a larger number of biomarkers, two or several narrowband lasers can be used simultaneously. Broadband multispecies detection can be achieved using optical parametric oscillators (OPOs), optical frequency combs (OFCs), fiber lasers, or external-cavity diode or quantum cascade lasers (ECDLs or EC-QCLs). In the latter approach, the laser is mounted in an optical cavity containing a wavelength-selective device, which provides a narrow laser linewidth and a broadband wavelength tuning range (up to 100 cm^{-1}). In principle, such laser sources can be used with all of the laser techniques that are described below. High-power lasers, e.g., OPO, EC-QCL, or CO_2 lasers, are most beneficial in photoacoustic spectroscopy, where the signal strength scales linearly with the optical power of the light source.

14.3.1 Infrared spectroscopy using thermal light sources

A basic NDIR spectroscopy setup consists of a broadband IR light source, such as a tungsten filament lamp or a light-emitting diode, a sample gas chamber, and a

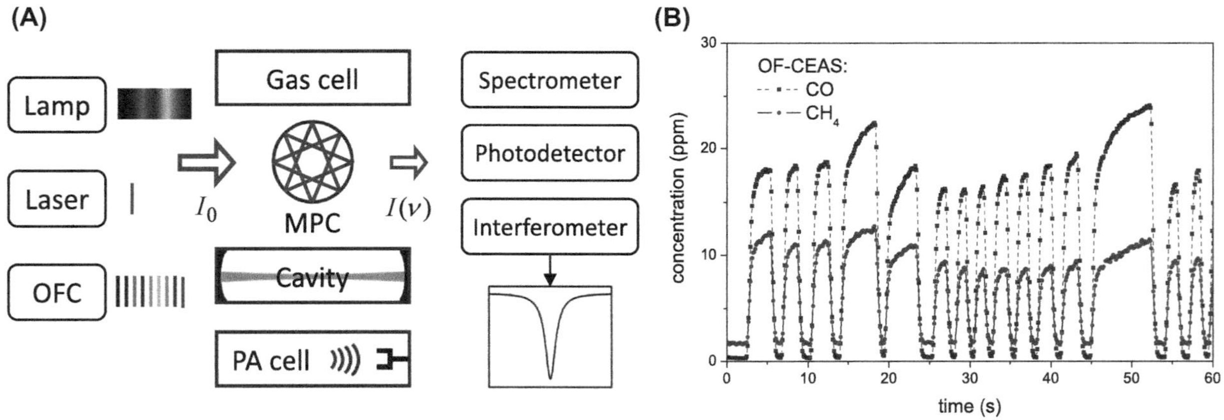

FIGURE 14.2

(A) Schematic drawing of experimental setups for absorption spectroscopy indicating the different types of light sources, sample cells, and detection systems. (B) Simultaneous, breath cycle–resolved (real-time) detection of carbon monoxide (CO) and methane (CH_4) with optical feedback cavity-enhanced absorption spectroscopy.[15]

I(ν), transmitted light intensity; I_0, incident light intensity; *OF-CEAS*, optical feedback cavity-enhanced absorption spectroscopy; *MPC*, multipass cell; *OFC*, optical frequency comb; *PA*, photoacoustic.

Table 14.1 Comparison between different spectroscopic techniques used in breath gas analysis.[a]

Technique	LOD/ precision	Selectivity	Real-time (sub-second)	Multi-species	Complexity/ Size/Cost
NDIR	ppm	Med	Yes	No	Low
FTIR	ppb–ppm	Med	No	Yes	High
TDLAS	ppb	High	Yes	1–3	Med
CEAS	ppt	High	Yes	1–3	Med–high
OFCS	ppt–ppb	High	Not yet[b]	Yes	High
LPAS	ppt	High	Yes	1–3	Med
FERS	ppm	Med	Yes	Yes	Med

[a] CEAS, *cavity-enhanced absorption spectroscopy;* FERS, *fiber-enhanced Raman spectroscopy;* FTIR, *fourier transform infrared;* LOD, *limit of detection;* LPAS, *laser photoacoustic spectroscopy;* Med, *medium;* NDIR, *nondispersive infrared;* OFCS, *optical frequency comb spectroscopy;* ppm, *parts per million;* ppb, *parts per billion;* ppt, *parts per trillion;* TDLAS, *tunable diode laser absorption spectroscopy.*

[b] *Sensitive and high-resolution real-time breath gas analysis has not been demonstrated with OFCS but is technically possible.*

photodetector (Fig. 14.2A). An optical filter in front of the detector eliminates all light outside the absorption band of the target molecule. An NDIR device can be made very compact and robust, but suffers from interference due to other species absorbing at the same wavelength, as well as from low sensitivity (the absorption path length is usually short). Thus, NDIR sensors are only suitable for the highly abundant breath species, such as CO_2 and H_2O. Application areas are capnography and $^{13}CO_2/^{12}CO_2$ isotope ratio analysis.[8,9]

Fourier transform infrared (FTIR) spectroscopy also employs a broadband thermal light source, but here the light is first directed through an interferometer, usually a Michelson interferometer. The photodetector signal, a so-called interferogram, is recorded as a function of the optical path length difference experienced by the two beams traveling through the two arms of the interferometer. The interferogram is then converted to a spectrum (absorption vs. wavenumber in cm^{-1}) by applying a Fourier transform.[4] While a wide absorption spectrum, including both narrowband and broadband absorbers, can be recorded with this technique, a typical FTIR does not provide the selectivity and sensitivity required for the detection of trace species in exhaled breath.

14.3.2 Tunable diode laser absorption spectroscopy

Tunable diode laser absorption spectroscopy (TDLAS) refers to absorption techniques employing narrowband semiconductor lasers that are tuned across the target absorption feature(s). This comprises direct absorption spectroscopy (DAS) and modulation techniques, such as wavelength modulation spectroscopy (WMS) in single-pass configuration or with MPCs, for example, Herriott, White, or ring cells.[10–13] Most TDLAS setups can be made compact and robust and are suitable for the targeted detection of one to three biomarkers with high selectivity and temporal resolution. TDLAS does not typically require frequent calibration to obtain quantitative results with high accuracy and precision. Real-time detection is possible for breath biomarkers in the ppb–ppm range, such as CO, CH_4, NH_3, and nitrous oxide (N_2O). TDLAS, and laser absorption spectroscopy in general, can also be combined with Faraday rotation spectroscopy,[14] which only works for paramagnetic species like NO but has the advantage of being insensitive to water.

14.3.3 Cavity-enhanced laser absorption spectroscopy

In cavity-enhanced absorption spectroscopy (CEAS), the breath gas sample is enclosed in an optical cavity consisting of two or more highly reflective mirrors (Fig. 14.2A). Depending on the reflectivity of the mirrors, the effective absorption path length can reach tens of kilometers, while the setup retains a compact physical size. The highest reflectivities are available in the near-IR region, where the mirror coating technology is advanced. In the mid-IR, the mirror reflectivity is normally lower, but this is compensated by the stronger absorption line strengths of the fundamental vibrational transitions. Similar to TDLAS, cavity-enhanced methods can make use of modulation techniques for noise reduction.

Cavity-enhanced absorption spectroscopy can be implemented in various configurations. One of the most common strategies is to record the time evolution of the laser intensity inside the optical cavity and extract the decay time constant, which depends on the level of absorption in the cavity. This approach is called cavity ring-down spectroscopy (CRDS) and has been employed in a number of breath investigations.[16–21] An alternative approach is integrated cavity output spectroscopy (ICOS), where the total amount of laser intensity that leaks out of the cavity is recorded.[22] The coupling of the laser light into the optical resonator can be performed in various ways, leading to techniques such as off-axis cavity-enhanced absorption spectroscopy (OA-CEAS)[10] and optical feedback cavity-enhanced absorption spectroscopy (OF-CEAS).[15,23] In intracavity absorption spectroscopy (ICAS), the sample is placed within the resonator of a fiber laser.[24] Typical absorption sensitivities achieved with CEAS are sufficient for the detection of almost all small biomarker molecules in exhaled breath, including those present at low ppb or ppt concentrations, such as NO, hydrogen cyanide (HCN), hydrogen sulfide (H_2S), carbonyl sulfide (OCS), and formaldehyde (CH_2O).

14.3.4 Optical frequency comb spectroscopy

An optical frequency comb combines the advantages of narrowband (single-mode) lasers with those of broadband light sources.[25] The spectrum of an OFC consists of hundreds of thousands of synchronized coherent modes, spanning a wide wavelength range (Fig. 14.1A). This feature makes the OFC an attractive light source for sensitive, broadband, and high-resolution spectroscopy, enabling high-speed, multispecies detection. Simultaneous detection of different biomarkers in human breath (CO, NH_3, and CH_4 in the presence of high concentrations of H_2O and CO_2) has been demonstrated in the near-IR wavelength range using cavity-enhanced optical frequency comb spectroscopy (CE-OFCS).[26] With OFC spectroscopy, it is thus possible to discriminate and characterize both narrow and broadband absorbers.

Optical frequency combs are traditionally generated from mode-locked lasers that cover the visible and near-IR wavelength ranges. Reliable mode-locked lasers emitting in the mid-IR are still under development.[27] Currently, the two main approaches to build mid-IR OFCs are nonlinear frequency conversion with OPOs or difference frequency generation (DFG) in combination with near-IR mode-locked lasers or direct generation from semiconductor lasers, ideally QCLs.[28] Compared to spectrometers that employ single-mode lasers, an OFC spectrometer is more complicated since the broadband spectrum must be spectrally analyzed to produce an absorption spectrum. This spectral analysis can be done using a diffraction grating or an FTIR. Both of these approaches have drawbacks when it comes to selectivity, measurement speed, and sensitivity. The most promising strategy for OFC spectroscopy is to employ two OFC sources in a so-called dual-comb spectroscopy setup.[29,30] Dual-comb spectroscopy provides the power of FTIR without mechanical elements in the interferometer that limit the spectral resolution and without a large number of photons needed to achieve a high signal-to-noise ratio.

14.3.5 Laser photoacoustic spectroscopy

Similar to the laser spectroscopic methods introduced above, laser photoacoustic spectroscopy (LPAS) is well-suited for sensitive and fast trace gas detection.[31] Unlike absorption spectroscopy, however, which measures a small decrease in transmitted light intensity, LPAS relies on the fact that molecules excited by the absorption of a photon release energy via collisions. This causes an increase in the temperature of the surrounding gas. By modulating the laser intensity, the gas temperature will change periodically, which gives rise to a periodical pressure change, and this, in turn, results in an acoustic wave that can be detected by a microphone (Fig. 14.2A).[32] Without absorbers, there is no acoustic wave at the modulation frequency; thus the signal is background-free.

While CEAS techniques derive their high sensitivity from long absorption path lengths, the LPAS signal increases with laser power. Therefore, in order to measure concentrations at ppb levels or below, LPAS employs lasers with relatively high-power ($>$100 mW) and resonant photoacoustic cells. LPAS allows a wide wavelength range to be accessed, limited only by the available light sources. This is in contrast to CEAS, in which the wavelength dependence of the highly reflective mirror coatings limits the spectral coverage of a given system. In addition, LPAS allows detection of gas samples in a very small volume (few mm^3), as demonstrated by quartz-enhanced photoacoustic spectroscopy (QEPAS).[33] Here, the microphone and the resonant cell are replaced by a quartz tuning fork—common elements used as frequency standard in clocks and smart phones. Another more complex approach is cantilever-enhanced photoacoustic spectroscopy (CEPAS), where the acoustic signal is detected by measuring the movement of a miniature silicon cantilever-type pressure sensor using a laser interferometer.[34] Portable LPAS sensors with low power consumption can be developed based on mid-IR ICL or QCL lasers.

14.3.6 Fiber-enhanced Raman spectroscopy

In Raman spectroscopy, the frequency of the detected light is shifted from the laser frequency due to inelastic scattering by the molecules. Since the Raman signal is intrinsically weak, it needs to be enhanced, usually by stimulated Raman spectroscopy. A novel strategy to achieve this is to improve the interaction of the laser light with the target molecules in an optical waveguide, such as an optical fiber. In general, simple waveguides suffer from high attenuation, but novel hollow-core photonic crystal fibers can guide the light with low attenuation, which dramatically increases sensitivity. This technique is called fiber-enhanced Raman spectroscopy (FERS) and has shown promising results for isotope-resolved breath gas detection.[35]

14.4 Selected applications

In the following, the impact of optical spectroscopy in breath gas analysis is exemplified by a few practical applications.

14.4.1 Lung function parameters

With the real-time capability and accuracy of optical spectroscopy, the possibility has emerged to measure lung function parameters. An important example is the recent work by Ciaffoni and coworkers, who have developed a sensor to determine oxygen consumption (VO_2) on a breath-by-breath basis at the mouth of intubated patients.[10] The compact device combines real-time detection of CO_2 and H_2O with DAS, and oxygen (O_2) with ICOS, all with a time response of 10 ms. This device enables clinical studies to explore the effects of a range of common interventions on VO_2 in patients in various degrees of shock, including the use of fluids, blood transfusions, elevated oxygen, and vasoactive drugs.

The highly precise, time-resolved data delivered by optical spectroscopy have made it possible to obtain information on lung inhomogeneity by use of an N_2 washout maneuver and a mathematical gas exchange model.[36] The shape of the N_2 washout reflects the width of the alveolar ventilation-to-volume distribution, while the CO_2 and O_2 exhalation profiles encode information about the anatomic deadspace and its distribution in the lung, as well as the distribution of ventilation-to-perfusion ratio in the lung. Key parameters that can be recovered from the model are the standard deviations for the distributions in deadspace, fractional lung compliance, and fractional vascular conductance. Significant differences in these parameters were found between chronic obstructive pulmonary disease (COPD) patients and healthy controls.

A detailed model of pulmonary gas exchange[37] was used by Ghorbani et al. to fit simulated CO exhalation profiles to experimental exhaled CO breath cycles recorded with multipass-enhanced TDLAS.[13] This method enables the extraction of end-tidal and alveolar CO, average lung diffusion capacity, and airway tissue CO concentration from single exhalations.

Sensitive breath cycle—resolved analysis of exhaled CO and CH_4 (Fig. 14.2B) was performed with OF-CEAS together with spirometry by Ventrillard et al. to map the levels of these gases during different ventilation phases[15] and to monitor bronchial inflammation. The role of exhaled CO and NO as biomarkers for the selection of lung grafts for transplantation was also investigated with the same optical technique. In a first demonstration, the gas from isolated pig lungs in an ex vivo lung perfusion device was analyzed to determine endogenous CO produced during the progressive rewarming of the lung after cold ischemia.[23] It was found that the level of CO production correlated with the severity of ischemia-reperfusion injury.

14.4.2 Oxidative stress and red blood cell lifetime

A commercial CO_2 laser—based LPAS analyzer with a detection limit of 300 ppt in 5 s (Sensor Sense, Netherlands) was used to assess lipid peroxidation by expired ethylene (C_2H_4) during cardiac surgery (Fig. 14.3A).[31] During the surgery, patients were undergoing the most common surgical valvular cardiopulmonary bypass and coronary artery bypass grafting on beating heart procedures. The study revealed, for the first time, real-time signatures of the surgical response triggered by oxidative stress and showed the significant contribution of diathermy and reperfusion events to overall oxidative stress. In addition, LPAS clearly revealed the effect of diathermy and sternotomy on breath C_2H_4 levels (Fig. 14.3B). This demonstrated that sensitive, noninvasive real-time analysis of lipid peroxidation via exhaled C_2H_4 is feasible in clinical and perioperative settings. The combination of fluid phase platforms and real-time analysis of lipid peroxidation could improve patient phenotyping prior to and during high-risk surgery.

An NDIR spectrometer has been used to demonstrate that the endocrine life span (the red blood cell lifetime) in humans can be accurately determined by measuring the alveolar CO concentration.[9] This implies a significant advantage compared to established methods, as it is much faster, does not involve radioactive substances, and allows for continuous noninvasive monitoring with sufficient accuracy.

14.4.3 Stable isotope tracer studies

Optical techniques play an important role as fast, low-cost alternatives or complements to isotope ratio mass spectrometry (IRMS) for stable isotope breath studies. This is clinically exploited with success in the $^{13}CO_2/^{12}CO_2$ urea breath test (with NDIR or CRDS) for *H. pylori* diagnosis[8] and for evaluating the activity of

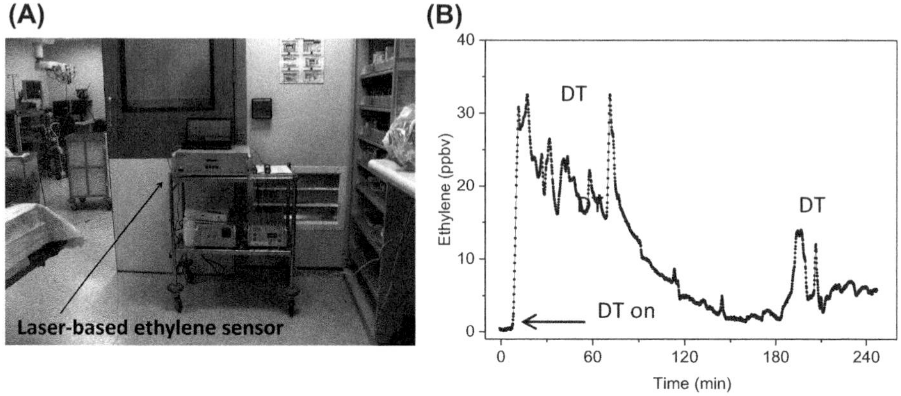

FIGURE 14.3

(A) Laser-based photoacoustic spectrometer in the cardiothoracic surgery department at Harefield Hospital, UK. (B) Continuous intraoperative real-time monitoring of ethylene by LPAS in patients undergoing coronary artery bypass grafting on beating heart. At t = 0 h, the patient was connected to the sampling line; *DT*, diathermy (electrocautery with high-frequency electric currents).

(A) Photo by S. Cristescu. (B) Adapted from[5].

drug-metabolizing enzymes using stable isotope-labeled substrates[38] (see Chapter 17). In another study, CRDS was used for real-time NO isotope detection.[21] Recently, stable carbon isotope ratio measurement on CH_4 with a commercial CRDS analyzer provided evidence that CH_4 is produced in all human beings.[17] CEAS was also used to measure ^{18}O and ^{13}C-isotopic fractionations of breath CO_2 in patients with end-stage renal disease undergoing hemodialysis.[22] The results suggest that $^{12}C^{18}O^{16}O$ and $^{13}C^{16}O^{16}O$ could be used as novel markers for rapid, noninvasive assessment of dialysis adequacy and for tracking the physiological status of individuals with end-stage renal disease or chronic kidney disease.

14.4.4 Biomarker physiology and in vitro studies

As is evident from the applications presented above, the high sensitivity, accuracy, and real-time capabilities of laser-based absorption spectroscopy open up avenues for detailed physiological and toxicological studies. Of special interest for the field is the investigation of biomarker origins and to scrutinize the dependence of breath concentrations on physiological, environmental, and technical factors, such as variations of breath sampling, external exposure, or exercise. Often, the ambient air concentration of the biomarker and exhaled CO_2 can be measured along with the target molecule, which can serve as a measure of the integrity of the breath sample.

Substantial work has been conducted with CRDS regarding the origin and physiology of exhaled NH_3, HCN, CH_4, and acetylene (C_2H_2).[16–20] Nose-exhaled HCN levels were found to correlate with systemic (blood) HCN concentrations.[19]

Furthermore, the HCN concentrations measured in the breath of cystic fibrosis (CF) patients were strongly correlated with the levels produced by *Staphylococcus aureus* strains isolated from sputum of these patients on the day of breath sampling.[39] Contrary to established belief, exhaled NH_3 mainly originates from NH_4^+/NH_3 in the oral cavity and saliva, not primarily from blood ammonia, and depends on saliva pH and the blood urea level.[16,20] Further, C_2H_2 was detected in the breath of healthy nonsmokers for the first time, with exhaled concentrations reflecting those of recently inhaled ambient air C_2H_2 concentrations.[18]

Optical spectroscopy is also a useful tool for in vitro studies of gases emitted by pathogenic bacteria.[39–41] Like humans, bacteria produce a wide variety of volatile organic and inorganic compounds as part of their metabolism (cf. Chapter 26). These metabolites may serve as biological fingerprints of bacterial presence directly in exhaled breath, urine, blood, saliva, sputum, or bronchoalveolar lavage (BAL) fluid. Several studies using LPAS and CRDS have investigated the production of specific target molecules, such as HCN and NH_3, by *Pseudomonas aeruginosa* in relation to CF and oral anaerobic pathogens.[40,41] Real-time measurements have shown the dynamics of these gases during all bacterial growth phases. The emission of HCN from *S. aureus*, for example, was continuously recorded during the first 24 h of in vitro growth using LPAS.[39] By comparing and contrasting the in vitro work with ex vivo or in vivo samples, knowledge on volatile biomarkers of clinically relevant species can be advanced toward the differentiation between antibiotic sensitive and resistant strains and polymicrobial metabolic interactions.

14.5 Future prospects of optical techniques in breath analysis

The sensitivity of an optical absorption measurement depends linearly on the line strength of the optical transition that is used to probe the analyte. When rovibrational transitions are used, the highest line strengths are generally associated with the fundamental transitions that occur in the mid-IR region. Accordingly, the trend in optical trace gas analysis is to move the instrumentation to work in the mid-IR. This enables detection limits down to sub-ppt levels in some cases, albeit with longer averaging times. Traditionally, the laser sources, detectors, and optics are cheaper and more developed in the near-IR region due to the use of that region for optical telecommunication. The current rapid development of the laser sources and other optical instrumentation, in particular in the mid-IR, ensures that the cost and size of laser-based trace gas analyzers will come down in the future.

One of the main limiting factors of current optical gas analyzers is the number of analytes that can be probed using a single laser. Normally, the analyzers are developed to measure one to three analytes simultaneously, and often only one of the species can be analyzed with high sensitivity. However, the development of OFC spectroscopy (see Section 14.3.4.) is bringing a paradigm shift to optical spectroscopy. Instead of a single monochromatic, tunable laser frequency, the OFC provides a hundred thousand or million such frequencies in a single beam. The best mid-IR OFCs can currently cover frequency ranges as wide as 3–5.5 μm[30] or 6–17 μm.[27] Such a broadband coverage enables, in principle, the simultaneous measurement of dozens or even hundreds of molecules that exhibit absorption features in this wavelength range. In addition to multispecies detection, the broad spectral coverage would enable the analysis of larger molecules that exhibit broad absorption features that cannot be recorded using lasers with a narrow scanning range. A more straightforward alternative to OFC spectroscopy is multiplexed spectroscopy, where several laser sources are combined in one analytical instrument to probe multiple analytes simultaneously.

Optical frequency comb spectroscopy comes with certain caveats, the first of which are cost and complexity. Frequency combs are still expensive and employ bulky light sources. In addition, the spectrum of the OFC must be spectrally analyzed to distinguish the absorption experienced by the individual comb lines. This is done either by using bulky dispersive spectrometers and FTIRs, or by employing dual-comb spectroscopy. Another caveat is sensitivity. To reach sub-ppb detection limits, an OFC needs to be coupled with cavity-enhanced detection (see Section 14.3.3) in a similar manner as with single-mode lasers. Even though this has ben demonstrated, the implementation of cavity-enhanced OFC spectroscopy is technically very demanding. A third practical obstacle is data analysis. The evaluation of broadband IR spectra of breath samples containing dozens of absorbing molecules, all at low concentrations, is not trivial. In addition, the analysis will be complicated by the presence of a few molecules (e.g., H_2O, CO_2) at very high concentrations and a few others (e.g., CH_4, NH_3, CO, acetone) at ppm and sub-ppm levels. The sharp absorption features of small molecules are well cataloged in various databases, but larger compounds with broad features that do not correspond to any analytical function are much less thoroughly documented. The practical analysis of such a broadband high-sensitivity spectrum of a breath sample will be complicated to perform, even more so automatically and in real time.

The unique features of optical spectroscopy (quantitative, fast, and sensitive) will continue to find use in breath analysis. For quantitative single-species analysis, the method could be even considered the best of all currently available techniques. As promising novel breathborne biomarkers are discovered using other techniques, new optical instruments can subsequently be developed to analyze these volatiles with cost-effective, robust, and compact devices suitable for routine use in the clinical practice. Whether the methodology will expand from the current single-species analysis of small molecules will largely depend on how far OFC spectroscopy will be developed in the future. Nevertheless, even with the currently available instrumentation, there are new frontiers in breath analysis to be explored using optical spectroscopy. One example is the high-speed measurement of trace species in breath that enables the extraction of pulmonary gas exchange parameters from individual expirograms.

14.6 Summary

Optical spectroscopy is a very powerful technique for exhaled breath analysis. It could be argued that for single-species real-time analysis of small molecules (two to five atoms), optical spectroscopy represents the gold standard for both breath analysis as well as for other trace gas applications. If the chemical identity of the breath biomarker is known and the molecule exhibits well-defined absorption features, in principle, it is quite straightforward to develop a robust optical analyzer that does not require an expert user. The cost, size, and complexity will ultimately depend on the required sensitivity. A small NDIR-based capnograph to measure CO_2 levels of 4%–5% will always be 10–100 times cheaper than a single-mode laser spectrometer based on cavity-enhanced spectroscopy that reaches a detection limit of 0.1 ppb.

Among the online analytical platforms used in breath gas analysis, optical spectroscopy lies somewhere between sensors and mass spectrometers. Optical spectrometers are generally more compact than most online analyzers based on MS. Laser-based analyzers can currently be manufactured down to shoe-box size, and NDIR instruments can be made even smaller. On the other hand, sensors can be miniaturized even further. Laser-based analyzers are, in general, more sensitive than sensors but cannot compete with most sensitive MS-based analyzers, except in special cases. Among online analysis techniques, the selectivity of optical spectroscopy is almost unparalleled, but this is the case only for relatively small molecules. Accurate quantification is easier than with probably any other technique used in breath research. Finally, optical analyzers can be built for use by nonexperts in a black box fashion. Commercial laser–based analyzers are available from several manufacturers. These instruments are currently marketed mainly for other applications than breath research (e.g., atmospheric research, soil flux measurements, petrochemical industry, exposure monitoring), but the same methodologies can be utilized also for breath analysis.

Optical spectroscopy is a powerful real-time analysis tool for accurate quantification of small breath biomarkers. Due to the robust and easy-to-use nature of optical analyzers, large-scale clinical studies can be realistically conducted by nonexpert users. Currently, optical instruments are not useful for nontargeted exploratory biomarker discovery, but with the rapid development of OFC spectroscopy, this might change in the near future.

References

1. Banwell CN, McCash EM. *Fundamentals of molecular spectroscopy*. New York: McGraw-Hill; 1994.
2. Gordon IE, Rothman LS, Hill C, Kochanov RV, Tan Y, Bernath PF, et al. The HITRAN2016 molecular spectroscopic database. *J Quant Spectrosc Radiat* 2017;**203**: 3–69.

3. . Linstrom PJ, Mallard WG. *NIST chemistry WebBook, NIST standard reference database number 69*. Gaithersburg MD: National Institute of Standards and Technology. p. 20899.
4. Maiti KS, Lewton M, Fill E, Apolonski A. Sensitive spectroscopic breath analysis by water condensation. *J Breath Res* 2018;**12**(4):046003.
5. Henderson B, Khodabakhsh A, Metsälä M, Ventrillard I, Schmidt FM, Romanini D, et al. Laser spectroscopy for breath analysis: towards clinical implementation. *Appl Phys B* 2018;**124**(8):161.
6. Metsälä M. Optical techniques for breath analysis: from single to multi-species detection. *J Breath Res* 2018;**12**(2):027104.
7. Wang CJ, Sahay P. Breath analysis using laser spectroscopic techniques: breath biomarkers, spectral fingerprints, and detection limits. *Sensors* 2009;**9**(10):8230–62.
8. Modak AS. Stable isotope breath tests in clinical medicine: a review. *J Breath Res* 2007; **1**(1):014003.
9. Zhang H-D, Ma Y-J, Liu Q-F, Ye T-Z, Meng F-Y, Zhou Y-W, et al. Human erythrocyte lifespan measured by Levitt's CO breath test with newly developed automatic instrument. *J Breath Res* 2018;**12**(3):036003.
10. Ciaffoni L, O'Neill DP, Couper JH, Ritchie GA, Hancock G, Robbins PA. In-airway molecular flow sensing: a new technology for continuous, noninvasive monitoring of oxygen consumption in critical care. *Sci Adv* 2016;**2**(8):e1600560.
11. Ghorbani R, Schmidt FM. Real-time breath gas analysis of CO and CO_2 using an EC-QCL. *Appl Phys B* 2017;**123**(5):144.
12. Ghorbani R, Schmidt FM. ICL-based TDLAS sensor for real-time breath gas analysis of carbon monoxide isotopes. *Optic Express* 2017;**25**(11):12743–52.
13. Ghorbani R, Schmidt FM. Fitting of single-exhalation profiles using a pulmonary gas exchange model—application to carbon monoxide. *J Breath Res* 2019;**13**(2):026001.
14. Wang Y, Nikodem M, Zhang E, Cikach F, Barnes J, Comhair S, et al. Shot-noise limited Faraday rotation spectroscopy for detection of nitric oxide isotopes in breath, urine, and blood. *Sci Rep* 2015;**5**:9096.
15. Ventrillard-Courtillot I, Gonthiez T, Clerici C, Romanini D. Multispecies breath analysis faster than a single respiratory cycle by optical-feedback cavity-enhanced absorption spectroscopy. *J Biomed Optic* 2009;**14**(6):064026.
16. Chen W, Metsälä M, Vaittinen O, Halonen L. The origin of mouth-exhaled ammonia. *J Breath Res* 2014;**8**(3):036003.
17. Keppler F, Schiller A, Ehehalt R, Greule M, Hartmann J, Polag D. Stable isotope and high precision concentration measurements confirm that all humans produce and exhale methane. *J Breath Res* 2016;**10**(1):016003.
18. Metsälä M, Schmidt FM, Skytta M, Vaittinen O, Halonen L. Acetylene in breath: background levels and real-time elimination kinetics after smoking. *J Breath Res* 2010;**4**(4): 046003.
19. Schmidt FM, Metsala M, Vaittinen O, Halonen L. Background levels and diurnal variations of hydrogen cyanide in breath and emitted from skin. *J Breath Res* 2011;**5**(4): 046004.
20. Schmidt FM, Vaittinen O, Metsälä M, Lehto M, Forsblom C, Groop P, et al. Ammonia in breath and emitted from skin. *J Breath Res* 2013;**7**(1):017109.
21. Halmer D, von Basum G, Horstjann M, Hering P, Murtz M. Time resolved simultaneous detection of 14NO and 15NO via mid-infrared cavity leak-out spectroscopy. *Isot Environ Health Stud* 2005;**41**(4):303–11.

22. Mandal S, Mukhopadhyay P, Ghosh C, Pal M, Banik GD, Chatterjee T, et al. Isotope-specific breath analysis to track the end-stage renal disease during hemodialysis. *J Breath Res* 2018;**12**(3):036019.
23. Maignan M, Briot R, Romanini D, Gennai S, Hazane-Puch F, Brouta A, et al. Real-time measurements of endogenous carbon monoxide production in isolated pig lungs. *J Biomed Optic* 2014;**19**(4). 047001−01.
24. Fjodorow P, Hellmig O, Baev VM. A broadband Tm/Ho-doped fiber laser tunable from 1.8 to 2.09 μm for intracavity absorption spectroscopy. *Appl Phys B* 2018;**124**(4):62.
25. Picqué N, Hänsch TW. Frequency comb spectroscopy. *Nat Photon* 2019;**13**(3):146−57.
26. Thorpe MJ, Balslev-Clausen D, Kirchner MS, Ye J. Cavity-enhanced optical frequency comb spectroscopy: application to human breath analysis. *Optic Express* 2008;**16**(4): 2387−97.
27. Butler T, Gerz D, Hofer C, Xu J, Gaida C, Heuermann T, et al. Watt-scale 50-MHz source of single-cycle waveform-stable pulses in the molecular fingerprint region. *Opt Lett* 2019;**44**(7):1730−3.
28. Villares G, Hugi A, Blaser S, Faist J. Dual-comb spectroscopy based on quantum-cascade-laser frequency combs. *Nat Commun* 2014;**5**:5192.
29. Coddington I, Newbury N, Swann W. Dual-comb spectroscopy. *Optica* 2016;**3**(4): 414−26.
30. Muraviev A, Smolski V, Loparo Z, Vodopyanov K. Massively parallel sensing of trace molecules and their isotopologues with broadband subharmonic mid-infrared frequency combs. *Nat Photon* 2018;**12**(4):209.
31. Cristescu SM, Kiss R, te Lintel Hekkert S, Dalby M, Harren FJ, Risby TH, et al. Real-time monitoring of endogenous lipid peroxidation by exhaled ethylene in patients undergoing cardiac surgery. *Am J Physiol-Lung C* 2014;**307**(7):L509−15.
32. Harren FJ, Cristescu SM. Photoacoustic spectroscopy in trace gas monitoring. In: Meyers RA, editor. *Encyclopedia of analytical chemistry: applications, theory and instrumentation*; 2019.
33. Patimisco P, Scamarcio G, Spagnolo V. Quartz-enhanced photoacoustic spectroscopy: a review. *Sensors* 2014;**14**(4):6165−206.
34. Peltola J, Vainio M, Hieta T, Uotila J, Sinisalo S, Metsälä M, et al. High sensitivity trace gas detection by cantilever-enhanced photoacoustic spectroscopy using a mid-infrared continuous-wave optical parametric oscillator. *Optic Express* 2013;**21**(8):10240−50.
35. Hanf S, Keiner R, Yan D, Popp Jr , Frosch T. Fiber-enhanced Raman multigas spectroscopy: a versatile tool for environmental gas sensing and breath analysis. *Anal Chem* 2014;**86**(11):5278−85.
36. Mountain JE, Santer P, O'Neill DP, Smith NMJ, Ciaffoni L, Couper JH, et al. Potential for noninvasive assessment of lung inhomogeneity using highly precise, highly time-resolved measurements of gas exchange. *J Appl Physiol* 2018;**124**(3):615−31.
37. Ghorbani R, Blomberg A, Schmidt FM. Modeling pulmonary gas exchange and single-exhalation profiles of carbon monoxide. *Front Psychol* 2018;**9**:927.
38. Modak AS. Point-of-care companion diagnostic tests for personalizing psychiatric medications: fulfilling an unmet clinical need. *J Breath Res* 2017;**12**(1):017101.
39. Neerincx AH, Linders YA, Vermeulen L, Belderbos RA, Mandon J, van Mastrigt E, et al. Hydrogen cyanide emission in the lung by *Staphylococcus aureus*. *Eur Respir J* 2016; **48**(2):577−9.

40. Chen W, Roslund K, Fogarty CL, Pussinen PJ, Halonen L, Groop P-H, et al. Detection of hydrogen cyanide from oral anaerobes by cavity ring down spectroscopy. *Sci Rep* 2016; **6**:22577.
41. Neerincx AH, Mandon J, van Ingen J, Arslanov DD, Mouton JW, Harren FJ, et al. Real-time monitoring of hydrogen cyanide (HCN) and ammonia (NH3) emitted by *Pseudomonas aeruginosa*. *J Breath Res* 2015;**9**(2):027102.

CHAPTER

Comprehensive gas chromatography-mass spectrometry

15

Delphine Zanella[1], Jean-François Focant[1], Jane E. Hill[2], Pierre-Hugues Stefanuto[1]

[1]*Organic and Biological Analytical Chemistry Group, MolSys Research Unit, University of Liège, Liège, Belgium;* [2]*Chemical and Biological Engineering, University of British Columbia, Vancouver, BC, Canada*

Chapter outline

15.1 Overview

Metabolomics studies aim at exploring the metabolites present in an organism. Exhaled breath analysis captures a subset of metabolites and the consequences of metabolism and is thus seen as a promising noninvasive and early diagnostic tool for infectious and noninfectious diseases in humans. Breath potentially contains valuable information about human health, while being relatively easy to collect. Nevertheless, as for most biological matrices, its composition is complex, and the untargeted screening of its composition requires powerful analytical techniques. Thanks to its enhanced resolution and its high sensitivity, comprehensive two-dimensional gas chromatography (GC×GC) coupled to time-of-flight mass spectrometry (TOFMS) is an ideal technique for the characterization of volatile organic compounds (VOCs) in exhaled breath. The power of GC×GC relies on the combination of two capillary columns of different stationary phases connected in series via a resampling "modulator" device. There, the production of narrow second dimension chromatographic pulses allows for the detection of up to 10-fold more compounds compared to classical single dimension (1D) GC.[1] The efficiency of the technique

Breathborne Biomarkers and the Human Volatilome. https://doi.org/10.1016/B978-0-12-819967-1.00015-3
Copyright © 2020 Elsevier B.V. All rights reserved.

has been demonstrated in numerous fields of research, including metabolomics,[2] environmental analysis,[3] and diverse medical applications.[4–7] In addition, the recent combination of GC×GC with high-resolution (HR) MS detectors, such as HRTOFMS, makes it a highly powerful tool for untargeted screening, enabling the accurate identification of compounds. Although increasing our understanding of the volatilome, these advanced analytical capabilities come with challenges, mainly in terms of method optimization and data management. Indeed, there is an ongoing need for the development of more user-friendly solutions to make GC×GC-HRTOFMS a key tool for exploring the human volatilome.

15.2 Gas chromatography; a required step for complex sample analysis

According to the International Union of Pure and Applied Chemistry (IUPAC) definition, chromatography is a physical method of separation in which the components to be separated are distributed between two phases, one of which is stationary (stationary phase) while the other (mobile phase) moves in a definite direction. According to their physicochemical properties (e.g., polarity, volatility, etc.), molecules will be retained on the column stationary phase to a lesser or greater degree. This differential migration generates separation of the different constituents of a complex mixture, i.e., chromatography.[8]

In GC, the mobile phase is an inert carrier gas—typically hydrogen (H_2) or helium (He)—and the stationary phase is generally a siloxane-based polymer, which can vary in polarity. GC can be hyphenated with a large variety of detectors, including a flame ionization detector (FID) or a mass spectrometry (MS), whereby the latter is most commonly used for untargeted analysis.[8]

Gas chromatography coupled to mass spectrometry (GC-MS) is the accepted gold standard for the analysis of complex volatile mixtures.[8] The chromatographic dimension allows separation of the compounds present in the sample prior to MS detection. This separation step provides a clearer view of the sample composition by warranting clean mass spectra, and the elution pattern provides an additional identification criterion (i.e., retention time).[8] The latter parameter is often corrected relative to a reference mixture—typically containing an *n*-alkane series—to establish a linear retention index (LRI; also known as the Kovats index), which caters for drift between runs or over time.[9]

15.3 Why add an extra chromatographic dimension?

In metabolomics, the major aim is to analyze the metabolome by characterizing and quantifying the small-molecule chemicals present in a biological sample. Such a characterization of the metabolome would enlarge the knowledge of metabolic pathways involved during the development of diseases, the presence of toxins, or the exposure to contaminants.

The idea of exhaled breath metabolomic profiling relies on the alteration of the breath profile when a pathological disorder appears. Moreover, the main working hypothesis states that this alteration can be detected and used for diagnosis and monitoring. The usefulness of VOC analysis in exhaled breath as a diagnostic tool has already been demonstrated for a wide variety of diseases, as reported in other chapters of this book.[4–6,10]

Gas chromatography-mass spectrometry remains the gold standard for breath analysis;[11] however, breath VOCs are usually present in large numbers, over a wide dynamic range, and at trace levels, making their detection and identification challenging (see Chapter 1). In such regard, to keep making progress and to gain knowledge in metabolomic processes, advanced and powerful analytical tools must be considered to detect, identify, and quantify unknown metabolites. Because of the large number of peak coelution in classical 1DGC, the richness of the breath volatilome is likely to be underestimated. As such, GC×GC is seen as a more suitable tool for the exhaustive detection of VOCs in complex samples, such as exhaled breath (Fig. 15.1). The enhanced peak capacity and the high sensitivity of GC×GC over classical GC makes it the method of choice for untargeted metabolomics applications,[12] especially when untargeted screening is required. In that context, major contributions of GC×GC have been reported in environmental analysis,[3,13] petroleum,[14] food and flavors,[15] and forensics.[16]

As shown in Fig. 15.2, the improved separation power of GC×GC is achieved by combining two capillary columns made of different stationary phases. This extra chromatographic dimension provides an increased peak capacity that allows a better resolution of the sample components. This high resolution is the keystone of untargeted screening. Indeed, it has been demonstrated that GC×GC is a powerful analytical technique for the study of VOCs in exhaled human breath. Typically, compared to classical GC-MS, GC×GC-MS enables the detection of 3- to 10-fold more compounds, extending knowledge on the human volatilome.[17] Therefore, the use of GC×GC in metabolomic applications will extend the boundaries of knowledge on

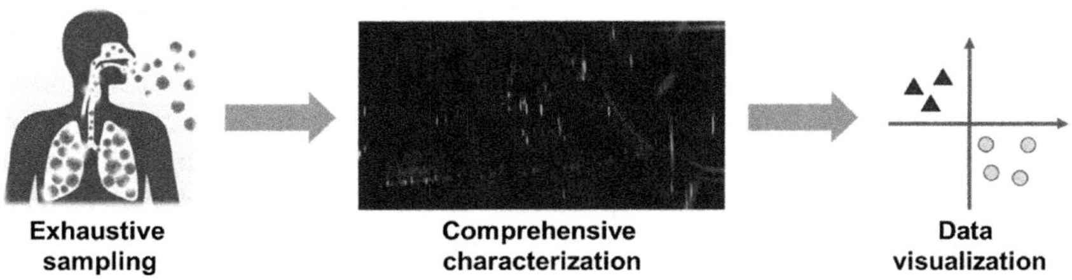

FIGURE 15.1

Analytical workflow representing the main steps of exhaled breath study with comprehensive gas chromatography (GC×GC), which provides a clear picture of the sample composition.

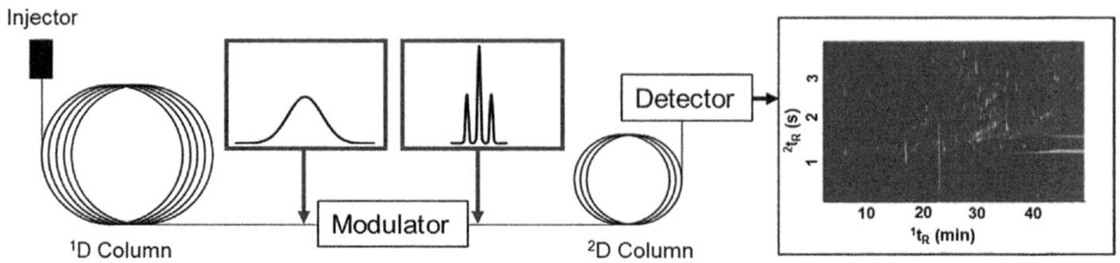

FIGURE 15.2

Schematic representation of a comprehensive gas chromatography (GC×GC) system.

pathological diseases. In such systems, the two dimensions are connected through a modulator. The modulator is the heart of the system: it controls the transfer of analytes between the two GC dimensions, and it traps, refocuses, and injects narrow pulses of the entire eluent from the first (^{1}D) into the second (^{2}D) column, allowing a comprehensive separation.[14] Moreover, the use of cryogenic modulators provides enhanced sensitivity due to mass conservation during the refocusing step.[18] Indeed, the cryogenic zone compression of the chromatographic peaks in the modulator provides better limits of detection for trace-level analytes.

The increase in peak capacity and the low limit of detection are complemented by a chemistry-based structuration of the chromatogram, which is considered as an additional identification criteria in GC×GC.[19] These are essential for comprehensive screening of the human volatilome and could reveal new potential biomarkers of diseases, leading to the development of improved diagnostic methods.

15.4 The impact of hyphenation with high-resolution mass spectrometry

The accurate description of GC×GC signals requires fast detectors capable of responding to the rapid ^{2}D separation process. Indeed, the resolution achieved in ^{1}D must be preserved in ^{2}D during the entire separation process. Therefore, late eluting analytes within a modulated pulse must exit ^{2}D before early eluting analytes of the next pulse of modulation. Such high-speed conditions in ^{2}D require the production of very narrow peaks (60–600 ms) that necessitate a detector response time in the range of 100 Hz. Flame ionization detectors (FID) were the first detectors used in combination with GC×GC as these offer acquisition frequencies of up to 300 Hz.[20] In the 1990s, the FID was the detector of choice for petrochemical analysis, which was one of the main applications of GC×GC.[14] Indeed, FID provides a response that is proportional to the number of carbon atoms, which enables both qualitative and quantitative analysis. However, such detectors do not provide structural information needed for compound identification. For such purpose, the use of MS is required.

The coupling of comprehensive GC×GC with MS offers an additional dimension of analytical separation and makes GC×GC-MS an efficient technique for complex sample characterization. GC×GC-MS provides a vast amount of information about sample composition. The fragments generated by the electron ionization (EI) source of the mass analyzer give valuable information about the chemical structure of the detected VOCs. This contributes to the identification of the VOCs present in exhaled breath, which offers insights into the biological processes occurring in the body and therefore about the health status of a patient.

Since the introduction of mass analyzers in GC×GC experiments, the most commonly used MS detectors were low-resolution time-of-flight mass spectrometers (LRTOFMS).[12] The attractiveness of TOFMS detectors is mainly related to the very high data acquisition speed, the nonscanning mode, accurate mass determination, and the high efficiency for deconvolution of MS signals.[12] In addition, TOFMS presents the ability to acquire full mass range spectra with no compromise on sensitivity, as all masses are collected simultaneously. The high acquisition speed of TOFMS enables the accurate definition of GC×GC peaks over a broad dynamic range (typically five decades, depending on the system). Therefore, the full spectra obtained are highly valuable for untargeted analysis.[7] A further strength of such analyzers is the mass spectral continuity generating consistent ion ratios across the chromatographic peaks. This characteristic enables the further deconvolution of coeluting peaks, if required. These combined abilities, i.e., to acquire full mass range spectra at high speed with high sensitivity, makes TOFMS a remarkable analyzer for both qualitative and quantitative analysis.

The combination of GC×GC with HRTOFMS makes it one of the most powerful analytical techniques for the analysis of complex samples, such as exhaled breath. HRTOFMS generates sensitive full-spectrum data in the same manner as LRTOFMS. In addition, HRTOFMS provides high mass resolution and accurate mass measures (see the next chapter). The possibility of having the accurate mass of the molecular ion is particularly valuable, since the elemental composition can be determined with increased confidence. Although retention times in both dimensions and the LRIs provide additional information on the chemical structure of VOCs, some compounds can only be tentatively identified by LRTOFMS. Indeed, relying on the MS libraries, such as the National Institute of Standards and Technology (NIST) library, presents some limitations. The main issues are encountered when being confronted with isomers that are not listed in such libraries or when identifying structural isomers of hydrocarbons that display high spectral similarities. Typically, GC×GC-HRTOFMS is used for untargeted or pretargeted analysis in the case of complex matrix elucidation.[4] The accurate full spectrum can also be searched a posteriori for the identification of compounds of interest; reference databases can aid in compound identification (see Chapter 39). Nevertheless, the necessity of the HR GC step combined with the HR mass analyzers is highly dependent on the sample complexity.[12]

The identification capacities of comprehensive GC coupled to HRMS for untargeted screening is further supported by the combination of orthogonal identification criteria, as chromatographic and spectral information offers the possibility to reach level 2 MSI (Metabolomics Standard Initiative) confidence in identification, even in the context of untargeted studies.[21] This confidence in compound identification is highly beneficial when aiming for complete validation of putative volatile markers.

15.5 A powerful and complex technique

The enhanced analytical capacities of multidimensional techniques, such as GC×GC, come with an increased complexity in method optimization and data management.[17] Indeed, the extra separation and the HR mass spectral dimensions require fine tuning in order to exploit the technique to its fullest extent. Even if optimization tools are available, expert users are still required to set up the analytical conditions, limiting the spread of the technique. In addition, to ensure reproducibility between the studies, quality control (QC) and validation of sampling and analysis methods must be performed and should be adequately reported.[4,22] As for most analytical techniques, because of the lack of robustness, results reported in the literature are highly dispersed and often not reproducible when tentatively identifying significant analytes for disease diagnosis. The constant improvement of the analytical techniques forces researchers to make a series of choices that can bias the selection toward different classes of molecules. Among these choices, the most important parameters include the breath VOC collection approach (see Chapter 2), the nature of the trapping material, the type of thermal desorption (TD) technique, and the chromatographic separation and detection parameters. Therefore, it is necessary to implement a minimum of reporting standard guidelines that describe every step of the analytical process, including the QC procedure, for the publication of robust GC×GC data. Even though exhaled breath analysis has a great potential for disease diagnosis and monitoring, this approach is not yet accepted in clinical research, demonstrating the necessity to implement standardized and controlled procedures.

The increase in size of the detectable volatilome provided by GC×GC also generates numerous challenges. Each collected data point contains two retention time values (summed as LRIs), a signal intensity value, and mass spectral information (made of both a fragmentogram and an accurate mass). This amount of information already implies the need to adopt powerful hardware and software mechanisms to store, process, visualize, and analyze those sets of generated data. This is of particular importance, especially during continuous long-term clinical studies.[4]

In addition, this information must be preprocessed to generated useable peak tables for statistical investigations. In general, robust alignment methods together with powerful machine learning algorithms are required to identify significant VOCs of pathological metabolism (see Chapter 38). However, powerful statistical methods are not enough to identify new potential markers. Excellent analytical skills remain indispensable to produce robust raw data, ensure their proper treatment, guarantee reproducibility, and avoid erroneous discoveries. To assess the robustness of the collected data, a platform for global data transparency could be established. Another challenge comes with the efficiency to gather and more importantly share the information with the clinic. A more comprehensive communication would increase the opportunity of these findings to play a major role in the exploration and discovery processes.[4,17,22] These problems are not only specific to GC×GC, but also appear with other untargeted screening methods. Much of the scientific initiatives organized by the International Association of Breath Research (IABR) and the Metabolomics Society are aimed at tackling those challenges.

Nowadays, more and more GC×GC user-friendly solutions are being developed by research groups and instrument providers. Nevertheless, the expansion of the technique requires integrated and simplified solutions. There are numerous opportunities to develop algorithms addressing data filtering, data transformation, feature extraction, and feature selection.

15.6 What has comprehensive chromatography brought to the field?

In 2002, Seeley and colleagues used a GC×GC system coupled to an FID for the first time to analyze healthy human breath collected onto multibed sorbent TD tubes.[23] The goal was to use different column sets to show their efficacy and specificity for breath analysis. This comprehensive analytical investigation of the breath matrix revealed the high complexity of its composition. For example, the authors demonstrated the importance of the second dimension to resolve coeluting alcohol isomers (e.g., primary vs. secondary), which are important compounds in breath research.[23] A few years later, Libardoni and Sanchez reported on a similar TD-GC×GC method coupled to FID and TOFMS to show the potential of GC×GC for exhaled breath analysis.[24,25] These proof-of-principle studies demonstrated the successful usage of GC×GC for exhaled breath, detecting compounds with higher sensitivity and specificity without increasing the time required to perform the analysis. These properties, together with the reduced compound misidentification that may have been induced by coelution of VOCs, and with the detection of a broader range of breath VOCs, will extend the accurate knowledge on the volatilome, leading to improved diagnostic methods. With the development of the technique and the availability of commercial instruments and software solutions, different successful applications of GC×GC for exhaled breath analysis have been conducted.

In 2006, Sanchez and colleagues published the first application of GC×GC-TOFMS in breath research for medical purposes.[25] The study demonstrated the ability of exhaled breath analysis to detect environmental and exposure contaminations. This proof-of-concept study identified markers of active smoking in exhaled breath (2,5-dimethylfuran, 2-methylfuran, and furan), which demonstrated the potential of breath research for noninvasive human biomonitoring.

Since then, GC×GC has mainly been used to evaluate the use of exhaled breath analysis as a diagnostic technique for several respiratory diseases. Asthma, tuberculosis (TB), and lung cancer have been the most widely studied.[4–7,26]

Asthma is a chronic lung disease that affects more 350 million people worldwide. Asthma is not curable, but proper monitoring helps to improve the quality of life of sufferers. The first study involving exhaled breath analysis by GC×GC for asthma diagnosis was performed by Caldeira et al. in 2012.[5] This study demonstrated the progress that could be made in health sciences using advanced analytical techniques, such as GC×GC, by exploring the exhaled breath analysis of 32 asthmatic children and 27 healthy children. The authors reported a combination of nine compounds comprising alkanes and aldehydes that could discriminate between the two sample classes. The partial least squares–discriminate analysis (PLS-DA) model had a classification rate of 96% and showed 98% sensitivity and 93% specificity, although those promising results are yet to be confirmed on a larger cohort. Several years later, Schleich et al., Ibrahim et al., and Wilde et al. used GC×GC to gain more insights into metabolic processes specific to asthma. Schleich et al. reported the first large-scale clinical study with the aim of phenotyping asthma by analyzing the breath of a cohort of 521 asthmatics patients.[4] The authors identified and validated five biomarkers to phenotype neutrophilic and eosinophilic asthma, with accuracies of 73% and 72%, respectively. In this study, the researchers reported optimized analytical methods, as well as the implementation of a strong QC procedure, to produce robust data that enabled reliable identification of the biomarkers. The biomarkers, hexane, 1-propanol, 2-hexanone, nonanal, and undecane, corroborated previous studies using GC-MS on asthma. The work of Schleich and colleagues also demonstrated the complementarity of multidimensional GC with classical GC-MS for exhaled breath analysis. Indeed, the discovery study was conducted using GC-MS and the validation using GC×GC-HRTOFMS. Ibrahim et al. and Wilde et al. both focused on the acute breathlessness symptoms associated with asthma, which are responsible for frequent hospital admission. In their studies, they reported optimized study designs, analytical methods, and QC procedures for conducting large-scale population screenings.[6,26]

In the area of infectious diseases, TB is one of the top 10 causes of death worldwide, with 1.5 million fatalities in 2018. This high mortality is mainly attributable to a lack of early screening tools to enable timely treatment. Exhaled breath is a promising solution to improve the situation.[27] In 2018, Beccaria et al. reported the lack of reproducibility of the numerous studies involving GC-MS for the identification of TB biomarkers in exhaled breath.[27] The study highlighted that the low concentration of volatile compounds released by infected patients required an advanced and sensitive analytical technique, such as GC×GC, to ensure the reproducibility of the studies. In a follow-up study, an optimized GC×GC method was used together with machine learning algorithms to highlight possible biomarkers associated with TB. A total of 23 discriminant features, mainly hydrocarbons, were highlighted using three different machine learning algorithms. Since no analytical validation study has been performed, the identity of the potential biomarkers was not revealed.

Indeed, as for any analytical techniques, an external validation study is of prime importance for biomarker identification in untargeted analysis. Further studies on TB are reported in Chapter 21.

In respiratory diseases, lung cancer is among the leading causes of death, affecting half a billion people worldwide.[28] Ma et al. performed qualitative and quantitative breath analysis by GC×GC for biomarker discovery of lung cancer.[29] Using this advanced technique, the authors reported detection limits below 1.0 ng/L, which is suitable for the analysis of trace level compounds found in breath. A few years later, Pesesse et al. highlighted the importance of a powerful analytical strategy together with robust statistical approach to extract significant information from breath VOCs. Indeed, using appropriate data preprocessing, the researchers could overcome the influence of exogenous VOCs and identify 37 possible biomarkers for lung cancer, although these have yet to be validated.[7]

Such studies demonstrate that this methodological approach using a widely available matrix together with powerful analytical instrumentation offers the possibility to broaden knowledge on metabolic pathways by increasing the spectra of detected and identified VOCs. This approach can lead to important discoveries in numerous fields of research. Nevertheless, these studies highlight the prime importance of good analytical practice when complex instrumentation is involved.

15.7 Future perspectives for comprehensive gas chromatography

Nowadays, the requirement of expert users for method development and data processing is the main challenge associated with GC×GC usage. The need of in-depth knowledge is slowing down the acceptance and development of such comprehensive techniques. Often, the need for rapid answers and the lack of resources are limiting the use of GC×GC, especially in the biomedical field, thus equally for exhaled breath analysis. However, the added value from the high-quality information acquired compensates for this complexity. The solution to these challenges lies in the development of fully integrated methods to make the use of GC×GC simpler and more straightforward. Currently, attempts are being made to make GC×GC more accessible. Indeed, the development of optimized, controlled, and reported processing tools, together with the implementation of robust statistical treatments, will lead to fully integrated solutions for sample analyses and data handling in the near future.

The coupling of GC×GC with HRTOFMS has further improved the identification capabilities and separation capacities, opening new routes for diverse applications. Indeed, this potential is pushing forward the use of GC×GC in clinical applications, where biomarker identification is of prime interest to implement targeted treatments. Nevertheless, there remains room for development. Indeed, biomarker identification for specific diseases should be reported only in cases where validation studies are carried out. This would ensure the robustness of clinical studies, putting more confidence in the significant analytes highlighted. It is expected that this will lead to a higher number of large-scale studies divided into two main parts, a discovery part that aims to identify compounds of interest, and a validation part to ensure their proper identification and their robustness using independent patient cohort data.

Evolving from classical GC to GC×GC in clinical research will be highly beneficial since it will expand knowledge on the composition of the human volatilome. In addition, the developed and optimized analytical methods will be easily applicable to other biological matrices, such as exhaled breath condensate (EBC; see Chapter 7), bronchoalveolar lavage fluid (BALF), or sputum, which are highly relevant in the clinical field. Therefore, despite some limitations, GC×GC, especially when coupled to HRTOFMS, will inevitably open new routes for innovative diagnosis methods.

15.8 Summary

The complex chemical composition of exhaled breath presents a real analytical challenge, especially for untargeted screening. Indeed, to ensure appropriate fingerprinting of a sample, high resolving power is required. For many years, 1DGC has been the go-to method for volatile mixture characterization. When untargeted resolution of complex samples is required, however, the resolution of the technique reaches its limits. Pushing forward the resolution limitations, GC×GC is the emerging solution. GC×GC relies on the combination of multiple chromatographic dimensions, which can be hyphenated with HRMS. The resulting system allows full sample resolution with a high confidence in compound identification.

Nevertheless, such analytical power comes with an increased complexity in method optimization and data handling. For each data point collected, two retention times and a full (HR) mass spectrum is collected at a rate of 100–200 data points per second. So, there is a growing importance in the use of powerful computer tools to assist optimization and high-dimensional data management. The recent efforts to overcome these complications are making GC×GC the method of choice for untargeted volatilomics.

In recent years, several successful studies using GC×GC have been conducted for different pathologies. In the complex field of cancer research, for example, the use of exhaled breath represents a real hope for large-scale population screenings. Different proof-of-concept and optimization studies have been successfully conducted in cancer research. The resolution power of GC×GC has been compared to classical 1DGC studies in different fields. For asthma characterization, different research projects are pushing the use of GC×GC to diagnose and understand lung inflammation. The first large-scale study combining the output of both GC-MS and GC×GC-HRTOFMS has been conducted for discovery and validation of breath volatiles. Moreover, much effort is being devoted to infectious disease detection, with the aim to complement time-consuming culture-based diagnosis.

The techniques of GC×GC-HRTOFMS and LRTOFMS are powerful analytical approaches, especially for untargeted screening. The method has demonstrated its added value for several breath-based research areas. The development of more user-friendly and integrated solutions, especially on the (automated) data processing front, will make GC×GC a key future tool in the field of breath research.

References

1. Ramos L, editor. *Comprehensive two dimensional gas chromatography.* 1st ed. Elsevier Science; 2009. p. 328.
2. Gruber B, Keller S, Groeger T, Matuschek G, Szymczak W, Zimmermann R. Breath gas monitoring during a glucose challenge by a combined PTR-QMS/GC×GC-TOFMS approach for the verification of potential volatile biomarkers. *J Breath Res* 2016;**10**(3): 36003.
3. Megson D, Focant J, Patterson DG, Robson M, Lohan MC, Worsfold PJ, et al. Can polychlorinated biphenyl (PCB) signatures and enantiomer fractions be used for source identification and to age date occupational exposure?. Vancouver *Environ Int* 2015:56–63.
4. Schleich FN, Zanella D, Stefanuto P, Bessonov K, Graff S, Moermans C, et al. Exhaled volatile organic compounds are able to discriminate between neutrophilic and eosinophilic. *Am J Respir Crit Care Med* 2019:1–44.
5. Caldeira M, Perestrelo R, Barros AS, Bilelo MJ, Morête A, Câmara JS, et al. Allergic asthma exhaled breath metabolome: a challenge for comprehensive two-dimensional gas chromatography. *J Chromatogr A* 2012;**1254**:87–97.
6. Ibrahim W, Wilde M, Cordell R, Salman D, Ruszkiewicz D, Bryant L, et al. Assessment of breath volatile organic compounds in acute cardiorespiratory breathlessness: a protocol describing a prospective real-world observational study. *BMJ Open* 2019;**9**(3):1–13.
7. Pesesse R, Stefanuto PH, Schleich F, Louis R, Focant JF. Multimodal chemometric approach for the analysis of human exhaled breath in lung cancer patients by TD-GC × GC-TOFMS. *J Chromatogr B Anal Technol Biomed Life Sci* 2019;**1114–1115**: 146–53.
8. McNair HM, Miller JM. Basic gas chromatography. *J Chem Educ* 1998;**75**:1094.
9. Kováts E. Gas-chromatographische Charakterisierung organischer Verbindungen. Teil 1: retentionsindices aliphatischer Halogenide, Alkohole, Aldehyde und Ketone. *Helv Chim Acta* 1958;**41**(7):1915–32.
10. Di Giovanni N, Meuwis M, Louis E, Focant J. Untargeted serum metabolic profiling by GC×GC-HRTOF-MS untargeted serum metabolic profiling by GC×GC-HRTOF-MS. *J Proteome Res* 2019:1–46.
11. Beauchamp J. Current sampling and analysis techniques in breath research - results of a task force poll. *J Breath Res* 2015;**9**(4):47107.
12. Tranchida PQ, Franchina FA, Dugo P, Mondello L. Introduction I. Comprehensive two-dimensional gas chromatography-mss spectrometry: recent evolution and current trends. *Mass Spectrom Rev* 2016;**35**:524–34.

13. de Vos J, Dixon R, Vermeulen G, Gorst-Allman P, Cochran J, Rohwer E, et al. Comprehensive two-dimensional gas chromatography time of flight mass spectrometry (GC×GC-TOFMS) for environmental forensic investigations in developing countries. *Chemosphere* 2011;**82**(9):1230–9.
14. Seeley JV, Seeley SK. Multidimensional gas chromatography: fundamental advances and new applications. *Anal Chem* 2013;**85**(2):557–78.
15. Cordero C, Liberto E, Bicchi C, Rubiolo P, Reichenbach SE, Tian X, et al. Targeted and non-targeted approaches for complex natural sample profiling by GC×GC-qMS. *J Chromatogr Sci* 2010;**48**(4):251–61.
16. Dekeirsschieter J, Stefanuto P, Brasseur C, Haubruge E. Enhanced characterization of the smell of death by comprehensive two-dimensional gas chromatography- time-of-flight mass spectrometry (GC×GC-TOFMS). *PLOS One* 2012;**7**(6).
17. Higgins Keppler EA, Jenkins CL, Davis TJ, Bean HD. Advances in the application of comprehensive two-dimensional gas chromatography in metabolomics. *TrAC Trends Anal Chem (Reference Ed)* 2018;**109**:275–86.
18. Patterson DG, Welch SM, Turner WE, Sjödin A, Focant JF. Cryogenic zone compression for the measurement of dioxins in human serum by isotope dilution at the attogram level using modulated gas chromatography coupled to high resolution magnetic sector mass spectrometry. *J Chromatogr A* 2011;**1218**(21):3274–81.
19. Dimandja JMD, Clouden GC, Colón I, Focant JF, Cabey WV, Parry RC. Standardized test mixture for the characterization of comprehensive two-dimensional gas chromatography columns: the Phillips mix. *J Chromatogr A* 2003;**1019**(1–2):261–72.
20. Adahchour M, Beens J, Vreuls RJJ, Brinkman UAT. Recent developments in comprehensive two-dimensional gas chromatography (GC×GC) II. *Modulation and detection* 2006;**25**(6).
21. Viant MR, Kurland IJ, Jones MR, Dunn WB. How close are we to complete annotation of metabolomes? *Curr Opin Chem Biol* 2017;**36**:64–9.
22. Purcaro G, Stefanuto P, Franchina FA, Beccaria M, Wieland-alter WF, Wright PF, et al. Analytica Chimica Acta SPME-GC×GC-TOF MS fi ngerprint of virally-infected cell culture: sample preparation optimization and data processing evaluation. *Anal Chim Acta* 2018;**1027**:158–67.
23. Seeley JV, Kramp FJ, Sharpe KS, Seeley SK. Characterization of gaseous mixtures of organic compounds with dual-secondary column comprehensive two-dimensional gas chromatography (GC × 2 GC). *J Separ Sci* 2002;**25**(1–2):53–9.
24. Libardoni M, Stevens PT, Waite JH, Sacks R. Analysis of human breath samples with a multi-bed sorption trap and comprehensive two-dimensional gas chromatography (GC × GC). *J Chromatogr B Anal Technol Biomed Life Sci* 2006;**842**(1):13–21.
25. Sanchez JM, Sacks RD. Development of a multibed sorption trap, comprehensive two-dimensional gas chromatography, and time-of-flight mass spectrometry system for the analysis of volatile organic compounds in human breath. *Anal Chem* 2006;**78**(9): 3046–54.
26. Wilde MJ, Cordell RL, Salman D, Zhao B, Ibrahim W, Bryant L, et al. Breath analysis by two-dimensional gas chromatography with dual flame ionisation and mass spectrometric detection – method optimisation and integration within a large-scale clinical study. *J Chromatogr A* 2019;**1594**:160–72.

27. Beccaria M, Bobak C, Maitshotlo B, Mellors TR, Purcaro G. Exhaled human breath analysis in active pulmonary tuberculosis diagnostics by comprehensive gas chromatography-mass spectrometry and chemometric techniques Exhaled human breath analysis in active pulmonary tuberculosis diagnostics by comprehensive gas chr. *J Breath Res* 2019;**13**:016005.
28. World Health Organization. [Internet]. [cited 2018 Jan 20]. Available from: www.who.int/respiratory/asthma/en.
29. Ma H, Li X, Chen J, Wang H, Cheng T, Chen K, et al. Analytical Methods patients and healthy controls with solid-phase comprehensive two-dimensional gas. *Anal Methods* 2014;**6**:6841–9.

CHAPTER

High-resolution mass spectrometry

16

M. Ariel Geer Wallace, James P. McCord

National Exposure Research Laboratory, United States Environmental Protection Agency, Research Triangle Park, NC, United States

Chapter outline

Breathborne Biomarkers and the Human Volatilome. https://doi.org/10.1016/B978-0-12-819967-1.00016-5
Copyright © 2020 Elsevier B.V. All rights reserved.

16.1 Overview

Many breath applications of mass spectrometry analysis have traditionally relied on low-resolution mass spectrometers, such as single quadrupoles and ion traps for targeted analyses of known compounds of interest. The implementation of high-resolution mass spectrometry (HRMS) in breath research has expanded the capabilities in data collection for both targeted and nontargeted analyses. High-resolution mass spectrometers can differentiate between two peaks with small *m/z* differences. Additionally, HRMS instruments often have high mass accuracy and can measure the mass of a molecule with an accuracy of 0.0001 Da or less (within 1 ppm) compared to the exact mass, in contrast to low-resolution mass spectrometers that report only integer mass measurements (~1 Da). Therefore, at its simplest, HRMS can be used to distinguish between molecules with the same integer masses but different exact masses, thereby improving feature identifications in nontargeted analyses. Chemical features can be observed as distinct masses detected in HRMS analyses that are representative of different compounds or compound fragments that have not been identified through targeted approaches. Data processing steps are necessary to assign chemical formulas to features, and many commercial software packages are now available for high-throughput processing of HRMS data.

Mass resolution has been defined by the International Union of Pure and Applied Chemistry (IUPAC) as $R = m/\Delta m$, where m is the measured mass of the peak and Δm is the peak width or the distance between two peaks of equal height, where the valley between the peaks is 10% or less of the maximum peak height.[1] The mass resolving power of an instrument is the ability to separate two peaks with small differences in *m/z* values (i.e., *m/z* 50.00 and 50.01). Mass resolving power varies with m and is reported as $m/\Delta m_{50\%}$, where m is the mass of the peak and $\Delta m_{50\%}$ is the peak width at the full width at half maximum height (FWHM) of the peak in mass units.[2] The performance characteristics of individual technologies are always improving, with both top-end resolving power and scan speed being chief targets for improvement. In general, low-resolution mass spectrometers, such as quadrupole filters and ion traps, have unit mass resolution (resolving power ~1000), while high-resolution time-of-flight (TOF) instruments have resolving powers in the tens of thousands. The resolving power of TOF instruments is dependent on molecular mass, with lower resolution observed for lighter compounds, which improves with increasing compound molecular mass. Orbitraps have mass resolving power on the order of hundreds of thousands, and Fourier-transform ion cyclotron resonance (FT-ICR) mass spectrometers reach resolving power of hundreds of thousands to millions.

The HRMS instruments for breath research applications use various sample introduction frontends, such as proton transfer reaction (PTR), gas chromatography (GC), liquid chromatography (LC), and thermal desorption (TD). The wide variety of instrumentation has opened the door for real-time breath sampling as well as exhaled breath sampling using canisters, sorbent tubes, and polymer bags (e.g., Tedlar) (see Chapter 2). While low-resolution mass spectrometry is capable of accurately identifying small molecules in targeted gas-phase analyses, larger molecules in exhaled breath condensate (EBC) and exhaled breath aerosol (EBA) samples benefit from HRMS analyses, which provide more accurate mass measurements that can be used to identify these organic molecules to a greater degree of certainty. The use of HRMS for these applications increases the amount of data that can be acquired for a single sample and improves the confidence of compound identifications during nontargeted analyses. This chapter will discuss some of the fundamental concepts of HRMS, provide an overview of common mass analyzers used for HRMS analyses, and highlight some current HRMS breath research applications.

16.2 Fundamentals of high-resolution mass spectrometry

16.2.1 Mass definitions

The precision of modern mass spectrometers requires careful specification of the "mass" of a molecule, as many different masses can be measured. Differentiating between the atomic mass, average mass, exact mass, monoisotopic mass, and accurate mass of a molecule can be confusing for data collection, interpretation, and reporting in mass spectrometry. The SI unit for mass is the kilogram (kg), but atomic/molecular scale measurements are commonly expressed in unified atomic mass units (u) or daltons (Da), which are defined as $1/12^{th}$ the mass of a neutral carbon 12 atom, with a mass of 1.66×10^{-27} kg.[3] Fig. 16.1 shows where the nominal, monoisotopic, and average masses can be found in the mass spectrum of an unknown compound.

Atomic mass

The atomic mass of an atom is an empirically measured property, which is equivalent to the sum mass of protons, neutrons, and electrons that make up the atom (with a small adjustment for nuclear binding energy). Isotopes have the same number of protons but different numbers of neutrons and thus have different atomic masses. Collections of atoms have a property called atomic *weight*, which incorporates all of the isotopes present.

Average mass

For most collections of atoms/molecules, the existence of isotopes contributes significantly to measurements of molecular mass. The average mass of an element, its atomic weight, is the average of all isotopes of the element weighted by their natural abundance. For molecules, the average mass is the sum of the underlying atomic weights. This is equivalent to the weighted sum of all isotopes of all atoms that make up the molecule. The average mass can be easily determined using the atomic weights of the elements found in the periodic table. The atomic masses of ^{12}C

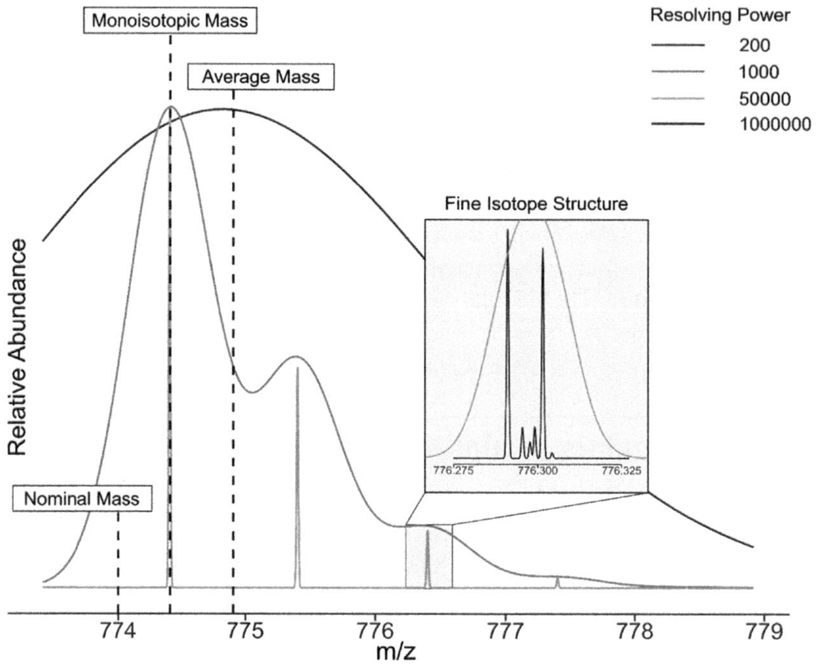

FIGURE 16.1

Simulation spectrum for an unknown compound at increasing mass resolving powers. Higher resolving powers readily distinguish isotope information and allow more accurate monoisotopic mass measurements. Very high resolving powers (inset) allow separation of isotopic fine structure, enabling more accurate formula prediction.

and ^{13}C, for example, are 12 Da and 13.003355 Da, respectively, while the atomic weight of carbon is 12.011 Da, which accounts for both the natural abundance and isotope together.

Exact mass

The exact mass of a compound is used when dealing with a single molecule or an isotopically pure collection of molecules. The exact mass is the calculated sum of the masses of the individual atomic isotopes that comprise the molecular formula. The exact mass of a molecule can be calculated from the sum of the atomic mass of each element in its chemical formula if the isotope composition of each element is specified. Table 16.1 lists the exact masses for common stable isotopes of elements found in HRMS analyses.

Monoisotopic mass

The monoisotopic mass is a special case of the exact mass, where the molecule contains only the single most abundant natural stable isotope of each element. For most small (organic) molecules, which contain primarily C, H, N, and O, the

Table 16.1 Relative abundance, exact mass, and mass defect of stable isotopes of select organic elements found in HRMS analyses. Isotopic composition fractions obtained from isotopic compositions of the elements.[4] Isotopic abundances are reported as intervals representing the variation found in natural materials.

Element	Isotope	Isotopic abundance intervals	Nominal mass (Da)	Exact mass (Da)	Mass defect (Da)
Hydrogen					
^{1}H	M	0.99972–0.99999	1	1.007825032	0.007825032
^{2}H	M+1	0.00001–0.00028	2	2.014101778	0.014101778
Carbon					
^{12}C	M	0.9884–0.9904	12	12.00000000	0.00000000
^{13}C	M+1	0.0096–0.0116	13	13.00335484	0.00335484
Nitrogen					
^{14}N	M	0.99578–0.99663	14	14.00307400	0.00307400
^{15}N	M+1	0.00337–0.00422	15	15.00010890	0.00010890
Oxygen					
^{16}O	M	0.99738–0.99776	16	15.99491462	−0.00508538
^{17}O	M+1	0.000367–0.000400	17	16.99913176	−0.00086824
^{18}O	M+2	0.00187–0.00222	18	17.99915961	−0.00084039
Fluorine					
^{19}F	M	1	19	18.998403162	−0.00159684
Phosphorous					
^{31}P	M	1	31	30.973761998	−0.02623800

Continued

Table 16.1 Relative abundance, exact mass, and mass defect of stable isotopes of select organic elements found in HRMS analyses. Isotopic composition fractions obtained from isotopic compositions of the elements.[4] Isotopic abundances are reported as intervals representing the variation found in natural materials.—*cont'd*

Element	Isotope	Isotopic abundance intervals	Nominal mass (Da)	Exact mass (Da)	Mass defect (Da)
Sulfur					
^{32}S	M	0.9441–0.9529	32	31.972071174	−0.02792883
^{33}S	M+1	0.00729–0.00797	33	32.971458910	−0.02854109
^{34}S	M+2	0.0396–0.0477	34	33.967867004	−0.03213300
^{36}S	M+4	0.000129–0.000187	36	35.96708071	−0.03291929
Chlorine					
^{35}Cl	M	0.755–0.761	35	34.96852682	−0.03147318
^{37}Cl	M+2	0.239–0.245	37	36.96590260	−0.03409740
Bromine					
^{79}Br	M	0.505–0.508	79	78.9183376	−0.08166240
^{81}Br	M+2	0.492–0.495	81	80.9162897	−0.08371030

monoisotopic mass is the most abundant mass. Therefore, the monoisotopic mass of a compound, such as carbon dioxide ($^{12}C^{16}O_2$), can be calculated using Table 16.1 as follows:

$$12.00000 + (2 \times 15.99491462) = 43.98982924.$$

The monoisotopic mass and the average mass for CO_2 calculated from the periodic table of the elements have low error:

$$12.011 + (2 \times 15.999) = 44.009.$$

For chlorinated molecules, however, the monoisotopic mass is not always the most abundant mass. The monoisotopic mass for $^{1}H^{35}Cl$, for example, can be calculated from Table 16.1 as follows:

$$1.007825032 + 34.96852682 = 35.97635185.$$

The average mass for HCl calculated from the periodic table of the elements is much larger than the monoisotopic mass of HCl:

$$1.008 + 35.45 = 36.458.$$

Accurate mass

Accurate mass is a common term used in the literature and marketing material that refers to an experimentally determined mass that has a degree of accuracy/precision necessary to predict the elemental composition of the underlying ion/molecule. Mass error for a measurement is reported as the difference between the accurate (experimental) and exact (theoretical) mass values. Mass error in HRMS is normally quite low and is commonly expressed in terms of the relative error in parts per million (ppm),[3] calculated as follows:

$$\frac{\text{accurate mass} - \text{exact mass}}{\text{exact mass}} * 10^6$$

16.2.2 Mass defect

Mass defect is the difference between the exact mass and the nominal integer mass of an atom or molecule, as shown in Table 16.1. The nominal mass requires knowledge of the formula to calculate but can be approximated by rounding the mass to the nearest integer number. A previous study of organic compounds in the DSSTox database (see Chapter 39) showed that only 1.31% of mass defects were incorrectly assigned using the rounding method, making this a reasonable technique to help discern the elemental composition of an unknown feature in HRMS.[5] The mass defect is positive for certain elements and negative for others, which contributes to the overall positive or negative mass defect of a compound, based on its elemental composition. Hydrocarbon compounds have significant positive mass defects, while the inclusion of halogens, sulfur, oxygen, and/or phosphorous causes the mass defect to trend toward negative. In HRMS, the mass defect can be used to predict the chemical class, as halogenated compounds (negative defect) are obviously distinct from nonhalogenated compounds (positive defect). Analysis of homologous compound classes can be accomplished with Kendrick mass defect plots, where related compounds form straight lines, which may aid in identifying compounds in complex breath samples.[6] Other mass defect plots can be constructed by comparison to reference molecules, which can help assign molecular class.[7]

16.2.3 Isotope ratios

The primary strength of HRMS is its ability to resolve individual combinations of isotopes with mass differences much smaller than 1 Da. The use of isotope peaks allows more accurate determination of the elemental composition of an unknown molecule than exact mass assignment alone. In fact, research has shown that knowledge of isotopic abundance patterns removes >95% of false identifications compared to exact mass matches alone, even at high mass accuracy (<1 ppm).[8] In an HRMS spectrum, isotopic peaks of compounds appear in clusters, weighted by the abundance of heavier (or lighter) isotopes. A spectrum is typically annotated with respect to the monoisotopic mass peak (A) based on the nominal number of 1 Da neutron shifts from the monoisotopic mass—A, A+1, A+2, etc. (see Fig. 16.1). As described in Pleil et al. (2018), carbon counting, sulfur counting, and halogen-counting techniques can be used to determine the molecular formula of a compound based on the isotopic abundance patterns observed in the mass spectrum and the known isotopic abundances of these elements.[9]

16.2.4 Fragmentation patterns

Fragmentation of molecules provides an important mechanism for structural identification. Fragmentation can occur either in source, as is the case with electron ionization (EI) sources, or in activation/collision cells in tandem mass spectrometers. Tandem mass spectrometry designs are used to isolate molecules of interest in the first stage, fragment them in a cell, and analyze the resulting fragments in a secondary mass spectrometry stage. Low-resolution mass spectrometers are familiar as the ubiquitous triple quadrupole (QqQ) mass spectrometer, where a single precursor is selected by the first quadrupole, fragmented in the collision cell, and a single product ion selected in the third quadrupole to yield an ion transition signal. Likewise, in-source fragmentation by EI is common in GC-MS with a single, low-resolution quadrupole analyzer. For high-resolution mass spectrometers, a quadrupole typically performs precursor selection similarly to low-resolution instruments, but the product ions are collected and scanned simultaneously in the high-resolution mass analyzer. This allows collection of a complete fragmentation spectrum in a single scan at high resolution. Formation of diagnostic ions can indicate the detection of a specific structural class, and tentative chemical identifications can be made by reconstructing the parent molecule from the constituent fragments. More accurate identification can be made by comparing fragmentation spectra to reference library spectra or against spectra collected from authenticated reference standards under the same conditions. While these techniques are applicable in low-resolution instruments, the increased mass accuracy of HRMS allows much more specific matching and structural prediction.[10]

16.3 Mass analyzers

The core of HRMS lies in the mass analyzer. Instrumental features of the mass analyzer include scan speed (duty cycle), dynamic range, mass resolving power (R), mass accuracy, and the capability of performing molecular fragmentation. To date, advancement in the major high-resolution mass analyzers, TOF, and Fourier-transform mass spectrometers (Orbitrap and FT-ICR) have highly differentiated platforms with a range of performance characteristics.

16.3.1 Time-of-flight mass spectrometers

Time-of-flight mass analyzers derive accurate mass measurements from the measured transit time for an ion with a known kinetic energy moving a known distance through an electric field. The TOF analyzer pulses a packet of ions using a known amount of energy and measures the arrival time of ions to determine their mass. Individual TOF spectra measurements (a transient) are rapid, on the order of microseconds, and multiple measurements are summed or averaged to produce a mass spectrum. Mass accuracy of the TOF is related to the accuracy of the time measurement, which requires nanosecond scale resolution. As consequence, the mass resolving power of a measurement is related to the mass of the measured analyte; more massive, slower moving analytes, where the time error is small relative to the flight time, have higher effective mass resolving power, while lower molecular weight compounds have less mass resolving power. The mass resolving power for the TOF analyzer is calculated as follows:

$$m/\Delta m_{50\%} = (\mathrm{T}/2\Delta \mathrm{t})$$

where T is the total ion flight time and Δt is the width of the mass spectral peak. The mass resolving power can be improved by using a reflectron plate, which extends the flight time of the ions.[2] The upper end mass range of the TOF analyzer is essentially unlimited, enabling easy analysis of proteins, long-chain lipids, glycans, and polymers.[11,12]

The greatest strength of TOF analyzers over other high-resolution analyzers is its fast acquisition speed. The high scan rate of TOF instruments allows more temporal resolution across a chromatographic run, allowing easier coupling to chromatographic systems with short peak widths or real-time sampling devices, such as in

PTR-TOF-MS (see Chapter 10). TOF resolving power and scan speed are largely fixed based on the platform, but resolving powers are normally on the order of tens of thousands, while scan speed is in the range of 1–50 Hz.[13]

16.3.2 Orbitrap mass spectrometers

Orbitrap mass analyzers are the descendants of trapping mass analyzer systems. The Orbitrap functions as both a mass analyzer and a detector, based on the configuration of its electrostatic trapping cell. Ions are injected as a packet into the analyzer and oscillate axially between an outer electrostatic charge plate and an internal electrode. Current generated by oscillating ions is measured by the device, and a Fourier transform is applied to determine the individual ion frequencies, which are proportional to the exact ion masses. The mass resolving power of Orbitrap instruments is related to the number of oscillation "beats" in the time transient, which is related to the mass of the measured analyte. Resolving power for small molecules is substantially higher as a result, with the effective resolving power more than halving between *m/z* 200 and *m/z* 1000. Orbitrap analyzers further have a dynamic mass resolving power based on the scan time, with the top-end resolving power based on the model. Reported top-end resolving power for Orbitrap instruments is 1,000,000 (FWHM @ *m/z* 200) at relatively slow scan speeds, although these can reach 40 Hz scan rates at reduced resolving power.[14]

The greatest strength of Orbitrap mass analyzers lies in their ability to obtain very high mass resolving powers and high mass accuracy with a benchtop footprint. Increased resolving power reduces the potential for mass collisions in a full scan spectrum, which allows for the detection of more distinct masses and isotopic clusters, as well as more accurate molecular assignments. Further, as trap-based instruments, Orbitraps can achieve very high sensitivities by lengthening ion collection times to collect more ions and generate higher quality MS and MS/MS spectra of very weak ion signals.

16.3.3 Fourier-transform ion cyclotron resonance mass spectrometers

Fourier transform ion cyclotron resonance (FT-ICR) mass spectrometry bases its measurements on the cyclic oscillation of a charged ion within a magnetic field (cyclotron frequency). The oscillation of the ions is detected as a time domain transient current consisting of the sum of all the ion motions and transformed to frequency measurements using a Fourier transform. Like Orbitraps, which are also FT-MS instruments, the oscillation frequency is related to the ion mass-to-charge and there is a proportional relationship between resolving power and scan speed, with top-end resolving powers on the order of millions with sub-ppm mass accuracy.[15]

Fourier transform ion cyclotron resonance instruments are specialty platforms requiring substantial infrastructure (high-field magnets), which place them outside common usage, although research studies using these platforms are still ongoing in national laboratories. Their primary benefit is the extremely high resolving power, which is capable of separating almost all isotopic mass collisions, and the unparalleled sub-ppm mass accuracy of FT-ICR-MS.

16.3.4 Collection of product ion spectra

While HRMS data are powerful for compound discovery and identification, the choice of data acquisition method influences how data can be processed and interpreted. The simplest type of data collection is to perform a high-resolution full-scan (MS1) experiment, where the instrument collects mass values across a wide scan range (e.g., *m/z* 100–2000). This dedicates 100% of the instrument scanning time to collecting MS information, at the cost of not obtaining product ion MS/MS scans that could yield more structural information. When using EI, fragmentation may still occur enough to determine chemical structure, but for softer ionization techniques, such as chemical ionization (CI) or electrospray ionization (ESI), the lack of fragmentation spectra will limit compound identification to formula predictions and accurate mass matches only.

A more thorough approach is to collect MS/MS scans as part of the experiment to obtain fragmentation information. Under typical conditions, the mass spectrometer is not fast enough to obtain MS/MS scans for every mass detected in an MS1 scan (especially not on the timescale of a fast GC or UPLC chromatographic peak); therefore some compromise approaches exist to maximize valuable MS/MS information. One can perform a full-scan experiment without MS/MS as a profiling experiment for semiquantification and perform follow-up experiments with MS/MS enabled for specific compounds to provide identification. With a sufficiently fast mass spectrometer and data processing software, it is common practice to interlace triggered MS/MS scans with MS1 scans in the same experiment.

The earliest, and still most common, approach to precursor and product spectra collection is data-dependent acquisition (DDA). In DDA experiments, MS1 scans are assessed according to a series of logical decisions to prioritize masses for subsequent MS/MS scans. The selected precursor masses are sequentially isolated by the instrument and fragmented to collect the fragment spectrum. In "Top N" mode, compounds of highest abundance are prioritized, with up to N MS/MS product scans per MS1 precursor scan (Fig. 16.2); N is determined by experimental factors, including the speed of the mass analyzer, abundance of ions, and the width of the chromatographic peak. Inclusion/exclusion lists can also be used, where masses are preferentially targeted or excluded from MS/MS collection either across the whole run or dynamically following collection of an MS/MS scan. Faster mass analyzers can provide deeper coverage with more MS/MS scans, but fast chromatographic separations with narrow peaks can still challenge the collection speed of the analyzer. In situations where sample quantity and instrument time are not limitations, sequential injections with exclusion lists can ensure almost complete coverage.[16]

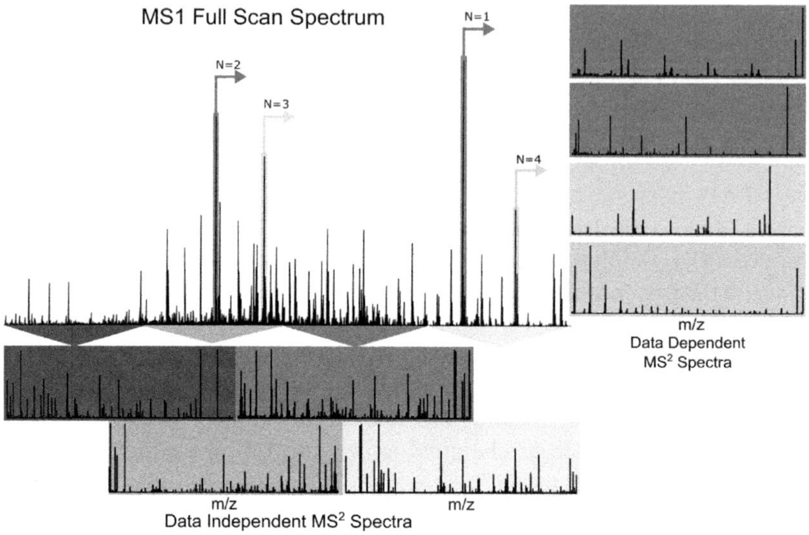

FIGURE 16.2

Selection of precursors for fragmentation from an MS1 scan under two fragmentation schemes. (Right) selection of Top four precursor ions for MS2 in data-dependent acquisition (DDA). (Bottom) isolation of four wide mass ranges for MS2 in data-independent acquisition (DIA).

Data-independent acquisition (DIA) methods, such as all-ion fragmentation (AIF), SWATH-MS,[a] MS^E, etc., offer an alternative to specific, triggered MS/MS scans. In this approach, no precursor fragmentation selection occurs; instead, a wide mass range (20+ Da) of incoming ions are fragmented simultaneously, without selection logic. This reduces any bias in selecting precursors, either from DDA scan logic or from experimental design, but has the downside of decoupling MS/MS information from precursor ion scans. The width of the ion selection range for scans can be tweaked to reduce the complexity of the resulting spectra, but it reduces the temporal resolution of MS/MS scans due to the increased number of scans, similarly to scaling the Top N number in DDA. Assignment of fragment ions/spectra to compounds relies on deconvolution algorithms and/or database searches to resolve the multiplexed fragment ions and assign them to parent compounds as reconstructed spectra.[17] Development of small molecule deconvolution and assignment software is still ongoing, compared to tools for protein analysis, and the acquisition speed necessary to support data analysis in this mode is exceptional, limiting the number of applicable platforms.

[a] Sequential Windowed Acquisition of All Theoretical Fragment Ion Mass Spectra.

16.4 High-resolution mass spectrometry applications for breath research

16.4.1 Techniques for different breath matrices

Introduction of breath samples to an HRMS analyzer is an important consideration when conducting nontargeted experiments. Traditionally, HRMS development has been driven by liquid coupled platforms (LC-MS), and many breath applications for HRMS have focused on liquid matrices, such as EBC. Alternative approaches, including GC-HRMS and real-time analyzers, such as PTR-TOF-MS, are a more recent development with rapidly evolving capabilities.

Gas-phase exhaled breath

Gas-phase exhaled breath consists of volatile organic compounds (VOCs) that originate from endogenous processes and exogenous exposures (see Chapter 1). Gas-phase exhaled breath may be analyzed in either real time or captured using a canister, polymer bag, or sorbent tube for laboratory analysis. Gas-phase breath samples are gathered as the subject exhales into the collection apparatus, capturing the end-tidal breath volume (see Chapter 2). PTR-TOF-MS instruments are amenable to real-time online exhaled breath analysis. PTR-TOF-MS instruments have high mass resolving power and are useful for gas-phase breath sampling because the instruments are smaller and more portable than conventional GC-MS and LC-MS instruments, making the PTR-TOF-MS more applicable for field work or studies in hospitals while retaining high mass resolving power. Such studies employing high-resolution PTR-TOF-MS analyses have been previously conducted in hospitals with mechanically ventilated patients.[18] Results of compounds in breath can be achieved immediately with real-time analysis, while sample collection and analysis by GC-MS may take several weeks or months for the sample to be delivered to the lab and analyzed. PTR-TOF-MS is treated in more detail in Chapter 10.

Gas-phase breath samples that are collected and stored can be analyzed off-line in an HRMS laboratory. Samples stored on sorbent tubes and canisters can be analyzed by TD with GC and PTR frontends. Any HRMS instrument, including GC-Q-TOF and GC-Orbitrap, that can be coupled to a TD system for analysis of sorbent tubes and canisters could be used for gas-phase exhaled breath analysis. While HRMS with GC separation is a more established technique for off-line analysis, in a recent study, TD was coupled to PTR-TOF-MS to analyze breath samples collected from patients prior to upper gastrointestinal endoscopy or surgery using a ReCIVA device (Owlstone Ltd., Cambridge, UK).[19] The use of HRMS in exhaled breath analyses using PTR makes it more likely to correctly identify compounds in real time without a compound separation step compared to GC-MS or LC-MS.

Exhaled breath condensate

Exhaled breath condensate is a liquid matrix consisting of water vapor and aerosol particles from within the lungs (see Chapter 7). EBC typically contains larger compounds than gas-phase exhaled breath, such as peptides, cytokines, leukotrienes, isoprostanes, particles, and exogenous compounds from the environment. EBC is collected as a subject exhales into a chilled breathing tube for several minutes to condense the aqueous fraction of exhaled air. As a liquid matrix, EBC is compatible with LC sample introduction to HRMS instruments, which is advantageous since HRMS instruments have been traditionally available with LC capabilities. More recently, some HRMS instruments have become available with GC frontends. EBC samples can be introduced to HRMS instruments through liquid injection using GC separation. In a recent study, a GC coupled to a Q Exactive Orbitrap MS was used to analyze EBC samples collected from cystic fibrosis patients to quantify thiocyanate levels.[20]

Exhaled breath aerosol

Exhaled breath aerosol consists of the aerosol fraction of EBC, including the larger molecules, such as fatty acids, proteins, viruses, and bacteria, and it may contain exogenous compounds originating from food, medications, and chemical exposures (see Chapter 7). EBA can be collected onto filters, mask materials, or solid surfaces and then extracted with solvents or buffers for LC-MS or GC-MS analysis.[21] Similar to EBC, HRMS is advantageous for providing greater confidence in identifications of the larger compounds in breath samples that originate from particulate matter.

Recent advances in HRMS instrumentation have provided a way to analyze EBA in real time. A chemical analysis of aerosol online (CHARON) particle inlet with a PTR-TOF-MS instrument can be used to collect and analyze particle phase, semivolatiles, and VOCs in gaseous samples. The CHARON-PTR-TOF-MS has been applied for atmospheric chemistry investigations, and this instrument shows promise for future use in breath research.[22]

16.4.2 Nontargeted analysis

Breath research using mass spectrometry has traditionally focused on targeted analysis of known compounds of interest, often using single quadrupole MS (e.g., GC-MS, PTR-MS) or QqQ-MS. These low-resolution mass spectrometers with single Da resolution are an excellent resource for identifying and quantifying known compounds in breath, such as acetone, isoprene, and nitrogen, as well as exogenous contaminants, including benzene, toluene, and ethylbenzene. However, the availability of HRMS instruments has opened the door for nontargeted analysis of compounds in breath that are not in chemical reference libraries. Identifying unknown compounds with low molecular mass (less than 200 Da) may be possible with low-resolution MS because there are only a limited number of possible molecular formulas.[8] As compounds increase in size, however, it becomes increasingly difficult, if not impossible, to make compound assignments without HRMS and knowledge of isotopic abundance patterns; the number of compounds with the same unit mass and/or similar spectra increases rapidly with mass.

Nontargeted analysis can be useful for assessing recent and long-term exposure to contaminants as well as to provide information on health state. Smolinska et al., for example, performed nontargeted GC-TOF-MS analysis of exhaled breath and fecal samples from individuals with Crohn's disease in order to correlate volatile metabolites with the microbiota composition.[23] The use of GC-TOF-MS in this study made it possible to perform nontargeted analysis with greater confidence in metabolite identification. Another important application for nontargeted analysis in breath has been for proteomic and metabolomic investigations. The fast scan speed of HRMS instruments along with the high mass resolution are ideal for omics investigations. In one study, 229 unique proteins, many of which were secreted from the respiratory tract, were detected in EBC samples using an LC-linear trap quadrupole (LTQ)-Orbitrap.[24] The hybrid LTQ-Orbitrap instrument takes advantage of the high resolution and mass accuracy of the Orbitrap and the high speed and sensitivity of the linear ion trap, making this a useful instrument for proteomics investigations. In this instrument, scan rates of 3 Hz have been achieved in MS/MS mode with accurate mass detection (low ppm).[14] LTQ analyzers have also been coupled to FT-ICR MS detectors with similar specifications as the LTQ-Orbitrap, achieving 100,000 mass resolution at m/z 400 with 2 ppm mass accuracy.[25] An LTQ-FT Ultra mass spectrometer has been employed to identify 119 proteins and 164 metabolites in EBC collected from intubated infants.[26] The use of HRMS instrumentation in these nontargeted breath research applications increased the number of proteins and metabolites that could be identified.

16.4.3 Limitations of high-resolution mass spectrometry for breath analysis

While HRMS represents a major frontier in breath research, it is important to consider the limitations of such work. The instrument platforms used in HRMS tend to be much more expensive than lower resolution instruments and require more upkeep and operator training to maximize their value, which may not be feasible for standard laboratories. HRMS instrumentation can generate large blocks of multidimensional data, which requires specialized analysis software and long periods of analysis time, as well as expertise in handling such datasets. Statistical analysis of HRMS data with large numbers of signals compared to the number of samples requires a different approach than targeted work, and interpretation and explanation of the results is non-trivial. Further, while HRMS can provide compound identifications, the ability to detect unknowns is limited by a combination of the quality of reference databases, availability of confirmatory reference standards, and the ability of an analyst to perform time-consuming de novo structural elucidation.

16.5 Summary

High-resolution mass spectrometry is a useful tool for analyzing complex sample matrices, such as exhaled breath, EBC, and EBA. The increased resolution of HRMS instrumentation makes it possible to distinguish isotopic distributions and generate fragmentation patterns, improving the accuracy of chemical formula prediction and library matching for compound identification. Mass analyzers, such as TOF, Orbitrap and FT-ICR, all have high mass resolving power that can be used to generate high-quality data. Nevertheless, HRMS instrumentation is not a replacement for standard low-resolution mass spectrometers found in many research laboratories. While HRMS instrumentation is advantageous for conducting nontargeted analyses when the compounds of interest are unknown, high resolution is not necessary for basic applications. HRMS instrumentation tends to be expensive and generates large quantities of highly complex data, which is not necessary for routine, targeted analyses of a limited subset of compounds that can be processed using low-resolution mass spectrometry. For such targeted analyses, single Da resolution methods—commonplace worldwide—are sufficient for detection and quantification.

Recent developments in HRMS instrumentation configurations have improved its usefulness for nontargeted analyses and applications in breath research. Real-time analyses for exhaled breath and EBA using PTR-TOF-MS and CHARON inlet filter will provide opportunities for on-site analysis in the field, laboratory, or medical facility. Real-time HRMS instrumentation will provide immediate feedback for a more accurate assessment of an individual's health state and environmental exposure. The availability in recent years of HRMS instruments with GC capabilities, such as GC-Orbitrap and GC-Q-TOF, has provided additional platforms for analyzing volatile and semivolatile compounds in breath. This will allow for breath samples to be more thoroughly assessed for a variety of compound types. A liquid EBC sample could be analyzed using both GC-HRMS and LC-HRMS to fully capture the volatile range of compounds in the sample in addition to the nonvolatile fraction. These HRMS capabilities will be used to increase the library of compounds known to be part of the human volatilome. As the accessibility of HRMS instrumentation improves, their use in breath research is expected to increase.

Disclaimer

This publication was reviewed and approved for publication by US Environmental Protection Agency (EPA). The views in this chapter are those of the authors and do not reflect policy of US EPA. References to products and trade names do not constitute endorsement for use.

References

1. McNaught A, Wilkinson A. IUPAC (1997) Compendium of chemical terminology. 2nd ed. (The "Gold Book"). In: Blackwell, Oxford (on-line corrected version: http://goldbook.iupac.org).
2. Xian F, Hendrickson C, Marshall AG. High resolution mass spectrometry. *Anal Chem* 2012;**84**:708–19.
3. Brenton AG, Godfrey AR. Accurate mass measurement: terminology and treatment of data. *J Am Soc Mass Spectrom* 2010;**21**:1821–35.
4. Meija J, Coplen TB, Berglund M, Brand WA, De Bievre P, Groning M, Holden NE, Irrgeher J, Loss RD, Walczyk T, Prohaska T. Isotopic compositions of the elements 2013 (IUPAC technical report). *Pure Appl Chem* 2016;**88**(3):293–306.
5. Pleil JD, Isaacs KK. High-resolution mass spectrometry: basic principles for using exact mass and mass defect for discovery analysis of organic molecules in blood, breath, urine and environmental media. *J Breath Res* 2016;**10**:012001.
6. Hughey CA, Hendrickson CL, Rodgers RP, Marshall AG, Qian K. Kendrick mass defect spectrum: a compact visual analysis for ultrahigh-resolution broadband mass spectra. *Anal Chem* 2001;**73**:4676–81.
7. Lerno Jr LA, German JB, Lebrilla CB. Method for the identification of lipid classes based on referenced Kendrick mass analysis. *Anal Chem* 2010;**82**(10):4236–45.
8. Kind T, Fiehn O. Metabolomic database annotations via query of elemental compositions: mass accuracy is insufficient even at less than 1 ppm. *BMC Bioinformatics* 2006;**7**:234.
9. Pleil JD, Wallace MAG, McCord J. Beyond monoisotopic accurate mass spectrometry: ancillary techniques for identifying unknown features in non-targeted discovery analysis. *J Breath Res* 2018;**13**:012001.
10. Kind T, Tsugawa H, Cajka T, Ma Y, Lai Z, Mehta SS, et al. Identification of small molecules using accurate mass MS/MS search. *Mass Spectrom Rev* 2018;**37**(4):513–32.
11. Boesl U. Time-of-flight mass spectrometry: introduction to the basics. *Mass Spectrom Rev* 2017;**36**(1):86–109.
12. Gale PJ, Vestal ML. *The development of time-of-flight mass spectrometry. The encyclopedia of mass spectrometry.* Elsevier; 2016. p. 32–42.
13. Niessen WMA, Falck D. Introduction to mass spectrometry, a tutorial. In: Kool J, Niessen WMA, editors. *Analyzing biomolecular interactions by mass spectrometry.* 1st ed. Wiley-VCH Verlag GmBH & Co. KGaA; 2015. p. 1–54.
14. Eliuk S, Makarov A. Evolution of orbitrap mass spectrometry instrumentation. *Annu Rev Anal Chem* 2015;**8**:61–80.
15. Brown SC, Kruppa G, Dasseux JL. Metabolomics applications of FT-ICR mass spectrometry. *Mass Spectrom Rev* 2005;**24**(2):223–31.
16. Koelmel JP, Kroeger NM, Gill EL, Ulmer CZ, Bowden JA, Patterson RE, et al. Expanding lipidome coverage using LC-MS/MS data-dependent acquisition with automated exclusion list generation. *J Am Soc Mass Spectrom* 2017;**28**(5):908–17.
17. Tsugawa H, Cajka T, Kind T, Ma Y, Higgins B, Ikeda K, et al. MS-DIAL: data-independent MS/MS deconvolution for comprehensive metabolome analysis. *Nat Methods* 2015;**12**(6):523.

18. Trefz P, Schmidt M, Oertel P, Obermeier J, Brock B, Kamysek Jr S, et al. Continuous real time breath gas monitoring in the clinical environment by proton-transfer-reaction-time-of-flight-mass spectrometry. *Anal Chem* 2013;**85**:10321–9.
19. Romano A, Doran S, Belluomo L, Hanna GB. High-throughput breath volatile organic compound analysis using thermal desorption proton transfer reaction time-of-flight mass spectrometry. *Anal Chem* 2018;**90**:10204–10.
20. Chandler JD, Horati H, Walker DI, Pagliano E, Tirouvanziam R, Veltman M, et al. Determination of thiocyanate in exhaled breath condensate. *Free Radic Biol Med* 2018;**126**:334–40.
21. Pleil JD, Wallace MAG, Madden MC. Exhaled breath aerosol (EBA): the simplest non-invasive medium for public health and occupational exposure biomonitoring. *J Breath Res* 2018;**12**:027110.
22. Eichler P, Müller M, D'anna B, Wisthaler A. A novel inlet system for online chemical analysis of semi-volatile submicron particulate matter. *Atmos Meas Tech* 2015;**8**: 1353–60.
23. Smolinska A, Tedjo DI, Blanchet L, Bodelier A, Pierik MJ, Masclee AA, et al. Volatile metabolites in breath strongly correlate with gut microbiome in CD patients. *Anal Chim Acta* 2018;**1025**:1–11.
24. Lacombe M, Marie-Desvergne C, Combes F, Kraut A, Bruley C, Vandenbrouck Y, et al. Proteomic characterization of human exhaled breath condensate. *J Breath Res* 2018;**12**: 021001.
25. Kononikhin AS, Starodubtseva NL, Chagovets VV, Ryndin AY, Burov AA, Popov LA, et al. Exhaled breath condensate analysis from intubated newborns by nano-HPLC coupled to high resolution MS. *J Chromatogr B* 2017;**1047**:97–105.
26. Makarov A, Denisov E, Kholomeev A, Balschun W, Lange O, Strupat K, et al. Performance evaluation of a hybrid linear ion trap/orbitrap mass spectrometer. *Anal Chem* 2006;**78**(7):2113–20.

PART

Breath tests

D

CHAPTER

^{13}C breath tests

17

Anil S. Modak, e-mail address: AModak2003@yahoo.com

Chapter outline

17.1 Overview

The carbon atom, often described as the *building block of life*, is a chemical element that is intrinsic to organic compounds. In its most abundant isotopic configuration, carbon contains six protons and six neutron, i.e., a nucleon number of 12, thus it is referred to as ^{12}C (or C-12); this isotope represents 98.9% of carbon atoms. The addition of one or two neutrons in the nucleus leads to the isotopes ^{13}C or ^{14}C, respectively. The former, ^{13}C, is a naturally occurring stable isotope that makes up approximately 1.1% of all carbon atoms; the latter, ^{14}C, also referred to as radiocarbon, is an unstable radionuclide that decays with a half-life of 5730 years, a phenomenon that is put to use in (radio) carbon dating of organic matter. It is the former, ^{13}C, which is at the focus of this chapter.

Antoine Lavoisier, the father of modern chemistry, and his colleague Pierre Simon Laplace were the first to analyze carbon dioxide (CO_2) in guinea pig breath

Breathborne Biomarkers and the Human Volatilome. https://doi.org/10.1016/B978-0-12-819967-1.00017-7
Copyright © 2020 Elsevier B.V. All rights reserved.

and then human breath, which was attributed as a product of combustion of oxygen and food. Carbon dioxide can easily be measured in exhaled breath, since it is present at approximately 5% by volume (see Chapter 1). The use of stable isotope ^{13}C-labeled compounds as diagnostic probes has been exploited in a research setting for almost half a century. The first report of using ^{13}C as a tracer was in a human metabolomics study in 1973, which followed the abundance of exhaled ^{13}C-carbon dioxide ($^{13}CO_2$) from metabolism of ingested ^{13}C-glucose.[1] Low-cost ^{13}C-substrates subsequently have become widely available, leading to their implementation as tracers in diverse applications, from nutritional to medical and veterinary research.

Broadly speaking, ^{13}C-labeled breath tests are based on monitoring $^{13}CO_2$ in exhaled breath after enzymatic breakdown of a labeled substrate. Specifically, a ^{13}C label is incorporated into a selected substrate at a site that is specifically designed to target an enzyme—either foreign or inherent—in a discrete metabolic pathway. A metabolite of the ensuing enzyme–substrate interaction is $^{13}CO_2$, which is subsequently expelled from the body during exhalation and can be detected in breath. Monitoring the abundance of $^{13}CO_2$ in breath over time offers an indirect measure of enzyme activity. The analysis of $^{13}CO_2$ in exhaled breath is typically achieved via spectroscopic methods (this technology is covered in detail in Chapter 14).

The viability of administering an oral dose of a ^{13}C-substrate and procuring metabolic or diagnostic information from its metabolism and conversion to $^{13}CO_2$ is, like other breath-based tests, attractive due to its noninvasive nature. The salient features of ^{13}C breath tests are that they are noninvasive, nonradioactive, safe, simple, and effective. Such breath tests have the potential to be companion diagnostic tests (CDx) for identifying responders and nonresponders for drugs metabolized by various polymorphic cytochrome CYP450 enzymes. Further, the scope of ^{13}C-stable isotope probes is currently being expanded to provide precise evaluations of the presence or absence of etiologically significant changes in metabolism due to a specific disease or the lack of a specific enzyme or drug–drug interaction.

Despite extensive research over the last four decades–and the obvious benefits of noninvasive, functional in vivo phenotyping–few breath tests have been approved by the Food and Drug Administration (FDA) in the United States or the European Medicines Association (EMA) in Europe, although a number are currently in a research and development phase. These breath tests can be subclassified into two groups, one that utilizes constitutive/endogenous volatile organic compounds (VOCs) generated via biological processes in the human body due to prevalent disease states, and the other that uses ^{13}C-substrates as in vivo probes for the generation of $^{13}CO_2$ due to specific enzymes in the human body. The paucity of approvals of more ^{13}C breath tests has been severely stymied by the lack of financial resources required for the development of these tests and their introduction into routine clinical practice for personalized medicine. This chapter introduces the concept of breath tests with ^{13}C-labeled probes and reviews the research and development of representative tests and their introduction into clinical practice.

17.2 Approved tests

There currently exist three FDA/EMA-approved human breath tests involving $^{13}CO_2$ as an in vivo biomarker as a diagnostic aid for diseases or infection after ingestion of ^{13}C-substrates. These are

^{13}C-Urea for *Helicobacter pylori* infection (Meretek Inc., approved in 1996)
^{13}C-Spirulina for gastroparesis (by Cairn Diagnostics, approved in 2015)
^{13}C-Methacetin for liver function (by Humedics GmbH, approved in 2017)

Each of these tests will be discussed in the following sections.

17.2.1 *Helicobacter pylori* infection

Helicobacter pylori is a global pathogen infecting approximately 50% of the world population.[2] *H. pylori* is a common pathogen that is able to alter host physiology and subvert host immune response. Thus, the infection can persist asymptomatic for the life of the host. Infection by this bacterium can lead to various gastroduodenal diseases, such as functional dyspepsia, mucosa-associated lymphoid tissue lymphoma (MALT), and distal gastric cancer. It is also the primary cause of peptic ulcers in 20% of the population and is subsequently responsible for gastric cancer if untreated with antibiotics. In the United States, the annual cost associated with peptic ulcer disease is estimated to be $6 billion, and globally gastric cancer kills over 700,000 people per year. The prevalence of *H. pylori* infection is extremely high (>50%) in much of the world, although the infection rates are dropping in most Western nations.

Before the noninvasive ^{13}C-urea breath test (UBT) was approved by the FDA in 1996, diagnosis of *H. pylori* infection was made via noninvasive stool antigen test, serology, or by invasive techniques, such as histology, rapid urease test, and culture tests. The first reported use of the ^{13}C-UBT was in 1987 by Graham et al., when it was used to determine the presence of *Campylobacter pylori* in the stomach in 26 individuals.[3] Since then, thousands of publications have appeared in the scientific and medical literature in relation to this test.

The premise of the ^{13}C-UBT is urease activity by the pathogen in the stomach, which converts the labeled urea—administered orally as a substrate—to $^{13}CO_2$-labeled that is excreted from the body via breath. Specifically, the test involves the ingestion of 75 mg ^{13}C-urea and the collection of breath samples, one immediately before (baseline) and then 20 min after administration. If an *H. pylori* infection is present, the urea will be hydrolyzed by the bacterium, generating $^{13}CO_2$ that is subsequently detectable in expired air. The results are expressed as delta over baseline (DOB). The cut-off for *H. pylori* infection is typically a DOB $> 2.4‰$.

The ^{13}C-UBT kit for diagnosing *H. pylori* appeared on the market in 1997, commercialized by Meretek Inc. The UBT for pediatric use by Otsuka Pharmaceuticals Co. Ltd., has subsequently received FDA approval (in 2012) and is in clinical practice throughout the United States, Europe, and other parts of the world. After the patent on the UBT expired in 2009, Exalenz Biosciences received FDA approval for the Exalenz BreathID Hp System in 2011. Subsequently in 2018, Exalenz Biosciences received FDA approval for the BreathID Hp point-of-care (PoC) system and the BreathID Hp Lab system for use in detecting *H. pylori* infection in children aged 3–17 years. More recently, in 2020, the Center for Devices and Radiological Health (CDRH) of the FDA approved ARJ Medical's premarket approval (PMA) application for the PyloPlus UBT System. The PyloPlus UBT system is intended for use in the qualitative detection of urease activity associated with *H. pylori* in the human stomach and is indicated as an aid in the initial diagnosis of *H. pylori* infection in adults of age 18 and over.

The clinical utility of the noninvasive and rapid (20 min) ^{13}C-UBT for reliably diagnosing *H. pylori* infection has been reviewed in the literature[4], as has its accuracy compared to other invasive diagnostic tests, such as gastric biopsies performed during endoscopy (rapid urease test, histology, culture).[5,6] Several studies have focused on the efficacy of eradication therapy, resistance to antibiotics therapy, and the selection of alternate therapies.[7,8] The test has proven to be clinically useful in diagnosis, as well as in evaluating the eradication of infection following antibiotic therapy.[9] A comprehensive review paper on *H. pylori* has recently been published in the literature.[10]

Unfortunately, ineffective, incompetent marketing of the ^{13}C-UBT, inadequate cost reimbursement compared to endoscopy, and a lack of patient and physician knowledge of the clinical utility of the test have resulted in poor acceptance in routine clinical practice in the United States. Despite this, the ^{13}C-UBT has been implicit in reduced endoscopy costs in clinical practice.

17.2.2 Gastric emptying

Gastroparesis, also known as delayed gastric emptying, is a muscular disorder that either slows or stops movement of food through the gastrointestinal (GI) tract. It commonly causes nausea, vomiting, heartburn, erratic blood glucose levels, early satiety, postprandial fullness, and upper gut symptoms. Normally, the smooth muscles of the GI tract move food down to the stomach and through the intestines, where it is excreted through the rectum. This process is controlled by the vagus nerve, a component of the parasympathetic nervous system. Gastroparesis is a form of gastric paralysis that causes significant morbidity. Weight loss, malnutrition, and dehydration may be prominent in severe cases and may require tertiary care.

The most common causes of gastroparesis are diabetes mellitus and surgery; it may also be of unknown cause. Diabetes accounts for almost one-third of cases of gastroparesis, with about 20%–40% of all diabetes patients experiencing this disorder. In diabetics, gastroparesis may adversely affect glycemic control. The etiology of gastroparesis is diverse; the main categories are diabetic, idiopathic, and postsurgical. Virtually any disease or condition that can induce neuromuscular dysfunction of the GI tract may cause gastroparesis.

Diagnosis is most often determined by performing endoscopy or using scintigraphy, a specialized radiologic imaging technique, to measure the rate at which radioactive food passes through the GI tract.[11] However, the technique is not generally available at major healthcare centers, and because it involves radiation exposure, this precludes use of the test in pregnant women and children. Breath tests using ^{13}C-substrates, such as ^{13}C-octanoic acid (^{13}C-OABT)[12] or the edible ^{13}C-enriched blue-green alga, *Spirulina platensis* for solid gastric emptying, or ^{13}C-sodium acetate for liquid gastric emptying,[13] have been extensively researched.[14] These substrates are rapidly absorbed in the proximal small intestine and then metabolized in the liver, with an accompanying fast generation and exhalation of $^{13}CO_2$ that exhibits low interindividual variability; $^{13}CO_2$ exhalation therefore reflects gastric emptying of food intake.[15] The most significant reason why noninvasive breath tests have not yet replaced the radioactive scintigraphy test for gastric emptying is the relatively long time required (240 min) for the breath test and the multiple breath samples required, although the current requirement of only five sampling points, typically at 45, 90, 120, 150, and 180 min postingestion of ^{13}C-substrate, makes the test more user-friendly.

A gastric emptying breath test (GEBT) using stable isotope–labeled ^{13}C-spirulina as a probe was developed by Cairn Diagnostics and approved by the FDA in 2015. Spirulina is a safe, edible, highly digestible blue-green microalga, which is primarily consumed in the United States as a health food due to its rich nutritional qualities. Spirulina was acknowledged as a "legally marketed" food by the FDA in 1981. In 2011, the Dietary Supplements Information Expert Committee (DSI-EC) of the United States Pharmacopeia (USP) performed a comprehensive review of information related to safety and toxicology of spirulina and subsequently unanimously voted to grant Class A safety assignment to *Spirulina platensis* and *Spirulina maxima*. Based on this decision, USP has classified spirulina as a dietary ingredient and is currently developing a quality monograph for spirulina. The recommended amount of spirulina as a dietary supplement is around 3–4.5 g/day. GEBT is easy to administer, thereby making it patient and nurse friendly, with the collection of only five to six breath samples over 3–4 h postingestion of a meal containing ^{13}C-spirulina. The dose of ^{13}C-spirulina—acquired by growing algae in a $^{13}CO_2$ atmosphere—in each GEBT meal is approximately 0.1 g.

17.2.3 Liver function

Hepatic dysfunction can have various etiologies, and the complexity of the liver, its diverse functions, and the severity of liver diseases make it difficult to use one specific ^{13}C-labeled substrate for determining liver metabolic function. There exist several blood tests to identify the severity of liver damage and to monitor patients with liver disease. These blood tests measure alanine aminotransferase (ALT), aspartate aminotransferase (AST), or bilirubin levels leaked into the bloodstream by dead liver cells. Scoring indexes, such as Pugh and MELD[a] are routinely utilized by physicians as complementary parameters to identify liver damage severity and to monitor

[a] Model for end-stage liver disease

patients. However, these tests reflect a static end point of hepatocellular damage but do not evaluate functional hepatic reserve. Therefore, while advanced liver disease can be diagnosed with confidence, early or marginal hepatic disease is poorly defined.

Breath tests using ^{13}C for identifying hepatic dysfunction are numerous and diverse; in fact, more ^{13}C-substrate probes have been studied in relation to liver diseases compared to tests for other medical conditions. A particular benefit of labeled substrate-based tests is that typically they deliver quantitative information on liver function. Over the last four decades, hepatologists have undertaken widespread research on ^{13}C breath tests for evaluating cytosolic, mitochondrial, and microsomal hepatic reserves (liver function).[16,17] Specific functional phenotype breath tests have been investigated for cytosolic (^{13}C-phenylalanine), microsomal (^{13}C-aminopyrine), and mitochondrial (^{13}C-methionine, ^{13}C-ketocaproic acid, ^{13}C-sodium octanoate, ^{13}C-pyruvate) liver functional capacity. In addition, breath tests for evaluation of CYP450 activity have been researched, such as ^{13}C-dextromethorphan for CYP2D6, ^{13}C-pantoprazole for CYP2C19, ^{13}C-methacetin for chronic liver disease and CYP1A2, and ^{13}C-caffeine for CYP1A2. Further ^{13}C-labeled substrates for exploring liver functional reserve include ^{13}C-galactose and ^{13}C-fructose. An overview of ^{13}C probes for targeting different enzymes in liver disorder tests is given in Table 17.1.

Overall, the aforementioned approaches provide information on diverse aspects of liver function. These include monitoring liver regeneration and functional capacity, monitoring liver graft and determining optimum transplantation timing, long-term liver disease prognosis prediction, efficacy evaluation of surgical intervention and/or therapy, and deficiencies in liver-related enzymes that lead to conditions such as phenylalanine hydroxylase deficiency (PKU), maple syrup urine disease (MSUD; see also Chapter 24), and propionic acidemia.

In terms of FDA approval, only one test for liver disease has passed this hurdle, namely one based on the use of ^{13}C-methacetin. Developed by Humedics GmbH and

Table 17.1 Overview of ^{13}C probes targeting different enzymes in a variety of liver function tests.

Liver function	^{13}C probe	Enzymes
Microsomal	Aminopyrine	CYP1A2, CYP2C9
	Phenacetin	CYP1A2, CYP2E1
	Methacetin	CYP1A2, CYP2E1, CYP2B6
	Caffeine	CYP1A2, CYP3A4
	Erythromycin	CYP3A4
	Diazepam	CYP2C19, CYP3A4
Cytosomal	Galactose	Galactokinase
	Phenylalanine	Phenylalanine hydroxyls
	Fructose	Fructokinase
	Alanine	Aminotransferase (ALT, GPT)
	Ornithine	Ornithine transcarboxylase
Mitochondrial	Methionine	Krebs cycle
	α-Ketoisocaproic acid	α-Keto acid dehydrogenase

named LiMAx - referring to maximum liver function capacity - the test involves the *intravenous* administration of ^{13}C-methacetin for an easy, yet invasive, method to evaluate liver function capacity.[18] After administration, ^{13}C-methacetin is metabolized by cytochrome CYP1A2 in the liver into paracetamol and $^{13}CO_2$ that is subsequently exhaled in breath. When the patient breathes into the LiMAx test mask, the ultrasensitive laser system continuously measures the ratio of $^{13}CO_2$ to $^{12}CO_2$ in every breath, which changes as a function of liver capacity. The results from the LiMAx test can be used to stratify a patient's liver function capacity into three different levels: (i) normal liver function, over 315 μg/kg/h; (ii) limited hepatic impairment, 140–314 μg/kg/h; and (iii) significant hepatic injury, 0–139 μg/kg/h.

The LiMAx test provides clinicians a diagnostic tool for a rapid estimation of liver function, thus improving clinical outcomes for patients undergoing liver surgery (transplantation or resection of cancerous tumors). A particularly beneficial feature of the system is that it is mobile and can therefore be used in intensive care units (ICUs), emergency rooms, and outpatient clinics. The test was approved by EMA in 2017–18 for clinical use in the United Kingdom, Germany, and Austria. One caveat of the test is the need for intravenous—rather than oral—administration of ^{13}C-methacetin. This will make the transition to the US market more difficult, as the manufacture of the ^{13}C-methacetin probe for intravenous use will face stringent good manufacturing practice (GMP) requirements, making it cost prohibitive, especially for obtaining FDA approval. By comparison, a ^{13}C-methacetin breath test with oral administration would make the test noninvasive, cheaper, and more patient friendly.

The race is on in the pharmaceutical industry to develop drugs to treat a form of fatty liver disease called nonalcoholic steatohepatitis (NASH). Industry experts estimate the global market for these new drugs to be \$35 billion. The Center for Disease Analysis (CDA) reports that the United States is spending \$5 billion annually in healthcare costs related to the disease, which include chemotherapy, transplants, tests, and hospitalizations. The National Institutes of Health (NIH) estimates as many as 12% of US adults (30 million people) have NASH, with healthcare costs rising to \$18 billion by 2030 if this disease continues to go unchecked. A number of the ^{13}C probes for evaluating liver function listed above could be used as companion diagnostics to monitor the efficacy of new drugs that are in the pipeline for the treatment of liver diseases like NASH, nonalcoholic fatty liver disease (NAFLD), and hepatitis C. Although most research in the last 6 years has been with ^{13}C-methacetin breath test, several other ^{13}C probes, viz. ^{13}C-aminopyrine, ^{13}C-galactose, ^{13}C-methionine, ^{13}C-α-ketoisocaproic acid, and ^{13}C-phenylalanine, have been researched for liver function.

17.3 Current nonapproved tests

Despite only three ^{13}C probe breath tests being approved by the FDA and/or EMA, several tests are currently in the research and development phase, or beyond. These tests will be reviewed in brief in this section.

17.3.1 Glucose

Insulin resistance is a metabolic syndrome that contributes to the pathophysiology of type 2 diabetes, obesity, and cardiovascular diseases. Therefore, quantifying insulin sensitivity and/or resistance in humans is of great significance for epidemiological studies, as well as clinical and basic science investigations for eventual use in clinical practice. Fasting blood glucose tolerance test (GTT), insulin tolerance test (ITT), insulin suppression test, and the hyperinsulinemic-euglycemic clamp technique are the typical invasive tests carried out in clinical practice. A ^{13}C-glucose breath test has been researched as a rapid, noninvasive alternative to either assess insulin resistance[19] or diagnosis of type-2 diabetes[20] or assess sucrose maldigestion with functional bowel disorder in adults and adolescents.[21]

17.3.2 Bacterial overgrowth

Bowel conditions including indigestion, bloating, gas, abdominal pain, and irregularity can manifest as a result of bacterial overgrowth in the gut, specifically the small intestine, which can arise through reduction in gastric acidity or digestive enzymes and an ensuing nonsterile environment. Intestinal obstructions can also result from Crohn's disease, adhesions, radiation damage, and lymphoma. Successful treatment of bacterial overgrowth significantly alleviates chronic symptoms such as diarrhea and abdominal pain in irritable bowel syndrome patients (IBS). Bacterial overgrowth in the gut produces methane and hydrogen, and detecting these gases is a sensitive measure to detect this condition. Several breath tests, namely utilizing ^{13}C-xylose, ^{13}C-D-sorbitol, or ^{13}C-glycocholic acid, represent potential alternatives to ascertain microbiological activity in the small bowel and thereby offer a practical way to monitoring this condition, which is often persistent throughout life.

17.3.3 Pancreatic insufficiency

Exocrine pancreatic insufficiency (EPI), sometimes shortened to pancreatic insufficiency, is the inability of the body to break down and digest food properly. Diseases involving the pancreas, such as cystic fibrosis (CF) and pancreatic cancer, lead to abnormal pancreatic function. Tests for these conditions have had a central role in the study of pancreatic disease for more than 70 years. The gold standard test for such conditions is the invasive secretin stimulation test, which measures the ability of the pancreas to respond to the hormone secretin. The small intestines produce secretin in the presence of partially digested food. Other tests for pancreatic function are noninvasive fecal elastase test, which measures the enzyme elastase, and invasive computed tomography (CT) scan with contrast dye, abdominal ultrasound, endoscopic retrograde cholangiopancreatography, endoscopic ultrasound, and magnetic resonance cholangiopancreatography. The ^{13}C-mixed triglyceride (1,3-distearoyl, 2-[^{13}C]-octanoyl glycerol) test (^{13}C-MTG-BT) is a noninvasive alternative for evaluating exocrine pancreatic function by comparing the $^{13}CO_2$ breath excretion with the lipase output in the duodenum.[22]

17.3.4 Cytochrome CYP450 enzymes

A majority of the drugs approved by the FDA are metabolized by CYP450 phase I enzymes in the liver. Over the past decade, stable isotope–labeled substrates have been used as probes for rapid, PoC, noninvasive, and user-friendly phenotype breath tests to evaluate activity of phase I drug-metabolizing enzymes. Several genotype tests have been approved by the FDA over the last 15 years for cytochrome P450 2D6, 2C19 enzymes; however, they have not been cleared for use in personalizing medications since they fall woefully short in identifying all nonresponders to drugs. CYP2D6 and CYP2C19 are among the most extensively studied drug metabolizing enzymes, involved in the metabolism of approximately 35% of FDA-approved drugs in clinical use. These enzymes are associated with large individual differences in medication efficacy or tolerability, essentially due to phenoconversion and/or drug–drug interactions.

The development and commercialization via FDA approval of the noninvasive, rapid (<60 min), PoC, in vivo, phenotype diagnostic tests to evaluate polymorphic CYP2D6 (^{13}C-dextromethorphan breath test)[23,24] and CYP2C19 (^{13}C-pantoprazole breath test)[25,26] enzyme activity by measuring exhaled $^{13}CO_2$ as a biomarker in breath will effectively resolve the currently unmet clinical need for individualized psychiatric, cancer, cardiovascular, and GI proton-pump inhibitor (PPI) drug therapy. These companion diagnostic breath tests could also be useful to monitor the progress of disease severity or medication efficacy or for in vivo study of the pharmacokinetics of xenobiotics with narrow therapeutic windows. The selection of initial and consequent maintenance dose-based on enzyme activity will help improve efficacy and reduce toxicity by maintaining the plasma levels of the active drugs within the therapeutic index after steady state is reached.

Currently, financial resources limit the progression of these promising breath tests through the regulatory process; as such, the transition from research to the clinic is on hold. Because all US, Canadian, and European patents on these ^{13}C breath tests (both issued and pending) have now been released by Cambridge Isotope Laboratories Inc/Otsuka Pharm, there is hope for future development from other companies.

17.3.5 Miscellaneous applications

Up to 30% of patients undergoing 5-fluorouracil therapy, either 5-FU (intravenous) or Xeloda (oral), experience severe toxicity. Dihydropyrimidine dehydrogenase (DPD) deficiency is responsible for 36%–61% of cases. Predicting toxicity is an unmet challenge. The ^{13}C-uracil breath test (Ura-BT) involves measuring $^{13}CO_2$ in exhaled breath after ingestion of 2-^{13}C-uracil to evaluate pyrimidine (and 5-FU) catabolism. The Ura-BT has the ability to identify individuals with DPD deficiency (partial or complete), who are likely to have severe toxicity after initiating the normal prescribed dose of 5-FU.[27]

Peripheral carbidopa (CD) levels directly impact on central dopamine (DA) production in Parkinson's disease (PD) through extracerebral inhibition of dopa decarboxylase (AADC), resulting in an increase in levodopa (LD) bioavailability.

The ^{13}C-L-DOPA breath test is a potential noninvasive diagnostic tool for evaluating AADC enzyme activity by measuring exhaled $^{13}CO_2$, offering a first step in personalizing CD doses for PD patients.[28]

Epidemic infection with the intestinal pathogen *Clostridium difficile* (*C. diff*) is becoming highly important worldwide, especially in the United States, with estimates of 300,000 cases of *C. diff.*—associated disease (CDAC) in 2005, with increasing incidence rates. Administration of a diagnostically effective amount of an isotopically labeled tyrosine and/or isotopically labeled ^{13}C-p-hydroxyphenylacetic acid to the patient and measuring the levels of $^{13}CO_2$ and/or an isotopically labeled p-cresol in exhaled breath by bacterial enzymes can evaluate the presence or absence of this bacterial infection.[29]

Naturally occurring stable ^{13}C isotopes are not equally distributed in body proteins, fats, and carbohydrates, and this phenomenon could be used as a noninvasive predictive marker for onset of infection in surgical ICU patients. The exhaled $^{13}CO_2/^{12}CO_2$ ratio, or breath delta value (BDV), changes during the onset and progression of infection and precedes physiological changes associated with infection. A BDV greater than 1.4‰ accurately differentiates subjects who develop infections from those who do not, and predicts the presence of infection up to 48 h before clinical confirmation. Earlier diagnosis of systemic inflammatory response syndrome (SIRS) could lead to earlier appropriate treatment and improve clinical outcomes.[30] The BDV correlates to workload only during energy balance and could alternatively serve as a biomarker for metabolic energy need in performance-trained athletes. Athletes and trainers can use BDV during performance workout trainings to understand and optimize the metabolic fuel sources being used to perform work and may provide immediate feedback on rest, recovery programs, and nutritional strategies during intense training.[31]

17.4 Regulatory aspects of ^{13}C breath tests

Breath tests utilizing ^{13}C probes as substrates and a spectrometer for analyzing breath samples have been deemed by the FDA as *Combination Products*. Combination product diagnostic breath tests consist of a drug component that is isotope-labeled, typically with ^{13}C, and the device component to capture and/or analyze exhaled breath for the detection of labeled CO_2 as a biomarker. The FDA has determined that the primary mode of action (PMOA) of such combination products is attributable to the role of the device components in the in vitro diagnosis of the disease or condition in question, while the drug component plays a secondary role in acting as the diagnostic substrate. In the United States, the CDRH has been assigned by the FDA to be the lead agency center for reviewing PMA applications. While the FDA is responsible for determining the safety and effectiveness of both the device and drug components of the combination product, the CDRH consults or collaborates with the Center for Drug Evaluation and Research (CDER) on issues pertaining to chemistry and GMP, pharmacology and toxicology, and clinical issues related to the ^{13}C-labeled drug component.[32]

The process of transitioning a ^{13}C breath test from research to clinical practice encounters many hurdles and pitfalls. Any breath test that has been developed must fulfill certain criteria in terms of accuracy, sensitivity, and specificity compared to the current gold standard, as well as return high true positives and negatives, and low false positives and negatives; these aspects in specific relation to ^{13}C breath tests is discussed in the literature and will not be covered in this chapter.[33]

17.5 The future of ^{13}C breath tests

Exhaled $^{13}CO_2$ can be a useful biomarker for phenotype breath tests to evaluate activity of various drug metabolizing polymorphic enzymes (CYP2D6, 2C19, 1A2) and thereby enable physicians to personalize medications with narrow therapeutic windows. The dearth of ^{13}C breath tests in routine clinical practice can be addressed by obtaining FDA approval through PMA applications for the promising diagnostic tests that are currently in the research and development stages. There are a number of scientific, developmental, regulatory, and marketing barriers to acceptance and integration of breath tests into routine clinical practice. Even after regulatory approval of diagnostic breath tests, the financial success will entirely depend on how well the test is marketed to physicians, healthcare organizations, payers (reimbursement), and insurance companies, and most importantly to patients, the eventual beneficiaries. Nevertheless, the biggest impediment for regulatory approval of these breath tests is the scarcity of financial resources from pharmaceutical and diagnostic companies. Several ^{13}C probe breath tests are currently under development, with the promise of joining current approved tests in the arsenal of noninvasive substrate tests for routine clinical practice.

17.5.1 Avisa breath test

The Avisa Breath Test is an easy-to-use, PoC device/drug combination product for the detection and monitoring of lung infections caused by pathogens like bacteria or fungi with urease enzyme activity that cause pneumonia or tuberculosis. The prominent bacteria with urease activity that cause lung infections include the so-called ESKAPE pathogens: *Enterococcus faecium*, *Staphylococcus aureus*, *Klebsiella pneumoniae*, *Acinetobacter baumannii*, *Pseudomonas aeruginosa*, and *Enterobacter* spp, as well as *Haemophilus influenzae*.[34]

Unlike other tests that require ingestion or intravenous administration, the Avisa Breath Test, developed by Avisa Pharmaceuticals, requires inhalation of ^{13}C-urea. This allows for the detection of live urease-containing bacteria, which convert the labeled urea to $^{13}CO_2$ and ammonia. The bacterial lung infection is determined by measuring the ratio of $^{13}CO_2$ to $^{12}CO_2$ in exhaled breath using a laser spectrometer,[35] thereby allowing rapid identification of patients with pneumonia whose disease is caused by bacteria with urease activity. This assists the physician in selection of personalized antibiotic therapy, and after initiating therapy, the breath test can be used to monitor the efficacy/resistance of the therapy.

17.5.2 Methacetin breath test

The LiMax liver function test based on ^{13}C-methacetin discussed in Section 17.2.3 requires *intravenous* administration of the labeled substrate. An alternative test is currently under development by Exalenz Bioscience in which labeled methacetin is administered orally.[36] The oral methacetin breath test (MBT) will have significant benefits over the EMA approved intravenous LiMAx, as an easy-to-use, PoC, noninvasive breath test for evaluation of liver function in patients, monitoring liver disease severity, and monitoring the efficacy of novel medications in a wide range of liver diseases like NASH, clinically significant portal hypertension (CSPH), and acute liver failure (ALF). Exalenz Bioscience is currently carrying out a series of clinical trials to establish the reliability and accuracy of the MBT for diagnosing and monitoring NASH, CSPH, and ALF to replace the need for invasive liver biopsy.

17.6 Summary

The ^{13}C probe–based breath tests, which generate exogenous $^{13}CO_2$ after ingestion of a ^{13}C-labeled substrate, are unique since they are highly specific to a certain enzyme and thereby offer an indirect measure of enzyme activity. Exhaled $^{13}CO_2$ in breath can be a useful biomarker for phenotype ^{13}C probe–based breath tests to evaluate activity of various drug-metabolizing polymorphic enzymes (CYP2D6, 2C19, 1A2) and enable physicians to personalize medications with narrow therapeutic windows, both initial doses and subsequent maintenance doses after steady state is reached. Three ^{13}C breath tests are currently approved as a diagnostic tool by the FDA in the United States and EMA in Europe, namely the ^{13}C UBT to ascertain *Helicobacter pylori* infection, the GEBT to determine gastroparesis, and a liver function test (LiMax) to establish liver disorders. Further ^{13}C-based tests have been developed—or are under development—but have not yet received regulatory approval. These include a glucose breath test to determine insulin intolerance or diagnose type 2 diabetes, a test for bacterial overgrowth in the gut, a test for EPI, and phenotyping tests for enzyme activity, amongst others. The PMA application is the regulatory pathway for the approval of ^{13}C probe–based breath tests, which constitute a drug (^{13}C probe) and a device (spectrometry or spectroscopy). A major impediment for regulatory approval of these breath tests is the reluctance of diagnostic and pharmaceutical companies to provide the financial resources required for their development, regulatory approval, and subsequent transition to the clinic. Despite these hurdles, the success of approved ^{13}C-based breath tests demonstrate their utility in aiding diagnosis in the clinical setting or as companion diagnostics in disease therapy. Certainly, current developments in the use of ^{13}C substrates for novel tests hold much promise. It is likely that further tests will receive regulatory approval, driving ^{13}C breath tests to be the most widely implemented, varied, and successful tests within the field of breath research.

References

1. Lacroix M, Mosora F, Pontus M, Lefebvre P, Luyckx A, Lopez-Habib G. Glucose naturally labeled with carbon-13: use for metabolic studies in man. *Science* 1973;**181**(4098): 445.
2. Peek RM, Blaser MJ. *Helicobacter pylori* and gastrointestinal tract adenocarcinomas. *Nat Rev Canc* 2002;**2**(1):28–37.
3. Graham D, Evans D, Alpert L, Klein P, Evans D, Opekun A, et al. Campylobacter pylori detected noninvasively by the ^{13}C-urea breat test. *Lancet* 1987;**329**(8543):1174–7.
4. Malfertheiner P. Diagnostic methods for *H. pylori* infection: choices, opportunities and pitfalls. *United European Gastroenterol J* 2015;**3**(5):429–31.
5. Calvet X, Sánchez-Delgado J, Montserrat A, Lario S, Ramírez-Lázaro MJ, Quesada M, et al. Accuracy of diagnostic tests for *Helicobacter pylori*: a reappraisal. *Clin Infect Dis* 2009;**48**(10):1385–91.
6. Nocon M, Kuhlmann A, Leodolter A, Roll S, Vauth C, Willich SN, et al. Efficacy and cost-effectiveness of the ^{13}C-urea breath test as the primary diagnostic investigation for the detection of *Helicobacter pylori* infection compared to invasive and non-invasive diagnostic tests. *GMS Health Technol Assess* 2009;**5**. Doc14–Doc.
7. Zullo A, Perna F, Ricci C, Hassan C, Tampieri A, De Francesco V, et al. ^{13}C-urea breath test values and *Helicobacter pylori* eradication. *Dig Dis Sci* 2008;**53**(2):370–4.
8. Safavi M, Sabourian R, Foroumadi A. Treatment of *Helicobacter pylori* infection: current and future insights. *World J Clin Cases* 2016;**4**(1):5–19.
9. Goderska K, Agudo Pena S, Alarcon T. *Helicobacter pylori* treatment: antibiotics or probiotics. *Appl Microbiol Biotechnol* 2018;**102**(1):1–7.
10. Crowe SE. *Helicobacter pylori* infection. *N Engl J Med* 2019;**380**(12):1158–65.
11. Abell TL, Camilleri M, Donohoe K, Hasler WL, Lin HC, Maurer AH, et al. Consensus recommendations for gastric emptying scintigraphy: a joint report of the American neurogastroenterology and motility society and the society of nuclear medicine. *J Nucl Med Technol* 2008;**36**(1):44–54.
12. Bruno G, Lopetuso LR, Ianiro G, Laterza L, Gerardi V, Petito V, et al. ^{13}C-octanoic acid breath test to study gastric emptying time. *Eur Rev Med Pharmacol Sci* 2013;**17**(2 Suppl):59–64.
13. Bertram F, Andresen V, Layer P, Keller J. Simultaneous non-invasive measurement of liquid gastric emptying and small bowel transit by combined ^{13}C-acetate and H_2-lactulose breath test. *J Breath Res* 2014;**8**(4):046007.
14. Sanaka M, Yamamoto T, Kuyama Y. Retention, fixation, and loss of the [^{13}C] label: a review for the understanding of gastric emptying breath tests. *Dig Dis Sci* 2008;**53**(7): 1747.
15. Ghoos YF, Maes BD, Geypens BJ, Mys G, Hiele MI, Rutgeerts PJ, et al. Measurement of gastric emptying rate of solids by means of a carbon-labeled octanoic acid breath test. *Gastroenterology* 1993;**104**(6):1640–7.
16. Modak AS. Stable isotope breath tests in clinical medicine: a review. *J Breath Res* 2007; **1**(1):014003.
17. Ilan Y. Review article: the assessment of liver function using breath tests. *Aliment Pharmacol Ther* 2007;**26**(10):1293–302.

18. Rubin TM, Heyne K, Luchterhand A, Jan B, Vondran F WR, Polychronidis G, et al. Kinetic validation of the LiMAx test during 10 000 intravenous ^{13}C-methacetin breath tests. *J Breath Res* 2017;**12**(1):016005.
19. Maldonado-Hernández J, Martínez-Basila A, Rendón-Macías ME, López-Alarcón M. Accuracy of the ^{13}C-glucose breath test to identify insulin resistance in non-diabetic adults. *Acta Diabetol* 2019;**56**(8):923–9.
20. Takemoto I, Kawagoe N, Kijima S, Sasaki Y, Watanabe T, Urita Y. ^{13}C-glucose breath tests: a non-invasive method for detecting early clinical manifestations of exogenous glucose metabolism in type 2 diabetic patients. *Acta Diabetol* 2019;**56**(4):449–56.
21. Opekun AR, Balesh AM, Shelby HT. Use of the biphasic ^{13}C-sucrose/glucose breath test to assess sucrose maldigestion in adults with functional bowel disorders. *BioMed Res Int* 2016;(7952891):12.
22. González-Sánchez V, Amrani R, González V, Trigo C, Picó A, de-Madaria E. Diagnosis of exocrine pancreatic insufficiency in chronic pancreatitis: ^{13}C-mixed triglyceride breath test versus fecal elastase. *Pancreatology* 2017;**17**(4):580–5.
23. Opdam FL, Modak AS, Gelderblom H, Guchelaar H-J. Breath tests to phenotype drug disposition in oncology. *Clin Pharmacokinet* 2013;**52**(11):919–26.
24. Opdam FL, Modak AS, Gelderblom H, Guchelaar HJ. Further characterization of a ^{13}C-dextromethorphan breath test for CYP2D6 phenotyping in breast cancer patients on tamoxifen therapy. *J Breath Res* 2015;**9**(2):026003.
25. Modak AS. Point-of-care companion diagnostic tests for personalizing psychiatric medications: fulfilling an unmet clinical need. *J Breath Res* 2017;**12**(1):017101.
26. Modak AS, Klyarytska I, Kriviy V, Tsapyak T, Rabotyagova Y. The effect of proton pump inhibitors on the CYP2C19 enzyme activity evaluated by the pantoprazole-^{13}C breath test. *J Breath Res* 2016;**10**(4):046017.
27. Cunha-Junior GF, De Marco L, Bastos-Rodrigues L, Bolina MB, Martins FL, Pianetti GA, et al. ^{13}C-uracil breath test to predict 5-fluorouracil toxicity in gastrointestinal cancer patients. *Canc Chemother Pharmacol* 2013;**72**(6):1273–82.
28. Modak A, Durso R, Josephs E, Rosen D. A rapid non invasive L-DOPA-^{13}C breath test for optimally suppressing extracerebral AADC enzyme activity – toward individualizing carbidopa therapy in Parkinson's disease. *J Parkinsons Dis* 2012;**2**(4):349–56.
29. Timmins G, Silks LP. *Inventors methods for diagnosing bacterial infections 2018*. June 19, 2018.
30. O'Rourke AP, Buckman SA, Evans DC, Kerwin AJ, Breunig EA, Bütz DE. Changes in exhaled $^{13}CO_2/^{12}CO_2$ breath delta value as an early indicator of infection in intensive care unit patients. *J Trauma Acute Care Surg* 2019;**86**(1):71–8.
31. Bütz DE, Schoeller DA, Breunig E, Brassil J, Chu J, Weidmann D. Exhaled $^{13}CO_2/^{12}CO_2$ delta value as a metabolic marker for metabolism in performance training. *Faseb J* 2017; **31**(1_Suppl.). 1036.18–18.
32. Modak AS. Regulatory issues on breath tests and updates of recent advances on [^{13}C]-breath tests. *J Breath Res* 2013;**7**(3):037103.
33. Modak AS. ^{13}C breath tests: transition from research to clinical practice. In: Amann A, Smith D, editors. *Breath analysis for clinical diagnosis and therapeutic monitoring*. 1st ed. Singapore: World Scientific; 2005. p. 457–78.

34. Bishai WR, Timmins GS. Potential for breath test diagnosis of urease positive pathogens in lung infections. *J Breath Res* 2019;**13**(3):032002.
35. Joseph DS, Massick SM, Peterson KA, Gomez AL, Silver JA. *Inventors Methods for detecting bacterial infections*. 2016. US 9,518,972 B2.
36. Stravitz RT, Ilan Y. Potential use of metabolic breath tests to assess liver disease and prognosis: has the time arrived for routine use in the clinic? *Liver Int* 2017;**37**(3):328–36.

CHAPTER

Breath monitoring in the intensive care unit

18

Jochen K. Schubert, Wolfram Miekisch

Department of Anesthesia and Intensive Care, Rostock University Medical Center, Rostock, Germany

Chapter outline

18.1 Overview

Life-threatening changes of vital conditions in critically ill patients can occur within minutes. Some of these changes translate into volatile organic compound (VOC) profiles in exhaled breath. Such changes can be monitored noninvasively and in real time by means of breath-resolved analysis, e.g., by direct mass spectrometry (MS). Other clinical conditions that take longer to develop can be monitored using offline sampling methods, such as solid-phase microextraction (SPME) or needle trap microextraction (NTME), with laboratory-based MS or other analytical techniques capable of VOC detection in the appropriate concentration range (nmol/L−pmol/L or ppb−ppt). In contrast to blood or tissue analysis, sampling breath is completely noninvasive and testing can be done repeatedly and frequently without any burden to the patient or risk to the staff collecting the samples.

Obtaining representative breath samples is a crucial issue in breath analysis (see Chapter 2). This can be particularly challenging in the clinical environment and

Breathborne Biomarkers and the Human Volatilome. https://doi.org/10.1016/B978-0-12-819967-1.00018-9
Copyright © 2020 Elsevier B.V. All rights reserved.

under mechanical ventilation in the intensive care unit (ICU) because of positive airway pressures.[1] Further, any system that is implemented for breath monitoring must be suitably compatible to be incorporated with the numerous devices already in place, as illustrated in Fig. 18.1. Substances detected in exhaled breath may have endogenous or exogenous origins. Some exhaled VOCs might reflect recent or past exposure and may continue to be exhaled for a long time after exposure. As substance concentrations found in exhaled breath are typically present in the range from μmol/L to fmol/L (ppm−ppt), analytical methods must be highly sensitive with low limits of detection.

This chapter outlines the diagnostic potential of endogenous VOCs in the breath of critically ill patients, including its application in monitoring patients in the ICU. Since many of these patients need ventilator support, aspects of breath analysis under mechanical ventilation are addressed. Analytical procedures, problems associated with the physiological meaning of breath markers, and future developments are discussed, with a focus on the analysis of VOCs in exhaled breath.

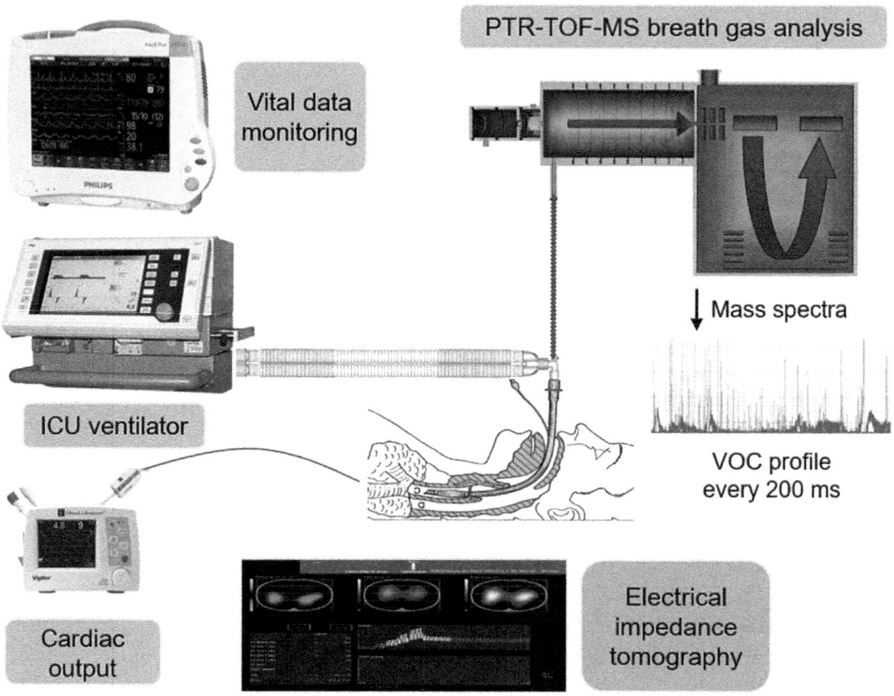

FIGURE 18.1

Example configuration for clinical breath analysis in a mechanically ventilated patient in the intensive care unit (ICU). Continuous analysis of volatile organic compounds (VOCs) by means of direct mass spectrometry, such as proton transfer reaction−time-of-flight−mass spectrometry (PTR-TOF-MS), and ICU vital data monitoring, including electrocardiogram, pulse oximetry, invasive blood pressure, cardiac output, and ventilation monitoring by means of electrical impedance tomography.

18.2 Volatile compound analysis in the intensive care unit

Modern mass spectrometric methods, such as proton transfer reaction—time-of-flight—mass spectrometry (PTR-TOF-MS; Chapter 10), selected ion flow tube—mass spectrometry (SIFT-MS; Chapter 9), secondary electrospray ionization—mass spectrometry (SESI-MS; Chapter 12), or other techniques, such as ion mobility spectrometry (Chapter 11), sensor systems (Chapter 13), and optical spectroscopy (Chapter 14), in principle all enable continuous breath-resolved analysis of volatile substances in breath. Time resolution and sensitivity of some of these methods still present problems to be resolved. Identification of the respiratory phases, as well as substance identification, must be unequivocal and reproducible.

Discontinuous, offline analysis of breath volatiles relies on breath sampling into a sampling receptacle and/or preconcentration of the volatile analytes. Considerable bias may be introduced if these methods are not chosen appropriately and in a well-adapted way to the problem under investigation (see Chapter 2).[2] The huge variation of substance concentrations reported in different studies is at least partly due to dilution and contamination of samples and losses of analytes during the sampling process.[2] It should be noted here that confounding factors can arise both physically, with respect to confounding VOCs (artifacts, environmental contaminants, etc.), and virtually, in relation to data overfitting during data processing (see Section 18.5).

Analysis of exhaled breath condensate (EBC) has been described as an easy and versatile alternative to VOC analysis in exhaled breath gas (see Chapter 7). While collection of EBC samples is simple, the actual quantification of substances in EBC with respect to blood concentrations still represents an unsolved issue. EBC formation and collection depends on various parameters, such as tidal volume, respiratory rate, or airflow, and reliable standardization is still lacking. Current developments aim at the detection of particles in breath rather than at analysis of the whole EBC fraction. Analysis of exhaled particles in healthy volunteers showed low dependency of results from pulmonary variables, such as expiratory flow or tidal volume and may, therefore, help to develop normalization procedures for EBC analysis.[4]

In this context, it is important to realize that a quantitative analysis of substances in breath or EBC does not automatically provide information on substance concentrations in blood or knowledge on the amount of a drug/substance that had been administered. Apart from these analytical and technical issues, physiology can have profound impact on results (see Chapter 2).[5–7] These effects may easily override any true effect produced by disease or therapy.

Direct MS methods such as PTR-TOF-MS represent an attractive tool for real-time analysis of VOC profiles in human breath, but sample gas matrix effects must be taken into account for accurate quantification of breath VOCs using such tools. Specifically, the performance of the instrument can be affected by the sample gas composition. Human breath not only contains several hundred VOCs at ppb—ppt concentrations but is also saturated with water and contains percentage levels of carbon dioxide (CO_2). Additionally, oxygen (O_2) concentrations may vary in a significant way under mechanical ventilation or when supplemental O_2 is supplied to patients via masks or (high-flow) cannulas.[8]

In PTR-TOF-MS, the mobility of ions in the reaction chamber is affected by sample humidity and CO_2 levels (see Chapter 10). The repercussions of this phenomenon have been studied systematically for exhaled breath matrices sampled in the ICU. The effects of CO_2 were found to be more pronounced in dry samples than in humid samples. Further, it was observed in dry samples that the signal intensities for aldehydes and ethanol decreased with increasing O_2 concentration in the sample, whereas those of acetonitrile and aromatic compounds increased with increasing O_2 concentrations. By comparison, isoprene, acetone, and 2-butanone signals showed only small O_2 dependencies. In humid samples, on the other hand, this effect was much less pronounced for aldehydes and ethanol, while acetone, acetonitrile, 2-butanone, and aromatic compounds did not vary with O_2 concentration.[9] It should be noted that such instrumental effects can be accommodated for with adequate calibration. Nevertheless, these results confirm that reliable quantification of VOCs in clinical studies is mandatory and that standardization of VOC research needs well-adapted calibration standards.

Recent developments and technical progress show a trend toward improved and simplified analytical techniques which, in principle, could be employed for routine use in the future. Sampling has become more reliable and reproducible owing to automated devices[2] or algorithms[10] that are able to separate the end-tidal phase from deadspace and inspiratory gas (see Chapter 2). Preconcentration of volatile substances has been simplified considerably by the introduction of microextraction techniques, such as SPME and NTME,[2] facilitating automatic preconcentration and desorption. Promising developments of real-time MS determination of volatile substances in terms of sensitivity, mass resolution, and speed continue (see Chapters 9–12). Nonspecific sensor array technologies may gain importance for diagnostic purposes when specific patterns of disease markers have been defined and interference factors can be excluded. Specific sensors for single substances hold promise if the presence or quantitative changes of these substances are in close relation with disease (see Chapter 13).

18.3 Diseased states

18.3.1 Ventilator-associated pneumonia

Ventilator-associated pneumonia (VAP) is a frequent and dreaded complication of mechanical ventilation. Diagnosis of hospital-acquired pneumonia remains challenging. One of the difficult issues in this context is the distinction between bacterial colonization and infection. A number of studies have addressed VAP by trying to identify volatile biomarkers that might be used to recognize infection and/or colonization in mechanically ventilated patients.[11–13] In a prospective, single-center, cross-sectional cohort study, breath from mechanically ventilated patients was analyzed using gas chromatography−mass spectrometry (GC-MS). Partial least square discriminant analysis (PLA-DA) classified patients with modest accuracy, with an area under the receiver operating characteristic curve (AUROC) value of 0.73 (95% confidence interval; CI: 0.57−0.88) after leave-one-out (LOO) cross-validation. For determining the colonization status of patients, the model had an AUROC of 0.69 (95% CI: 0.57−0.82) after LOO cross-validation. 1-Propanol and hexafluoroisopropanol showed the highest AUROC values, with further substances discriminating healthy from diseased patients being sevoflurane and other halogenated compounds, acetone, carbon disulfide, 2-ethoxy-2-methyl-propane, cyclohexane, and methyl-isobutyl-ketone.

Although numerous substances have been proposed as potential markers for hospital-acquired pneumonia, sensitivity and specificity for diagnostic testing remain unsatisfactory, and none of these markers have been confirmed in prospective, blinded, and randomized trials, or have made it into clinical practice. This is not surprising, as substances of exogenous origin and metabolites present at largely varying concentrations in breath, such as acetone, were used as discriminative markers.

18.3.2 Ischemia/reperfusion

Ischemia/reperfusion (IR) is a prevalent cause of morbidity and mortality. Following complete or partial interruption of oxygen supply, which can occur during shock, coronary heart disease, transplantation, or sepsis, reactive oxygen species (ROS) are generated when perfusion is restored in the tissue. These reactive molecules can destroy cellular structures, including DNA and RNA, and can cause single or multiple organ damage. IR injury caused through sepsis often induces multiorgan dysfunction syndrome (MODS) or multiorgan failure (MOF). According to the National Institute of General Medical Sciences (NIGMS), each year at least 1.7 million adults in North America develop sepsis and nearly 270,000 die as a result.

Coronary artery disease (CAD) is the most common type of heart disease, affecting about 18.2 million adults worldwide and killing over a third of a million people in 2017 (CDC/WHO). Approximately one in five CAD deaths occur in adults less than 65 years old. Clinical symptoms, such as angina, are not always recognizable during myocardial ischemia, especially in the critically ill. Diabetic patients with a high risk of CAD can experience myocardial infarction even without any clinical symptoms. Sensitivity and specificity of current diagnostic methods are not satisfactory in terms of early recognition and prevention of IR injury. Conventional laboratory parameters (e.g., creatinine kinase [CK and CK-MB], troponin-T, lactate), vital data (electrocardiogram, blood pressure, heart rate, cardiac output) or histological findings (biopsy) are not sufficiently sensitive or are too slow to provide early recognition of IR. To date, no diagnostic method has been established that is able to provide information on onset or presence of organ hypoperfusion and IR online and in real time. As a consequence, diagnosis of IR is typically too late, when organ damage cannot be prevented. IR injury is a significant contributor to morbidity and mortality in critical care. As such, the outcome of patients with sepsis, systemic inflammatory response syndrome (SIRS), CAD, or following shock or transplantation might be expected to be markedly improved through early recognition of IR via breath biomarkers to enable immediate subsequent induction of therapy. Pilot studies in humans and animals have demonstrated that IR elicits specific VOC profiles in exhaled breath. During cardiac surgery or spinal ischemia, increased concentrations of metabolic markers (acetone, isopropanol) and substances linked to oxidative stress (pentane, hexane) could be observed, thus providing promising indications that a breath-based test for IR is potentially viable.[14–16]

18.3.3 Acute respiratory distress syndrome

Acute respiratory distress syndrome (ARDS) is a life-threatening lung disorder that can be precipitated by various pathophysiological conditions that affect the lung, either directly or indirectly, such as pneumonia, thoracic trauma or surgery, inhalation injury, mass transfusion, multiple trauma, sepsis, or use of extracorporeal circulation. The incidence of ARDS is high, ranging between 3.7 and 9.6 per 100,000 people in Western nations, and up to 40% of patients with this condition die. Inflammation and ventilation/perfusion mismatch play pivotal roles in disease progression, and this can be observed via changes in exhaled VOC profiles when ARDS is present.

The first observations of changes in exhaled peroxidation markers and isoprene in association with ARDS were made more than two decades ago.[17,18] Interestingly, exhaled isoprene concentrations were found to be lower in ARDS than in the healthy population, which was associated with reactions of isoprene with inflammation-derived reactive species in the alveoli. Numerous studies have since been published proposing different biomarkers for ARDS diagnosis.[19,20] Common problems in these studies were tendencies to overfit data using elaborate statistics on a large number of parameters originating from a comparably low number of independent measures (see Section 18.5). Nevertheless, reported sensitivities and specificities remain poor and—more critically—diagnosis of ARDS is not the primary problem, as this condition is not a disease but a syndrome, with well-established clinical criteria (Berlin definition) and cornerstones of therapy treating the underlying conditions and lung-protective (mechanical) ventilation.

18.3.4 End-stage renal disease

Patients with end-stage renal disease (ESRD), as well as (acute) renal failure, are at risk of numerous complications. In a clinical study, exhaled alveolar gas from 30 ESRD patients during 4 h thrice-weekly hemodialysis (HD) was analyzed by means of headspace SPME-GC-MS. Corresponding venous blood samples were taken to determine conventional serum parameters. Exhaled concentrations of isoprene (10–589 ppb) were observed to drop at initiation of HD, then increased at the end of the procedure. Further, isoprene concentration changes mirrored changes in serum lactate dehydrogenase (LDH) activities. By comparison, exhaled pentane (0.3−12 ppb) increased at onset of HD and then returned to baseline levels after completing the procedure. Ethanol and dimethyl sulfide concentrations remained unchanged during HD. This trial provided evidence of the utility of breath analysis follow hemolysis during HD, offering promise for noninvasive breath testing for renal conditions. In particular, breath tests could be used to monitor patients with renal failure under HD or other renal replacement therapies commonly used in the ICU, such as continuous venovenous hemo(dia)filtration (CVVH(D)F), and thereby help to prevent patients from being affected by associated well-known detrimental side effects.[21]

18.4 Monitoring in the intensive care unit

18.4.1 Mechanical ventilation

Many critically ill patients need ventilatory support, making mechanical ventilation monitoring a vital issue in critical care. It is important when considering potential monitoring procedures that aspects of gas exchange, recruitment of atelectases, and adverse effects of mechanical ventilation are addressed. Exhaled gas from mechanically ventilated patients was analyzed continuously following cardiac surgery in a clinical trial.[22] Acetone and isoprene were monitored by PTR-TOF-MS and CO_2 was measured via capnography. During a lung recruitment maneuver, isoprene and acetone concentrations were observed to decrease by just over half or a quarter, respectively, whereas CO_2 decreased by only 13%. After the maneuver, endogenous acetone had decreased by 14%, whereas isoprene and CO_2 respectively increased by 4% and 9% compared to at-rest concentrations.[22] Further, relationships were observed between changes in ventilation in dorsal lung areas and acetone and isoprene concentrations. These observations reflect the principal dependency of exhaled VOC concentrations on minute ventilation. The extent of these correlations depends on the physicochemical properties of the compounds, as well as their total quantities in the body. Following the notion of ventilation perfusion matching, an increase in ventilation in well-perfused dorsal lung areas improves ventilation/perfusion matching and leads to a decrease in VOC concentrations during the recruitment maneuver, when substances are exhaled at greater rates.

18.4.2 Medication

Critically ill patients typically receive a variety of medications, such as analgetic, hypnotic, antibiotic, and/or antimycotic drugs, muscle relaxants, catecholamines, antiarrhythmic or blood pressure lowering drugs, or antidiabetics. Some of these drugs are sufficiently volatile to be detectable in exhaled breath; volatile metabolites might similarly be present. Monitoring drug levels in patients without the need to draw blood samples is an attractive prospect in the critical care environment. Apart from improved therapy control, a breath-based approach would be a step toward personalized medicine. Drug detection in breath can be either qualitative or quantitative, depending on the matrix that is analyzed and the general goal of the assay. As drugs and their metabolites tend toward larger, less volatile compounds, technologies such as SESI-MS or analysis of particulate matter might represent promising approaches.

Fentanyl, a synthetic opioid, is widely applied as an intravenous analgesic and is mainly eliminated by hepatic biotransformation, with only a small portion being eliminated unchanged. The latter has been detected in both urine and feces, as well as in patients' respiratory circuits and in the ambient air of the operating theater. Detection limits of 0.01 ng/mL for fentanyl in exhaled breath have been achieved by SPME-GC-MS. In a clinical study, SPME was used to sample breath from respiratory circuits of mechanically ventilated patients. Following fentanyl administration, concentrations in breath were detected in the range 6.00–20.89 pg/mL; the broad range was attributed to concentrations associated with variations in pulmonary uptake and distribution over time. Peak concentrations of fentanyl in exhaled breath were reached 15–20 min after administration, despite plasma concentrations already peaking after 30 s.

A series of (illicit) drugs and medications has been shown to be detectable in exhaled breath by sensitive real-time MS, including narcotics (fentanyl, norfentanyl, morphine) and naloxone, cocaine, nicotine, and γ-hydroxybutyrolactone.[14] Tramadol, an analgesic, and its metabolite O-desmethyltramadol (ODMT) was detected in exhaled breath by means of liquid chromatography–tandem MS (LC-MS/MS) analysis. The study participants received a single oral dose of tramadol, and their breath was analyzed repeatedly over 48 h. Tramadol was detectable in breath up to 48 h after intake in all but one subject, while ODMT was partially detectable up to 24 h. The ratio of ODMT to tramadol was lower in breath compared to plasma.[4] The detection of drug intake via breath analysis is the subject matter of Chapter 30.

Propofol is a hypnotic agent that is frequently used for sedation of patients in intensive care. Since propofol is administered intravenously, in contrast to inhalation anesthetics a noninvasive method for continuous monitoring of propofol plasma concentrations is not available. In clinical practice, administration of propofol takes into account the patient's age, body weight, and preexisting diseases. Overdosage of propofol can induce hemodynamic instability, whereas low propofol concentrations increase the risk of awareness; a potentially life-threatening condition termed "propofol infusion syndrome" may occur, depending on the total amount of drug infused. Propofol is volatile enough to be detected in the gas phase, thus also in exhaled breath. Experimental and clinical studies have demonstrated the presence of propofol in exhaled breath after intravenous administration. Concentrations in breath typically ranged from 0.04 to 0.5 nmol/L (ppb range), whereas concentrations in whole blood ranged from 2 to 212 μmol/L (ppm range).[4] Only loose correlations between breath and blood concentrations have been determined; this was on account of expiratory propofol concentrations only reflecting blood concentrations under steady-state conditions. Pharmacological factors, such as compartmental distribution, pulmonary metabolism, and elimination of the drug, can affect the relationship between breath and blood concentrations. Even interindividual differences in partial pressure ratios of propofol in blood and in exhaled breath were reported.

Physiology plays an important role in the exhalation of VOCs. Changes in cardiac output, pulmonary perfusion, alveolar and deadspace ventilation, or changes in the ventilation/perfusion ratio and intrapulmonary shunt, can affect exhalation of xenobiotics like propofol. Such effects are poorly understood, yet nevertheless clinically relevant. Propofol is injected intravenously, is bound >98% to proteins and erythrocytes, and is transported to the lungs and other tissues via blood flow, which depends on cardiac output. The free fraction of propofol crosses the alveolar–capillary membrane and is exhaled in end-tidal breath. Disturbances in all sections can influence the relationship between blood and alveolar concentrations of propofol. Patients undergoing lung surgery showed different blood: breath propofol correlation coefficients compared to patients with nonthoracic surgery.[23] The significant impact on the correlation between blood and breath propofol concentrations can be attributed to an impaired ventilation/perfusion ratio of the lung, whereby an intrapulmonary shunt can lead to consistently lower propofol concentrations in exhaled breath in such patient groups despite identical blood concentrations.

Changes in cardiac output lead to considerable changes in organ perfusion and distribution-related changes of propofol blood and breath concentrations. Hemodynamic instability and changes in cardiac output occur frequently in critical care, especially in older and morbid patients. Increase in cardiac output, e.g., due to infusion of catecholamines, causes opposite alteration of propofol blood and breath levels. In an animal model, decreasing pulmonary blood flow and cardiac output did not significantly affect the relationship between propofol concentrations in breath and blood, whereas increasing cardiac output due to dobutamine infusion caused a deterioration of this relationship (Fig. 18.2).[24] Propofol breath gas analysis does not reflect its blood levels adequately during phases of distinct and rapid changes in cardiac output or during states of ventilation/perfusion mismatch. Propofol breath monitoring therefore might not be suitable to assess anesthetic depth in such cases. In addition, alveolar propofol concentrations can be affected by impaired diffusion characteristics of the alveolar–capillary membrane due to pulmonary edema, fibrosis, or emphysema. Furthermore, alveolar ventilation and thus changes of respiratory adjustment in relation to tidal volume and respiratory rate might influence propofol exhalation.

Online monitoring of propofol in breath could be advantageous if relevant influences of pharmacology and pathophysiology on propofol exhalation are considered. The aforementioned challenges must be addressed, however, before propofol breath analysis can be introduced into routine clinical practice. Clinical conditions and disease states that can cause impairment of the propofol breath/blood correlation and thus reduce reliability and safety of the method require examination. Additionally, the purpose of application must be clearly defined, since these have different requirements: simple disconnection monitoring of propofol delivery is a significantly

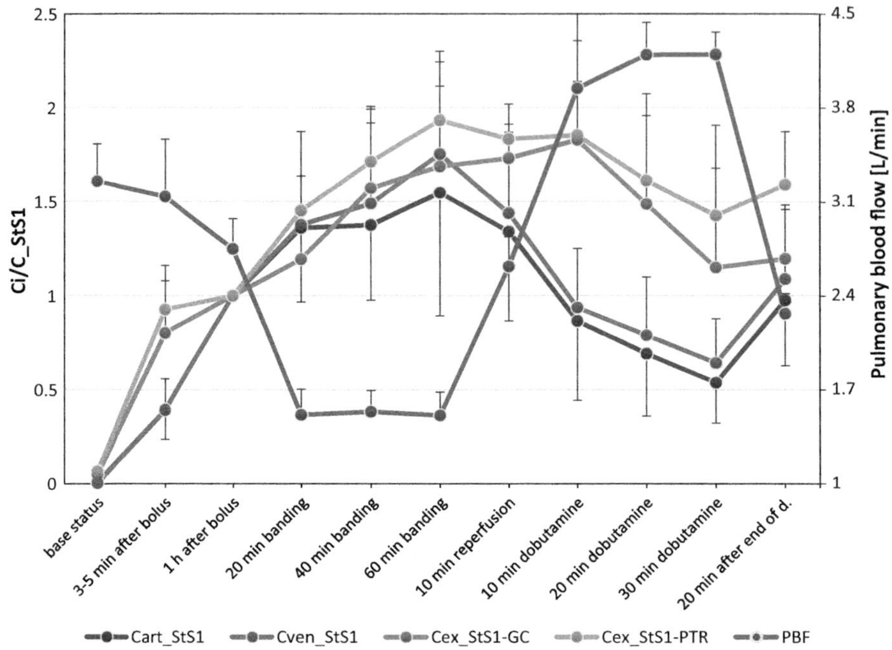

FIGURE 18.2

Breath propofol concentrations (Cex_StS1-GC and Cex_StS1-PTR), arterial (Cart_StS1), and mixed venous (Cven_StS1) concentrations of propofol in relation to pulmonary blood flow (PBF; L/min). Breath and blood concentrations were normalized to values at steady state.[24]

different application to breath monitoring intended for indirect estimation of anesthetic depth.

In terms of current developments in sensor technology, propofol monitoring could be achieved by sensor systems or spectroscopic devices. Despite these breath sensor technologies being promising, however, clinical implementation has yet to be realized. Any detection device for point-of-care (PoC) propofol breath monitoring has to be validated via large multicentered studies, including the use of target controlled infusion (TCI) and effect-site monitoring, e.g., bispectral index (BIS), are imperative for further evaluation. It is mandatory that these requirements are met and that limitations are taken into account before noninvasive, real-time propofol monitoring in exhaled breath is adopted in the ICU, for example, integrated into ICU ventilators, as a future vision.

18.4.3 Metabolism

Metabolic disorders and starvation play an important role in critical care medicine. Protein catabolism and anaerobic metabolic pathways cause additional morbidity in the critically ill. Markers for these pathological states that are both easy to measure and reliable markers are still lacking. Several breath volatiles are associated with carbohydrate, lipid, and protein metabolism. Acetone (dimethyl ketone) is formed by decarboxylation of acetoacetate, which derives from lipolysis or lipid peroxidation. Acetone is one of the most abundant compounds in human breath. Ketone bodies like acetone are oxidized in peripheral tissue via the Krebs cycle. Ketone bodies in blood (including acetoacetate and β-hydroxybutyrate) are increased in ketonemic subjects, in times of fasting, starving, or during diet. Breath acetone concentrations are elevated in patients with (uncontrolled) diabetes mellitus. As acetone is produced by spontaneous decarboxylation of acetoacetate, it is impossible to quantify the fraction that arises from lipid peroxidation. 2-Propanol is postulated to be a product of an enzyme-mediated reduction of acetone.

Volatile amines are generated during protein metabolism and show immediate changes in exhaled concentrations after protein ingestion.[25] Critically ill patients frequently suffer from respiratory or metabolic acid—base imbalances; here, it is important to bear in mind that exhalation of ammonia and amines is pH-dependent, e.g., ammonia (NH_3) reacts with water (H_2O) as a Lewis base: $NH_3 + H_2O \rightleftharpoons NH_4^+ + OH^-$. According to mass action law, concentration ratios of ammonia, ammonium ions (NH_4^+), hydroxyl ions (OH^-), and water are constant for a given temperature and pressure. From this, it can be derived that $[NH_4^+]:[NH_3] = 63:1$ at pH 7.45 and 79:1 at pH 7.35, i.e., $[NH_4^+]:[NH_3]$ changes by 33% when pH is changed by only 0.1.[22]

As energy production and fuel consumption also translate into VOC profiles,[26] monitoring VOCs in exhaled breath has the potential to assess aspects of nutrition, starvation, and energy production in the critically ill in a noninvasive manner.

18.5 Data mining pitfalls

Most—if not all—potential diagnostic applications in breath research involve changes in biomarker concentrations, rather than unique breath markers that are present only in the diseased state. Hence, data interpretation is a crucial step in breath analysis. Mathematical treatment and data mining approaches are covered in other chapters of this book (see Chapters 37 and 38), but a brief discourse on the pitfalls of inappropriate data handling is offered in the following.

To avoid artificial significance of breath-based data, every effort should be made to implement method validation, data cross-testing, and statistical validation along the entire process.[27] The most common problems related to data analysis in breath research can be classified into three groups, namely confounding variables, which have a real correlation with both the diseased state and a breath marker, but lead to the erroneous conclusion (VCs) that there is a causal relationship between disease and breath; "voodoo" correlations, which can be understood as statistically true correlations that arise coincidentally as a result of the vast number of measured variables; and statistical misconceptions in the study design (SMSD).[13]

Typical confounding variables are environmental exposure and medical history, host factors, such as sex, age, weight, etc., and parameters that could affect the quality of breath data, including subject breathing mode, effects of breath sampling, or effects of the analytical technique itself. Disinfectants and volatiles off-gassing from medical equipment, such as endotracheal tubes, respiratory circuits, or intravenous-tubing (IV-tubing), must be recognized and taken into account in breath analysis undertaken in the ICU (as well as in other environments). Other factors that might impact exhaled VOCs in the ICU are (par)enteral feeding and numerous medications commonly given to the critically ill.

In cases when the number of measured variables overwhelms the number of independent samples, the chances of finding coincidental "voodoo" correlations grow proportionally. Such correlations can be typically expected in the following scenarios: insufficient number of patients, (too) many measurement variables, use of advanced statistical data mining methods (overfitting), or data validation using nonindependent datasets. Statistically, this problem is due to the lack of correction for multiple comparisons and the nonindependence problem.[28–30] The latter refers to the inflation of cluster-wise statistical estimates when the constituent data are selected using the same statistical measure.[31]

Nonprospective, nonblinded, and nonrandomized trials, a priori biased study populations, or group selection with unrealistically high disease prevalence typically represent misconceptions of study design. Further discussion on study design is offered in Chapter 35.

18.6 Summary

Analysis of volatile substances in exhaled breath gas can be considered a noninvasive window into biochemical processes in the lung, the airways, and the whole body. As such, breath analysis represents a promising alternative to serum chemistry and other invasive diagnostic procedures for intensive care patients. Continuous monitoring of exhaled volatiles by means of direct MS or spectroscopy seems especially attractive for the dynamical state of critically ill patients, while sensor systems offer opportunities for bedside (PoC) analyses.

Countless volatile substances have been suggested as biomarker candidates for specific conditions, yet many of these observations and results could never be reproduced, and none of those proposed biomarkers have transitioned into clinical practice. Correlations between breathborne VOCs and diseases are often an inevitable outcome when too many parameters are matched with too few distinct measurements.

Requirements for clinical applications of exhaled breath analysis include time-resolved, onsite, and PoC applicability. Modern analytical methods such as sensor technologies potentially meet these requirements, yet many sensors and analytical methods do not offer sufficient resolution to achieve breath-by-breath resolved analysis. Additional effort is therefore necessary to provide well-defined and reproducible breath samples for measurement devices. Substances in exhaled breath have varied pathways: some substances are bloodborne and originate from the alveoli; others, such as NO and CO, are generated in the airways; many compounds represent contaminants from ambient air; and other VOCs may be generated in the oral cavity. This complexity dictates a necessity for thoroughly controlled sampling to ensure reliable analysis of breath biomarkers. In order to introduce sensor technology into clinical breath analysis, automated, controlled, and adjustable sampling devices are required. These devices must be easy to handle, inexpensive, and adaptable in terms of sample volume, flow, and origin (spontaneously breathing or mechanically ventilated individuals). For reasons of standardization, relevant data, such as sampling flow rate, total sampled volume, or CO_2 partial pressures during sampling, must be recorded continuously.

Despite a number of very promising observations of exhaled markers offering potential diagnostic properties, the analysis of volatile compounds in the breath of the critically ill has not yet been introduced into clinical practice. The main obstacles hindering this transition are technical problems involving sampling, preconcentration, and analysis, as well as basic methodological issues, such as normalization, expression of data, and the problem of background contamination. With the exception of NO, such standards do not exist. For this reason, considerable effort has been expended in establishing reproducible and reliable analytical methods and in understanding how various compounds end up in exhaled breath.

Three broad questions challenge researchers studying volatile markers in the breath of critically ill patients: How should breath samples be collected? Given the low concentrations of VOCs in breath, how can they be detected and quantified in a fast and reliable way? And finally, is there a clear link between certain volatile substances and particular diseases?

In terms of longitudinal studies, breath biomarkers represent promising alternatives for metabolic profiling, drug monitoring, and optimization of mechanical ventilation. When breath analysis is used for cross-sectional investigations, researchers must be aware of crucial issues, such as high natural variations of breath biomarkers, pitfalls of statistics, confounding variables, and clinical relevance. Biomarkers play an important role for staging purposes and for detection of disease progression rather than for primary diagnosis. Strong points of modern breath analysis are its noninvasiveness, speed and continuity, potential bedside applicability without any burden to the patient, and unlimited repeatability. In a first step towards clinical applications, these features must offer optimal conditions to gather additional and complementary information in metabolomics, control of disease progression, pharmacodynamics, and kinetics.

References

1. Miekisch W, et al. Impact of sampling procedures on the results of breath analysis. *J Breath Res* 2008;**2**:026007.

2. Trefz P, Rosner L, Hein D, Schubert JK, Miekisch W. Evaluation of needle trap micro-extraction and automatic alveolar sampling for point-of-care breath analysis. *Anal Bioanal Chem* 2013;**405**:3105–15.
3. Beauchamp J, Herbig J, Gutmann R, Hansel A. On the use of Tedlar (R) bags for breath-gas sampling and analysis. *J Breath Res* 2008;**2**.
4. Trefz P, et al. Drug detection in breath: non-invasive assessment of illicit or pharmaceutical drugs. *J Breath Res* 2017;**11**:024001.
5. Sukul P, et al. FEV manoeuvre induced changes in breath VOC compositions: an unconventional view on lung function tests. *Sci Rep* 2016;**6**:28029.
6. Sukul P, Trefz P, Schubert JK, Miekisch W. Immediate effects of breath holding maneuvers onto composition of exhaled breath. *J Breath Res* 2014;**8**:037102.
7. Sukul P, Trefz P, Kamysek S, Schubert JK, Miekisch W. Instant effects of changing body positions on compositions of exhaled breath. *J Breath Res* 2015;**9**:047105.
8. Nishimura M. High-Flow Nasal Cannula Oxygen Therapy in Adults: Physiological Benefits, Indication, Clinical Benefits, and Adverse Effects. *Respir Care* 2016;**61**:529–41.
9. Trefz P, Schubert JK, Miekisch W. Effects of humidity, CO_2 and O_2 on real-time quantitation of breath biomarkers by means of PTR-ToF-MS. *J Breath Res* 2018;**12**:026016.
10. Schwoebel H, et al. Phase-resolved real-time breath analysis during exercise by means of smart processing of PTR-MS data. *Anal Bioanal Chem* 2011;**401**:2079–91.
11. van Oort PM, et al. Exhaled breath metabolomics reveals a pathogen-specific response in a rat pneumonia model for two human pathogenic bacteria: a proof-of-concept study. *Am J Physiol Lung Cell Mol Physiol* 2019;**316**:L751–6.
12. Schnabel R, et al. Analysis of volatile organic compounds in exhaled breath to diagnose ventilator-associated pneumonia. *Sci Rep* 2015;**5**.
13. May AK, et al. A Pilot Study of the Noninvasive Assessment of the Lung Microbiota as a Potential Tool for the Early Diagnosis of Ventilator-Associated Pneumonia. *Chest* 2015;**147**:1494–502.
14. Püschel A, et al. Can Recognition of Spinal Ischemia Be Improved? Application of Motor-Evoked Potentials, Serum Markers, and Breath Gas Analysis in an Acutely Instrumented Pig Model. *Ann Vasc Surg* 2018;**49**:191–205.
15. Pabst F, Miekisch W, Fuchs P, Kischkel S, Schubert JK. Monitoring of oxidative and metabolic stress during cardiac surgery by means of breath biomarkers: an observational study. *J Cardiothorac Surg* 2007;**2**:37.
16. Szűcs S, et al. Detection of Intestinal Tissue Perfusion by Real-Time Breath Methane Analysis in Rat and Pig Models of Mesenteric Circulatory Distress. *Crit Care Med* 2019;**47**:e403–11.
17. Schubert JK, Miekisch W, Geiger K, Nöldge-Schomburg GFE. Breath analysis in critically ill patients: potential and limitations. *Expert Rev Mol Diagn* 2004;**4**:619–29.
18. Schubert JK, Müller WP, Benzing A, Geiger K. Application of a new method for analysis of exhaled gas in critically ill patients. *Intensive Care Med* 1998;**24**:415–21.
19. Zhou M, et al. Rapid breath analysis for acute respiratory distress syndrome diagnostics using a portable two-dimensional gas chromatography device. *Anal Bioanal Chem* 2019;**411**:6435–47.
20. Bos LDJ, Schultz MJ, Sterk PJ. Exhaled breath profiling for diagnosing acute respiratory distress syndrome. *BMC Pulm Med* 2014;**14**:72.
21. Goerl T, et al. Volatile breath biomarkers for patient monitoring during haemodialysis. *J Breath Res* 2013;**7**:017116.

22. Brock B, et al. Monitoring of breath VOCs and electrical impedance tomography under pulmonary recruitment in mechanically ventilated patients. *J Breath Res* 2017;**11**: 016005.
23. Miekisch W, Fuchs P, Kamysek S, Neumann C, Schubert JK. Assessment of propofol concentrations in human breath and blood by means of HS-SPME-GC-MS. *Clin Chim Acta Int J Clin Chem* 2008;**395**:32–7.
24. Kamysek S, et al. Drug detection in breath: effects of pulmonary blood flow and cardiac output on propofol exhalation. *Anal Bioanal Chem* 2011;**401**:2093–102.
25. Pugliese G, Trefz P, Brock B, Schubert J, Miekisch W. Extending PTR based breath analysis to real-time monitoring of reactive volatile organic compounds. *Analyst* 2019.
26. McCue MD, Pollock ED. Measurements of substrate oxidation using (13)CO 2-breath testing reveals shifts in fuel mix during starvation. *J Comp Physiol [B]* 2013;**183**: 1039–52.
27. Miekisch W, Herbig J, Schubert JK. Data interpretation in breath biomarker research: pitfalls and directions. *J Breath Res* 2012;**6**:036007.
28. Vul E, Harris C, Winkielman P, Pashler H. Puzzlingly High Correlations in fMRI Studies of Emotion, Personality, and Social Cognition. *Perspect Psychol Sci* 2009;**4**:274–90.
29. Vul E, Pashler H. Voodoo and circularity errors. *NeuroImage* 2012;**62**:945–8.
30. Fiedler K. Voodoo Correlations Are Everywhere-Not Only in Neuroscience. *Perspect Psychol Sci J Assoc Psychol Sci* 2011;**6**:163–71.
31. Kriegeskorte N, Simmons WK, Bellgowan PSF, Baker CI. Circular analysis in systems neuroscience: the dangers of double dipping. *Nat Neurosci* 2009;**12**:535–40.

CHAPTER

Chronic obstructive pulmonary disease and smokers

19

Andras Bikov[1], Zsofia Lazar[2]

[1]*North West Lung Centre, Manchester University NHS Foundation Trust, Manchester, United Kingdom;* [2]*Department of Pulmonology, Semmelweis University, Budapest, Hungary*

Chapter outline

Breathborne Biomarkers and the Human Volatilome. https://doi.org/10.1016/B978-0-12-819967-1.00019-0
Copyright © 2020 Elsevier B.V. All rights reserved.

19.1 Overview

Chronic obstructive pulmonary disease (COPD) is currently the fourth leading cause of death worldwide. It is characterized by progressive airflow limitation and by persistent symptoms of breathlessness and usually productive cough.[1] It is a multifactorial disease induced by chronic exposure to noxious particles, especially cigarette smoke, in susceptible individuals. Other factors, such as premature birth, frequent childhood infections, asthma, or passive smoking, can also lead to impaired lung growth and subsequently to the development of fixed airway obstruction and respiratory symptoms. The diagnosis of COPD is based on three criteria: symptoms (breathlessness or productive cough), suggestive medical history (exposure to noxious particles), and fixed airway obstruction (ratio of postbronchodilator forced expiratory volume in 1 s to forced vital capacity; $FEV_1/FVC < 0.70$). Apart from diagnosis, lung function is used for staging (Table 19.1). Although widely recognized as a progressive disease, the activity of disease varies largely between patients. Around half of COPD patients have a rapid FEV_1 decline ($\geq$50 mL/year loss)[2] and around 30% are prone to acute exacerbations, which are major events that lead to health deterioration and are associated with high healthcare burden and mortality.[3] The pharmacological treatment for COPD depends on symptoms and exacerbation burden (Figs. 19.1 and 19.2). The nonpharmacological treatment includes smoking cessation, pulmonary rehabilitation, vaccinations, oxygen therapy, and ventilatory support.

Chronic obstructive pulmonary disease comprises several distinct phenotypes, with various underlying mechanisms contributing to disease development.[4] Noxious particles, especially cigarette smoke, induce airway inflammation in susceptible individuals, which leads to the development of chronic bronchitis, small airway narrowing (bronchiolitis), and emphysema. These are usually present in every COPD patient, but the extent of each pathological change varies largely between patients. Emphysema usually affects the upper lobes and induces considerable ventilation heterogeneity in COPD. Moreover, during normal expiration, airways in the basal segments collapse, contributing to a more pronounced heterogeneity. Due to the enhanced collapsing potential of the small airways in COPD, airway closure occurs

Table 19.1 The Global Initiative for Chronic Obstructive Lung Disease (GOLD) COPD grading system based on airflow limitation.

FEV_1 (% predicted)	Grade
≥80	GOLD 1 (mild)
50–79	GOLD 2 (moderate)
30–49	GOLD 3 (severe)
<30	GOLD 4 (very severe)

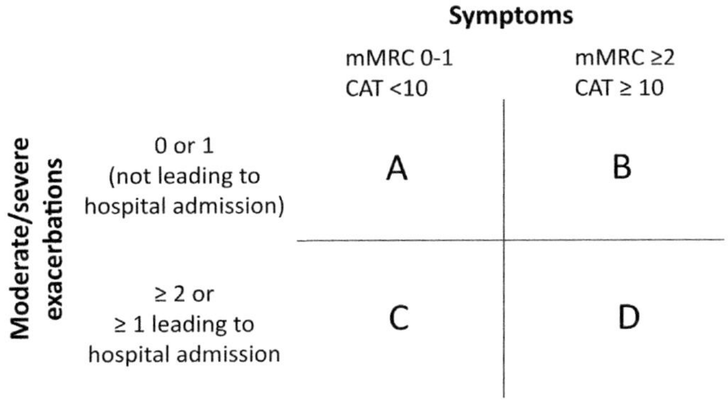

FIGURE 19.1

The Global Initiative for Chronic Obstructive Lung Disease (GOLD) grouping system to categorize patients into A, B, C, and D groups based on symptoms and exacerbation burden. This grouping help in choosing the initial treatment (Fig. 19.2.).

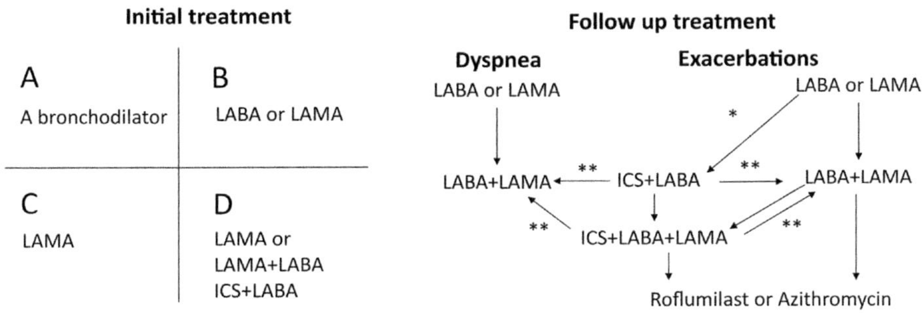

FIGURE 19.2

Pharmacological management of stable COPD. Initial management depends on the grouping system (left panel). If the patient is symptomatic or has persistent exacerbations despite initial treatment, the following algorithm is suggested (right panel): *, consider if blood eosinophils $\geq 300 \times 10^9/L$ or $\geq 100 \times 10^9/L$ and ≥ 2 moderate exacerbations/1 hospitalization. **, consider deescalation of inhaled corticosteroids if pneumonia, inappropriate original indication or lack of response to inhaled corticosteroids. LABA: long-acting β2-agonist; LAMA: long-acting muscarinic antagonist; ICS: inhaled corticosteroid.

at the level of the more proximal airways compared to in healthy individuals. This can potentially contribute to the trapping of a fraction of molecules during a single exhalation maneuver.

Chronic bronchitis is characterized by goblet cell hyperplasia, excessive mucus production, epithelial and ciliary dysfunction, enhanced bronchial circulation, microbial colonization, and influx of inflammatory cells (mainly macrophages and neutrophils and in lesser extent lymphocytes and potentially eosinophils). Small airway disease is featured by airway narrowing due to remodeling and loss of elastic recoil, as well as increased airway inflammation. Finally, emphysema is related to the destruction of the alveolar walls, leading to impaired diffusion capacity and vascular remodeling.

In many subjects, smoking causes airway inflammation, symptoms, and lung structure damage without reaching the diagnostic criterion of fixed airway obstruction. If smoking persists, it is very likely that these people will develop COPD eventually, therefore non-COPD smokers with respiratory symptoms are often labeled as "early COPD."

19.2 Detection of chronic obstructive pulmonary disease in breath

Hundreds of articles have investigated exhaled breath in COPD, to date. Most of these studies employed a case–control design aimed at identifying discriminative/diagnostic biomarkers. However, COPD diagnosis is already possible by the simple, cheap, reliable, and widely used pulmonary function test. Beneficial new (breath-based) tests would offer (1) phenotyping patients based on their inflammatory profile and exacerbation etiology, (2) treatment response prediction, or (3) lung function deterioration and exacerbations prognoses.

In this chapter, different collection and analytical techniques in relation to target compounds are discussed, including exhaled breath condensate (EBC; see Chapter 7), particles (see Chapter 8), volatile organic compounds (VOCs), nitric oxide (NO; see Chapters 4 and 5), carbon monoxide (CO; see Chapter 6), and exhaled breath temperature (EBT). These methods have different clinical potential in COPD (Table 19.2). Rather than discussing specific molecules, the focus is on methodological considerations that may influence exhaled biomarker results in COPD.

19.3 Exhaled breath condensate

19.3.1 Methodological limitations and confounding factors of exhaled breath condensate measurements

Mediator concentrations in EBC are around the detection limit of commercially available assays, which is one of the reasons for poor intrasubject repeatability of most markers (detailed in Ref. 5). This can be improved by either sample enrichment or a better choice of detection methods. For instance, the detection rate of cytokines in EBC from patients with COPD by an enzyme-linked immunoassay (ELISA) can be improved by sample enrichment using lyophilization, and the between-day variability of malondialdehyde (MDA) concentration is markedly lowered when high-pressure liquid chromatography (HPLC) is combined with mass spectrometry (MS) during measurements. In line with this, the highly sensitive signals captured by nuclear magnetic resonance (NMR) spectroscopy show good within-days stability in COPD.

The effect of cigarette smoking on EBC biomarker readings is ambiguous. Increased levels of hydrogen peroxide (H_2O_2), 8-isoprostane, and nitrotyrosine are observed in smokers. Acute cigarette smoking induces a transient increase in EBC nitrate concentrations, but it does not influence pH values, interleukin 1 beta (IL-1β), interleukin 8 (IL-8), tumor necrosis factor-alpha (TNFα), or MDA

Table 19.2 The potential clinical applications of exhaled biomarkers in COPD.

Exhaled breath condensate (fingerprint analyses)	Exhaled particles	Exhaled VOCs	Exhaled nitric oxide	Exhaled carbon monoxide	Exhaled breath temperature
Aiding disease phenotyping	Aiding disease phenotyping	Aiding disease phenotyping	Prediction of exacerbations	Monitoring smoking cessation	Prediction of disease progression
Aiding differential diagnosis	Aiding differential diagnosis	Aiding differential diagnosis	Prediction of response to inhaled and systemic steroid treatment		Prediction of disease exacerbations
		Detecting microbial colonization	Diagnosing eosinophilic exacerbations		Diagnosing exacerbations with increased airway neutrophilia
		Diagnosing infectious exacerbations	Simultaneously monitoring the effects of antiinflammatory drugs on central and peripheral airway inflammation		

concentrations. Furthermore, physiological variations can also modulate EBC biomarker concentrations as H_2O_2 concentration shows diurnal variability in patients with COPD and it is strongly affected by the respiratory pattern and exercise. However, the effect of gastroesophageal reflux disease, a common comorbidity in COPD, on EBC pH and pepsin concentration has not been conclusive.

In support, the recent technical standard document on exhaled biomarkers endorsed by the European Respiratory Society (ERS) details the technical and patient-related difficulties of the method and discusses the problems arising during data interpretation.[6] The expert panel concluded that the EBC technique should be standardized for individual markers, usually requiring the adjustment of various steps of the measurement protocols.

19.3.2 Clinically relevant exhaled breath condensate biomarkers in smokers and patients with chronic obstructive pulmonary disease

A detailed description of biomarkers in COPD are provided in Ref. 5. EBC pH is the most studied marker, but results are inconsistent regarding changes in COPD compared to control nonsmoking or smoking subjects. Both lower and similar EBC pH values were reported in COPD. Importantly, an association of EBC pH to disease characteristics of COPD, including lung function, sputum inflammatory profile, emphysematous phenotype, current relapse, or frequent exacerbations, is not proven. This can be partly explained by the high within- and between-day variations in EBC pH of stable COPD patients, with no apparent relationship between the degree of variability and clinical factors.

The measurements of H_2O_2, 8-isoprostane, and MDA in EBC reflect the burden of airway oxidative stress. None of these markers has been shown unambiguously to be increased in COPD or to be linked with variables and activity of the disease. Of note, current smoking is a significant confounder when monitoring these mediators. Markers of nitrative stress, including nitrate, nitrite, and nitrosothiols, have been detected in EBC of patients with COPD, but data do not support the usefulness of these markers to monitor disease severity. More sophisticated detection methods, such as capillary electrophoresis, might provide more reliable readings.

Among the arachidonic acid metabolites in EBC, leukotriene B_4 (LTB_4), and cysteinyl leukotrienes are the most studied mediators. EBC LTB_4 levels have consistently been reported to be elevated in COPD, which seems to be a marker for monitoring airway inflammation, although salivary origin is a potential source of mediator contamination.

The detection of single proteins in EBC, including cytokines, chemokines, and circulating receptors, seems unreliable with immunoassays, as contradictory findings have been reported not only in control smokers but also in patients with COPD. Proteomic analysis that returns a profile of multiple proteins might be a step forward, as the fingerprint generated by this technique could distinguish patients with different phenotypes, including those with and without emphysema or subjects with emphysema associated with alpha-1-antitrypsin deficiency.

A similar method, metabolomics, which measures metabolic compounds, can differentiate patients with COPD from controls and patients with asthma, and it differentiates COPD with emphysema, but these new techniques need further validation and their reproducibility should be proven (these techniques and recent results are reviewed in Ref. 7).

Finally, it should be emphasized that at present no EBC biomarkers have been validated to identify clinical phenotypes or to predict disease relapse or progression in COPD.

19.4 Exhaled particles

19.4.1 Exhaled particles in smokers

Only one small study has investigated exhaled particles in smokers that included some patients with airway obstruction.[8] No difference in the amount of exhaled particles was observed between smokers and nonsmokers, but particles collected from smokers were found to contain more sodium and potassium salts, as well as protonated and sodiated phosphocholine adducts and a relatively lower quantity of phosphocholine and phosphocholine potassium adducts. These differences were related to smoking history.

19.4.2 Exhaled particles in chronic obstructive pulmonary disease

The formation and origin of exhaled particles is described in Chapter 8 in detail. Concerning COPD, maneuvers necessary to yield the highest particle amount (i.e., exhalation to residual capacity followed by inhalation to total lung capacity and measured exhalation) may be challenging to some patients.[9] In addition, as physiological airway closure during exhalation happens at the levels of the more proximal airways in COPD patients compared to healthy individuals, the anatomical region sampled may be slightly different. In addition, some patients with COPD have increased mucus production in the conducting airways, which can also be a potential source of particles. As the temperature of exhaled breath affects particle formation, this may also be a confounding factor to consider in COPD (see Section 19.6).

The number of studies in COPD that have analyzed exhaled particle impaction is very low. It seems that the amount of exhaled particles in COPD is lower than in healthy subjects. A potential explanation could be that exhalation to residual volume can be challenging for many patients with COPD. Analyzing specific molecules, the levels of surfactant protein A normalized to the amount of particles was observed to be lower in COPD, while there was no difference in albumin. The effect of inhaled medications on exhaled particles has not been investigated to date, but bronchodilators can potentially improve airway closure leading to altered particle levels and composition.

19.4.3 Volatile organic compounds

Factors associated with altered exhaled kinetics in COPD

Endogenous exhaled VOCs originate either from the airway wall or from the pulmonary circulation by diffusing through the alveolocapillary barrier.[10,11] In the conducting airways, VOCs are either released as products of the local cellular metabolism, inflammation, and oxidative stress, or they diffuse from the bronchial circulation. For instance, expiratory flow dependency of exhaled ethanol suggests that it partly originates from the bronchial circulation. As the bronchial wall is thickened in COPD, this could contribute to altered levels. During acute exacerbation, however, when the bronchial circulation is accelerated, the release of VOCs can be augmented. In addition, COPD is often characterized by an acidic mucus hypersecretion, hence acidophilic, water-soluble VOCs may trap in the airway lining fluid. Secondly, ventilation heterogeneity and airflow limitation in COPD, which may affect the exhalation kinetics, make the concentrations of VOCs unpredictable and variable between each breath. To minimize this confounding effect, the multiple breathing technique is preferred over the single exhalation method in breath sampling.

Emphysema can be present and may lead to impaired diffusion capacity even in early disease stages. Reduction in diffusion capacity, together with vascular remodeling, may affect the excretion of bloodborne VOCs (i.e., acetone, isoprene) into exhaled breath.

In summary, changes in exhaled VOCs in COPD may represent acceleration in airway metabolism, inflammation, and oxidative stress, but it can also be related to alteration in exhalation kinetics (Fig. 19.3).

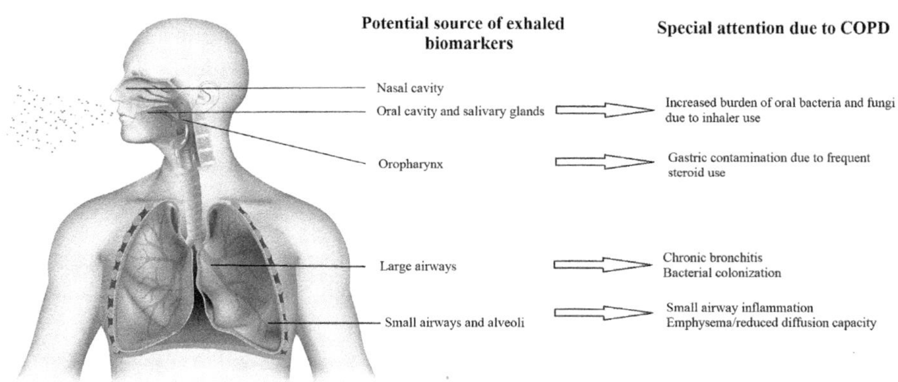

FIGURE 19.3

Potential sources of exhaled biomarkers with a special attention to COPD.

The role of microbial colonization on exhaled VOCs in COPD

Due to the aberrant and partly paralyzed airway immunity, and frequent steroid and antibiotic use, the airways of patients with COPD are often colonized by different bacteria and fungi that are associated with a unique VOC pattern.[12] More specifically, *Streptococcus pneumoniae*, which often colonizes the airways of COPD patients, releases acetaldehyde, isoprene, formaldehyde, butadine, and isopentanol. In addition, *Pseudomonas aeruginosa*, which can colonize patients with very severe COPD, is associated with a unique pattern of VOCs. Although this has a confounding effect in stable disease, during acute exacerbations, VOCs can potentially help with diagnosing infectious etiology and guide antimicrobial treatment.

Effect of smoking

Smoking can increase the levels of exhaled pentane, benzene, and acetonitrile and alter the exhaled volatile compound pattern.[11] Interestingly, smoking can affect the discrimination potential of carbon polymer electronic noses to detect COPD, while gold nanoparticle–based instruments were not affected by this confounding factor.

Exhaled VOCs in COPD

Exhaled VOCs have been analyzed in COPD in a number of studies.[10,11] Most studies employed a case–control design and only few found correlations of exhaled VOCs with disease activity, respiratory mechanics (i.e., lung function and diffusion capacity), or disease-specific pheno/endotypes. Some studies specifically identified exhaled VOCs, whereas others only tentatively reported molecules, and some used unspecific methods, such as electronic noses.

In the case–control studies, alterations in the levels of exhaled organic acids, alcohols, aldehydes, and hydrocarbons were reported. Most of the studies failed to reproduce the results of others. The direction of differences was also variable, suggesting that variances in exhaled kinetics and bacterial load discussed above may have a larger impact on exhaled VOC results than airway inflammation itself.

Investigating disease endotype, some VOCs were related to airway eosinophilia and a history of frequent exacerbations.[13] Stratifying on disease phenotype, analytical signal responses have been associated with emphysema, airway obstruction, and impaired diffusion capacity, yet compound identification was not made. The correlations between exhaled VOC pattern, airway eosinophilia, and lung function were demonstrated in electronic nose studies, too. Moreover, an electronic nose could differentiate COPD and asthmatic patients with fixed airway obstruction. The application of sensor systems to breath research is considered in detail in Chapter 13.

The effect of medications and comorbidities on exhaled VOCs in COPD

Some medications are metabolized by liver enzymes and their metabolites are excreted via exhaled breath. As patients with COPD usually suffer from multimorbidities, the effect of medications cannot be neglected. Regarding inhaled corticosteroids (ICS), no difference was reported between ICS-user and ICS-naïve patients, but no study has investigated the effect of inhaled medications on VOCs in COPD longitudinally.

Due to common etiologies (e.g., age and smoking), systemic inflammation and hypoxemia in COPD, and frequent systemic corticosteroid use for exacerbations, COPD is often accompanied by cardiovascular (i.e., hypertension, coronary artery disease, arrhythmia, heart failure) and metabolic (i.e., osteoporosis and diabetes) disease. These are associated with distinctive patterns of exhaled VOCs. In addition, heart failure can alter pulmonary blood flow, which may influence the exhalation kinetics of bloodborne VOCs.[11]

19.5 Exhaled nitric oxide

19.5.1 Confounding factors and methodological considerations of exhaled nitric oxide measurements

Several patient-related factors can influence exhaled nitric oxide (NO) concentrations (see Chapters 4 and 5). The fractional concentration of exhaled NO (F_ENO) is reduced by cigarette smoking and is lower in ex-smokers than in never-smoking controls. Although the inducible NO synthase (iNOS) is upregulated in the airways under inflammation, such as in smoking-related chronic bronchitis or COPD, NO reacts very rapidly with reactive oxygen species (ROS) in the airways, and the lack of elevation in F_ENO may be a result of the coexistence of increased oxidative and nitrative burden. Furthermore, inhaled drugs, including short- and long-acting β2 agonists and ICS, can influence F_ENO,[6] but the administration of a short-acting β2 agonist to patients with COPD did not modify alveolar NO concentration (*C*ANO).[14] Not specific to COPD, it is known that lung function maneuvers and a nitrate-rich diet alter F_ENO, and a trend in F_ENO in relation to age exists (at least in nonsmoking subjects).[6]

The extended NO analysis requires measurements at multiple exhalation flows. The recently published ERS technical task force document recommends the use of at least three flows, both in the linear and nonlinear models, ranging between 20 and 400 mL/s (the required flow rates depend on the mathematical model in use; see Chapter 4).[6] However, patients with airflow obstruction have difficulties generating technically acceptable exhalation maneuvers for F_ENO measurements, especially at very low and high flow rates, which are required in the nonlinear equations. Therefore, the linear model has been proposed as a more feasible method, with exhalations flows between 100 and 250 mL/s.[15]

19.5.2 Exhaled nitric oxide parameters in smokers and patients with chronic obstructive pulmonary disease

The change in F_ENO in patients with COPD compared to smoking-matched controls is ambiguous, but it is mostly reported to be unchanged. Importantly, persistently elevated F_ENO ($\geq$20 ppb) is associated with an increased risk of exacerbation and a shorter time to the first moderate or severe relapse, but it is not related to blood eosinophil counts.[16] F_ENO is decreased by ICS treatment, with a higher reduction in ex-smoking patients than in smoking subjects with COPD.

Elevated F_ENO is measured in severe COPD exacerbations compared to stable states,[15] which is reduced by systemic steroid therapy, but it also predicts a higher improvement in FEV_1 by the end of hospitalization. Moreover, a significant relationship exists between sputum eosinophil percentage and F_ENO during a relapse. An eosinophilic type exacerbation of COPD may be related to viral infections, which induces an increase in F_ENO.

The most studied parameters generated by the extended NO analysis is the total flux of NO in the conducting airway compartment ($J_{aw}NO$) and *C*ANO (see Chapter 4). $J_{aw}NO$ is decreased in smokers and ex-smokers compared to nonsmokers, but it remains unchanged in patients with stable COPD in relation to smoking-matched controls. Similar to F_ENO, it increases at the onset of an exacerbation and responds to a course of systemic steroids.[15]

In contrast to F_ENO and $J_{aw}NO$, the measurement of alveolar NO concentration holds the potential to monitor inflammation in the peripheral airways, the major site of pathological alterations in COPD. *C*ANO is increased in smoking patients with COPD compared to nonsmokers and ex-smoking patients or current smokers with no COPD. Interestingly, a negative association might exist between current smoking and *C*ANO in control smokers. Using the nonlinear model, no difference in *C*ANO exists among the clinical phenotypes of stable COPD, including patients with emphysema, chronic bronchitis, frequent exacerbations, and disturbed body composition. Additionally, exacerbated patients showed increased *C*ANO compared to smoking controls but not to stable patients. In stable COPD, a short course of either inhaled or systemic steroids does not change *C*ANO, which might be linked to the known corticosteroid resistance of airway inflammation in COPD. In line with this, alveolar NO is not modulated by systemic steroid therapy during an exacerbation either.[15]

Current clinical guidelines of COPD do not recommend the measurement of exhaled NO parameters during disease monitoring. This is due to the lack of systematic studies, such as those assessing the long-term stability of NO parameters of the central and peripheral airway inflammation in patients with COPD. Furthermore, the value—either as a single marker or an adjunct to blood eosinophil count—of exhaled NO readings to predict clinical response to ICS has not been evaluated, to date. In addition, current evidence shows that alveolar NO concentration might be a marker of peripheral airway inflammation in both stable and exacerbated states, but the clinical benefit of its potential pharmacological modulation is not known.

19.6 Exhaled carbon monoxide

19.6.1 Origin and detection

Carbon monoxide (CO) is endogenously produced in the human body under physiological and pathophysiological circumstances. Most of the CO in the body is generated during hemoglobin degradation, when the heme oxygenases (HO) 1 and

2 react with heme, and biliverdin is formed. The HO-1 expression is induced in response to systemic inflammation and oxidative stress, which are associated with increased heme metabolism. CO binds to hemoglobin in the circulation forming carboxyhemoglobin (COHb), and the gas molecule is released into the alveolar space and exhaled. The biological background of CO generation is treated in detail in Chapter 6.

In airway disease, such as asthma and COPD, HO-1 is also upregulated in pulmonary resident and inflammatory cells, directly releasing CO in the exhaled breath. Hence, measuring exhaled CO ($_{E}CO$) provides information about the systemic and pulmonary inflammatory states of the subject.

The inhalation of CO at a low concentration can also exert antiinflammatory effects, as mainly shown in animal models of acute lung injury, endotoxemia, ischemia/reperfusion injury, or graft rejection, as previously reviewed in detail.[17] CO is a ligand and activator of the soluble guanylate cyclase, but it also induces the mitogen-activated protein kinase and phosphatidylinositol 3-kinase pathways to abrogate inflammatory cytokine production by macrophages, to prevent apoptosis of endothelial cells, and to mitigate proliferation of vascular smooth muscle cells.

Detection of $_{E}CO$ can be made by a number of handheld portable devices currently available on the market. These are usually based on electrochemical technology, with sensitivity and accuracy in the low ppm range (see Chapter 13). Higher sensitivity is achieved by more sophisticated techniques, including gas chromatography and spectroscopy (see Chapter 15), although these are often suitable only for experimental purposes.[17]

CO from environmental sources is a major confounder during exhaled measurements (see Chapter 6). It is found in high concentrations in cigarette smoke, and smoking one cigarette can increase $_{E}CO$ approximately threefold. Hence, monitoring $_{E}CO$ concentration is a well-known tool to detect active smoking of patients and to follow-up subjects in a smoking cessation program. A cutoff value of 6.0−6.5 ppm was suggested to distinguish smokers from nonsmokers.[17] Importantly, the incomplete burning of organic matter, such as wood, coal, and gas, leads to increased ambient CO concentration; therefore, nonsmoking subjects in heavily urbanized areas can exhibit $_{E}CO$ at similarly high concentrations as current smokers, which underlies the importance of the simultaneous detection of background CO levels in certain settings.

19.6.2 Exhaled carbon monoxide in smokers and chronic obstructive pulmonary disease patients

The concentration of $_{E}CO$ has been observed to be higher in ex-smoking patients with COPD than in nonsmoking controls, and it is further elevated in currently smoking patients, although no correlation was found between exhaled NO and CO concentrations, and $_{E}CO$ was not related to lung function values in patients.[18]

Exhaled CO levels are elevated in smoking patients with an acute exacerbation of COPD compared to nonsmoking exacerbated subjects.[19] This difference is also noted during convalescence after systemic steroid treatment, with CO concentrations found to be similar at the onset and recovery in both subgroups, and no observed relation to the severity of lung function limitations. Similarly, $_ECO$ was not altered in exacerbations with bacterial etiology. As expected, $_ECO$ and blood COHb concentration showed a significant correlation in smoking patients. These data prove that although airway inflammation and oxidative stress are elevated at the onset of a COPD exacerbation, no associated change in $_ECO$ concentrations arise.

19.7 Exhaled breath temperature

19.7.1 Origin and detection

Measurement of body temperature has been a tool for centuries to monitor disease processes. Since the transfer of thermal energy from the core to the body surface is influenced by tissue interfaces, body temperature readings can also provide information about the functions of these organs. The temperature of exhaled breath measured at the mouth arises from the transfer of the thermal energy from the alveolar capillaries to the alveolar air, but it is also changed by heat exchange in the lower and upper airways during exhalation. Based on this phenomenon, the measurement of EBT can provide useful insights into alveolar and airway pathologies.

Pathologies of the alveolocapillary unit can modulate the transfer of the core temperature to the alveolar space. Degradation of alveolar walls and capillaries, as seen in emphysema, results in decreased thermal exchange between the alveolar wall and alveolar air. Moreover, hypoxemia, present in many respiratory diseases, can alter pulmonary and bronchial blood flow. In addition, ventilation-perfusion mismatch can also influence the thermal exchange in the alveolar space.

The bronchial arteries originate from the systemic circulation, more specifically from the aorta and intercostal arteries. The bronchial circulation supplies the extra- and intrapulmonary airways where the arteries form a peribronchial and a submucosal plexus. On the one hand, inflammatory mediators, such as bradykinin, prostaglandins, neuropeptides, and platelet activating factor, increase local bronchial blood flow and EBT in airway disease. On the other hand, inflammation-induced airway remodeling manifested as thickened basement membrane, reduced vascular bed, and mucous hyperproduction, is observed in airway pathologies, including COPD, and hampers heat and water exchange in the airways. Inflammatory mediators, including endothelins, induce vasoconstriction of the bronchial vessels, while hyperinflation might also contribute to reduced bronchial blood flow. Hence, a net effect of bronchial vascularity and local pathophysiological processes will determine the heat exchange in the bronchial tree, which can be assessed by EBT measurements.

There are currently two main strategies to measure EBT: the single controlled exhalation technique and the multiple tidal breathing method. Both methods can distinguish smokers from nonsmokers and from patients with COPD. No relationship between EBT and body temperature could be observed in smoking subjects or patients with COPD, proving that this parameter is a thermal signal from the airways rather than a readout of systemic processes.

The single exhalation technique is based on recording the increment of the temperature rise and the plateau temperature during a flow- and pressure-controlled expiration. This requires patient training and bulky instrumentation, and results are influenced by ambient conditions, including room temperature and air pressure, exhalation flow rate, and the position of the thermocouple in relation to the mouthpiece.[20]

By comparison, the multiple tidal breathing method can be performed using a handheld device, where the expired thermal energy is collected into an insulated vessel containing a heat sink. EBT recordings are less influenced by ambient parameters, but measurements last for a couple of minutes (typically 3–7 min in patients with COPD). EBT values are not influenced by ambient temperature in the range of 20–25°C, but values are decreased when environmental temperature during the measurements is low (3–10°C). Importantly, the length of measurement does not affect peak EBT values in smoking controls and patients with COPD.[21]

Sex can influence EBT readings using the tidal breathing method, with nonsmoking males tending to have higher EBT, although this has not been observed in smokers or patients with stable or exacerbated COPD.[21] A weak correlation between peak EBT and age was noted in elderly subjects[22] but not in middle-aged adult smokers or patients with COPD.[21] Of note, the smoking of a single cigarette induces an acute rise in peak EBT in control smokers.[22]

19.7.2 Exhaled breath temperature in smokers and chronic obstructive pulmonary disease patients

The rise of EBT in stable COPD is slower than in nonsmoking controls, as measured with the single controlled exhalation technique, but no difference was recorded in plateau temperatures.[20] The increment of EBT was related to F_ENO in COPD. Notably, the inhalation of a short-acting β2 agonist albuterol, a known vasodilator, resulted in an acute rise in the EBT slope. These data prove that this parameter reflects a change in bronchial blood flow triggered by airway inflammation.

No difference in peak EBT measured with the multiple tidal breathing method was observed between nonsmoking and smoking control subjects, but patients with stable COPD presented a lower peak EBT than smokers.[21] This might be caused by pathological processes, including decreased alveolocapillary surface, ventilation-perfusion mismatch, reduced bronchial vascularity, and airway remodeling. Importantly, acute severe exacerbation of COPD is associated with an increase in EBT, which is lowered by the systemic steroid treatment of the relapse. This observation suggests that vasoactive mediators accumulate in the airways during exacerbation, which results in the transient rise in bronchial blood flow and EBT. Interestingly, EBT during disease worsening is related to airway neutrophilia, but not to systemic inflammatory markers, lung function variables, the type of inhaled medication in use, the severity of the COPD, or the patient's health status.

Peak EBT is elevated after smoking a single cigarette in control smokers and in patients with mild COPD. Importantly, the magnitude of this increase is predictive for disease progression on a 2-year time frame in patients and in smokers with newly diagnosed COPD.[22]

19.8 Summary

Despite the numerous studies on COPD, exhaled breath measurement did not reach its potential for this disease and its clinical value is currently very limited. Exhaled carbon monoxide and fractional exhaled nitric oxide measurements are the most widely applied breath analyses in COPD. Exhaled CO can potentially detect active smoking; however, its specificity (higher exhaled levels due to pollution) and sensitivity (rapid decrease in >6 h of smoking abstinence) limit its value. Although F_ENO is often used to differentiate asthma from COPD, this routine is not endorsed by guidelines. There is a direct but only moderately strong relationship between F_ENO and airway eosinophilia in asthma, COPD, and health. As assessment of blood eosinophils is currently part of the Global Initiative for Chronic Obstructive Lung Disease (GOLD) recommendations to tailor pharmaceutical treatment, F_ENO can be a potential surrogate, although this is yet to be investigated in clinical trials. EBC, exhaled particles, exhaled VOCs, and EBT have served well to better understand the pathophysiology of COPD, but the results are conflicting without any single biomarker being concordantly shown to have a discriminative, prognostic, or predictive value. This could be explained by the variable designs of these studies using different techniques to collect and analyze breath samples. COPD is also a broad term comprising several endo- and phenotypes, complicated by various medications and comorbidities, which could all contribute to the variety of findings reported. Most importantly, large follow-up studies are lacking, and the temporal variability of exhaled biomarkers is as yet unexplored.

Microbiome colonizing the airways of COPD patients has achieved significant interest from the scientific and clinical community recently. It contributes to persistent airway inflammation and progressive airway and alveolar destruction, and it can be responsible for acute flare-ups. The high number of articles associating specific VOCs with airway microorganisms is especially encouraging to design studies that focus on this aspect of COPD.

References

1. Vogelmeier CF, Criner GJ, Martinez FJ, Anzueto A, Barnes PJ, Bourbeau J, et al. Global strategy for the diagnosis, management, and prevention of chronic obstructive lung disease 2017 report: GOLD executive summary. *Eur Respir J* 2017;**49**(3). pii: 1700214.
2. Lange P, Celli B, Agusti A, Boje Jensen G, Divo M, Faner R, et al. Lung-function trajectories leading to chronic obstructive pulmonary disease. *N Engl J Med* 2015;**373**(2): 111–22.
3. Hurst JR, Vestbo J, Anzueto A, Locantore N, Mullerova H, Tal-Singer R, et al. Susceptibility to exacerbation in chronic obstructive pulmonary disease. *N Engl J Med* 2010; **363**(12):1128–38.
4. Barnes PJ, Burney PG, Silverman EK, Celli BR, Vestbo J, Wedzicha JA, et al. Chronic obstructive pulmonary disease. *Nat Rev Dis Primers* 2015;**1**:15076.

5. Lázár Z, Horváth I, Vestbo J, Bikov A. Exhaled breath condensate in chronic obstructive pulmonary disease: methodological challenges and clinical application. *Minerva Pneumol* 2018;**57**(2):42–56.
6. Horvath I, Barnes PJ, Loukides S, Sterk PJ, Hogman M, Olin AC, et al. A European Respiratory Society technical standard: exhaled biomarkers in lung disease. *Eur Respir J* 2017;**49**(4).
7. Maniscalco M, Fuschillo S, Paris D, Cutignano A, Sanduzzi A, Motta A. Clinical metabolomics of exhaled breath condensate in chronic respiratory diseases. *Adv Clin Chem* 2019;**88**:121–49.
8. Bredberg A, Josefson M, Almstrand AC, Lausmaa J, Sjovall P, Levinsson A, et al. Comparison of exhaled endogenous particles from smokers and non-smokers using multivariate analysis. *Respiration* 2013;**86**(2):135–42.
9. Bake B, Larsson P, Ljungkvist G, Ljungstrom E, Olin AC. Exhaled particles and small airways. *Respir Res* 2019;**20**(1):8.
10. Bos LD, Sterk PJ, Fowler SJ. Breathomics in the setting of asthma and chronic obstructive pulmonary disease. *J Allergy Clin Immunol* 2016;**138**(4):970–6.
11. Bikov A, Lazar Z, Horvath I. Established methodological issues in electronic nose research: how far are we from using these instruments in clinical settings of breath analysis? *J Breath Res* 2015;**9**(3):034001.
12. Bos LD, Sterk PJ, Schultz MJ. Volatile metabolites of pathogens: a systematic review. *PLoS Pathog* 2013;**9**(5):e1003311.
13. Basanta M, Ibrahim B, Dockry R, Douce D, Morris M, Singh D, et al. Exhaled volatile organic compounds for phenotyping chronic obstructive pulmonary disease: a cross-sectional study. *Respir Res* 2012;**13**:72.
14. Amer M, Cowan J, Gray A, Brockway B, Dummer J. Effect of inhaled beta2-agonist on exhaled nitric oxide in chronic obstructive pulmonary disease. *PloS One* 2016;**11**(6): e0157019.
15. Lazar Z, Kelemen A, Galffy G, Losonczy G, Horvath I, Bikov A. Central and peripheral airway nitric oxide in patients with stable and exacerbated chronic obstructive pulmonary disease. *J Breath Res* 2018;**12**(3):036017.
16. Alcazar-Navarrete B, Ruiz Rodriguez O, Conde Baena P, Romero Palacios PJ, Agusti A. Persistently elevated exhaled nitric oxide fraction is associated with increased risk of exacerbation in COPD. *Eur Respir J* 2018;**51**(1).
17. Ryter SW, Choi AM. Carbon monoxide in exhaled breath testing and therapeutics. *J Breath Res* 2013;**7**(1):017111.
18. Montuschi P, Kharitonov SA, Barnes PJ. Exhaled carbon monoxide and nitric oxide in COPD. *Chest* 2001;**120**(2):496–501.
19. Antus B, Drozdovszky O, Barta I. Assessment of exhaled carbon monoxide in exacerbations of chronic obstructive pulmonary disease. *Phys Int* 2016;**103**(2):211–9.
20. Paredi P, Caramori G, Cramer D, Ward S, Ciaccia A, Papi A, et al. Slower rise of exhaled breath temperature in chronic obstructive pulmonary disease. *Eur Respir J* 2003;**21**(3): 439–43.
21. Lazar Z, Bikov A, Martinovszky F, Galffy G, Losonczy G, Horvath I. Exhaled breath temperature in patients with stable and exacerbated COPD. *J Breath Res* 2014;**8**(4): 046002.
22. Labor M, Vrbica Z, Gudelj I, Labor S, Juric I, Plavec D. Exhaled breath temperature as a novel marker of future development of COPD: results of a follow-up study in smokers. *COPD* 2016;**13**(6):741–9.

CHAPTER

Clinical phenotyping

20

Waqar Ahmed[1], Paul Brinkman[2], Stephen Fowler[1,3]

[1]*School of Biological Sciences, Faculty of Biology, Medicine and Health, University of Manchester, Manchester, United Kingdom;* [2]*Department of Respiratory Medicine, Amsterdam UMC, University of Amsterdam, Amsterdam, The Netherlands;* [3]*Manchester Academic Health Science Centre and NIHR Manchester Biomedical Research Centre, Manchester University Hospitals NHS Foundation Trust, Manchester, United Kingdom*

Chapter outline

20.1 Overview

Most published studies that consider future clinical applications of breath analysis have focused on diagnosis, seeking to discover potential exhaled biomarkers that could identify the disease of interest. Because of the complexity of breath sampling and analysis, these have thus far mostly been conducted on a relatively small scale in a single center, and external validation (either within the same study or across studies and research groups) is rare. The opportunity for discovering genuine disease-specific markers that can be used in future diagnostic pathways may be further reduced by two factors: study design (specifically the choice of control groups) that do not reflect the "real world" clinical scenario and pathophysiological heterogeneity of the disease in question.

The concept introduced here proposes the use of biomarkers, or their patterns, to distinguish among disease phenotypes. Shortness of breath, for example, could be a sign of asthma, but certain expressed biomarkers could be used to discern

Breathborne Biomarkers and the Human Volatilome. https://doi.org/10.1016/B978-0-12-819967-1.00020-7
Copyright © 2020 Elsevier B.V. All rights reserved.

occupational, nonallergic, inflammatory, bronchoconstrictive, exercise-induced, or other subsets of the disease that could more quickly guide the appropriate intervention.

To deal briefly with control groups and clinical pathways, any potential biomarker(s) must perform within the clinical scenario in which it will be required. Taking asthma (the exemplar disease that will be used throughout this chapter), the physician's perspective must be considered, usually in primary care. The physician sees people of all ages who present with symptoms such as cough, chest tightness, and breathlessness and performs a test that will identify (and/or rule out) asthma in such patients. Thus, the appropriate control group for biomarker discovery would comprise such patients who turn out not to have the disease, not "healthy controls." Such is the complexity of the breath matrix that it seems unlikely that studies using healthy controls will generate clinically useful data, beyond proof of principle that breath is generally "different" in the disease under study.

The second issue, that the situation does not present a unique and uniform pathophysiology, applies to many diseases, including asthma. Asthma is defined by the presence of symptoms together with physiological evidence of variable narrowing of the airways.[1] Underlying are myriad inflammatory and structural abnormalities that may be absent, or present to any degree, in a given patient. Most commonly, the inflammatory pathways implicated in asthma are associated with upregulation of type 2 cytokines (including interleukin-4 [IL-4], IL-5, and IL-13) and the associated eosinophilia and airway mastocytosis. However, such a typical pattern is by no means always present; similar physiological abnormalities are found in the absence of any clear sign of airway "inflammation," or there may be apparent upregulation of non-type 2 pathways, driven by IL-8 and with associated airway neutrophilia.[2,3] A further important factor to consider is the airway microbiome, with or without overt "infection."[4]

Approaching asthma from the diagnostic perspective also allows consideration of the multiple pathophysiological abnormalities that could lead to its defining feature, namely variable airflow obstruction. These include contributors to airway wall thickness (inflammation, hypervascularity, smooth muscle hypertrophy), hypersensitivity (inflammation, sensory neuronal activation), hyperreactivity (as for airway wall thickness, plus motor neuronal upregulation), and intraluminal factors (epithelial damage, mucus hypersecretion, and hyperviscosity). Exhaled volatiles associated with asthma are likely to arise from, and/or be altered by, every aspect of the inflammatory, structural, microbial, and airflow abnormalities described above, to a greater or lesser degree in any individual patient. Hence, finding an "asthma breath signature" in studies comprising limited patient numbers seems unlikely; however, as outlined below, it may become possible to use biomarker signatures to tease out specific subtypes of the disease.

20.2 Overlapping paradigms

The potentially overwhelming complexity of asthma described above has led to the development of three overlapping paradigms, which can be used to help clinicians and scientists (and patients) formulate strategies for research and treatment: phenotypes, endotypes, and treatable traits.[3,5]

Phenotypes used in this context—and often described as "clinical phenotypes," in order to distinguish them from the original biological definition—may be best defined as "identifiable manifestations of disease."[2] In asthma, terms such as "eosinophilic asthma" and "neutrophilic asthma" are used to describe the pattern of inflammation, which may be seen in blood and/or airway samples.

Endotypes are linked to specific pathways (such as those in the type 1 or 2 inflammatory pathways above) and may be especially useful in identifying targets for the development of novel therapies.[3]

Treatable traits is a newer, holistic, and clinically focused process for categorization, which aims to illuminate possible personalized strategies for helping the patient.[5,6] Hence, they include not only "pulmonary" (such as airflow obstruction and inflammation) but also "extrapulmonary" (such as obesity and osteoporosis) and "psychosocial" (smoking, anxiety, and depression, etc.) traits. Drug selection, monitoring, and toxicity may also be considered under this framework.

Considering these classification systems for asthma may well lead to better design of biomarker discovery studies, especially in omic studies such as exhaled breath. It would be more straightforward to attribute mechanistic importance to volatile organic compounds (VOCs) if they are discovered in well-conducted studies that are able to carefully describe the structural, inflammatory, and/or microbial status of the patients. Such studies may also minimize the problem of a suitable control group: the population can be defined as the disease under study, with the "cases" being those with the pheno- or endotypes of interest, and the "controls" being those without.

20.3 Volatiles for phenotyping airways diseases

While the field of breathomics is moving toward questions in the spirit of personalized medicine, a common stated primary aim in literature is to investigate potential breath biomarkers for the use of diagnosis of asthma. Thereby, the comparisons are mostly made with healthy controls or those with chronic obstructive pulmonary disease (COPD; see also Chapter 19).[7] A systematic review and metaanalysis by Rufo et al. showed that exhaled VOCs have a sensitivity and specificity, respectively, of 87% and 86%, for the diagnoses of asthma.[8] When reviewing outcomes of these "traditional asthma diagnosis" papers and other exhaled breath asthma papers through the lens of phenotyping, a series of topics regarding potential mechanistic pathways of importance and personalized treatment come to light.

20.3.1 Targeted patient treatment and exacerbations

Most patients with asthma and COPD experience exacerbations, defined as an acute increase in respiratory symptoms. Exacerbations have a negative impact on patients' quality of life, accelerate lung function decline, and represent a huge economic burden to society. Objective assessment of an exacerbation might allow for early intervention, thereby improving patient-related outcomes and reducing costs.

Production of reactive oxygen species (ROS), which stimulates lipid peroxidation, is widely recognized as an important component of airway inflammation in asthma.[9] Based on that, exhaled pentane (a product of polyunsaturated membrane fatty acid peroxidation) was tested by Olepade et al. as an exhaled marker of active inflammation in asthma by the use of gas chromatography (GC). With the conclusion that exhaled pentane levels are increased in patients with acute asthma and decreased significantly once a stable phase of the disease is reached again, this paper can be considered as one of the first papers focusing on the development of an exhaled biomarker for asthma (exacerbations).[10] While Olepade et al. followed up acute asthma patients presenting at the emergency department, Brinkman et al. used a different model to test whether profiles of VOCs could discriminate between clinically stable and unstable episodes of asthma. During a prospective steroid withdrawal study ($n = 23$), exhaled metabolites were sampled at baseline, during loss of control, and after recovery, and analyzed by GC mass spectrometry (GC-MS) and electronic nose (eNose) technology. Based on the obtained breath profiles, accuracies from 68% to 95% were reached. Furthermore, associations between exhaled metabolites versus sputum eosinophils, lung function PbFEV1%, and fraction of exhaled nitric oxide (F_ENO) were observed.[11] One of the exhaled compounds reported as potentially relevant for the discrimination between different stable and unstable periods of asthma in this study was methanol, a molecule that is also reported as possibly relevant for the distinction between stable and unstable phases within cystic fibrosis.[12]

As with asthma, a biomarker for the early detection of an exacerbation is an unmet need for COPD. So far, mostly cross-sectional studies are reported on this topic. According to a pilot study by Pizzini et al., patients with stable COPD can be discriminated from those having an exacerbation by analysis of four VOCs, namely n-butane, 2-pentanone, cyclohexanone, and 4-heptanone. While the metabolic sources and the pathways related to these VOCs remain uncertain, a cautious suggestion has been made concerning a bacterial origin.[13] A recent data study using real-time secondary electrospray ionization—high-resolution mass spectrometry (SESI-MS) showed differences in exhaled metabolites during clinically stable episodes between COPD patients with frequent and nonfrequent exacerbations, whereby again a connection was drawn with oxidative stress and lipid peroxidation.[14] These findings during stable episodes of COPD suggest that not only the exacerbations themselves but also the phenotype of frequent exacerbations is associated with distinguishable exhaled metabolomic profiles. A longitudinal VOC analysis during and after COPD exacerbations using eNose technology resulted in a success rate of 72% when aiming for a differentiation of phases of diseases within patients.[15] Similar numbers were reported by van Velzen et al., who found significant differences in breath acquired during baseline, exacerbation, and recovery, as analyzed by both GC-MS and eNose. The four reported compounds that drive the discriminative GC-MS signal were acetone, 1,2-pentadiene, toluene, and limonene. These compounds could suggest that bacterial infections are involved in the VOC signal during exacerbations of COPD, although the authors emphasize that these compounds are common in exhaled air.[16]

20.3.2 Airways disease cell types

Since asthma and COPD are considered as complex conditions with different underlying inflammatory mechanisms, phenotyping according to airway inflammation could allow the identification of individuals who are more likely to respond to targeted therapy. Induced sputum is currently the best available noninvasive technique to assess bronchial inflammation. Based on this technique, four distinct inflammatory subtypes have been classified: (1) eosinophilic, (2) neutrophilic, (3) mixed granulocytic, and (4) paucigranulocytic. Unfortunately, (induced) sputum collection is complex, time-consuming, and not widely applicable. Considering the noninvasive and easy character of breathomics, it is often tested in studies aiming to develop a clinical tool for defining a classic airway inflammation profile.

One of the studies that assessed the ability of breath profiles to discriminate between bronchial inflammation−based phenotypes was conducted by Plaza et al.[17] They analyzed breath samples using an eNose sensor system (Cyranose 320), whereby three sputum based inflammatory phenotypes had been classified: eosinophilic (>3% eosinophils), neutrophilic (>61% neutrophils), and paucigranulocytic (<61% neutrophils and <3% eosinophils). Differences in eosinophilic compared with both neutrophilic and paucigranulocytic phenotypes were found, with accuracies of 73% and 74%, respectively. The comparison of "neutrophilic" and "paucigranulocytic" breathprints resulted in an accuracy of 89%. With slightly different cutoffs (eosinophilia ≥2 and neutrophilia ≥40%), accuracies of 83% for eosinophilic versus noneosinophilic inflammation and 72% for neutrophilic versus nonneutrophilic were reported by Ibrahim et al. after performing principal component analysis (PCA) combined with logistic regression-based modeling of GC-MS−driven breath analysis.[18] Also, an inflammatory phenotyping study in COPD reported associations between exhaled compounds and sputum cell counts. Among other compounds, pentane and toluene were found as molecules that correlated with neutrophilic inflammation, whereas 2-butanone and 5-nonanol were linked to eosinophilic inflammation.[19] A prospective study among 521 patients with asthma recruited from the University Asthma Clinic of Liege had similar aims.[20] During a two-phase study, patient groups were split into a training and a validation set. During the discovery study (n = 276), seven exhaled VOCs were marked as potential surrogate biomarkers for recognition of sputum inflammatory phenotypes. During the validation phase (n = 245), two markers for eosinophilic (hexane and 2-

hexanone) and three for neutrophilic inflammation (nonanal, 1-propanol, and hexane) could be confirmed. The identification of the neutrophilic asthma phenotype resulted in an area under the curve (AUC) of the receiver operating characteristic (ROC) curve of 73% for the validation set. During an extreme analysis (sputum neutrophils >76% and sputum cell counts greater than the 95th percentile), an AUC of 97% was reached.

Although the GC-MS compounds of interest reported by different authors often vary due to instrument variation and limited chemical validation, the presented compounds have structural similarities. To avoid this lack of confirmation from becoming a hampering factor as it concerns valorization of breathomics, new breathomics studies concentrating on surrogate markers for bronchial inflammation would be well advised to apply a strategy in which a targeted analysis based on current findings is applied.

20.3.3 Unbiased phenotyping

Apart from the option to use exhaled metabolites for the detection of established airway inflammation profiles, there is also the option to use this noninvasive technology for the identification of new molecularly driven phenotypes within asthma and COPD. Through the application of unbiased clustering strategies, different series of potentially clinically relevant subgroups have been delineated.

In an attempt to capture clinical/inflammatory new phenotypes in patients with chronic airway disease using an eNose technology, a multicenter cross-sectional study among asthma and COPD patients was performed in a training/validation setting (n = 321 and validation n = 114) by de Vries et al.[21] Through PCA followed by unsupervised hierarchical clustering, the authors obtained five significant asthma/COPD clusters with clinically characteristic differences regarding ethnicity, systemic eosinophilia and neutrophilia, body mass index, $F_{E}NO$, atopy, and exacerbation rates. Using a similar analysis strategy, but including four different brands of eNoses, the U-BIOPRED[a] consortium[22] performed a cluster analysis within severe asthma patients, whereby longitudinal within-patient stability was assessed.[23] Three eNose-driven clusters were discovered, showing differences in circulating eosinophil and neutrophil percentages and usage of corticosteroids. Longitudinal within-patient cluster stability, based on 12–18 months follow-up visit, was linked to deviations in sputum eosinophil percentages. In an additional exploratory subanalysis, VOC profiles (n = 28) were compared with the U-BIOPRED sputum transcriptomic phenotypes by Kuo et al.,[24] resulting in an accuracy of 93%. Furthermore, using hierarchical clustering, but now in a combination of clinical characteristics and four exhaled VOCs (all four detectable in asthma patients but not in healthy control subjects), Meyer et al. described seven asthma subgroups and reported that some had different VOC profiles, but similar clinical characteristics,

[a] Unbiased BIOmarkers in PREDiction of respiratory disease outcomes

while other clusters had similar VOC profiles, but distinct clinical characteristics.[25] The distinct clinical characteristics that were observed concerned atopy, disease severity, medication, lung function, and symptom scores.

While there is a difference in study population, the reported clinical characteristics of the described clusters among these three unbiased cluster analyses studies partly overlap. Specifically, the usage of medication, atopic, or nonatopic and inflammation profiles seem prominent characteristics, whereby a link to (potential) pathophysiological pathways to these features is known. Although that does not prove the relevance of unbiased cluster analysis, it does show its potential.

20.3.4 Treatment monitoring and stratification

Good adherence and adequate inhalation technique are associated with improved outcomes, but it is difficult for clinicians to monitor appropriate medication intake and drug effectiveness in individual patients. Furthermore, and in line with personalized medicine, it is important to prescribe the right drug to a patient, in order to avoid wasteful treatment or potentially harmful side effects. A series of studies has been performed on both "drug related" topics.

In a study concerning monitoring of (changes in) the metabolites after usage of inhaled bronchodilator therapy, exhaled levels of metabolites were measured in a group of patients ($n = 50$) at baseline and 10 and 30 min after inhalation of salbutamol and compared with the exhaled features of a control group ($n = 48$) after inhaling a placebo spray. The salbutamol-related metabolites salbutamol-4-O-sulfate and formoterol were found to be generally increased in salbutamol users as compared to the placebo group. This suggests that real-time breath analysis (e.g., by high-resolution mass spectrometry, such as SESI-MS) is a useful method for noninvasive therapeutic drug monitoring (see Chapter 12). In a real-life study conducted within the U-BIOPRED consortium, the association between urinary levels of salbutamol and oral corticosteroids (OCS) and exhaled VOCs (analyzed by GC-time-of-flight-MS; GC-TOF-MS) in severe asthma patients was investigated in a training ($n = 48$), replication ($n = 30$), and validation ($n = 30$) setting. Baseline AUCs were 82.1 for salbutamol and 78.8 for OCS, whereby outcomes could adequately be replicated and validated. Finally, in a study regarding steroid responsiveness concerning steroid-free patients with established asthma, Van der Schee et al. showed that through exhaled breath analysis by eNose technology, steroid responsiveness was predicted with greater accuracy by VOC analysis ($AUC = 88.3$) than F_ENO ($AUC = 54.5$) or sputum eosinophils ($AUC = 61.0$).[26]

Metabolomics of exhaled breath appears to be associated with markers of recent drug usage and even in the prediction of steroid responsiveness (see Chapter 30). These outcomes should be extended toward time series analysis, which argues for pharmacokinetic pathway and dose-ranging studies on asthma medication.

20.3.5 Mechanistic pathways

Studies frequently associate exhaled breath compounds with anabolic and catabolic biological pathways. With regard to asthma, the most commonly accepted breath marker of inflammation is nitric oxide (NO), which is synthesized from the amino acid L-arginine through (inducible) nitric oxide synthase (iNOS). Nitric oxide subsequently acts as a signaling molecule in several proinflammatory pathophysiological mechanisms including vasodilation (see Chapters 4 and 5). Proinflammatory molecules such as ROS and prostaglandins are also released as a defense mechanism in immune system activation.

Oxidative stress can occur due to an imbalance of ROS resulting in cell damage. Lipids that are exposed to oxidative conditions have increased concentrations of derivatives such as aldehydes. Studies investigating inflammatory lung diseases have identified increased aldehydes and aliphatic hydrocarbons in exhaled breath. The aldehyde hexanal is one such example, identified in the breath of patients with COPD.[27] The eicosanoid 8-isoprostane is a well-established marker of lipid peroxidation and is found in exhaled breath condensate (EBC) of patients with asthma (see Chapter 7). Such lipid peroxidation products are reported in the volatile fraction of breath where studies have observed changes in aldehyde and hydrocarbon levels from patients with asthma.

Compounds with a phenyl and/or halogen group are regarded as exogenous compounds originating from environmental pollutants or synthetic drugs, for example, in β2-adrenergic agonists which are commonly used to manage asthma and contain one or more benzene ring. Such compounds, or their metabolized products, are most likely excreted through xenobiotic metabolism pathways and therefore show potential for treatment or exposure monitoring.

Isoprene and acetone are the most abundant endogenous VOCs found in breath. Although they are highly regarded as nonspecific, some studies suggest changes in exhaled breath concentration by activated inflammatory pathways. Acetone is produced through the decarboxylation of acetoacetate, which derives from lipid peroxidation. Acetoacetate is also a component in acetyl coenzyme A (CoA), which is the starting metabolite of the mevalonate pathway. Isoprene units make up the backbone of terpenoids (e.g., cortisol and cholesterol) in the mevalonate pathway, which have essential roles in cellular development and regulation of antiinflammatory mechanisms as steroid hormones. By inhibiting 3-hydroxy-3-methyl-glutaryl-CoA (HMG-CoA) reductase in the mevalonate pathway using statins, breath isoprene levels have been shown to decrease (see Chapter 3). Studies have also found changes in abundance of isoprene with circadian rhythm in asthmatic patients.[28] Table 20.1 lists VOCs reported by two or more studies with relevance to clinical phenotyping of asthma and overlap with COPD, and summarizes their possible role in disease metabolic activity.

Table 20.1 List of VOCs associated with asthma phenotypes from two or more cross-sectional and/or longitudinal studies.

Compound name	Classification	Possible role in metabolism	Healthy versus asthma	Exacerbation	Early diagnosis	Cellular phenotype	Overlap with COPD
Nonanal[a]	Aldehyde	Lipid peroxidation	(−) Caldeira 2012[29]	(+) Van vliet 2017[30]		(Neutrophilia) Schleich 2019[20]	(+) Basanta 2012[31]
2,4-Dimethylheptane	Alkane		(−) Caldeira 2011[32] (+) Dragonieri 2007[33]		(+) Smolinska 2014[34]		
2,6,10-Trimethyldodecane[a]					(−) Smolinska 2014[34]	(Eosinophilla) Ibrahim 2011[18]	
4-Methyloctane			(+) Dragonieri 2007[33] (−) Caldeira 2011[32]				Berkel 2010[35]
Decane			(+) Caldeira 2011[32] (+) Caldeira 2012[29]				(+) Cazzola 2015[27]
Dodecane[a]			(+) Caldeira 2011[32] (+) Caldeira 2012[29]				
Ethane			(+) Paredi 2000[36] (+) Lärstad 2007[37]				
Tetradecane			(+) Caldeira 2011[32] (+) Caldeira 2012[29]				
Undecane			(−) Dallinga 2010[38]			(Paucigranulocytic) Schleich 2019[20]	Berkel 2010[35]
Isoprene	Alkene	Terpenoid biosynthesis	(−) Lärstad 2007[37] (+) Dragonieri 2007[33] (+) Caldeira 2011[32]				Philips 2012[39]

Continued

Table 20.1 List of VOCs associated with asthma phenotypes from two or more cross-sectional and/or longitudinal studies.—*cont'd*

Compound name	Classification	Possible role in metabolism	Healthy versus asthma	Exacerbation	Early diagnosis	Cellular phenotype	Overlap with COPD
p-Xylene[a]	Aromatic	Xenobiotic excretion	(–) Dallinga 2010[38]	(+) Robroeks 2013[40]			
1,4-Dichlorobenzene			(+) Gahleitner 2013[41]		(–) van de kant 2013[42]		
Acetone	Ketone	Decarboxylation of ketone bodies	(+) Dragonieri 2007[33] (–) Caldeira 2011[32]		(–) Smolinska 2014[34]		

[a] *VOCs were also found using a targeted approach in U-BIOPRED breath sample data from asthma patients.*[44]

Adapted from [7, 8, 43]

20.4 Summary

For airways diseases at least, some of the most promising findings in breathomics to date are from studies designed to report phenotype-specific biomarkers. Such biomarkers, related to well-defined subpopulations, still require validation and further tests before clinical implementation. There is already a clear clinical need and relevance to biomarkers that relate to specific (and targetable) pathways and drugs.

Beyond asthma, a similar approach may well reap rewards before nonspecific "diagnostic" discovery studies. Taking lung cancer as an example, it seems unlikely that nontargeted searches for markers will be successful if no account is taken of cell type, location, dissemination, and systemic effects. Likewise for infection, given the complexity of the microbiome, it seems likely that only longitudinal studies that use patients as their own controls and perhaps include an intervention (most likely antimicrobials) to change the baseline state will be able to help determine which volatiles are from the host and which are from the microbes of interest.

Future work would also benefit from consideration of combining in vitro and in vivo studies, a "bottom-up" approach where volatile combinations reported from cell and microbial cultures are then prospectively sought in well-designed patient cohorts. Finally, as for all studies in breath volatiles, study designs that utilize or develop standardized sampling and analytical methods that are suitable for reproduction across centers are recommended; only this will allow confidence in external validation that can then lead to the required interventional studies that will utilize breath volatiles as companion diagnostics.

Using biomarkers to screen for diseases from the general population is a daunting task due to the huge variability in pathophysiology within- and between-human systems. Once a disease cohort is identified by symptomology, however, the use of biomarkers to "fine-tune" disease screening to the phenotype level could become a valuable tool.

References

1. Global Initiative for Asthma (GINA). *Guidelines for the diagnosis and management of asthma*. 2019.
2. Wenzel SE. Asthma phenotypes: the evolution from clinical to molecular approaches. *Nat Med* 2012;**18**(5):716–25.
3. Lotvall J, Akdis CA, Bacharier LB, Bjermer L, Casale TB, Custovic A, et al. Asthma endotypes: a new approach to classification of disease entities within the asthma syndrome. *J Allergy Clin Immunol* 2011;**127**(2):355–60.
4. Huang YJ, Marsland BJ, Bunyavanich S, O'Mahony L, Leung DY, Muraro A, et al. The microbiome in allergic disease: current understanding and future opportunities-2017 PRACTALL document of the American academy of allergy, asthma & immunology and the European academy of allergy and clinical immunology. *J Allergy Clin Immunol* 2017;**139**(4):1099–110.

5. Agusti A, Bel E, Thomas M, Vogelmeier C, Brusselle G, Holgate S, et al. Treatable traits: toward precision medicine of chronic airway diseases. *Eur Respir J* 2016;**47**(2):410–9.
6. Simpson AJ, Hekking PP, Shaw DE, Fleming LJ, Roberts G, Riley JH, et al. Treatable traits in the European U-BIOPRED adult asthma cohorts. *Allergy* 2019;**74**(2):406–11.
7. Azim A, Barber C, Dennison P, Riley J, Howarth P. Exhaled volatile organic compounds in adult asthma: a systematic review. *Eur Respir J* 2019;**54**(3).
8. Rufo JC, Madureira J, Fernandes EO, Moreira A. Volatile organic compounds in asthma diagnosis: a systematic review and meta-analysis. *Allergy* 2016;**71**(2):175–88.
9. Barnes PJ. Reactive oxygen species and airway inflammation. *Free Radic Biol Med* 1990; **9**(3):235–43.
10. Olopade CO, Zakkar M, Swedler WI, Rubinstein I. Exhaled pentane levels in acute asthma. *Chest* 1997;**111**(4):862–5.
11. Brinkman P, van de Pol MA, Gerritsen MG, Bos LD, Dekker T, Smids BS, et al. Exhaled breath profiles in the monitoring of loss of control and clinical recovery in asthma. *Clin Exp Allergy : J British Soci Allergy Clin Immunol* 2017;**47**(9):1159–69.
12. Montuschi P, Paris D, Melck D, Lucidi V, Ciabattoni G, Raia V, et al. NMR spectroscopy metabolomic profiling of exhaled breath condensate in patients with stable and unstable cystic fibrosis. *Thorax* 2012;**67**(3):222–8.
13. Pizzini A, Filipiak W, Wille J, Ager C, Wiesenhofer H, Kubinec R, et al. Analysis of volatile organic compounds in the breath of patients with stable or acute exacerbation of chronic obstructive pulmonary disease. *J Breath Res* 2018;**12**(3):036002.
14. Gaugg MT, Nussbaumer-Ochsner Y, Bregy L, Engler A, Stebler N, Gaisl T, et al. Real-time breath analysis reveals specific metabolic signatures of COPD exacerbations. *Chest* 2019;**156**(2):269–76.
15. Shafiek H, Fiorentino F, Merino JL, Lopez C, Oliver A, Segura J, et al. Using the electronic nose to identify airway infection during COPD exacerbations. *PloS One* 2015; **10**(9):e0135199.
16. van Velzen P, Brinkman P, Knobel HH, van den Berg JWK, Jonkers RE, Loijmans RJ, et al. Exhaled breath profiles before, during and after exacerbation of COPD: a prospective follow-up study. *COPD* 2019;**16**(5–6):330–7.
17. Plaza V, Crespo A, Giner J, Merino JL, Ramos-Barbon D, Mateus EF, et al. Inflammatory asthma phenotype discrimination using an electronic nose breath analyzer. *J Invest Allergol Clin Immunol* 2015;**25**(6):431–7.
18. Ibrahim B, Basanta M, Cadden P, Singh D, Douce D, Woodcock A, et al. Non-invasive phenotyping using exhaled volatile organic compounds in asthma. *Thorax* 2011;**66**(9): 804–9.
19. Fens N, de Nijs SB, Peters S, Dekker T, Knobel HH, Vink TJ, et al. Exhaled air molecular profiling in relation to inflammatory subtype and activity in COPD. *Eur Respir J* 2011; **38**(6):1301–9.
20. Schleich FN, Zanella D, Stefanuto PH, Bessonov K, Smolinska A, Dallinga JW, et al. Exhaled volatile organic compounds are able to discriminate between neutrophilic and eosinophilic asthma. *Am J Respir Crit Care Med* 2019;**200**(4):444–53.
21. de Vries R, Dagelet YWF, Spoor P, Snoey E, Jak PMC, Brinkman P, et al. Clinical and inflammatory phenotyping by breathomics in chronic airway diseases irrespective of the diagnostic label. *Eur Respir J* 2018;**51**(1).
22. Shaw DE, Sousa AR, Fowler SJ, Fleming LJ, Roberts G, Corfield J, et al. Clinical and inflammatory characteristics of the European U-BIOPRED adult severe asthma cohort. *Eur Respir J* 2015;**46**(5):1308–21.

23. Brinkman P, Wagener AH, Hekking PP, Bansal AT, Maitland-van der Zee AH, Wang Y, et al. Identification and prospective stability of electronic nose (eNose)-derived inflammatory phenotypes in patients with severe asthma. *J Allergy Clin Immunol* 2019;**143**(5):1811−20. e7.
24. Kuo CS, Pavlidis S, Loza M, Baribaud F, Rowe A, Pandis I, et al. T-helper cell type 2 (Th2) and non-Th2 molecular phenotypes of asthma using sputum transcriptomics in U-BIOPRED. *Eur Respir J* 2017;**49**(2).
25. Meyer N, Dallinga JW, Nuss SJ, Moonen EJ, van Berkel JJ, Akdis C, et al. Defining adult asthma endotypes by clinical features and patterns of volatile organic compounds in exhaled air. *Respir Res* 2014;**15**:136.
26. van der Schee MP, Palmay R, Cowan JO, Taylor DR. Predicting steroid responsiveness in patients with asthma using exhaled breath profiling. *Clin Exp Allergy : J British Soci Allergy Clin Immunol* 2013;**43**(11):1217−25.
27. Cazzola M, Segreti A, Capuano R, Bergamini A, Martinelli E, Calzetta L, et al. Analysis of exhaled breath fingerprints and volatile organic compounds in COPD. *COPD Res Practice* 2015;**1**(1):7.
28. Wilkinson M, Maidstone R, Loudon A, Blaikley J, White IR, Singh D, et al. Circadian rhythm of exhaled biomarkers in health and asthma. *Eur Respir J* 2019;**54**(4).
29. Caldeira M, Perestrelo R, Barros AS, Bilelo MJ, Morete A, Camara JS, et al. Allergic asthma exhaled breath metabolome: a challenge for comprehensive two-dimensional gas chromatography. *J Chromatogr A* 2012;**1254**:87−97.
30. van Vliet D, Smolinska A, Jobsis Q, Rosias P, Muris J, Dallinga J, et al. Can exhaled volatile organic compounds predict asthma exacerbations in children? *J Breath Res* 2017;**11**(1):016016.
31. Basanta M, Ibrahim B, Dockry R, Douce D, Morris M, Singh D, et al. Exhaled volatile organic compounds for phenotyping chronic obstructive pulmonary disease: a cross-sectional study. *Respir Res* 2012;**13**:72.
32. Caldeira M, Barros AS, Bilelo MJ, Parada A, Camara JS, Rocha SM. Profiling allergic asthma volatile metabolic patterns using a headspace-solid phase microextraction/gas chromatography based methodology. *J Chromatogr A* 2011;**1218**(24):3771−80.
33. Dragonieri S, Schot R, Mertens BJ, Le Cessie S, Gauw SA, Spanevello A, et al. An electronic nose in the discrimination of patients with asthma and controls. *J Allergy Clin Immunol* 2007;**120**(4):856−62.
34. Smolinska A, Klaassen EM, Dallinga JW, van de Kant KD, Jobsis Q, Moonen EJ, et al. Profiling of volatile organic compounds in exhaled breath as a strategy to find early predictive signatures of asthma in children. *PloS One* 2014;**9**(4):e95668.
35. Van Berkel JJ, Dallinga JW, Moller GM, Godschalk RW, Moonen EJ, Wouters EF, et al. A profile of volatile organic compounds in breath discriminates COPD patients from controls. *Respir Med* 2010;**104**(4):557−63.
36. Paredi P, Kharitonov SA, Leak D, Ward S, Cramer D, Barnes PJ. Exhaled ethane, a marker of lipid peroxidation, is elevated in chronic obstructive pulmonary disease. *Am J Respir Crit Care Med* 2000;**162**(2 Pt 1):369−73.
37. Larstad MA, Toren K, Bake B, Olin AC. Determination of ethane, pentane and isoprene in exhaled air–effects of breath-holding, flow rate and purified air. *Acta Physiologica (Oxford, England)* 2007;**189**(1):87−98.
38. Dallinga JW, Robroeks CM, van Berkel JJ, Moonen EJ, Godschalk RW, Jobsis Q, et al. Volatile organic compounds in exhaled breath as a diagnostic tool for asthma in children. *Clin Exp Allergy : J British Soci Allergy Clin Immunol* 2010;**40**(1):68−76.

39. Phillips CO, Syed Y, Parthalain NM, Zwiggelaar R, Claypole TC, Lewis KE. Machine learning methods on exhaled volatile organic compounds for distinguishing COPD patients from healthy controls. *J Breath Res* 2012;**6**(3):036003.
40. Robroeks CM, van Berkel JJ, Jobsis Q, van Schooten FJ, Dallinga JW, Wouters EF, et al. Exhaled volatile organic compounds predict exacerbations of childhood asthma in a 1-year prospective study. *Eur Respir J* 2013;**42**(1):98–106.
41. Gahleitner F, Guallar-Hoyas C, Beardsmore CS, Pandya HC, Thomas CP. Metabolomics pilot study to identify volatile organic compound markers of childhood asthma in exhaled breath. *Bioanalysis* 2013;**5**(18):2239–47.
42. van de Kant KD, van Berkel JJ, Jobsis Q, Lima Passos V, Klaassen EM, van der Sande L, et al. Exhaled breath profiling in diagnosing wheezy preschool children. *Eur Respir J* 2013;**41**(1):183–8.
43. Christiansen A, Davidsen JR, Titlestad I, Vestbo J, Baumbach J. A systematic review of breath analysis and detection of volatile organic compounds in COPD. *J Breath Res* 2016;**10**(3):034002.
44. Ahmed WM, Brinkman P, Weda H, Knobel HH, Xu Y, Nijsen TM, et al. Methodological considerations for large-scale breath analysis studies: lessons from the U-BIOPRED severe asthma project. *J Breath Res* 2018;**13**(1):016001.

CHAPTER

Breath analysis for respiratory infections

21

Cristina E. Davis[1], Jane E. Hill[2], Matthias Frank[3,4], Mitchell M. McCartney[1], Michael Schivo[5,6,7], Heather D. Bean[8]

[1]*Department of Mechanical and Aerospace Engineering, University of California, Davis, CA, United States;* [2]*Chemical and Biological Engineering, University of British Columbia, Vancouver, BC, Canada;* [3]*Lawrence Livermore National Laboratory, Livermore, CA, United States;* [4]*Department of Biochemistry and Molecular Medicine, University of California, Davis School of Medicine, Sacramento, CA, United States;* [5]*Center for Comparative Respiratory Biology and Medicine, University of California, Davis, School of Medicine, Davis, CA, United States;* [6]*Department of Internal Medicine, University of California, Davis, School of Medicine, Sacramento, CA, United States;* [7]*VA Northern California Health Care System, Mather, CA, United States;* [8]*School of Life Sciences, Arizona State University, Tempe, AZ, United States*

Chapter outline

21.1 Overview

Respiratory pathogen diagnostics are an important area of current breath research, and researchers have taken various tactics to approach studies on this topic.[1] Any studies of infectious diseases in humans have relied on subjects in preidentified cohorts who tested positive for the diseases under normal standards of care, using gold-standard confirmatory methods. Breath samples, collected with informed consent, have been concentrated and/or directly analyzed in a variety of ways. Room air analytical controls are sometimes used as a best practice. A "control" cohort of subjects is usually employed as a comparison and typically is comprised of uninfected age-matched and gender-matched healthy individuals, although in more ideal circumstances, this would be a group that was suspected of having the infection but was later ruled out (see Chapter 35 for further details on clinical study design). Indeed, recruiting those with the same symptoms in the control groups, including noninfectious disease subjects, such as sarcoidosis patients in the case of breath

Breathborne Biomarkers and the Human Volatilome. https://doi.org/10.1016/B978-0-12-819967-1.00021-9
Copyright © 2020 Elsevier B.V. All rights reserved.

sampling for tuberculosis, is an increasingly essential parameter in a study design. Once breath samples are obtained by an appropriate sampling method (see Chapter 2), a common theme has been to use different forms of mass spectrometry (MS) analysis of the breath samples to provide complete metabolomic chemical identification of the sample contents. Using an untargeted approach to statistical analysis, researchers then applied various data analysis techniques to try to identify candidate breath biomarkers of that specific pulmonary infection.

Parallel animal model studies have provided insight into human breath biomarker studies. The animals may be chosen from natural disease occurrence, similar to the human studies, or laboratory animals in carefully controlled studies that target a specific pathogen at different doses of inoculation. These animal models should be—and often are—of high human relevance and have included macaques, mice, ferrets, and swine. Animal model studies have played a key role in developing and maturing biomarker identification, as well as providing fundamental information, such as how biomarkers change over an infection time course postinoculation.

Breath biomarkers of infection from human and animal studies can arise from the host, generated by the immune system[2] or by cellular responses of infected tissues, or they can arise from the microbial pathogen as a function of virulence or as a consequence of normal cell cycles and proliferation. Thus, there is also a role for in vitro cell culture models, which can produce translatable biomarkers or be used to understand the origin of biomarkers during disease progression (see Chapter 26).

Together, this triad of approaches (human, animal, and in vitro cell culture studies) has allowed researchers to identify candidate breath biomarkers that can be carried forward into larger studies. A summary of these advances is outlined in this chapter.

21.2 Bacterial infections

Tuberculosis (TB) is the largest infectious disease killer in the world and the biggest killer of people with human immunodeficiency virus (HIV) infection and/or acquired immune deficiency syndrome (AIDS). TB is one of the respiratory diseases for which breath analysis studies have been performed in both humans and animal models. Traditional means of TB detection, still used in many countries,[3] are slow and require weeks-long sputum culturing to confirm diagnosis and assess antibiotic resistance. Nucleic acid amplification approaches generate useful diagnostic data faster, typically in three days, but this timeline is problematic as many patients do not return for their result and thus are lost to follow up. And, children and immunodeficient patients generally do not produce sputum. Thus, breath diagnostics are poised to provide a much more rapid timeline for physicians, especially to allow patient treatment to begin rapidly.

Breath studies to diagnose bacterial infections caused by *Mycobacterium tuberculosis* (*Mtb*) are more likely to succeed than any other approach due to the extensive groundwork done by the Belgian-Tanzanian group APOPO (Anti-Persoonsmijnen

Ontmijnende Product Ontwikkeling) who have evaluated the volatiles from tens of thousands of sputum samples using trained giant Gambian rats.[4] Morozov and colleagues performed a proof-of-principle study aimed at developing noninvasive diagnostics for pulmonary TB based on potential TB biomarkers in microdroplets exhaled by TB patients and collected on electrospun fibers.[5] This study involved 42 TB patients (including recent and chronic TB cases) and 13 healthy volunteers. Samples were tested for the presence of *Mtb* cells, *Mtb* DNA, and protein biomarkers. While no *Mtb* cells or *Mtb* DNA could be detected, an ultrasensitive immunoassay was developed that detected immunoglobulin A (IgA) in >90% of samples from both TB patients and at higher rates than healthy volunteers. Antigen-specific IgA was detected at higher rates in the patient samples compared with the healthy control samples. Although the authors report and comment on the relatively low sensitivity and specificity, they suggest that expanding this method to include inflammation-specific biomarkers in addition to the TB-specific antibodies promises to increase the level of discrimination in the future.

Beccaria and colleagues conducted two studies evaluating the use of human breath collected and stored on thermal desorption tubes and analyzed by comprehensive gas chromatography–time-of-flight mass spectrometry (GC×GC-TOFMS) to diagnose active TB in subjects with confirmed *Mtb* infection.[6,7] The control samples used in each study included room air, as well as either a group presenting to the clinic with the same symptoms (South Africa) or in subjects with nonrespiratory infectious diseases (Haiti). About 84 subjects were evaluated, of which 34 were patient controls. Both studies present a chemometric pathway, utilizing statistical and machine learning tools to translate thousands of analytical signals to a putative biomarker set in the context of an underpowered study design.

Hill and colleagues also investigated the potential of breath analysis for detecting mycobacterial infections using a murine model and *Mycobacterium bovis* bacillus Calmette–Guerin (BCG),[8] as well as *Mtb* in macaques.[9,10] This was the first report of breath being collected from mice or nonhuman primates infected with a pathogen from the *Mycobacterium tuberculosis* complex (MTBC). A related study on mice used breath collected at two time points for eight mice infected with *M. bovis* BCG and eight healthy mouse controls exposed to instilled phosphate-buffered saline (PBS). Room air samples were collected as controls. The breath of ventilated mice was collected in Tedlar bags and then concentrated onto thermal desorption tubes. Analysis of the breath samples was performed using GC×GC-TOFMS. Statistical analysis using random forest identified 23 features in the data that discriminated between the breath samples of infected mice and the controls. Tentative identification of these 23 compounds was provided by mass spectral matches to a reference library. Four of these markers were also reportedly found in previous breath studies on animals with MTBC infections. This study showed that this overall methodology can differentiate between healthy and infected mice using breath analysis. It also indicated that a mouse breath model might be useful to study TB pathogenesis and evaluate preclinical drug regimen efficacy.

A design for the collection of breath in gently anesthetized macaques in a biological safety level three laboratory was published by Mellors et al., in 2017.[8] A chemometric process for evaluating the repeatability of breath samples in terms of chemical composition was given, and a pilot level determination of 37 putative biomarkers to distinguish *Mtb* lung infections from healthy animals using breath was generated. In a follow-up study,[10] the same group looked for and found substantial synergy between cell culture volatile molecules of *Mtb* in the breath of nine cynomolgus macaques with infection by the same strain (Erdman). Thirty-seven molecules found in culture could distinguish the breath of infected and healthy macaques with an area under the receiver operating characteristic curve (AUROC) of 87%. The authors point out, however, that the origin of the molecules in the breath of the macaques was unknown; therefore, the translation from culture to breath should be considered precarious unless supported by unique metabolism, as can be found in some fungal infections (see Section 21.4).

Bergmann and colleagues investigated the in vivo volatile organic compound (VOC) signatures of *Mycobacterium avium* ssp. *paratuberculosis* (MAP) in goats[11] in a study that involved 26 MAP-inoculated animals and 16 healthy controls. Volatile molecules were collected from the animals' breath using an automated alveolar sampling device and needle trap microextraction, as well as from feces using solid-phase microextraction (SPME), with sampling performed over various time points spanning up to 48 weeks after inoculation. All samples were analyzed by GC-MS, with tentative substance identification accomplished using a reference library. In parallel, blood was sampled and analyzed for MAP-specific antibodies and MAP-specific interferon-γ response to correlate measured breath molecule patterns to measurements of these blood-based biomarkers. The observed VOC patterns differed between MAP-inoculated and noninoculated animals. Although the measured VOC patterns evolved during the infection, notable differences in observed VOCs between inoculated and noninoculated animals remained throughout the course of infection. A total of 28 VOCs were identified as potential markers for MAP infection in vivo, some of which have been found in previous in vitro studies in the headspace of MAP cultures.[12] Some of the observed VOC markers in breath were attributed to host response, and the authors cautioned that results from in vitro VOC studies (bacterial headspace sampling) may not simply transfer to in vivo conditions. The results of this work also indicated the potential usefulness of VOC profiles from feces to detect the presence of mycobacteria in the gut of ruminants. Further details of this and other studies with ruminants are given in Chapter 27.

Several human studies have targeted chronic *Pseudomonas aeruginosa* (PA) in subjects with cystic fibrosis. Cystic fibrosis (CF) is a genetic disease that results in polymicrobial lung infections, with the presence of PA, in particular, being linked to poorer outcomes.[13] Extricating a breathprint from the milieu of complex microbial ecology and host comorbidities is one of the most complicated undertakings in breath research. Gilchrist and colleagues evaluated hydrogen cyanide (HCN) in breath using selected ion flow tube-mass spectrometry (SIFT-MS) as an early biomarker for PA infection in 233 patients over 2 years (2055 total breath samples).[14] Grounded in microbiology confirmation, which revealed 71 PA-positive cultures, the specificity of the approach was estimated to be 99%, although the low sensitivity (estimated at 33%) makes the current test inadequate as a screening tool. Španěl and colleagues evaluated the breath of 20 CF patients with confirmed *P. aeruginosa* infection and contrasted those with 38 CF patients who were PA-free. Using a panel of 16 molecules obtained from a linear logistic model, they achieved an AUROC of 0.91. Further details of these studies with SIFT-MS are given in Chapter 9. Nasir and colleagues evaluated a lung sample, bronchoalveolar lavage (BAL), from patients with CF, seeking potential biomarkers for *P. aeruginosa* and *Staphylococcus aureus*.[15] Using 154 samples, they identified breathprints for each infection type, with good negative predictive values for *P. aeruginosa* (0.92) and good positive predictive values for *S. aureus* (0.86).

Two mouse studies also provided insights into lung infections by bacterial pathogens. Purcaro and colleagues evaluated a core breathprint for PA in mouse lung infections, the contribution of the host to the breathprint, and explored which molecules could be used to distinguish mice infected with the major clades of *P. aeruginosa* (PAO1, PA14, PAK, PA7).[16] Sham infection controls (PBS) were employed. Nine compounds from the headspace of those cultures were sufficiently consistent and frequently observed, and were thereby considered core in the in vivo model at 24 and 48 h infection time points. A breathprint containing ten molecules for mice infected with different strains could somewhat reflect the phylogenetic groups to which the strains belong. Barbour and colleagues developed a method to sample nose-only breath from mice and performed VOC analysis of breath from mice with systemic bacterial infections caused by *Borrelia hermsii*.[17] While inflammatory and antiinflammatory cytokines were found to be elevated in infected mice compared with uninfected controls, no significant difference was found in the profiles of 72 different VOCs from nasal sampling, although carbon monoxide (CO) was notably elevated in the breath of infected mice. Control experiments of headspace sampling from the cultures of *B. hermsii* indicated that bacteria themselves did not produce CO and thus were unlikely to be the source of elevated CO in mouse breath; however, in vivo verification is still required to confirm this hypothesis. The findings of this work indicate the potential utility of CO concentration in exhaled breath during systemic infection and inflammation. Exhaled CO in inflammatory pulmonary diseases is treated at length in Chapter 6.

21.3 Viral infections

Numerous viruses can infect the human airway, with common examples being influenza viruses and respiratory syncytial virus (RSV). Most respiratory viruses cause similar symptoms of a fever and aches, but can be quite severe for patients of certain ages or who are immunocompromised, leading to more developed maladies, such as pneumonia or bronchitis. RSV leads to nearly 14,000 deaths annually in US adults, a number that is likely underestimated.

As with bacteria, there have been significant advancements in breath tests for viral infections. While bacteria and fungi can generate their own metabolite signatures, viruses rely on the metabolic machinery of their host to propagate. Thus, viral detection often looks for direct evidence of the virus, accomplished by hunting for nucleic acid segments specific to a virus in a bodily fluid sample, such as blood or mucous. This technique of nucleic acid screening, known as polymerase chain reaction (PCR), is very sensitive but often includes a lag time with samples being shipped to offsite laboratories and additional time for the analysis to be conducted. More rapid tests for viral infections are needed, as correct diagnoses by even a day sooner could save lives. In addition, specifying between viral versus bacterial infection is rather important for treatment, as prescription of antibiotics is ineffective against viruses. Finally, methods to collect bodily fluids for PCR can be quite invasive, especially for respiratory infections. Bronchoscopies are a common technique, in which an instrument is threaded through the mouth or nose and down into the lungs, for instance, to collect liquid biosamples for analysis, as in BAL. As breath is collected noninvasively, it is a much more attractive technique, especially for those already suffering from pulmonary disorders.

Around the turn of the millennium, pigs in industrial farm settings were contracting Aujeszky's disease, caused by Suid herpesvirus 1 (SHV-1). The disease starts with respiratory symptoms that often become severe, leading to death. Mortality rates were as high as 100% for piglets, and at the time it was not understood how the virus spread. Gillespie et al. posited that as infected pigs coughed or sneezed, they would produce aerosols that could travel long distances, carrying the virus and infecting nearby pigs. In their study, they inoculated a set of pigs with Aujeszky's disease. Once they showed symptoms, an infected pig was paired with a healthy "recipient" pig. A mouthpiece was placed over each pig's snout, which was then connected with tubing. This allowed the recipient pig to inspire the exhaled breath of the infected pig, without physical contact between animals. Afterward, nasal mucus was collected from all pigs. Each infected pig had detectable levels of SHV-1 in their mucus samples, but no virus was detected from the recipient pigs. Yet, all recipient pigs developed clinical symptoms of Aujeszky's disease[a] The authors made two conclusions, namely (1) that viruses can be carried by respiratory aerosols and infect others nearby and (2) that pigs are much more sensitive indicators of airborne SHV-1 virions than laboratory aerosol detection methods at the time.[18] This work demonstrates what is commonly known today—that respiratory viruses can be spread through exhaled aerosols.

Influenza infection is a major cause of seasonal morbidity and mortality in humans worldwide, and efforts to detect influenza viruses and virus infections rapidly and specifically are evolving. Phillips and colleagues conducted a before-and-after study looking at exhaled metabolites emitted by subjects receiving live attenuated influenza vaccine (LAIV).[19] Confounding biologic

[a] Since this study was first reported, a vaccine has been developed for Aujeszky's disease.

variables were likely minimized because each subject was his/her own control. LAIV breath metabolites were measured and compared pre- and postvaccination. The authors hypothesized that oxidative stress products (e.g., 2,8-dimethyl-undecane and 4,8-dimethyl-undecane) were seen in post−LAIV-treated subjects because influenza virus characteristically causes increased reactive nitrogen oxide species. Indeed, oxidative stress products were observed with changing concentrations over time post-LAIV inoculation, and these products did not appear in the LAIV headspace. The breath test performances in classifying subjects as pre- or post-LAIV inoculation ranged from a C-statistic of 0.82 (day 2) to 0.95 (day 7) to 0.95 (day 14), thereby demonstrating the feasibility of this approach. This study provides an important framework for future studies assessing influenza virus infection in its elegant design.

Aksenov and colleagues assessed metabolites from several different influenza strains that infected cultured human B-lymphoblastoid cells.[20] Lymphoblastoid cells were used to minimize the contribution of apoptosis to metabolite production. Several metabolite patterns were increased significantly over background and were used to identify specific influenza virus strains (e.g., H1N1 vs. H6N2) and strength of virus inoculation (e.g., H1N1 multiplicity of infection [MOI] 1 vs. MOI 10). This proof-of-concept study has many implications for influenza detection at both an individual level and for epidemiologic studies (consider pandemic influenza strain testing). Importantly, there were many overlapping alkene and other compounds seen in the Aksenov and Phillips studies. This provides phenomenological significance in the metabolites generated from influenza-infected cells, despite not knowing the exact mechanism of metabolite production.

Zoonotic reservoirs are common for influenza A viruses, and efforts to determine breath metabolites from animals may shed light on understanding emerging pandemic influenza strains. Traxler and colleagues assessed breath metabolites from swine during a complete H1N1 infection cycle.[21] The study reported six metabolites that classified swine as infected, as measured against standard diagnostic testing. Although the animals showed biochemical evidence of infection, they did not show clinical evidence of infection, such as decreased food or water intake, lethargy, and so on. This suggests that breath metabolite analysis may identify subclinical influenza A infection in animals, which may ultimately become more virulent when they bridge from animals to humans. As mentioned, a limitation of these studies is that they do not address the biologic origin or mechanism(s) of metabolite production, which will be important considerations of future studies.

Lower respiratory tract infections, such as pneumonia, are more challenging to diagnose compared with upper respiratory infections. Doctors are required to order invasive tests for the patients with suspect infections, including oropharyngeal swabs, induced sputum collection, and BAL. Toward less invasive techniques for lower respiratory tract diagnoses, PneumoniaCheck is a commercially available tool to collect aerosol particles from exhaled breath patients. Users simply cough into a handheld device, which diverts away breath exhaled from the upper respiratory tract and allows air from the lower respiratory tract to collect onto a filter.

The filter can then undergo molecular and biochemical analyses for disease detection. While previously demonstrated as a successful tool for bacterial infections, Patrucco et al. investigated its utility for viral diagnostics. Using PneumoniaCheck, samples were collected from persons with pneumonia infections, in addition to BAL. Using PCR, both sets of specimens were screened for a panel of common respiratory viruses, such as influenza, rhinovirus, and bocavirus. A high concordance was observed among the paired samples collected by BAL and PneumoniaCheck. Specificity rates were 100%, but sensitivity was 66%. Thus, while more investigation needs to be done, this work shows that this approach has high potential for noninvasive respiratory viral diagnostics.[22]

Other ways in which breath could be used as a viral detection matrix have also been explored. Etiological studies have recently demonstrated a relationship between certain viruses and the development of lung cancers. Specifically, two herpes viruses have been of great interest. The Epstein–Barr virus (EBV), a herpes virus, is a recognized carcinogen. Another herpes virus, cytomegalovirus (CMV), has been associated with maladies of the brain, lung, and colon. Early diagnosis of these viral infections could lead to lifestyle changes to reduce the associated cancer risk. Carpagnano et al. investigated breath as a sample matrix for detection of EBV and CMV. PCR was used to investigate if EBV and/or CMV DNA were present in exhaled breath condensate (EBC) from 110 patients, among which 70 had some type of lung cancer and 40 were healthy controls. These studies demonstrated that these two viruses were directly detectable in breath samples and increased in positivity among lung cancer patients, especially those in advanced stages. This technique could be a huge relief from the current gold-standard approaches, whereby viral nucleic acid is screened from bronchial brushings during fiber-optic bronchoscopies.[23]

Detection paradigms using breath have shown specificity to certain pathogens. Bacterial infections of PA, for example, are especially problematic to patients when co-infected with RSV. Purcaro et al. modeled the human airway by culturing human epithelial cells, a technique covered more in depth in Chapter 26. Cultures were inoculated into four classes: (1) PA only, (2) RSV only, (3) coinfection with PA and RSV, and (4) no infection (healthy controls). Detection of VOCs released by the cells was used to create a volatile profile of each category. Using higher statistics, the study demonstrated that a coinfection of the virus did not inhibit the ability to identify cells with the bacterial infection. In this particular case, however, cells inoculated with just RSV did not yield a sufficient change to the volatile profile for accurate diagnoses, providing evidence that each respiratory viral infection may have to be independently evaluated for its ability to be detected directly in breath samples.[24]

21.4 Fungal and other eukaryotic infections

Over 150 million people worldwide have serious fungal disease and more than 1.6 million people die from fungal infections annually, a rate similar to TB.[25]

Risk factors for fungal infections include comorbidities, such as HIV/AIDS, TB, cancer, asthma, and chronic obstructive pulmonary disease (COPD), as well as immunosuppressive treatments for chronic conditions, such as transplant rejection and rheumatoid arthritis.[25] As the incidences of these risk factors and comorbidities are rising globally, so too are the incidences of fungal infections.[25]

In the clinic, the diagnosis of a fungal infection is usually made based on fungal growth on selective media or the detection of fungal antigens or antibodies in patient specimens. While these tests can be highly sensitive, if a good-quality lung specimen is obtained, e.g., via biopsy, bronchoscopy, or sputum induction, these procedures are contraindicated in many of the patients who are at high risk for fungal pneumonias. Therefore, the detection of fungal infections via biomarkers in breath or EBC would significantly enhance the surveillance and detection of mycoses.

One of the most prevalent causes of pulmonary fungal infections, globally, is *Aspergillus* spp.,[25] which can cause invasive disease in immunosuppressed individuals. Koo and colleagues designed a study to identify putative breath biomarkers of invasive aspergillosis (IA) by taking an "in vitro—informed" approach to selecting classes of secondary metabolites to analyze in their clinical experiments.[26] The majority of cases of IA are caused by *Aspergillus fumigatus*, and therefore they focused their efforts on identifying VOCs that discriminate *A. fumigatus* from other *Aspergillus* spp., first by culturing the fungi in vitro. The study design included two important steps for the in vitro experiments to enhance the translation of results to clinical samples. First, oxygen and temperature conditions that favor hyphal growth (the dominant *Aspergillus* infection morphology) over conidia formation (the dominant environmental morphology) were established and benchmarked by comparing the in vitro transcriptome to the murine lung infection transcriptome. Second, *Aspergillus* spp. VOCs were characterized in a wide range of different media, from an all-purpose rich fungal medium (yeast extract peptone dextrose (YPD) broth) to media that are low iron, low nitrogen, alkaline stress, and *Aspergillus* minimal media. *A. fumigatus*—specific VOCs that were produced in most of these conditions were identified to enhance robustness of the approach, with several terpenes named as putative *A. fumigatus* IA biomarkers.

To validate the IA biomarkers, Koo et al. prospectively collected breath from 64 patients with suspected IA and analyzed the samples by GC-MS, focusing on the detection and identification of the terpenes observed in the in vitro studies. Blinded to the breath sample data, two clinicians reviewed the clinical data to classify the subjects as IA (having proven or probable aspergillosis, based on signs and symptoms) or non-IA (having possible IA, or another cause of invasive fungal disease), resulting in 34 IA and 30 non-IA breath samples. Several of the *Aspergillus* in vitro terpenes were observed in the breath of patients with or without IA, which were thus eliminated as possible biomarkers. Nevertheless, four terpenes and terpenoids specific to in vivo samples were identified that distinguished IA from non-IA with 94% sensitivity and 93% specificity. In one case, of the two false-positive samples, one was from a patient whose respiratory specimens were fungal culture negative and antigen negative, but upon autopsy, pulmonary nodules were recovered that stained positive for *Aspergillus*. This not only highlights the difficulty in benchmarking new biomarkers against a clinical diagnostic with poor sensitivity but also underscores the potential impact of adding breath tests to the diagnostic armamentarium.

Fungi can also play a significant role in noninfectious respiratory disease. Fungi are implicated in more than 11 million annual cases of allergic asthma and 12 million annual cases of rhinosinusitis globally,[25] but studying the causal relationships between fungal colonization and disease severity is impeded by respiratory specimen access. Carpagnano and colleagues sought to determine whether fungi could be cultured from EBC from asthma patients, with the ultimate goal of supplanting sputum induction, which is not advised for asthmatics with moderate-to-severe disease.[27] EBC was collected from 28 atopic (i.e., allergic) asthma patients, 19 nonatopic asthma patients, and 20 controls, and these samples were paired with induced sputum, where clinically permissible. In addition, data on disease severity, anthropometric variables (body mass index, age, sex, etc.), and fraction of exhaled nitric oxide (F_ENO) were collected. Using fungal growth on selective media as a positive result, 70% of EBC from asthmatic subjects were found to be positive for fungi, whereas none of the control subjects were fungal-positive. Where a paired sputum sample was available, the same culture results were obtained from induced sputum and EBC, yielding a 100% sensitivity for the EBC test. Interestingly, EBC fungal positivity correlated with disease severity, with 100% of moderate-to-severe asthma patients culturing fungi versus 90% of mild and 63% of intermittent asthmatics, but there was no correlation to F_ENO. A weakness in this study was the lack of gold-standard specimens (induced sputum) in the moderate-to-severe cases of disease, but this again highlights the significant clinical need that EBC can fill for tracking fungal disease and colonization.

While not specific to the airway, malaria is a well-known viral infection, with hundreds of millions of worldwide cases each year. Early diagnosis is crucial to prevent death, but diagnosis is still contingent on a century-old test: visualization of the parasite in stained blood films. Berna et al. collected breath samples from subjects involved in controlled malaria studies for the purposes of developing vaccines and treatments.[28] Astonishingly, nine VOCs were not only elevated in the breath of the malaria cohort but also the concentration of these volatiles tracked with parasitemia levels measured in their blood. When subjects began their antimalarial treatments 8 days after being infected, the levels of these VOCs decreased, still correlating with the measured level of blood parasites. Thus, the interaction of the parasite with the body produced volatile compounds in human breath that could be used for malaria diagnostics. Detection of malaria infection via breath is covered in depth in the next chapter.

21.5 Summary

Over the last decade, there have been tremendous advances in breath analysis techniques applied to respiratory infection and colonization, which represents one of the most clinically relevant applications for the field of breath diagnostics and monitoring. The studies outlined in this chapter present a paradigm of untargeted biomarker discovery, when study design is employed to uncover breath compounds that are statistically correlated with specific clinical conditions. As with all breath studies, pitfalls exist. In most human studies in this area, there have been limited subject numbers, and the breath biomarkers have not been confirmed in independent cohorts. Currently, there is no adequate understanding of how confounding factors or other unrelated clinical conditions may affect biomarker profiles. It is possible that breathborne biomarker signals may change over time and may vary with pathogenicity and microbial load. It is also possible that animal studies could be performed to inform pilot human studies, but that is not yet a common practice, and there have not yet been exhaustive comparisons of animal and human trials to cross-correlate and confirm results between the two.

There is a tremendous amount of ongoing parallel research to elucidate biomarker signals that emanate from in vitro cell culture models—in addition to the human/animal studies work (as outlined in Chapter 26). Most of this in vivo and in vitro work has been accomplished using various types of confirmatory analytical chemistry methods (e.g., mass spectrometry). While bringing novel sensors into this research area will eventually be an important step toward clinical translation, unequivocal chemical identification is critical during this untargeted exploratory phase of biomarker discovery. The field is now on the cusp of profiling candidate breath biomarkers of infection that can be pursued further. The best clinical studies to date have compared human/animal breath data paired with studies of in vitro organism cell models, especially those that examined biomarkers from the actual microbial clinical isolates involved in the clinical work.

Applications to aid modern emerging infectious disease epidemics or pandemics truly represent a holy grail of breath research—the chance to provide meaningful early, real-time, noninvasive diagnostics and distinguish trivial conditions from grave infectious pathogens. Given that so many respiratory infections result in nonspecific clinical presentations (e.g., elevated body temperature, malaise, etc.), it would be extremely valuable to differentiate patients who are critically ill with a specific circulating pathogen of interest. It would also allow clinicians to focus on truly ill patients and distinguish them from the "worried well" in western nations who may descend on hospital emergency departments during an infectious disease outbreak. In the far future, it would also allow public health officials to use potential technology to help limit or prevent disease spread and could provide information to help during quarantine endeavors in extreme situations. For epidemiologists, asymptomatic biomarker detection of infection could be very valuable while tracing contacts during an outbreak. While the breath research community works toward these goals, more research is needed before any of these scenarios is possible. Ideally, researchers would have immediately available high-fidelity in vivo cell culture and animal models to generate candidate breath biomarkers without the need for parallel human discovery efforts. This would save the most time during an infectious disease outbreak or in the unlikely event of intentional or engineered biological pathogen release. None of this is possible at present, but as in vitro cell culture models advance and begin to increase in fidelity, it is possible this could ultimately occur. This paradigm would have been very beneficial three times since the turn of the century, during the 2002 severe acute respiratory syndrome (SARS) pandemic, the 2009 swine influenza H1N1 pandemic, and the 2019 coronavirus disease (COVID-19) pandemic currently underway.

References

1. van Oort PM, Povoa P, Schnabel R, Dark P, Artigas A, Bergmans DCJJ, et al. The potential role of exhaled breath analysis in the diagnostic process of pneumonia—a systematic review. *J Breath Res* 2018;**12**(2):024001.
2. Bean HD, Jiménez-Díaz J, Zhu J, Hill JE. Breathprints of model murine bacterial lung infections are linked with immune response. *Eur Respir J* 2015;**45**(1):181.
3. Pai M, Nicol MP, Boehme CC. Tuberculosis diagnostics: state of the art and future directions. *Microbiol Spectr* 2016;**4**(5).
4. Poling A, Mahoney A, Beyene N, Mgode G, Weetjens B, Cox C, et al. Using giant African pouched rats to detect human tuberculosis: a review. *The Pan African Medical Journal* 2015;**21**:333–8.
5. Morozov VN, Nikolaev AA, Shlyapnikov YM, Mikheev AY, Shlyapnikova EA, Bagdasaryan TR, et al. Non-invasive approach to diagnosis of pulmonary tuberculosis using microdroplets collected from exhaled air. *J Breath Res* 2018;**12**(3):036010.
6. Beccaria M, Bobak C, Maitshotlo B, Mellors TR, Purcaro G, Franchina FA, et al. Exhaled human breath analysis in active pulmonary tuberculosis diagnostics by comprehensive gas chromatography-mass spectrometry and chemometric techniques. *J Breath Res* 2018;**13**(1):016005.
7. Beccaria M, Mellors TR, Petion JS, Rees CA, Nasir M, Systrom HK, et al. Preliminary investigation of human exhaled breath for tuberculosis diagnosis by multidimensional gas chromatography – time of flight mass spectrometry and machine learning. *J Chromatogr B* 2018;**1074-1075**:46–50.
8. Franchina FA, Mellors TR, Aliyeva M, Wagner J, Daphtary N, Lundblad LKA, et al. Towards the use of breath for detecting mycobacterial infection: a case study in a murine model. *J Breath Res* 2018;**12**(2):026008.
9. Mellors TR, Blanchet L, Flynn JL, Tomko J, O'Malley M, Scanga CA, et al. A new method to evaluate macaque health using exhaled breath: a case study of *M. tuberculosis* in a BSL-3 setting. *J Appl Physiol* 2017;**122**(3):695–701.
10. Mellors TR, Nasir M, Franchina FA, Smolinska A, Blanchet L, Flynn JL, et al. Identification of *Mycobacterium tuberculosis* using volatile biomarkers in culture and exhaled breath. *J Breath Res* 2018;**13**(1):016004.
11. Bergmann A, Trefz P, Fischer S, Klepik K, Walter G, Steffens M, et al. Vivo volatile organic compound signatures of *Mycobacterium avium* subsp. paratuberculosis. *PloS One* 2015;**10**(4):e0123980.
12. Trefz P, Koehler H, Klepik K, Moebius P, Reinhold P, Schubert JK, et al. Volatile emissions from *Mycobacterium avium* subsp. paratuberculosis mirror bacterial growth and enable distinction of different strains. *PloS One* 2013;**8**(10):e76868.

13. Nixon GM, Armstrong DS, Carzino R, Carlin JB, Olinsky A, Robertson CF, et al. Clinical outcome after early *Pseudomonas aeruginosa* infection in cystic fibrosis. *J Pediatr* 2001;**138**(5):699–704.
14. Gilchrist FJ, Belcher J, Jones AM, Smith D, Smyth AR, Southern KW, et al. Exhaled breath hydrogen cyanide as a marker of early *Pseudomonas aeruginosa* infection in children with cystic fibrosis. *ERJ Open Res* 2015;**1**(2).
15. Nasir M, Bean HD, Smolinska A, Rees CA, Zemanick ET, Hill JE. Volatile molecules from bronchoalveolar lavage fluid can 'rule-in' *Pseudomonas aeruginosa* and 'rule-out' *Staphylococcus aureus* infections in cystic fibrosis patients. *Sci Rep* 2018;**8**(1):826.
16. Purcaro G, Nasir M, Franchina FA, Rees CA, Aliyeva M, Daphtary N, et al. Breath metabolome of mice infected with *Pseudomonas aeruginosa*. *Metabolomics* 2019; **15**(1):10.
17. Barbour AG, Hirsch CM, Ghalyanchi Langeroudi A, Meinardi S, Lewis ER, Estabragh AS, et al. Elevated carbon monoxide in the exhaled breath of mice during a systemic bacterial infection. *PloS One* 2013;**8**(7):e69802.
18. Gillespie RR, Hill MA, Kanitz CL, Knox KE, Clark LK, Robinson JP. Infection of pigs by Aujeszky's disease virus via the breath of intranasally inoculated pigs. *Res Vet Sci* 2000;**68**(3):217–22.
19. Phillips M, Cataneo RN, Chaturvedi A, Danaher PJ, Devadiga A, Legendre DA, et al. Effect of influenza vaccination on oxidative stress products in breath. *J Breath Res* 2010;**4**(2):026001.
20. Aksenov AA, Sandrock CE, Zhao W, Sankaran S, Schivo M, Harper R, et al. Cellular scent of influenza virus infection. *Chembiochem* 2014;**15**(7):1040–8.
21. Traxler S, Bischoff A-C, Saß R, Trefz P, Gierschner P, Brock B, et al. VOC breath profile in spontaneously breathing awake swine during Influenza A infection. *Sci Rep* 2018;**8**(1): 14857.
22. Patrucco F, Gavelli F, Ravanini P, Daverio M, Statti G, Castello LM, et al. Use of an innovative and non-invasive device for virologic sampling of cough aerosols in patients with community and hospital acquired pneumonia: a pilot study. *J Breath Res* 2019;**13**(2):021001.
23. Carpagnano GE, Lacedonia D, Natalicchio MI, Cotugno G, Zoppo L, Martinelli D, et al. Viral colonization in exhaled breath condensate of lung cancer patients: possible role of EBV and CMV. *Clin Res J* 2018;**12**(2):418–24.
24. Purcaro G, Rees CA, Melvin JA, Bomberger JM, Hill JE. Volatile fingerprinting of *Pseudomonas aeruginosa* and respiratory syncytial virus infection in an in vitro cystic fibrosis co-infection model. *J Breath Res* 2018;**12**(4):046001.
25. Bongomin F, Gago S, Oladele RO, Denning DW. Global and multi-national prevalence of fungal diseases-estimate precision. *Journal of Fungi* 2017;**3**(4).
26. Koo S, Thomas HR, Daniels SD, Lynch RC, Fortier SM, Shea MM, et al. A breath fungal secondary metabolite signature to diagnose invasive aspergillosis. *Clin Infect Dis* 2014; **59**(12):1733–40.
27. Carpagnano GE, Malerba M, Lacedonia D, Susca A, Logrieco A, Carone M, et al. Analysis of the fungal microbiome in exhaled breath condensate of patients with asthma. *Allergy Asthma Proc* 2016;**37**(3):41–6.
28. Berna AZ, McCarthy JS, Wang RX, Saliba KJ, Bravo FG, Cassells J, et al. Analysis of breath specimens for biomarkers of plasmodium falciparum infection. *J Infect Dis* 2015; **212**(7):1120–8.

CHAPTER

Volatile biomarkers of malaria infection

22

Amalia Z. Berna[1], Justin J. Miller[1], Audrey R. Odom John[1,2,3]

[1]*Department of Pediatrics, Washington University School of Medicine, Saint Louis, MO, United States;* [2]*Department of Molecular Microbiology, Washington University School of Medicine, Saint Louis, MO, United States;* [3]*The Children's Hospital of Philadelphia, Philadelphia, PA, United States*

Chapter outline

22.1 Overview

In 2018, an estimated 228 million cases of malaria occurred worldwide, resulting in approximately 405,000 deaths. Aggressive global efforts to control malaria over the last 15 years have been highly successful, with 20 million fewer cases in 2017 than in 2010. However, progress has stalled alarmingly over the last several years.[1] Among the more pressing current challenges to malaria control is the need for additional effective diagnostic tools to detect both symptomatic and asymptomatic infections.

Malaria is caused by infection by protozoan parasites in the genus *Plasmodium*. While several *Plasmodium* spp. infect humans, the majority of severe and life-threatening malaria is due to *P. falciparum* infection. Transmitted person-to-person by *Anopheles* mosquitoes, symptomatic *P. falciparum* malaria is characterized by repeated cycles of asexual replication within mature human erythrocytes. A small proportion of asexual bloodstream parasites undergo sexual differentiation. Ongoing malaria transmission requires consumption of these sexual-stage parasites, termed gametocytes, during blood meals by female *Anopheles*. Malaria diagnostics used for point-of-care (PoC) clinical diagnostic testing of symptomatic individuals in malaria-endemic areas must be highly sensitive to the presence of

Breathborne Biomarkers and the Human Volatilome. https://doi.org/10.1016/B978-0-12-819967-1.00022-0
Copyright © 2020 Elsevier B.V. All rights reserved.

asexual bloodstream *Plasmodium* infection. In contrast, a diagnostic test that reflects gametocytemia would impact public health strategies to identify individuals at risk of malaria transmission.

For more than a century, microscopic evaluation of capillary blood samples has been used to identify bloodstream malaria parasites.[2] Highly sensitive and specific nucleic acid–based tests have been also developed[3,4] but are not readily available in low- and middle-income countries where malaria is endemic. Conversely, lateral flow–based rapid diagnostic tests (RDTs) are sensitive and require little training, making them overwhelmingly the method of choice to evaluate for PoC diagnosis in malaria-endemic areas. In 2018, 412 million RDTs were sold, and an estimated 64% of diagnostic testing in sub-Saharan Africa was performed by RDT.[1]

The current generation of RDTs largely relies on the detection of a distinct *P. falciparum*–specific protein, histidine rich protein 2 (HRP2), but HRP2-based RDTs possess critical weaknesses. False-positive RDTs are common as HRP2 can be detected up to 1 month after malaria clearance, making it impossible to distinguish acute from recent infections.[5] More critically, HRP2-based RDTs are at risk due to the recent emergence and spread of *P. falciparum* strains that lack HRP2. *Hrp2*-null parasites were first reported in 2010 in South America,[6] but gene deletions have subsequently been found in several locations in Africa.[7–10] In a recent study in Eritrea, up to 80% of all patients were infected with *hrp2*-null parasites.[7] Computational modeling projects a dramatic rise in RDT-undetectable parasites, as widespread use of RDTs has maintained ongoing selective pressure against parasites that still express HRP2.[11]

Other challenges in the diagnosis of malaria include addressing the large asymptomatic reservoir, as nearly 75% of individuals infected with *Plasmodium* spp. are asymptomatic. Asymptomatic individuals constitute a major source of ongoing transmission; these individuals are more likely to be bitten by mosquitoes than parasite-free individuals, but they do not present for care (and are therefore not diagnosed or treated) and are often mobile, increasing their potential for malaria transmission and geographic spread.[12] Most human-infecting *Plasmodium* species do not have a zoonotic reservoir acting as a barrier to malaria eradication. *P. knowlesi,* predominantly a simian parasite, can transmit to humans; however, the prevalence of *P. knowlesi* malaria is not widely known, as microscopy-based diagnostic tests often misdiagnose *P. knowlesi*.

Growing evidence suggests that vector mosquito species can differentiate between malaria-infected and malaria-uninfected individuals based on odor. This finding has inspired recent work to address the possibility of diagnosing malaria via volatile biomarkers emitted by breath and/or skin. This chapter examines the changes that *Plasmodium* spp. infection imparts on host odors and the resulting impacts on vector behavior. Recent studies on human malaria that have evaluated volatiles from the skin and breath of *Plasmodium*-infected individuals are reviewed.

22.2 Mosquito attraction to malaria-infected hosts

Female *Anopheles* mosquitoes require mammalian blood meals to mature eggs; however, mosquito biting behavior is highly complex and context-dependent. Mosquitoes detect human hosts using a combination of cues, with olfactory cues being undoubtedly the most important group of external stimuli—affecting mosquito behavior.[13] From a distance, female mosquitoes sense carbon dioxide (CO_2) and preferentially migrate to areas of higher CO_2. As they move closer to the target blood meal, they sense host heat, skin odors, and potentially the breath volatiles of the target host.[14] Increasing evidence, reviewed below, indicates that host volatiles—and, as a result, mosquito behavior—are exploited by mammalian *Plasmodium* spp. parasites, including *P. falciparum*, to increase likelihood of transmission (Fig. 22.1).

Humans infected with *P. falciparum* may be more mosquito attractive than uninfected humans. For example, investigators evaluated the attraction of *Anopheles gambiae* to 5- to 12-year-old Kenyan children of variable infection status and their uninfected classmates, using a dual-choice olfactometer (a setup which allows testing preference of one odor against another).[15] *Plasmodium*-infected children were treated with antimalarials and, following parasite clearance, the attraction of mosquitoes to these children was reassessed. Investigators found that *A. gambiae* mosquitoes were nearly twice as attracted to children carrying high burdens of gametocytes (microscopically visible) than parasite-free individuals, individuals infected with only asexual stages, or individuals carrying submicroscopic levels of gametocytes.[15] As expected, treatment with antimalarials reduced asexual and sexual-stage parasite levels below the limits of molecular detection. Importantly, following antimalarial treatment, *Plasmodium*-infected children were no longer preferentially attractive to *A. gambiae* mosquitoes. This is in

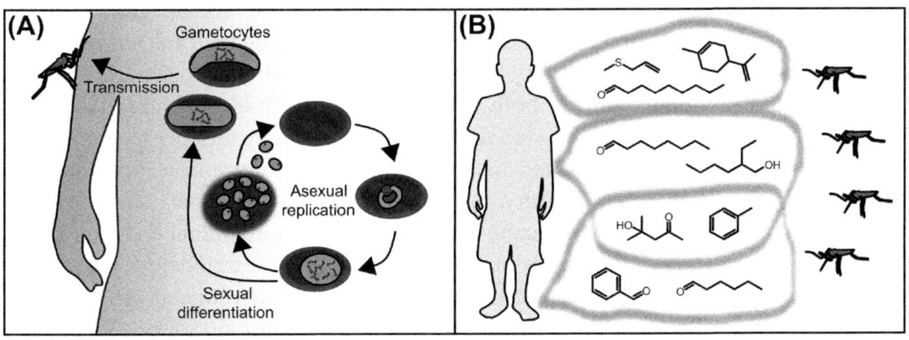

FIGURE 22.1

(A) Life cycle of *Plasmodium falciparum*. (B) Human body odor is altered upon malaria infection and attracts *Anopheles* spp. mosquitoes.

agreement with a previous study by Lacroix et al.[16] in which mosquito attractiveness was seen to be approximately double in gametocyte-positive children, relative to uninfected children or those with asexual-stage parasitemia.

While there is increasing evidence that *P. falciparum* infection alters *A. gambiae* host-seeking behavior, it is not clear which volatile compounds may be responsible. To address this question, Robinson et al.[17] assessed the behavioral response of *A. gambiae* to the foot odors of 5- to 12-year-old Kenyan school children, before and after antimalarial treatment. Foot odors of asymptomatic *P. falciparum*–infected and *P. falciparum*–uninfected children were collected on socks over 20 h and extracted for mass spectrometry (MS) analysis. For infected individuals, odors were collected after administration of the first dose of treatment with the antimalarial. Following confirmed parasite clearance (21 days later), odor samples were collected in the same manner from the same children. *A. gambiae* mosquitoes were offered the choice of either Day 1 or Day 21 odor samples from the same child, in a dual-choice cage assay. Investigators found that mosquitoes did not differentiate between Day 1 and Day 21 odor samples from uninfected children but were more attracted to the Day 1 samples from children harboring asexual or gametocyte-stage *Plasmodium* parasites. Somewhat unexpectedly, researchers did not observe the gametocyte-specific effect that was previously described.[15,16] The authors suggest that imperfect detection of low densities of gametocytes could play a role in this discrepancy. Alternatively, gametocyte-specific attraction may be communicated through a distinct mechanism.

Nonhuman animal studies also provide evidence for increased mosquito attractiveness of *Plasmodium*-infected individuals. De Moraes et al.[18] for example, evaluated whole-body volatiles from healthy and *P. chabaudi*–infected mice throughout the course of infection. Investigators found that *A. stephensi* were preferentially attracted to infected mice, relative to control mice, during the period in which mice harbored relatively high levels of gametocytes. The investigators also observed increased attraction to gametocyte-positive versus gametocyte-negative individuals during this period. Alongside these behavioral studies, the investigators observed distinct body odor profiles between healthy and *Plasmodium*-infected individuals, during both acute and chronic stages of infection. The characteristic components of *Plasmodium*-infected odor profiles were identified by MS and subsequently confirmed to increase mosquito attraction individually, when added to the odor of healthy mice. Specifically, hexanoic acid, 2- and 3-methyl butanoic acid, and tridecane were increased in abundance in the odor profiles of malaria-infected animals and displayed a direct relationship with mosquito attractiveness. Conversely, benzothiazole, present in reduced amounts in chronically infected mice, was inversely related to mosquito attraction.

While several studies have focused on identifying the changes in odor profiles resulting from *Plasmodium* infection, fewer studies have focused on the origin of these changes. Three hypotheses currently exist for the basis of *Plasmodium*-dependent odor changes: (1) *Plasmodium* infection leads to changes in the skin microbiome that indirectly change the host odor profile; (2) *Plasmodium* infection stimulates endogenous host changes that alter the host odor profile; and (3) *Plasmodium* spp. directly generate and release malaria-associated volatile (odoriferous) compounds.

The direct production of *Plasmodium*-volatile compounds by *Plasmodium* garnered attention as Kelly et al.[19] identified the plant-like terpenes α-pinene and limonene as arising from cultured *P. falciparum*–erythrocytes. Apicomplexan parasites, including *Plasmodium* species, contain an apicoplast, an organelle with a similar endosymbiotic evolutionary origin to plant chloroplasts, which synthesizes isoprenoids (such as α-pinene and limonene) via the methylerythritol phosphate (MEP) pathway. This pathway is not present in animals, though a parallel metabolic route, the mevalonate pathway, does exist. Kelly et al.[19] hypothesized that *P. falciparum* parasites might utilize the MEP pathway to produce terpenes and indeed observed that inhibition of the pathway via the MEP pathway–specific inhibitor (fosmidomycin) ablated accumulation of α-pinene and limonene. Interestingly, Emami et al.[20] found that, even in the absence of *P. falciparum* parasites, introduction of the isoprenoid precursor (E)-4-hydroxy-3-methyl-but-2-enyl pyrophosphate (HMBPP) altered erythrocyte headspace volatile profiles and increased mosquito attractiveness. Specifically, HMBPP-treated erythrocytes produced higher headspace levels of aldehydes (octanal, nonanal, and decanal) and monoterpenes (α-pinene, β-pinene, and limonene). This result remains surprising, as there is no clear biosynthetic route to monoterpenes in human erythrocytes and the metabolic origin of these compounds in erythrocytes is enigmatic. There is some overlap in the types of compounds (terpenes and aldehydes) found in *P. falciparum*–infected and HMBPP-treated erythrocytes. Indeed, as noted below, several of these compounds have been also detected in the odor profiles of malaria-infected human subjects.[21–24]

Given the important role of host odor profiles on *Anopheles* spp. mosquito attraction, the finding that *Plasmodium* spp. infection consistently alters mosquito attraction provides an important proof of concept that malaria parasites reproducibly alter host volatile profiles. The current literature on breath and skin odor changes during asymptomatic and symptomatic *Plasmodium* spp. infection in humans is reviewed below.

22.3 Breath odor profiles in asymptomatic malaria

While acute, uncomplicated *P. falciparum* infection in children is most often characterized by fever, a large proportion of *P. falciparum* infections in semi-immune individuals in highly endemic regions will be asymptomatic. As individuals with asymptomatic infections do not present for care, asymptomatic infections are an important public health concern as they represent a hidden reservoir that contributes to persistent malaria transmission. Detecting submicroscopic infections requires sensitive molecular diagnostic methods, such as polymerase chain reaction (PCR) or loop-attenuated isothermal amplification (LAMP). These molecular diagnostic tests have shown that even in regions of low endemicity, asymptomatic *Plasmodium* spp. infection is common, representing up to 75% of positive individuals in community surveys.[25]

To identify biomarkers of asymptomatic malaria, odor profiles of breath and skin have been characterized both under natural field conditions in malaria-endemic regions and during controlled human malaria infection (CHMI). During CHMI, volunteers receive a direct intravenous inoculation of *Plasmodium*-infected erythrocytes, followed by close monitoring and treatment. CHMI trials are increasingly being used to aid vaccine and drug development.

The first study on breath biomarkers of *P. falciparum* CHMI volunteers was published by Berna et al.[26] One liter of breath was collected from each volunteer as malaria infection progressed and additional samples were taken following antimalarial administration. Exhaled breath was collected using sampling bags, and volatiles were then transferred from the bags to sorbent tubes (Tenax/Unicarb) via a pump. The authors identified nine compounds whose concentrations varied significantly over the course of malaria infection: carbon dioxide, isoprene, acetone, benzene, cyclohexanone, and four thioethers (allyl methyl sulfide, 1-methylthio-propane, (Z)-1-methylthio-1-propene, and (E)-1-methylthio-1-propene) (Table 22.1). The malaria-associated thioethers were of particular interest as potential disease biomarkers because they have not been previously associated with any pathological condition and because their concentrations changed during infection for all individuals. Machine learning methods were further applied, which accurately classified all samples into "active infection" and baseline/post-*P. falciparum* clearance on the basis of thioester levels. Of note, due to the nature of CHMI, parasite counts were quite low and gametocytes, which take approximately 2 weeks to develop, were never detected. Berna et al.[26] attempted to detect thioethers in in vitro cultures of *P. falciparum* but did not find any appreciable levels. While the metabolic origin of breath thioethers found during CHMI is not known, these results suggest that interplay between host and parasite metabolic pathways may be required. In a follow-up study, these same researchers[21] found that thioether concentrations in breath exhibit a diurnal cyclical pattern and, in general, thioether levels are significantly higher in *P. falciparum* CHMI volunteers compared to healthy control individuals. Moreover, the authors demonstrated that breath volatiles have a time-of-day variation that impacts the ability to predict *P. falciparum* infection using the thioethers. Additionally, this study found that terpenes (α-terpinene, m-cymene, limonene, and terpinolene) were elevated in the breath of *P. falciparum*—infected individuals. Of these terpenes, limonene had been previously reported to be associated with cultured asexual *P. falciparum.*[19] The top two compounds with the highest classification accuracies (healthy vs. *P. falciparum*) were m-cymene (92.7% correct classification), followed by terpinolene (87.7% correct classification).

Malaria-associated volatile biomarkers have also been investigated in skin samples. De Boer et al.[23] collected skin odor samples from two cohorts of *P. falciparum* CHMI volunteers. Skin odors were collected by placing one foot of the volunteer

Table 22.1 Summary of studies on individuals with asymptomatic *Plasmodium* spp. infection.

Sample type	Compounds	Study population	Reference
Breath	(E)-1-Methylthio-1-propene ↑ (Z)-1-Methylthio-1-propene ↑ 1-Methylthio-propane ↑ Allyl methyl sulfide ↑ α-Terpinene ↑ m-Cymene ↑ Limonene ↑ Terpinolene ↑	Controlled human malaria infection (malaria-naïve adults)	Berna et al.[21,26]
Skin	Unidentified sesquiterpene ↑ 1-Dodecene ↑ 2-Methylbutanal ↑ Dodecanal ↑	Controlled human malaria infection (malaria-naïve adults); foot odors	de Boer et al.[23]
	Foot **Nonanal ↓** **2-Ethylhexan-1-ol ↓** Benzaldehyde ↓ 1-Ethyl-3-methylbenzene ↓ **Toluene ↓** **4-Hydroxy-4-methylpentan-2-one ↓** **Ethylbenzene ¦** **Hexanal ↓** Arm **2-Ethylhexan-1-ol ↓** **Ethylbenzene ↑** **Toluene ↓** Dodecane ↓ **Octanal ¦** Octane ¦ 2,4-Dimethylhept-1-ene ↓	Naturally infected school children; foot/arm odors	De Moraes et al.[24]
	(E)-2-Decenal ↑	Naturally infected school children; foot/arm odors	Robinson et al.[17,*]

Bold text: volatiles were also found in subjects with symptomatic Plasmodium *infection (Table 22.2).* ↑ *indicates levels of volatile increased with infection;* ↓ *indicates volatile levels decreased with infection;* ¦ *indicates direction of change was not provided.* [*]*Volatiles reported in this table are based on comparisons relative to parasite-free individuals versus all individuals with detectable* Plasmodium *spp. parasitemia.*

into a clean bag, and volatiles from the bag were then transferred to Tenax and Porapak filters. Skin odor samples were collected 2 days prior to parasite challenge, during *Plasmodium* infection, and posttreatment. Foot odor profiles were distinct in *P. falciparum*–challenged individuals. Table 22.1 lists those volatile compounds that showed significant differences "before treatment" versus "during malaria infection" (qPCR-positive for *P. falciparum*). Several compounds (e.g., an unidentified sesquiterpene, 1-dodecene, 2-methyl butanal, and dodecanal) increased significantly upon infection. Aldehydes have been previously found in mammalian skin odors (see Chapter 25) and are well recognized as host attractants for hematophagous insects. The authors hypothesize that the increased aldehyde levels found in malarious samples originate from lipid peroxidation, caused by oxidative stress induced by *P. falciparum*. Alternatively, the authors postulate that the aldehydes might be produced directly by *Plasmodium* parasites, as aldehydes are emitted by HMBPP-treated erythrocyte cultures.[20] In parallel with compound identification, researchers performed dual-choice olfactometer experiments to determine whether changes in body odor during *P. falciparum* infection affect mosquito behavior. Unexpectedly, they found reduced attractiveness of parasite-positive participants in one cohort and no significant effect of *P. falciparum* infection in a second group. This discrepancy was attributed to the use of different parasite strains in both cohorts. Mature gametocytes were not detected in any of the participants during any portion of this study.

A more recent study also aimed to identify malaria-associated skin odor biomarkers, through evaluation of more than 400 primary school children (aged $\leq$12 y) in Western Kenya.[24] In this study, investigators profiled arm and foot volatiles from each individual, using a portable volatile collection system. Collection from both sites occurred in parallel, prior to treatment of malaria-positive individuals, and odors were stored on HayeSep adsorbent polymer prior to gas chromatography−MS (GC-MS). Machine learning was employed to identify volatile patterns that predicted malaria infection (Table 22.1). Of these, only 2-ethylhexan-1-ol and ethylbenzene were found in both foot and arm samples. Critically, predictive models successfully identified asymptomatic infections with 100% sensitivity in foot samples and 75% sensitivity in arm samples. The majority of the identified foot volatiles increased in concentration during infection, whereas the direction of the change was not specified for the majority of skin volatiles. The authors indicate that all identified compounds have either been previously reported from human volatile collections or have known mechanisms of natural production from humans or potentially human-associated microbes. Interestingly, none of the volatiles reported in this study were also reported from skin volatile profiling from asymptomatic CHMI volunteers.[23]

To explore the molecular basis of odor manipulation by malaria, Robinson et al.[17] both quantitatively and qualitatively compared the volatile compounds emitted from the feet of asymptomatic infected children in Western Kenya. This work was done in parallel with the mosquito attraction work mentioned earlier in this chapter. Of note is that the study site was the same as De Moraes[24] and Busula.[15] For each child, one foot was placed in a plastic bag, and volatiles were collected in Porapak filters (similar to de Boer et al.[23]) and sampled for 100 min (at 500 mL/min). Detailed infection status (uninfected, infected with low parasitemia, high parasitemia, or infected with gametocytes) was collected using an 18S-based qPCR for *P. falciparum* asexual stages and QT-NASBA qPCR for quantifying gametocytes.

(E)-2-Decenal was the only compound that showed significant differences if individuals were categorized simply as *Plasmodium*-positive or parasite-free (Table 22.1). The analysis also revealed higher abundance in the levels of the aldehydes heptanal, octanal, nonanal, (E)-2-octenal, and (E)-2-decenal by infected asymptomatic individuals compared to control (solvent) or empty bag. A positive trend on the levels of these VOCs was also observed when associated to the parasite densities (low and high). Additionally, the ketone 2-octanone was found to be associated with the presence of microscopic gametocytes. Although a similar collection technique for volatile organic compounds (VOCs) was used by de Boer et al.[23] (from the same research group), none of the compounds identified were reported in this work. Inconsistencies could be due to the low parasitemia present in de Boer et al.,[23] as well as the absence of gametocytes.

Taken together, the results from this collection of recent studies on volatiles emitted from skin and breath clearly establish that malaria infection is associated with changes in volatile profiles. Of the malaria-associated VOCs reported (Table 22.1), only nonanal was detected in two independent studies of individuals with asymptomatic parasitemia. However, the two studies problematically disagree about the direction of change with infection: in one study, nonanal levels increase with malaria infection[17]; in the other, nonanal levels decrease.[24] A possible source of this discrepancy may be the variability in volatile collection methods, absorbent materials, and/or analytical techniques employed. While diet is thought to influence breath and skin volatile profiles, both studies in question recruited from the same locality in Western Kenya. More broadly, the use of malaria-naïve individuals (as in CHMI) versus naturally infected individuals (in endemic areas) also presents challenges for comparison purposes, as study participants in field studies are likely to have had previous malaria exposure.

22.4 Breath odor profiles in symptomatic *Plasmodium* infection

The first report of candidate diagnostic biomarkers in the breath of symptomatic *P. falciparum*—infected children from a typical malaria-endemic clinical setting was published recently by Schaber et al.[22] Samples were collected in Malawi from children 3—15 years old presenting for care for fever. In the study, 1 L of exhaled breath was collected in a sample bag and transferred to a sorbent tube (Tenax/Carbograph/Carboxen). Investigators found global differences in breath VOC composition based on infection status. In addition, six breath volatiles were highly correlated with infection status and together yielded a classification accuracy of 83%: methyl undecane, dimethyl decane, trimethyl hexane, nonanal, isoprene, and tridecane (Table 22.2). Of these six VOCs, methyl undecane and dimethyl decane levels increased with malaria infection. The only compound that had previously been associated with *Plasmodium* infection was nonanal.

Table 22.2 Summary of studies on individuals with symptomatic *Plasmodium* spp. infection.

Sample type	Compounds	Patients/ volunteers	Reference
Breath	Dimethyl decane ↑ Isoprene ↓ Methyl undecane ↑ **Nonanal ↓** Tridecane ↓ Trimethyl hexane ↓ 3-Carene ↑ α-Pinene ↑	Naturally infected children (age 3–15)	Schaber et al.[22]
Skin	Foot **Hexanal ↓** **4-Hydroxy-4-methylpentan-2-one ↓** **Toluene ↓** Ethylcyclohexane ↑ **Ethylbenzene ¦** 1-Ethyl-3-methylbenzene ↑ Arm **Octanal ↓** **2-Ethylhexan-1-ol ↓** m-Xylene (or p-xylene) ↓ **4-Hydroxy-4-methylpentan-2-one ↓** **Toluene ↓** Ethylcyclohexane ↑ **Ethylbenzene ¦** 2,4-dimethylhept-1-ene ¦	Naturally infected school-aged children; foot/arm odors	De Moraes et al.[24]

Bold text: volatiles were also found in subjects with asymptomatic Plasmodium *infection (Table 22.1). ↑ indicates levels of volatile increased with infection; ↓ indicates volatile levels decreased with infection; ¦ indicates direction of change was not provided.*

Plasmodium-infected individuals have decreased levels of skin-emitted nonanal, when arm volatiles were sampled.[24] In contrast, skin-emitted nonanal from the feet of *Plasmodium*-infected individuals was elevated compared to controls.[17]

Schaber et al.[22] also found significantly increased breath levels of two terpenes, α-pinene ($P = .04$, with 20% higher mean), and 3-carene ($P = .01$, with a 28% higher mean). α-Pinene has been observed reproducibly to arise during *P. falciparum* asexual infection of erythrocytes in culture.[19] In addition, this terpene is a direct, potent, and specific activator of *A. gambiae* odorant receptors.[19] Both α-pinene and the related 3-carene are among the volatiles produced by mosquito-preferred nectar-providing plant species.[27] The investigators postulate that because malaria-induced volatiles are chemically identical to those produced by mosquito-preferred plants, parasites might produce or induce production of these volatiles in order to hijack mosquito behavior and increase transmission.

Of note, the results of this research group were somewhat distinct from the previous breath metabolite findings reported by Berna et al.[26] from a population of naïve healthy adults undergoing CHMI *P. falciparum* infection. Thioethers were not detected, suggesting that parasite densities, parasite stage, or age of host might induce a range of physiological changes in the human body that manifests in the breath and body odor. Alternatively, *P. falciparum* may produce volatiles in a density- or stage-specific manner. Prior parasite exposure may also be required for host-generated volatiles produced during *P. falciparum* infection. The different sorbent material and storage conditions used in both studies may also contribute to an absence of thioethers in the work of Schaber et al.[22]

De Moraes et al.[24] also examined symptomatic *Plasmodium*-infected children at a primary school located in Western Kenya. In this work, researchers collected foot and arm volatiles and employed machine learning algorithms to develop predictive models for infection status. The following volatiles were found to be key predictors of malaria: 4-hydroxy-4-methylpentan-2-one, toluene, ethylcyclohexane, and ethylbenzene (Table 22.2). Toluene is notable in that it has previously been reported to be produced by *P. falciparum* in vitro[19] and has also been found to be associated with human skin. De Moraes et al.[24] suggest that toluene could be produced by *Clostridium* spp. residing in the human microbiome.[28] Most of the volatiles found by De Moraes et al.[24] in arm and foot samples were in lower abundance in *Plasmodium*-infected individuals. Importantly, seven of the volatiles reported in symptomatic malaria cases were also detected in samples from asymptomatic *Plasmodium*-infected individuals (VOCs in bold text—Tables 22.1 and 22.2). Predictive models using foot volatiles exhibited greater sensitivity (91%) than using arm volatiles (89%). It is important to note that these sensitivities were achieved using samples that were collected over 3 years across 41 schools in Western Kenya. Additionally, some subjects were coinfected with multiple *Plasmodium* species and, in some cases, were coinfected with other organisms, including human immunodeficiency virus (HIV) and intestinal helminths.

No volatiles were found in common among studies with symptomatic patients, possibly due to variations in collection methods and/or materials.

22.5 Summary

Despite substantial global investment, malaria remains a serious global health problem. Young infants and children are particularly at risk, with more than 400,000 deaths each year. Over the past decade, RDTs have transformed malaria diagnosis and have been instrumental to malaria control efforts. Nevertheless, there remains an urgent need to develop new malaria diagnostics. Because parasite proteins can persist in the bloodstream long after treatment, "false-positive" tests are common in children who do not have malaria, meaning that nonmalaria infections can be missed and untreated. Even more ominously, variant parasite strains have emerged, such that in some parts of the world, 80% of parasites are no longer detected by current tests. There is a pressing need for highly sensitive and specific malaria diagnostics that are also simple and affordable. For public health campaigns, noninvasive testing would also represent a major advance, since all current malaria tests require blood samples.

There is mounting evidence that malaria parasites affect the behavior of *Anopheles* mosquito vectors and hosts in ways that increase the contacts between them to favor parasite transmission. Such changes in attractiveness have been demonstrated in both animal and human malaria systems, as well as in other vector-borne disease systems. Body odor, comprising the volatile compounds emitted from the skin and breath of vertebrates, is the most important cue used by *Anopheles* for host location. While increased attractiveness of *Plasmodium*-infected individuals has been demonstrated in a malaria-endemic setting, remarkably very few studies have investigated the chemical ecology underlying this phenomenon. Volatiles emitted from human skin, breath, and cultured human red blood cells are all altered in the presence of *Plasmodium* spp. infection. Of note, the identified volatile biomarkers have been highly variable from study-to-study. These differences may reflect variability in volatile collection methods, absorbent materials, and/or analytical techniques employed. In addition, study populations were distinct with respect to prior exposure to malaria and duration of infection, which may influence the volatile profiles due to presence/absence of gametocytes and parasite densities. Nevertheless, the reproducible finding that malaria induces volatile changes provides compelling hope for a future malaria diagnostic that identifies both asexual parasitemia and the presence of gametocytes. Future work should include research in different geographical regions using collection methods and analytical techniques similar to those used in previous studies. In addition, a compelling question is how diet, age of host, genetic, and environmental factors affect the volatile fingerprint of malaria-infected individuals. The biological origin of those volatile compounds induced by malaria are yet unknown and remains an outstanding question of importance in understanding the specificity of these volatiles as biomarkers to be used in a noninvasive diagnostic for malaria.

References

1. WHO. *World health organization, global malaria programme*. Geneva: World malaria report; 2019.
2. Tangpukdee N, Duangdee C, Wilairatana P, Krudsood S. Malaria diagnosis: a brief review. *Kor J Parasitol* 2009;**47**(2):93–102.
3. Kobayashi T, Gamboa D, Ndiaye D, Cui LW, Sutton PL, Vinetz JM. Malaria diagnosis across the international centers of excellence for malaria research: platforms, performance, and standardization. *Am J Trop Med Hyg* 2015;**93**(3):99–109.

4. Walk J, Schats R, Langenberg MCC, Reuling IJ, Teelen K, Roestenberg M, et al. Diagnosis and treatment based on quantitative PCR after controlled human malaria infection. *Malar J* 2016;**15**.
5. Wilson ML. Malaria rapid diagnostic tests. *Clin Infect Dis* 2012;**54**(11):1637−41.
6. Gamboa D, Ho MF, Bendezu J, Torres K, Chiodini PL, Barnwell JW, et al. A large proportion of *P. falciparum* isolates in the Amazon region of Peru lack pfhrp2 and pfhrp3: implications for malaria rapid diagnostic tests. *PLoS One* 2010;**5**(1).
7. Berhane A, Anderson K, Mihreteab S, Gresty K, Rogier E, Mohamed S, et al. Major threat to malaria control programs by *Plasmodium falciparum* lacking histidine-rich protein 2, Eritrea. *Emerg Infect Dis* 2018;**24**(3):462−70.
8. Beshir KB, Sepulveda N, Bharmal J, Robinson A, Mwanguzi J, Busula AO, et al. *Plasmodium falciparum* parasites with histidine-rich protein 2 (pfhrp2) and pfhrp3 gene deletions in two endemic regions of Kenya. *Sci Rep UK* 2017;**7**.
9. Cheng Q, Gatton ML, Barnwell J, Chiodini P, McCarthy J, Bell D, et al. *Plasmodium falciparum* parasites lacking histidine-rich protein 2 and 3: a review and recommendations for accurate reporting. *Malar J* 2014;**13**.
10. Koita OA, Doumbo OK, Ouattara A, Tall LK, Konare A, Diakite M, et al. False-negative rapid diagnostic tests for malaria and deletion of the histidine-rich repeat region of the hrp2 gene. *Am J Trop Med Hyg* 2012;**86**(2):194−8.
11. Watson OJ, Slater HC, Verity R, Parr JB, Mwandagarlirwa MK, Tshefu A, et al. Modelling the drivers of the spread of *Plasmodium falciparum* hrp2 gene deletions in sub-Saharan Africa. *Elife* 2017;**6**.
12. Peprah S, Tenge C, Genga IO, Mumia M, Were PA, Kuremu RT, et al. A cross-sectional population study of geographic, age-specific, and household risk factors for asymptomatic *Plasmodium falciparum* malaria infection in western Kenya. *Am J Trop Med Hyg* 2019;**100**(1):54−65.
13. Takken W, Knols BGJ. Odor-mediated behavior of afrotropical malaria mosquitoes. *Annu Rev Entomol* 1999;**44**(1):131−57.
14. Raji JI, DeGennaro M. Genetic analysis of mosquito detection of humans. *Current opinion in insect science* 2017;**20**:34−8.
15. Busula AO, Bousema T, Mweresa CK, Masiga D, Logan JG, Sauerwein RW, et al. Gametocytemia and attractiveness of *Plasmodium falciparum*-infected Kenyan children to *Anopheles gambiae* mosquitoes. *J Infect Dis* 2017;**216**(3):291−5.
16. Lacroix R, Mukabana WR, Gouagna LC, Koella JC. Malaria infection increases attractiveness of humans to mosquitoes. *PLoS Biol* 2005;**3**(9):e298.
17. Robinson A, Busula AO, Voets MA, Beshir KB, Caulfield JC, Powers SJ, et al. Plasmodium associated changes in human odor attract mosquitoes. *Proc Natl Acad Sci Unit States Am* 2018;**115**(18):E4209−18.
18. De Moraes CM, Stanczyk NM, Betz HS, Pulido H, Sim DG, Read AF, et al. Malaria-induced changes in host odors enhance mosquito attraction. *Proc Natl Acad Sci USA* 2014;**111**(30):11079−84.
19. Kelly M, Su C-Y, Schaber C, Crowley JR, Hsu F-F, Carlson JR, et al. Malaria parasites produce volatile mosquito attractants. *mBio* 2015;**6**(2).
20. Emami SN, Lindberg BG, Hua S, Hill SR, Mozuraitis R, Lehmann P, et al. A key malaria metabolite modulates vector blood seeking, feeding, and susceptibility to infection. *Science* 2017;**355**(6329).

21. Berna AZ, McCarthy JS, Wang XR, Michie M, Bravo FG, Cassells J, et al. Diurnal variation in expired breath volatiles in malaria-infected and healthy volunteers. *J Breath Res* 2018;**12**(4).
22. Schaber C, Katta N, Bollinger LB, Mwawi M, Mlotha-Mitole R, Trehan I, et al. Breathprinting reveals malaria-associated biomarkers and mosquito attractants. *J Infect Dis* 2018;**217**(10):1553–60.
23. de Boer JG, Robinson A, Powers SJ, Burgers SLGE, Caulfield JC, Birkett MA, et al. Odours of *Plasmodium falciparum*-infected participants influence mosquito-host interactions. *Sci Rep UK* 2017;**7**.
24. De Moraes CM, Wanjiku C, Stanczyk NM, Pulido H, Sims JW, Betz HS, et al. Volatile biomarkers of symptomatic and asymptomatic malaria infection in humans. *Proc Natl Acad Sci USA* 2018;**115**(22):5780–5.
25. Bousema T, Okell L, Felger I, Drakeley C. Asymptomatic malaria infections: detectability, transmissibility and public health relevance. *Nat Rev Microbiol* 2014;**12**(12): 833–40.
26. Berna AZ, McCarthy JS, Wang RX, Saliba KJ, Bravo FG, Cassells J, et al. Analysis of breath specimens for biomarkers of *Plasmodium falciparum* infection. *J Infect Dis* 2015; **212**(7):1120–8.
27. Nyasembe VO, Tchouassi DP, Kirwa HK, Foster WA, Teal PEA, Borgemeister C, et al. Development and assessment of plant-based synthetic odor baits for surveillance and control of malaria vectors. *PLoS One* 2014;**9**(2).
28. Peng G, Hakim M, Broza YY, Billan S, Abdah-Bortnyak R, Kuten A, et al. Detection of lung, breast, colorectal, and prostate cancers from exhaled breath using a single array of nanosensors. *Br J Canc* 2010;**103**(4):542–51.

CHAPTER

Lifestyle applications

23

Sebastian Abegg[1], Andreas T. Güntner[1,2]

[1]*Particle Technology Laboratory, Department of Mechanical and Process Engineering, ETH Zurich, Zurich, Switzerland;* [2]*Department of Endocrinology, Diabetes, and Clinical Nutrition, University Hospital Zurich, Zurich, Switzerland*

Chapter outline

23.1 Overview

Winning the fight against obesity and related metabolic diseases, such as diabetes and nonalcoholic fatty liver disease (NAFLD), is one of the biggest challenges modern society currently faces. The prevalence of obesity (i.e., body mass index (BMI)$\geq$ 30 kg/m^2) worldwide approximately tripled over the past three decades from 4.5% in 1975 to 13% in 2016, with an estimated 1.9 billion adults and 340 million children being classified as obese (WHO). Despite a growing diagnostic toolset and a plethora of preventative and therapeutic interventions to tackle the consequences of an obesogenic environment, no real success has been demonstrated, to date. Energy expenditure by physical activity and exercise, and dietary intervention, is

Breathborne Biomarkers and the Human Volatilome. https://doi.org/10.1016/B978-0-12-819967-1.00023-2
Copyright © 2020 Elsevier B.V. All rights reserved.

simply too outbalanced by energy intake, causing an epidemic growth of overweight and obese populations.

Monitoring physical activity and nutrition is still a challenge due to a lack of point-of-care (PoC) user feedback. In particular, the complex interaction and strongly varying metabolic reactions of physical activity and nutrition make it currently impossible to guide the user in a personalized way. Nevertheless, continuous monitoring of outcome parameters during treatment is highly important to assess success and provide guidance during therapeutic interventions. Thus, dependable tools for daily, personal, and convenient (thus noninvasive) monitoring of metabolic status are highly desirable. Such technologies must be portable and inexpensive to offer widespread and affordable application.

So far, the use of continuous or repetitive self-monitoring by the layperson is mostly restricted to simple anthropometric (e.g., weight, body composition), blood circulatory (e.g., heart rate, pulse oximetry), or lifestyle measures (e.g., food intake, pedometers). Direct measurement of metabolic parameters (e.g., β-hydroxybutyrate [BOHB] or glucose) in everyday life is only possible by blood measurement via fingerpricking or by subcutaneously implanted glucose sensors. These are invasive and expensive, thus are primarily used in risk groups or the diseased population. Also, measurement of urinary acetoacetate (AcAc) is inconvenient, not always accessible, and rather inaccurate.[1] New technologies that allow the routine and personal analysis of metabolic data inexpensively, conveniently, and noninvasively, but accurately, are therefore desirable.

Nowadays, the focus lays on the analysis of noninvasively accessible body fluids like sweat, saliva, or tears. Exhaled breath is promising as it is rich in physiological information, always accessible, and least invasive compared to blood assays. Breath is an ideal medium for easy and routine measurements in lifestyle applications. The capacity of breath analysis to recognize metabolic alterations immediately by tracing related molecules, i.e., breath markers, in real time can guide obesity treatment by providing PoC feedback. This chapter discusses state-of-the-art knowledge of breath markers and the use of breath monitoring in exercise and dieting and highlights advances and challenges of suitable breath analysis technologies. A particular focus is given to breath acetone, a volatile by-product of lipolysis. Breath studies on dieting and physical activity are discussed and compared to established blood ketone markers and indirect calorimetry. Finally, an overview of selected portable technologies with proven performance in human breath analysis is presented. Therein, challenges and advances toward integration into portable breath analyzers for guiding exercise and diet are highlighted.

23.2 Breath markers, biochemical pathways, and sampling

Exhaled human breath is composed of nitrogen, oxygen, carbon dioxide, water, inert gases, and a small fraction of over 800 volatile organic compounds (VOCs) (see Chapter 1). The latter are present at low concentrations, ranging from parts per trillion (ppt) to parts per million (ppm). To monitor metabolic conditions, endogenous substances are particularly attractive, as discussed in the following.

23.2.1 Oxygen and carbon dioxide

Monitoring exercise and diet via breath is an established methodology as an indirect estimate of the carbohydrate and fat metabolism, known as indirect calorimetry. Here, the respiratory gas exchange ratio (RER) is measured, which is defined as the ratio of carbon dioxide exhalation (VCO_2) to oxygen uptake (VO_2).[2] The human body converts nutritional substrates like carbohydrates, fat, and proteins into the energy-intermediate adenosine triphosphate (ATP), whereby the amounts of consumed O_2 and produced CO_2 differ depending on the substrate.[2] The stoichiometry of the complete oxidation of glucose ($C_6H_{12}O_6$) is given as[2]

$$C_6H_{12}O_6 + 6\ O_2 \rightarrow 6\ CO_2 + 6\ H_2O$$

Accordingly, RER for glucose oxidation is parity. A similar RER is also obtained for other carbohydrates, e.g., sucrose. For oxidation of a typical fat,[2] the reaction stoichiometry is

$$C_{55}H_{104}O_6 + 78\ O_2 \rightarrow 55\ CO_2 + 52\ H_2O$$

The corresponding RER for fat is about 0.7; for proteins, RER is approximately 0.8.[2]

Different indirect calorimetry systems are available. These include (1) confinement systems, where the subject is enclosed in a chamber; (2) closed-circuit systems that apply CO_2 and moisture adsorbers/absorbers, and typically measure only O_2 consumption; (3) total collection systems, where the exhaled breath is sampled in a bag to measure its volume and composition; and (4) open-circuit systems that assess the air flow and measure the composition of a subject breathing ambient air. Handheld indirect calorimeters to measure the resting energy expenditure are available nowadays, but they suffer from limited accuracy compared to benchtop devices.

23.2.2 Acetone

Exhaled acetone is one of the most abundant breath VOCs and is related to fat metabolism. Specifically, in the hepatic mitochondria, fatty acids are converted through β-oxidation to acetyl coenzyme A (acetyl-CoA), which can be transformed into AcAc,[3] as shown in Fig. 23.1A. From the latter, the ketones acetone and BOHB are formed through spontaneous decarboxylation and enzymatic degradation, respectively.[3] The coefficient of determination between breath acetone and blood BOHB averaged over several studies is $R^2 = 0.77$.[3] A key advantage of acetone over BOHB and AcAc is its high volatility. As a result, it can be measured in exhaled human breath. Breath acetone concentrations range from 0.15 to 2.7 ppm in healthy, nonfasting individuals[4] and can increase up to several hundred ppm during diabetic

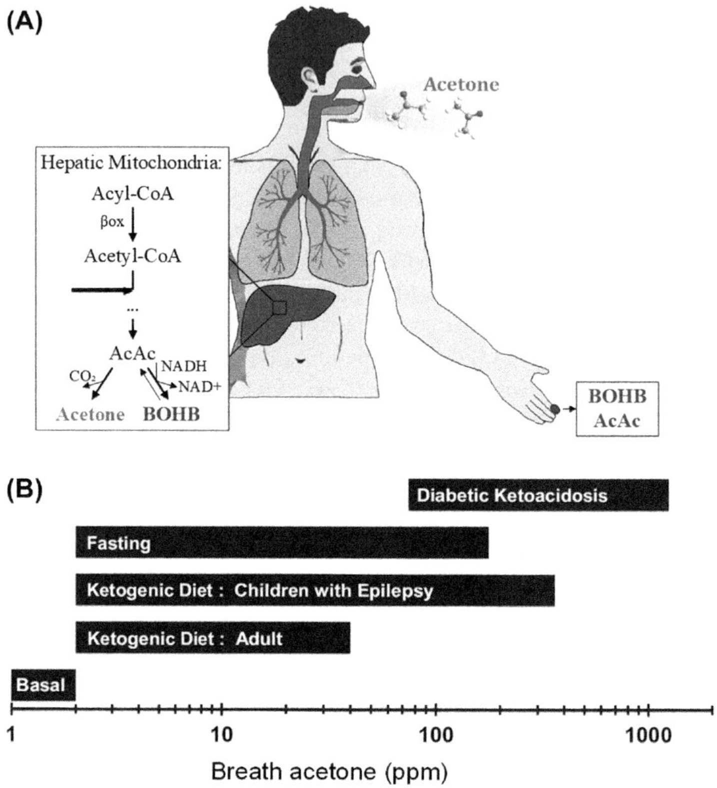

FIGURE 23.1

(A) Metabolic pathway of acetone and β-hydroxybutyrate (BOHB) formation in the hepatic mitochondria (box). Acetone is volatile and can be measured in breath while BOHB and AcAc can be measured in the blood. (B) Range of breath acetone concentrations for different conditions.

(A) Adapted from [5]; (B) Reproduced from [3].

ketoacidosis.[3] During ketogenic diets, concentrations typically range from 2 to 40 ppm in adults.[3] They might be higher in children (up to 360 ppm) and during fasting (up to 170 ppm), as had been mapped by Anderson et al.[3] in Fig. 23.1B.

23.2.3 Isoprene

Exhaled isoprene might be a by-product of the cholesterol biosynthetic pathway.[6] Typical breath concentrations range from 22 to 234 ppb,[7] with lower levels found in young adults.[8] Most interestingly, a correlation between breath isoprene and blood cholesterol was proposed for patients treated with lovastatins[6] that lower cholesterol. Also, an extrahepatic production site in muscle tissue had been suggested based on observations that breath isoprene levels rapidly increase during muscle activation[9] (see Chapter 3) but not in muscle dystrophy patients.[10]

23.3 Breath analysis in diet monitoring and exercise

23.3.1 Dieting

Dieting is an effective way to lose weight by establishing a negative energy balance. In the ketogenic diet, for instance, carbohydrate intake is restricted, forcing the body to use fat as a primary energy source. It is important to monitor the status of ketosis during a ketogenic diet to ensure its effectiveness. Breath acetone has been proposed as a noninvasive surrogate for the currently applied plasma BOHB and urinary AcAc measurements.[1] Higher breath acetone concentrations were observed for participants undergoing high-fat (79% and 90%) compared to low-fat (29%) diets,[11] reflecting increased fat metabolism during ketogenic diet.

The onset of ketosis during the first 36 h of a ketogenic diet was studied in relation to breath acetone and plasma BOHB concentrations (protocol shown in Fig. 23.2A).[5] During the first 12 h, breath acetone concentrations typically increased by a factor of three (Fig. 23.2B), in agreement with previous reports.[1] Thereafter, breath acetone increased more markedly, for some volunteers up to 60 ppm after 36 h (volunteer #5 in Fig. 23.2B). By comparison, blood BOHB (Fig. 23.2C) indicated that most individuals entered nutritional ketosis (0.5−3 mM) after 9 h, with some exhibiting mild ketosis (2−7 mM) after 36 h of the ketogenic diet. Most interestingly, the volunteers responded quite differently to the same ketogenic diet conditions, emphasizing the need for personalized guidance by repetitive and longitudinal monitoring. Furthermore, breath acetone and blood BOHB showed some differences in their dynamics (e.g., volunteer #1 in Fig. 23.2) despite similar origin from the β-oxidation of fatty acids in the hepatic mitochondria (Fig. 23.1A). Therefore, further studies on their biochemical relationship are required to scrutinize the usability of breath acetone as a marker for ketosis.

During mid- and long-term ketogenic diet, breath acetone concentrations increased up to 60 ppm $\pm$ 14 ppm in ten children when dieting for at least 1 month.[12] This was significantly elevated compared to controls without ketogenic diet (0.46 $\pm$ 0.22 ppm).[12] These breath acetone levels should be rather stable, as observed over 2 weeks in epileptic children on a ketogenic diet for several months (8−64 months) before testing.[13] Also, breath acetone concentrations of children on a ketogenic diet for at least 1 month did not change after the consumption of a ketogenic diet meal, in contrast to nondieting controls.[12]

Few studies on dieting have correlated the changes in ketone concentrations (i.e., acetone and BOHB) to fat oxidation rates determined by indirect calorimetry. For instance, fat oxidation rates during fasting (1.13 $\pm$ 0.06 mg/kg/min) and low

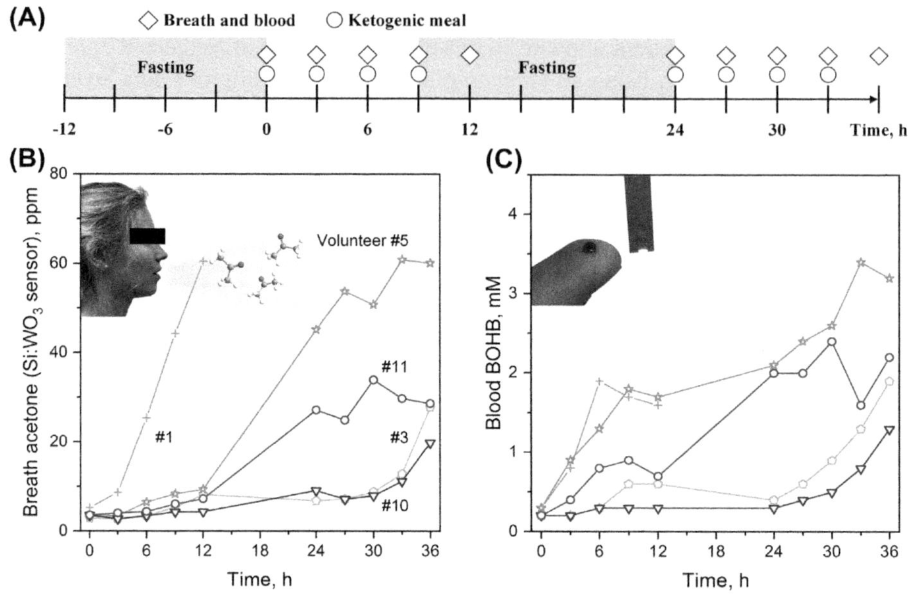

FIGURE 23.2

(A) Protocol of a 36 h ketogenic diet with regular sampling of breath and blood (diamonds) and intake of the ketogenic meals (circles). (B) Breath acetone and (C) blood β-hydroxybutyrate (BOHB) profiles shown for five volunteers.

Reproduced from[5].

carbohydrate diet (1.16 ± 0.06 mg/kg/min) were significantly higher compared to a high carbohydrate diet (0.94 ± 0.04 mg/kg/min).[14] This trend was in agreement with breath acetone concentrations, which increased during fasting and low carbohydrate diet but decreased during a high carbohydrate diet.[14] Another study investigated changes in RER in a volunteer when lowering the fat content of the diet from 82% to 75% by increasing the amount of protein supplements.[15] An increasing fat oxidation, as determined through a decreased RER from 0.84 to 0.7, was accompanied by an increase in acetone concentration at a rate of 0.3 ppm/h.[15] This motivates further studies to investigate correlations between these markers.

23.3.2 Exercise

Regular exercise is an effective way to lose and maintain a healthy weight by increasing the total energy expenditure. As with dieting, breath acetone has been suggested as a noninvasive marker to monitor the dynamics of lipolysis.[3] Significant increases in breath acetone concentrations have been observed after walking, especially in type 1 diabetic subjects.[16] Fig. 23.3 shows the cycling power (A), heart rate

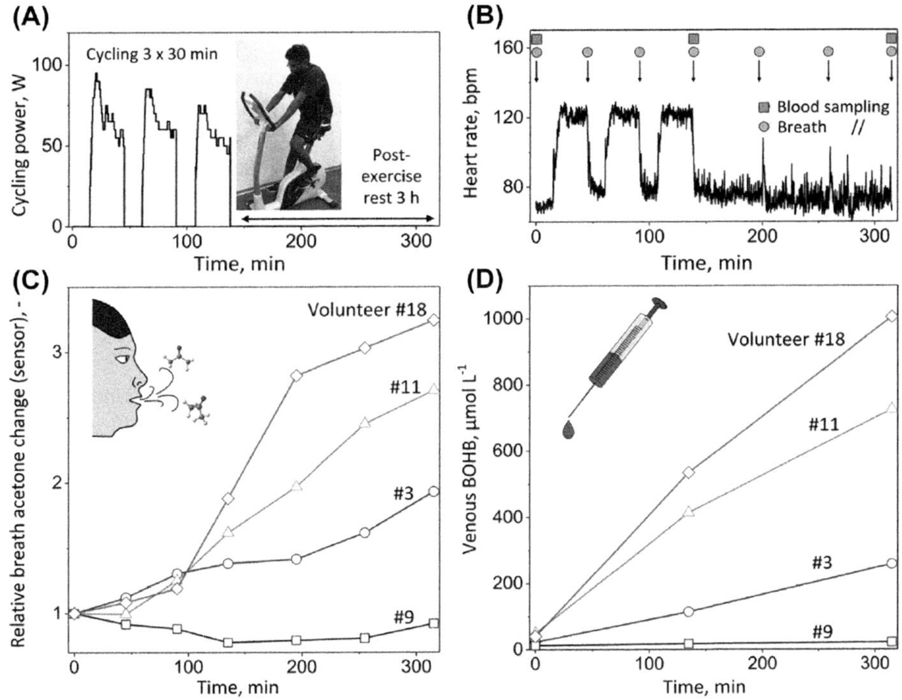

FIGURE 23.3

Individual monitoring during exercise after overnight fasting. Cycling power (A), heart rate (B), breath acetone (C), and blood β-hydroxybutyrate (BOHB) (D) dynamics of selected volunteers during 3 × 30 min cycling and subsequent resting for 3 h.

Reproduced from [17].

(B), breath acetone (C), and blood BOHB (D) dynamics in overnight-fasted subjects during 3 × 30 min of cycling and 3 h postexercise rest.[17] Volunteer #18 (diamonds) showed a threefold increase in breath acetone concentrations, with the strongest increase after exercise, which was in good agreement with blood BOHB. In contrast, volunteer #9 (squares) hardly showed any change, highlighting that the metabolic response can vary quite strongly between individuals, even when following the same protocol.

Exhaled acetone increased continuously during graded[18] and nongraded[17] exercise. Further, acetone concentration optima were observed during graded exercise, occurring when fuel preference shifted from fat and carbohydrate to carbohydrate-only, as measured by indirect calorimetry.[19] Such differences are probably related to the different exercise protocols, for instance, the total intensity during nongraded exercise or intensity increase rate during graded exercise. Portable devices that individually optimize the fat oxidation rate are therefore desirable.

Breath isoprene also has potential for metabolic monitoring, but its biochemical pathway and clinical utility must first be clarified. In pilot studies, breath isoprene concentrations peaked at the beginning of exercise and decreased to about 20% of the base levels.[19] This behavior was related to an extrahepatic isoprene storage in the muscles that is released rapidly during activity.[19] This theory is supported by the observation that such isoprene peaks did not occur in muscle dystrophy patients.[10] Ventilation-perfusion effects on exhaled isoprene levels have also been suggested.[9]

23.4 Portable breath analyzers

Portable and inexpensive breath analyzers that do not require trained personnel hold potential for widespread monitoring during exercise and dieting. Nevertheless, the detectors need to fulfill key criteria for the integration into portable devices and their application as lifestyle or even medical products. In breath analysis, relevant markers (e.g., acetone or isoprene) occur typically at ppb to ppm levels, therefore detectors must exhibit sufficient sensitivity and limit of detection for the target analyte. Further, the chosen breath marker must be selectively detectable and quantifiable in real time over the other hundreds of breath compounds at high relative humidity. Additionally, small size and low power consumption are necessities for integration into portable or even handheld, battery-driven devices. Especially the latter is crucial for affordable and widespread application. The detector performance should be stable over several months or even years, ideally without the need for recalibration. Finally, clinical relevance needs to be demonstrated if a medical application is targeted.

The development of portable detectors initially involves extensive testing and optimization of the analytical performance under laboratory conditions. To prove the applicability in everyday life, however, the breath analyzer must also be tested rigorously in clinical settings in studies involving a sufficiently large cohort size to assess accuracy, reliability, and repeatability.[20] Furthermore, parallel cross-validation with reference analytical methods, such as mass spectrometers and/or blood measurements, is essential. Finally, medical specificity and sensitivity for the intended use must be evaluated.[20] Different portable technologies with proven performance in the detection of metabolic markers in human breath are discussed below.

23.4.1 Metal oxide sensors

Chemoresistive semiconducting metal oxide (MOx) gas sensors offer simple application, high miniaturization potential, low power consumption, minimal production cost, and fast response times for real-time applications (see Chapter 13). From that perspective, MOx sensor systems represent the ideal detector type for handheld breath analyzers. In fact, state-of-the-art micromachined sensor substrates are only a few mm^2 in size. With suspended heater design to reduce thermal mass, such sensors feature power consumptions as low as few tens of mW, thereby enabling battery-driven operation in a handheld device.

Importantly, such sensors are capable of detecting VOCs down to ppb concentrations; thus, these detection limits are sufficiently low for monitoring metabolism.[20] A key drawback, however, is their limited selectivity.[20] This may be overcome, to some extent, by tuning the material characteristics (such as composition and morphology).[20] For instance, selective acetone detection down to 20 ppb was enabled by doping WO_3 with Si and thereby stabilizing the metastable monoclinic ε-phase.[21] Selectivity can be further enhanced by applying filters (e.g., microporous membranes or packed beds[22]) to exploit the molecular size and/or sorption properties of the target analyte. Alternatively, such sensors can be combined as arrays, and statistical response analysis is applied for analyte prediction. Ideally, such sensors in an array feature *orthogonal* selectivity, offering enhanced estimation stability and improved accuracy.[23] Nearly-orthogonal arrays were already successfully applied to detect entrapped humans by detecting acetone, ammonia, and isoprene, among others, with high accuracies of 19, 21 and 3 ppb, respectively (see Chapter 31).

Doped SnO_2 and Pt WO_3 sensors have previously been integrated into a simple handheld acetone analyzer featuring wireless communication to a smartphone.[24] Using a straw as a mouthpiece, a pressure sensor detected the start of exhalation. The system was validated by gas chromatography (GC) over 2 weeks, measuring the breath of 17 volunteers (238 samples, $R^2 = 0.9$), although variations in the exhalation flow from different users caused errors. Also, Si-doped WO_3 sensors showed excellent agreement with proton transfer reaction time-of-flight mass spectrometry (PTR-TOF-MS) analysis when tested on 20 volunteers during and after exercise (280 breath samples, $R^2 = 0.94$).[17] A buffered end-tidal (BET) breath sampler with controlled flowrate and CO_2 sensor for controlled and reliable breath analysis was used, and the sampler−sensor system was applied successfully to monitor breath acetone during ketogenic diet between 1 and 66 ppm (105 samples, $R^2 = 0.95$ with a quadrupole mass spectrometer).[5] Other studies applied a frontend GC on a ZnO gas sensor to enhance selectivity, whereby a miniature pump and solenoid valve were used to guide the breath sample through the column to the sensor, with results immediately reported on a display for near real-time detection of breath acetone.[25] This system was tested with three volunteers undergoing ketogenic and normal diets over 3 days, however, no cross-validation with benchtop analytical instrumentation was performed.[25]

23.4.2 Enzyme-based biosensors

Enzyme-based biosensors feature high miniaturization potential and can be fabricated at low cost. While such sensors are widely established in metabolic monitoring for plasma glucose detection in diabetes management, breath sensors are currently in the development phase. For instance, secondary alcohol dehydrogenase (s-ADH) catalyzes acetone in the presence of nicotinamide adenine dinucleotide

(NADH).[26] The latter can be detected from its fluorescent emission after ultraviolet (UV) excitation.[26] Acetone has been detected by these means within a range from 7 ppb to 10 ppm, depending on the molarity of the NADH buffer solution.[26] Mixtures of multiple enzymes have also been used in biosensors. For instance, a reaction pathway starting from acetone and using the enzymes s-ADH, lactate dehydrogenase, and pyruvate oxidase, was used to extract electrons, resulting in a measurable current.[27] This handheld system detected acetone in breath samples in the range from 0.18 to 10.2 ppm in excellent agreement with GC-MS ($R^2 = 0.971$, 38 breath samples).[27]

23.4.3 Mass spectrometers

Mass spectrometers exhibit exceptionally high sensitivity to detect analytes down to ppt levels with high selectivity to interferents. Real-time mass spectrometric techniques, such as selected ion flow tube mass spectrometry (SIFT-MS) or proton transfer reaction mass spectrometry (PTR-MS) have been largely applied for breath analysis (see Chapters 9 and 10, respectively), as has secondary electron spray ionization mass spectrometry (SESI-MS; see Chapter 12). Such devices can be considered as benchtop methods for real-time breath analysis and are indispensable for marker identification and cross-validation of other portable/handheld sensor systems.

There are several challenges in miniaturizing MS-based real-time detectors, for instance, the ionization of the sample and the transport of the resulting ions from ambient pressure to high vacuum. Recent advances for such challenges have been summarized by Snyder et al.[28] Portable MS that weigh ~2 kg are available but are currently predominantly employed in safety applications (e.g., detection of chemical warfare agents).

23.4.4 Ion mobility spectrometers

Ion mobility spectrometers (IMS) can selectively detect gases down to ppt levels (e.g., 50 ppt acetone) in real time (see Chapter 11). Such instruments are operated at ambient pressure, which is beneficial for miniaturization as vacuum pumps and other auxiliary systems are avoided. Handheld systems with small dimensions ($33 \times 10.5 \times 11.5$ cm^3) and weight (3.2 kg) have been reported for the detection of chemical warfare agents,[29] and a portable, backpack sized IMS was used to detect skin and breath emitted VOCs (see Chapter 11). Such systems are often combined with GC to improve sensitivity.

23.4.5 Optical spectroscopy

Optical spectroscopy has already been successfully implemented into commercial, portable breath analyzers, e.g., for CH_4 detection with infrared lasers (see Chapter

14). Acetone has been selectively detected with an optical cavity ring-down spectrometer in breath samples with a detection limit of 57 ppb, as validated by GC-MS ($R^2 > 0.99$).[30] Such devices can be portable with small dimensions ($65 \times 40 \times 20\ cm^3$) and low weight (15 kg).[30]

23.5 Summary

Breath analysis is a promising procedure to continuously monitor the human metabolic state and recognize rapid changes in an entirely noninvasive manner. Next to established indirect calorimetry, the current primary focus is on VOCs. Acetone, a by-product of lipolysis, is particular promising as a breath marker that has been investigated primarily during dieting and physical activity in small cohort pilot studies. Other markers (e.g., isoprene) hold potential, but further investigations are needed to clarify their biochemical pathway and clinical utility.

Currently, a key limitation of breath analysis for lifestyle applications is the scarce availability of portable and dependable breath detectors to monitor metabolic markers at home. Established MS exhibits outstanding sensitivity and selectivity and there are strong efforts to miniaturize such instruments. Most promising are probably chemical sensors where handheld, battery-driven and cost-effective devices already exist. With an appropriate sensor material, filter, and/or array design, such detectors feature sufficient sensitivity to detect specific compounds down to relevant ppb to ppm concentrations with sufficient selectivity in human breath. Nevertheless, their usability needs to be expanded to other metabolic markers, while reliability and stability concerns need to be addressed. Meeting these criteria can result in a new generation of personal and noninvasive metabolic monitors to guide dieting and physical activity for individual use at home.

Acknowledgments

This research was primarily supported by the ETH Particle Technology Laboratory and in part by the ETH Research Grants ETH-21 18−1 and ETH-05 19−2 and the Swiss National Science Foundation (project grant 175754).

References

1. Musa-Veloso K, Likhodii SS, Cunnane SC. Breath acetone is a reliable indicator of ketosis in adults consuming ketogenic meals. *Am J Clin Nutr* 2002;**76**(1):65−70.
2. Frayn KN. Calculation of substrate oxidation rates in vivo from gaseous exchange. *J Appl Physiol Respir Environ Exerc Physiol* 1983;**55**(2):628−34.
3. Anderson JC. Measuring breath acetone for monitoring fat loss. *Obesity* 2015;**23**(12): 2327−34.

4. Turner C, Spanel P, Smith D. A longitudinal study of ammonia, acetone and propanol in the exhaled breath of 30 subjects using selected ion flow tube mass spectrometry, SIFT-MS. *Physiol Meas* 2006;**27**(4):321–37.
5. Güntner AT, Kompalla JF, Landis H, Theodore S, Geidl B, Sievi N, et al. Guiding ketogenic diet with breath acetone sensors. *Sensors* 2018;**18**(11):3655.
6. Stone BG, Besse TJ, Duane WC, Evans CD, DeMaster EG. Effect of regulating cholesterol biosynthesis on breath isoprene excretion in men. *Lipids* 1993;**28**(8):705–8.
7. Španěl P, Davies S, Smith D. Quantification of breath isoprene using the selected ion flow tube mass spectrometric analytical method. *Rapid Commun Mass Spectrom* 1999;**13**(17):1733–8.
8. Smith D, Španěl P, Enderby B, Lenney W, Turner C, Davies SJ. Isoprene levels in the exhaled breath of 200 healthy pupils within the age range 7–18 years studied using SIFT-MS. *J Breath Res* 2009;**4**(1):017101.
9. King J, Kupferthaler A, Unterkofler K, Koc H, Teschl S, Teschl G, et al. Isoprene and acetone concentration profiles during exercise on an ergometer. *J Breath Res* 2009;**3**(2):027006.
10. King J, Mochalski P, Unterkofler K, Teschl G, Klieber M, Stein M, et al. Breath isoprene: muscle dystrophy patients support the concept of a pool of isoprene in the periphery of the human body. *Biochem Biophys Res Commun* 2012;**423**(3):526–30.
11. Prabhakar A, Quach A, Zhang H, Terrera M, Jackemeyer D, Xian X, et al. Acetone as biomarker for ketosis buildup capability-a study in healthy individuals under combined high fat and starvation diets. *Nutr J* 2015;**14**(1):41.
12. Musa-Veloso K, Rarama E, Comeau F, Curtis R, Cunnane S. Epilepsy and the ketogenic diet: assessment of ketosis in children using breath acetone. *Pediatr Res* 2002;**52**(3):443.
13. Musa-Veloso K, Likhodii SS, Rarama E, Benoit S, Liu YMC, Chartrand D, et al. Breath acetone predicts plasma ketone bodies in children with epilepsy on a ketogenic diet. *Nutrition* 2006;**22**(1):1–8.
14. Bovey F, Cros J, Tuzson B, Seyssel K, Schneiter P, Emmenegger L, et al. Breath acetone as a marker of energy balance: an exploratory study in healthy humans. *Nutr Diabetes* 2018;**8**.
15. Prabhakar A, Quach A, Wang D, Zhang H, Terrera M, Jackemeyer D, et al. Breath acetone as biomarker for lipid oxidation and early ketone detection. *Glob J Obes Diabetes Metab Syndr* 2014;**1**(1):12–9.
16. Samudrala D, Lammers G, Mandon J, Blanchet L, Schreuder TH, Hopman MT, et al. Breath acetone to monitor life style interventions in field conditions: an exploratory study. *Obesity* 2014;**22**(4):980–3.
17. Güntner AT, Sievi NA, Theodore SJ, Gulich T, Kohler M, Pratsinis SE. Noninvasive body fat burn monitoring from exhaled acetone with Si-doped WO_3-sensing nanoparticles. *Anal Chem* 2017;**89**(19):10578–84.
18. Yamai K, Ohkuwa T, Itoh H, Yamazaki Y, Tsuda T. Influence of cycle exercise on acetone in expired air and skin gas. *Redox Rep* 2009;**14**(6):285–9.
19. Schubert R, Schwoebel H, Mau-Moeller A, Behrens M, Fuchs P, Sklorz M, et al. Metabolic monitoring and assessment of anaerobic threshold by means of breath biomarkers. *Metabolomics* 2012;**8**(6):1069–80.
20. Güntner AT, Abegg S, Königstein K, Gerber PA, Schmidt-Trucksäss A, Pratsinis SE. Breath sensors for health monitoring. *ACS Sens* 2019;**4**(2):268–80.
21. Righettoni M, Tricoli A, Pratsinis SE. Thermally stable, silica-doped epsilon-WO_3 for sensing of acetone in the human breath. *Chem Mater* 2010;**22**(10):3152–7.

22. van den Broek J, Guüntner AT, Pratsinis SE. Highly selective and rapid breath isoprene sensing enabled by activated alumina filter. *ACS Sens* 2018;**3**(3):677−83.
23. Pineau NJ, Kompalla JF, Güntner AT, Pratsinis SE. Orthogonal gas sensor arrays by chemoresistive material design. *Microchim Acta* 2018;**185**(12):563.
24. Toyooka T, Hiyama S, Yamada Y. A prototype portable breath acetone analyzer for monitoring fat loss. *J Breath Res* 2013;**7**(3):036005.
25. Jung H, Cho W, Yoo R, Lee H-s, Choe Y-S, Jeon JY, et al. Highly selective real-time detection of breath acetone by using ZnO quantum dots with a miniaturized gas chromatographic column. *Sensor Actuator B Chem* 2018;**274**:527−32.
26. Ye M, Chien P-J, Toma K, Arakawa T, Mitsubayashi K. An acetone bio-sniffer (gas phase biosensor) enabling assessment of lipid metabolism from exhaled breath. *Biosens Bioelectron* 2015;**73**:208−13.
27. Landini BE, Shane T. Bravard, Breath acetone concentration measured using a palm-size enzymatic sensor system. *IEEE Sensor J* 2009;**9**(12):1802−7.
28. Snyder DT, Pulliam CJ, Ouyang Z, Cooks RG. Miniature and fieldable mass spectrometers: recent advances. *Anal Chem* 2015;**88**(1):2−29.
29. Yamaguchi S, Asada R, Kishi S, Sekioka R, Kitagawa N, Tokita K, et al. Detection performance of a portable ion mobility spectrometer with 63 Ni radioactive ionization for chemical warfare agents. *Forensic Toxicol* 2010;**28**(2):84−95.
30. Sun M, Jiang C, Gong Z, Zhao X, Chen Z, Wang Z, et al. A fully integrated standalone portable cavity ringdown breath acetone analyzer. *Rev Sci Instrum* 2015;**86**(9):095003.

PART

Beyond human breath

E

CHAPTER

Biomarkers in urine and stool

24

Ben de Lacy Costello, Oliver Gould, Norman M. Ratcliffe
Centre for Research in Biosciences, University of the West of England, Frenchay, United Kingdom

Chapter outline

Breathborne Biomarkers and the Human Volatilome. https://doi.org/10.1016/B978-0-12-819967-1.00024-4
Copyright © 2020 Elsevier B.V. All rights reserved.

24.1 Overview

Analysis of small molecules is a facile method for assessing stool and urine. Gut chemistry is exceedingly complex and the analysis of volatile organic compounds (VOCs) offers a simplified picture of the chemical soup that is in the gut. Much remains to be understood of the chemical compounds in the gut and their mode of synthesis, and the difficulties are compounded by the diverse food intake of humans and the very dynamic nature of the biotransformations that continuously occur within the gut.

Much of the published work in this area has been undertaken using small studies, and there is a risk that any correlations found between "biomarkers" and disease may not hold out in larger studies. It could be questioned why some researchers are assessing VOCs for, say, cancer diagnosis in urine, when there are as yet no clear biosynthetic routes established for specific cancer biomarker volatile compounds. The answer must be that this endeavor is worthwhile because the present test(s) to identify some cancers, particularly nonbenign prostate cancer and bowel cancer, are inadequate and

i. successful tests based on biomarker(s) could potentially save a very large number of lives, reduce suffering, and reduce the (cost) burden on health services, and
ii. urine/stool volatile analysis is a noninvasive procedure and potentially low cost.

It is the first reason that is especially important: it is worthwhile to invest in what might be a long shot if the potential returns are so significant.

Urine and stool are rich in volatiles, as is immediately apparent when encountered by the human sensory detection apparatus, the nose. As a major excretion pathway for xenobiotics and metabolites, urine and stool are potentially valuable sources of biomarkers. In particular, the presence of (odoriferous) VOCs in these biological specimens might be exploited to determine pathogen infections or physiological imbalances. This is already practiced in relation to odor of urine, which is a known indicator of specific conditions. In several cases, the disease synonym even reflects this, such as the fishy smell of urine—caused by the presence of trimethylamine—in "fish odor syndrome" (or trimethylaminuria), or the sweet "aroma" of maple syrup—relating to keto acids—in "maple syrup urine disease" (MSUD) (or branched-chain ketoaciduria). Certainly also odorless volatiles in urine or stool have potential as markers for disease, but identifying compounds and establishing screening practices requires objective analytical procedures, which might complement subjective odor perception in some cases.

Recent years have seen an increased interest in the human microbiome, much of which resides in the gut, and how this might be adjusted to treat infectious diseases such as *C. difficile*, and also disease more widely. Therefore, the study of stool VOCs provides a potential window on the current state of the gut microbiome and a method to monitor interventions. Stool VOCs are abundant and often at high concentrations when compared to urine and breath samples, so their study and causal link to disease states is also important in order that biomarkers can be translated to other media (breath/urine), the collection of which are better tolerated by patients. This chapter reviews current knowledge on VOCs that are present in urine and stool and that might represent biomarkers of certain illnesses and conditions. Ultimately, analytical methods that exploit the presence of volatile excretion products in these biological media to indicate disease present a desirable opportunity for noninvasive (early) screening in a painless manner that could increase patient uptake and result in early treatment and reduced mortality.

24.2 Urine volatiles and disease

The first major study of VOCs in urine was undertaken by Mills and Walker[1] using solid-phase microextraction (SPME) in conjunction with gas chromatography–mass spectrometry (GC-MS) to investigate urine headspace. The study reported 103 volatile compounds associated with urine taken from five patients with metabolic conditions. A later review of volatiles from urine expanded on the numbers of compounds by compiling data from six studies, including samples from elderly, apparently healthy individuals.[2] Nine compounds were commonly reported in all studies, namely propanone, 2-butanone, 2-pentanone, 2-heptanone, 3-hexanone, 4-heptanone, 2,5-dimethylfuran, 2-ethyl-5-methylfuran, and toluene, and so must be considered to be present in urine with a very high degree of certainty. The source of some or all of these compounds may be the gut. A study of 4-heptanone in urine strongly suggests that its presence originates at least in part from in vivo oxidation of the plasticizer component, 2-ethylhexanoic acid. Propanone, 2-butanone, 2-pentanone, and 2-heptanone were also found ubiquitously in the headspace of stool from a large number of samples donated from apparently healthy individuals, while a significant number of individuals were found to have 2,5-dimethylfuran, 2-ethyl-5-methylfuran, 3-hexanone, and 4-heptanone.[3] A list of compounds detected in urine and stool from healthy individuals—as well as breath, saliva, blood, sweat, and breast milk—has been compiled in an expansive table published by de Lacy Costello et al. in 2014.[4] The large number of ketones in urine probably at least partially arises from bacterial action in the gut, maybe by decarboxylation from the corresponding oxoacids, since ketones have been found at much lower concentrations in the urine of "germ-free" rats.[1] A significant number of terpenes are described, likely derived from foods.

Fewer urinary VOCs have been reported compared to stool, with 279 identified in the review of VOCs from healthy humans[4] and an update to this review increasing this number to >300 compounds, which is still much fewer than in stool or breath. This may be in part due to a lack of effort in characterizing urine and stool volatiles versus breath, and also indicates a relative lack of studies that focus on identifying VOCs in healthy individuals, as much of the data extracted are from controls in clinical studies. The overriding focus of most published studies has been the diagnosis of disease, as outlined below. Further, many of these studies do not identify the markers of disease, but rather rely on extracting spectral features or combining sensor outputs and using pattern recognition and classification techniques.

Few data exist on quantitative measurements of VOCs in urine. Concentrations of phenol and p-cresol (typically excreted in urine at 10 mg/day and 52 mg/day, respectively) have been reported to increase in urine with increasing protein intake. Their formation is considered to be due to gut microflora acting on tyrosine; anaerobic bacteria in the left colon produces phenol, and aerobic bacteria in the ileum/cecum produces p-cresol. The relationship is complicated by fiber intake, whereby high intake with high protein results in a smaller increase in concentration due to decreased transit time, as demonstrated in a study that was motivated by phenols being implicated in bladder and colon cancer.[5] Normal ranges reported for alcohols are 0–46 mg/24 h for ethanol, 0–300 μg/24 h for n-propanol, and 0–18 μg/24 h for n-butanol, which approximately mirror blood serum levels.[6] Trimethylamine (TMA) and 4-heptanone have been quantified in urine at 0.5–20 μg/mL and 40–800 ng/mL, respectively.[7] The highest concentrations in a range of volatile short-chain fatty acids (SCFAs) and semivolatile acids found in urine were hippuric>glycolic>benzoic>ethanoic>2-ketoglutaric>2-hydroxyisovaleric>lactic, and 2-hydroxyisobutanoic>oxaloacetic>pyruvic acids, with propionic, isobutanoic, butanoic, and 2-methylbutanoic acids as minor components.[8] Interestingly, there was poor correlation between urine and serum levels, with lactic>oleic>acetic>palmitic>3-hydroxybutanoic acids as the five most abundant acids in blood serum. It has been suggested that methylamine and other short-chain aliphatic amines may play a significant role in the central nervous system disturbances observed during hepatic and renal disease.[9] The average output of this compound was 11 mg/day, with a range of 1.7–62 mg/day; diet was found to have only a small effect on concentrations and the source was considered mainly endogenous. Gut bacteria are likely to be implicated in the production of methylamine (probably from creatinine), as less than half the output was observed in rats with no gut bacteria.[9] The average daily output for dimethylamine was about 17 mg, with values for the majority of the population lying within the range 0.68–35.72 mg/day.[10] Healthy young adults excrete about 1 mg/day TMA and 40 mg/day trimethylamine N-oxide, although these levels are markedly influenced by diet, particularly when containing marine fish, resulting in the excretion of several hundred mg/day trimethylamine N-oxide.[11]

There is a very large number of compounds found in low concentrations in urine.[4] The concentrations of most of these compounds is unknown and their association with health is either little known or unknown.

24.2.1 Alcoholic hepatitis

Alcoholic hepatitis, as the name suggests, is induced from prolonged alcohol consumption. In a very small study of a patient with alcoholic-induced liver failure, dimethyl disulfide (DMDS) was found to be the main component in urine, as detected via headspace analysis.[1] The presence of DMDS was considered to be due to impaired methionine catabolism and decreased clearance of sulfur compounds from compounds biosynthesized by intestinal microorganisms. In this context, it should be noted that homocystinuria, an inherited disease adversely affecting the catabolism of methionine, also produces large amounts of DMDS in urine headspace.

24.2.2 Alcohols and short-chain fatty acids

Alcohols present in blood can have differing origins. An increase in blood ethanol following consumption of glucose and accompanied by no methanol increase and only slight increases in other alcohols has been suggested to result from yeast overgrowth. In contrast, an increase in concentration of a range of alcohols, but not ethanol, has been suggested to be due to a bacterial dysbiosis in the small intestine, possibly due to malabsorption. Similarly, elevated levels of SCFAs, with normal or nearly normal blood alcohols, suggest increased bacterial fermentation in the colon, probably due to small intestinal malabsorption.

The origin of urinary acids must in part be from the gut. It is well known that SFCAs are extensively synthesized by intestinal microbacteria from carbohydrates and amino acids, and bovine urine acid concentrations showed a marked dependence on acid concentration changes in the gastrointestinal (GI) tract.[12] The hydrogen breath test to assess malabsorption and possible microbial overgrowth could clearly be used to aid differentiation between bacterial and fungal overgrowth, as fungi are not known to produce hydrogen. Carboxylic acid profiling in urine and plasma has been successfully undertaken for the diagnosis and study of inherited metabolic disorders using GC-MS.[8] Branched- and/or straight-chain SCFAs, such as acetic, n-butanoic, isobutanoic, and n-octanoic acid, for instance, have been observed as being prominent in the headspace of acidified urine for different acyl-CoA dehydrogenase deficiencies, which are inherited diseases (see also isovaleric acidemia, below).[1]

24.2.3 Ammonia, amines, and trimethylaminuria

Most ammonia in the body forms when protein is broken down by bacteria in the intestines. The liver normally converts ammonia into urea, which is then eliminated in urine. Ammonia levels in the blood rise when the liver is not able to convert ammonia to urea. This may be caused by cirrhosis or severe hepatitis. The diagnostic symptoms of confusion, excessive sleepiness, coma, or hand tremor, and the neuro psychological condition hepatic encephalopathy, may be at least partially due to damage to the liver and brain caused by ammonia poisoning. GI bleeding may cause elevated blood ammonia levels as red blood cells contain a high concentration of protein. One of the features of Reye's syndrome, which leads to symptoms similar to those seen with hepatic encephalopathy, and even seizures and coma in severe cases, is increased blood ammonia levels. There are tests for measuring the amount of ammonia in the blood, but it appears that there are no reports on ammonia levels in urine headspace for conditions other than urinary tract infections (UTIs).

It is well known that trimethylamine can be formed in the healthy human body, mainly from dietary choline, of which eggs, liver, and soybeans are the main sources. A rapid and simple assay procedure employing headspace GC analysis has been developed for the routine quantification of volatile methylamines and trimethylamine N-oxide present in human urine. The method was developed to offer rapid screening and diagnosis of patients with fish odor syndrome (trimethylaminuria, an inborn error of oxidative metabolism), where an odor similar to decaying fish is produced.[11] The very malodorous TMA is a by-product of the breakdown by bowel bacteria of carnitine (biosynthesized from lysine and methionine) and choline that are in high concentrations in marine fish, egg yolks, certain beans, and liver. TMA is normally absorbed from the gut and reoxidized in the liver to the nonsmelling N-oxide, which, together with small amounts of nonoxidized TMA, is excreted through the urine. In some individuals, the N-oxidation process in the liver is impaired, resulting in reduced conversion of TMA to trimethylamine-N-oxide. Trimethylaminuria often has very negative psychological effects on sufferers: social exclusion that can lead to depression, low self-esteem, and anxiety.

24.2.4 Cystinuria

Cystinuria is a rare genetic disease that is characterized by the accumulation of cystine in the body. People with cystinuria cannot properly reabsorb cystine into their bloodstream during the filtering process in the kidneys. Most of the time, this excess cystine is simply excreted in the urine. The urine may have a characteristic "rotten egg" odor.

24.2.5 Diabetes

When the pancreas releases insufficient amounts of insulin or no insulin at all, the body cannot utilize glucose. The body then switches to muscle, especially fat, to burn as fuel. Ketone bodies are the by-product of this process, which consist chemically of three substances, namely β-hydroxybutanoic acid, acetoacetic acid, and acetone. When ketone bodies are released in the body, they enter the bloodstream, acidify the blood, and are eventually excreted, mostly in urine, which exhibits a fruity smell, mainly due to acetone. Rhodes et al. analyzed urine from rats with diabetes and found that levels of a number of aliphatic ketones, predominantly 4-heptanone and 2-heptanone, were increased in the urine, which corroborated previous studies in humans.[13,14] A ketogenic diet, which is high in fats and low in protein and carbohydrates, or fasting, raises the level of ketone bodies. Home tests kits are available to check ketone levels using dipstick technology, and it is unlikely that VOC detector systems will have much of a role to play in this area. Abnormally increased alcohol levels, particularly ethanol, n-propanol, isobutanol, n-butanol, and isopentanol, have also been reported in association with diabetes mellitus.[14]

24.2.6 Gastrointestinal conditions

Very little research has been undertaken to assess GI diseases in relation to VOCs in urine. Urine is a much more pleasant biological material to deal with and easier to collect than stool. However, more research must be undertaken to correlate stool chemistry with urine chemistry, especially as the linking of volatiles in the gut to diseases is, itself, a nascent area. A series of urine and fecal samples have been obtained from patients with ulcerative colitis, Crohn's disease, and diabetes, and analyzed using an electronic nose. The study was small, but showed promise in differentiating disease groups, although further work is needed to establish these findings.[15]

24.2.7 Isovaleric acidemia

Isovaleric academia—isovaleric acid CoA dehydrogenase deficiency—is a genetic disease that leads to an increase in 3-hydroxyisovaleric acid in urine; the acid has a cheesy, sweaty feet odor. The signs and symptoms of this disorder become apparent within a few days after birth—or may appear during childhood and may come and go over time—and include poor feeding, vomiting, seizures, and lack of energy that can progress to coma. Symptoms are often triggered by an infection or by eating an increased amount of protein-rich foods.

24.2.8 Maple syrup urine disease

As the name implies, the urine in patients with MSUD exhibits a strong smell of maple syrup, which is associated with the presence of keto acids and is often a first indicator in the diagnosis of this condition (ketoacidosis). Symptoms of MSUD include mental retardation. The disease is caused by a deficiency in enzyme activity that catalyzes decarboxylation of 2-oxocarboxylic acids, as a step in the metabolism of the branched-chain amino acids, valine, leucine, and isoleucine. As a result, there is build-up of the acids in tissues and urine: 2-oxoisocaproic acid from leucine, 2-oxo-3-methylvaleric acid from isoleucine, and 2-oxoisovaleric acid from valine (plus their reduced products). Patients with MSUD most commonly have $<2\%$ of the enzyme and will die unless treated. The second most common form of MSUD is with enzyme activity between 8% and 15% of normal. Children with intermittent MSUD may not show any signs of the disorder until they are 12–24 months of age, usually in response to an illness or a rapid increase in protein intake. During episodes of illness or other metabolic stress, the child may develop seizures. Successful diagnosis, preferably soon after birth, followed by a restrictive diet, minimizing branched-chain amino acids, minimizes symptoms.

24.2.9 Methionine malabsorption syndrome

Methionine malabsorption—also termed Oast-house syndrome or Smith–Strang disease—is a rare genetic disease that results in reduced absorption of methionine in the small intestine. This leads to the amino acid becoming available for breakdown in the large intestine by intestinal bacteria, with the resulting α-hydroxybutyric acid causing a beer-like or yeasty odor to urine. Symptoms can be convulsions, white hair, and mental retardation, among others.

24.2.10 Ovulation

Acetone and ammonia levels in the headspace above urine from a healthy woman have been observed to vary on a daily basis over the course of three separate menstrual cycles.[16] Concurrent with the time of ovulation, a 3- to 12-fold increase in the level of acetone in the urine headspace was observed. The successive peaks in acetone level and the subsequent return to baseline values were mirrored by similar increases in the ammonia levels, albeit out of phase by 1 day. Following this pilot study, a larger cohort of seven women were studied, whereby several-fold increases in urine headspace acetone levels were observed in five women 2–3 days after the predicted day of ovulation.[17]

24.2.11 Phenylketonuria

Phenylketonuria is a genetic disease resulting in reduced ability to convert phenylalanine (Phe) to tyrosine in the liver. The accumulated Phe is converted to phenylacetate in the body, which imparts a musty odor to urine. Other urinary metabolites of this condition include phenylacetate, phenylpyruvate, and phenethylamine. Interestingly, the major volatile component in the headspace of urine samples in this disease is benzaldehyde—probably derived from benzoic acid—which exhibits an approximate 10-fold increase compared to urine from healthy individuals. If the condition is left untreated, it can cause problems with brain development, leading to progressive mental retardation, brain damage, and seizures.

24.2.12 Prerenal azotemia

Prerenal azotemia is the most common form of kidney failure in hospitalized patients. It can be caused by any condition that reduces blood flow to the kidney. The inability of the kidneys to filter blood results in accumulation of nitrogen-rich compounds in blood, resulting in a very strong concentrated ammonia-like urine smell.

24.2.13 Tuberculosis

An estimated 1.5 million people died from tuberculosis (TB) in 2018. Improving the early and effective diagnosis of TB is very important and the development of a rapid, low-cost, noninvasive method, with the potential to be administered in the field, remains a challenge. Although TB is a disease of the respiratory system, researchers have studied[18] VOCs in human urine from TB patients and healthy controls. A significant increase in the abundance of o-xylene and isopropyl acetate and decreased levels of 3-pentanol, dimethylstyrene, and cymol were observed in TB patients compared to controls. This study suggests a possibility of using urinary VOCs for the diagnosis of human TB, although further work is required to confirm these results and to ascertain whether other conditions lead to false positives. The detection of TB via breathborne biomarkers is discussed in Chapter 21.

24.2.14 Cancer

The search for a relationship between VOCs and cancer is presently adventurous research. Work showing promise, even of a pilot study nature, gives confidence that this is a worthwhile area of research to pursue. This is the case so far with urine analyses. The impressive ability of dog olfactory senses (see Chapter 34) has been tested for assessing urine for bladder cancer detection. The first study involving canines used six different dogs trained with urine from patients with bladder cancer and from patients who were either healthy or had some other disease.[19] The dogs were subsequently tested for their ability to pick out one bladder cancer urine sample from six controls; 54 samples (27 from cancer patients) were used in training and 27 for testing. The dogs correctly selected urine from patients with bladder cancer on 22 out of 54 occasions, yielding a mean success rate of 41% compared with 14% expected by chance alone. This suggests that tumor-related VOCs are present in urine, imparting a characteristic odor signature.

Another proof-of-principle study was undertaken with a Belgian Malinois shepherd dog trained to recognize urine of people with prostate cancer.[20] After a learning phase and a training period of 24 months, the dog's ability to discriminate prostate cancer and control urine was tested in a double-blind procedure. Urine was obtained from 66 patients that had been referred to a urologist for elevated prostate-specific antigen or abnormal digital rectal examination. All patients underwent prostate biopsy and two groups were considered: 33 patients with cancer and 33 controls presenting negative assessments. During each test, the dog was required to signal a cancer urine among six samples containing only one cancer urine and five randomly selected controls. The dog completed all the tests and correctly designated the cancer samples in 30 of 33 cases. Of the three cases incorrectly classified as cancer by the canine, one patient was rebiopsied and prostate cancer was subsequently diagnosed. The sensitivity and specificity were both 91%. This study shows that dogs can be trained to detect prostate cancer by smelling urine with a significant success rate. It also suggests that prostate cancer gives an odor signature to urine.

Gas sensor arrays in combination with pattern recognition algorithms have been used to assess detection of volatile bladder cancer biomarkers in urine. The best diagnostic performance that was obtained with independent test data provided by healthy volunteers and bladder cancer patients was 70% overall accuracy (70% sensitivity and 70% specificity).[21] When the data of patients suffering from other noncancerous urological diseases were added to those of the healthy controls, the classification accuracy fell to 65% with 60% sensitivity and 67% specificity. While this is not sufficient for a diagnostic test, it is significantly better than random chance, leading the authors to conclude that there was useful information in the urine headspace but that a more informative analytical technique, such as MS, was required to exploit this approach fully. Although the specificity achieved using this method was less than that of conventional urine cytology, sensitivity was appreciably higher.

There have been limited MS-based studies published on volatile analyses and cancer diagnoses from urine samples. An early study implicated formaldehyde as a putative biomarker,[22] but this work specifically targeted only one compound, rather than the full range of VOCs present. A later study assessed the pattern of VOCs in urine from patients known to have prostate cancer compared to controls; differences were observed, but since this was a small study, caution must be expressed.[23] The metabolic composition of urine metabolites has been further shown to be consistently altered in lung cancer patients ($n = 71$) relative to control subjects ($n = 54$), suggesting a metabolic signature for lung cancer in urine.[24] The compounds showing higher importance for this discrimination were mainly aldehydes (e.g., hexanal, acetaldehyde, decanal) and ketones (e.g., pentane-2,4-dione, hexan-3-one, 2,4-dimethylpentan-3-one).

Urinary volatile organic metabolites have also been investigated as potential cancer biomarkers by SPME-GC-MS.[25] A significant increase in 2-methyl-3-phenyl-2-propenal, p-cymene, anisole, 4-methyl-phenol, and 1,2-dihydro-1,1,6-trimethyl-naphthalene in the urine of cancer patients (oncological group: 14 leukemia, 12 colorectal and seven lymphoma) versus 21 healthy individuals was observed, and the authors found, on average, statistically significant lower abundances of DMDS in cancer patients.

Urinary VOC analysis has been used to diagnose colorectal cancer (CRC)[26,27,28] as the simplicity of urine collection and patient acceptability compared with stool makes it an excellent potential target for a screening test. Di Lena et al.,[26] have reviewed the use of VOCs for the detection of CRC. This review discusses multiple studies that have investigated both fecal and urinary VOCs using a variety of instrumentation, including GC-MS, field-asymmetric ion-mobility spectrometry (FAIMS), and selected ion flow tube−mass spectrometry (SIFT-MS).[26] McFarlane et al.[27] analyzed the fecal microbiome in combination with urinary VOCs for the detection of CRC and found that FAIMS could be used to differentiate CRC patients from noncancer controls with an area under the curve (AUC) of the receiver operating characteristic (ROC) curve (i.e., AUROC) of 0.72. Mozdiak et al.[28] also analyzed urinary VOCs for the detection of CRC using FAIMS and GC-IMS and found that FAIMS gave better results with an AUROC of 0.98, while the GC-IMS yielded an AUROC of 0.83.

Advancements in the use of urinary VOCs for disease detection has not been limited to CRC; a 2019 review by Gao and Lee[29] describe studies investigating the utility of urinary VOCs for malignancies including bladder, prostate, and renal cancer. This review also included general information about analytical techniques and favored real-time techniques of SIFT-MS and proton transfer reaction−mass spectrometry (PTR-MS) (see Chapter 9 and 10, respectively) over the more traditional chromatographic techniques, such as GC-MS.[29] A shift toward these types of techniques has the benefit of not requiring preconcentration of samples and results being delivered quickly (in under 5 min). Recent advances in design and user-friendliness have led to these instruments being operable by semiskilled (rather than highly skilled) staff, yet these tools remain bulky and expensive. Other studies have focused on using FAIMS, or hyphenated IMS or sensor techniques, such as GC-FAIMS[29] or GC-sensor,[30] which can be more easily miniaturized and are lower cost. Further advancements to these instruments will aid their ability for clinical VOC analysis, for example, increased automation, or the addition of thermal desorption functionality, e.g., to SIFT-MS, which allows for clinical sample preconcentration storage, transport, and rapid quantification of VOCs linked to disease states.

A recent study used GC coupled to a quadrupole-time-of-flight (QTOF) mass spectrometer (GC-QTOF-MS) to analyze urinary VOCs to look at localized and metastatic breast cancer models.[31] Models of this type may be useful in translating markers to human studies and establishing reliable data on the efficacy of VOCs to follow the stages of cancer development.

In a recent review focusing on innovative diagnostic methods for prostate cancer via urine analysis, Bax et al.,[32] reported on VOC analysis using dogs and "senso-instrumental" analysis (i.e., electronic-nose, or e-nose). The variability of the dogs studies reported were very wide, ranging from 13% to 100% sensitivity and 36%−98.7% specificity, which is attributable to poor reproducibility from different dogs and different sample sets (see Chapter 34). By comparison, the review reports a sensitivity of the e-nose analysis ranging from 71.4% to 82%, and an associated specificity of 67%−92.8%. As such, substituting large and costly mass spectrometers for low-cost sensors or sensor arrays might lead to the development of an accurate test that can be performed rapidly, potentially even at point-of-care (PoC). Other studies using coupled GC separation with sensor systems have demonstrated good sensitivity and specificity, such as that reported on a combined GC−metal oxide sensor for detecting bladder cancer and prostate cancer based on urine VOCs, which offered >90% sensitivity and specificity.[30] Sensor systems for biomarker detection are the subject of Chapter 13.

24.2.15 Bacteria in urine

Urinary tract infections are a significant cause of morbidity, with an estimated 92 million cases worldwide.[33] Thirty-one percent of nosocomial infections in medical intensive care units (ICUs) are attributable to UTI, and it is estimated that 20% of females aged 20–65 years suffer at least one episode per year. Furthermore, UTI is the most frequent cause of infection among nursing home residents and the most commonly documented source of bacteremia in the elderly population.[34] UTI is also the most common bacterial infection in infants and young children worldwide, and newborn infants are at greatest risk of UTI, with 10% of those affected developing bacteremia.[35] Prompt diagnosis and early treatment are highly desirable. An important aspect of testing is for asymptomatic bacteriuria, which occurs in 5%–10% of pregnancies and is associated with risks of preterm birth.[36] Fortunately, the prescription of the correct antibiotic(s) is normally effective in rapidly reducing bacteria levels in urine and reducing symptoms.

Bladder urine is normally sterile, thus the presence of pathogenic bacteria in urine is indicative of UTI. When voided naturally, urine from men may contain small levels of bacteria from the urethra or prepuce and with women from the urethra, vulva, or vagina, but this is typically less than or equal to 200 organisms/mL. Levels of 105 or greater are diagnostic of UTI in a symptomatic patient.[34] Levels as low as 103 may be clinically significant in symptomatic patients suspected of UTI. Urine culture remains the reference standard for diagnosis of UTI, but at least 18 h are needed for detection of bacterial growth on culture media by standard microbiological techniques, which can result in delayed or inappropriate treatment. Urine culture is also expensive and needs a microbiological laboratory with skilled technicians.

Rapid urine tests do not identify the pathogenic organism or suitable antibiotic treatment for UTIs; therefore, they cannot be used as a replacement test for urine culture. Because of the clinical importance of early diagnosis and delays with the gold standard microbial culturing method, however, rapid urine tests, such as urine dipsticks and microscopy of urine for white cells and bacteria, are used widely. The nitrite test-strip method, which is based on a color change when nitrite is present, is very rapid. Most bacteria responsible for UTI (e.g., Enterobacteriaceae) are able to reduce nitrate in the diet to nitrite, but this test will not detect the species of bacteria (including *Enterococcus*, *Pseudomonas*, and *Staphyloccocus*) that do not reduce nitrate. Approximately 80% of uncomplicated UTI are caused by *Escherichia coli* and 20% by enteric pathogens, such as *Enterococci*, *Klebsiellae*, *Proteus* spp., *coagulase (−) Staphylococci*, and fungal opportunistic pathogens, such as *Candida albicans*.[37] Guidelines for the management of UTIs recommend microscopy for both white cells and unstained bacteria in children younger than 3 years of age, and dipstick testing for children older than 3 years. However, an analysis of data from 95 studies in 95,703 children revealed that microscopy for white cells should not be used for the diagnosis of UTI because microbiology laboratory facilities are needed, so the results are considerably delayed compared with dipstick results, and its accuracy is no better than that of dipstick.[38]

Clearly, there remains a requirement for a low-cost, PoC rapid test to preferably identify UTI and desirably the particular bacteria in urine with a very high specificity and sensitivity. There have been several approaches to tackle the problem of making a successful volatile analyzer for UTI. Bacterial urease enzymes break down urea to ammonia and this can be used to advantage as a biomarker for the presence of bacteria in urine. The potential utility of this has been established by GC-MS studies that have shown that ammonia and acetic acid can occur in the urine headspace from UTI-infected patients.[37] Urine samples spiked with various microorganisms (*Proteus mirabilis*, *Pseudomonas aeruginosa*, *Staphylococcus aureus*, *Staphylococcus faecalis*, *C. albicans*, *Escherichia coli*, and *Klebsiella* spp.) were analyzed in comparison to clinical urine samples; negative samples (defined as having $<10^5$ cfu/mL) averaged 3 ppm acetic acid, while positives (defined as being $>10^5$ cfu/mL) ranged from 8 to 500 ppm, with an average value of approximately 100 ppm.[37]

24.3 Stool volatiles and disease

There are many GI conditions where it would be advantageous if PoC, rapid analyses could be undertaken. Currently, gut-derived hydrogen is measured on breath to assess small intestinal bacterial overgrowth and sugar malabsorption syndromes.[39] Presently, it can take days for a microbiological assessment of stool to be undertaken, e.g., for *Clostridium difficile* infection. Rapid diagnosis of inflammatory bowel disease (IBD) and differentiating between Crohn's disease, ulcerative colitis, and irritable bowel syndrome (IBS) would be beneficial for clinicians. A rapid, accurate screening test for CRC would be especially desirable and has been a target for many studies.

The analysis of volatile metabolites from stool offers the possibility of rapid, low-cost diagnoses of a range of disease conditions, in a noninvasive manner. Increased knowledge of the complex chemistry and microflora of the gut from healthy individuals and those with GI conditions would enable more rapid progress to be made in correlating chemical metabolites and disease state. The following section reviews the volatile compounds that have been found in the headspace of stool, which is mainly qualitative in nature.

Overall, there are import questions to be asked with regard to the clinical use of fecal VOCs, which despite a number of advances in analysis still remain. There is still no standardized method to collect fecal samples; the standardization of sample collection remains a pressing topic for VOC analysis in general (see Chapter 2). Then, there is the issue of participant uptake; in general, people are more reluctant to donate stool samples than breath and urine donations, which are considered much more palatable. In addition to the issue of uptake is the issue of compliance; following instructions that involve the careful handling (and possibly even storage) of fecal matter in a person's home are at best unappealing.

24.3.1 Volatiles in stool

The first report of gas analysis from stool was in 1861 when Ruge reported that human rectal gas contained hydrogen, carbon dioxide, and methane, in addition to other unidentified gases.[40] 130 years later, Tomlin drew similar conclusions,[41] and in recent years, Levitt has led the field in the study of flatus, publishing a seminal paper on its composition in 1997.[42] The paper confirmed Ruge's observation and went on to quantify the most abundant gases in the flatus of 16 volunteers. Typically, about 70% of gas in flatus was reported to be produced intraluminally, with flatus considered to be a mixture of hydrogen (0%–50%), nitrogen (5%–90%), oxygen (0%–10%), carbon dioxide (10%–30%), and methane (0%–10%). Methane production occurs in about 50% of the healthy population, with some producing higher levels than others; methane production is correlated with methanogenic bacteria. Similarly, sulfate-reducing bacteria are responsible for the generation of pungent sulfides.[43] Thus, evidence shows that methane[44] and sulfide production[45] can be directly related to intestinal flora.

It has been shown that methanogens use hydrogen to synthesize methane by reduction of carbon dioxide, and as a result high hydrogen levels and high methane levels are considered to occur in flatus. Thus, to assess bacterial overgrowth in the small intestine, both methane and hydrogen breath levels would be ideally measured.

The 2014 review paper by de Lacy Costello et al.[4] reports 381 compounds identified from fecal samples from healthy humans. More recently, further compounds have been identified from several publications including Ref. [46,57], giving an updated total in excess of 400 compounds. An online searchable database has been compiled to include all compounds identified from breath, with the other matrices including stool, urine, etc., currently being added (see Chapter 39); certainly, this database will be a valuable resource for researchers looking to crosscheck and validate their findings and has additional utility in assisting to map the metabolic basis for the appearance of VOCs.

In terms of healthy fecal samples, some efforts have been made to quantify their volatile constituents. Gould et al. used ^{13}C-labeled compounds to calculate the mass of compounds retained in the stool.[46] This work quantified dimethyl sulfide (26–25626 ng/g), acetone (442–3006 ng/g), ethyl butanoate (39–2468 ng/g), ethyl 2-methylbutanoate (0.3–180 ng/g), DMDS (35–1303 ng/g), 1-octen-3-one (12 ng/g), dimethyl trisulfide (10–410 ng/g), 1-octen-3-ol (0.4–58 ng/g), ethanoic acid (672–12963 ng/g), butanoic acid (2493–11553 ng/g), 3-methylbutanoic acid (64–8262 ng/g), pentanoic acid (88–21886 ng/g), indole (290–5477 ng/g), and 3-methyl indole (37–3483 ng/g). Additionally, 15 compounds were identified for the first time in stool, including TMA, which required the feces to be made alkali with sodium hydroxide prior to analysis.[46] This study also set out to examine the differences in fecal volatile profiles between South American and UK participants, revealing five compounds to be significantly different, namely ethyl butanoate, 1-octen-3-ol, ethanoic acid, butanoic acid, pentanoic acid, and indole. A discriminant model based on ethyl butanoate and butanoic acid allowed the two populations to be distinguished with an AUROC of 0.937. Conversely, UK participants on an omnivore and vegetarian diet could not be distinguished using the same approach, highlighting that microbiome differences may be more significant than diet in inducing large shifts in fecal VOCs. Understanding these differences in ethnic groups is critical to the development of volatile-based clinical tests, as understanding what constitutes healthy is what allows disease states to be identified.

Another study used selective reagent ionization time-of-flight mass spectrometry with NO^+ as the reagent ion (SRI-TOF-MS(NO^+)) in conjunction with GC and headspace SPME to quantify 16 VOCs in the urine of 19 healthy volunteers.[47] Quantification was made on ten ketones (acetone, 2-butanone, 3-methyl-2-butanone, 2-pentanone, 3-methyl-2-pentanone, 4-methyl-2-pentanone, 2-hexanone, 3-hexanone, 2-heptanone, and 4-heptanone), three volatile sulfur compounds (dimethyl sulfide, allyl methyl sulfide, and methyl propyl sulfide), and three heterocyclic compounds (furan, 2-methylfuran, 3-methylfuran). The median concentrations of the species studied varied between 0.55 nmol/L for allyl methyl sulfide and 11.6 μmol/L for acetone. Limits of detection (LODs) ranged from 0.08 nmol/L for allyl methyl sulfide to 1.0 nmol/L for acetone and furan (with relative standard deviations [RSDs] ranging from 5% to 9%).

Patients and healthcare professionals have observed that feces often smell abnormal during GI disease, thus identification of VOC biomarkers from stool offers the potential for developing a method for rapid diagnosis of GI diseases. Surprisingly, little is known about VOCs found in the gut. The diagnostic and health implications of most of these compounds remain to be explored. Significant concentrations of a range of SCFAs,[48] branched-chain fatty acids (BCFAs), indoles,[49] and phenols[50] have been observed. Fermentation of carbohydrates in the gut produces ethanoic, propionic, butanoic, pentanoic, and hexanoic acids, particularly by *Bacteroides*.[51] In vitro studies have provided evidence that proteinaceous foods also produce SCFAs via the action of bacteria, such as *Clostridia* spp.[52]; BCFAs, such as 2-methylbutanoic acid and methylpropionic acids, are principally produced by gut microbial action on proteins via the respective branched amino acid. Volatiles, such as methanethiol and ammonia, are considered derivable from methionine by the action of bacteria, such as *Clostridium sporogenes*.[53] Hydrogen sulfide and methanethiol can be damaging to the large intestinal epithelium and can be generated from sulfur-containing substances in the diet.[54] Similarly, fermentation of tyrosine and tryptophan in stool has been shown to produce phenol and indole, respectively.[53] Phenol and p-cresol are considered to be produced by aerobic intestinal microflora acting on tyrosine, and the latter by anaerobic organisms.[55] Fig. 24.1 describes some of the compounds shown to be produced by incubating amino acids with fresh stool.

Two independent studies state that a total of 297 and 135 different VOCs have been identified in the headspace of stool from apparently healthy individuals on an ad libitum diet.[3,56] These two studies show many similarities as well as some differences. Typically for each donor the number of VOCs ranged from 78 to 125 (median = 101). Interestingly, 44 compounds were stated to be common to 80% of the cohort samples.[3] Thus, many compounds were shared by asymptomatic donors.

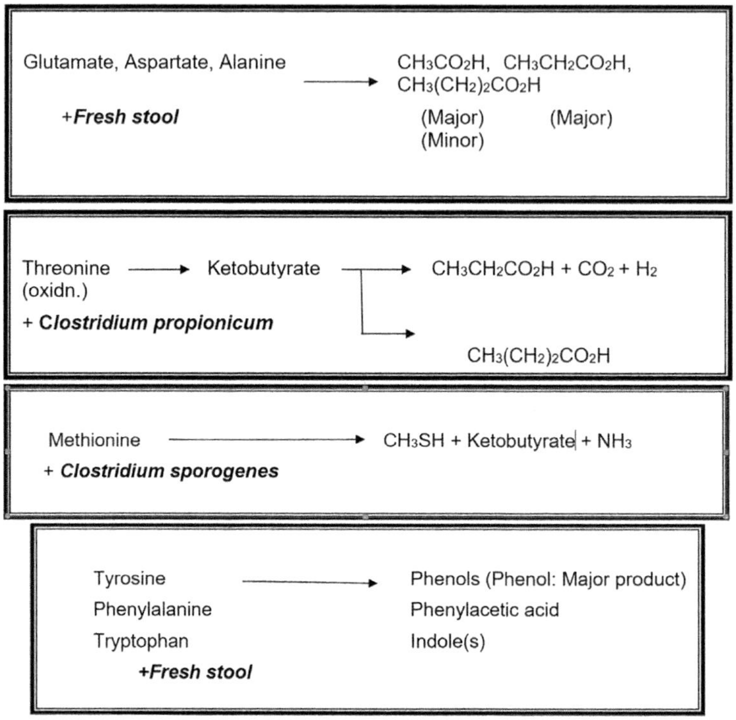

FIGURE 24.1

Origins of volatiles in the headspace of stool by incubating fresh stool/microorganisms with amino acids.

Headspace SPME has greatly facilitated the isolation and analysis of VOCs from human stool. Pioneering human fecal VOC metabolomic investigations have utilized a single SPME fiber type for analyte extraction and analysis. However, Couch et al. hypothesized that the varied functionality of the metabolites in the headspace of stool dictated the use of several diverse SPME fiber coatings for more comprehensive metabolomic coverage.[55] They evaluated eight different commercially available SPME fibers in combination with GC (coupled to flame ionization detection, FID, and MS). This approach appears very promising; 267 peaks were found via GC-FID, although not all compounds could be identified.

24.3.2 Acids, alcohols, and esters

Short-chain fatty acids in stool have been reported to be very common in several studies[4] and shown in Table 24.1A. SCFAs arise from metabolism of undigested carbohydrate, such as dietary fiber, by colonic bacteria.[51] BCFAs are also known to be abundant in feces, and their presence can be explained from the dissimilation of amino acids in some cases.[53] SCFA concentrations are also present in appreciable

Table 24.1A A description of straight-chain alcohols, short-chain fatty acids, aldehydes, alkanes, and methyl ketones found in the headspace of stool and urine of apparently healthy volunteers.

Class	Carbon number (n)																			
	1	2	3	4	5	6	7	8	9	10	11	12	13	14	15	16	17	18	19	20
Alcohol	SU*	S	S	S	S	S	S	S	S	S		S	S			S				
$H\text{-}(CH_2)_n\text{-}OH$		U	U	U	U	U		U												
Acid	S	S	S	S	S	S	S	S	U											
$H\text{-}(CH_2)_{n-1}\text{-}CO_2H$	U	U	U	U	U			U												
Aldehyde		S	S	S	S	S	S	S	S	S	S	S	S	S	S	S		S		
$H\text{-}(CH_2)_{n-1}\text{-}CHO$		U	U		U	U	U	U	U											
Alkane	S				S	S		S		S	S	S	S	S	S	S	S	S	S	S
$H\text{-}(CH_2)_n\text{-}H$						U								U						
Methyl ketone		S	S	S	S	S	S	S	S	S										
$H\text{-}(CH_2)_{n-2}\text{-}CO\text{-}CH_3$		U	U	U	U	U	U	U	U											

S, *found in stool;* U, *found in urine;* U*, *not detected in urine in healthy nondrinkers, easily detectable in urine of alcoholic beverage drinkers.*

Table 24.1B A description of straight-chain esters found in the headspace of stool and urine of apparently healthy volunteers.

Ester		Acid component carbon number (n)								
		0	1	2	3	4	5	6	7	8
Methyl	$H\text{-}(CH_2)_n\text{-}CO_2\text{-}CH_3$	S	S	S	S	S	S	S		
Ethyl	$H\text{-}(CH_2)_n\text{-}CO_2\text{-}C_2H_5$		S U	S	S	S	S	S		
Propyl	$H\text{-}(CH_2)_n\text{-}CO_2\text{-}C_3H_7$		S	S	S	S	S		S	
Butyl	$H\text{-}(CH_2)_n\text{-}CO_2\text{-}C_4H_9$	S	S	S	S U	S	S			
Pentyl	$H\text{-}(CH_2)_n\text{-}CO_2\text{-}C_5H_{11}$	S	S	S						
Hexyl	$H\text{-}(CH_2)_n\text{-}CO_2\text{-}C_6H_{13}$		S	S						
Heptyl	$H\text{-}(CH_2)_n\text{-}CO_2\text{-}C_7H_{15}$	S								

S, *found in stool;* U, *found in urine.*

concentrations, for instance, circa 37, 12 and 13 mmol/L of stool for acetate, propionate, and butanoate, respectively.[55] Methanoic acid is not commonly observed, nor is methyl esters. Of the other acids, benzoic acid has received some attention in quantitative rather than headspace studies of stool due its ability to inhibit the growth of fungi and some bacteria (by inhibiting glucose fermentation). Its origin is considered to be from microbial activity in the gut, but mainly from diet, particularly from fruit and coffee containing quinic acid.

Alcohols were thought uncommon in adult feces,[58] but it is now known that at least 45 different alcohols can be present (see Table 24.1A and the literature).[4] Ethanol is very commonly observed and has been linked to yeast overgrowth, as well as bacterial synthesis by the Embden–Meyerhof pathway. Interestingly, specific long-chain aliphatic alcohols (C_6–C_{20}), which can inhibit the growth of various bacteria and fungi,[59] can be also found. It is likely that these compounds play a role in providing an ecological niche for some bacteria by inhibiting others.

Esters were found to represent the largest group of compounds; 59 esters have been detected in stool, of which four contained sulfur, as listed in Table 24.1B and in the literature.[4] An interesting readily observed feature of Table 24.1B is the similarity of the higher molecular weight compounds; these either possess a long-chain acid and short-chain alcohol or a short-chain acid and long-chain alcohol. This suggests that the number of the esters described is not a true picture of what is present in the stool and reflects limitations of the method, i.e., the low volatility of the esters and the associated difficulty in sampling them. It is very likely that a more sensitive method or better preconcentration would significantly increase the number of compounds observed. An in vitro study using carbon-labeled butanoic acid has shown the origin of some esters by stool-mediated syntheses, as depicted in Fig. 24.2.

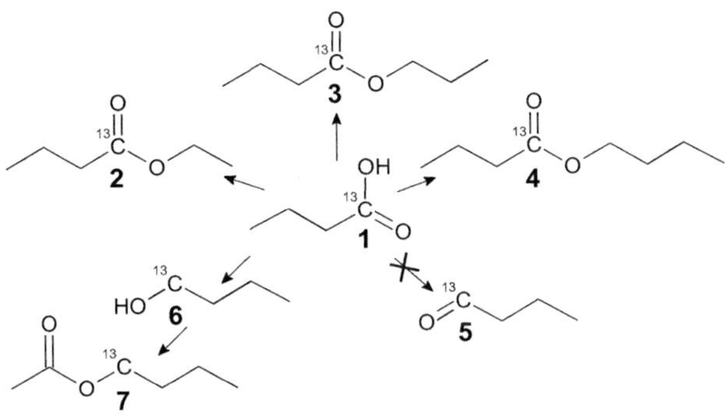

1 Butanoic acid, **2** Butanoic acid ethyl ester, **3** Butanoic acid propyl ester
4 Butanoic acid butyl ester, **5** Butanal, **6** 1-Butanol, **7** Ethanoic acid butyl ester

FIGURE 24.2

The products found from the conversion of [1-^{13}C]butanoic acid on incubation with fresh stool from an asymptomatic donor.[3]

24.3.3 Benzenoid and heterocyclic compounds

A diverse range of 56 aromatic compounds have been reported in stool, including mono-, di-, tri-, and tetra-substituted benzenoids, mono- and di-substituted furans, and nitrogen-containing derivatives of pyridine, pyrrole, and indole.[3] Most of these have been only relatively recently reported in feces, although it has been established that phenolic and indole compounds arise from the metabolism of aromatic amino acids by gut bacteria.[52] It is considered that aerobic bacteria in the ileum tend to produce phenol from tyrosine, while anaerobic bacteria in the left colon produced p-cresol; production rates are estimated at 10 mg/day and 52 mg/day, respectively.[55] A wide range of furan derivatives, mainly aromatic, have been found, possibly of plant origin,[60] although this is doubtful, as furans are acid labile and would not be expected to survive the digestive process. There are many publications reporting observations of alkyl furans being produced by fungi, yet in contrast there is a paucity of publications relating to furan biosynthesis to bacteria. Fungi are well known to be commensal organisms in the gut, which could explain the origins of furans in stool, possibly from the metabolism of fructose. Some benzenoid compounds, such as dimethylbenzenes, ethylbenzene, and toluene (constituents of petrol), may arise from air pollution.

24.3.4 Aldehydes and ketones

The aldehydes ethanal, propanal, butanal, and hexanal have been reported in significant numbers of individuals.[3] Ethanal is of particular interest due to its abundance and its purported role in promoting mutagenesis[61] associated with bowel cancer; the toxic effects of higher aldehydes have received much less attention. The origins of some aldehydes may be dietary, for instance, 2-methylpropanal, 3-methylpropanal, hexanal, nonanal, decanal, and benzaldehyde are found in potato tubers and hexanal in carrots. However, it is doubtful that these compounds would remain unchanged through the digestive system, thus biosynthesis by microorganisms in the gut appears more likely. Ketones have been reported to be the second largest class of VOC found in stool and include four of the most common compounds, of which acetone and 2-butanone probably arise from fatty acid and carbohydrate metabolism.[62] Methyl ketones can be produced by many species of bacteria, as well as by fungi from the respective alkanoic acid. The 3- and 4-ketones—and undoubtedly other ketonic compounds—can also be synthesized by bacteria. Some compounds of this group may have dietary origins, for example, cheddar cheese contains 2-propanone, 2-butanone, and 2-heptanone, potatoes contain acetone and 2-heptanone, and 6-methyl-5-hepten-2-one is an abundant volatile in carrots. The universal presence of 2,3-butanedione is interesting, since it may have health implications by impacting on the growth of some bacteria and yeasts.[63] This group of compounds, as indeed other groups, is not normally the end-products of metabolism by microorganisms; therefore, their concentrations would be expected to be continually changing in the gut.

24.3.5 Alkanes, alkenes, and alicyclic compounds

Methane is a product from bacterial fermentation of monosaccharides, reduction of carbon dioxide, or from acetic acid. Pentane has been reported in feces, but there is little evidence of other short-chain species, which could be due to the analytical procedures used. Longer-chain species have been found in small numbers in stool. Isoprene has been extracted from feces and has been investigated as a breath marker for disease.[64] Isoprene in the gut may be the result of cholesterol biosynthesis,[65] and it is considered the most common hydrocarbon in the human body, thus it would be expected to be found in stool. Many alkenes/terpenoid compounds found are well documented as naturally occurring plant products. Limonene has been reported as the most abundant of the terpenoid compounds. Most of the terpenes identified are found in vegetable foodstuffs and do not originate from animal products.

24.3.6 Ether and chloro compounds

At least seven ether compounds have been isolated from the headspace of stool, although it is very unlikely that chlorinated compounds found are of biological origin; consumption of contaminated food or water is the likely source of these compounds. Chloroform may arise as a stool VOC component from several sources: it is an air contaminant and has been detected in foodstuffs. Chlorination for disinfection of drinking water is another source resulting in the production of chloroform and halogenated methanes. Methylene chloride is not known to occur naturally, although it has been detected in the ambient air in some areas and can enter the aquatic environment from industrial wastewater.

24.3.7 Nitrogen and sulfur compounds

Many nitrogen compounds have been reported in stool, and they likely arise from diet; for instance, methylpyrazine, pyridine, and pyrrole are constituents of coffee. In contrast, pyrrole readily polymerizes with acid and, therefore, its presence in feces is unlikely to be dietary, as it would be unlikely to survive the journey through the stomach. Ammonia results from microorganism activity and increasing the amount of protein in the diet from 63 g/day to 136 g/day has been found to increase the amount of fecal ammonia from 15 to 30 mmol/L; but increasing the amount of fiber to the high protein diet was reported to not alter the ammonia concentration.[5]

A diverse range of sulfur compounds has been identified, including methanethiol and dimethyl sulfide (DMS); the former is, at least in part, considered to be produced from methionine by *Clostridia* in the gut.[53] Methanethiol has a toxicity approaching cyanide and the factors controlling its concentration and biosynthesis might bear more examination. Methanethiol and DMS may also be produced by methylation of hydrogen sulfide (H_2S) as a detoxification mechanism by mucosal thiol S-methyltransferase.[66] DMDS and dimethyl trisulfide (DMTS) have both been commonly reported in stool.[3,42,67] Hydrogen sulfide is probably most likely to occur due to metabolism of sulfate by sulfate-reducing bacteria.[68] Sulfate, which is poorly absorbed in the small bowel, is naturally present in cruciferous vegetables (cabbage, broccoli) and nuts and as an additive in bread and beer. The main sulfur-containing flatus components in healthy individuals have been quantified (stated as mean values with standard error of the mean in parentheses): hydrogen sulfide (1.06 (0.2) μmol/L), followed by methanethiol (0.21 (0.04) μmol/L) and DMS (0.08 (0.01) μmol/L).[61] The authors were concerned about the social aspect of pungent flatus and found in their study that hydrogen sulfide and methanethiol appeared to be principally responsible, and not indole-based compounds, as previously thought.

24.3.8 Metabolism of acid in stool

Long-chain alcohols may be made by reduction of the corresponding acid, for which extensive homologous series are found in the headspace of stool. This hypothesis was investigated by incubating feces with labeled butanoic acid, whereby it was found that labeled butanol was synthesized (in increasing amounts as incubation time progressed).[3] No conversion to [1-^{13}C]butanal was observed. Esterification of acids was also found to readily occur in stool, as evidenced by the formation of [1-^{13}C]butanoic acid ethyl ester. Conversion of [1-^{13}C]butanoic acid to [1-^{13}C] butanoic acid propyl ester after 3 h incubation was also confirmed by the same method (Fig. 24.1). It is likely that microbially mediated reduction of other fatty acids also occurs.

24.4 Gastrointestinal disease and cancer

Headspace screening of fluid obtained from the gut during colonoscopy has been undertaken as a novel approach in the diagnosis of GI diseases, but no significant differences were found between patients with IBS and IBD and healthy volunteers.[69]

A large study of 205 individuals demonstrated the potential of pattern analysis of fecal VOCs in patients with IBD, as determined by GC-MS.[70] This may provide a noninvasive method in the diagnosis of IBD. A further development has been the use of a more sensitive GC-MS analysis method using a time-of-flight MS (GC-TOF-MS), which has enabled breath to be analyzed for bowel conditions.[71] It was found that six to eight compounds could be used to classify (with 95%–100% accuracy) between remission and active disease for Crohn's disease and ulcerative colitis; the sensitivity to distinguish IBD patients from controls was 100%.[71]

A review by Chan et al. offers a good overview of the diagnosis of GI illnesses using fecal headspace analysis. This review outlines the various studies undertaken to explore and diagnose CRC, IBD, IBS, and various forms of infectious diarrhea.[72] The review also details less commonly studied diseases, including celiac disease, necrotizing enterocolitis, and nonalcoholic fatty liver disease (NAFLD). Many of the studies quote sensitivity and specificity values >90% for a range of diseases, which highlights the potential of fecal VOCs to diagnose GI conditions. Nevertheless, many of the studies have relatively small numbers and only perform within-sample cross validation, highlighting a need for larger studies to validate any findings. Analysis of volatiles from the stool of neonates demonstrated that there is promise in the early detection of necrotizing enterocolitis, which is the main reason for death in premature babies.[73] The analyses of VOCs in the stools from patients with ulcerative colitis, *C. jejuni* and *C. difficile*, has also been undertaken and resulted in the characterization of 149, 183, and 145 compounds, respectively.[3] Group-to-group differences were clearly observed, as well as some interesting differences among samples from the various groups. Butanoic acid, for example, was common in all groups except *C. difficile* samples, where it was present in just 41% of samples. By comparison, butanol was ubiquitous in *C. difficile* samples. *Clostridia* are known to produce alcohols, such as ethanol, isopropanol, and butanol, at the expense of producing less butanoate.[74] Such observations fit with the results of the previously reported study.[3] Another example of the utility of stool VOC analyses is that stool from cholera sufferers could be differentiated clearly from noncholera controls by their VOC signatures.[75]

There is an ever-increasing interest in how the microbiome is altered in disease states and how this will directly impact the fecal volatile metabolites that evolve. A study from 2013 explored fecal microbiota in relation to VOC profiles from the same cohort of participants.[76] NAFLD patients were compared to healthy controls, whereby there were significantly more *Lactobacillus* and some phylum firmicutes species in NAFLD samples, which corresponded to a significant increase in ester compounds in the NAFLD patients versus the control group.[76]

Couch et al.[77] investigated the link between the fecal microbiome and the volatilome. This work focused on the difference in volatile profile and microbiome associated with sample collection, specifically with regard to the differences between samples obtained during colonoscopy and those passed naturally (collected from home). There was little change in the microbiome between the two collections methods, but significant differences in the VOCs emitted were noted.[77] Further, differences relating to extraction time of the sample headspace (using headspace SPME) have been discerned by the authors of this chapter, which highlights the importance and influence of the analytical method in VOC analysis.

Hydrogen production in the gut is well known and has been related to bacterial overgrowth in the small intestine, and excess hydrogen production has been used for the diagnosis of lactose and fructose malabsorption. Since in this case hydrogen is measured in breath and not from stool, it is not covered in this chapter.

Several studies in recent years have used fecal VOCs to explore their potential clinical utility. Bond et al., for example, have shown the potential for fecal VOCs as markers of CRC.[78] This group showed that a person is six times more likely to have CRC if each of propan-2-ol, hexan-2-one, and ethyl 3-methyl-butanoate are all present. CRC markers are an excellent target for VOC analysis, as the current gold standard diagnosis is the highly invasive colonoscopy; reducing the number of unnecessary colonoscopies would not only be beneficial from the patient perspective, but also decrease the burden on healthcare providers.

A review by Bosch et al. examined the potential of using VOCs for the detection of CRC; the primary finding of this review was that while there was a lack of standardization of fecal sample analysis, there were reoccurring features, with CRC patients tending to have higher levels of amino acids and SCFAs, with reductions in bile acids and polyol alcohols.[79]

24.5 Summary

The most notable feature of a comparison of VOCs found in urine and stool is that many more compounds have been reported in stool compared to urine. More compounds may exist in urine, and may simply be at too low a level to be measured, when compared to the concentrations found in stool. The most prominent difference in chemical class were the esters, which represent the largest class in stool, with 59 compounds found compared to two compounds in urine. The extraction and sample treatment method must be considered in the data analysis; hydrolysis of esters would certainly be expected if alkaline work-up of the sample is undertaken. Almost certainly, the source of many urinary compounds is the gut, but in most cases there is no definitive published evidence (e.g., by using labeled compounds to prove the relationship). Some compounds are undoubtedly decreased in concentration from their journey from the gut to the bladder, e.g., benzoic acid (which occurs naturally in most berries) is derivatized (in the liver) and excreted in urine as the less volatile hippuric acid (benzoylaminoethanoic acid). Although representatives of many chemical classes were found in the headspace of urine, in general shorter-chain species were found in the headspace compared to stool. Again, this is likely to be due to their presence at lower concentrations. One of the most notable differences between stool and urine is for straight-chain hydrocarbons, with 15 compounds reported in stool but only two compounds in urine, strongly suggesting that the hydrocarbons were ingested and remained unchanged through the digestion process, their low water solubility preventing their detection in urine.

Gut chemistry is very complex and this applies to omnivores in particular. The huge number of chemicals synthesized by plants enter the gut with chemical modification to a lesser or greater extent by the digestive system. The gut microorganisms are small chemical factories themselves and then act on undigested food, shed intestinal cells, bile fluids, etc. The analysis of volatiles from the headspace of both urine and stool offers a simplified assessment of the microbiome function and dysbiosis caused by disease. There is the potential for a range of disease diagnoses, in a noninvasive manner, via VOC analyses from urine and/or stool, yet more work must be undertaken, with larger studies, to determine the viability of such tests.

The ever-increasing sensitivity of instruments, such as multidimensional or tandem-MS GC systems (GC-GC-TOF-MS, GC × GC-TOF-MS, or GC-MS-MS) and real-time techniques (SIFT-MS and PTR-MS) has opened new doors to the exploration of both urinary and fecal VOCs. Despite the work carried out by research groups all over the world, however, there is no urinary or fecal VOC clinical test in widespread use, to date. This is due to a lack of standardization across research groups in terms of sample collection, storage, preprocessing, and analysis, making it difficult for research groups to have their findings supported—or even validated—by other groups. Moreover, there is a lack of large-scale studies, which are needed to make accurate inferences about how VOC profiles in health and disease equate to the general population.

Despite developments in instrumentation, there are still relatively few prospective studies to determine the exact VOCs emanating from stool and urine. Studies on large cohorts and diverse populations are required if the diagnostic potential of VOC analysis is to be realized, by providing a basis for a better understanding of the metabolic processes and the role of the microbiome in disease development. Such studies should be designed to maximize knowledge of VOCs, qualitatively and quantitatively, as it is unlikely that single biomarkers of disease states exist. The use of the most sensitive and up-to-date hyphenated MS techniques should yield many more VOCs, which should ideally be confirmed with standards (given the likelihood that gut chemistry and metabolism produce a range of VOCs that are not currently represented in comprehensive mass spectral databases). Longitudinal studies are also needed to understand variation in populations with time. Realistically, these endeavors can only be achieved by large partnership projects to establish a comprehensive picture of the human volatilome.

Note on this chapter

The previous incarnation of this chapter (in the first edition of this book) aimed to list all VOCs found up until that point in stool and urine samples. This chapter itself inspired the idea of a compendium of VOCs emanating from the human body, which was subsequently published as a review in 2014 in *Journal of Breath Research* (ref. [4] in this chapter). Rather than replicate the associated tables here, we refer the interested reader to the original chapter, the subsequent review (and imminent update), and the EPA database on the human volatilome (see Chapter 39).

Acknowledgments

We must acknowledge our friend Anton Amann for his part in our joint venture to catalog the volatiles from healthy humans.

References

1. Mills GA, Walker V. Headspace solid-phase microextraction profiling of volatile compounds in urine: application to metabolic investigations. *J Chromatogr B* 2001; **753**(2):259–68.
2. Smith S, Burden H, Persad R, Whittington K, DeLacy Costello B, Ratciffe NM, Probert CS. A comparative study of the analysis of human urine headspace using gas chromatography-mass chromatography. *J Breath Res* 2008;**2**(10pp):037022.
3. Garner CE, Smith S, DeLacy Costello B, White P, Spencer R, Probert CS, Ratcliffe NM. Volatile organic compounds from feces and their potential for diagnostics of gastrointestinal disease. *FASEB (Fed Am Soc Exp Biol) J* 2007;**21**:1675–88.
4. de Lacy Costello B, Amann A, Al-Kateb H, Flynn C, Filipiak W, Khalid T, et al. A review of the volatiles from the healthy human body. *J Breath Res* 2014;**8**(1):014001.
5. Bone E, Tamm A, Hill M. The production of urinary phenols by gut bacteria and their possible role in the causation of large bowel cancer. *Am JClin Nutr* 1976;**29**(12): 1448–54.
6. Undo Liebich HM, Buelow HJ, Kallmayer R. Quantification of endogenous aliphatic alcohols in serum and urine. *J Chromatogr A* 1982;**239**(30):343–9.
7. Wahl HG, Hofmann A, Luft D, Liebich HM. Analysis of volatile organic compounds in human urine by headspace gas chromatography mass spectrometry with a multipurpose headspace sampler. *J Chromatogr, A* 1999;**847**(1–2):117–25.
8. Schatowitz B, Gercken G. Simulatneous determination of C2-C22 non-esterified fatty acids and other metabolically relevant carboxylic acids in biological material by gas chromatography of their benzyl esters. *J Chromatogr* 1988;**425**:257–68.
9. Mitchell SC, Zhang AQ. Methylamine in human urine. *Clin Chim Acta* 2001;**312**(1–2): 107–14.
10. Zhang AQ, Mitchell SC, Smith RL. Dimethylamine in human urine. *Clin Chim Acta* 1995;**233**(1–2):81–8.

11. Zhang AQ, Mitchell SC, Ayesh R, Smith RL. Determination of trimethylamine and related aliphatic amines in human urine by head-space gas chromatography. *J Chromatogr B* 1992;**584**(2):141–5.
12. Hradecky P. Volatile fatty acids in urine and vaginal secretions of cows during reproductive cycle. *J Chem Ecol* 1986;**12**(1):187–96.
13. Rhodes G, Holland ML, Wiesler D, Novotny M, Moore SA, Peterson RG, Felton DL. Excretion of urinary volatile metabolites in response to alloxan induced diabetes of short duration in rats. *J Chromatogr B* 1982;**228**:33–42.
14. Leibich HM, Al-Babbili O, Zlatkis A, Kyoung K. Gas-chromatographic and mass-spectrometric detection of low molecular weight aliphatic alcohols in urine of normal individulas and patients with Diabetes Mellitus. *Clin Chem* 1975;**21**(9):1294–6.
15. Quraishi MN, Harrison L, Wedlake L, Ouaret N, Maxim J, Nwokolo C, Andreyev J, Bardhan KD, Covington JC, Arasaradnam RP. Sniffing gastrointestinal diseases with an electronic E-nose–Is this the future for diagnostics? *Absts 10th Congr Euro Fed Int. Med./Eur J Intern Med* 2011;**22**(Suppl. 1):S78.
16. Diskin AM, Spanel P, Smith D. Increase of acetone and ammonia in urine headspace and breath during ovulation quantified using selected ion flow tube mass spectrometry. *Physiol Meas* 2003;**24**(1):191.
17. Smith D, Khaled MK, Ismail DAM, Chapman G, Magnay JL, Španěl P, O'Brien S. Increase of acetone emitted by urine in relation to ovulation. *Acta Obstet Gynecol Scand* 2006;**85**(8):1008–11.
18. Banday KM, Pasikanti KK, Chan EC, Singla R, Rao KV, Chauhan VS, Nanda RK. Use of urine volatile organic compounds to discriminate tuberculosis patients from healthy subjects. *Anal Chem* 2011;**83**(14):5526–34.
19. Willis CM, Church SM, Guest CM, Cook WA, McCarthy N, Bransbury AJ, Church MRT, Church JCT. Olfactory detection of human bladder cancer by dogs: proof of principle study. *BMJ* 2004;**329**:712.
20. Cornu JN, Cancel-Tassin G, Ondet V, Girardet C, Cussenot O. Olfactory detection of prostate cancer by dogs sniffing urine: a step forward in early diagnosis. *Eur Urol* 2011;**59**(2):197–201.
21. Weber CM, Cauchi M, Patel M, Bessant C, Turner C, Britton LE, Willis CM. Evaluation of a gas sensor array and pattern recognition for the identification of bladder cancer from urine headspace. *Analyst* 2011;**136**(2):359–64.
22. Spanel P, Smith D, Holland TA, Singay WA, Elder JB. Analysis of formaldehyde in the headspace of urine from bladder and prostate cancer patients using selected ion flow tube mass spectroscopy. *Rapid Commun Mass Spectrom* 1999;**13**(14):1354–9.
23. Smith S, White P, Redding J, Ratcliffe NM, Probert C. Application of similarity coefficients to predict disease using volatile organic compounds. *Sensors Journal IEEE* 2010; **10**(1):92–6.
24. Carrola J, Rocha CM, Barros AS, Gil AM, Goodfellow BJ, Carriera IM, Bernardo J, Gomes A, Sousa V, Carvolho DIF. Metabolic signatures of lung cancer patients in biofluids: NMR based metabonomics of urine. *J Proteome Res* 2011;**10**(1):221–30.
25. Silva CL, Passos M, Câmara JS. Investigation of urinary volatile organic metabolites as potential cancer biomarkers by solid-phase microextraction in combination with gas chromatography-mass spectrometry. *Br J Canc* 2011;**105**:1894–904.
26. Di Lena M, Porcelli F, Altomare DF. Volatile organic compounds as new biomarkers for colorectal cancer: a review. *Colorectal Dis* July 2016;**18**(7):654–63.

27. McFarlane M, Millard A, Hall H, Savage R, Constantinidou C, Arasaradnam R, Nwokolo C. Urinary volatile organic compounds and faecal microbiome profiles in colorectal cancer. *Colorectal Dis* November 2019;**21**(11):1259–69.
28. Mozdiak E, Wicaksono AN, Covington JA, Arasaradnam RP. Colorectal cancer and adenoma screening using urinary volatile organic compound (VOC) detection: early results from a single-centre bowel screening population (UK BCSP). *Tech Coloproctol* 2019;**23**(4):343–51.
29. Gao Q, Lee WY. Urinary metabolites for urological cancer detection: a review on the application of volatile organic compounds for cancers. *Am j clin exp urol* 2019;**7**(4):232.
30. Aggio RB, de Lacy Costello B, White P, et al. The use of a gas chromatography-sensor system combined with advanced statistical methods, towards the diagnosis of urological malignancies. *J Breath Res* 2016;**10**(1):017106. Published 2016 Feb 11.
31. Woollam M, Teli M, Angarita-Rivera P, et al. Detection of volatile organic compounds (VOCs) in urine via gas chromatography-mass spectrometry QTOF to differentiate between localized and metastatic models of breast cancer. *Sci Rep* 2019;**9**:2526.
32. Bax C, Taverna G, Eusebio L, Sironi S, Grizzi F, Guazzoni G, Capelli L. Innovative diagnostic methods for early prostate cancer detection through urine analysis: a review. *Cancers* 2018;**10**(4):123.
33. Global Burden of Disease Study 2013 Collaborators. Global, regional, and national incidence, prevalence, and years lived with disability for 301 acute and chronic diseases and injuries in 188 countries, 1990-2013: a systematic analysis for the Global Burden of Disease Study 2013. *Lancet* August 22, 2015;**386**(9995):743–800.
34. Barnett BJ, Stephens DS. Urinary Tract infection: an overview. *Am J Med Sci* 1997; **314**(4):245–9.
35. Pantell RH, Newman TD, Bernweig J, Bergman DA, Takyama JI, Segel M, Finch CA, Wasserman RC. Management and outcomes of care of fever in early infancy. *J Am Med Assoc* 2004;**291**(10):1203–12.
36. Lumbiganon P, Laopaiboon M, Thinkhamrop J. Screening and treating asymptomatic bacteriuria in pregnancy. *Curr Opin Obstet Gynecol* 2010;**22**(2):95–9.
37. Persaud KC, Pisanelli AM, Evans P, Travers PJ. Monitoring urinary tract infections and bacterial vaginosis. *Sensor Actuator B* 2006;**116**(1–2):116–20.
38. Williams GJ, Macaskill P, Chan SF, Turner RM, Hodson E, Craig JC. Absolute and relative accuracy of rapid urine tests for urinary tract infection in children: a meta analysis. *Lancet Infect Dis* 2010;**10**(4):240–50.
39. Bate JP, Irving PM, Barrett JS, Gibson PR. Benefits of breath hydrogen testing after lactulose administration in analysing carbohydrate malabsorption. *Eur J Gastroenterol Hepatol* 2010;**22**(3):318–26.
40. Ruge E. Beitrag zur kenntness der darmgase. *Sitzber Kaiser Aka* 1961;**44**:739–62.
41. Tomlin J, Lowis C, Read NW. Investigation of normal flatus production in healthy volunteers. *Gut* 1991;**32**(6):665–9.
42. Suarez F, Furne J, Springfield J, Levitt M. Insights into human colonic physiology obtained from the study of flatus composition. *Am J Physiol Gastrointest Liver Physiol* 1997;**272**(5):G1028–33.
43. Gibson GR, Macfarlane GT, Cummings JH. Sulphate reducing bacteria and hydrogen metabolism in the human large intestine. *Gut* 1993;**34**:437–9.

44. Miller TL, Wolin MJ. Methanogens in animal and human intestinal tracts. *Syst Appl Microbiol* 1986;**7**:223–9.
45. Willis CL, Cummings JH, Neale G, Gibson GR. Nutritional aspects of dissimilatory sulphate reduction in the human colon. *Curr Microbiol* 1997;**35**:294–8.
46. Gould O, de Lacy Costello B, Smart A, Jones P, Macmaster A, Ransley K, Ratcliffe N. Gas chromatography mass spectrometry (GC-MS) quantification of metabolites in stool using ^{13}C labelled compounds. *Metabolites* October 31, 2018;**8**(4):75.
47. Mochalski P, Unterkofler K. Quantification of selected volatile organic compounds in human urine by gas chromatography selective reagent ionization time of flight mass spectrometry (GC-SRI-TOF-MS) coupled with headspace solid-phase microextraction. *HS-SPME) Analyst* 2016;**141**:4796–803.
48. Cummings JH. Short chain fatty acids in the human colon. *Gut* 1981;**22**:763–79.
49. Yokoyama MT, Carlson JR. Microbial metabolites of tryptophan in the intestinal tract with special reference to skatole. *Am J Clin Nutr* 1979;**32**(1):173–8.
50. Gostner A, Blaut M, Schaeffer V, Kozianowski S, Theis S, Klingeberg M, Dombrowski D, Martin S, Ehrhardt D, Taras A, Schwiertz A, Kleessen H, Luhrs H, Schauber J, Dorbath D, Menzel T, Scheppach W. Effect of isomalt consumption on faecal microflora and colonic metabolism in healthy volunteers. *Br J Nutr* 2006;**95**:40–50.
51. Backhed F, Ley RE, Sonnenberg JL, Peterson DA, Gordon JI. Host-bacterial mutualism in the human intestine. *Science* 2005;**307**:1915–20.
52. Mortensen PB, Clausen MR. Short-chain fatty acids in human colon: relation to gastrointestinal health and disease. *Scand J Gastroenterol* 1996;**31**:132–48.
53. Smith EA, Macfarlane GT. Dissimilatory amino acid metabolism in human colonic bacteria. *Anaerobe* 1997;**3**(5):327–37.
54. Pitcher MCL, Beatty E, Cummings JH. The contribution of sulphate reducing bacteria and 5-aminosalicylic acid to faecal sulphide in patients with ulcerative collitus. *Gut* 2000;**46**:64–72.
55. Cummings JH, Hill MJ, Bone ES, Branch WJ, Jenkins DJA. The effect of meat protein and dietry fiber on colonic function and metabolism. *Am J Clin Nutr* 1979;**32**:2094–101.
56. De Preter V, Van Staeyen G, Esser D, Rutgeerts P, Verbeke K. Development of a screening method to determine the pattern of fermentation metabolites in faecal samples using on-line purge-and-trap gas chromatographic-mass spectrometric analysis. *J Chromatogr A* 2009;**1216**(9):1476–83.
57. Dixon E, Clubb C, Pittman S, Ammann L, Rasheed Z, Kazmi N, Keshavarzian A, Gilevet P, Rangwala H, Couch R. Solid-phase microextraction and the human fecal VOC metabolome. *PLoS One* 2011:0018471.
58. Lington AW, Bevan C, Clayton GD, Clayton FE. In: *Patty's industrial hygiene and toxicology*. 4th ed. New York: John Wiley & Sons; 1994. p. 2585–760.
59. Elgaali H, Hamilton-Kemp TR, Newman MC, Collins RW, Yu K, Archbold DD. Comparison of long chain alcohols and other volatile compounds emmitted from food-borne related gram positive and gram negative bacteria. *J Basic Microbiol* 2002;**42**(6):373–80.
60. Perez Locas C, Yaylayan VA. Origin and mechanistic pathways of formation of the parent furans, a food toxicant. *J Agric Food Chem* 2004;**52**(22):6830–6.
61. 61Seitz HK, Matsuzaki S, Yokoyama, Homann N, Vakevainen S, Dong Wang X. The role of acetaldehyde in alcohol-associated cancer of the gastrointestinal tract. *Alcohol Clin Exp Res* 2001;**25**(S1):137S–43S.

62. Decombaz J, Arnaud MJ, Milon H, Moesch H, Philipposian G, Thelin AL, Howald H. Energy metabolism of medium-chain triglycerides versus carbohydrates during exercise. *Eur J Appl Physiol Occup Physiol* 1983;**52**(1):9–14.
63. Ott A, Germond JE, Chaintreau A. Vicinal diketone formation in yoghurt: 13C precursors and and effect of branched-chain amino acids. *J Agric Food Chem* 2000;**48**:724–31.
64. Salerno-Kennedy R, Cashman KD. Potential applications of breath isoprene as a biomarker in modern medicine; a concise overview. *Wien Klin Wochenschr* 2005; **117**(5–6):180–6.
65. Deneris ES, Stein RA, Mead JF. Acid -catalysed formation of isoprene from a mevalonate-derived product using a rat liver cytosolic fraction. *J Biol Chem* 1985; **260**(3):1382–5.
66. Weiseger RA, Pinkus LM, Jakoby WB. Thiol S-methyltransferase: suggested role in detoxification of intestinal hydrogen sulphide. *Biochem Pharmacol* 1980;**29**(20): 2885–7.
67. Moore JG, Jessop LD, Osborne DN. Gas-chromatographic and mass-spectromeric analysis of the odor of human feces. *Gastroenterology* 1987;**93**(6):1321–9.
68. Suarez FL, Springfield J, Levitt MD. Identification of gases responsible for the odour of human flatus and evaluation of a device purported to reduce this odour. *G*ut 1998;**43**: **10**0–4.
69. Lechner M, Colvin HP, Ginzel C, Lirk P, Rieder J, Tilg H. Headspace screening of fluid obtained from the gut during colonoscopy and breath analysis by proton transfer reaction-mass spectrometry: a novel approach in the diagnosis of gastro-intestinal diseases. *Int j mass spectrosc* 2005;**243**(2):151–4.
70. Ahmed I, Probert CS, Smith S, Ratcliffe N, Greenwood R. Diagnostic potential of volatile organic compounds in inflammatory bowel disease. *Gastroenterology* 2010; **138**(5):S763–4.
71. Bodelier A, Pierik MJ, Dallinga JW, Berkel JV, Moonen E, Hameeteman W, Jonkers D, Masclee A, Van Schooten J. Analysis of volatile organic compounds in exhaled air as a non invasive biomarker of inflamatry bowel disease. *Gastroenterology* 2009;**13**(5):A35.
72. Chan DK, Leggett CL, Wang KK. Diagnosing gastrointestinal illnesses using fecal headspace volatile organic compounds. *World J Gastroenterol* 2016;**22**(4):1639–49.
73. Garner CE, Ewer AK, Elasouad K, Power F, Greenwood R, DeLacy Costello B, Probert CS, Ratcliffe NM. Analysis of faecal volatile organic compounds in preterm infants who develop necrotising enterocolitis: a pilot study. *J Pediatr Gastroenterol Nutr* 2009;**49**(5):559–65.
74. May T, Mackie RI, Fahey GC, Cremin JC, Garleb KA. Effect of fibre source on short-chain fatty acid production and on the growth and toxin production by *Clostridium difficile*. *Scand J Gastroenterol* 1994;**29**:916–22.
75. Garner CE, Smith S, Bardhan PK, Probert CS, Ratcliffe NM. A pilot study of faecal volatile organic compounds in faeces from cholera patients in Bangladesh to determine their utility in disease diagnosis. *Trans r soci trop med hygeine* 2009;**103**(11):1171–3.
76. Raman M, Ahmed I, Gillevet PM, Probert CS, Ratcliffe NM, Smith S, Greenwood R, Sikaroodi M, Lam V, Crotty P, Bailey J. Fecal microbiome and volatile organic compound metabolome in obese humans with nonalcoholic fatty liver disease. *Clin Gastroenterol Hepatol* July 1, 2013;**11**(7):868–75.

77. Couch RD, Navarro K, Sikaroodi M, Gillevet P, Forsyth CB, Mutlu E, Engen PA, Keshavarzian A. The approach to sample acquisition and its impact on the derived human fecal microbiome and VOC metabolome. *PloS One* 2013;**8**(11).
78. Bond A, Greenwood R, Lewis S, Corfe B, Sarkar S, O'Toole P, Rooney P, Burkitt M, Hold G, Probert C. Volatile organic compounds emitted from faeces as a biomarker for colorectal cancer. *Aliment Pharmacol Therapeut* April 2019;**49**(8):1005–12.
79. Bosch S, Berkhout DJ, Larbi IB, de Meij TG, de Boer NK. Fecal volatile organic compounds for early detection of colorectal cancer: where are we now? *J Canc Res Clin Oncol* January 22, 2019;**145**(1):223–34.

CHAPTER

Volatile emissions from skin

25

Leena A. Nylander-French[1], Jonathan D. Beauchamp[2], Joachim D. Pleil[1]

[1]*Department of Environmental Sciences and Engineering, Gillings School of Global Public Health, Univeristy of North Carolina, Chapel Hill, NC, United States;* [2]*Fraunhofer Institute for Process Engineering and Packaging IVV, Freising, Germany*

Chapter outline

25.1 Overview

A primary function of the skin is to form a barrier to external exposures, preventing the entry of toxicants, pathogens, and allergens, while limiting the loss of moisture. The principal barrier is the epidermis, which consists of layers of differentiated keratinocytes resulting in the stratum corneum (SC) (see Fig. 25.1 for skin structure). In addition, the skin serves as the body's thermostat through changes in capillary perfusion and perspiration. Due to its complex structure and functions, many events can take place in the skin, which are related to its physical, chemical, and biochemical constitution, as well as to physicochemical behaviors of the compounds that come into contact with skin and are available for percutaneous absorption and/or evaporation from its surface. As such, the skin is second only to breath as a source of volatile organic compound (VOC) emissions through direct perfusion of gases from the surface capillaries, as well as being a source of human scent via perspiration and activities of surface microbiome. Much of the scientific work to date has focused upon skin barrier functions and absorption of chemicals

Breathborne Biomarkers and the Human Volatilome. https://doi.org/10.1016/B978-0-12-819967-1.00025-6
Copyright © 2020 Elsevier B.V. All rights reserved.

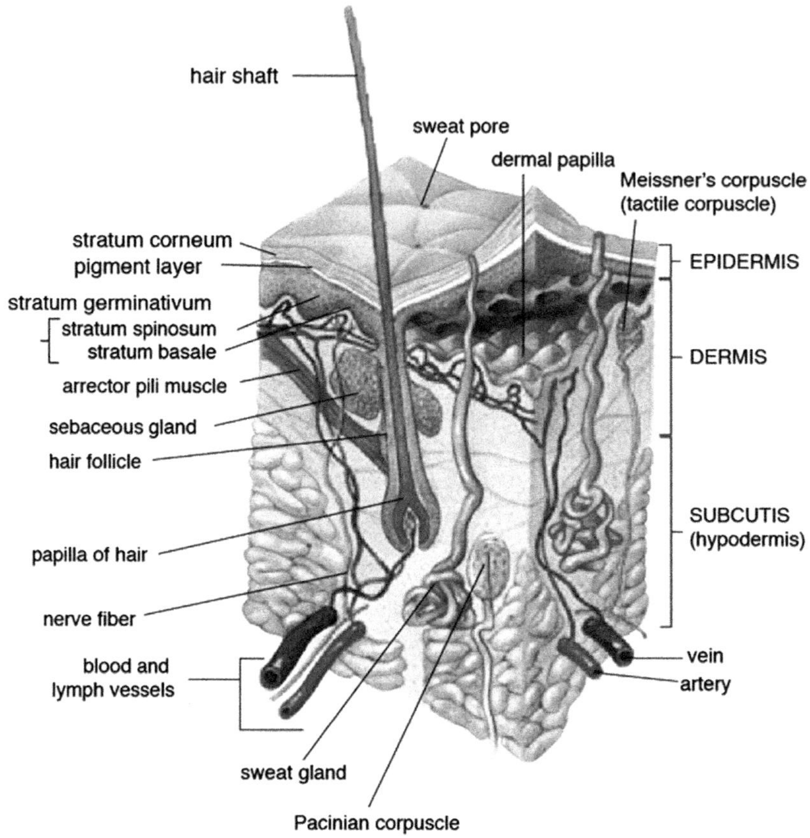

FIGURE 25.1

The structure of human skin. Sweat glands and pores, hair shafts, and blood and lymph vessels can all contribute to chemical releases through the skin.

Reproduced from [4].

into and through the skin, and the potential for these to cause adverse health effects. During the past decade, however, the emissions of VOCs from the human skin have been given an increased attention, both in scientific[1,2] and clinical[3] settings.

Like breath, skin VOCs can be of endogenous or exogenous origin, whereby the former can reflect general biochemical processes within the body or potentially indicate an imbalance associated with compromised health, and the latter mirror environmental exposures or intake of xenobiotics, including food constituents. Unlike breath, however, the exterior nature of skin and its complex surface composition presents an additional facet in relation to volatile markers. Skin lipids, which are abundant across the entire surface—albeit with localized quantitative variations—are highly susceptible to peroxidation through ultraviolet (UV) radiation and atmospheric radicals, leading to a complex and varying surface chemistry. Further, the skin microbiota contributes to volatile emissions at the surface. At the very least, these additional sources of VOCs require consideration in terms of their potential to confound datasets targeting emissions through the skin surface from within the body, but equally they represent phenomena that are worthy of investigation in their own right.

This chapter describes the exchange of chemicals through the skin, serving as both source and sink of environmental chemicals in the body, and addresses skin surface chemistry as an additional feature of exploring volatile emissions. In addition, this chapter reviews the sampling and analytical methods for measuring skin emanations and the interpretation of these data for assessing environmental exposures and health state.

25.2 Skin structure and volatilome

As indicated above, the principal function of the skin is twofold, namely to act as a barrier to external exposures while reducing moisture losses from within the body. The epidermis, which is the principal barrier, is represented by the SC. The SC contains enucleated cells connected in an integrated matrix of specialized proteins and lipids and provides the structural barrier protecting the skin and the underlying tissues from heat, UV radiation, water loss, and other environmental insults. Physiologically, skin participates directly in thermal regulation, metabolism, electrolyte and hormonal regulation, and immune regulation as follows:

- *Thermal regulation*: regulation of blood flow, hair growth, and perspiration
- *Metabolism*: keratin, collagen, melanin, lipids, vitamin D synthesis, respiration, and biotransformation
- *Electrolyte and hormonal regulation*: apocrine, eccrine, and sebaceous secretions, as well as endocrine functions
- *Immune regulation*: sensitization

The physical, chemical, and biochemical constitution of the skin elicit complex events at its surface. Skin is a highly metabolic organ and epidermis is the major site for metabolism of xenobiotics, steroids, and vitamins. Readily absorbed chemicals, such as solvents, can permeate human skin at rates exceeding 1 $mg/cm^2/h$, while very lipophilic compounds, for example, flame retardants, are absorbed at much slower rates, typically ng—$pg/cm^2/h$. Most environmental xenobiotics are lipophilic (i.e., they can penetrate lipid membranes), thus they represent substrates for biotransforming enzymes in the skin. Human skin expresses myriad transport and metabolic enzymes that provide active uptake and biotransformation of many xenobiotics, including drugs, solvents, and carcinogens. Like in other organs in the human body, endogenous and exogenous compounds are commonly metabolized by phase I and phase II detoxification pathways in the skin. These metabolic functions have a significant effect on both local and systemic transdermal delivery and provide the critical pathway for volatile and nonvolatile metabolites to be eliminated through skin, breath, or urine.

Adult human skin has a large surface area (1.5–2.0 m^2) and is perfused by ample supply of blood carrying plentiful endogenous and exogenous compounds, including VOCs, from other tissues. In a comprehensive review of VOCs identified in healthy individuals, of the 1849 compounds reported, 504 compounds related to emissions from skin (compared to 874 compounds in breath).[5] While some VOCs are specific to a body fluid or tissue, and others are common to emanation sources, very little is known about the tissue of origin or production mechanisms for the major portion of these VOCs. Common skin VOCs include aldehydes, hydrocarbons, ketones, esters, and alcohols.[6] Many of the skin VOCs are derived from food, consumer products, or as contaminants from the environment (e.g., ambient air).

Apocrine, eccrine, and sebaceous gland secretions are the major sources for human skin VOCs, i.e., metabolism and transformation of sweat and sebum.[1,2,7] Additional skin VOCs can originate from blood (e.g., internal hormonal or metabolic processes) or result from metabolism by skin microbiota and their use of skin gland secretions.[1] Many of the skin volatiles are odorless, but odoriferous VOCs derived from skin microbial activity are the primary source of human body odor. Any disturbance in the homeostatic balance of skin functions, e.g., due to bacterial infection, can induce changes in both the quality and quantity of VOCs emitted from the skin. Skin VOCs have been associated with a number of diseases and infections and also used as diagnostic biomarkers for infectious and metabolic diseases, as well as genetic disorders.[3,7]

25.3 Volatile compounds emitted by skin

A comprehensive review of the VOC emanations from human skin is beyond the scope of this chapter, and the interested reader is directed to the literature.[1–3] Skin VOCs mainly arise from secretions from the eccrine, sebaceous, and apocrine glands but are also derived from the metabolic functions of skin microbiome, interaction of the microbiome with the excreta from these glands, and from internal hormonal, metabolic, or thermoregulatory processes carried by blood from other tissues. The glands in the skin are distributed to a varying extent across the human body, and different VOC profiles have been observed between skin regions. Eccrine glands are the major sweat glands in human skin, and their main function is thermoregulation. Their highest density occurs on palms, soles, axillae, and forehead, producing odorless sweat containing mostly water, sugars, glycoproteins, lactic acid, amino acids, electrolytes, and sialomucin.[8] Apocrine glands are mostly located in axillae and genital regions. These produce lipids, proteins, steroids, and sialomucin, and their secretion is stimulated by adrenaline. The function of apocrine glands is unclear, but they may represent scent glands. Sebaceous glands, which secrete sebum and lipids, are distributed all over the body but are concentrated on the upper part of the body (head, scalp, face, forehead, chest, and back).[9] These glands contain cholesterol, cholesterol esters, long-chain fatty acids (FAs), squalene, and triglycerides. Skin microbiome is rich in these skin areas, and these compounds provide an optimal growth environment for these microbes. Skin microbiome activity with sebaceous secretions causes emission of acids, alcohols, and aldehydes from the skin.[10] Human body odors emitted by different skin areas reflect the distribution of these glands on the skin surface, as well as the composition of skin microbiome, which greatly differs between individuals.

The chemical profile of an individual's skin VOCs has been observed to vary due to skin area, age, gender, ethnicity, hormonal status, enzyme expression, diet, and various pathological conditions.[1–3] Gallagher et al.[11] observed significant quantitative and qualitative differences in VOCs between upper back and forearm, even though many compounds were in common. No observations of significant differences in skin VOCs relating to sex were made, but small differences related to age were observed, particularly for the compounds dimethyl sulfone, benzothiazole, and nonanal in older individuals. The differing distribution of apocrine, eccrine, and sebaceous glands in the human body could be the most likely reason for variation of VOCs in different skin areas. The axillary region, for example, can produce a variety of VOCs due to the high densities of apocrine, eccrine, and sebaceous glands and a cohabitation of variable microbiome. Two main odoriferous compounds, (*E*)-3-methylhex-2-enoic acid and 3-hydroxy-3-methylhexanoic acid, in axilla samples have been reported.[12] Further, a series of carboxylic acids have been observed to be emitted at the axilla.[13] Lipophilic corynebacteria have been observed to be responsible for the production of malodorous VOCs on the axillary region.[12]

In other regions of the body, Bernier et al.[14] observed hundreds of VOCs in skin secretions collected from the palms and backs of hands and reported that most of these compounds were organic C_{16} and C_{18} saturated, monounsaturated, and diunsaturated acids. Interestingly, these organic acids are not volatile at body temperature. The most commonly identified VOCs in the hands are aldehydes and ketones, and particularly 6-methyl-5-hepten-2-one (6-MHO), nonanal, decanal, undecanal, and geranylacetone ((E)-6,10-dimethyl-5,9-undecadien-2-one). These same compounds have also been identified as major VOCs in forearms, along with some alkanes and carboxylic acids.[11,15] Curran et al.[16] observed six compounds to be frequent VOC emanations from hands, namely furan-2-carbaldehyde, furan-2-ylmethanol, phenol, nonanal, decanal, and dimethyl hexanedioate. Other studies, however, have reported only nonanal, decanal, and some carboxylic acid-methyl esters to have been isolated regularly from hands.[17,18]

Volatile organic compounds emitted from feet have received little attention in comparison to compounds derived from other human body regions. Skin microbiome is dense on the feet and responsible for the production of many malodorous VOCs.[19] Microbiome on the feet has also been reported to vary between feet exhibiting low and high levels of odor.[19] Common foot VOCs are carboxylic acids.

Ara et al.[20] reported several short-chain FAs in solvent extracts of foot sweat and identified 3-methylbutanoic acid (isovaleric acid) as the likely culprit for strong foot odor. Qiu and colleagues[21], on the other hand, did not detect such compounds when analyzing foot odors from headspace extracts collected using Tenax sorbents. Caroprese et al.[22] detected FAs from solid-phase microextraction (SPME) fibers exposed to cotton samples collected from the interdigital areas of the human feet. Isolation of short-chain FAs from foot skin emanations is probably dependent on both the sampling method used and the skin area sampled. The production of foot VOCs may be modified by the variation of microbiota in the feet or inhibited by the use of fragrances or sweat reducing agents. Identification of VOCs from feet has had a great commercial interest for the development of devices and/or compounds that can reduce foot malodor[20,22] and for detection of olfactory cues that mediate host-seeking behavior of blood-sucking insects.[1,21]

Skin VOCs emitted by the whole human body were investigated in two studies by placing human volunteers within individual aluminumized plastic bags, with only their heads outside.[23,24] Logan et al.[24] identified 24 volatile compounds using coupled gas chromatography–electroantennographic detection (GC-EAD), and six of the compounds, namely 6-MHO, octanal, nonanal, decanal, and geranylacetone, were observed to be strongly attractive to anthropophilic insects. Similarly, Harraca et al.[23] isolated four C_7-C_{10} aldehydes (heptanal, octanal, nonanal, decanal), 6-MHO, and geranylacetone using gas chromatography (GC).

The current literature shows that the skin site is a major contributor to the observed differences in skin VOC emanations and that the chemical composition of these VOCs is varied. Nevertheless, only a few chemical groups are prevalent; these include carboxylic acids, aldehydes, alkanes, alcohols, esters, and ketones. The most commonly reported compounds are nonanal, decanal, 6-MHO, and geranylacetone. Gallagher et al.[11] also observed that many of the identified skin VOCs could be attributed to exogenous sources, like fragrances, shampoos and soaps, and other skin care products.

25.4 Skin surface chemistry

The extensive area of skin and its rich abundance of lipids predispose it to complex surface chemistry, initiated through exposure to solar radiation or atmospheric radicals. While reaction chemistry products are not biomarkers in the strictest sense, they can indicate exposure—as is relevant when exploring the human exposome (see Chapter 1)—or represent potential confounders that should be taken into account when endogenous skinborne biomarkers are targeted. Skin surface chemistry is complex. A thorough review is beyond the scope of this chapter, thus this section will only scratch the surface, so to speak (pun intended), by introducing aspects of this topic in relation to key studies that have explored volatile emissions as a result of environment stressors.

25.4.1 Ultraviolet radiation

Exposure of the skin to solar radiation, specifically in the UV range at wavelengths from 200 to 400 nm—as similarly encountered in sunbed use for skin "tanning" purposes—induces a production of reactive oxygen species (ROS) and an ensuing peroxidation of skin lipids, specifically unsaturated FAs, primarily linoleic and palmitic acid. During lipid peroxidation, volatiles are produced as by- or end products of the reaction process, and these can be detected at the skin surface. Typical peroxidation products in skin are saturated and unsaturated aldehydes. Different methods of sampling and analyzing skin volatiles are presented and discussed below, but typically headspace samples are extracted for offline analysis, such as via gas chromatography−mass spectrometry (GC-MS). This approach allows a comprehensive picture of the volatile fingerprint of skin to be taken, but collecting and analyzing discrete and intermittent samples is not particularly conducive to exploring the rapid skin chemistry induced by UV radiation.

An alternative methodology is to employ an analytical platform that is capable of online analysis, i.e., continual, real-time detection. One such technology is proton transfer reaction−mass spectrometry (PTR-MS; see Chapter 10), which has been used to explore the kinetics of UV radiation−induced peroxidation of skin lipids in vivo. The study in question utilized a cuvette system to sample gas emanations directly from the skin surface before, during, and after UV irradiation.[25] Acetaldehyde had the greatest abundance during exposure, followed by propanol and several other instrumental signals of unknown identity, one of which was tentatively attributed to pentanal or 3-methylbutanal. Concomitant detection of ethane (not detectable by PTR-MS) was made with a laser-based photoacoustic system, whereby a similar profile of production to the aldehydes was observed. Notably, the real-time analyses revealed a marked increase in aldehydes immediately (within seconds) upon irradiation, with a plateau in concentrations after just over 3 min; likewise, an immediate drop in aldehyde production was observed when the UV lamp was switched off. A further interesting observation of this study was that the abundance of compounds generated was proportional to the intensity of the radiation source. Taken together, the results of such online irradiation studies shed light on prospective peroxidation products and their kinetics of production, which have potential utility in exploring the protective function of the skin against UV radiation.

25.4.2 Ozone-induced reactions

As a barrier, the skin is constantly exposed to gaseous ambient constituent, including inorganic and organic compounds, and particulate matter. Although many of these components have the potential to react at the skin surface, it is the highly reactive compounds that induce the most notable reactions. Among the reactive species are tropospheric radicals, with ozone (O_3) being the most relevant for skin. The degree of exposure to tropospheric O_3 varies with season and location, but typically ambient levels range from around 20–40 ppb in remote forest or marine environments, 50–150 ppb in rural settings, and 100–400 ppb in suburban and urban locations. Unlike UV (notwithstanding intentional irradiation in a sunbed), exposure to O_3 is not limited to outdoor settings, since this radical can be transported indoors through passive or active ventilation systems.

As outlined above, the surface of the skin is rich in lipids, among which is squalene, which constitutes 10%–12% of the surface lipids.[9] Squalene is the skin lipid that is most susceptible to ozonolysis. Like with UV-induced changes, skin surface reactions arising from O_3 exposure are rapid; as such, online analytical platforms, such as PTR-MS, are ideally suited for exploring the kinetics of these processes. A seminal study on the volatile compounds generated during ozonolysis of squalene using PTR-MS revealed the production of several compounds previously not reported due to the limitations in conventional analytical techniques. In addition to acetone, the exposure of the forehead (rich in squalene) to 50 ppb O_3 led to the generation of predominantly 6-MHO, decanal, and 4-oxopentanal, among other compounds.[26] Steady-state mixing ratios of these reaction products were observed to scale with O_3 concentration. As with the UV irradiation study, rapid analyses of reaction products during exposure offer the opportunity to characterize the kinetics of surface chemistry, thereby providing insights into human–environment interactions.

25.5 Measuring skin volatiles

The profile of skin VOCs varies greatly among the published studies, and the use of different sampling procedures and analytical chemistry methods is responsible for some of these variations. To increase our knowledge on skin VOC emanations, further studies are needed to compare and evaluate the efficiency of the sampling and analytical methods to quantify different VOCs produced by human skin. It is possible that due to the diversity of skin VOCs, two or more different methods must be simultaneously applied to characterize the complete VOC profile. A further significant need is to develop convenient methods to be used in field conditions. The ensuing section reviews sampling procedures and analytical platforms for the determination of skin VOCs.

Current analytical chemistry techniques allow rapid analyses of VOCs in human tissues and excreta, including skin. Skin VOC analyses require high sensitivity and specificity because these compounds are emitted at very low concentrations from the skin. The most commonly used methods are soft chemical ionization mass spectrometry (SCIMS) techniques, such as selected ion flow tube–mass spectrometry (SIFT-MS; see Chapter 9), PTR-MS (see Chapter 10), ion mobility spectrometry (IMS; see Chapter 11), and GC-MS (see Chapter 15). A variety of sampling techniques for skin VOCs have been reported, including liquid extraction, adsorbent materials applied in direct contact with the skin, and headspace sampling.

Shirasu and Touhara[3] documented progress in sampling and analytical techniques that provided rapid analyses of VOCs derived from skin, breath, and other bodily fluids. Similar to biomarkers derived in commonly sampled human media, such as blood and urine, recent progresses in sampling and analytical techniques allow analyses of xenobiotics, including VOCs, from skin surface. Samples for analyses are readily obtained by tape stripping or wiping/sabbing the skin with a liquid (e.g., organic solvent with low toxicity, such as acetone) or by collecting emitted VOCs from the skin directly into an absorbent material (e.g., filter or fiber).

Jiang et al.[27] developed a noninvasive thin film headspace sampling method for in vivo skin VOC sampling coupled with GC-MS analysis, which was considered to be an improvement over direct sampling methods. The headspace membrane sandwich (film between metal grids) significantly reduced background signal intensity, minimized contamination from the skin, and was reproducible. Soini et al.[28] used a rolling stir bar coated with polydimethylsiloxane (PDMS) for quantitative profiling of human skin VOCs with thermal desorption GC-MS analysis. The authors concluded that this approach would be suitable for field sampling because the skin VOCs and standards were stable under cool storage conditions for at least 14 days.

Gallagher et al.[11] identified a total of 92 compounds in the skin emanations using SPME and solvent extracts combined with GC-MS analysis. They reported that 58 of these compounds were identified in SPME samples and 49 in solvent extracts, although only 15 compounds were common in both sample sets. This observation led the authors to conclude that the two sampling approaches are complementary and both should be utilized in order to obtain a comprehensive profile of skin VOCs. Mochalski et al.[29] quantified 64 C_4–C_{10} compounds from the skin of healthy volunteers using SPME and GC-MS analysis. Aldehydes and hydrocarbons were found to be the predominant compounds, with not only acetaldehyde but also acetone and 6-MHO observed to be emitted from the skin at high emission rates (100 $fmol/cm^2/min$).

Duffy et al.[30] identified 37 compounds using SPME and GC-MS analysis, with aldehydes (hexanal, nonanal, decanal), acids, and hydrocarbons (squalane and squalene) dominating the VOC profile. The authors also observed that tape stripping affected the VOC profile; several compounds derived from sebaceous components and their oxidation products were altered following barrier disruption (squalane, squalene, octanal, and nonanal).

The technique of SPME-GC-MS has emerged as the method of choice for analyses of skin volatiles due to the sensitivity of mass spectrometry and ease of performing headspace sampling with SPME. In particular, SPME is suitable for low molecular weight compounds that are volatile at body temperature and is also portable and thereby appropriate for field sampling. Headspace sampling has become the preferred method for skin VOC analysis because direct contact sampling methods, like tape strips and sampling pads, can also capture nonvolatile compounds (e.g., lipids) or cause loss of VOCs during sampling.[27] Tape-strip sampling has become a preferred method when analyzing both absorbed and metabolized and/or adducted compounds in the skin.[31,32]

Research involving the comparison of relative abundances of VOCs emanating from human skin is currently limited by the methodology used for sample collection and preconcentration. The use of silicone rubber PDMS passive sampling devices constructed in the form of bracelets and anklets has been explored for the nontargeted collection and quantification of skin VOCs (reviewed in Duffy and Morrin[2]). Compounds collected encompassed a wide range of compound classes. Several compounds identified, notably cyclic ketones, have not been previously reported in skin VOC literature. Comparison of normalized unique mass peak area signals has revealed relative quantitative differences and similarities between the samples collected from two individuals' wrists, as well as between an individual's wrist and ankle. The sampling method was evaluated based on its ability to provide many candidate compounds for potential biomarker discovery. The reported results show the ability of this new easy-to-use and economical sampling method for augmenting the current knowledge on human skin VOC emissions.

25.6 Skin biomarkers of disease

Skin VOCs are of great interest to researchers in medical and forensic fields, as well as to biologists studying the ecology of blood-feeding insect vectors of human disease. Human skin is an organ of high metabolic activity where a rich variety of biomarkers are secreted from the SC. Skin VOCs contain the imprint of cellular activities and thus offer unique insights into the intricate processes of cutaneous physiology. Shirasu and Touhara[3] noted that hundreds of VOCs are emitted from the human body, which includes those in skin, exhaled breath, and other bodily fluids. The components of VOCs usually reflect the metabolic condition of an individual, including underlying health conditions and diseases. Specifically, contracting an infectious or metabolic disease often results in a change in body odor–related VOCs. Disease-specific VOCs can be used as diagnostic olfactory biomarkers of infectious diseases, metabolic diseases, genetic disorders, and other diseases. Duffy and Morrin[2] examined the growing body of research on skin VOC markers relating to skin physiology and pathology, whereby variations in skin, including both intrinsic and microbial metabolic processes, together result in unique VOC profiles. Their review summarizes the current knowledge on chemical and clinical aspects of body-derived VOCs and provides a brief consideration of the future of olfactory diagnosis. Elucidation of pathophysiological mechanisms underlying production of disease-specific VOCs may provide novel insights into therapeutic approaches for treatments for various diseases.

Broza et al.[7] noted that endogenous metabolism may result from oxidative stress that oxidizes nucleic acids, polyunsaturated fatty acids (PUFAs), and proteins. Volatile hydrocarbons are produced from peroxidation of PUFAs. Ethane and pentane are produced from peroxidation of FAs. Cytochrome P450 enzymes can upregulate ROS production in cancers. Alcohols are absorbed from the gastrointestinal track into venous blood and aldehydes arise from metabolized alcohols. Ketones are produced in the liver and are influenced by specific diets (e.g., acetone). Exogenous VOCs may result from the metabolism of drugs, nutrients, and food additives, among other sources. VOCs may also result from specific gland metabolism and their microbiota emitting VOCs into extracellular fluid, which are transferred into blood and ultimately to skin VOC emanations.

Duffy et al.[30] observed VOCs in normal skin and tape-stripped skin that induced a barrier impairment. This approach provided valuable information in the scientific and clinical fields of dermatological disorders associated with compromised skin barrier, including chronic conditions, such as atopic dermatitis and psoriasis. Further work is ongoing in this regard and is characterized by extension of VOC biomarkers into the dermatology clinic and the use of VOCs as biomarkers of skin disorders associated with impaired barrier function. This approach may permit identification of diagnostic biomarkers and/or improve our understanding of the pathologic etiology of impaired barrier disorders.

Emerging evidence for unique VOC biomarkers linked to skin perturbations and skin cancer is described for melanoma as a special case.[2] Melanoma and basal cell carcinoma have been anecdotally reported to emit specific odors. Canine olfactory detection of these skin cancers has motivated the training of dogs for use as a diagnostic tool (see Chapter 34). Trained canines were able to detect melanomas hidden beneath patient bandages and were correct in identifying 75%–86% of the cancers. The results suggest the existence of melanoma-specific VOC profiles that can be distinguished from other skin lesions. Duffy and Morris[2] summarized four different studies of melanoma biopsy material that detected significant levels of similar VOC profiles. Microbial-derived VOCs have also been investigated as prospective diagnostic markers, and their potential to indicate the composition of the local skin microbiota, and consequently the status of cutaneous health.

Numerous studies have investigated the chemical composition of skin odors, and various sampling methods have been used for this purpose (reviewed by Dormont et al.[1]). This area of science is of interest for the development of diagnostic tools, perfumes, deodorants, and for investigations of vectorborne human diseases and forensic science. Many studies have also been aimed at describing the chemical composition of skin odors and VOCs, finding human pheromones, and understanding the chemical transformation of perfumes caused by skin, as well as the development and evaluation of devices to reduce malodorous VOCs from human skin. Perhaps the most significant area of human odor science, however, is the identification of olfactory stimulants and the development of methods for vector control because insects are responsible for transmission of infectious diseases, including malaria and dengue. Further studies of the olfactory cues that mediate host-seeking by anthropophilic mosquitoes will have to consider the possible changes in human body odors in people with infectious diseases. People infected with transmissible stages of malaria were reported to be more attractive to anopheline mosquitoes.[33] Whether such modifications of host odor following infection may be of adaptive significance, for either the parasite or the host, is under debate.[34,35] Whether humans infected and uninfected by malaria show differences in their VOC profiles is not known, and further investigations on skin VOCs emitted by subjects infected by malaria are required to better understand the chemical ecology of human/vector/pathogen interaction on skin VOCs. The subject of volatile biomarkers of malaria infection in relation to both skin and breath volatiles is discussed in depth in Chapter 22.

Finally, a recent groundbreaking study on skin biomarkers is worthy of especial mention in the context of this chapter. Parkinson's disease (PD) is a neurodegenerative disorder affecting the central nervous system that disrupts motor functions, leading to hand tremors, among other symptoms, and ultimately dementia. At the University of Manchester, UK, researchers encountered anecdotal evidence of a woman who claimed that her husband, who had been diagnosed with PD, exhibited a unique body odor.[36] The woman, Joy Milne—referred to as a "super smeller"—was challenged with identifying skin swabs from PD sufferers compared with healthy controls in a blinded trial and demonstrated high accuracy. Subsequently, she was invited to participate in trials to identify the odorous molecules responsible for this distinct smell. Having determined that the odor is predominantly expressed in areas of high sebum production, such as the upper back and forehead, samples collected from these regions were analyzed using GC-MS/olfactometry (GC-MS/O).[37] In GC-MS/O, the compounds eluting from the chromatographic capillary are directed to the MS and simultaneously to an olfactory detection (or sniffing) port, the latter allowing for human odor perception assessment of individual compounds as they elute through the GC run. Mrs. Milne was trained at GC-O evaluation and flagged when odors similar to the familiar PD were present, allowing for concomitant MS analysis for compound identification purposes. Research is still ongoing, but the compounds perillic aldehyde and eicosane have been named as potential volatile biomarkers of PD diagnosis.

An unmet research need is a direct comparison of VOC analysis from skin and exhaled breath to determine agreement of the analytical methods employed to date. In addition, scientific criteria need to be established for basic methodology required in skin VOC analysis. Skin VOCs that represent normal homeostasis conditions and those that represent an individual's dysregulation of homeostasis and resultant disease have not been critically investigated. For example, it is important to determine whether VOCs identified from skin emanation and/or exhaled breath analysis appear to be dependent or independent of any underlying disease conditions due to differences in methodology and statistical interpretation. Observed outcomes (i.e., measurement of VOCs associated by different analytical methods) may have statistical type 1 or 2 errors impacting diagnostic implications influenced by age, body site, sex, or other interindividual differences. Future research focused on determination of the relative contribution and variance related to heritability and environmental contributions will be of significant interest and value to understanding skin VOCs as biomarkers of health and disease.

25.7 Summary

This chapter develops the concepts that the skin is a major source of human VOCs and that the chemical composition and quantity of skin emissions depend on the glandular skin, resident microbiota, blood flow, and metabolic processes within the individual. Skin VOCs are of significant interest for their potential contribution to assessment of environmental and occupational exposures, medical diagnosis and/or therapeutic monitoring, and investigation of the physiology and pathophysiology of the skin as an integral organ of the human body. Similar to those found in exhaled breath, skin VOCs complement the overall use of volatile and semivolatile biomarkers for environmental and occupational exposure assessment, as well as health diagnostics. Although this chapter focused solely on the role of chemicals emanating from the skin, it is important to note that the skin is perfused with blood that contacts every other cell in the body. As such, integration of VOCs emitted from the skin and those from exhaled breath and other biological media (e.g., blood, urine, or saliva) may be best evaluated together.

Analysis of VOCs is a promising approach for noninvasive, rapid, and potentially inexpensive diagnostics, much like using breath sampling rather than blood sampling. This is particularly important for future public health and other population-wide scoping studies wherein streamlining infield logistics of collecting samples from a large number of individuals is important. Incorporating skin emissions into an overall biomarker diagnostic strategy could ultimately provide a more comprehensive picture of VOC-related pathological changes than relying on VOC quantitation from single media specimens. The qualitative and quantitative methods described herein offer a variety of options for environmental and/or occupational exposure studies, mapping interindividual differences in metabolic properties, novel genome-wide association studies (GWAS), and preclinical experimental animal models to aid etiology and mechanistic research.

References

1. Dormont L, Bessiere JM, Cohuet A. Human skin volatiles: a review. *J Chem Ecol* 2013; **39**(5):569–78.
2. Duffy E, Morrin A. Endogenous and microbial volatile organic compounds in cutaneous health and disease. *Trac Trends Anal Chem* 2019;**111**:163–72.
3. Shirasu M, Touhara K. The scent of disease: volatile organic compounds of the human body related to disease and disorder. *J Biochem* 2011;**150**(3):257–66.
4. U.S. National Institutes of Health. SEER training manual: skin cancer (melanoma), anatomy of the skin.

5. de Lacy Costello B, Amann A, Al-Kateb H, Flynn C, Filipiak W, Khalid T, et al. A review of the volatiles from the healthy human body. *J Breath Res* 2014;**8**(1):014001.
6. Amann A, Costello Bde L, Miekisch W, Schubert J, Buszewski B, Pleil J, et al. The human volatilome: volatile organic compounds (VOCs) in exhaled breath, skin emanations, urine, feces and saliva. *J Breath Res* 2014;**8**(3):034001.
7. Broza YY, Mochalski P, Ruzsanyi V, Amann A, Haick H. Hybrid volatolomics and disease detection. *Angew Chem Int Ed Engl* 2015;**54**(38):11036–48.
8. Sato K, Kane N, Soos G, Sato F. The eccrine sweat gland; basic science and disorder of eccrine sweating. In: Moshell AN, editor. *Progress in dermatolog*. Evanston, IL: Dermatology Foundation; 1995. p. 1–11.
9. Nicolaides N. Skin lipids: their biochemical uniqueness. *Science* 1974;**186**(4158):19.
10. Leyden JJ, Nordstrom KM, McGinley KJ. Cutaneous microbiology. In: Goldsmith LA, editor. *Physiology, biochemistry, and molecular biology of the skin*, vol. 1. New York, NY: Oxford University Press; 1991. p. 1403–24.
11. Gallagher M, Wysocki CJ, Leyden JJ, Spielman AI, Sun X, Preti G. Analyses of volatile organic compounds from human skin. *Br J Dermatol* 2008;**159**(4):780–91.
12. Natsch A, Derrer S, Flachsmann F, Schmid J. A broad diversity of volatile carboxylic acids, released by a bacterial aminoacylase from axilla secretions, as candidate molecules for the determination of human-body odor type. *Chem Biodivers* 2006;**3**(1):1–20.
13. von Hartungen E, Wisthaler A, Mikoviny T, Jaksch D, Boscaini E, Dunphy PJ, et al. Proton-transfer-reaction mass spectrometry (PTR-MS) of carboxylic acids: determination of Henry's law constants and axillary odour investigations. *Int J Mass Spectrom* 2004;**239**(2–3):243–8.
14. Bernier UR, Kline DL, Barnard DR, Schreck CE, Yost RA. Analysis of human skin emanations by gas chromatography/mass spectrometry. 2. Identification of volatile compounds that are candidate attractants for the yellow fever mosquito (*Aedes aegypti*). *Anal Chem* 2000;**72**(4):747–56.
15. Syed Z, Leal WS. Acute olfactory response of *Culex* mosquitoes to a human- and bird-derived attractant. *Proc Natl Acad Sci U S A* 2009;**106**(44):18803–8.
16. Curran AM, Prada PA, Furton KG. The differentiation of the volatile organic signatures of individuals through SPME-GC/MS of characteristic human scent compounds. *J Forensic Sci* 2010;**55**(1):50–7.
17. Kusano M, Mendez E, Furton KG. Comparison of the volatile organic compounds from different biological specimens for profiling potential. *J Forensic Sci* 2013;**58**(1):29–39.
18. Prada PA, Curran AM, Furton KG. The evaluation of human hand odor volatiles on various textiles: a comparison between contact and noncontact sampling methods. *J Forensic Sci* 2011;**56**(4):866–81.
19. Marshall J, Holland KT, Gribbon EM. A comparative study of the cutaneous microflora of normal feet with low and high levels of odour. *J Appl Bacteriol* 1988;**65**(1):61–8.
20. Ara K, Hama M, Akiba S, Koike K, Okisaka K, Hagura T, et al. Foot odor due to microbial metabolism and its control. *Can J Microbiol* 2006;**52**(4):357–64.
21. Qiu YT, Smallegange RC, Hoppe S, van Loon JJ, Bakker EJ, Takken W. Behavioural and electrophysiological responses of the malaria mosquito *Anopheles gambiae* giles sensu stricto (Diptera: Culicidae) to human skin emanations. *Med Vet Entomol* 2004;**18**(4):429–38.
22. Caroprese A, Gabbanini S, Beltramini C, Lucchi E, Valgimigli L. HS-SPME-GC-MS analysis of body odor to test the efficacy of foot deodorant formulations. *Skin Res Technol* 2009;**15**(4):503–10.

23. Harraca V, Ryne C, Birgersson G, Ignell R. Smelling your way to food: can bed bugs use our odour? *J Exp Biol* 2012;**215**(Pt 4):623−9.
24. Logan JG, Birkett MA, Clark SJ, Powers S, Seal NJ, Wadhams LJ, et al. Identification of human-derived volatile chemicals that interfere with attraction of *Aedes aegypti* mosquitoes. *J Chem Ecol* 2008;**34**(3):308−22.
25. Steeghs MML, Moeskops BWM, van Swam K, Cristescu SM, Scheepers PTJ, Harren FJM. On-line monitoring of UV-induced lipid peroxidation products from human skin in vivo using proton-transfer reaction mass spectrometry. *Int J Mass Spectrom* 2006; **253**(1−2):58−64.
26. Wisthaler A, Weschler CJ. Reactions of ozone with human skin lipids: sources of carbonyls, dicarbonyls, and hydroxycarbonyls in indoor air. *Proc Natl Acad Sci USA* 2010;**107**(15):6568−75.
27. Jiang R, Cudjoe E, Bojko B, Abaffy T, Pawliszyn J. A non-invasive method for in vivo skin volatile compounds sampling. *Anal Chim Acta* 2013;**804**:111−9.
28. Soini HA, Bruce KE, Klouckova I, Brereton RG, Penn DJ, Novotny MV. In situ surface sampling of biological objects and preconcentration of their volatiles for chromatographic analysis. *Anal Chem* 2006;**78**(20):7161−8.
29. Mochalski P, King J, Unterkofler K, Hinterhuber H, Amann A. Emission rates of selected volatile organic compounds from skin of healthy volunteers. *J Chromatogr B* 2014;**959**: 62−70.
30. Duffy E, Jacobs MR, Kirby B, Morrin A. Probing skin physiology through the volatile footprint: discriminating volatile emissions before and after acute barrier disruption. *Exp Dermatol* 2017;**26**(10):919−25.
31. Fent KW, Jayaraj K, Ball LM, Nylander-French LA. Quantitative monitoring of dermal and inhalation exposure to 1,6-hexamethylene diisocyanate monomer and oligomers. *J Environ Monit* 2008;**10**(4):500−7.
32. Kang-Sickel JC, Butler MA, Frame L, Serdar B, Chao YC, Egeghy P, et al. The utility of naphthyl-keratin adducts as biomarkers for jet-fuel exposure. *Biomarkers* 2011;**16**(7): 590−9.
33. Lacroix R, Mukabana WR, Gouagna LC, Koella JC. Malaria infection increases attractiveness of humans to mosquitoes. *PLoS Biol* 2005;**3**(9):e298.
34. Lefvre T, Koella JC, Renaud F, Hurd H, Biron DG, Thomas F. New prospects for research on manipulation of insect vectors by pathogens. *PLoS Pathog* 2006;**2**(7):e72.
35. Prugnolle F, Lefevre T, Renaud F, Moller AP, Misse D, Thomas F. Infection and body odours: evolutionary and medical perspectives. *Infect Genet Evol* 2009;**9**(5):1006−9.
36. Morgan J. Joy of super smeller: sebum clues for PD diagnostics. *Lancet Neurol* 2016; **15**(2):138−9.
37. Trivedi DK, Sinclair E, Xu Y, Sarkar D, Walton-Doyle C, Liscio C, et al. Discovery of volatile biomarkers of Parkinson's disease from sebum. *ACS Cent Sci* 2019;**5**(4): 599−606.

CHAPTER

Cell cultures as in vitro models for breath research 26

Cristina E. Davis[1], Mitchell M. McCartney[1], Matthias Frank[2,3], Michael Schivo[4,5,6]

[1]Department of Mechanical and Aerospace Engineering, University of California, Davis, CA, United States; [2]Lawrence Livermore National Laboratory, Livermore, CA, United States; [3]Department of Biochemistry and Molecular Medicine, University of California, Davis, School of Medicine, Sacramento, CA, United States; [4]Center for Comparative Respiratory Biology and Medicine, University of California, Davis, School of Medicine, Davis, CA, United States; [5]Department of Internal Medicine, University of California, Davis, School of Medicine, Sacramento, CA, United States; [6]VA Northern California Health Care System, Mather, CA, United States

Chapter outline

26.1 Overview

As the field of breath research has expanded, researchers have explored in vitro cell culture models of breath signatures. Many of the same approaches and tools used in breath research have been applied to these studies. While clinical breath biomarkers appear in both the volatile and nonvolatile domain, most in vitro cellular studies thus far have focused on gas-phase volatiles. These volatile organic compounds (VOCs) are carbon-containing chemicals with low molecular mass and low boiling points and vapor pressures at normal atmospheric conditions. This "volatilomics" approach parallels those in clinical breath studies. Over the last decade, significant work has amassed to examine these compounds as they emanate from cell cultures—bacterial, fungal, and mammalian. A wide array of sampling approaches have been employed to capture and analyze cellular VOCs. These sampling techniques are similar to those being currently used in the clinical breath analysis community. Nevertheless, there are some differences noted due to the specific nature of cell-based experiments.

Breathborne Biomarkers and the Human Volatilome. https://doi.org/10.1016/B978-0-12-819967-1.00026-8
Copyright © 2020 Elsevier B.V. All rights reserved.

The nature and origin of the chosen cell cultures to be studied is not insignificant. Most studies have focused on cultures of a single cell type in laboratory based conditions, as is described in this chapter. It can be simplistic to consider monocultures alone, but it is very challenging to try and truly replicate the cellular composition from the human respiratory tract in an incubator. Many different cell types are present, and the field has not yet expanded to consider a fully representative in situ model of the lung. It is expected that the trajectory of the research will move towards more realistic in situ models coupled with more sophisticated methodologies and analyses.

Great care must be taken in in vitro experiments, and they are technically challenging to execute. For example, living cell cultures that are actively proliferating must be afforded appropriate gas exchange needed for cellular respiration. This ypically involves allowing cultures in liquid or on solid media to be in active gas exchange in a sterile incubator environment, and gases actively diffuse into the media or exchange directly with the cells on the media support. Thus, a consistent oxygen environment must be continually passed over the cultures, and then sampled for potential VOC emissions. Reasonable flow rates needed for this gas exchange can potentially have a negative effect of diluting the already trace VOCs emitted by the cells, and it is challenging to capture, preconcentrate, and chemically analyze these compounds while maintaining cell health. Also important to note, the medium "food" for the cells typically has a VOC signature itself, and this may change over time as nutrients are depleted and cellular waste by-products are created. The choice of medium may also have a significant impact on the specific VOC emissions, too. Nonetheless, despite these challenges, researchers have reported several successful experimental systems, described in the papers cited within this chapter. When coupled with traditional chemoanalytical systems, such as mass spectrometry or other approaches, this becomes an excellent starting point to build bottom-up models of biomarkers observed in clinical studies.

26.2 Bacteria cultures

It has long been established that bacteria produce VOCs as either products or by-products of their metabolism. For instance, the degradation of fatty acids through β-oxidation produces precursors for volatiles like alkanes or alkenes. Cheese-ripening bacteria produce a variety of gaseous sulfur compounds through catabolism of L-methionine amino acid.

Researchers have shown that the bacterial VOC profile, the sum of all emitted volatiles, can be species-specific, either through detection of unique compounds only found in one type of bacteria or through relative comparisons of compound abundances across bacteria types, i.e., a VOC "fingerprint" or "signature." Furthermore, the VOC emissions of a bacterium change during its growth cycles. Shnayderman et al. grew separate cultures of *Escherichia coli, Bacillus subtilis, Bacillus thuringiensis*, and *Mycobacterium smegmatis*, and measured VOCs at three growth cycle stages: lag, exponential, and stationary. The overall VOC profiles demonstrated variation during growth: within *M. smegmatis* samples, the team identified compounds that only appeared during the lag growth phase and also identified compounds that did not appear until the exponential phase. Using bioinformatics pattern recognition algorithms, these VOC profile measurements could be used to accurately distinguish volatile samples collected from the four bacterial cultures and at their different growth time points, with upward of 90% accuracy.[1] Küntzel et al. cultured three strains of *Mycobacterium avium* ssp. *paratuberculosis* (MAP), measuring VOCs at different incubation periods. The team detected 43 total MAP volatiles, with eight compounds showing significant abundance differences between the three strains. The entire VOC composition changed with increasing bacterial density, indicating that growth stage and metabolism impact the volatile profile.[2] These two studies demonstrate that incubation duration and growth periods must be standardized within a bacterial study to ensure conclusions are void of such confounding effects. Additionally, they suggest that a VOC detection platform could be used to predict bacterial density. Finally, a latest paper by Reese et al. shows this is possible with clinically important bacterial strains that are relevant to national security concerns.[3]

The VOC profile of bacteria cultures is not just dependent on strain or concentration but also on the growth medium. The medium itself is a source of VOCs, and researchers must analyze blank samples of pure medium to remove these signal artifacts from cultured samples. Bacteria can actively interact with media in several ways, such as consuming media volatiles, or be influenced by the media to engage or disengage certain volatile biochemical production pathways. Dryahina et al. performed headspace sampling experiments on four different bacteria cultured in three different media. Their multivariate statistical analysis of VOCs from *Pseudomonas aeruginosa*, *Staphylococcus aureus*, *Stenotrophomonas maltophilia* and the *Burkholderia cepacia* complex revealed that the four taxa can be well discriminated by differences in the headspace VOC concentration profiles in spite of effects introduced by the culture media used.[4] In a study by Chippendale et al., a single strain of *E. coli* was cultured in two different media. Ethanol, acetaldehyde, and hydrogen sulfide were major components of samples cultured in Dulbecco's modified eagle medium, whereas ammonia was the major component of the volatile profile from the same cultures grown in lysogeny broth (LB).[5] Bean et al. evaluated the effects of four growth media types on the ability to use VOCs to distinguish *Staphylococcus aureus* and *S. epidermidis*. Regardless of media type, the two species had unique profiles that were easily distinguished. Within one species, the volatile samples could further be distinguished by media type. The number of emitted volatiles widely varied—*S. epidermidis* grown in tryptic soy broth released 66 compounds not found in *S. epidermidis* samples of the other three media types.[6]

With the case made that growth conditions and media types influence bacterial VOC emissions, it can be reasonably deduced that bacteria might have a different volatile profile while proliferating in the human body in vivo relative to laboratory cultures in vitro. Furthermore, it is plausible that the host response to infection—the interaction of the airway with the pathogen—might generate a separate set of unique VOCs. Thus, under the context of understanding which volatile compounds in breath might indicate bacterial infection, it is ideal to culture bacteria in conditions as close to "native" as possible. To accomplish this in a laboratory setting, blood can be added to typical growth media, or even be used as the growth medium. Umber et al. grew *E. coli* in pure LB broth and in whole blood. They reported six unique VOCs—dimethyl disulfide, carbon disulfide, ethanol, acetaldehyde, methyl butanoate, and one unidentified—that were observed only in *E. coli*-inoculated whole blood and not found in pure LB broth, *E. coli*–inoculated broth, or even in sterile whole blood. This indicates that these six compounds were only generated during the interaction of *E. coli* and blood.[7] In hospital settings, critically ill patients are at risk for bloodstream infections of *Klebsiella pneumoniae*. Rees et al. comprehensively profiled the volatile metabolites generated by *K. pneumoniae* cultured in human blood during exponential and stationary growth phases. Twenty-two volatiles were found more abundant in inoculated blood relative to sterile blood; nine had not been previously reported to be produced by *K. pneumoniae* cultures.[8] These compounds, created during the interaction of host and pathogen, could serve as biomarkers for infection.

There have been a number of researchers looking at respiratory-specific bacterial infections. Among these, Professor Jane E. Hill is a notable example. Her laboratory has focused on using advanced chemical detection platforms to define the volatile profile of many respiratory bacteria with extremely high resolution. This is accomplished using a technique called comprehensive gas chromatography time-of-flight mass spectrometry (GC×GC-TOF-MS), which provides high levels of sensitivity and specificity in chemical separation and detection (see Chapter 15). Hill's studies include mycobacteria, infectious bacteria that lead in worldwide mortality, with most infections caused by *Mycobacterium tuberculosis*. Nontuberculous mycobacteria (NTM) also have high prevalence of death in North America. Diagnostic specificity is crucial with regard to patient treatment, as NTMs are unresponsive to antituberculosis therapies. Hill and researchers analyzed volatile compounds emitted by 18 strains of four mycobacteria (three NTM and one *M. tuberculosis*). Of the 217 total detected VOCs, 58 were enough to discriminate the four mycobacteria. This first-of-kind study demonstrates that bacterial VOCs could be used to reduce time-to-diagnosis of certain bacterial infections.[9] In another of their work, Hill's team collected 24 clinical isolates of *Pseudomonas aeruginosa* from patients' respiratory tracts, throat, blood, skin, and elsewhere. Using techniques familiar to comparative genomics, the researchers conducted a comparative analysis of the volatile metabolite profile, or *volatilome*, which was based on a total of 391 unique compounds. Of these, 70 compounds were present in all clinical isolates, representing the "core" components of the *P. aeruginosa* VOC profile. Variation between isolates was driven by the nearly 250 volatiles that were present in various subsets of isolates, with little similarity among strains collected from the similar infection sites.[10]

While many studies have demonstrated that bacteria cultures can be used to discover biomarkers of infection, researchers must also show that the investigated bacterium has the biochemical means, or genetic capacity, to produce such volatiles. Furthermore, understanding these metabolic pathways could be useful toward development of treatment, therapies, and vaccines. Researchers can highlight the exact transduction pathways of active VOC production by feeding bacteria cultures isotopically labeled compounds. Isotopically labeled compounds are compounds in which an atom or atoms of a molecule have been replaced with another isotope of the same atom. Thus, uptake of these altered compounds can be traced during the metabolic process to verify the ability of the microbe to produce certain volatile compounds. (This concept is similarly exploited for human breath analysis by ingestion of labeled substrates, as is covered in Chapter 17.) Phan et al., for example, noted that cystic fibrosis patients had detectable levels of 2,3-butanedione in their breath. Speculation was made that this compound must be from microbial origin, as it is not a known human metabolite. Three bacterial strains were fed isotopically labeled glucose (D-glucose-13C_6) and water (^2H_2O). After sampling the volatile headspace above in vitro cultures, *Rothia mucilaginosa* and *Streptococcus salivarius* were confirmed to have taken up labeled glucose and water, metabolically converting these compounds into isotopically labeled 2,3-butanediol. This confirmed that the observation of 2,3-butanediol in breath has microbial origin.[12]

Occasionally, the field of breath research blurs its boundaries, extending its investigations into volatile compounds emitted through orifices other than the mouth and nose. In social animals, scent markers have a critical role in communication within and across species. Domestic cats, for example, use scent secretions from their anal sacs for individual recognition. Yamaguchi et al. have demonstrated that bacteria can play a significant role in the volatile composition of anal sac secretions. The authors isolated three taxa of bacteria from the anal sac of a Bengal cat, in addition to swabbing the cat's anal sac with a cotton ball. The cotton ball was chemically analyzed, providing a volatile profile of the entire anal sac aroma. When comparing the volatile composition of the anal sac with the volatiles produced by cultures of the three isolated bacteria, there was a large overlap: 51 volatile compounds were both found in the anal sac bouquet and produced by at least one of the three isolated taxa.[13] This work shows the importance of local commensal microbes on scent-based animal communication.

26.3 Fungus cultures

Fungal respiratory infections can be just as severe as bacterial. As fungi fall within a distinct kingdom from humans and animals, their genetically coded metabolism has profound differences to bacteria, and thus can generate different types and ratios of VOCs. This feature can be exploited in breath research to develop rapid tests to distinguish fungal and bacterial infections. Rees et al. measured the volatile profile of clinically isolated *Candida* spp. yeast and nine bacteria. Comparison of the fungal profile with the nine bacterial profiles was quite distinct. Generally, only about 4%–5% of volatiles emitted by the bacteria overlapped with volatiles produced by *Candida*.[11]

Like bacteria, VOCs emitted by cultured fungi can depend on growth conditions. Tejero Rioseras et al. used secondary electrospray ionization–high-resolution mass spectrometry (SESI-HRMS) to make volatile measurements every minute from cultured baker's yeast (*Saccharomyces cerevisiae*). The cultures were initially in glucose minimal medium. After injection of isotopically labeled glucose ($^{13}C_6$-glucose) into the media, the researchers tracked the corresponding production of ethanol and acetic acid by the yeast, which continued until all the glucose had been consumed. Then, the team compared wild type to mutant strains of *S. cerevisiae* and found stark contrasts between produced volatiles. One mutant strain emitted methyl ketones not found in the other mutant strains or the wild type; the wild type produced twice as much ethanol as the mutant strains.[14]

Volatile organic compound changes related to growth conditions can have clinical implications. Fungi in the genus *Aspergillus* cause life-threatening diseases. *Aspergillus* lesions in the human pulmonary system are often low-oxygen microenvironments, which cause the fungus to undergo gene expression changes to survive these hypoxic conditions. Rees et al. cultured *A. fumigatus* in oxygen replete (normoxia) and deplete (hypoxia) environments. Using only 19 of the detected volatiles, the authors were able to completely distinguish samples based on their oxygen-growing conditions. In conjunction with transcriptome data, the team posited the metabolic pathways that could account for observed VOC changes.[15]

Interestingly, fungal "breath" emissions can be used in applications outside of human studies. For instance, chicken eggshells have pores that can be contaminated with microbes, putting human health at risk during consumption. Cumeras et al. measured the volatile profile of eggs with likely internal fungal infections and fresh eggs. Afterward, the eggs were cracked open and cultures were made from collected microbes, including two fungi (*Cladosporium macrocarpum* and *Botrytis cinereal*). Compounds produced by the fungi, such as trichloromethane, appeared in volatile measurements of intact, infected eggs.[16]

26.4 Mammalian cell cultures

As the breath research field begins to understand the origins of VOCs in human tissue, it clearly is important to understand the background signals that arise from normal mammalian cell types. Mammalian cells in culture are frequently liquid suspension monocultures of a single-cell type, and like microorganism cultures require suitable media to be healthy, grow, and proliferate. Many commercially available cell types were originally derived from a single individual, and then were taken through a series of steps to "immortalize" them so that they do not follow a normal senescence sequence. Thus, the cells can be cultured almost indefinitely in the laboratory and represent a single genotype of the original cell donor if they are not contaminated. There is some recurring controversy in the cell biology field about mammalian cell culture identity, and recent best practices include periodically commercially genotyping the cultures used in research labs to ensure they have not been compromised with other invasive mammalian cell types, such as HeLa. In contrast, there are also "primary" cell cultures that are obtained from unique human subjects and are cultured and proliferated until the cells senesce and are discarded. These cell types are genetically unique to each subject, and they are free from artificial cell cycle alterations that immortalized cells have. Finally, stem cell cultures can develop into many different cell types and can also be a model for in vitro breath biomarker studies. By culturing cells in enclosed or semienclosed environments, researchers from many groups have been able to confirm specific VOCs that are generated in the headspace above the cells.

Some of the early work in this area of mammalian cell culture VOCs centered on the attractive idea of understanding how transformed cell lines could function as engines to generate volatile biomarkers that may be circulated in the blood and subsequently detected in the exhaled breath.[17,18] These in vitro models were thought to potentially provide a way to clinically preidentify or confirm observed breath biomarkers in cancer patients. Lung cancer was an obvious choice as a potential model, and authors of a 2008 study used a common commercially available immortalized small-cell lung cancer cell line CALU-1 and found that these produced up to 60 VOCs, as measured via gas chromatography—mass spectrometry (GC-MS).[17] Other researchers investigated VOCs produced by colon cancer cells SW 480 and SW 1116 in comparison to the normal colon cell line NCM460.[18] These efforts were notable for their early entry into this research topic. However, the researchers also recognized the potential pitfalls with these cell culture models. Cancer rarely has one single genotype, and a great heterogeneity of mutations is observed within the same clinical phenotype. While transformed cell lines are easier to work within research settings, they are only partial models of actual primary tissue types. More recent work in the field has sought to explore mutation-specific VOCs by examining multiple cell lines.[19] By using statistical analysis, the researchers were able to tease apart the VOCs putatively associated with various genotypes across 19 commercially available cell lines each containing one of three major oncogene mutations. By grouping the datasets for conserved genetic mutations, the authors were able to speculate specific VOC profiles that seemed associated with each one. Other researchers also looked at VOCs released from transformed and normal lung epithelial cell lines, and these were then compared to breath biomarkers from human clinical studies.[20] Together all of this work has helped set the stage for identifying cancer-related VOCs from in vitro cell culture models. By observing mutations, VOCs could start to be identified in a flexible model to accommodate genetic information of tumors from patients.

Since the original work in this area has advanced, several groups have attempted to create new model systems and protocols to measure VOC headspace above active mammalian cell cultures. Specifically, epithelial cells from the respiratory tract are unique in that they grow at the liquid—air interface in the lung, and several groups have explored VOC sampling from this cell type.[21–23] Importantly, these begin to increase the in vitro model complexity to illustrate potential origins of breath biomarker compounds. In 2014, Schivo et al. showed that VOCs could be generated from primary human tracheobronchial (TBE) cells being cultured on Transwell plates.[21] Once the TBE cells proliferated to confluence, they were transferred to air–liquid interface (ALI) culture conditions, and VOCs could be captured using solid-phase microextraction (SPME) fibers and identified by GC-MS. By carefully controlling headspace capture above the cells, trace biomarkers observed in actual human breath samples were also identified in this in vitro model. In 2018, Traxler et al. noted that there was not a standardized setup to chemically monitor VOCs from these types of cultures, and they championed a novel platform that could be used to make online and offline cell headspace measurements.[22] By using inert materials and standard universal sampling ports, they were able to make real-time VOC measurements using proton transfer reaction time-of-flight mass spectrometry (PTR-TOF-MS). They also showed it was possible to employ needle-trap microextraction (NTME) coupled with GC-MS. They employed a mouse embryonic multipotent stem cell line, and they observed VOCs that were also reported in human breath samples in the literature. That same year, Yamaguchi et al. illustrated an analogous system that improved on the Schivo study by using headspace sorptive extraction (HSSE) to capture the released VOCs.[23] This improved collection system increased the VOC recovery, resulting in 97 more compounds detected from the same cell types that were used by Schivo et al. As an application of this experiment system, the team published a second paper in 2018 that detailed exogenous effects on cell biomarkers.[24] The in vitro system was used to study oxidative stress impacts using hydrogen peroxide (H_2O_2) and cigarette smoke extract (CSE) exposure directly to the cells. Resulting VOC releases were monitored at various times postexposure, and the chemicals were putatively identified using GC-MS. Some aldehyde chemicals were observed, which is consistent with prior human breath studies showing those biomarkers were typical of oxidative stress in clinical settings. In a similar study, researchers showed that genotoxic carcinogens could also alter cellular VOCs.[25]

One particularly compelling application of VOC monitoring from mammalian cells is process control for bioreactors. Increasingly, cell cultures are at the forefront of biologic therapeutics that include stem cell therapies, vaccine development, and regenerative medicine approaches to cancer treatments. There are very few "inline" sensors that can monitor the health of bioreactors in real-time, such as pH and dissolved oxygen sensors. Currently, detailed assessments require drawing an aliquot of inoculated media out of the bioreactor, which introduces risk for accidental contamination. Luchner et al. demonstrated that volatile measurements can be taken from the exhaust line of bioreactors as a potential tool for cell culture health monitoring. Using proton transfer reaction–mass spectrometry (PTR-MS), a tool used for continuous VOC readings (see Chapter 10), the group confirmed that volatiles emanating from *E. coli* in bioreactors could be measured online, although only a few compounds were identifiable.[26] The work of McCartney et al. took this idea further, by collecting downstream samples in the gas exhaust lines of CHO and primary T-cell reactors daily for 8 days. Using GC-MS, these researchers profiled the emitted volatiles and confirmed that bioreactor emissions not only differed between CHO and T cell cultures, but also within one cell type evolved during 8 days of cellular growth. Further, statistical models could be trained to accurately predict cell density inside the bioreactor based on the VOC emissions, which could be a very useful tool for noninvasive monitoring. Another interesting observation from this study was that certain volatiles, confirmed to be artifacts of the culture medium, were found to deplete over time, even with active media exchange in the bioreactor. This suggests that the cultures were consuming these compounds, but depleting them faster than they were replaced. Thus, VOCs have a potential as markers to ensure culture media integrity and nutrient availability in reactors.[27]

Similarly, emerging stem cell therapies in human medicine have gained momentum for use in clinical settings, but assessing culture health can be time-consuming and requires destruction of precious and expensive cells. Work by Bischoff et al. found that murine pluripotent stem cells off-gas VOCs, such as thiirane and methyl-methoxy-hydroxy-methyl-amine. These compounds correlated with cell abundance, providing an opportunity for noninvasive monitoring of cellular expansion.[28] The ability to monitor cellular health during stem cell differentiation would be further beneficial. In the work by Klemenz et al., parallel cultures of nondifferentiating and differentiating mesenchymal stromal/stem cells (ASCs) were incubated for 21 days, with VOC measurements made weekly. Certain volatiles, such as acetaldehyde, increased in abundance during the 3 weeks of differentiation, but remained at relatively lower, stable concentrations in the nondifferentiating cells. Conversely, other compounds, such as benzaldehyde, remained in higher concentrations in nondifferentiating cells relative to differentiating cells. This work demonstrates that VOCs emitted by stem cells are affected by metabolic activity and cellular state.[29]

In 2012, Aksenov et al. explored whether VOCs could be useful for monitoring cell culture models that partially mimic the human immune system.[30] A suspension culture of immortalized human lymphocyte B cells was shown to generate baseline VOCs into the headspace above the culture, and the chemicals were identified using GC-MS. These cells are important in the response to pathogens and are also associated with human body odor and mate selection. The cell lines had the genes at the human leukocyte antigen (HLA) class IA and B loci deleted from a parent cell line, and daughter cell lines were transfected to express specific HLA gene products to allow for specific replacement of gene sequences against the whole genome background that remained the same. The researchers found that certain HLA alleles specifically influenced the cell VOC emissions, which potentially demonstrates a novel avenue to explain human scent markers that are related to immune status. This was the first time that researchers had shown that specific single gene products could be associated with distinct VOC signature changes in vitro.

Unlike bacteria or fungi, viruses are not living and do not generate their own VOCs that could be used for their detection. Instead, volatiles released by the host during infection are the only means for VOC-based viral detection. In mammalian cell cultures, baseline measurements can be made of healthy cells, and after inoculation with a virus, and researchers can monitor effects on mammalian VOC production as a function of time after infection and/or the quantity of viral load. Aksenov et al. inoculated immortalized human lymphocyte B cells with three live influenza virus subtypes: H9N2 (avian), H6N2 (avian), and H1N1 (human). As the host B cells responded to infection, their volatile profile changes over time were unique for each virus subtype and followed specific time courses. Metabolites of oxidative stress, such as esters, were particularly noted.[31]

In another study, Schivo et al. made models of the human airway using primary human TBE cells, which were inoculated with either live human rhinovirus (HRV) or heat-killed HRV. Similar to the findings of Aksenov et al. with influenza, HRV lead to volatile emission changes of airway cells related to oxidative damage, with release of aliphatic alcohols and branched hydrocarbons. These biomarker changes were not observed in the heat-killed HRV-infected cells, suggesting these biomarkers are specific to infection.[21] Purcaro et al. compared the volatile emissions of human laryngeal cancer cells infected with either syncytial virus or influenza A virus. In their publication, the authors posted a comprehensive table of all detected volatiles emitted by virus-infected cells. Using three statistical model types, the authors presented the cellular volatiles that could discern between syncytial and influenza A infection. These volatile differences were observed as quickly as 24 h after inoculation, demonstrating that VOC techniques could diagnose viral infections much faster than a patient might present symptoms.[32]

Certain viral infections, such as influenza A, weaken a person's immune system significantly, allowing for coinfections of opportunistic pathogens. How does the presence of a secondary infection affect VOC production of mammalian cells? Traxler et al. examined pharynx carcinoma (Detroit) cells that were either monoinfected with influenza A virus, monoinfected with the bacteria *Streptococcus pyogenes*, or coinfected with both the virus and the bacteria. Among the monoinfected cells, there were distinct biomarkers of each pathogen: production of acetaldehyde and propanal were upregulated in bacteria-infected cells, whereas acetone and n-propyl acetate were upregulated in the virus-infected cells. Both the bacteria biomarkers and the viral biomarkers were also upregulated in the coinfected cells.[33] This result is rather impactful, as it demonstrates that secondary infections may not diminish the effectiveness of a volatile-based diagnostic platform.

With demonstration that pathogens can either release volatile compounds or cause mammalian cells to release volatiles, a question is asked about the effects of VOC exposure on airway cells. For instance, during infection, bacteria can generate cytotoxic volatiles, harming surrounding mammalian cells. Also, the airway is routinely exposed to toxicants through inhalation of pollutants, such as vehicle emissions and smog. Wichmann et al. developed a model system to investigate immunomodulating effects of toxicant VOCs. A culture of immune cells, human peripheral blood mononuclear cells (PBMC), was grown. Cells were exposed to known concentrations of toluene, a known carcinogen, by doping cell media. There was a direct relationship between the concentration of doped toluene with the quantity of toluene emitted by the cells. No concentration of toluene, from 22 to 725,000 mg/m^3, influenced cell proliferation and vitality. The authors also measured a panel of cytokines, a group of signaling molecules secreted by immune cells to mediate and regulate inflammation, and it was found that toluene in concentrations above 72,500 mg/m^3 caused immune cells to decrease cytokine secretion and increase the proinflammatory marker tumor necrosis factor alpha (TNF-α).[34] This work shows how mammalian cultures can be used to understand the biochemical processes by which toxic volatiles affect health.

26.5 Algal cultures

Although research on VOCs produced by algae may not be directly relevant to breath analysis, this is also an active field of research that should be briefly mentioned here. The research methods and approaches for studying VOCs from cell cultures can easily be adapted to study metabolism of algae and their interactions with other organisms. Early work on the "odor of seaweed," the volatiles from marine algae, was summarized by R.E. Moore in 1977.[35] Since then, algal VOC production has been associated not only with metabolism of algae but also with intra- and interspecies communication, allelopathy, semiochemical production, and predator deterrence (for a recent review, see Zuo[36]). In recent years, the use of algae for industrial biomass production in large-scale algal ponds has renewed the interest in exploring the use of algal VOC emissions for monitoring the health status of algal ponds and as early indicators for culture crashes caused by pathogens or predator infestation. Recent work by Reese et al., for example, investigated volatiles emitted from cultures of *Microchloropsis salina* infected with a predator rotifer *Brachionus plicatilis*. The addition of *B. plicatilis* to healthy cultures of *M. salina* resulted in decreased algal cell numbers relative to uninfected controls, and generated characteristic VOCs attributed to carotenoid degradation.[37] In the future, VOCs from algal cultures may provide a noninvasive tool to continuously monitor and control algal biomass production.

26.6 Summary

Clearly, many advances have been made in the area of cellular respiration and VOC releases as an in vitro model for breath analysis, and future breakthroughs will no doubt be forthcoming in this area. Future work is likely to expand these models to include mixed cultures, and more expansive media choices. Those models are likely to include multiple mammalian cell types, which are found in vivo in the human or animal respiratory system. For example, a future system could be envisioned that utilizes a mixture of fully ciliated functional epithelial cells connected with cells forming culture-based blood vessels that could be connected with interstitial tissue cell types and immune cells representing the native lung structure. Current technology does not allow this to be achieved yet, but advances in tissue engineering are likely to make this possible one day. Such a system could also be coupled with cocultures of both commensal organisms and pathogenic organisms to challenge the cells in a way that mimics pulmonary infections in vivo. This also must eventually be expanded to include cells that are observed throughout the respiratory tract, including the oral, nasal and pharyngeal, down to the trachea and bronchi. The muscular cells and structures that underlay these cell types and exchange signaling chemicals and mechanical stimuli should also be considered. Finally, these culture models have not yet been challenged in situ by the environmental pollutants (low or high level) that all human subjects in clinical trials are regularly exposed to.

In future work, it will be critically important to include cell models that make use of more primary cultures. While immortalized cells and those pathogen bacteria commonly used in cell culture facilities have been a good initial start to this important work, only primary cells and a wide range of clinical pathogen isolates truly begin to replicate the biological complexity observed in real human systems. Unfortunately, sampling and culturing many respiratory cells from humans is still technically challenging, and not all cell types have yet been sampled and proven in culture to exactly mimic their in vivo functions. Additionally, primary cells only genetically represent the single individual human they were extracted from. As a research field, the available cell types must be expanded to include a broader array of genotypes that truly represent the human population. Only then will an in vitro model be viably used to predict exhaled breath biomarkers before they are identified in a clinical setting.

Finally, as the field of breath research continues to expand to include nonvolatile breath biomarkers, cell models must simultaneously expand to examine these compounds, too. Some recent research described in this chapter has already begun to do this. Continued observations of the liquid effluent of cell cultures for potentially large molecules that may be present in exhaled breath condensate or aerosols should be made. These future developments will aid in compiling the most complete "bottom-up" in vitro model of breath biomarker origins.

References

1. Shnayderman M, Mansfield B, Yip P, Clark HA, Krebs MD, Cohen SJ, et al. Species-specific bacteria identification using differential mobility spectrometry and bioinformatics pattern recognition. *Anal Chem* 2005;**77**(18):5930–7.
2. Küntzel A, Fischer S, Bergmann A, Oertel P, Steffens M, Trefz P, et al. Effects of biological and methodological factors on volatile organic compound patterns during cultural growth ofMycobacterium aviumssp. paratuberculosis. *J Breath Res* 2016; **10**(3). 037103.

3. Reese KR, Avila JR, Jones AD, Frank M. Metabolic profiling of volatile organic compounds (VOCs) emitted by the pathogens *Francisella tularensis* and *Bacillus anthracis* in liquid culture. *Sci Rep* 2020 [in review].
4. Dryahina K, Sovová K, Nemec A, Španěl P. Differentiation of pulmonary bacterial pathogens in cystic fibrosis by volatile metabolites emitted by their in vitro cultures:*Pseudomonas aeruginosa, Staphylococcus aureus, Stenotrophomonas maltophilia* and the *Burkholderia cepacia* complex. *J Breath Res* 2016;**10**(3). 037102.
5. Chippendale TWE, Španěl P, Smith D. Time-resolved selected ion flow tube mass spectrometric quantification of the volatile compounds generated by *E. coli* JM109 cultured in two different media. *Rapid Commun Mass Spectrom* 2011;**25**(15):2163–72.
6. Jenkins CL, Bean HD. Influence of media on the differentiation of *Staphylococcus* spp. by volatile compounds. *J Breath Res* 2019;**14**(1). 016007.
7. Umber BJ, Shin H-W, Meinardi S, Leu S-Y, Zaldivar F, Cooper DM, et al. Gas signatures from *Escherichia coli* and *Escherichia coli*-inoculated human whole blood. *Clin Transl Med* 2013;**2**(1):13.
8. Rees CA, Smolinska A, Hill JE. The volatile metabolome of *Klebsiella pneumoniae* in human blood. *J Breath Res* 2016;**10**(2). 027101.
9. Mellors TR, Rees CA, Wieland-Alter WF, von Reyn CF, Hill JE. The volatile molecule signature of four mycobacteria species. *J Breath Res* 2017;**11**(3). 031002.
10. Bean HD, Rees CA, Hill JE. Comparative analysis of the volatile metabolomes of *Pseudomonas aeruginosa* clinical isolates. *J Breath Res* 2016;**10**(4). 047102.
11. Rees CA, Burklund A, Stefanuto P-H, Schwartzman JD, Hill JE. Comprehensive volatile metabolic fingerprinting of bacterial and fungal pathogen groups. *J Breath Res* 2018; **12**(2). 026001.
12. Phan J, Meinardi S, Barletta B, Blake DR, Whiteson K. Stable isotope profiles reveal active production of VOCs from human-associated microbes. *J Breath Res* 2017; **11**(1). 017101.
13. Yamaguchi MS, Ganz HH, Cho AW, Zaw TH, Jospin G, McCartney MM, et al. Bacteria isolated from Bengal cat (*Felis catus* × Prionailurus bengalensis) anal sac secretions produce volatile compounds potentially associated with animal signaling. *PloS One* 2019; **14**(9). e0216846.
14. Tejero Rioseras A, Garcia Gomez D, Ebert BE, Blank LM, Ibáñez AJ, Sinues PML. Comprehensive real-time analysis of the yeast volatilome. *Sci Rep* 2017;**7**(1):14236.
15. Rees CA, Stefanuto P-H, Beattie SR, Bultman KM, Cramer RA, Hill JE. Sniffing out the hypoxia volatile metabolic signature of *Aspergillus fumigatus*. *J Breath Res* 2017;**11**(3). 036003.
16. Cumeras R, Aksenov AA, Pasamontes A, Fung AG, Cianchetta AN, Doan H, et al. Identification of fungal metabolites from inside Gallus gallus domesticus eggshells by non-invasively detecting volatile organic compounds (VOCs). *Anal Bioanal Chem* 2016;**408**(24):6649–58.
17. Filipiak W, Sponring A, Mikoviny T, Ager C, Schubert J, Miekisch W, et al. Release of volatile organic compounds (VOCs) from the lung cancer cell line CALU-1 in vitro. *Canc Cell Int* 2008;**8**(1):17.
18. Zimmermann D, Hartmann M, Moyer MP, Nolte J, Baumbach JI. Determination of volatile products of human colon cell line metabolism by GC/MS analysis. *Metabolomics* 2007;**3**(1):13–7.

19. Peled N, Barash O, Tisch U, Ionescu R, Broza YY, Ilouze M, et al. Volatile fingerprints of cancer specific genetic mutations. *Nanomed Nanotechnol Biol Med* 2013;**9**(6):758–66.
20. Filipiak W, Filipiak A, Sponring A, Schmid T, Zelger B, Ager C, et al. Comparative analyses of volatile organic compounds (VOCs) from patients, tumors and transformed cell lines for the validation of lung cancer-derived breath markers. *J Breath Res* 2014; **8**(2). 027111.
21. Schivo M, Aksenov AA, Linderholm AL, McCartney MM, Simmons J, Harper RW, et al. Volatile emanations from in vitro airway cells infected with human rhinovirus. *J Breath Res* 2014;**8**(3). 037110.
22. Traxler S, Bischoff A-C, Trefz P, Schubert JK, Miekisch W. Versatile set-up for non-invasive in vitro analysis of headspace VOCs. *J Breath Res* 2018;**12**(4). 041001.
23. Yamaguchi MS, McCartney MM, Linderholm AL, Ebeler SE, Schivo M, Davis CE. Headspace sorptive extraction-gas chromatography–mass spectrometry method to measure volatile emissions from human airway cell cultures. *J Chromatogr B* 2018;**1090**: 36–42.
24. Yamaguchi MS, McCartney MM, Falcon AK, Linderholm AL, Ebeler SE, Kenyon NJ, et al. Modeling cellular metabolomic effects of oxidative stress impacts from hydrogen peroxide and cigarette smoke on human lung epithelial cells. *J Breath Res* 2019;**13**(3). 036014.
25. Fijten RRR, Smolinska A, Shi Q, Pachen DM, Dallinga JW, Boots AW, et al. Exposure to genotoxic compounds alters in vitro cellular VOC excretion. *J Breath Res* 2018;**12**(2). 027101.
26. Luchner M, Gutmann R, Bayer K, Dunkl J, Hansel A, Herbig J, et al. Implementation of proton transfer reaction-mass spectrometry (PTR-MS) for advanced bioprocess monitoring. *Biotechnol Bioeng* 2012;**109**(12):3059–69.
27. McCartney MM, Yamaguchi MS, Bowles PA, Gratch YS, Iyer RK, Linderholm AL, et al. Volatile organic compound (VOC) emissions of CHO and T cells correlate to their expansion in bioreactors. *J Breath Res* 2019;**14**(1). 016002.
28. Bischoff A-C, Oertel P, Sukul P, Rimmbach C, David R, Schubert J, et al. Smell of cells: volatile profiling of stem- and non-stem cell proliferation. *J Breath Res* 2018;**12**(2). 026014.
29. Klemenz A-C, Meyer J, Ekat K, Bartels J, Traxler S, Schubert KJ, et al. Differences in the emission of volatile organic compounds (VOCs) between non-differentiating and adipogenically differentiating mesenchymal stromal/stem cells from human adipose tissue. *Cells* 2019;**8**(7).
30. Aksenov AA, Gojova A, Zhao W, Morgan JT, Sankaran S, Sandrock CE, et al. Characterization of volatile organic compounds in human leukocyte antigen heterologous expression systems: a cell's "chemical odor fingerprint". *ChemBioChem* 2012;**13**(7): 1053–9.
31. Aksenov AA, Sandrock CE, Zhao W, Sankaran S, Schivo M, Harper R, et al. Cellular scent of influenza virus infection. *Chembiochem* 2014;**15**(7):1040–8.
32. Purcaro G, Rees CA, Wieland-Alter WF, Schneider MJ, Wang X, Stefanuto P-H, et al. Volatile fingerprinting of human respiratory viruses from cell culture. *J Breath Res* 2018;**12**(2). 026015.
33. Traxler S, Barkowsky G, Saß R, Klemenz A-C, Patenge N, Kreikemeyer B, et al. Volatile scents of influenza A and S. pyogenes (co-)infected cells. *Sci Rep* 2019;**9**(1):18894.

34. Wichmann G, Mühlenberg J, Fischäder G, Kulla C, Rehwagen M, Herbarth O, et al. An experimental model for the determination of immunomodulating effects by volatile compounds. *Toxicol In Vitro* 2005;**19**(5):685–93.
35. Moore RE. Volatile compounds from marine algae. *Acc Chem Res* 1977;**10**(2):40–7.
36. Zuo Z. Why algae release volatile organic compounds—the emission and roles. *Front Microbiol* 2019;**10**(491).
37. Reese KL, Fisher CL, Lane PD, Jaryenneh JD, Moorman MW, Jones AD, Frank M, Lane TW. Chemical profiling of volatile organic compounds in the headspace of algal cultures as early biomarkers of algal pond crashes. *Sci Rep* 2019;**9**:13866.

CHAPTER

Ruminants

27

Petra E. Reinhold[1], Peter Gierschner[2], Anne Küntzel[1], Jochen K. Schubert[2], Wolfram Miekisch[2], Heike U. Köhler[1]

[1]*Institute of Molecular Pathogenesis at 'Friedrich-Loeffler-Institut' (Federal Research Institute for Animal Health), Jena, Germany;* [2]*Department of Anesthesia and Intensive Care, Rostock University Medical Center, Rostock, Germany*

Chapter outline

27.1 Overview

Among the roughly 200 species of living ruminants, the most important domestic ones are bovines, goats, and sheep. Beside farming and veterinary medicine, those species may also serve as large animal models in translational biomedical research. With respect to respiratory diseases and chronic infections, numerous large animal

Breathborne Biomarkers and the Human Volatilome. https://doi.org/10.1016/B978-0-12-819967-1.00027-X
Copyright © 2020 Elsevier B.V. All rights reserved.

models offer high biological relevance due to size, lung volumes, and anatomical similarities to humans.[1] Spontaneous or induced pathological processes can be studied under biologically relevant conditions, while the phenotypic characteristics of the diseases to be examined resemble the same phenomenon in other species of animals or even in humans. Pertinent animal models are still indispensable to understand characteristic features of pathogenesis, i.e., mechanisms, and pathways that underlie the observed phenotypic alterations. Compared to human clinical studies, one of the strengths of large animal models is a high level of standardization with respect to nutrition, housing conditions, ambient conditions, and medication. Thus, confounding factors resulting from personal lifestyle in humans (including diet, smoking behavior, medication) and various environmental exposures can be either excluded or studied regarding their unique effects.

Assessment of potential biomarkers in exhaled breath is a promising tool in human and veterinary medicine because collection of breath is completely noninvasive and repeatable at any time; it just requires spontaneous breathing and is applicable to conscious animals. Methodologically, exhaled breath analysis comprises several approaches: (i) analysis of exhaled breath condensate (EBC), (ii) analysis of exhaled particles (PExs), and (iii) analysis of exhaled volatile organic compounds (VOCs). In any approach, appropriate sampling and analytical expertise is mandatory to assess reliable and repeatable results (see Chapter 2). This chapter summarizes the current knowledge with respect to the different approaches of exhaled breath sampling and analysis in ruminants. VOC analysis has emerged as the most promising strategy, thus it bears thematic priority. Whether volatile compounds that are exhaled or emitted are robust enough to withstand as diagnostically relevant biomarkers in a clinical context is yet to be elucidated.

27.2 Exhaled breath condensate

Per definition, exhaled breath condensate (EBC) is obtained by cooling exhaled breath through contact with a cold surface or condenser. Samples are collected as fluid or frozen material and analyzed immediately or later for volatile and nonvolatile macromolecules (see Chapter 7).

Encouraging work has been performed in standardizing collection and analysis of EBC samples in numerous species of animals, including ruminants.[2,3] Besides a number of homemade collection systems, commercial systems for EBC collection are available that can be applied to animals via tightly fitting facemasks, taking into account that nose breathing is obligatory for most of the large animal species. Components of interest identified in EBC of large animals were predominantly related to markers of oxidative stress, inflammation, or infection.[3,4]

It has been challenging to compare EBC results for different animal species published by different groups due to sampling differences and repeatability issues. In light of current knowledge of the field, this is not surprising. After two decades of EBC research, the initial enthusiasm among EBC researchers was replaced by caution in relation to pitfalls and limitations. Overall, even 1 mL EBC may not be a suitable sample volume for measuring and quantifying potential biomarkers, depending on the analyte(s) and the measurement assay. Besides the efficacy of the condensation system, the individual spontaneous pattern of breathing contributes significantly to the amount and composition of EBC, as demonstrated exemplarily in pigs and calves.[5] There is no clear information on how EBC is influenced by the environment (inhalation) or by water dilution. Furthermore, steps of processing EBC until analysis, as well as preservation techniques and analytical methods, differ between working groups. Due to this level of methodological inhomogeneity, highly variable and even conflicting results can be found in the literature.

For EBC research in the future, standardization of collection, processing, and analysis, as well as data interpretation, would be necessary before EBC data are interpretable in a biological sense. Most likely, these different standardization steps need to be adjusted to each potential biomarker of interest. To overcome the problem of dilution in EBC by condensed water, exhaled particles could be of interest, and precise determination of exhaled aerosols could possibly be a useful prerequisite to standardize the assessment of nonvolatile mediators in EBC (by normalization to the aerosol emission rate).

27.3 Exhaled particles

Per definition, particles in exhaled breath (PExs) are endogenously generated droplets that contain water and nonvolatile material originating from the respiratory tract lining fluid (see Chapter 8). The theory behind aerosol formation is based on the bronchiole fluid film burst model, also called the "reopening model." Since the very peripheral airways are "closed" at the end of exhalation (i.e., the airway walls are attached to each other), the respiratory fluid film within these airways bursts while airways are "reopened" during the next inhalation. The resulting aerosol droplets are drawn into the alveoli until they are expelled with airflow during the next exhalation.

It is well known from human studies that emission of PEx differs significantly between subjects, and this intersubject variability is strongly associated with lung physiology and individual respiratory patterns.[6] Tidal breathing generates only a low number of extremely small particles whose mass is negligible. Consequently, breathing maneuvers are required to force airway opening following airway closure in order to generate more particles and increasing the number of PEx substantially.

In animal models, exhaled aerosols have been studied in the context of aerogen transmission of pathogens to understand aerogen transfer routes of infections. A study in calves assessed the quantity of *Chlamydia psittaci* exhaled for a defined time by ensuring that all exhaled breath was directed through a dust sampler equipped with a gelatin filter.[7] Although excretion and/or exhalation of the pathogen was sufficient to transmit the infection to naïve sentinel calves cohoused with the infected animals, the presence of the pathogen in exhaled breath was only demonstrated sporadically. This finding was not surprising, taking the limitations known from humans into account, particularly when PEx assessment is based on tidal breathing.

Systematic studies on PEx in large animals are neither available nor promising. First, standardized breathing maneuvers cannot be demanded in spontaneously breathing large animals. Second, nose breathing is the obligatory route of respiration for most of the large animal species. Compared to the adult human whose nasal cavity is relatively short and high, the shape of the nasal cavity in *Artiodactyla* (even-toed ungulate animal species) is stretched and flat. Functionally, the two respiratory turbinates, located directly in the main part of the nasal airflow, are most important for the mandatory contact between breathing air and the respiratory mucosa. Besides its respiratory function that comprises moistening and filtering of the inspired air, the nose most effectively ensures thermoregulation and odor recognition in many animal species, including sheep and goats. The anatomical structure of nostrils and nasal interior is linked with nasal airflow resistance, thus the airflow runs nearly completely turbulent through the nose. Taking all these anatomical and functional peculiarities of nose breathing together, the chance of using PEx in spontaneously breathing large animals (including ruminants) for diagnostic purposes seems very limited.

27.4 Volatiles in exhaled breath

27.4.1 Methodology of collection and analysis

A schematic overview of the principal analytical strategies for the detection of volatile biomarkers is given in Fig. 27.1. Among all the possibilities depicted, the following three methodologies have been proven applicable to analyze volatiles in the exhaled breath of ruminants: (I) pattern recognition of VOC mixtures, (II) sampling of preconcentrated exhaled breath for later analysis in specialized laboratories, and (III) VOC analysis in real time.

I. ***Pattern recognition:*** Different technologies of electronic noses (E-noses) and differential ion mobility spectrometry (DMS/IMS) have been used in ruminants (bovines and goats) to identify animals with defined infections (Table 27.1). E-noses are commonly based on nonselective absorption-based or electrochemical sensors to obtain a fingerprint of the condition assessed. The DMS uses mobility differences of chemicals as they interact with tightly controlled electric fields to quantify and characterize analytes of interest. In both cases, the data outputs can be analyzed using pattern recognition for chemical identification. Although some of these technologies have been demonstrated to be able to distinguish between the infected groups of ruminants and the noninfected ones in controlled studies, the individual volatile substances generating the fingerprint of interest are not explored in the reported studies.[8–12] The biochemical backgrounds of the recognized patterns of interest and the robustness of this methodological approach have yet to be elucidated.

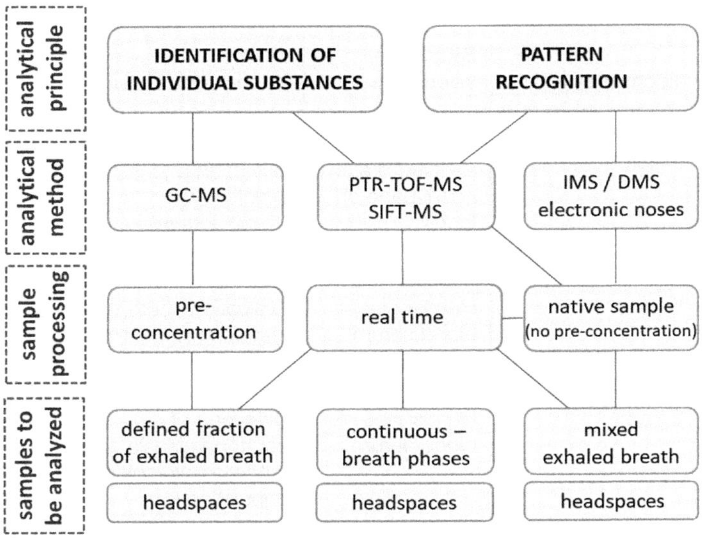

FIGURE 27.1

Analytical strategies for volatile biomarker detection.
DMS, differential ion mobility spectrometry; *GC-MS*, gas chromatography—mass spectrometry; *IMS*, ion mobility spectrometry; *PTR-TOF-MS*, proton transfer reaction—time of flight—mass spectrometry; *SIFT-MS*, selected ion flow tube—mass spectrometry.

II. ***Sampling of preconcentrated exhaled breath and laboratory-based analysis:*** The technical setup for exhaled breath collection needs to be adjusted to the fraction of exhaled gas to be analyzed. A controlled sampling system, e.g., using exhaled carbon dioxide (CO_2), is an indispensable prerequisite in order to collect predominantly alveolar gas, and the latter is most likely representative for blood-borne volatiles. If all fractions of exhaled breath are collected, a mixture between a significant amount of deadspace gas and alveolar gas will be analyzed. Compared to the human lung, the proportion of deadspace volume per breath is much greater in large animals due to the anatomy of extrathoracic airways (i.e., nasal and laryngeal passages, and trachea). While anatomical or physiological deadspace, respectively, comprises about 30% of tidal volume in a healthy human adult, the proportion of deadspace volume per breath in cattle and sheep is about 60% or even higher. Including deadspace gas in exhaled breath analysis may be helpful if volatiles derived from the airway epithelium are of interest (for example, under conditions of airway inflammation). For metabolic markers, however, any deadspace gas included in the analysis will dilute any markers that are present in alveolar gas.

The preanalytical phase between sampling and analysis is influenced by storage material, time of storage, transport conditions, etc., and each factor may impact the analytical results (see Chapter 2). To avoid these confounding influences,

Table 27.1 Overview of VOC analyses performed in ruminants.

Animal species	Technology	Biological specimen	Focus of the study	Ref.
Pattern recognition				
Goats	DMS	Exhaled breath, headspace of feces	Detection of chronic infection with *Mycobacterium avium* subspecies *paratuberculosis* (MAP)	[8]
Cattle (calves)	E-nose	Headspace of blood serum	Evaluating acute-phase response (experimentally induced, *Mannheimia haemolytica*)	[9]
Cattle (adult)	E-nose	Headspace of blood serum	Detection of paratuberculosis and brucellosis in cattle on population level	[10]
Cattle (calves) Badgers	E-nose	Headspace of blood serum	Detection of *Mycobacterium bovis* infection	[11]
Cattle	E-nose	Headspace of blood serum	Detection of *Mycobacterium bovis* infection (bovine tuberculosis)	[12]
Laboratory-based analysis				
Goats	GC-MS (NTD)	Alveolar fraction of exhaled breath	Impact of food intake on exhaled breath	[16]
Goats	Breath: GC-MS (NTD) Feces: GC-MS (SPME)	Alveolar fraction of exhaled breath, headspace of feces	Effects of changes in nutrition from milk-based food to plant-based food and somatic growth	[15]
Goats	Breath: GC-MS (NTD) Feces: GC-MS (SPME)	Alveolar fraction of exhaled breath, headspace of feces	Detection of *Mycobacterium avium* subsp. *paratuberculosis* infection	[14]
White-tailed deer	GC-MS (SPME)	Headspace of feces	Detection of chronic wasting diseases	[21]
White-tailed deer	GC-MS (SPME)	Headspace of feces	Detection of *Mycobacterium bovis* infection	[22]
Cattle (calves)	GC-MS (SPME)	Headspace of feces	Discrimination between nonvaccinated and BCG-vaccinated cattle prior to	[23]

Table 27.1 Overview of VOC analyses performed in ruminants.—*cont'd*

Animal species	Technology	Biological specimen	Focus of the study	Ref.
			and after *Mycobacterium bovis* challenge	
Cattle (calves, adult)	GC-MS (Tenax)	Exhaled breath gas inside the facemask	Detection of *Mycobacterium bovis* infection	[26,27]
Cattle	GC-MS (SPME)	Exhaled breath gas	Characterization of clinical morbidity	[34]
Cattle (adult)	GC-MS (SPME)	Rumen headspace gas	Composition of rumen gas; effects of feeding	[20]
Cattle (adult)	SIFT-MS	Rumen headspace gas	Composition of rumen gas; effects of feeding	[19]
Cattle (adult)	GC-MS (polyterephthalic ester bags)	Exhaled breath	Detection of ketosis in dairy cows	[35]
Bison	GC-MS (Tenax)	Exhaled breath	Detection of VOC in *Brucella abortus* —seropositive animals	[36]
Real-time analysis				
Cattle (adult)	PTR-TOF-MS	Exhaled breath	Introducing a new technical setup for breath analysis in bovines	[17]
Cattle, goats, sheep	PTR-TOF-MS	Exhaled breath	Effect of eructation on exhaled VOC profiles in ruminants	[18]
Cattle (adult)	PTR-TOF-MS	Stable air	VOC emissions from dairy cows (herd monitoring)	[24]
Cattle (adult)	PTR-MS	Environmental chamber	VOC emissions from dairy cows (predominantly urine and feces)	[37]
Cattle (calves)	Tunable diode laser absorption spectroscopy	Exhaled breath (lower airways)	Measurement of exhaled NO	[38]
Cattle (heifer)	Infrared laser detection of CO, N_2O, and CO_2	Exhaled breath	Evaluation of exhaled CO, exhaled N_2O, and exhaled CO_2 in bovine respiratory disease	[39]

DMS, *differential ion mobility spectrometry;* E-nose, *electronic nose;* GC-MS, *gas chromatography —mass spectrometry;* NTD, *needle trap devices (used for preconcentration);* PTR-MS, *proton transfer reaction—mass spectrometry;* PTR-TOF-MS, *proton transfer reaction—time of flight—mass spectrometry;* SIFT-MS, *selected ion flow tube—mass spectrometry;* SPME, *solid-phase microextraction (used for preconcentration);* Tenax, *glass cartridge containing sorbent material (used for preconcentration).*

preconcentration of the gas mixture during the sampling process already has been considered as one step of standardization. Preconcentration steps using needle trap devices (NTD), for example, have been proven to reduce sampling time and to improve detection limits, stability, and reproducibility of the results of exhaled breath gas analysis.[13] A CO_2-controlled device combining automated sampling of the alveolar fraction of exhaled breath and preconcentration in NTD has been successfully applied to goats and cattle.[14–17] Based on predefined flow rates and sampling volumes, this system generates reproducible alveolar gas samples in ruminants ranging from juvenile goats (body mass less than 10 kg) to adult dairy cows (body mass of about 500 kg) and even at high respiratory rates (until 80 breaths per minute) that may occur in nonphysiological conditions (e.g., during heat stress).

Laboratory-based analyses of preconcentrated breath gas by gas chromatography—mass spectrometry (GC-MS) ensure precise identification and quantification of the volatiles present in the sample. However, some logistical restrictions must be taken into account. First, sampling at the animal site and analysis occur separately from each other at different times and places. Thus, results can only be interpreted retrospectively. Second, collecting predefined volumes of breath gas in a cumulative way over time provides samples that represent mixtures of gas fractions over a number of breathing cycles. The latter, however, vary markedly between animals due to their individual breathing patterns and create a certain interindividual biological variability.[5,17,18] Third, ruminants have the peculiarity of highly developed forestomachs where complex gas mixtures, primarily CO_2 and methane (CH_4), are produced due to microbial-based fermentative digestion. These gases must be removed rapidly to prevent rumen distension. When released by eructation, they contribute to exhaled breath. Since some of the eructation episodes are "silent" (not detectable visually or acoustically during the measurements), lung-derived breath gas and rumen-derived gases can hardly be separated from each other in pooled samples collected over numerous exhalations.[18–20]

III. ***Real-time analysis:*** VOC analysis in real time has advantages of being fast and allowing continuous monitoring of the complete exhaled VOC profiles in a breath-by-breath time-resolved way. Using a proton transfer reaction—time of flight—mass spectrometer (PTR-TOF-MS), both practical applicability and usefulness of breath gas analysis have been demonstrated in cattle, sheep, and goats.[17,18] Regardless of respiratory rate, age, or body condition score, a clear discrimination between inspiratory and expiratory phases of the breathing cycles is possible, even over a wide range of respiratory rates (<15 to about 50 breaths/min). Instantly available, repeatable, and reliable results have been gained in each of the three ruminant species tested.[17] Despite its sophistication, PTR-TOF-MS can be implemented in a robust setup appropriate for field trials. Particularly in ruminants, real-time analysis of breath gas is a crucial requirement for the detection and elimination of confounding signals that derive predominantly from ruminal gases.[18]

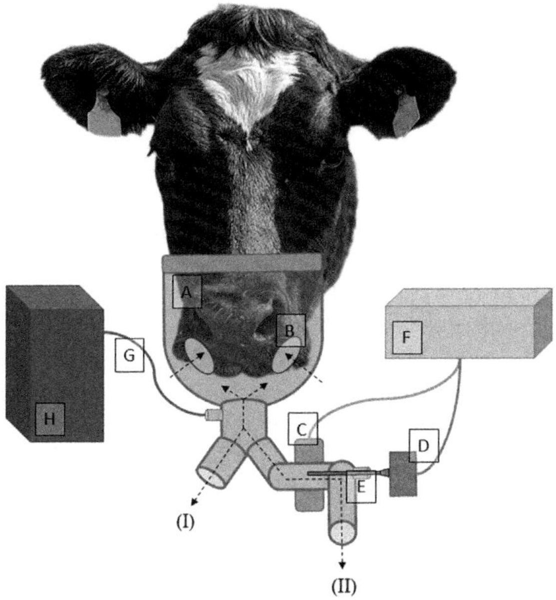

FIGURE 27.2

Setup for the combined collection of alveolar gas and real-time breath analysis in cattle. A: tightly fitting facemask, B: inspiratory valve, C: CO_2 sensor, D: CO_2 triggered flow valve, E: needle trap device, F: sampling device for alveolar gas, G: heated transfer line, H: real-time analyzer, vents: (I) Ø 25 mm (respiratory flow) and (II) Ø 10 mm.

A combination of the two different methodologies, i.e., (II) collection of alveolar gas (based on capnography and combined with preconcentration followed by GC-MS analysis) for identification and quantification of volatile compounds and (III) breath-by-breath analysis in real time to distinguish between pure lung-derived exhaled breath and contaminations of ruminal gases, is currently the most suitable approach for research directed on exhaled biomarkers in any ruminant species (Fig. 27.2). This complex methodology is still an indispensable prerequisite to identify marker profiles for health and certain diseases and to evaluate biomarker candidates regarding their robustness, sensitivity, and specificity. Later simplification toward sensor techniques could be the next step, but it is still a future vision in veterinary medicine.

27.4.2 Forestomach gases must be separated from exhaled breath in ruminants

Ruminants acquire nutrients from plant-based feed by using a specialized forestomach system (reticulum, rumen, and omasum) where both degradation of complex plant material and microbial protein synthesis occur. Degradation of polysaccharides, such as cellulose, is ensured by a variety of microbes, mainly obligate or facultative anaerobes, and is assisted by the process of rumination, i.e., rechewing of partly fermented, regurgitated ingesta. After hydrolysis of cellulose, the ruminal microbiome further degrades carbohydrates (and nitrogen) to short-chain fatty acids, mainly acetate, lactate, propionate, and butyrate. Gases are by-products of this fermentation process. Eructation is the principle route of gas elimination from the forestomachs, and the main stimulus for eructation is gas pressure in the dorsal part of the rumen. Regular eructation (physiologically about once or twice per minute) is of vital importance for ruminants and is coordinated by afferent and efferent fibers of the *nervus vagus* as a vagovagal reflex circuit. Eructation gas predominantly consists of CO_2 (about 60%) and CH_4 (about 35%), but includes at least 50 other gases and volatiles, such as several alcohols, nitrogen (N_2), ammonia (NH_3), traces of oxygen (O_2) and hydrogen (H_2), sulfide components, and phenolic compounds, as well as products of fermentation, i.e., volatile fatty acids, acetone, butanol, etc.[19,20] When eructation occurs, the ruminal gas is released into the esophagus, where it travels up at a mean velocity of about 2 m/s. Physiologically, the majority of eructated gas is first inspired into the lungs before it contributes to expiration later on.

A recent study revealed, for the first time, how eructation significantly alters exhaled breath composition. About 20–30 exhaled VOCs increased markedly after one episode of eructation in goats, sheep, and cattle, and the number of breathing cycles required before levels returned to initial levels ranged from 1 to about 20 (maximum: 36 breaths). In particular, compounds with high water solubility and unsaturated hydrocarbons, such as 2-methylbutene, showed the highest increase after eructation and needed longer to reequilibrate.[18] Taking this new knowledge into account is a mandatory requirement for further research regarding breathborne biomarkers in ruminants and will strengthen the validity of results. A custom-made algorithm, the so-called "Burp Tracker", recently published by Oertel et al.,[18] now enables an automated identification of eructation episodes in breath gas analysis of ruminants based on comparing the difference of the intensity of two consecutive exhalations with respect to mean intensity over all measured breaths for defined VOCs[a].

Retrospectively, discriminating between breath gas and contamination by ruminal volatiles has led to a better understanding of variations in exhaled breath data published in the literature, to date. In a prospective view, pure breath data and exhaled eructation gases could be analyzed separately, providing distinguished real-time information about the exhalation of bloodborne volatiles (alveolar fraction), volatiles produced in the airway epithelium (deadspace fraction), and fermentation-derived volatiles from digestive processes in the forestomach system, particularly rumen. This new research strategy requires further investigation and interdisciplinary collaboration.

[a] stacks.iop.org/JBR/12/036014/mmedia

27.4.3 Sources of biological variability of exhaled volatiles

From a biological perspective, there are many factors, conditions, and influences that determine the total VOC profile of an organism. It is ancient knowledge that different species of animals smell differently and that male sheep or goats present a characteristic odor compared to females of the same species. Beside species and sex, there are numerous sources of biological variability, and those evaluated in scientific studies particularly for ruminants are discussed below.

Development from a juvenile animal to a functional ruminant and resultant somatic growth

For the first months of life, young lambs and calves are functionally monogastric mammals. After birth, all of them depend on milk. While suckling, a reflex (closing muscular folds to a channel, i.e., esophageal groove) ensures direct transport of milk to the abomasum while the forestomachs are bypassed. Although the forestomachs are anatomically present, they are not functionally developed yet (preruminant phase). With the change in feed composition toward plant-based products that include coarse-grained concentrates and roughage (hay, straw), the rumen will be colonized by microbes (bacteria, protozoa, fungi), which start fermentation, and the system of forestomachs develops further. This transition phase from milk-based feedstuff to solid feed materials, i.e., until ruminants become pure herbivores, takes a few months in cattle, goats, and sheep. During this period, glucose is replaced as the primary source of energy by short-chain fatty acids as the products of ruminal fermentation. The manifold dynamic changes in metabolism are clearly associated with changes in the metabolome assessed by VOC analyses.

One long-term study over 1 year assessed the physiological changes in exhaled VOC profiles that occurred due to the replacement of the predominant milk intake (in goat kids) by an increasing percentage of plant products through to pure plant-based nutrition. This was performed in healthy goats under standardized conditions with respect to their nutrition regime, husbandry, and environmental influences. In addition to the expected changes due to degradation of different feedstuffs, somatic growth was associated with changing concentrations and compositions of blood proteins due to maturing processes in liver function and the immunological system. Under these circumstances, exhaled aromatic hydrocarbons, aldehydes, and ketones changed significantly over time and became much less variable as soon as the animals became full ruminants (digesting only plant-based feed for the rest of their life).[15]

This long-term study in a biologically relevant large animal model revealed the critical importance of a well-founded knowledge regarding metabolic effects of nutrition and somatic growth effects as "background information" that will contribute significantly to the composition of all VOC profiles emitted at different sites of the body in any animal species.

Intra and intersubject variability of exhaled volatiles

Animal models bear the advantage of a high level of standardization regarding housing conditions, nutrition regime, and medications. Defined experimental conditions and a high degree of homogeneity within a uniform cohort of animals (breed, sex, age, body weight, genetic background, etc.) assure that intersubject variability can be kept as low as possible. Thus, interindividual variations of any parameter observed in well-controlled animal studies reflect the *minimum* of biological variability that has to be taken into account. A much larger intersubject variability will appear under field conditions where populations are more heterogeneous and surrounding conditions are less standardized.

Compared to intersubject variability, intrasubject variations are usually much lower (about 20%–30% of total variability) but still may become significant when considering the reliability and the repeatability of consecutive measurements. Variations over time in the same animal have been linked to many factors, including intake of feedstuff, circadian rhythm, or habituation effects. Animals that provided consecutive exhaled breath samples on different days adapted to the collection procedures and handling (i.e., wearing a facemask, short-time fixation, peculiarities of the technology used, etc.); this effect was comparable to "training," resulting in lower variability in VOCs over time when measurements were performed at the same time of day.[17] Over the course of the day, however, diurnal or circadian influences are associated with the progression of VOC concentrations for many substances, as measured exemplarily in exhaled breath of young cattle.[17] Regardless of the animal species (goats or cattle), the lowest exhaled VOC concentrations were constantly observed in the morning, i.e., before the first feeding of the day.[16,17] After intake of a concentrated mixture of plants and plant products, particularly the postprandial concentrations of both saturated and unsaturated hydrocarbons showed a concentration pattern that was characterized by an initial increase, a peak about 60 min after food intake, and a subsequent gradual decrease over the next hours.[16] With respect to standardization of protocols for exhaled VOC analyses, any feeding or spontaneous intake of feedstuff generates physiological effects on exhaled VOC concentrations that need to be taken into consideration.

27.5 Volatiles from other biological samples obtained from ruminants

27.5.1 Volatiles emitted from feces

While exhaled VOCs are more representative of bloodborne variables, fecal samples have had direct contact with the inside of the gut and might therefore reflect intestinal and metabolic processes. Methodologically, the headspace above fecal samples has been studied in ruminants (particularly in goats and white-tailed deer) using both pattern recognition techniques and GC-MS.[8,14,15,21–23] The questions raised so far were related to either physiological[15] or infectious[8,14,21–23] conditions (Table 27.1).

Physiological influences resulting from intake of feed and digestion

As mentioned previously, the development from the preruminant calf, lamb, or goat kid (with milk in the composition of feedstuff) to a full ruminating adult is accompanied by significant changes in metabolism and microbiota colonizing the digestive tract, particularly the forestomachs. Consequently, consistency and composition of feces change over time, and so does the composition of VOCs emitted from the feces. In goats, concentrations of ketones, aliphatic hydrocarbons, furans, and sulfur-containing compounds were much higher in the headspace of feces during the milk-suckling phase and decreased to lower concentrations in the advanced ruminant, whereas emissions of other VOCs, such as isoprene and methyl acetate, increased with age.[15] These findings show exemplarily the extent to which different nutrition regimes significantly influence intestinal homeostasis and, consequently, the VOC profiles emitted. This knowledge strengthens the need for standardized feeding and nutritional management in research studies in order to demonstrate the robustness of potential volatile biomarkers.

27.5.2 Volatiles assessed in stable air (herd monitoring)

Based on the knowledge that VOCs are emitted from feces, urine, skin, and via exhalation, and based on the hypothesis that infected and/or diseased animals emit diagnostically relevant VOCs in a pattern that will differ from the physiological one, it is postulated here that a mixture of indicative volatile markers will be present in the air of the stable that houses the animals. Further, it is postulated that the concentration of biologically relevant volatile markers depends on the percentage of animals infected and/or diseased within the herd. Following this hypothesis, stable air collected or analyzed close to the animals might be an interesting diagnostic sample for screening at the herd level, without the necessity of directly interacting with any individual animal (Fig. 27.3).

The suitability of this novel application has recently been demonstrated in cattle.[24] Employing real-time analysis by PTR-TOF-MS, groups of dairy cows from one herd were monitored consecutively within a defined area while waiting to be milked. The groups differed with respect to their metabolic conditions (i.e., gravidity, time after calving, milk yield) and health status (lameness, infections). In this pilot study on approximately 600 animals, a clear discrimination was possible between the VOC profiles emitted by the different groups. Changes in VOC emissions were related to different average milk yield per group and the time of the day (morning vs. noon milking time). Particularly in early lactation, dairy cows with high milk production run into a deficit of energy that results in mobilization of body fat and the endogenous formation of ketone bodies. The corresponding metabolic situation, i.e., mostly subclinical ketosis or acetonemia, is associated with increased concentrations of acetone in exhaled breath. In contrast, acetone emissions from cows in the late lactation phase and/or with lower milk yield were reduced compared to fresh-milking cows and/or cows with high milk yields.[24] Due to these encouraging first results, the novel research strategy of herd monitoring deserves further attention.

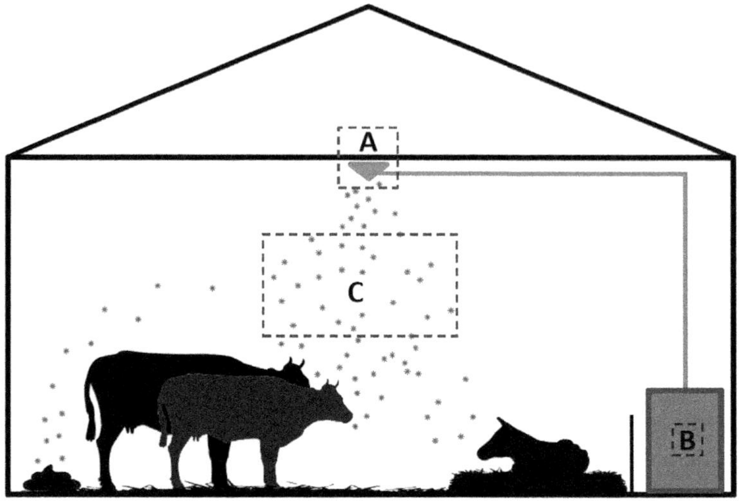

FIGURE 27.3

Principle of stable air measurement (herd monitoring) in flocks of ruminants. Red/gray stars indicate VOCs emitted from animals by exhalation, feces, etc. A: sensor or sampling inlet, B: analytical device, and C: mixture of VOCs from various sources in stable air. Red (gray in print version) stars indicate VOCs emitted from animals by exhalation, feces, etc. A: sensor or sampling inlet, B: analytical device, and C: mixture of VOCs from various sources in stable air.

27.6 Association between volatile markers and infections

Using VOCs as indicators for the presence of certain infections has gained interest in human and veterinary medicine. Infectious processes within the airways and the lung may generate diagnostically relevant VOCs at the site of infection, thus the chance of using exhaled breath for diagnosing pulmonary infections is quite high (see also Chapter 21). Using exhaled breath to detect extrapulmonary or systemic infections is based on the hypothesis that diagnostically relevant volatile markers—generated elsewhere in the body—are transported by blood to the lungs, where they equilibrate between blood and alveolar gas. Because the lungs receive the total cardiac output, the large endothelial surface of the pulmonary vascular bed is very important for and very efficient in removing many substances from the blood that were previously produced at different sites of the body.

In ruminants, infections with mycobacteria have been the predominant focus of interest because these chronic infections are economically important for affected farms and bear a zoonotic risk for humans. While *Mycobacterium bovis* (*M. bovis*) is the main agent of bovine tuberculosis, with many sites of infection, including the lungs, *Mycobacterium avium* subspecies *paratuberculosis* (MAP) primarily causes granulomatous enteritis called paratuberculosis. Worldwide, the two pathogens or diseases, respectively, can be found in domestic and wild ruminants, and eradication programs depend on early diagnosis in living animals. The latter is still a challenge because infected animals remain clinically unsuspicious for many months or even years. During this long period of clinical silence, the infected animals hardly produce antibodies and shedding of the pathogen is intermittent.[25] To overcome the current diagnostic dilemma, VOC analysis as a potential diagnostic tool has gained interest, and different biological specimens have been included in the search for diagnostically relevant VOC marker sets (Table 27.1).

The first encouraging data were published in 2005, where headspace of blood sera from cattle infected with *M. bovis* and from noninfected controls were differentiated using an E-nose.[11] In later studies, headspace from cattle feces and a mixture of exhaled breath collected from the deadspace of the facemask were analyzed with GC-MS for the same diagnostic purpose.[26,27] Further evidence for using VOC analyses diagnostically resulted from a caprine model of paratuberculosis in which DMS/IMS was capable of identifying goats infected with MAP using both exhaled breath and fecal headspace.[8] Other in vitro studies have investigated specific VOC profiles emitted from the different species of mycobacteria.[28–30]

It is less likely that a VOC profile emitted from a pathogen under artificial or standardized growth conditions can be rediscovered in living subjects. On the contrary, host-specific reactions to the pathogen or pathophysiological and immunological responses by the host animal, respectively, will either contribute to the VOC profile emitted by the pathogen or even generate a unique pattern of VOCs.

Since none of the in vivo studies mentioned so far differentiated between VOC markers attributed to the pathogen itself and VOC markers resulting from the underlying host–pathogen interactions, two different caprine models representing two different clinical onsets of mycobacterial infections were exploited. A model of paratuberculosis, induced by MAP, represents the initial phase of a chronic granulomatous enteritis that remains clinically nonapparent in most of the animals for several months.[31] In contrast, *Mycobacterium avium* subspecies *hominissuis* (MAH) induced an acute and even life-threatening disease comparable to systemic inflammatory response syndrome (SIRS) in about 50% of the animals challenged.[32]

Recent knowledge gained from the two different models induced by the different mycobacterial species revealed, for the first time, that different biological specimens used for VOC analysis represent different pathophysiological backgrounds:

Exhaled volatiles

Based on alveolar gas samples and blood-based variables of host response, significant associations have been shown to exist between exhaled VOC biomarkers and systemic immunological responses or acute-phase reactions by the host.[14] Thus, there is strong evidence that VOC profiles assessed in exhaled breath are more closely linked to the host response than to the presence of the bacterium. Interestingly, a number of exhaled VOCs relating to the MAP model were at lower concentrations in the infected animals compared to the noninfected controls.[14,33]

Fecal volaitles

In contrast, VOCs assessed in the headspace above fecal samples seem to be more strongly linked with the intestinal mycobacterial infection itself.[14,33] The latter is unsurprising because the intestinal system was the main location of infection. Despite these promising findings with respect to intestinal processes, it has yet to be elucidated which VOC profile from feces specifically represents the presence of the pathogen within the gut (as postulated by Bergmann et al.)[14] and which pattern of volatiles derives from the locally induced pathogen-specific inflammatory reaction.

Volatiles in stable air

Herd monitoring in a dairy farm provided evidence for the presence of animals with paratuberculosis (i.e., a chronic granulomatous enteritis caused by MAP) and optimism for its detection within the herd under field conditions. A few substances reported in connection with MAP elsewhere (tentatively identified as 2-butanone and methyl acetate) expressed different concentrations between groups of cows containing MAP-infected animals and noninfected groups.[24]

Summarizing the current knowledge, the novel insights regarding both detection and evaluation of potential volatile marker sets clearly demonstrate that much more research is required in order to elucidate which volatiles are associated with the pathogen/infection itself, and which VOC biomarkers result from host responses. Furthermore, sensitivity, specificity, and robustness of potential marker sets remain to be defined. To clarify these multiple issues, large animal models provide a suitable basis for future research.

27.7 Summary

Volatile organic compounds as indicators of (patho)physiological conditions or to identify disease states are a promising novel diagnostic aid for veterinary applications, including large animal models. These exhaled or emitted VOCs can be monitored continuously and noninvasively in conscious animals. In ruminants, VOC analyses have not only focused on exhaled breath but also include VOCs in the headspace of biological specimens, or even in ambient stable air.

As principle analytical strategies, identification of individual compounds (using laboratory-based separation techniques or real-time analysis with mass spectrometric detectors), as well as pattern recognition of VOC mixtures, have been utilized so far. For laboratory-based analysis, a controlled preconcentration of the gas to be analyzed and a parallel collection of inspired air are an indispensable prerequisite for reliable results. CO_2-control of breath sampling enables the collection of alveolar gas samples related to bloodborne markers. Real-time mass spectrometry allows parallel identification of inspiratory and expiratory phases of tidal breathing and identification of eructation phases. Once markers (patterns) have been unequivocally identified and quantified by those techniques, less selective techniques (such as E-noses, IMS/DMS, or species-selective sensor systems) may be applied in field studies.

Careful evaluation of physiological and diurnal sources of variability is mandatory before VOC analysis can be used for diagnostics of metabolic conditions or infectious diseases. Particularly in ruminants, two main biological peculiarities need to be taken into account. First, the young lambs, goat kids, and calves are functionally not ruminants. During the preruminant phase in the first months of life, they depend on milk, with glucose as a primary source of energy. In parallel, they start to acquire the ability to gain nutrients from the fibrous material of plants. With the change in feed composition toward pure plant-based products, the forestomach system develops, and short-chain fatty acids become the main products of ruminal fermentation. Such age-dependent changes in metabolism are clearly associated with changes in the metabolome assessed by VOC analyses. Second, gas mixtures produced physiologically in the forestomach system and released by regular eructation contribute to exhaled breath as confounding gases. Thus, separation between breath gas and ruminal volatiles is indispensable for the concept of a diagnostic use of exhaled VOCs. Further known effects of feed intake (i.e., composition of feedstuff and time of feeding), circadian influences, and housing and handling of the animals complement the knowledge about physiological variability of VOC composition and strengthen the importance of standardizing biological conditions and operating procedures to improve data quality.

Since mycobacteria are relevant pathogens in ruminants worldwide, considerable efforts have been made on studying associations between VOC markers and infections caused by either *Mycobacterium bovis* or *Mycobacterium avium* subspecies *paratuberculosis*. Results have revealed that VOC profiles assessed in the infected animals differed markedly from VOC patterns of the pathogens emitted under artificial and standardized growth conditions. Most likely, the pathophysiological and immunological host responses and the induced inflammatory processes generate unique patterns of VOCs. Large animal models and further studies in ruminants provide a suitable basis for future research that will help to evaluate sensitivity, specificity, and robustness of diagnostically relevant VOC marker sets that indicate infections with mycobacteria or other pathogens and allow assessment of host–pathogen interactions.

References

1. Reinhold P, Menge C, Ladwig-Wiegard M, Thöne-Reineke C. Increasing importance of large animal models in biomedical research – an overview. *Berl Münch Tierärztl Wochenschr* 2018;**131**(7/8):341–50.

2. Reinhold P, Knobloch H. Exhaled breath condensate: lessons learned from veterinary medicine. *J Breath Res* 2010;**4**(1):017001.
3. Cathcart MP, Love S, Hughes KJ. The application of exhaled breath gas and exhaled breath condensate analysis in the investigation of the lower respiratory tract in veterinary medicine: a review. *Vet J* 2012;**191**(3):282–91.
4. Reinhold P, Deaton C, Marlin D. Potential for and limitations of exhaled breath analysis in animal models. In: Amann A, Smith D, editors. *Breath analysis for clinical diagnosis and therapeutic monitoring*. World Scientific Publishing Co. Pte. Ltd.; 2005. p. 503–14.
5. Reinhold P, Jaeger J, Schroeder C. Evaluation of methodological and biological influences on the collection and composition of exhaled breath condensate. *Biomarkers* 2006;**11**(2):118–42.
6. Schwarz K, Biller H, Windt H, Koch W, Hohlfeld JM. Characterization of exhaled particles from the healthy human lung–a systematic analysis in relation to pulmonary function variables. *J Aerosol Med Pulm Drug Deliv* 2010;**23**(6):371–9.
7. Ostermann C, Rüttger A, Schubert E, Schrödl W, Sachse K, Reinhold P. Infection, disease, and transmission dynamics in calves after experimental and natural challenge with a Bovine *Chlamydia psittaci* isolate. *PloS One* 2013;**8**(5):e64066.
8. Purkhart R, Köhler H, Liebler-Tenorio E, Meyer M, Becher G, Kikowatz A, Reinhold P. Chronic intestinal *Mycobacteria* infection: discrimination via VOC analysis in exhaled breath and headspace of feces using differential ion mobility spectrometry. *J Breath Res* 2011;**5**(2):027103.
9. Knobloch H, Schroedl W, Turner C, Chambers M, Reinhold P. Electronic nose responses and acute phase proteins correlate in blood using a bovine model of respiratory infection. *Sens Actuators B* 2010;**44**:81–7.
10. Knobloch H, Köhler H, Commander N, Reinhold P, Turner C, Chambers M. Volatile organic compound (VOC) analysis for disease detection: proof of principle for field studies detecting *paratuberculosis* and brucellosis. In: Pardo M, Sberveglieri G, editors. *Olfaction and electronic nose: proceedings of the 13th international symposium*. Ann Arbor, MI, USA: AIP (American Institutes of Physics) Publishing; 2009. p. 195–7 [ISBN: 978-0-7354-0674-2/09].
11. Fend R, Geddes R, Lesellier S, Vordermeier HM, Corner LA, Gormley E, et al. Use of an electronic nose to diagnose *Mycobacterium bovis* infection in badgers and cattle. *J Clin Microbiol* 2005;**43**(4):1745–51.
12. Cho YS, Jung SC, Oh S. Diagnosis of bovine tuberculosis using a metal oxide-based electronic nose. *Lett Appl Microbiol* 2015;**60**(6):513–6.
13. Trefz P, Rösner L, Hein D, Schubert JK, Miekisch W. Evaluation of needle trap micro-extraction and automatic alveolar sampling for point-of-care breath analysis. *Anal Bioanal Chem* 2013;**405**(10):3105–15.
14. Bergmann A, Trefz P, Fischer S, Klepik K, Walter G, Steffens M, et al. In vivo volatile organic compound signatures of *Mycobacterium avium* subsp. *paratuberculosis*. *PloS One* 2015;**10**(4). e0123980.
15. Fischer S, Trefz P, Bergmann A, Steffens M, Ziller M, Miekisch W, et al. Physiological variability in volatile organic compounds (VOCs) in exhaled breath and released from faeces due to nutrition and somatic growth in a standardized caprine animal model. *J Breath Res* 2015;**9**(2):027108.
16. Fischer S, Bergmann A, Steffens M, Trefz P, Ziller M, Miekisch W, et al. Impact of food intake on in vivo VOC concentrations in exhaled breath assessed in a caprine animal model. *J Breath Res* 2015;**9**(4):047113.

17. Küntzel A, Oertel P, Trefz P, Miekisch W, Schubert J, Köhler H, et al. Animal science meets agricultural practice: preliminary results of an innovative technical approach for exhaled breath analysis in cattle under field conditions. *Berl Münch Tierärztl Wochenschr* 2018;**131**(9/10):444–52.
18. Oertel P, Küntzel A, Reinhold P, Köhler H, Schubert JK, Kolb J, et al. Continuous real-time breath analysis in ruminants: effect of eructation on exhaled VOC profiles. *J Breath Res* 2018;**12**(3):036014.
19. Dewhurst RJ, Evans RT, Mottram TT, Spaněl P, Smith D. Assessment of rumen processes by selected-ion-flow-tube mass spectrometric analysis of rumen gases. *J Dairy Sci* 2001;**84**(6):1438–44.
20. Cai L, Koziel JA, Davis J, Lo YC, Xin H. Characterization of volatile organic compounds and odors by in-vivo sampling of beef cattle rumen gas, by solid-phase microextraction, and gas chromatography-mass spectrometry-olfactometry. *Anal Bioanal Chem* 2006; **386**(6):1791–802.
21. Ellis CK, Volker SF, Griffin DL, VerCauteren KC, Nichols TA. Use of faecal volatile organic compound analysis for ante-mortem discrimination between CWD-positive, -negative exposed, and -known negative white-tailed deer (*Odocoileus virginianus*). *Prion* 2019;**13**(1):94–105.
22. Stahl RS, Ellis CK, Nol P, Waters WR, Palmer M, VerCauteren KC. Fecal volatile organic ccompound profiles from white-tailed deer (*Odocoileus virginianus*) as indicators of *Mycobacterium bovis* exposure or *Mycobacterium bovis* Bacille Calmette-Guerin (BCG) vaccination. *PloS One* 2015;**10**(6). e0129740.
23. Ellis CK, Rice S, Maurer D, Stahl R, Waters WR, Palmer MV, et al. Use of fecal volatile organic compound analysis to discriminate between non-vaccinated and BCG-Vaccinated cattle prior to and after *Mycobacterium bovis* challenge. *PloS One* 2017; **12**(7). e0179914.
24. Gierschner P, Küntzel A, Reinhold P, Köhler H, Schubert JK, Miekisch W. Crowd monitoring in dairy cattle-real-time VOC profiling by direct mass spectrometry. *J Breath Res* 2019;**13**(4):046006.
25. Köhler H, Gierke F, Möbius P. *Paratuberculosis* – current concepts and future of the diagnosis. *Magyar Allatorvosok Lapja (Hungarian Veterinary Journal)* 2008;**130**(Suppl. I):67–9.
26. Peled N, Ionescu R, Nol P, Barash O, McCollum M, VerCauteren K, et al. Detection of volatile organic compounds in cattle naturally infected with *Mycobacterium bovis*. *Sens Actuators B* 2012;(s171–172):588–94.
27. Ellis CK, Stahl RS, Nol P, Waters WR, Palmer MV, Rhyan JC, et al. A pilot study exploring the use of breath analysis to differentiate healthy cattle from cattle experimentally infected with *Mycobacterium bovis*. *PloS One* 2014;**9**(2):e89280.
28. Mellors TR, Rees CA, Wieland-Alter WF, von Reyn CF, Hill JE. The volatile molecule signature of four mycobacteria species. *J Breath Res* 2017;**11**(3):031002.
29. Küntzel A, Oertel P, Fischer S, Bergmann A, Trefz P, Schubert J, et al. Comparative analysis of volatile organic compounds for the classification and identification of mycobacterial species. *PloS One* 2018;**13**(3). e0194348.
30. Küntzel A, Weber M, Gierschner P, Trefz P, Miekisch W, Schubert JK, et al. Core profile of volatile organic compounds related to growth of *Mycobacterium avium* subspecies *paratuberculosis* – a comparative extract of three independent studies. *PloS One* 2019;**14**(8). e0221031.

31. Köhler H, Soschinka A, Meyer M, Kather A, Reinhold P, Liebler-Tenorio E. Characterization of a caprine model for the subclinical initial phase of *Mycobacterium avium* subsp. *paratuberculosis* infection. *BMC Vet Res* 2015;**11**:74.
32. Schinköthe J, Möbius P, Köhler H, Liebler-Tenorio EM. Experimental infection of goats with *Mycobacterium avium* subsp. *hominissuis*: a model for comparative tuberculosis research. *J Comp Pathol* 2016;**155**(2–3):218–30.
33. Kasbohm E, Fischer S, Küntzel A, Oertel P, Bergmann A, Trefz P, et al. Strategies for the identification of disease-related patterns of volatile organic compounds: prediction of *paratuberculosis* in an animal model using random forests. *J Breath Res* 2017;**11**(4): 047105.
34. Spinhirne JP, Koziel JA, Chirase NK. Sampling and analysis of volatile organic compounds in bovine breath by solid-phase microextraction and gas chromatography-mass spectrometry. *J Chromatogr A* 2004;**1025**(1):63–9.
35. Dobbelaar P, Mottram T, Nyabadza C, Hobbs P, Elliott-Martin RJ, Schukken YH. Detection of ketosis in dairy cows by analysis of exhaled breath. *Vet Q* 1996;**18**(4):151–2.
36. Bayn A, Nol P, Tisch U, Rhyan J, Ellis CK, Haick H. Detection of volatile organic compounds in *Brucella abortus*-seropositive bison. *Anal Chem* 2013;**85**(22):11146–52.
37. Shaw SL, Mitloehner FM, Jackson W, Depeters EJ, Fadel JG, Robinson PH, et al. Volatile organic compound emissions from dairy cows and their waste as measured by proton-transfer-reaction mass spectrometry. *Environ Sci Technol* 2007;**41**(4):1310–6.
38. Roller CB, Holland BP, McMillen G, Step DL, Krehbiel CR, Namjou K, et al. Measurement of exhaled nitric oxide in beef cattle using tunable diode laser absorption spectroscopy. *Appl Optic* 2007;**46**(8):1333–42.
39. Burciaga-Robles LO, Holland BP, Step DL, Krehbiel CR, McMillen GL, Richards CJ, et al. Evaluation of breath biomarkers and serum haptoglobin concentration for diagnosis of bovine respiratory disease in heifers newly arrived at a feedlot. *Am J Vet Res* 2009; **70**(10):1291–8.

CHAPTER 28

Breath analysis in marine mammals

Laura C. Yeates[1], Eva Borras[2], Raquel Cumeras[3,4,5], Cristina E. Davis[2]

[1]Olivet Nazarene University, Bourbonnais, IL, United States; [2]Department of Mechanical and Aerospace Engineering, University of California, Davis, CA, United States; [3]Metabolomics Platform, Institut d'Investigació Sanitària Pere Virgili (IISPV), CERCA, Universitat Rovira i Virgili (URV), Reus, Spain; [4]Biomedical Research Centre in Diabetes and Associated Metabolic Disorders (CIBERDEM), ISCIII, Madrid, Spain; [5]Genome Center, University of California, Davis, CA, United States

Chapter outline

28.1 Overview

While the field of breath analysis has primarily focused on human health since its inception, there are other communities with urgent need of noninvasive diagnostics. Marine animals have an interesting parallel to human physiology and can benefit from exhaled breath testing. Skin puncture for routine testing compromises their main physiological barrier to the environment, and repeated blood draws for frequent testing are not prudent. Also, routine blood sampling does not test for contaminants, such as environmental pollutants, petroleum products, plastics, or bio-products of algal blooms, that affect the health of the animals, particularly long term. Thus, noninvasive breath testing is attractive for animals under managed human care. For free-ranging animals, this is even more attractive in order to minimize animal stress during health assessments taking place for population health estimates.

For animals under managed human care, early diagnosis of health aberrations is very important. Dolphins are particularly good at masking disease symptoms, and clinical signs often do not appear until the disease has become advanced, requiring

Breathborne Biomarkers and the Human Volatilome. https://doi.org/10.1016/B978-0-12-819967-1.00028-1
Copyright © 2020 Elsevier B.V. All rights reserved.

longer or more aggressive treatment. These animals are also particularly susceptible to pneumonia. Finally, dolphins can have advanced age-associated susceptibilities to pneumonia—similar to humans.[1,2]

Sampling marine mammal exhaled breath has been explored and achieved largely through three capture mechanisms: (1) the chemical content of the breath gases in the vapor phase can be directly assessed; (2) exhaled breath can be condensed into a liquid and analyzed; and (3) normal exhaled "blow" the animals expel when they surface above the waterline can be captured for offline analysis. Blow does not have a human analog but appears to be a mixture of endogenously produced aerosol droplets, seawater, and exhaled gases. For context, marine mammal "chuff" is a mucus mixture purposefully expelled from the respiratory system, as the animal analog to human sputum. All of these potential sample types have a rich milieu of biochemical metabolites that can be measured.

The research community has shown that there are thousands of compounds in both the volatile and nonvolatile phases of dolphin breath that may serve as noninvasive indicators for early disease. The volatile fraction includes small organic molecules, such as short- to medium-chain alcohols and aldehydes, small amines, and branched hydrocarbons, as well as exogenous compounds metabolized in the body (e.g., petroleum products, pharmaceuticals). The nonvolatile fraction detectable in exhaled breath condensate (EBC) of marine mammals has been shown to include glycans, lipids, peptides, carbonic acids, and pharmaceuticals. There is a continuing need to identify the chemical basis of specific disease conditions (e.g., inflammation and infection) and to evaluate the contributing factors to the total metabolic "fingerprint" of an animal as reflected by breath exhalate.

28.2 Cetacean lung physiology

Cetaceans have evolved several physiological and anatomical adaptations that make them uniquely suitable for using exhaled breath as a tool to measure pulmonary disease as well as overall health status. Unlike terrestrial mammals, their respiratory tract allows for separation of resources, such as oxygen and food intake—they consume food through their mouth and inhale/exhale through a blowhole on the top of their head. Thus, the separation of oral and nasal cavities allows the movement of air from the blowhole to the upper respiratory tract and lungs, which is separated from the digestive tract.

In terrestrial mammals, including humans, the nasal cavity and the mouth meet at the pharynx at the back of the nose and mouth. The pharynx is part of the digestive system as well as the respiratory system because it carries both food and air. The two systems share a common pathway allowing for incidental contamination of the breath chemicals by odors from the digestive system, e.g., during eructation. The larynx extends up into the nasal cavity rather than stopping at the opening into the throat. By comparison, the cetacean trachea is short, with a large diameter to allow small cetaceans like *Tursiops* to be “explosive” breathers, which means they can complete a full respiratory cycle in less than 0.3 s. During this breath maneuver, they exchange up to 75%–90% of their lung capacity.[3]

Total lung volume of dolphins is in the range of approximately 50–70 mL/kg, and researchers have measured an expired volume of up to 6 L and estimated lung volumes of 7 L[4] for an approximately 138 kg *Tursiops*, and 11 L in a 200 kg dolphin. Peak airflow of 24 L/s is also possible, depending on the animal’s body size. Although expired air volume in *Tursiops* is 5–6 times that of humans, their total lung volume is similar to average terrestrial mammals. There may be considerable variation in lung volume between different cetacean species and perhaps even between individuals of the same species from different geographic areas.

Additionally, unlike other marine mammals (pinnipeds, sea otters, polar bears), cetaceans do not have nasal turbinates, nasal hairs, or cilia in the upper respiratory tract to protect them from inspiring potential contaminants or pathogens.[4] In general, cetacean lungs also contain more alveoli than human lungs. Dolphin lungs are made up of two layers of capillaries, and this structural arrangement increases the efficiency of gas exchange since most mammals have only one layer of capillary. Thus, the lung surface area is increased, and gas exchange can occur rapidly.

With this unique respiratory physiology, dolphins are much more susceptible to airborne pathogens that may be present at the water surface. Dolphins have high rates of lung infections, including viral and bacterial pneumonias. Consequently, maintenance of pulmonary health in both wild and managed populations of bottlenose dolphins is an important focus for ecologists, conservationists, and marine mammal veterinarians. Currently, standard-of-care pulmonary assessments in small cetaceans include procedures such as ultrasound, bronchoscopy, cytology from forced exhaled breath content chuffs, and thoracic radiography. For all of these reasons, breath analysis is a reasonable avenue to begin exploring animal and population health of these important species.

28.3 Exhaled respiratory gasses and volatile organic compounds

While most mammals have traditional gas exchange at atmospheric pressure, a landmark paper in 1969 showed how marine mammals manage their breathing and gas exchange while at sea level and subsequently while diving.[5] Over the last several decades, much work has been placed on understanding gas exchange in diving mammals, most commonly oxygen (O_2), carbon dioxide (CO_2), nitrogen (N_2), and nitric oxide (NO). Ranges of these breath gasses can vary based on health status of the animal and also diving and breath-hold profiles.

An important publication in 2013 by Yeates et al. provided a summary of NO concentrations in exhaled dolphin breath under various conditions.[6] The authors reported the average NO concentrations after animals performed breath-holds for 30,

60, 90, and 120 s, but no apparent differences were observed. One animal with illness presented an interesting case study, exhibiting higher NO levels than healthy animals. NO levels were also recorded pre- and postfasting, and healthy dolphins had higher concentrations after feeding. The authors noted that there were high standard deviations of NO measurements in their study, and additional work was needed.

A thought piece by Dr. Sam Ridgway in 2013 begins to tie together some fascinating ideas about dolphin carbohydrate metabolism.[7] At a normal physiological state, dolphins mimic phenomena resembling human hyperthyroid diabetics; they have high thyroid hormone turnover and high plasma glucose levels while fasting. After feeding, the plasma glucose levels remain elevated for some time and can exceed 300 mg/dL (twice the human threshold) when glucose appears in the urine. As a potentially natural animal model of human diabetes, the author hypothesized that breath studies in the animal could yield information that could translate to the human disease state.

Likewise in 2013, another paper by Schivo et al. proposed that Atlantic bottlenose dolphins may be a gold standard animal model of human diabetes.[8] In human studies, both lipid peroxidation products and acetone have been studied to understand their response to altered carbohydrate metabolism in disease states. The authors speculated that future studies of diet manipulation in those animals coupled with breath testing could provide insight on human clinical trials.

While respiratory gases are a primary concern in the vapor phase of marine animal breath, other compounds, such as volatile organic compounds (VOCs), are also present.[9] When scaling exhaled breath collection to larger cetaceans, research also faces different issues and the need for novel sampling approaches to assess health in large, less accessible cetaceans. In a report published in 2014, the authors sought to measure respiratory VOCs from a large cetacean in a brackish inland waterway. The authors reached the animal with water craft and adapted breath sampling tools to be placed near the blowhole as the whale reached the surface to breathe. This method can be challenging given the size of some whale species. The authors used a telescopic bamboo sampling pole with a Tedlar sampling bag to sample the whale breath exhalations from a gray whale (*Eschrichtius robustus*). The gray whale traveled 7 km up the Klamath River (Northern California, USA) in June 2011, sampling was done between July 31 and August 2, and the whale died of septicemia on August 16. The bamboo pole was found to be sufficiently light and flexible to allow the whale's breath to be collected as it swam in the river. Collected whale breath exhalations were immediately transferred onto Tenax-TA traps (one exhalation per trap) through a custom manifold system and then returned to the laboratory where they were analyzed using solid-phase microextraction (SPME) and gas chromatography-mass spectrometry (GC-MS).

The whale breath showed a rich diversity of VOCs, once field sample blank artifacts were removed.[9] The authors observed 70+ chemicals in the breath samples, with 58 of them identified. Some of the common breath VOC chemical classes were alcohols, aliphatics, amines, aromatic hydrocarbons, substituted alkane and alkenes, esters, sulfonated compounds, terpenes, ketones, benzenoid compounds, halogenated hydrocarbons, and polycyclic aromatic hydrocarbons (PAHs). While some of these compounds are likely endogenous and generated from the animal itself, some compounds clearly had environmental origins of exposure. A total of 44% of the reported compounds had been previously reported in human breath samples.

28.4 *Tursiops truncatus* health assessments via exhaled breath condensate

Exhaled breath condensate has been widely explored in human studies and provides the unique possibility of quantifying both volatile and nonvolatile metabolites (see Chapter 7). EBC can be standardized by sample volume and has high reproducibility in human studies. Various mass spectrometry methods have been developed and refined to determine the broadest possible metabolite profiles.[10]

Using a custom EBC sampling device adapted for marine mammals, initial EBC sampling was achieved in bottlenose dolphins under managed human care.[11] A soft inflatable donut-shaped first responder mask was used as an interface between the sampler and the dolphins head and blowhole. After training, the dolphins allowed the device to rest above their blowhole, and they exhaled through the device on command. The interior of the EBC sampler had a glass tube around which dry ice was packed to cool the breath as it transited the tube. During exhalation, the expired breath passed through the chilled tube to condense, while during inhalation a one-way valve allowed incoming ambient air to bypass the chilled tube. This avoided having the animals exposed to extreme temperature differences in inhaled air temperatures. The animals did not exhibit discomfort during the collection. The initial device dimensions were chosen based on tidal lung volume of average sized animals and typical peak exhaled flow rates. Over 5 min, breath sampling produced EBC volumes of ~0.5 mL. The condensed breath was removed and stored at −80°C for later chemical analysis. The EBC was then lyophilized and resuspended in a small volume of solvent, and both GC-MS and liquid chromatography-mass spectrometry (LC-MS) analysis were performed. A wide range of endogenous and exogenous compounds were measured in the dolphin EBC.

Especially volatile compounds were observed in the dolphin breath, and these had also been previously reported in human studies. Examples include various aliphatic and aromatic alcohols, hydrocarbons, and carbonyls. Other volatile small amines, presumably associated to the dolphin's "fishy" breath odor, such as 1-pentanamine, bis (2-hydroxypropyl) amine, and 5-(hydroxymethyl)-2-pyrrolidinone, were also detected and have not been reported in humans. The nonvolatile fraction included amino acids, peptides, lipids such as steroids, phospholipids, prostaglandins, carbohydrates, and small molecules such as carbonic acids, amines, and pharmaceuticals. Additionally, phthalates (plasticizers) and other water and environmental contaminants were detected and attributed to environmental sources. Dolphins spend their lives in water and are potentially exposed to contaminants over long periods, especially if they are close to urbanized coastal areas or other industries. Those external compounds could be absorbed through the skin as well as inhaled or ingested, metabolized, and subsequently exhaled.[11] In this first study, the authors provided a survey of compounds of interest in the animal's breath. Comparisons were also made between the metabolomic fingerprint between specific populations of animals, such as differences between those under managed human care and wild dolphin groups. In general, higher variability was observed in the wild animal population compared to the EBC profiles provided by those under managed human care. This difference is likely due to increased variability of the environment and food sources for the free-ranging wild animals.

Sampling temperatures are known to have an effect on the metabolite content in EBC.[12] While this is not dramatic, it does inform choices of how to best standardize EBC collection from marine mammals in field settings. Given that temperatures can rise rapidly in the EBC collection tube of the first generation device, a second device was redesigned to ensure appropriate sample collection.[13] This device had several key improvements: the device was designed to collect larger EBC sample volumes by increasing the inner tube length and inner diameter of the glass tube; it provided more space around the glass tube for crushed dry ice to bring down the sampler temperature for efficient condensate formation; and finally, heat transfer and breath flow models helped determine the optimal length of the device to extract as much EBC as possible for the relatively high exhaled breath flow rates of the animals. The device had enhanced performance and sampling reproducibility, and it could be used on both land and for animals in the water in field conditions.

These improvements in the EBC collection device led to a follow-on study to examine associations between known disease states in cetaceans and their breath metabolites profiles. Complementary orthogonal physiological information was collected to characterize hematological and serum biochemical parameters, and these testing results could be correlated with EBC breath metabolites. One initial study applied this approach to animals from the largest marine oil spill in US history—the 2010 Deepwater Horizon oil spill in the Gulf of Mexico.[14] The EBC metabolome was used to identify correlations between lung abnormalities in cetaceans in the exposed region. Metabolites consistent with chronic inflammation were found, and products of lung epithelial cellular breakdown and arachidonic acid cascade metabolites were associated with sonographic evidence of lung consolidation commonly observed with active pulmonary infections. Serum hormone concentrations (cortisol and aldosterone), hepatobiliary enzyme levels, white blood cell counts, and iron homeostasis were also correlated with the EBC profile.

A second longitudinal assessment study aimed to monitor health conditions and detect early asymptomatic stages of a disease in dolphins under managed human care.[15] EBC samples and paired blood samples were regularly collected over a 14-month period. The metabolomic EBC breath profile was used to help identify biomarkers related to specific disease development in five dolphins followed up over that year. Three of the animals provided promising case study information about biomarkers related to common conditions in cetaceans, such as cutaneous infections, respiratory infections, dental disease, or hormonal changes due to pregnancy.

The relevant results obtained with the previous studies supported the hypothesis that breath sampling could be of value for assessing the health of cetaceans when compared to conventional health measures used as a routine clinical diagnostic. That is the case of type 2 diabetes and the metabolic syndrome, prevalent in epidemic proportions, that can result in significant comorbid disease. Since Atlantic bottlenose dolphins (*Tursiops truncatus*) have been proposed as a good model for human diabetes, the use of noninvasive breath testing of exhaled compounds can help to understand this health condition.[8] By studying the metabolic and cholesterol responses of dolphins to dietary modifications, researchers may gain insight into human diabetes, and breath testing in this population may shed important light on how to design human clinical studies.

28.5 Blow exudate, chuff metabolites, and biomarkers

The first published report of measuring the metabolic content (i.e., not respiratory gases or VOCs) from dolphin exhaled breath using modern analytical chemistry mass spectrometry approaches was published in 2005 by Hogg et al.[16] The authors developed a noninvasive way of collecting saliva and exhaled exudate to measure testosterone in bottlenose dolphins. Saliva was collected onto gauze, and blow samples were collected into polypropylene tubes held inverted over the blowhole. Dolphins were trained to chuff into the tube, and a preservative was added to the tubes to prevent degradation of the testosterone and ensure the preservation of blow mucus. The samples were analyzed using LC-MS. While groundbreaking, it was difficult to determine the physical origin for each sample type: mucus secretions around the blowhole, expectorant (phlegm), or aerosolized droplets from the deep lung that appear in exhaled gas-phase breath. In addition, the LC-MS data were criticized because the chemical identity of the testosterone was based on the presence of a single ion, presumed to be the protonated molecule of interest, rather than multiple diagnostic *m/z*'s. Nevertheless, this was clearly a critical study that helped to inform the community and set the stage for further breath metabolite analysis in marine mammals.

In 2009, the same researchers followed up their original study by expanding it to examine whale blow for the presence of reproductive hormones.[17] The authors again used LC-MS to analyze collected samples. The sampling itself involved either cotton gauze or nylon stocking stretched over a ring that was held via an extended pole that was moved into the whale blow cloud as the animal surfaced. Subjective scoring was needed to estimate the proximity of the collection device to the blow cloud for each sample. The samples were pretreated with an inhibitor mix and then prepared for LC-MS analysis. Two different species of free-ranging whales were sampled opportunistically in field environments ($n = 35$ humpback whales, *Megaptera novaeangliae*; and $n = 18$ North Atlantic right whales, *Eubalaena glacialis*). Of the total animals, 12 had detectable testosterone and 15 animals had progesterone measured. While not standardized, with variable quality between samples, this was the first time that reproductive hormones had been assessed by such a noninvasive method.

Contemporarily, other research groups were also exploring blow for free-ranging marine mammals.[18–26] These groups made progress using respiratory blow samples for endocrine assessments of bottlenose dolphins.[22] The potential for noninvasive sampling makes breath analysis a very intriguing concept to study reproductive or adrenal assessments in free-ranging marine mammals. In this study, the authors were able to adequately sample blow using the methods described earlier. They could quantify steroid hormones, and they compared favorably with gold standard blood measurements. Specifically, they examined levels of progesterone, testosterone, oestradiol, and cortisol. Blubber steroid levels correlated with other factors, such as age, sex, reproductive status, and time of year of sampling, but the blow measurements varied widely, and the authors noted difficulties encountered with adequate sampling. Specifically, seawater contamination had a major effect on sample volumes. Also, nylon sampling materials led to variable sample volumes, and this influenced concentration measurements. These challenges led the authors to attempt standardization of the blow measurements using two novel approaches. First, they normalized their blow steroid hormone measurements against the urea concentration measured in blood. This idea of an "endogenous standard" is very appealing when blood samples can be taken simultaneously. Second, the authors examined the ratios of steroid hormones relative to each other in the same blow sample. Unfortunately, neither of these normalization approaches had an effect on their analysis, and the authors noted a need for additional work on the sampling methods before blow analysis can be used for large-scale studies.

Most recently, the blow proteome was analyzed in aquarium-based bottlenose dolphin and was compared to physiological data.[25] The authors correlated the proteins identified in blow with those in plasma of the same animals. In this case, a simple collection technique was used by holding an inverted 50 mL Falcon tube above the blowhole for 10 sequential exhalations. This resulted in reduced volumes collected (60–122 μL) compared with EBC sample collection with a cold ice tube device (~500 μL). From the identified proteins, some of them were categorized as protein binding (26%), cytoplasm constituents (16%), and immune response (16%) that dominated the molecular function, cellular component location, and biological process domains, respectively. Plasma correlation showed that approximately 5% of the proteins were common in the two matrices. Almost 50% were related to the immune system (e.g., complement- and immunoglobulin-related proteins) and the remainder of the proteins was associated with various other biological systems.

28.6 Aerial blow sampling and animal behavior

As breath and blow sampling has gained traction as a way to assess health in the marine mammal community, aerial sampling via drones was thought to be a viable option to extend the reach of this technology. In 2017, Dominguez-Sanchez et al. reported using this type of autonomous vehicle to sample blow in free-ranging whales[a,18,27]. Behavior of blue whales was observed via drone during blow capture, in the absence of visible whale-watching water craft in the area. Individual animal behavior was attributed using biometric identification. The authors flew the drone into a blow cloud as animals surfaced and were able to quantify the animal behaviors in each encounter (e.g., surface and diving time, blows per surfacing, blow interval, full cycle length, blow rate, and surface time proportion). The authors concluded that there were no statistical differences in animal behavior due to drone activity above the water surface. While pilot skill, aircraft noise, and other factors could affect animals in some cases, this study provided a proof of concept that remote sampling of marine mammal exhaled breath holds great promise.

28.7 Summary

There is a unique opportunity to utilize exhaled breath metabolites to assess health for the marine mammal community. This method could be useful for single animal health tracking or as community health monitoring for conservation purposes, as well as be useful in stranding events when large numbers of dolphins beach themselves for unknown reasons. The range of breath metabolites reported from the research community to date mirrors results from human studies. There are unique breath sampling differences between the marine mammal and human analogs, and each sample type has niche areas where breath analyte measurements can be useful. A wide range of biochemical breath biomarkers have been identified in the breath of these marine mammals. While some biomarkers in these animals have been previously reported in humans, some new biomarkers are reported in this community for the first time. Like human studies, breath compounds appear to originate both endogenously and from exposure to environmental contaminants (see Chapter 1).

Breath vapor contains biological metabolites and traditional respiratory gases. These may include O_2, CO_2, N_2, and NO. Some of the common breath VOC chemical classes found in ceteacens were alcohols, aliphatics, amines, aromatic hydrocarbons, substituted alkane and alkenes, esters, sulfonated compounds, terpenes, ketones, benzenoid compounds, halogenated hydrocarbons, and PAHs. EBC contains some of these VOCs, along with additional nonvolatile biomarkers including (but not limited to) cortisol, therapeutic drugs an animal was prescribed, amino acids, steroids, lipophilic compounds, terpenoids, and others. Blow is a sample type with no human analog and is composed of seawater, respiratory aerosols and gasses, and mucoids from the respiratory tract. The major compounds of interests to date have been cortisol and reproductive hormones, such as testosterone and

[a] The reader might be interested to note that this research was awarded the prestigious Ig Nobel prize for Engineering in 2010; https://www.improbable.com/ig-about/winners/#ig2010.

progesterone. Finally, the chuff sample type is analogous to human sputum. Animals under managed human care can be taught to expel chuff on demand, and it represents another rich sample type. Most recently, investigators have explored the proteomic content of this sample for diagnostic purposes.

Standardization is still an issue that is actively being addressed, and tighter controls on sampling may aid this effort. Specifically, EBC sampling in marine animals has been achieved with novel devices and is fairly reproducible. Blow sampling could be more normalized by potentially videoing the sampling episode. Subsequently, the distance and percentage of sample capture could be quantified and then used for normalization of the breath metabolomic datasets. Finally, there is interest in using endogenous compounds, such as urea, as a normalization metric. Futher work should examine the marine mammal breath metabolome to observe if there are appropriate benchmarking compounds that achieve this function.

In the future, autonomous vehicles could help sampling in free-ranging wild animal populations. Given that the technology does not appear to influence animal behavior, this is a very promising way to gain proximity to the large animals in their native environment. It is envisioned that this could be used widely to assess population animal health of protected and/or endangered species.

References

1. Venn-Watson S, Daniels R, Smith C. Thirty year retrospective evaluation of pneumonia in a bottlenose dolphin *Tursiops truncatus* population. *Dis Aquat Org* 2012;**99**(3): 237–42.
2. Venn-Watson S, Smith CR, Jensen ED, Rowles T. Assessing the potential health impacts of the 2003 and 2007 firestorms on bottlenose dolphins (*Tursiops trucatus*) in San Diego Bay. *Inhal Toxicol* 2013;**25**(9):481–91.
3. Irving L, Scholander PF, Grinnell SW. The respiration of the porpoise, *Tursiops truncatus*. *J Cell Comp Physiol* 1941;**17**(2):145–68.
4. Ridgway S. *Mammals of the sea: biology and medicine. Homeostasis in the aquatic environment*. 1972. p. 590–747.
5. Ridgway SH, Scronce BL, Kanwisher J. Respiration and deep diving in the bottlenose porpoise. *Science* 1969;**166**(3913):1651.
6. Yeates LC, Carlin KP, Baird M, Venn-Watson S, Ridgway S. Nitric oxide in the breath of bottlenose dolphins: effects of breath hold duration, feeding, and lung disease. *Mar Mamm Sci* 2014;**30**(1):272–81.
7. Ridgway S. A mini review of dolphin carbohydrate metabolism and suggestions for future research using exhaled air. *Front Endocrinol* 2013;**4**(152).
8. Schivo M, Aksenov AA, Yeates LC, Pasamontes A, Davis CE. Diabetes and the metabolic syndrome: possibilities of a new breath test in a dolphin model. *Front Endocrinol (Lausanne)*. 2013;**4**:163.
9. Cumeras R, Cheung WH, Gulland F, Goley D, Davis CE. Chemical analysis of whale breath volatiles: a case study for non-invasive field health diagnostics of marine mammals. *Metabolites* 2014;**4**(3):790–806.

10. Aksenov AA, Zamuruyev KO, Pasamontes A, Brown JF, Schivo M, Foutouhi S, et al. Analytical methodologies for broad metabolite coverage of exhaled breath condensate. *J Chromatogr B Analyt Technol Biomed Life Sci* 2017;**1061–1062**:17–25.
11. Aksenov AA, Yeates L, Pasamontes A, Siebe C, Zrodnikov Y, Simmons J, et al. Metabolite content profiling of bottlenose dolphin exhaled breath. *Anal Chem* 2014;**86**(21): 10616–24.
12. Zamuruyev KO, Borras E, Pettit DR, Aksenov AA, Simmons JD, Weimer BC, et al. Effect of temperature control on the metabolite content in exhaled breath condensate. *Anal Chim Acta* 2018;**1006**:49–60.
13. Zamuruyev KO, Aksenov AA, Baird M, Pasamontes A, Parry C, Foutouhi S, et al. Enhanced non-invasive respiratory sampling from bottlenose dolphins for breath metabolomics measurements. *J Breath Res* 2016;**10**(4). 046005.
14. Pasamontes A, Aksenov AA, Schivo M, Rowles T, Smith CR, Schwacke LH, et al. Noninvasive respiratory metabolite analysis associated with clinical disease in cetaceans: a deepwater horizon oil spill study. *Environ Sci Technol* 2017;**51**(10):5737–46.
15. Borras E, Aksenov AA, Baird M, Novick B, Schivo M, Zamuruyev KO, et al. Exhaled breath condensate methods adapted from human studies using longitudinal metabolomics for predicting early health alterations in dolphins. *Anal Bioanal Chem* 2017; **409**(28):6523–36.
16. Hogg CJ, Vickers ER, Rogers TL. Determination of testosterone in saliva and blow of bottlenose dolphins (*Tursiops truncatus*) using liquid chromatography-mass spectrometry. *J Chromatogr B Analyt Technol Biomed Life Sci* 2005;**814**(2):339–46.
17. Hogg CJ, Rogers TL, Shorter A, Barton K, Miller PJO, Nowacek D. Determination of steroid hormones in whale blow: it is possible. *Mar Mamm Sci* 2009;**25**(3):605–18.
18. Acevedo-Whitehouse K, Rocha-Gosselin A, Gendron D. A novel non-invasive tool for disease surveillance of free-ranging whales and its relevance to conservation programs. *Anim Conserv* 2010;**13**(2):217–25.
19. Burgess EA, Hunt KE, Kraus SD, Rolland RM. Get the most out of blow hormones: validation of sampling materials, field storage and extraction techniques for whale respiratory vapour samples. *Conserv Physiol* 2016;**4**(1). cow024-cow.
20. Hunt KE, Moore MJ, Rolland RM, Kellar NM, Hall AJ, Kershaw J, et al. Overcoming the challenges of studying conservation physiology in large whales: a review of available methods. *Conserv Physiol* 2013;**1**(1).
21. Hunt KE, Rolland RM, Kraus SD. Detection of steroid and thyroid hormones via immunoassay of North Atlantic right whale (*Eubalaena glacialis*) respiratory vapor. *Mar Mamm Sci* 2014;**30**(2):796–809.
22. Mingramm FMJ, Dunlop RA, Blyde D, Whitworth DJ, Keeley T. Evaluation of respiratory vapour and blubber samples for use in endocrine assessments of bottlenose dolphins (*Tursiops* spp.). *Gen Comp Endocrinol* 2019;**274**:37–49.
23. Richard JT, Robeck TR, Osborn SD, Naples L, McDermott A, LaForge R, et al. Testosterone and progesterone concentrations in blow samples are biologically relevant in belugas (*Delphinapterus leucas*). *Gen Comp Endocrinol* 2017;**246**:183–93.
24. Thompson LA, Spoon TR, Goertz CEC, Hobbs RC, Romano TA. Blow collection as a non-invasive method for measuring cortisol in the beluga (*Delphinapterus leucas*). *PloS One* 2014;**9**(12). e114062.

25. Bergfelt DR, Lippolis J, Vandenplas M, Davis S, Miller BA, Madan R, et al. Preliminary analysis of the proteome of exhaled breath condensate in bottlenose dolphins (*Tursiops truncatus*). *Aquat Mamm* 2018;**44**(3):256–66.
26. Burgess EA, Hunt KE, Kraus SD, Rolland RM. Quantifying hormones in exhaled breath for physiological assessment of large whales at sea. *Sci Rep* 2018;**8**(1):10031.
27. Domínguez-Sánchez CA, Acevedo-Whitehouse KA, Gendron D. Effect of drone-based blow sampling on blue whale (*Balaenoptera musculus*) behavior. *Mar Mamm Sci* 2018; **34**(3):841–50.

PART

Field applications

F

CHAPTER

Breath analysis in law enforcement

29

Alan W. Jones

Department of Clinical Pharmacology, University Hospital, University of Linköping, Linköping, Sweden

Chapter outline

29.1 Overview

The smell of alcohol on a person's breath as detected with the human nose as a chemical sensor represents an early, albeit primitive, test for overconsumption of alcohol and drunkenness. The first quantitative measurement of ethanol in breath is credited to British scientist Dr. Francis Anstie (1833–74), who found that only a small fraction of the total amount of alcohol a person consumed could be recovered unchanged in the expired air.[1] Dr. Anstie determined ethanol in breath by oxidation with chromic acid (Cr^{6+}) reagent, which changed color from orange-yellow to green-blue (Cr^{3+}) after reduction by ethanol. When reporting the results of his attempts to determine the breath-alcohol concentration (BrAC), Anstie wrote

> *"It must not be tried during at least the first quarter of an hour after a dose had been taken for the mouth retains the characteristic smell even if the most moderate dose of any of the stronger smelling drinks for fully this time"*

This "mouth alcohol effect" has been verified in many subsequent studies and is reflected in a mandatory requirement of a 15 min observation/deprivation time

Breathborne Biomarkers and the Human Volatilome. https://doi.org/10.1016/B978-0-12-819967-1.00029-3
Copyright © 2020 Elsevier B.V. All rights reserved.

before the police administer an evidential breath-alcohol test. Waiting for this period avoids the risk of contaminating the breath sample with alcohol present in the oral mucosa from recent consumption of an alcoholic beverage.[2] Furthermore, modern instruments for breath-alcohol analysis are equipped with algorithms that monitor the shape of the exhalation profile of ethanol. Finding an unusual or irregular profile might suggest an eructation episode just prior to exhalation, leading to a falsely elevated result. However, the reliability of slope detectors on some instruments is questionable, and a 15 min waiting time after the last drink is still an important quality assurance measure for legal purposes.[3]

This chapter reviews the use of breath-alcohol analysis in law enforcement as evidence for apprehending and prosecuting drunken drivers. This represents an early application of breath analysis in forensic science dating back to the late 1930s.[4] The Breathalyzer instrument was developed in 1954 by Robert Borkenstein of Indiana University and was approved and used by police forces in Australia, Canada, and the United States.[5]

29.2 Physiological principles of breath-alcohol testing

The physiological principles governing the excretion of ethanol and other volatile organic compounds (VOCs) in breath and the existence of a blood/breath ratio (BBR) of alcohol concentration were formulated in a classic paper published in 1930[6] entitled *"Über die Ausscheidung des Alkohols mit der Exspirationsluft"* or *"On the excretion of alcohol in the expired air."*

After healthy volunteers drank alcohol or received the dose by rectal administration, samples of venous blood and breath were analyzed at various times postdosing. The blood and breath-alcohol profiles showed good agreement within the same subjects, and the BBR was determined to be 2000:1. These in vivo results agreed fairly well with in vitro blood/air ethanol distribution studies conducted at the same temperature as exhaled breath. The high solubility of ethanol in blood and other body fluids means that only a small fraction of the dose ingested (2%–3%) is expelled unchanged via the lungs.

Scores of human alcohol dosing studies verify a high statistical correlation and a close agreement between BrAC and BAC over a wide range of ethanol doses and patterns of drinking.[7] This further supports the use of breath-alcohol analysis as an indirect way to determine a person's BAC, which is then used as a proxy for the level of intoxication.

29.3 First methods for breath-alcohol analysis

Because of the US constitution—particularly fourth amendment rights—a person is protected against unreasonable searches, such as drawing samples of blood for analysis in criminal prosecutions.[8] Taking samples of breath for analysis is much less invasive and avoids penetrating the skin with a needle. It might also be argued that the suspect had voluntarily discarded the required sample of breath during normal processes of lung ventilation.

The Breathalyzer instrument was designed to capture the end portion of a prolonged exhalation (52.5 mL), and the ethanol content was determined by oxidation with chromic acid reagent contained in a small glass ampoule.[5] After allowing the reaction to proceed for 90 s, the end point was determined by visible photometry at 450 nm. The Breathalyzer instrument was precalibrated to give a readout of the person's BAC assuming a 2100:1 BBR of alcohol; hence, the measured BrAC $\times$ 2100 = BAC.

By the 1980s, new and improved technologies were developed and incorporated into instruments for breath-alcohol analysis, including gas chromatography,[9] infrared (IR) spectroscopy,[10] and electrochemical oxidation.[11] The use of an electrochemical sensor (fuel cell) led to the manufacture of small handheld instruments that were ideally suited for use as a roadside screening test for alcohol influence.[12] After a positive roadside test, the suspect was arrested and taken to a police station where an evidential breath-alcohol test was made or venous blood was taken for toxicological analysis.

Table 29.1 summarizes historical developments in the methods used for the determination of ethanol in breath, along with names of various breath-testing instruments used for legal purposes in different countries. Since the 1980s, it is no longer necessary to convert BrAC into the coexisting BAC assuming a constant BBR because most countries now enforce statutory concentration limits of ethanol in breath.

29.4 Alcohol in the body

When people consume an alcoholic beverage, the ethanol it contains becomes rapidly absorbed from the gastrointestinal tract to reach the portal venous blood and from there it gets transported to the liver where there are oxidative enzymes that begin to metabolize the drug. Ethanol molecules escaping first-pass metabolism in the gut and/or the liver enter the hepatic vein and become mixed with deoxygenated blood in the inferior vena cava. Blood then enters the right side of the heart and is pumped via pulmonary veins into the lungs, where CO_2 is released and O_2 absorbed across the alveolar–capillary membranes.

Fig. 29.1 illustrates the disposition and fate of ethanol in the body and the relative fraction of the dose that undergoes oxidative metabolism or becomes excreted unchanged in the breath and the urine. From the lungs, oxygenated blood is returned to the heart and pumped to the brain and throughout the entire systemic circulation.

Ethanol and other VOCs in pulmonary blood diffuse across the alveolar–capillary membrane and are eliminated from the body during respiration. Less

Table 29.1 Development of analytical methods for measuring ethanol in exhaled breath and the various instruments approved for legal purposes.

Period	Analytical methods for determination of ethanol in breath and the instruments that incorporated these principles
1870s–1930s	Sampling of mixed expired breath or the end part of a prolonged exhalation (alveolar air). Ethanol was measured by oxidation with chromic acid or some other oxidizing agent. The quantitative analysis depended on the number of breaths required to decolorize the chemical reagent or a visual comparison of the change in color with standard solutions or some type of volumetric analysis.
1940s–1950s	In the 1940s, the first instruments for breath analysis were developed in the United States and utilized chemical oxidation with potassium permanganate (Drunkometer and Intoximeter) or oxidation with iodine pentoxide (Alcometer). The proportion of alveolar air in the breath was determined by measuring CO_2 content and assuming that alveolar air was 5.5% by volume CO_2. Later versions of the Drunkometer instrument determined ethanol after the initial exhalation had been rebreathed five times.
1960s	The Breathalyzer instrument was developed in 1954 and was designed to capture the last portion of a prolonged exhalation. The ethanol content was determined by chemical oxidation with a mixture of potassium dichromate and sulfuric acid, and photometric analysis was used to detect the end point. Similar instruments were also developed, such as the Ethanograph and the Photo-electric intoximeter.
1970s	Gas chromatography utilizing a thermal conductivity detector (AlcoAnalyzer) or a flame ionization detector (GC intoximeter) was applied to breath-alcohol analysis in small compact benchtop instruments.
1970s	Single wavelength infrared absorptiometry at 3.39 μm (Intoxilyzer 4011).
1970s–1980s	Handheld electrochemical (fuel cell) analyzers such as Alcolmeter SD-2, or 400, Alco-Sensor III and IV, and various models of the Alcotest instrument.
1980s	Infrared spectroscopy at a single wavelength (3.4 μm) and a Taguchi semiconductor sensor for added selectivity (Intoximeter 3000) or three infrared wavelengths 3.39–3.42 μm, C–H stretching frequency (Intoxilyzer 5000 and DataMaster), or at 9.5 μm for (C–O) bond stretching.
1990s	Dual sensor technology incorporating both electrochemical oxidation (fuel cell) and infrared absorption at 9.5 μm: Alcotest 7110, 7310, and 9510.
2010s	Contact-free quantitative analysis of ethanol in breath using water vapor as an internal standard enabling rapid analysis without the need for making a prolonged exhalation or attaching a mouthpiece to the instrument (Servotek).

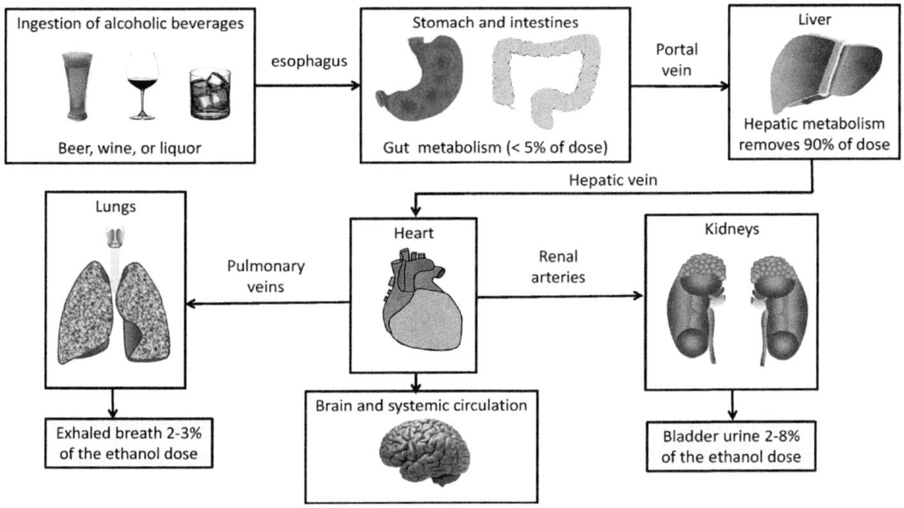

FIGURE 29.1

Illustration of the disposition and fate of ethanol in the body showing the organ systems involved and the approximate percentage of the dose eliminated by metabolism (gut and liver) and excreted unchanged in breath and urine.

than 10% of the ingested dose of ethanol is eliminated from the body unchanged, in breath, urine, and perspiration.[13] Elimination of ethanol from the body mainly occurs through metabolism in the liver by oxidative enzymes, first to acetaldehyde, then acetic acid, and the latter is eventually converted into the end products CO_2 and H_2O in extrahepatic tissues.

During normal inhalation and exhalation, the alveolar air, which is initially at body temperature (37°C), cools during its passage through the upper airways and when it leaves the mouth its temperature has dropped to about 34.5°C.[14] Owing to its high solubility in water, ethanol undergoes a continuous reequilibration between inspired and expired air and the mucus surfaces covering the upper airways and mouth.[15] The concentration of ethanol in exhaled breath will depend on many factors, including the volume exhaled in relation to the person's vital capacity,[16] the temperature and humidity of the ambient air breathed, and also the subject's breathing pattern just prior to exhalation.[17]

29.5 Arteriovenous blood ethanol differences

After drinking alcoholic beverages, the ethanol they contain becomes distributed throughout the total body water compartment, which represents about 50%–60% of a person's body weight. The ethanol in oxygenated arterial blood is transported to all the body fluids and tissues and diffuses across capillaries into muscular tissue, which contains ~80% water, thus acting as a reservoir for ethanol. This results in the development of an ethanol concentration gradient between the arterial blood circulation and the venous side of the tissue returning deoxygenated blood to the heart.[18]

Fig. 29.2 shows concentration—time profiles of ethanol in arterial (A) and venous (V) blood in one male subject who drank a moderate dose of ethanol on an empty stomach. Arterial—venous (A-V) differences in concentration are evident with A-BAC being higher than V-BAC for about 2 h postdrinking and thereafter V-BAC is always higher than A-BAC. The presence of ethanol in the venous blood returning to the heart from tissue can be detected for about 1 h longer than in the arterial blood circulation. The time course of ethanol concentration in breath runs more closely to the arterial BAC rather than venous BAC, which is the specimen analyzed in forensic casework.[19,20]

Arterial-venous differences in concentration of ethanol are evident during absorption—distribution and the elimination phase of the alcohol curve. During the absorption—distribution phase, the A-V difference is positive, then decreasing to zero by 1—2 h postdosing, and is thereafter negative at all later times.[19] The existence of A-V difference in ethanol concentrations are accounted for by the initial uptake of ethanol into muscle tissue and after absorption and distribution are complete the continuous clearance of ethanol from central arterial blood by metabolism in the liver. Ethanol in the tissue water compartments gets returned via capillaries to the venous side of the systemic circulation, such that the V-BAC is slightly higher than A-BAC in the postabsorptive state.[19]

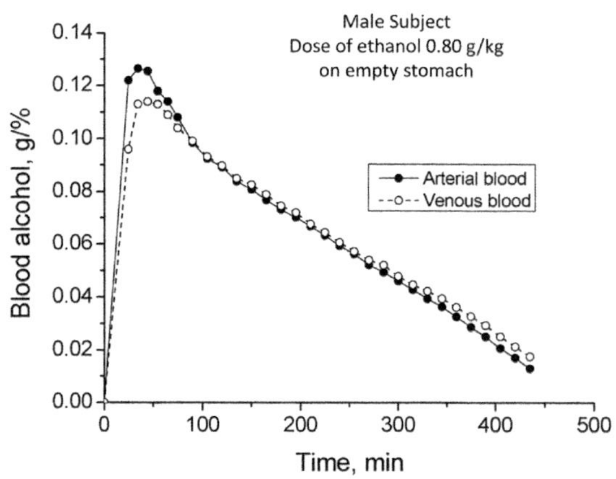

FIGURE 29.2

Comparison of blood-alcohol concentration—time profiles in venous blood (cubital vein) and arterial blood (radial artery) in one male subject who drank a moderate dose of ethanol (0.80 g/kg) on an empty stomach. Ethanol was determined in blood by gas chromatography.

Arterial-venous differences in ethanol concentration have implications when BBRs of alcohol are calculated and compared.[20] In forensic science practice, venous blood is the sample used for toxicological analysis and not blood obtained by arterial puncture. The V-BAC never reaches an equilibrium with ethanol at the pulmonary–capillary membrane of the lungs, which makes the venous BAC/BrAC ratio a moving target. The V-BAC reflects the concentration of ethanol in the body tissue being drained. The A-BAC and pulmonary–capillary BrAC (and exhaled BrAC) are more closely related during absorption, distribution, metabolism, and excretion (ADME) of ethanol.[20]

Nevertheless, the venous BAC and BrAC are highly correlated for different patterns of drinking and various doses of ethanol. Studies have shown that venous BAC/BrAC ratios are lower than the generally accepted value of 2100:1 during absorption–distribution of ethanol in the body, becoming higher than 2100:1 in the postabsorptive phase of ethanol metabolism.[21] The arterial BAC/BrAC ratio shows much less variation in relation to sampling time after end of drinking and one study reported this ratio to be 2251:1 on average.[22] Accordingly, BrAC x 2251:1 gives a close approximation of the arterial BAC throughout absorption, distribution, and elimination phases of the BAC curve.

29.6 Traffic law enforcement

Excessive drinking impairs normal body functioning in a dose-dependent manner, which has serious consequences when skilled tasks, such as driving, are performed.[23] After drinking moderate amounts of alcohol, such as 1–2 units per day (8–16 g), ethanol typically has no ill effects on a person's performance and behavior. However, overconsumption of alcohol and drunkenness impairs both cognitive and psychomotor functioning, thus increasing the risk of road traffic crashes as the BAC reaches 50–80 mg%.

The dangers of drunken driving for traffic safety were recognized early after motor vehicles appeared on the roads.[8] The first punishable BAC limits for driving were introduced in Norway in 1936 (0.50 g/kg or ~50 mg%), followed by Sweden in 1941 (0.80 g/kg or ~80 mg%). The corresponding legal limit in England and Wales was enacted in the Road Traffic Act of 1967 and was set at 80 mg% (~0.80 g/kg), where it still remains today.

Table 29.2 summarizes the statutory BAC and BrAC limits for driving in various countries. Note the fourfold variation (20–80 mg%), which probably is more a reflection of local political forces and activity of public pressure groups rather than results of traffic safety research or studies of crash risk. There is abundant evidence that risk of being involved in a traffic crash increases as a driver's BAC passes 20 mg% and becomes statistically significant at BAC of 50–80 mg% compared with sober drivers.[24]

Table 29.2 also lists the BBR of alcohol adopted in the various countries when statutory BrAC limits for driving were derived from the existing BAC limits. As is

Table 29.2 Statutory blood-alcohol concentration (BAC) and breath-alcohol concentration (BrAC) limits in various countries around the world and the blood/breath ratios (BBR) of alcohol used when the statutory BrAC limit was derived from the preexisting BAC.

Country/Nation[a,b]	Statutory BAC[c]	Statutory BrAC[c]	Assumed BBR
Most European Nations	0.50 mg/mL (g/L)	0.25 mg/L	2000:1
Netherlands	0.50 mg/mL (g/L)	220 μg/L	2300:1
Norway, Sweden, Poland	0.20 mg/g (g/kg)[d]	0.10 mg/L	2100:1[d]
Finland	0.50 mg/g (g/kg)[d]	0.22 mg/L	2400:1
United States of America[e]	0.08 g/100 mL	0.08 g/210 L	2100:1
England, Wales[f]	80 mg/100 mL	35 μg/100 mL	2300:1
Ireland, Scotland	50 mg/100 mL	22 μg/100 mL	2300:1
Japan	0.30 mg/mL	0.15 mg/L	2000:1
India	30 mg/100 mL	Screening test only	2100:1
China	0.20 mg/mL	Screening test only	2100:1
Canada	0.08 g/100 mL	0.08 g/210 L	2100:1
Australia	0.05 g/100 mL	0.05 g/210 L	2100:1
South Africa	0.05 g/100 mL	0.22 mg/L	2300:1
New Zealand	50 mg/100 mL	250 μg/L	2000:1

[a] *Some countries also enforce a lower limit for novice or provisional drivers; for those under 21 year in the United States, the limit is 0.02 g% and 0.04 g% for operators of commercial vehicles.*
[b] *Most countries also have a higher statutory alcohol limit corresponding to a more serious traffic offense, often referred to as aggravated drunken driving, with more severe penalties, including incarceration.*
[c] *Statutory BAC and BrAC values are reported in the units stated in legislation of the respective country/nation.*
[d] *Because density of whole blood is 1.055 g/mL on average, 0.20 g/kg is the same as 0.21 g/L, which means that the blood/breath ratio of ethanol in Norway and Sweden is 2100:1.*
[e] *The state of Utah lowered its statutory BAC limit for driving to 0.05 g% in December 2018.*
[f] *Also enforced are statutory concentration limits of alcohol in urine of 107 mg/100 mL (England and Wales) and 67 mg/100 mL (Scotland and Ireland), hence a urine/blood ratio of 1.34:1.*

apparent from the table, there was no international consensus about the most appropriate BBR, which ranged from 2000:1 to 2400:1 in different jurisdictions.

The punishable BAC and BrAC limits for driving are increasingly being lowered, with a limit of 0.20 g/kg (~20 mg%) enforced in China, Sweden, Norway, and Poland. Scotland lowered its legal BAC limit from 80 to 50 mg% in 2016. Most countries also enforce heavier penalties for drivers apprehended with higher BAC or BrAC, which represents a more serious offense known as aggravated drunken driving. Besides monetary fines and withdrawal of the driving permit, a person risks a mandatory prison sentence if found guilty of aggravated drunken driving. Such penalties and sanctions, even for a first offense, underscore the need for using highly reliable analytical methods to determine ethanol in blood or breath samples for legal purposes.

29.7 Classification of instruments for breath ethanol analysis

The instruments available for breath-alcohol analysis can be subdivided into different categories depending on their intended purpose or application, such as

(i) Quantitative evidential breath-alcohol instruments, the results of which are used for prosecuting traffic offenders if the BrAC exceeds a statutory limit for driving.
(ii) Preliminary breath-test devices, such as handheld instruments used by police officers at the roadside to test driver sobriety. The results produced serve as presumptive evidence that a person is in breach of the law.
(iii) Contact-free passive alcohol sensors are held in close proximity to a person's mouth as they are engaged in conversation. This type of instrument is useful when large-scale testing is required, to obviate using plastic mouthpieces. Such instruments are used by the police to test drivers apprehended at roadblocks or at border crossings or alcohol screening of spectators attending sporting events.
(iv) Personal breath-alcohol analyzers intended for purchase and use by the general public before they decide to drive home after consumption of alcoholic beverages. These instruments are similar in design and performance as the roadside breath-screening instruments used by the police authorities.
(v) Breath-alcohol detectors are increasingly being fitted to the ignition switch in private vehicles—referred to as ignition interlock devices. The aim is to prevent a drunken driver from being able to start a vehicle if alcohol is detected above a certain threshold value. Vehicles owned by convicted drunken drivers have interlocks fitted as a condition for them being relicensed and as a way to prevent recidivism.
(vi) Breath-alcohol instruments are used in emergency medicine to test patients smelling of alcohol or when they are admitted for treatment of skull trauma or other injuries sustained when drunk.
(vii) Random breath-alcohol tests are performed on individuals required to refrain from drinking alcohol, either because of ongoing treatment for an alcohol use disorder or as a condition of employment or participating in safety-sensitive work or public transportation.

29.8 Preliminary breath-alcohol screening tests

The first roadside breath-alcohol screening test was developed in the mid-1950s, and this consisted of a single-use chemical tube and plastic bag.[25] The alcohol in a person's breath was oxidized with chromic acid adsorbed on crystals of silica gel, which changed color from yellow to green after the redox reaction was complete. The test person was required to blow through the chemical tube to inflate a plastic bag (~1 L) attached to the other end of the tube. The heat generated when water vapor in the breath reacted with the sulfuric acid helped to promote the oxidation reaction. Without any alcohol in the person's blood or breath, the gold-yellow color of the chemicals turned slightly darker. With alcohol present, the length of the green color stain increased as a function of ethanol concentration in the breath sample.[26] If the green coloration extended up to a graduation mark on the tube, this indicated that the person's BAC exceeded the legal limit for driving.

In the 1970s, electrochemical oxidation methods were developed for breath-alcohol analysis that made use of a fuel cell detector incorporated into handheld devices.[27] After drinking alcoholic beverages, the only alcohol excreted in human breath of any significance is ethanol, which meant that the bulky chromatographic separation stage was unnecessary.[28] The first commercially available handheld instrument with an electrochemical detector was developed in 1973 and was known as Alcolmeter pocket model.[29] Today, a large number of instruments are available from various manufacturers that incorporate electrochemical oxidation with a fuel cell detector.[30] An example of the output from such a detector is shown in Fig. 29.3.

Quantitative analysis is achieved by measuring the height of the peak response or the area under the response curve, as indicated on the graph. The latter approach is useful as a test of specificity because the rate of oxidation varies for different aliphatic alcohols, which allows distinguishing ethanol from methanol and isopropanol.

Fig. 29.4 (left) shows a blood–breath scatter plot when handheld breath-alcohol screening instruments were used to test apprehended drivers at the roadside. The venous blood samples were always taken after the roadside testing was completed, sometimes several hours later. The statistical analysis shows a

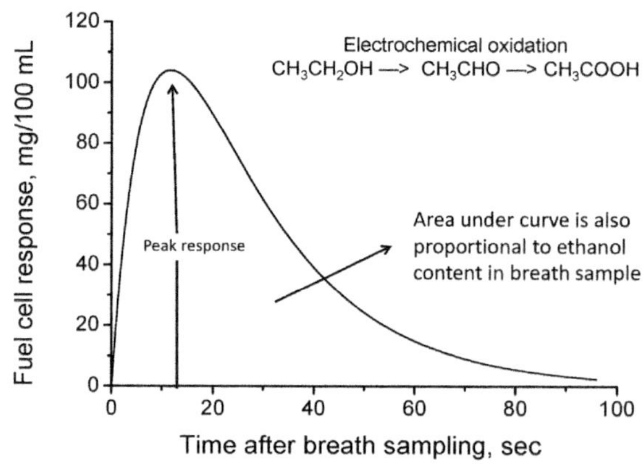

FIGURE 29.3

Time-response curve from a handheld electronic breath analyzer incorporating an electrochemical fuel cell sensor for determination of ethanol.

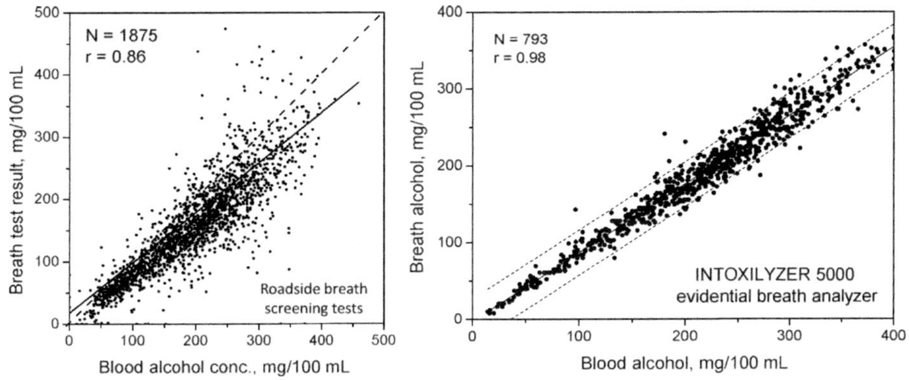

FIGURE 29.4

Scatter plots of results of breath-alcohol analysis using handheld roadside screening instruments (left) fitted with electrochemical sensors compared with venous blood-alcohol concentration (BAC) determined by gas chromatography. The scatter plot to the right shows the results when apprehended drivers were tested with a quantitative IR analyzer (Intoxilyzer 5000) and the corresponding venous BAC by gas chromatography.

high correlation ($r = 0.85$), although scatter of points around the regression line was appreciable, indicating that BAC should not be estimated from BrAC results. Fig. 29.4 (right) compares venous BAC with BrAC determined using an evidential breath analyzer (Intoxilyzer 5000).

29.9 Evidential breath-alcohol analysis

The quantitative determination of ethanol in breath is necessary for evidence quality results and prosecution of traffic offenders. The legal consequences for a person found guilty of a drunken driving charge necessitates the use of accurate, precise, and specific analytical methods that are fit for their intended purpose and subjected to method validation, quality assurance procedures, and regulation.[31]

The Breathalyzer instrument appeared in 1954, although this is now obsolete, being superseded by more modern breath-alcohol technology and instruments with electronic microprocessor control of the entire sampling and analysis process. In the 1970s, methods for the analysis of ethanol in breath based on gas–liquid chromatography, IR spectroscopy, and electrochemical oxidation became available.[32] Modern instruments incorporate several new features compared with the Breathalyzer, including automated sampling of the breath specimen at end-exhalation, calibration control of the instrument in conjunction with a subject being tested using air–ethanol vapor standards, the analysis of a room–air blank sample, monitoring the exhalation profile of ethanol and setting standards for exhalation time, and discarded breath volume before sampling.

Table 29.3 Example of quality assurance features in conjunction with the use of an evidential breath-alcohol analyzer for law enforcement purposes.

Test number	Type of performance test according to approved protocol and microprocessor control	Accepted results and/or tolerance requirement
1	Analysis of room air blank check	±0.05 mg/L
2	Internal standard check[a]	±5%
3	Obtain new baseline reference	±0.05 mg/L for 6 s
4	External standard check (0.50 mg/L)[b]	Range accepted: 0.47–0.53 mg/L
5	Analysis of room air blank check	±0.05 mg/L
6	Obtain new baseline reference	±0.05 mg/L for 6 s
7	First subject test result	Minimum exhalation time: 6 s
8	Analysis of room air blank check	±0.05 mg/L
9	Obtain new baseline reference	±0.05 mg/L for 6 s
10	Second subject test[c] result	The second subject test is done after a minimum time of 10 min after test one
11	Analysis of room air blank check	±0.05 mg/L
12	Obtain new baseline reference	±0.05 mg/L for 6 s
13	External standard check at 0.50 mg/L[2]	Range accepted: 0.47–0.53 mg/L
14	Analysis of room air blank check	±0.05 mg/L
15	Compare results from the two subjects test[c]	If test one minus test two is < 0.11 mg/L, this is deemed acceptable
16	Calculate mean of the two results	Mean breath-alcohol content
17	Deduction made for uncertainty	0.07 mg/L
18	Subject's breath-alcohol concentration at the time of testing and used for prosecution[d]	The suspect's breath-alcohol concentration was not less than the figure shown with 99.9% confidence

[a] *The internal standard is an optical test of the infrared signal amplitude from each of three filter channels at selected frequencies.*

[b] *The external standard is from the analysis of ethanol generated from a wet bath simulator at a constant temperature of 34°C producing a target effluent concentration 0.50 mg/L.*

[c] *For an approved evidential test, the absolute difference between the two subject tests should not exceed 0.11 mg/L.*

[d] *The mean of the two subject tests is calculated and truncated to two decimal places before subtracting 0.07 mg/L as a statistically calculated allowance for uncertainty to give the prosecution result with 99.9% confidence.*

Most evidential instruments for breath-ethanol analysis employ IR absorption spectroscopy or electrochemical oxidation or both analytical principles within the same instrument.[33] Those instruments using IR detectors monitor the absorption of light by ethanol molecules at several different wavelengths to enhance the selectivity for identification of ethanol. Two main regions in the IR spectrum are utilized, namely 3.4 μm and/or 9.4 μm, which correspond to the C—H stretching vibrations and O—H stretch in the molecules, respectively.

Examples of the rigorous quality assurance requirements necessary when breath-alcohol analyzers are used for evidential purposes are listed in Table 29.3, which are based on regulations approved by police authorities in Sweden. This sequence of tests include room air blank checks, calibration control of the accuracy of ethanol measurement results, and precision as reflected in differences between the duplicate determinations of BrACs. Finally, from the mean of two determinations, a deduction of 0.07 mg/L is made to allow for analytical uncertainty of the alcohol measurements. The mean ethanol concentration after subtracting 0.07 mg/L gives a certain advantage to the person (or suspect) being tested. After making the deduction, it can be stated that the prosecution BrAC is not less than the true concentration of ethanol in breath with a high degree of statistical probability (99.9%). The information in Table 29.3 can be provided to the court as evidence when a case goes to trial.

Fig. 29.5 shows a typical exhalation profile of ethanol in breath determined with an Evidenzer instrument approved for use by police in Sweden. For determination of ethanol, this instrument utilizes absorption of IR at five wavelengths between 3.3 and 3.5 μm. The trace in Fig. 29.5 shows how the BrAC increases during a prolonged exhalation. It is apparent that 77% of the final end-exhaled result is reached after just 2 s of starting to exhale into the instrument. So even the worst possible breath sample gives a good indication of the final end-expired BrAC.

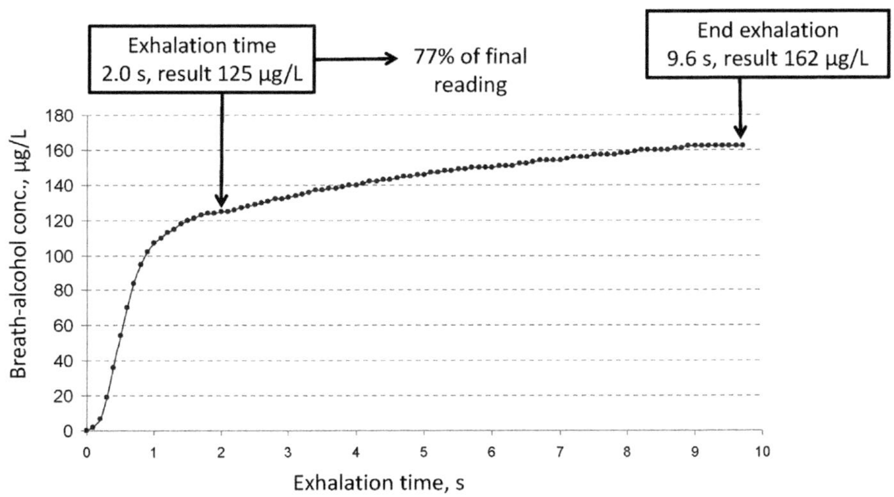

FIGURE 29.5

Concentration–time profile of ethanol in breath during a continuous exhalation into an Evidenzer breath-alcohol analyzer that uses IR absorption spectroscopy for ethanol analysis. Note that the BrAC after just 2 s of exhalation is 77% of the final end-exhaled concentration.

29.10 Advantages and limitations of breath- versus blood-alcohol analysis

There are both advantages and limitations with sampling and analysis of ethanol in breath compared with drawing blood samples for forensic toxicological analysis.[34] The most important considerations when comparing and contrasting the two alternatives are listed below:

1. Results of a breath-alcohol test are obtained within a few minutes after starting the analysis, whereas results of BAC determinations can take several days or longer to obtain from a forensic laboratory. Much depends on circumstances, such as location of the arrest, workload, and availability of an accredited laboratory for the analytical work.
2. The concentrations of VOCs in breath reflect the concentration in deep lung alveolar air in equilibrium with pulmonary arterial blood. During exhalation, there is a cooling of the breath and reequilibration of ethanol with the watery mucus surfaces of the upper airways. Nevertheless, BrAC is highly correlated with BAC over a wide range of ethanol concentration and drinking conditions.
3. When an evidential breath-alcohol analysis is performed, the suspect is identified from driving permit or fingerprints, whereas blood samples need to be labeled, packaged, and shipped to a laboratory and registered and stored prior to analysis, which necessitates precautions, such as documenting the chain of custody of specimens.
4. The sampling of breath is a noninvasive procedure and thus less objectionable to the person providing the sample compared with penetrating a vein with a needle and drawing a suitable blood specimen.
5. Breath-alcohol instruments are simple to operate by trained police officers, whereas laboratory personnel trained in analytical chemistry, often university graduates, are involved with the analysis of blood samples.
6. An evidential breath-alcohol test is usually conducted at a police station closer in time to the drunken driving offense, whereas blood for toxicological analysis is usually taken 1–2 h after the time of driving.
7. Although it is possible to save a breath sample for later confirmatory analysis, this is technically more difficult compared with reanalysis of blood, which might be necessary if integrity of the result is challenged. However, replicate (at least duplicate) samples of breath are routinely analyzed at the time a subject is tested.
8. Some cooperation is necessary from the arrested driver to obtain an approved sample of breath, by making a continuous forced exhalation into the instrument. This is not the case with blood samples, which can be taken from an unconscious person or by force if necessary (e.g., uncooperative subjects).
9. Approximately 15 min must elapse after an arrest to ensure that the sample of breath analyzed is not contaminated with alcohol in the oral mucosa from a

recent drink. Blood can be drawn at any time after an arrest is made, although swabbing the skin with an alcohol disinfectant should be avoided.

10. Other impairing drugs are easily determined in blood samples, although this is more difficult to accomplish in real time using breath. Methods have been developed for trapping breath for later analysis of nonalcohol drugs, although the analytical results are not so easy to interpret as for ethanol intoxication.
11. Several biomarkers of heavy drinking (e.g., carbohydrate-deficient transferrin) and abuse of alcohol (various liver enzymes) can be determined in blood samples, but such testing is not yet feasible using exhaled breath.

The various methods and instruments for breath-alcohol analysis are among the oldest and most well-established analytical techniques used in forensic science and legal medicine. The advantages of breath over blood are difficult to deny and new technological advances and microprocessor control prompted many European nations to reevaluate their drunken-driving laws and introduce statutory limits of BrAC (see Table 29.2).

29.11 Summary

The analysis of VOCs in exhaled breath furnishes a noninvasive way to obtain information about various metabolic processes and pathways in the body, such as in connection with clinical and diagnostic medicine as biomarkers for various disease states. However, the most well-established application of breath analysis, besides capnography, is for law enforcement purposes as a test for overconsumption of ethanol and unfitness to drive safely.[33] Immediate results are obtained after a breath-alcohol test is administered, which allows preventative action, such as confiscation of the driving permit, incarceration until the person sobers up, or impounding the vehicle.

The concentration of ethanol in exhaled breath is strongly correlated with its concentration in arterial blood, and BrAC is therefore a good reflection of exposure of the brain to impairment effects of this depressant drug.[8] The analysis of ethanol in blood or breath represents a more objective way to establish overconsumption of alcohol and drunkenness compared with looking for the usual clinical signs and symptoms, such as the person's general appearance and behavior, smell of alcohol on the breath, slurred speech, or ability to walk in a straight line.

With modern instruments for breath-ethanol analysis, the accuracy, precision, and selectivity of the results are comparable with those obtained from the determination of ethanol in blood and urine samples by gas chromatography.[35] All analytical methods used in forensic science must be carefully regulated and strict quality control procedures followed to ensure meeting the high standards required when used as evidence in a criminal prosecution.[31]

The first instruments for breath-alcohol analysis were developed in the United States, and the results were used for legal purposes as an indirect way to estimate the person's BAC assuming a constant BBR of 2100:1. However, this important ratio varies both within and between subjects depending on many biological factors, such as the time after drinking when samples are taken, or the breathing pattern prior to exhaling into the instrument. To avoid legal challenges directed against biological variations in the BBR, the drunken-driving laws in most nations now include a statutory definition of drunken driving in terms of a certain threshold BrAC (Table 29.2). This makes it unnecessary to convert the BrAC into BAC when apprehended drivers are prosecuted.

Improvements in breath-testing technology, especially the use of more specific analytical methods for identifying ethanol, and microprocessor control of the testing procedures, have led to worldwide acceptance of breath-alcohol instruments for legal purposes. The noninvasive nature of sampling breath compared with taking venous blood samples to determine ethanol concentration is a major advantage of breath-alcohol testing. Handheld breath analyzers are used at the roadside as screening tests for alcohol intoxication, whereas more sophisticated benchtop instruments provide the evidence necessary for prosecuting traffic offenders if the statutory BrAC limit for driving is exceeded.

Small and compact breath analyzers are available for purchase by the general public for personal use, such as after a night out at a bar or restaurant before driving home or early in the morning before driving to work. One problem with the use of personal breath testers is the need to ensure correct calibration and that it remains stable for a long time. Such self-testing should not be done within the first 15 min after the last drink, owing to a mouth-alcohol effect, and it is advisable to make repeated tests at 15 min intervals to ensure the BrAC is no longer increasing before starting to drive a vehicle.

In some countries, breath-alcohol analyzers are fitted to public transport vehicles, such as buses, trains, and taxis, to ensure the operators are alcohol-free before taking charge of the vehicles. Similar breath-alcohol devices are connected to the ignition switch in privately owned vehicles, which prevents the engine from starting if there is alcohol in the person's breath above some preset value. In some jurisdictions, interlock devices have become a mandatory requirement for people arrested multiple times for drunken driving as one of the conditions for being granted a new driving permit. Overall, the future looks bright for continued innovations and applications of breath-alcohol analysis in law enforcement. Certainly, the use of breath-alcohol instruments as a test for intoxication and control of driver sobriety has become so well established worldwide that this forensic science application is here to stay.

References

1. Anstie FE. Prognosis and treatment of certain diseases. *Lancet* 1867;**90**:385–7.
2. Sterling K. The rate of dissipation of mouth alcohol in alcohol positive subjects. *J Forensic Sci* 2012;**57**:802–5.

3. Labianca DA. Non-foolproof nature of slope detection technology in the Dräger Alcotest 9510. *Forensic Tox* 2018;**36**:222–4.
4. Harger RN, Lamb EB, Hulpieu HR. A rapid chemical test for intoxication employing breath a new reagent for alcohol and a procedure for estimating the concentration of alcohol in the body from the ratio of alcohol to carbon dioxide in the breath. *J Am Med Assoc* 1938;**110**:779–85.
5. Borkenstein RF, Smith HW. The Breathalyzer and its applications. *Med Sci Law* 1961;**2**: 13–22.
6. Liljestrand G, Linde P. Über die Ausscheidung des Alkohols mit der Exspirationsluft. *Skand Arch Physiol* 1930;**60**:273–98.
7. Taylor MD, Hodgson BT. Blood/breath correlations: intoxilyzer 5000C, alcotest 7110 and breathalyzer 900A breath alcohol analyzers. *Can Soc Forensic Sci J* 1999;**28**: 153–64.
8. Jones AW, Morland JG, Liu RH. Driving under the influence of psychoactive substances - a historical review. *Forensic Sci Rev* 2019;**31**:103–40.
9. Jones AW. The precision and accuracy of a gas chromatograph intoximeter breath alcohol device part II–in-vivo experiments. *J Forensic Sci Soc* 1978;**18**:81–7.
10. Harte RA. An instrument for the determination of ethanol in breath in law-enforcement practice. *J Forensic Sci* 1971;**16**:493–510.
11. Jones AW. Electrochemical measurement of breath-alcohol concentration: precision and accuracy in relation to blood levels. *Clin Chim Acta* 1985;**146**:175–83.
12. Kriikku P, Wilhelm L, Jenckel S, Rintatalo J, Hurme J, Kramer J, et al. Comparison of breath-alcohol screening test results with venous blood alcohol concentration in suspected drunken drivers. *Forensic Sci Int* 2014;**239**:57–61.
13. Kalant H. Pharmacokientics of ethanol: absorption, distribution, and elimination. In: Beglieter H, Kissin B, editors. *The pharmacology of alcohol and alcohol dependence*. New York: Oxford University Press; 1996. p. 15–58.
14. Cowan JM, Burris JM, Hughes JR, Cunningham MP. The relationship of normal body temperature, end-expired breath temperature, and BAC/BrAC ratio in 98 physically fit human test subjects. *J Anal Toxicol* 2010;**34**:238–42.
15. Hlastala MP, Anderson JC. Alcohol breath test: gas exchange issues. *J Appl Physiol* 2016;**121**:367–75.
16. Anderson JC, Hlastala MP. The alcohol breath test in practice: effects of exhaled volume. *J Appl Physiol* 2019;**126**:1630–5.
17. Jones AW. How breathing technique can influence the results of breath-alcohol analysis. *Med Sci Law* 1982;**22**:275–80.
18. Martin E, Moll W, Schmid P, Dettli L. The pharmacokinetics of alcohol in human breath, venous and arterial blood after oral ingestion. *Eur J Clin Pharmacol* 1984;**26**:619–26.
19. Jones AW, Lindberg L, Olsson SG. Magnitude and time-course of arterio-venous differences in blood-alcohol concentration in healthy men. *Clin Pharmacokinet* 2004;**43**: 1157–66.
20. Lindberg L, Brauer S, Wollmer P, Goldberg L, Jones AW, Olsson SG. Breath alcohol concentration determined with a new analyzer using free exhalation predicts almost precisely the arterial blood alcohol concentration. *Forensic Sci Int* 2007;**168**:200–7.
21. Jones AW. Variability of the blood:breath alcohol ratio in vivo. *J Stud Alcohol* 1978;**39**: 1931–9.

22. Lindberg L. A review of basic physical properties and physiological mechanisms involved in alcohol airway exchange processes and the alcohol breath test. *Blutalkohol* 2018;**55**:395–415.
23. Kalant H. Research on tolerance: what can we learn from history? *Alcohol Clin Exp Res* 1998;**22**:67–76.
24. Blomberg RD, Peck RC, Moskowitz H, Burns M, Fiorentino D. The Long Beach/Fort Lauderdale relative risk study. *J Saf Res* 2009;**40**:285–92.
25. Grosskopf K. Die Atemalkohol Bestimmung als analytische Aufgabe. *Angew Chem* 1954;**66**:295.
26. Prouty RW, O'Neill B. *An evaluation of some qualitative breath screening tests for alcohol*. Washington DC: Insurance Institute for Highway Safety; 1971.
27. Jones TP, Jones AW, Williams PM. Some recent developments in breath alcohol analysis. *The Police Surgeon* 1974;**6**:80–6.
28. Cremer E, Gruber HL, Huck H. Electro-chemical detector for the gas chromatographic determination of alcohols and aldehydes. *Chromatographia* 1969;**2**:197–203.
29. Jones AW. Evaluation of breath alcohol instruments. II. In vivo experiments with alcolmeter pocket model. *Forensic Sci* 1978;**12**:11–23.
30. Zuba D. Accuracy and reliability of breath alcohol testing by handheld electrochemical analysers. *Forensic Sci Int* 2008;**178**:e29–33.
31. Dubowski KM. Quality assurance in breath-alcohol analysis. *J Anal Toxicol* 1994;**18**: 306–11.
32. Dubowski KM. *The technology of breath alcohol analysis*, vol. 38. Rockville, MA: US Department of Health and Human Services; 1991. NIAAA.
33. Wigmore JG, Langille RM. Six generations of breath alcohol testing instruments: changes in the detection of breath alcohol since 1930. An historical overview. *Can Soc Forensic Sci J* 2009;**42**:276–83.
34. Jones AW. Medicolegal alcohol determination - blood- or breath-alcohol concentration? *Forensic Sci Rev* 2000;**12**:23–47.
35. Grubb D, Rasmussen B, Linnet K, Olsson SG, Lindberg L. Breath alcohol analysis incorporating standardization to water vapour is as precise as blood alcohol analysis. *Forensic Sci Int* 2012;**216**:88–91.

CHAPTER

Drugs in breath

30

Olof Beck

Department of Clinical Neuroscience, Karolinska Institute, Stockholm, Sweden

Chapter outline

30.1 Overview

The main applications for measuring drugs in human biological specimens include clinical trials, therapeutic drug monitoring, and clinical and forensic toxicology in relation to (illicit) drug misuse or abuse. The predominant media currently used for this are blood and urine. Blood plasma or serum are the most common specimens for measuring therapeutic drug concentrations and drug intoxication, and in forensics for postmortem and drugged driving investigations. Urine is the standard specimen for clinical and forensic drug testing aimed at monitoring illicit drug use. For most of these applications, there is strong interest for alternative specimens that offer convenience, better protection against adulteration, and less intrusion of privacy. Exhaled breath is a potential medium that offers these advantages.

Breathborne Biomarkers and the Human Volatilome. https://doi.org/10.1016/B978-0-12-819967-1.00030-X
Copyright © 2020 Elsevier B.V. All rights reserved.

In a text from 1976 entitled "*Toxicology in the next 25* years," the late British forensic toxicologist Alan S. Curry stated: "*One biological medium that has only been partly exploited is breath. Everyone is aware of breathtesting tubes for alcohol ... but is it too much to hope that in the not too distant future the concept of a super detector coupled to a miniature computerized mass spectrometer by the bedside may not be a reality? Breath samples are so easily available.*"[1]

It is without any doubt that the successful and widely accepted application of alcohol testing in breath (see Chapter 29) has triggered a continued interest in toxicology for its application to other drugs of abuse.

30.1.1 Early developments

In the early 1980s, some exploratory work on the possibility of using breath for toxicology applications was undertaken.[2] Detector signals relating to amphetamine and secobarbital in collected breath samples following controlled intake were observed, but the presence of drugs in breath was deemed to be "very questionable" and further work was discontinued. Despite this, and shortly thereafter in 1983, Manolis and coworkers published a study that demonstrated the detection by gas chromatography−mass spectrometry (GC-MS) of tetrahydrocannabinol (THC) in breath 12 min after smoking marihuana.[3]

Despite these promising early investigations, the next apparent development was only almost two decades later in 2002 when a US patent application was filed by R.J. Melker and coworkers describing an apparatus based on electronic sensor technology for detecting illicit drugs in breath.[4] The system was exemplified with γ-butyrolactone (a γ-hydroxybutyric acid, GHB, precursor), but it was claimed to be universal for illicit drugs. The field of drug detection in breath then again remained very silent in the scientific literature, until 10 years ago, when interest gradually grew.

30.1.2 Later developments

In 2010, a publication reported the detection of amphetamine and methamphetamine in breath collected from intoxicated drug users following recovery from intoxication.[5] The study was initiated after a discussion with scientists involved in measuring explosives in the ambient air with a portable detector based on surface-enhanced Raman spectroscopy (SERS) who wanted to know if this technology could be used for narcotics detection in exhaled air. A pilot study was performed to explore this proposition, since at that time drug detection in exhaled breath was not established. The study comprised sampling from two patients being treated for drug misuse, one of whom had admitted intake of amphetamine the evening before sampling. Exhaled breath was sampled on a solid-phase extraction (SPE) cartridge normally used for extraction of amphetamine from liquid specimens. Analysissing liquid chromatography−tandem mass spectrometry (LC-MS/MS) elicited signals for amphetamine in the sample from the subject who had admitted recent intake, while the control sample produced no signals (Fig. 30.1).

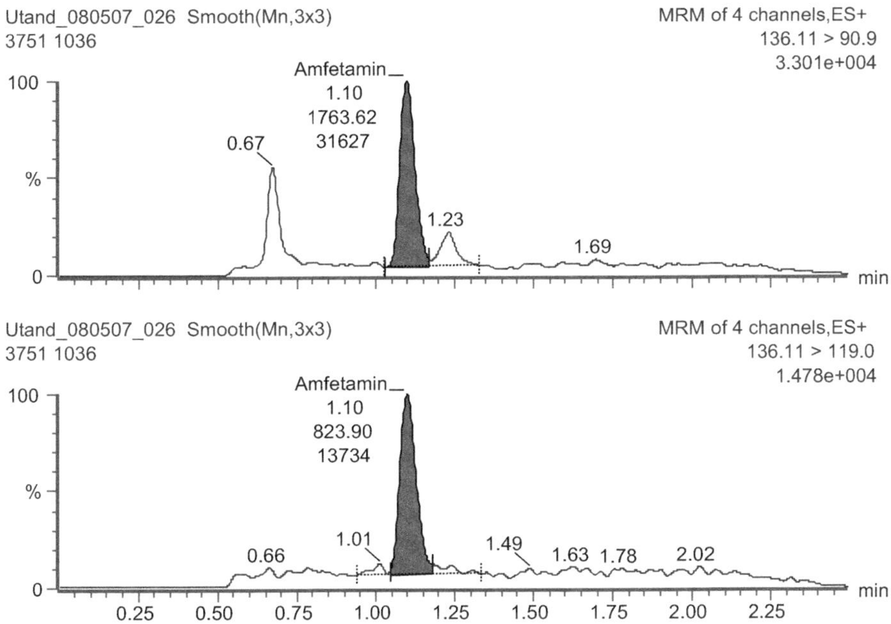

FIGURE 30.1

The chromatograms obtained from the first observation of amphetamine detection in exhaled breath following drug intake. The patient admitted amphetamine intake the evening before sampling the next morning. The analyses were performed using liquid chromatography–tandem mass spectrometry (LC-MS/MS) in selected reaction monitoring mode. The correct retention time and the two transitions monitored for amphetamine confirmed its identification.

This promising observation led to a more comprehensive study of the presence of amphetamine and methamphetamine in breath from patients recovering from acute intoxications.[5] This yielded the first demonstration of the possibility to detect a drug not administered through smoking or inhaling as detectable in exhaled breath, and opened the field for further work on this topic. In the study, 12 patients were recruited following recovery from acute intoxication and after admitting recent intake of amphetamine. In most cases, sampling was carried out several hours (and all within 24 h) after drug intake. Analyses of plasma and urine samples confirmed recent intake of amphetamine and/or methamphetamine. By comparison, amphetamine or methamphetamine was not detected in the exhaled breath of healthy controls. Breath was sampled via a facemask through an SPE cartridge for 10 min. An unexpected finding was the presence of methamphetamine in breath, blood, and urine in many cases. In at least seven (out of 12) cases, the data strongly indicated intake of methamphetamine either alone or with amphetamine. This result was surprising, as it had long been established in Sweden, where the study took place, that amphetamine was the primary stimulant of drug users, but at the time of the study the police authority reported that drug dealers in Stockholm had begun selling methamphetamine as an amphetamine substitute. This observation provided extra confidence in the analytical findings of the study, as well as confidence in breath as a suitable sampling medium. The identification of amphetamine and/or methamphetamine followed commonly accepted forensic criteria. The observed amounts of these drugs in exhaled breath—measured as excretion rates—were in the range 0.2–139 pg/min.[5] To measure such small amounts of analytes is challenging, even for mass spectrometry, and this might explain why Alan Curry's prediction took longer than expected before it could be realized.

30.2 Sampling biological specimens

30.2.1 Exhaled breath gas and aerosols

As reported above, early research demonstrated the potential of exhaled breath as a sampling medium for the detection of drug intake. Besides these early studies on amphetamine and methamphetamine, the presence of methadone in exhaled breath of patients receiving this medication as part of a substitution treatment for heroin addiction was investigated. The aim of these studies was to complement and confirm the amphetamine/methamphetamine findings with another substance, administered under controlled conditions, and to establish experimental settings for future developmental work. Using a modified sampling procedure, the presence of methadone in exhaled breath was successfully confirmed by sampling 8–60 min after taking a daily dose.[6] The newly developed sampling procedure used an SPE filter with larger diameter, a mouthpiece, and a shorter distance between the mouth and the filter (see Fig. 30.2). Methadone was detected in all subjects with a mean excretion rate of 1.8 ng/min.

This newly designed sampling device for methadone was constructed in a way that allowed the filter to be exchanged in order to test alternatives. The goal was to find a sampling material with a low flow resistance so that the pump support was no longer needed. Several options were tested, including particle filters.[7,8]

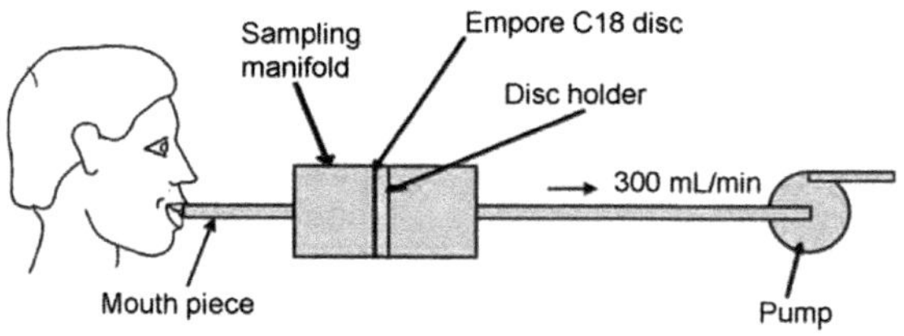

FIGURE 30.2

Schematic diagram of the device used to sample exhaled breath on an Empore C18 disc. The exhaled breath was sampled for 10 min.

Reproduced from.[6]

One configuration used a glass fiber particle filter to collect airborne particles with a pore size of 1 μm. Sample collection via this system demonstrated that the particle filter was more effective in collecting methadone from exhaled breath than the SPE filter. In addition, a polymer filter was alternatively tested for collection of particles from ambient air, and it was observed that this filter was almost as effective as the glass filter. These results were taken as evidence that methadone is carried out of lung within the aerosol fraction of breath.[8] This conclusion was supported in a later study using a better-defined particle fraction that had already been shown to contain components from the airway lining fluid.[9] Two other sampling techniques that did not require pump support were found to collect methadone from exhaled breath: one used Amberlite XAD-2 (a hydrophobic polymer material) beads and the other used a solid-phase microextraction (SPME) fiber, but both approaches were found to be much less effective than the reference SPE filter.[8] The best option for collecting methadone in exhaled breath—and avoiding the need for a pump—was therefore the polymer filter, which was commercially available.

Based on these findings, a simple commercial product was developed with the aim of supporting further progress in the field of capturing and detecting nonvolatiles in breath. The product was commercialized under the name SensAbues (see Fig. 30.3; top). This simple, commercial device was subsequently used to explore the utility of exhaled breath aerosol (EBA) for detecting drug exposure.

Although the SensAbues device was instrumental for advancement in the field of breath drug testing, it suffered from some drawbacks. Specifically, the device had to be assembled before use, it allowed only one specimen to be collected (making reanalysis impossible), and the filter material itself often disturbed the performance of the analytical method. To overcome these issues, a second generation of a simple collection device for nonvolatiles in exhaled breath was developed.[10] The new device enabled the collection of three parallel samples, without the need for prior assembly, and was constructed from materials free of interfering components (see Fig. 30.3; bottom).

30.2.2 Exhaled breath condensate

Besides exhaled breath gas, exhaled breath condensate (EBC) is a medium of interest for the detection of endogenous or exogenous compounds (see Chapter 7). Research on EBC has been ongoing for almost three decades and has focused on finding disease biomarkers rather than measuring drugs.[11] Nevertheless, the detection of drugs in EBC has been demonstrated, for example, for methadone. As part of the development of a simple collection device for measuring drugs of abuse, methadone was used as a model substance because of the readily available patients from whom to collect samples. Methadone was detected in EBC, but at lower concentration than in specimens collected by membrane filters.[12] More recently, methadone in EBC was studied using a newly developed liquid chromatography–UV method and a new patented sampling procedure.[13] The study confirmed the presence of methadone in EBC but found no correlation to serum concentrations.

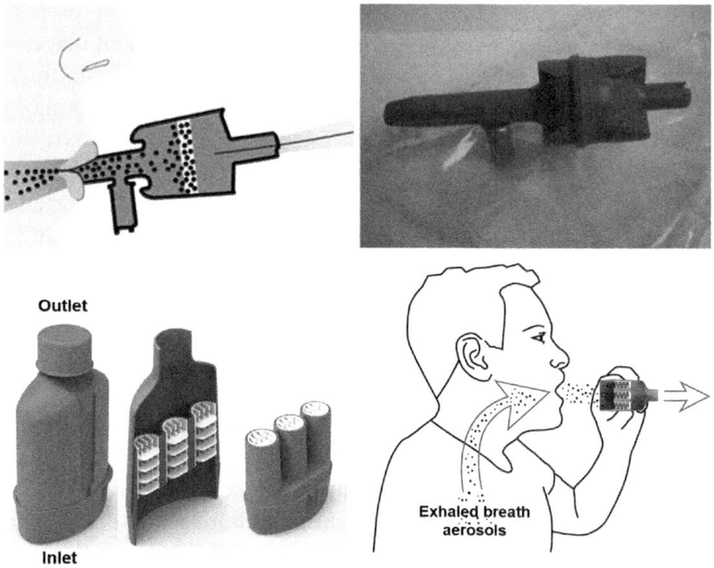

FIGURE 30.3

Top left: a schematic of the SensAbues device for sampling of exhaled breath aerosols (EBAs); the sampling duration is 1–3 min, with the plastic bag filled when the sampled exhaled gas reaches about 30 L. Top right: an image of the SensAbues device; the plastic bag and mouthpiece are discarded and the manifold holding the filter is capped and used for the analytical investigation. Bottom left: the second generation sample device for collecting EBA, housing three filter shafts; the product is constructed from medical grade plastic and collects three parallel samples. Bottom right: the subjects exhales through the device without restriction.

Top figures reproduced from[19] and bottom figures adapted from.[10]

Recently, however, a study that focused on the presence of morphine and hydromorphone in EBC successfully detected several metabolites of these drugs that exhibited good correlations with their associated concentrations in blood.[14]

30.2.3 Saliva

When collecting a breath specimen via the oral cavity, as is the case with breath gas or EBC sampling, it is an obvious possibility that components from saliva might contaminate the sample. EBC collection can be carried out with equipment constructed with saliva protection, where it is considered that saliva contamination is minimized.[15] Saliva contamination was already considered in the aforementioned early work with methadone[6] when oral fluid was collected in parallel with exhaled breath. The study concluded that although contamination from the oral cavity can occur, this is a rare event. A later study using codeine to contaminate the oral cavity also demonstrated this possibility, but similarly concluded that this was a random event.[9] This conclusion was supported by the fact that methadone was detected in the particle fraction known to originate from the lower airways.[9]

Recently, the issue of saliva-related contamination was reinvestigated more extensively.[16] Using measurements of abundant salivary proteins (e.g., α-amylase) with sensitive mass spectrometry, it was shown that contamination from saliva was not observable when sampling was made using a device with a saliva protection configuration. As surfactant contains a unique composition of phospholipids, this feature was exploited to characterize the specimen collected in the aerosol fraction by measuring the phosphatidylcholines (PCs) 16:0/16:0 and 16:0/18:1.[17] The mean ratio between these PCs was two in the collected breath specimens, while in oral fluid it was an order of magnitude lower, at 0.2. Nevertheless, the results from some individuals indicated a possible contribution from oral fluid, as was evident from the low ratio value observed.

In general, when collecting EBA, contamination from saliva in the oral cavity is uncommon but cannot be ruled out in specific cases. The ratio between the two PCs could be used as a sample validation parameter. Interestingly, the amount of PC 16:0/16:0 correlates with the sample size measured as particle mass.[18]

30.3 Drug detection

Many drugs or metabolites have been demonstrated to be detectable in breath or breath-based specimens, to date, as summarized in Table 30.1. This section highlights notable studies in this field.

30.3.1 A cocktail of drugs

In a comprehensive study from 2013, 47 drug users with ranging substance addiction were recruited to provide samples after recovering from acute intoxication at an emergency ward.[19] Blood, urine, and exhaled breath were collected, as was self-reported recent drug use. The analytical investigation using the SensAbues sampling device and mass spectrometric detection targeted a wide range of drugs, namely methadone, amphetamine, methamphetamine, 6-acetylmorphine, morphine, benzoylecgonine, cocaine, diazepam, oxazepam, alprazolam, buprenorphine, and THC. Analytes associated with all of the 12 drugs could be detected among the patient cohort. Since the sampling device featured a sample volume indicator to allow for "standardized" samples to be collected, which was equivalent to ~30 L of air (1–3 min sampling time), the unit of quantitation used was pg/filter, as opposed to pg/min used prior to that study. Among the samples, recent methadone intake was evident by blood and urine analyses and was detected in the breath samples in 11 cases, with methadone levels observed to be lower than the predose levels in patients receiving methadone maintenance treatment. In the first studies on methadone, sampling was carried out shortly after dose intake,[6,12] but later studies documented the detection of methadone 24 h after dose intake.[7]

Table 30.1 Drugs and metabolites that have been detected in exhaled breath.

Substance	Note
6-Acetylmorphine	Metabolite of heroin
7-Aminoclonazepam	Metabolite of clonazepam
Alprazolam	
Amphetamine	
Benzoylecgonine	Metabolite of cocaine
Bisoprolol	
Budesonide	
Buprenorphine	
Caffeine	
Cannabidiol	Cannabis constituent
Cannabinol	Cannabis constituent
Clonazepam	
Cocaine	
Codeine	
Dehydrochloromethyltestosterone	
Diazepam	
2-Ethylidene-1,5-dimethyl-3,3-diphenylpyrrolidine	EDDP, metabolite of methadone
Meldonium	
Meropenem	
Methadone	
Methamphetamine	
3,4-Methylenedioxymethamphetamine	MDMA, ecstasy
Methylhexanamine	
Methylphenidate	
Morphine	
Nicotine	
O-desmethyltramadol	Metabolite of tramadol
Oxazepam	
Piperacillin	
Pseudoephedrine	
Propofol	
Ritalinic acid	Metabolite of methylphenidate
Tetrahydrocannabinol	THC
Tramadol	

Within the same study outlined above, buprenorphine was detected in either urine and/or blood in seven cases, but was detected in exhaled breath in each related case. Concentrations were well above the quantitation limit in four of these cases, with values in the range 120–567 pg/filter. Amphetamine was detected in either blood or urine or both, in 13 cases, but again in all cases in breath. By comparison, methamphetamine was detected in either blood or urine in ten cases, but in breath it was only detectable in four cases. The concentrations of amphetamine and methamphetamine were between 20 and 4700 pg/filter. Cocaine and/or its metabolite benzoylecgonine were detected in blood and/or urine in four cases, and in breath in each case. The concentrations of cocaine in breath were between 64 and 13,000 pg/filter and were always higher than those of benzoylecgonine.

For opiates (heroin/morphine), eight cases indicated recent intake, with six of those cases additionally exhibiting the presence of morphine and/or the heroin metabolite 6-acetylmorphine in breath. These results demonstrated the importance of detecting 6-acetylmorphine, which was a better target analyte than morphine for detecting heroin intake. Three benzodiazepines—oxazepam, diazepam, and alprazolam—were included in the study and the intake of any of these was indicated in 28 cases through urine analyses. The detection of oxazepam, diazepam, and alprazolam in breath was possible in three, eight, and seven cases, respectively, and in each case was associated with high concentrations in blood and urine.

The main conclusion from this comprehensive study was that although breath sampling was often carried out around 24 h after drug intake, the detection rate was high for most substances. In the case of the benzodiazepines, increased analytical sensitivity is needed to improve detection rates. Overall, this study made a strong case for the use of exhaled breath as a new matrix in toxicology applications.

30.3.2 Detection of cannabis

The detection of cannabis intake is one of the more important applications of drug testing and is presently especially in focus due to ongoing legalization of this drug in many countries (see also Section 30.3.6). THC, a constituent of cannabis, is metabolized to the corresponding carboxylic acid (THCCOOH), but this metabolite is long-lived in the body so its detection might not necessarily indicate recent cannabis use. In the extensive drug study cited above,[19] this metabolite of THC was detected in urine in 20 cases. By comparison, THC was detected in blood samples in nine cases, indicating a more recent intake, and was also detected in breath in eight of those cases.

Several studies have focused on the detection of cannabis use in breath. Using the prototype SPE filter–based sampling procedure described above, an early study confirmed that THC could be detected in breath shortly after smoking cannabis,[20] as had been similarly demonstrated in the earlier work of Manolis and coworkers.[3] A follow-up controlled study involving 24 cannabis smokers who had smoked a standard formulation of pure THC in cigarette form further investigated its detection in breath.[21] The participants smoked one cigarette ad libitum over a 10 min period and THC was subsequently detected in exhaled breath in all but one subject, whereby concentrations were observed to be higher in chronic cannabis users compared to occasional smokers. The detection was typically between 0.5 and 2 h after smoking, but the method employed had rather limited sensitivity.

Another study of illicit cannabis consumption observed higher concentrations of THC in breath, which was detected in all 13 subjects 3 h after consumption.[22] Interestingly, the time curve of THC in breath coincided with the physiological measurements of pulse rate and pupil diameter, which provided evidence that THC in breath might relate to impairment. Finally, a fourth study of controlled intake demonstrated that THC could be detected in the breath of four occasional cannabis smokers 6 h after intake.[23]

Overall, detecting cannabis use by the detection of THC in breath shortly after intake has been demonstrated in several studies, whereby the exact detection time was dependent on the sensitivity of the method employed. Clearly, THC is the primary analyte in exhaled breath for determining cannabis use.

30.3.3 Drug testing in the workplace

Breath-based methods have been employed for workplace drug tests in parallel with urine and oral fluid. The two main applications for this are in random and preemployment testing in heavy industry, and in recruitment of candidates for military and police training. Analyses in these scenarios have been made by mass spectrometry according to the published method[19] and following forensic guidelines. Table 30.2 reports the incidences of sample positives for drugs in workplace drug tests in breath compared with findings in oral fluid and urine.

30.3.4 Therapeutic drugs

There is an overlap between substances used as therapeutic drugs and those subjected to abuse and misuse, e.g., opioids, because of effects on the central nervous system. Methadone, buprenorphine, methylphenidate, and amphetamine are drugs that are used in substitution treatment or for medical treatment of specific afflictions, such as attention deficit hyperactivity disorder (ADHD). For these treatments, exhaled breath testing can offer a means for compliance monitoring without the need for time-consuming and inconvenient urine collection.[24–26]

Table 30.2 Analytical findings in routine workplace drug tests using exhaled breath as compared with oral fluid and urine.

	Positive rates (‰) in		
Substance	**Breath (n = 2295)**	**Oral fluid (n = 18,095)**	**Urine (n = 35,142)**
Amphetamine/methamphetamine	1.7	1.1	2.7
Cannabis	0.9	3.4	12.7
Cocaine/benzoylecgonine	0.9	0.3	0.5
Opiates[a]	0.9	2.4	11.0
Benzodiazepines	1.3	0.1	15.5
∑ Positive rate (‰)	5.2	7.3	35.5

[a] *Most of the positives in oral fluid and urine were from codeine intake.*

Tramadol is an opioid used for pain management. A small clinical trial comprising 12 healthy volunteers was performed to compare concentrations of this drug in plasma, oral fluid, and exhaled breath.[27] After administering an oral dose of 50 mg, tramadol was detected in exhaled breath for up to 48 h in 11 of the 12 subjects. The concentration profiles in breath followed those of plasma and oral fluid, albeit with a small time shift (Fig. 30.4), which might reflect a slower turnover of the airway lining fluid compared to blood and oral fluid. The tramadol metabolite O-desmethyltramadol was also detected in breath samples.

The detection in exhaled breath of therapeutic drugs for asthma or antibiotics is also of potential interest. In a study on antibiotics, piperacillin and meropenem were present in the particle fraction of exhaled breath collected from critically ill patients, although no correlation with plasma concentrations could be determined.[28] In a preliminary study, budesonide was detected in the exhaled breath of patients prescribed this substance for treatment of asthma.

30.3.5 Doping control

Exhaled breath has also found interest in the field of sports doping control and was evaluated and proposed as a possible complementary matrix in a review paper in 2016.[29] A study using 12 model compounds subsequently demonstrated the usefulness of exhaled breath for sports doping control by successfully detecting stimulants, an anabolic steroid, meldonium, and a beta-blocker after controlled administration.[30] The study concluded that exhaled breath is a viable alternative for sports doping control, but pointed out the need for highly sensitive analytical methods to make this possible.

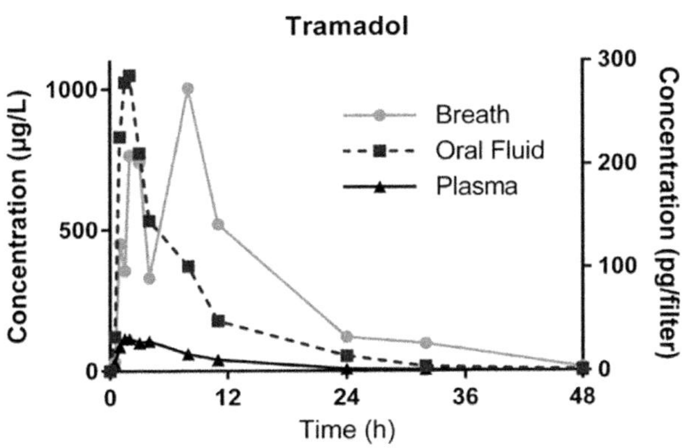

FIGURE 30.4

Transient concentrations of tramadol in breath, plasma, and oral fluid after intake of 50 mg tramadol. Data represent mean values for 12 individuals.

Reproduced from.[27]

30.3.6 Law enforcement applications

The successful implementation in law enforcement of a breathalyzer for alcohol impairment (see previous chapter) has been a driver in recent years for similar developments in relation to drug impairment, with a particular focus on cannabis use.[21] Collecting a blood sample presents a practical problem in law enforcement, since this cannot be performed at the roadside and therefore requires that the suspect is taken to a medical facility for professional blood sampling, which is both time-consuming and costly.

A very recent study in relation to law enforcement included the collection of exhaled breath in parallel with forensic analysis of blood and urine to investigate driving after drug use.[31] The analysis demonstrated similar success rates for exhaled breath as for the established forensic methods (blood and urine), with, respectively, 45 and 46 drivers out of 67 cases identified as being under the influence of drugs. The promising results lay the foundations for further work to explore this possible application. The roadside collection of a valid forensic specimen for drug testing is not only attractive because it could save resources, but also because it adds reliability to the investigation, since the sample can be taken immediately, without delay. This latter aspect is especially important for detecting cannabis intake, since blood levels decrease to below the reporting limits very rapidly after intake.[32]

30.3.7 Developments toward real-time detection

In both the clinical and forensic fields, there is sometimes a need for on-site and real-time drug measurement. One example is the possibility to measure cannabis intoxication at roadside using a breathalyzer, as discussed above. Another obvious application is to monitor anesthetic drug concentrations during surgical interventions. Propofol is a common anesthetic agent that has received great interest for this purpose due to the need to monitor administration in individual patients during anesthesia.[33] Since blood concentrations cannot be measured in real time, breath has been considered as a possible surrogate specimen. Despite the obvious pharmacokinetic differences between blood and breath, the latter has nevertheless been demonstrated to be clinically useful.[34] Online analytical technologies enable the detection of certain anesthetic drugs in the exhaled breath of anesthetized patients, as has been demonstrated using proton transfer reaction—mass spectrometry (PTR-MS; see Chapter10) and ion mobility spectrometry (IMS; see Chapter 11). The monitoring of propofol in breath during anesthesia is further discussed in Chapter 18.

30.4 Summary

Interest in detecting drugs in exhaled breath has emerged from an associated interest in the lung as the target organ for drug effect or administration, as well as compliance monitoring of other therapeutic drugs. Most drugs are nonvolatile and as such are not readily present in the gas phase of exhaled breath, meaning that early interest in exhaled breath for this application waned. The more recent discovery that drugs are carried out from lung in the form of aerosol particles sparked a renewed interest in breath as a potential biological specimen for drug detection and monitoring applications.

In the fields of drug monitoring in pharmacology and toxicology, it is common practice to measure drug exposure by analytical investigations of body fluids. Blood sampling is invasive and requires skilled health personnel. Urine sampling has the drawback of needing a special facility with supervised urination, as well as being time-consuming. Exhaled breath sampling has the advantage of being noninvasive, readily available, and easy from many aspects. Work in the past decade has demonstrated that drugs, in addition to nonvolatile biomarkers, can be measured in exhaled breath. A large number of illicit and therapeutic drugs are now documented to be detectable in breath. Simple devices for the collection of EBA and EBC under field conditions are available.

It is well known in the field of breath analysis that sampling procedures are a key factor in making a breath test clinically useful. Currently, no generally accepted procedure exists for sampling nonvolatiles collected in aerosols. More knowledge is needed for a better understanding of the relationship between EBA and the measured nonvolatile components. Better documentation of the origin of the particle fraction, as well as the introduction of standardized collection procedures, would greatly improve success in the field.

Exhaled breath gas, EBC or EBA, represent biological specimens that have great potential for determining drug use or misuse. Possible applications of breath-based drug testing include

- Quantification of therapeutic drugs, especially those that target the lung for effect or administration
- Compliance control of therapeutic drugs in general
- Clinical drug testing where urine and blood collection is problematic
- Clinical toxicology in emergency medicine
- Workplace drug testing to offer a more convenient and safer collection procedure
- Roadside testing of suspected drugged drivers
- Accidents and incidents for legal purposes
- Criminal justice
- Research on drug prevalence in populations where urine or blood collection is not feasible

Similar to the development of many other medical tests, it is likely that the introduction of drugs measurements via exhaled breath will need time to develop to its full potential. The scientific achievements that have been made in the field to date will offer a solid platform for the future implementation and success in this promising area of application.

References

1. Curry AS. Toxicology in the next 25 years. In: Heyndrickx A, editor. *Human toxicology Proceedings of the International Congress of the International Association of Forensic toxicologists. European meeting, held at the State University of Ghent, Department of Toxicology. August 26–27, 1976*. Ghent: European Press; 1978.
2. Jeffcoat AR. *Analysis for drugs in saliva and breath*. US Department of Transportation Report; September 1981. Report number DOT-HS-806194.
3. Manolis A, McBurney LJ, Bobbie BA. The detection of Δ9–tetrahydrocannabinol in the breath of human subjects. *Clin Biochem* 1983;**16**:229–33.
4. Melker RJ, Goldberger BA, Gold M. *Method and apparatus for detecting illicit substances*. US patent application US 2002/0177232 A1. November 2002.
5. Beck O, Leine K, Palmskog G, Franck J. Amphetamines detected in exhaled breath from drug addicts: a new possible method for drugs-of-abuse testing. *J Anal Toxicol* 2010; **34**(5):233–7.
6. Beck O, Sandqvist S, Eriksen P, Franck J, Palmskog G. Method for determination of methadone in exhaled breath collected from subjects undergoing methadone maintenance treatment. *J Chromatogr B Analyt Technol Biomed Life Sci* 2010;**878**(24):2255–9.
7. Beck O, Sandqvist S, Böttcher M, Eriksen P, Franck J, Palmskog G. Study on the sampling of methadone from exhaled breath. *J Anal Toxicol* 2011;**35**(5):257–63.
8. Beck O, Sandqvist S, Franck J. Demonstration that methadone is being present in the exhaled breath aerosol fraction. *J Pharmaceut Biomed Anal* 2011;**56**(5):1024–8.
9. Ljungkvist G, Ullah S, Tinglev Å, Stein K, Bake B, Larsson P, et al. Two techniques to sample non-volatiles in breath-exemplified by methadone. *J Breath Res* 2017;**12**(1): 016011.
10. Seferaj S, Ullah S, Tinglev Å, Carlsson S, Winberg J, Stambeck P, et al. Evaluation of a new simple collection device for sampling of microparticles in exhaled breath. *J Breath Res* 2018;**12**(3):036005.
11. Davis MD, Montpetit AJ. Exhaled breath condensate: an update. *Immunol Allergy Clin* 2018;**38**(4):667–78.
12. Beck O, Sandqvist S, Eriksen P, Franck J, Palmskog G. Determination of methadone in exhaled breath condensate by liquid chromatography-tandem mass spectrometry. *J Anal Toxicol* 2011;**35**(3):129–33.
13. Khoubnasabjafari M, Ansarin K, Jouyban-Gharamaleki V, Panahi-Azar V, Hamidi S, Azarmir Z, et al. Methadone concentrations in exhaled breath condensate, serum and urine of patients under maintenance treatment. *Iran J Pharm Res* 2017;**16**(4):1621–30.
14. Borras E, Cheng A, Wun T, Frank M, Schivo M, Davis CE. Detecting opioid metabolites in exhaled breath condensate (EBC). *J Breath Res* 2019;**13**:046014.
15. Montuschi P. Analysis of exhaled breath condensate in respiratory medicine: methodological aspects and potential clinical applications. *Ther Adv Respir Dis* 2007;**1**(1):5–23.
16. Garzinsky AM, Walpurgis K, Krug O, Thevis M. Does oral fluid contribute to exhaled breath samples collected by means of an electret membrane? *Drug Test Anal* 2019; **11**(11–12):1764–70.
17. Ullah S, Sandqvist S, Beck O. Measurement of lung phosphatidylcholines in exhaled breath particles by a convenient collection procedure. *Anal Chem* 2015;**87**(22): 11553–60.

18. Tinglev ÅD, Ullah S, Ljungkvist G, Viklund E, Olin AC, Beck O. Characterization of exhaled breath particles collected by an electret filter technique. *J Breath Res* 2016;**10**(2):026001.
19. Beck O, Stephanson N, Sandqvist S, Franck J. Detection of drugs of abuse in exhaled breath using a device for rapid collection: comparison with plasma, urine and self-reporting in 47 drug users. *J Breath Res* 2013;**7**(2):026006.
20. Beck O, Sandqvist S, Dubbelboer I, Franck J. Detection of Δ9-tetrahydrocannabinol in exhaled breath collected from cannabis users. *J Anal Toxicol* 2011;**35**(8):541–4.
21. Himes SK, Scheidweiler KB, Beck O, Gorelick DA, Desrosiers NA, Huestis MA. Cannabinoids in exhaled breath following controlled administration of smoked cannabis. *Clin Chem* 2013;**59**(12):1780–9.
22. Coucke L, Massarini E, Ostijn Z, Beck O, Verstraete AG. Δ(9)-Tetrahydrocannabinol concentrations in exhaled breath and physiological effects following cannabis intake - a pilot study using illicit cannabis. *Clin Biochem* 2016;**49**(13–14):1072–7.
23. Kintz P, Mura P, Jamey C, Raul JS. Detection of Δ9-tetrahydrocannabinol in exhaled breath after cannabis smoking and comparison with oral fluid. *Forensic Toxicol* 2017;**35**:173–8.
24. Skoglund C, Hermansson U, Beck O. Clinical trial of a new technique for drugs of abuse testing: a new possible sampling technique. *J Subst Abuse Treat* 2015;**48**(1):132–6.
25. Carlsson S, Olsson R, Lindkvist I, Beck O. Application of drug testing using exhaled breath for compliance monitoring of drug addicts in treatment. *Scand J Clin Lab Invest* 2015;**75**(2):156–61.
26. Kintz P, Mathiaux F, Villéger P, Gaulier JM. Testing for drugs in exhaled breath collected with Exabreath in a drug dependence population: comparison with data obtained in urine after liquid chromatographic-tandem mass spectrometric analyses. *Ther Drug Monit* 2016;**38**(1):135–9.
27. Meyer MR, Rosenborg S, Stenberg M, Beck O. First report on the pharmacokinetics of tramadol and O-desmethyltramadol in exhaled breath compared to plasma and oral fluid after a single oral dose. *Biochem Pharmacol* 2015;**98**(3):502–10.
28. Herregodts J, Van Vooren S, Deschuyteneer E, Dhaese SAM, Stove V, Verstraete AG, et al. Measuring antibiotics in exhaled air in critically ill, non-ventilated patients: a feasibility and proof of concept study. *J Crit Care* 2019;**51**:46–50.
29. Thevis M, Geyer H, Tretzel L, Schänzer W. Sports drug testing using complementary matrices: advantages and limitations. *J Pharmaceut Biomed Anal* 2016;**130**:220–30.
30. Thevis M, Krug O, Geyer H, Schänzer W. Expanding analytical options in sports drug testing: mass spectrometric detection of prohibited substances in exhaled breath. *Rapid Commun Mass Spectrom* 2017;**31**(15):1290–6.
31. Beck O, Ullah S, Kronstrand R. First evaluation of the possibility of testing for drugged driving using exhaled breath sampling. *Traffic Inj Prev* 2019;**20**(3):238–43.
32. Hartman RL, Brown TL, Milavetz G, Spurgin A, Gorelick DA, Gaffney GR, et al. Effect of blood collection time on measured Δ9-tetrahydrocannabinol concentrations: implications for driving interpretation and drug policy. *Clin Chem* 2016;**62**:367–77.
33. Trefz P, Kamysek S, Fuchs P, Sukul P, Schubert JK, Miekisch W. Drug detection in breath: non-invasive assessment of illicit or pharmaceutical drugs. *J Breath Res* 2017;**11**(2):024001.
34. Colin P, Eleveld DJ, van den Berg JP, Vereecke HEM, Struys MMRF, Schelling G, et al. Propofol breath monitoring as a potential tool to improve the prediction of intraoperative plasma concentrations. *Clin Pharmacokinet* 2016;**55**(7):849–59.

CHAPTER

Urban search and rescue 31

Veronika Ruzsanyi[1], Helmut Wiesenhofer[1], Dahlia Salman[2], C. L. Paul Thomas[2], Peter Hamm[3], Paweł Mochalski[1,4], Christopher A. Mayhew[1,5]

[1]*Institute for Breath Research, University of Innsbruck, Dornbirn, Austria;* [2]*Centre for Analytical Science, Department of Chemistry, Loughborough University, Loughborough, United Kingdom;* [3]*University Hospital for Anaesthesia and Intensive Care, Medical University of Innsbruck, Innsbruck, Austria;* [4]*Institute of Chemistry, Jan Kochanowski University, Kielce, Poland;* [5]*Molecular Physics Group, School of Physics and Astronomy, University of Birmingham, Birmingham, United Kingdom*

Chapter outline

31.1 Overview

Urban search and rescue (USaR) is needed for a variety of emergency incidences, including earthquakes, floods, tornados, terrorist attacks, and accidents. Earthquakes are one of nature's most powerful and dangerous forces. They belong to the most devastating natural disasters, occurring unexpectedly and often causing many deaths and traumatic injuries. More than a million earthquakes occur worldwide every year, causing approximately 1.5 million deaths and directly affecting another two billion people.[1] The latter figures are likely to rise, given the rapid pace of socioeconomic development and urbanization, an increasing global population, and the construction of taller and more complex buildings.

Regardless of origin—whether from a natural or manmade disaster—victims frequently become trapped under complex debris, especially in urban areas, often for days. Survival depends on the health conditions of the victims, the injuries

Breathborne Biomarkers and the Human Volatilome. https://doi.org/10.1016/B978-0-12-819967-1.00031-1
Copyright © 2020 Elsevier B.V. All rights reserved.

sustained, and the availability of food and water. The accessibility to water is more important than food, because the human body can only survive 3–5 days without water, but many weeks without food. The chance of survival depends on the duration a victim is trapped under debris. Hence, search and rescue operations are most effective if trapped people can be found within hours of an event. The fast location of victims is paramount. Unfortunately, in the event of an incident that leads to people being trapped under collapsed buildings, emergency search and rescue efforts often struggle, especially in the initial stages of a disaster, owing to the absence of fast sensing devices, readily available rescue equipment, and personnel with emergency response skills.

Another pertinent issue is that earthquakes, accidents, or terrorist attacks can lead to the release of hazardous toxic chemicals or radioactive materials into the environment, all of which need to be detected quickly to ensure the safety of both the victims and first responders.

The detection of hidden people is equally relevant and crucial in relation to human smuggling or trafficking. This transnational crime is a growing problem, with the illegal movement of people across borders affecting many countries, notably the United States and in the European Union (EU). In recent years, the trafficking and smuggling of people to Europe have reached epidemic proportions. This not only places a major strain on European resources but also puts at risk the health and lives of the people being trafficked or smuggled. Criminal networks promote migrant smuggling, offering transportation in trucks, containers, and specially reconstructed vehicles; this "service" frequently commands very high prices, despite the inhumane conditions often endured. These victims barely receive food or drink, and often no information about their location is given. Security personnel are unable to cope with the high influx of people illegally crossing borders. In this context, an early detection and interception of smuggled or trafficked people is of particular importance, not just to protect borders but also to save people from life-threatening situations. On highways with high traffic rates, the time of inspection plays a critical role because of the flux of trucks and containers crossing borders daily, such as between many EU countries, thereby limiting the time for any disruption caused by vehicle and container searches. To be viable, an inspection of any truck or container should be completed within 2–3 min, and preferably without opening any doors or breaking any customs import/export seals that are in place.

31.2 Locator systems

The capability of fast reaction and deployment in a disaster scenario requires well-trained and specially equipped rescuers, good communications, and reliable human locator systems. There is therefore a demand for the development of sensitive and portable detectors that can help to discover and locate people, which could be used either to complement or to replace other locator procedures, such as search and rescue dogs.

Many search and rescue teams rely increasingly on specialized technical tracking systems. In Europe, search operations are currently carried out primarily with the help of carbon dioxide (CO_2) sensors and infrared (IR) (heat) cameras. Both technologies have limitations. IR imaging is not effective where there are only very small temperature differences between the body and the surroundings, as can occur in the summer when ambient temperatures can reach 35°C or higher, or when metal shielding is present, thus can only be used in trucks with tarpaulins (no closed structures). Carbon dioxide sensors suffer from high false-positive rates in the presence of elevated CO_2 emissions from organic materials, such as wood, vegetables, and fruits, as has been reported by police officers conducting controls at the Austrian–Italian border crossing points. Other techniques used in USaR operations are X-rays (only occasionally applied), or drones equipped with cameras and heartbeat detectors that use microwave radiation to detect the chest movements associated with heartbeats and breathing.[2] Other standalone technologies available to locate people are based on optical or acoustic approaches. While useful, these similarly have limitations. The use of optical cameras, for example, requires unobstructed views of the emergency site. The use of acoustic devices generally requires victims to be conscious and these systems suffer a large number of false alarms due to interference from other noise sources.

Search and rescue dogs provide first responders with a system that is capable of rapidly locating victims. Thus, dogs are still the preferred choice for the quick detection of entrapped people.[3] Nevertheless, these equally have limitations, such as risk of injury, fatigue of the dog after only about 30 min of intensive search, and stressed and frustrated behavior if they are unsuccessful at locating people.[4] Another issue is the significant cost in the training/handling and boarding of dogs. For the interested reader, the use of canine olfaction in the detection of human volatilome emissions in relation to disease detection—but not search and rescue applications—is treated at length in Chapter 34.

Imitating dogs by sniffing the characteristic scents of humans by technological means represents a potentially fast method to detect the presence of people in emergency scenarios. Consequently, the chemical analysis of air in small, enclosed spaces without extensive air exchange can provide tangible indications of entrapped or hidden humans. An artificial human scent detector has the potential to improve significantly the success rate of any USaR and for use at border controls to find people hiding in containers or trucks.

31.3 Tracing volatile signals

The human body releases hundreds of trace volatile organic compounds (VOCs) through breath, sweat, and skin,[5,6] thereby representing a continuous emission source that can be exploited for various applications. In addition, volatile substances excreted by blood, urine,[7] and feces[8] (see also Chapter 24) can also be characteristic of human proximity, although the presence of these specimens is transient.

The combination of volatiles released from these bodily fluids produces a human chemical signature that can be detected in the immediate environment using sensitive analytical techniques.

The trace volatiles from breath, skin, urine, saliva, blood, sweat, and feces collectively make up the human volatilome (see Chapter 1). Fig. 31.1 depicts a schematic overview of the volatile compound classes emitted by the human body, which can be used as a chemical fingerprint for the presence of people.[9]

31.3.1 Continuous emission sources: breath and skin volatiles

Breath is an important source of emissions of gas-phase compounds into the environment from live victims. Breath-specific volatiles can be therefore used by default to find trapped or hidden people who are still alive. Although the release of breathborne compounds is continuous, they might vary in quantity, concentration, or type, depending on the activity (possible movements), health state, level of satiety or starvation, etc., of the victim.

Breath fluxes of the most important volatiles have been estimated by Mochalski et al.[9] Calculated fluxes were based on values measured during sleep that were applied to a simple model mimicking the situation of victims in an entrapped environment. After CO_2, which has an estimated breath emission rate of 66.8×10^2 μmol/min/person, the highest emissions of volatiles were found to be carbon monoxide (CO), followed by ammonia (NH_3), acetone, and methanol, with breath fluxes being between 524 and 45 nmol/min/person. The majority of other volatile compounds detectable in exhaled breath exhibited fluxes below 1 nmol/min/person.

In addition to the continuous exhalation of VOCs from breath, skin is a major source of volatiles (see Chapter 25). Skin is the largest organ of the body, but its composition and thickness varies over all parts of the body. As such, measured values of volatiles emitted by selected body parts (mainly peripheries have been investigated) cannot be interpolated to the whole body surface.[5] Furthermore, other factors, such as age, sex, race, health condition, skin temperature, or environmental parameters (e.g., ambient temperature and humidity), can change the intensity of perspiration, sebum production, and VOC release from skin.[10] Sebum is a complex mixture of lipids that include squalene, esters of glycerol, wax, cholesterol, and fatty acids. Oxidation processes in sebum can lead to the production of many volatile metabolites, such as aldehydes, ketones, and esters.[11]

Studies using body plethysmograph chambers provide the capability to investigate whole body VOC emissions, either for the separate determination of skin or the combination of both breath and skin, mimicking an entrapped environment. Using such a chamber, the following volatiles have been reported to provide characteristic contributions to the human scent: seven aldehydes (acrolein, 2-methylpropanal, 3-methylbutanal, 2-ethacrolein, n-hexanal, n-heptanal, and benzaldehyde), three ketones (acetone, 2-pentanone, and 4-methyl-2-pentanone), five esters (ethyl formate, ethyl propionate, vinyl butanoate, butyl acetate, and

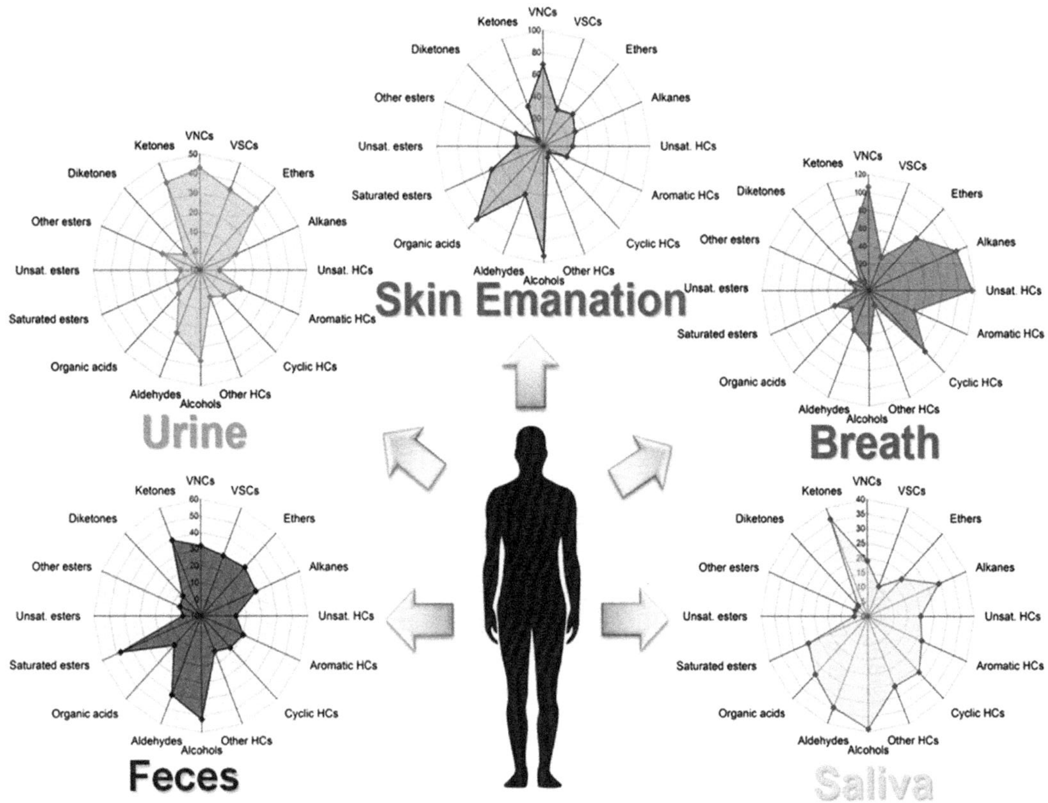

FIGURE 31.1

Volatile emission profile of the human body representing the human volatilome; each body fluid emits characteristic classes of chemical compounds. *HCs*, hydrocarbons; *VNCs*, volatile organonitrile compounds; *VSCs*, volatile sulfur compounds.

ethyl isovalerate), one alcohol (2-methyl-1-propanol), and one organic acid (acetic acid). The observed median emission rates ranged from 0.28 to 44.8 nmol/person/min.[5,12]

Recently, Güntner et al.[13] applied three nanostructured metal oxide sensors (Si-doped WO_3, Si-doped MoO_3, and Ti-doped ZnO) for highly sensitive and selective detection of three major species emitted by the human body: acetone, ammonia, and isoprene. The study involving a body plethysmograph chamber confirmed the usefulness of these species as potential markers of human presence using sensors with a high level of confidence.

31.3.2 Transient emissions sources: blood, urine, and feces

Urine, feces, and blood are transient emissions sources that exhibit changing composition and concentrations over time. Therefore, their emitted VOCs are not always present and thus cannot be considered fully eligible to locate living victims. Nevertheless, the occurrence of injuries, such as fractures and bleeding, during natural disasters are frequent; therefore, bloodborne volatiles might be of use to help locate injured victims within a relatively short time after becoming entrapped. Besides ammonia, the most important chemical classes of urineborne VOCs are ketones, aldehydes, and sulfur compounds.[7,14] Changes in the health conditions, especially kidney function of entrapped or hidden people, will considerably influence the production and types of volatiles present in urine (depending on the volatility and solubility of compounds), and hence complicate the use of these volatiles to locate people.

31.3.3 Volatiles from confounding sources

Another complicating factor in locating humans based on their volatile emissions is the often complex chemical matrix of the environment in which people are trapped or hidden. This can affect all volatiles. Organic and inorganic materials release volatiles into the surrounding atmosphere, which may confound the volatiles emitted by humans. It is therefore crucial to select volatile markers that provide a unique signature of human presence. Based on this requirement, a shortlist of 11 metabolites has been proposed as being suitable for providing the characteristics for detecting people with sufficient concentrations for current sensitive and selective analytical techniques, namely CO_2, ammonia, acetone, 6-methyl-5-hepten-2-one, isoprene, n-propanal, n-hexanal, n-heptanal, n-octanal, n-nonanal, and acetic acid.[12]

This use of key volatile biomarkers for locating or tracking people opens up new possibilities for different search and rescue scenarios, such as after natural catastrophes (earthquakes), explosions, or terrorist attacks, as well as in searching for humans being trafficked or smuggled across borders.

31.4 Portable field–deployable volatile compound detectors

Most volatile compounds of interest in relation to the human volatilome are emitted only at trace concentrations, which places high demands on available analytical technologies for their accurate detection. Many ultrasensitive analytical techniques currently exist that can easily detect concentrations down to parts per billion (or better) within seconds in real time. These include instrumentation such as selected ion flow tube mass spectrometry (SIFT-MS) or proton transfer reaction mass spectrometry (PTR-MS) (see Chapters 9 and 10, respectively), but these instruments are not portable. As such, although they are ideal for monitoring key volatiles, they cannot be easily transported to a disaster site and would be of limited use in the field. Portable and high-sensitivity analytical technologies that can be easily carried or handheld by emergency personnel are essential for on-site field applications. Other platforms, such as drones, offer potential widespread detection coverage through rapid sweeping of large areas. Regardless of platform, however, it is especially important to identify technologies with adequate selectivity to reduce false positives and false negatives, especially in view of confounding environmental factors, as discussed above. In addition to the chemical background, debris and dust, as is often encountered in natural disasters, can equally compromise the performance of sensitive detection systems.

Keeping in mind these relevant aspects, the gas sensors mentioned earlier have considerable potential for USaR and chemical, biological, radiological, and nuclear (CBRN) attacks owing to their small size, low cost, and fast response time. Numerous companies manufacture sensor systems, often comprising several gas sensors for multiple analyte detection to monitor indoor air pollutants or toxic chemicals. The idea of using these for human "scent" detection is relatively new but presents a challenge due to background signals from complex chemical environments, thereby making the development of a commercial sensor product difficult.

There are many different types of gas sensors that could be appropriate for detecting human volatile emissions, including electrochemical sensors,[15] semiconducting metal oxide sensors,[16] or multiwalled carbon nanotubes,[17] to name a few (see Chapter 13). Many of these exhibit good selectivity and sufficient sensitivity for selected compounds or compound groups in the higher ppb to ppm range, making them useful for detecting compounds such as CO_2, acetone, isoprene, and ammonia.[13] However, achieving sensitivities in the lower ppb range while maintaining good selectivity is still a major challenge for the detection of the majority of trace volatiles.

Ion mobility spectrometry (IMS) is in widespread use in homeland security applications to detect warfare agents and explosives, but has equally proven useful in civilian applications in different areas, including food quality control, medical science, and industrial processing (see Chapter 11). Given its small size and high analytical performance, IMS is a particularly promising analytical device for use in USaR and CBRN. IMS is a robust technique that has the advantage of portability and high sensitivity for rapid (seconds) and direct (without sample preparation) detection of VOCs. Further, chromatographic preseparation can be applied to enhance the selectivity for the parallel detection of selected VOCs at trace quantities in complex samples.[18] Although the use of a preseparation stage presents a disadvantage due to the increased analysis time of up to several minutes, systems can still be made portable by using isothermal separation and a controlled flow program to reduce peak tailing.[19]

In research undertaken by the Institute for Breath Research at the University of Innsbruck, a portable backpack human odor tracking system that employed IMS was developed. The device combines a state-of-the-art aspiratory IMS (aIMS) (ChemproDM, Environics, Oy, Finland) with a dual-channel CO_2 sensor module (Telaire, range: 0–5000 ppmv, Amphenol Advanced Sensors, Pforzheim, Germany) and an aldehyde sensor (IT Dr. Gambert GmbH, Wismar, Germany). The detection of CO_2 in conjunction with VOCs enhances the analytical capabilities of the instrument. Exhaled CO_2 is easily detected, given its high concentrations in human breath, and can be used as a possible sign of life. CO_2 emissions can therefore be searched for first, and if detected, the more sophisticated aIMS detector can be used to search for volatiles specific to humans. The combination of CO_2 and volatile detectors therefore represents a highly specific analytical device for first responders or security personnel to detect the presence of people with a higher level of confidence. The short analysis times of the aIMS allow for frequent and continuous sample collection, with VOCs such as ketones and amines being detectable with high sensitivity (down to pg/L concentrations).[20]

The aforementioned device is currently a working prototype that is still being developed and field-tested. The housing containing the key components of the aIMS is water resistant and is combined with a backpack carry system (Fig. 31.2A–C). To operate the device, a gas sample is collected from the end of a thin (1/8th inch diameter) and inert flexible tube. The narrow diameter and flexibility of the sampling probe makes it an appropriate size for inserting through small holes (e.g., in closed places under debris or in sealed containers). The gas sample is transported through a specially designed sampling system—including inert tubing, dust separator, and pump—to the CO_2 sensor, an aldehyde sensor, and the aIMS.

The portable system has been tested in the laboratory with biological samples, as well as in field trials. A protocol for data collection was refined during field trials, with a simple experimental setup to investigate the parameters needed for an adequate number (several hundreds to thousands) of chemicals detected for machine learning algorithms. After the training procedure, the specificity and sensitivity of the device were found to be both around 90%. However, the high diversity of background volatile signals of transported goods resulted in

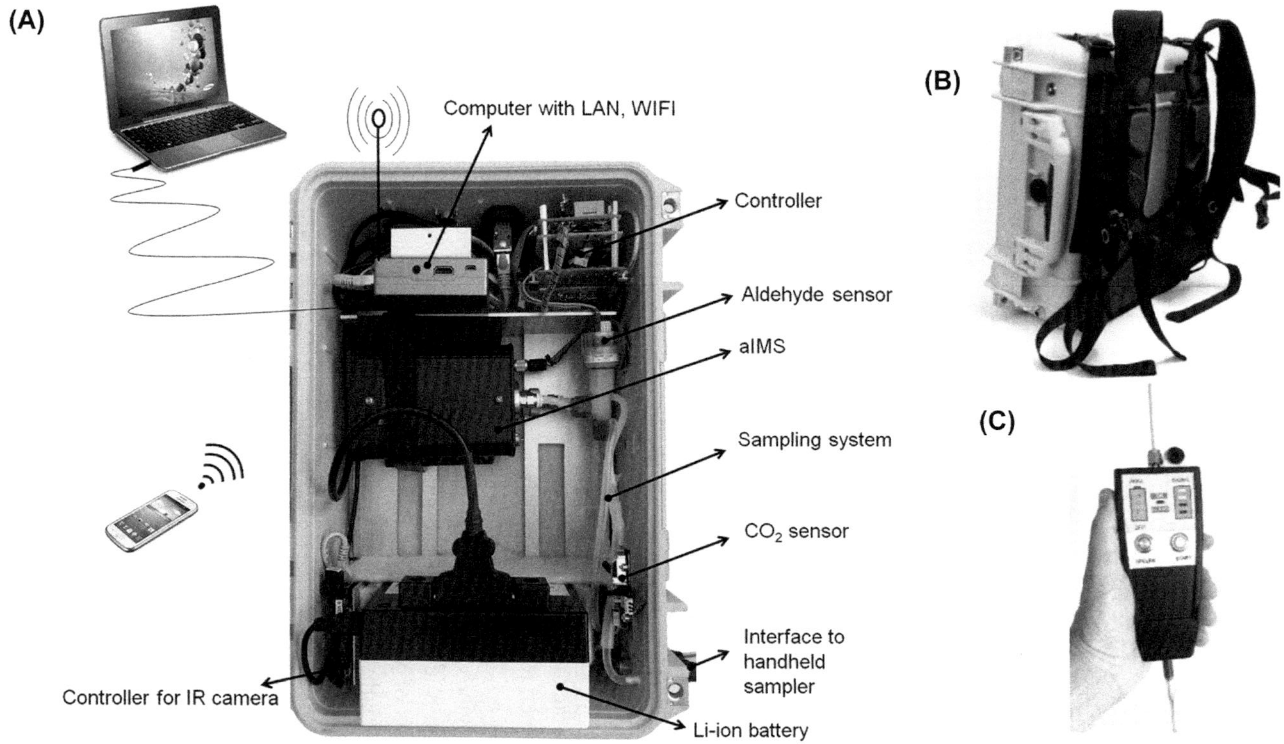

FIGURE 31.2

(A) Setup of portable locator that combines CO_2 and aldehyde detection via sensors with an aspiratory ion-mobility spectrometer (aIMS) for detection of breathborne volatiles, (B) backpack carry system, and (C) handheld sampler.

confounding signals; therefore, further training scenarios need to be investigated before this instrument advances beyond a prototype tool.[21]

Such research endeavors, as outlined here, require an interdisciplinary collaborative team to ensure that a developed device is designed according to user requirements and can be readily adopted for use in the field by end users. Ideally, national security authorities or sponsors should participate in the device specification/design phase of work to specify further requirements and appraise suitability.

31.5 Chemical, biological, radiological, and nuclear attacks

Terrorist attacks are usually less devastating than natural disasters but can still be significant. The global threat of terrorist attacks has peaked to unprecedented levels since the 9/11 World Trade Center atrocity of 2001 and the rise of Da'esh in 2014. There is a heightened risk of CBRN attacks similar to those that occurred in Syria (2015), Tokyo (1995), and Halabja (1988) involving the use of banned nonconventional weapons by rogue groups or regimes. In fact, the recent use of the nerve agent Novichok on UK soil has made the threat from CBRN attacks more real and ever closer to European citizens than before. Therefore, there is an urgent need for rapid detection of CBRN materials along with the identification and monitoring of critically exposed victims in triage incidents.

TOXI-triage was the name of a major EU Horizon 2020 funded multidisciplinary project that brought together 18 teams from eight different countries spanning university academics, defense agencies, and emergency and health services.[22] The project aimed to improve the preparedness of emergency services in the unfortunate event of a CBRN attack in Europe. A key aim of the project was to equip emergency services with advanced detection tools for rapid deployment during a CBRN incident to establish type and exposure levels of chemicals involved.[22]

The analysis of VOCs in exhaled breath offers a promising approach to underpin metabolic pathways associated with CBRN injuries and their toxicities, and permits the use of biomarkers as noninvasive bedside or field tests.[23] TOXI-triage used clinical data from toxic injuries in routine community medical care to discover and fingerprint CBRN-related metabolomes, and translate those into rapid sensors for detection and real-time monitoring of triage incidents. The selected CBRN breath markers aim to help first responders to accurately identify people who have been exposed to a chemical, biological, or radiological attack, and to predict the probability of that person suffering from related toxicities.

The first critical step toward achieving the goal of CBRN detection and accurate dose quantification through breath analysis is the ability to map changes in VOC profiles before and after CBRN injury, and to understand the effects of CBRN injuries on metabolic pathways over time. As part of TOXI-triage, clinical trials based in Loughborough University, University of Edinburgh, Oslo University

Hospital, and Peradeniya Hospital involved collecting breath samples from cancer patients before and after radiotherapy treatment, as well as from patients who were admitted to the hospital following pesticide or toxic alcohol poisoning. After identifying CBRN-related VOC biomarkers in breath using gas chromatography–mass spectrometry (GC-MS),[24] these were translated for use into a rapid online IMS detection system.[25]

A related clinical trial focused on the identification of markers of radiation exposure. Patients with deep body tumors were recruited and breath samples were collected before and after radiotherapy.[26] A range of methylated alkanes, alkenes, ketones, furans, and sulfur-containing compounds were identified as radiation markers. Sulfur-containing compounds have not been previously reported in the breath of people who have received radiation exposure, and this observation was derived from an unsupervised multivariate treatment of the data. Tracking the development of a radiation injury by following specific VOC profiles was also demonstrated, which is of interest as it can potentially benefit emergency services in determining the progression of radiation exposure. These volatile compounds can be successfully detected by gas chromatography–IMS (GC-IMS) and potentially translated into an online, fast detection method that could be used in scenarios requiring triage.

31.6 Summary

In the immediate aftermath of major disasters it is crucial to locate entrapped people in the rubble very quickly, owing to the rapid decrease in survival rates with time. Similarly, the detection of people being smuggled or trafficked is crucial for the victims and for the protection of national security. These aspirational goals are challenging. The use of sensitive, selective, and fast analytical techniques to detect volatile biomarkers of human life has a promising future. Current developments in this field have been highlighted in this chapter. The detection of CO_2 and trace volatile compounds emitted by humans can provide evidence of a unique chemical signature of life. Small and lightweight devices are being developed that could be integrated on various platforms, including a backpack or mounted on drones, allowing for rapid sweeps of large areas. The aim of such developments is to provide a novel low-cost and lightweight field deployable instrument for use by first responders and security forces. There are still many important scientific, technical, and operational issues and challenges to be addressed and overcome for both the sensor arrays and IMS-based systems. In particular, validated signatures of human volatiles in the presence of complex chemical and dusty environments and at different altitudes, humidities, and temperatures are needed to ensure adequate specificity and improve victim location. Importantly, the direct input and requirements of first responders are crucial to device design and deployment.

References

1. Bartels SA, VanRooyen MJ. Medical complications associated with earthquakes (vol 379, pg 748, 2012). *Lancet* 2012;**379**(9817):712.
2. Donelli M. A rescue radar system for the detection of victims trapped under rubble based on the independent component analysis algorithm. *Prog Electroma Res M* 2011;**19**: 173–81.
3. Ferworn A. *Canine ergonomics: the science of working dogs*. Boca Raton: CRC Press; 2009.
4. Wong CR J. Document number 207771. In: *(NIJ) FEMAFatNIoJ*; 2004.
5. Mochalski P, King J, Unterkofler K, Hinterhuber H, Amann A. Emission rates of selected volatile organic compounds from skin of healthy volunteers. *J Chromatogr B* 2014;**959**: 62–70.
6. Ruzsanyi V, Mochalski P, Schmid A, Wiesenhofer H, Klieber M, Hinterhuber H, et al. Ion mobility spectrometry for detection of skin volatiles. *J Chromatogr B Analyt Technol Biomed Life Sci* 2012;**911**:84–92.
7. Mochalski P, Krapf K, Ager C, Wiesenhofer H, Agapiou A, Statheropoulos M, et al. Temporal profiling of human urine VOCs and its potential role under the ruins of collapsed buildings. *Toxicol Mech Methods* 2012;**22**(7):502–11.
8. de Lacy Costello B, Amann A, Al-Kateb H, Flynn C, Filipiak W, Khalid T, et al. A review of the volatiles from the healthy human body. *J Breath Res* 2014;**8**(1).
9. Mochalski P, Unterkofler K, Teschl G, Amann A. Potential of volatile organic compounds as markers of entrapped humans for use in urban search-and-rescue operations. *Trac Trends Anal Chem* 2015;**68**:88–106.
10. Boer M, Duchnik E, Maleszka R, Marchlewicz M. Structural and biophysical characteristics of human skin in maintaining proper epidermal barrier function. *Postep Derm Alergol* 2016;**33**(1):1–5.
11. Ruzsanyi V, Fischer L, Herbig J, Ager C, Amann A. Multi-capillary-column proton-transfer-reaction time-of-flight mass spectrometry. *J Chromatogr A* 2013;**1316**:112–8.
12. Mochalski P, Wiesenhofer H, Allers M, Zimmermann S, Guntner AT, Pineau NJ, et al. Monitoring of selected skin- and breath-borne volatile organic compounds emitted from the human body using gas chromatography ion mobility spectrometry (GC-IMS). *J Chromatogr B Analyt Technol Biomed Life Sci* 2018;**1076**:29–34.
13. Guntner AT, Pineau NJ, Mochalski P, Wiesenhofer H, Agapiou A, Mayhew CA, et al. Sniffing entrapped humans with sensor arrays. *Anal Chem* 2018;**90**(8):4940–5.
14. Smith S, Burden H, Persad R, Whittington K, Costello BD, Ratcliffe NM, et al. A comparative study of the analysis of human urine headspace using gas chromatography-mass spectrometry. *J Breath Res* 2008;**2**(3).
15. Obermeier J, Trefz P, Wex K, Sabel B, Schubert JK, Miekisch W. Electrochemical sensor system for breath analysis of aldehydes, CO and NO. *J Breath Res* 2015;**9**(1).
16. Staerz A, Weimar U, Barsan N. Understanding the potential of WO3 based sensors for breath analysis. *Sensors (Basel)* 2016;**16**(11).
17. Ramalingame R, Udhayakumar N, Torres R, Neckel IT, Müller C, Kanouna O. MWCNT-PDMS nanocomposite based flexible multifunctional sensor for health monitoring. *Proc Eng* 2016;**168**:1775–8.
18. Kanu AB, Hill HH. Ion mobility spectrometry detection for gas chromatography. *J Chromatogr A* 2008;**1177**(1):12–27.

19. Contreras MD, Arroyo-Manzanares N, Arce C, Arce L. HS-GC-IMS and chemometric data treatment for food authenticity assessment: olive oil mapping and classification through two different devices as an example. *Food Control* 2019;**98**:82–93.
20. Utriainen M, Karpanoja E, Paakkanen H. Combining miniaturized ion mobility spectrometer and metal oxide gas sensor for the fast detection of toxic chemical vapors. *Sensor Actuator B Chem* 2003;**93**(1–3):17–24.
21. Ruzsanyi V, Wiesenhofer H, Ager C, Mayhew CA. *Detecting people being trafficked or smuggled via the sensing of trace volatiles. IDIMT-2017 Digitalization in Management, Society and Economy.* Trauner Verlag; 2017: 127–133.
22. TOXI-triage.
23. Lourenco C, Turner C. Breath analysis in disease diagnosis: methodological considerations and applications. *Metabolites* 2014;**4**(2):465–98.
24. Das S, Pal S, Mitra M. Significance of exhaled breath test in clinical diagnosis: a special focus on the detection of diabetes mellitus. *J Med Biol Eng* 2016;**36**(5):605–24.
25. Chouinard CD, Wei MS, Beekman CR, Kemperman RH, Yost RA. Ion mobility in clinical analysis: current progress and future perspectives. *Clin Chem* 2016;**62**(1):124–33.
26. Exposure, evidence for glutathione derived thiols in breath following therapeutic radiation. In: Salman D, Eddleston M, Darnley K, Nailon W, McLaren D, Hadjithelki A, editors. *International Association Breah Research Conference*; 2018. Maastrict.

CHAPTER

Breath analysis in occupational medicine

32

Silvia Ranzieri, Massimo Corradi

Department of Medicine and Surgery, University of Parma, Parma, Italy

Chapter outline

32.1 Overview

Modern environmental and clinical biomonitoring, as well as epidemiological study of disease, traces their roots to observations from occupational medicine. Workers are well-defined groups, generally with very specific chemical or physical stressors, making any adverse health outcomes relatively straightforward to detect. Given the many environmental and occupational pollutants encountered in everyday life, and the fact that the respiratory system is an important route of their entry into the body, respiratory diseases remain one of the major causes of morbidity and mortality throughout the world. Occupational respiratory diseases are caused by materials inhaled in the workplace; the traditionally prevalent incidences of silicosis and asbestosis have now been replaced by asthma and other allergic respiratory disorders, whose frequency and relative proportion among occupational lung diseases are increasing, at least in Western nations.

Before discussing occupational exposure, it is important to emphasize the distinction between the terms "exposure" and "dose," which can be defined as follows:

Breathborne Biomarkers and the Human Volatilome. https://doi.org/10.1016/B978-0-12-819967-1.00032-3
Copyright © 2020 Elsevier B.V. All rights reserved.

Exposure is the pollutant concentration in the air at the point of contact between the body and the external environment.

Dose is the amount of the pollutant that actually crosses one of the body's boundaries and reaches a target tissue.

Inflammation is a hallmark of the pathophysiology cascade that leads to respiratory diseases, such as asthma or chronic obstructive pulmonary disease (COPD) (see Chapter 19). As such, it is important to understand the biological mechanisms and biomarkers of inflammatory and oxidative stress at the target tissue level. Such pathophysiological processes can be identified by means of a wide range of diagnostic tools, which still represent the gold standard in epidemiological and clinical settings, including bronchoscopy or bronchoalveolar lavage (BAL), or less invasive induced sputum, to sample specific burden and assess the involvement of inflammatory and structural cells (Fig. 32.1). Nevertheless, these tools cannot be extensively applied in routine clinical practice because of their relative invasiveness, which hampers their use in screening programs, exposure monitoring, or the follow-up of workers exposed to hazardous pollutants.

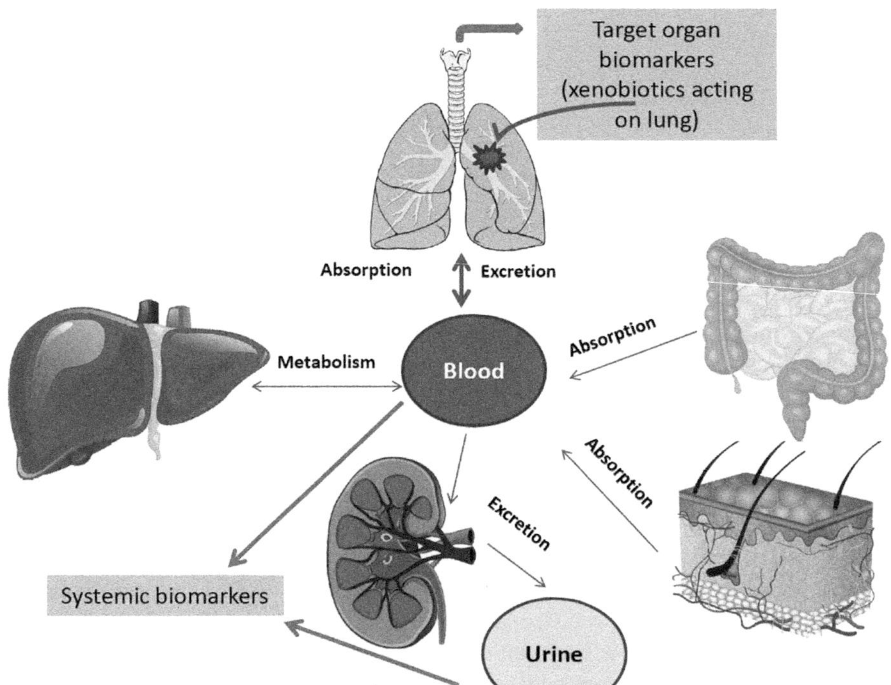

FIGURE 32.1

Kinetics of xenobiotics.

In the last decade, there has been an increased application of exhaled breath analysis, either considering exhaled gases or exhaled condensates, in occupational exposure research. Exhaled breath analysis has enormous potential as an easy, noninvasive means of monitoring inflammation and oxidative stress in the airways, particularly in nondiseased subjects, with widespread application in exploring the exposome of individuals (see Chapter 1). Breath can be collected without the need of unpleasant stimulation of the airways, as is necessary in sputum induction or BAL sampling.

While exhaled gas containing volatile organic compounds (VOCs) has been examined extensively, exhaled breath condensate (EBC), which is obtained by cooling exhaled breath under conditions of spontaneous breathing, is a relatively unexplored biological medium that could provide an assessment of lung pathobiology. EBC is a biological fluid mainly constituted by water and also containing small droplets of airway lining fluid. The interest in EBC is justified by the fact that its collection is entirely noninvasive and does not cause any discomfort or risk to examined subjects (see Chapter 7).

Exposure assessment calls for noninvasive procedures to screen presumably healthy subjects, such as solvent-exposed workers or selected groups of the general population exposed to environmental pollutants. Exposure—dose relationships and biomonitoring values for several substances have been established and can be easily converted into their ambient air equivalent. Indeed, much of current knowledge on breath analysis in respiratory medicine derives from years of experience gained in occupational settings and, more recently, from studies addressing the health effects of chemicals polluting the general environment.

This chapter reviews the main exhaled biomarkers used in clinical and experimental research for the study of occupational diseases and/or exposures. Priority is given to those biomarkers of exposure, susceptibility, and effect whose lung specificity is the highest.

32.2 Biomarkers of exposure

Conventional occupational exposure studies are based on the analysis of workplace air (ambient monitoring) and the determination of safe exposure level limits to prevent acute health effects. Despite these exposure assessments being relatively simple in occupational evaluation, it can be very complex for general conditions where low levels of exposure may occur via multiple pathways. Thus, simply monitoring ambient air levels may not be sufficient to assess health effects in nonacute conditions. Furthermore, it must be stressed that exposure and dose are two distinct phenomena, as defined above.

The most specific exposure biomarker is the direct measurement of the chemical of interest in the body, whereby much of the knowledge on this aspect derives from toxicological and occupational risk assessment perspectives, or measuring target tissue dose of pneumotoxic substances, especially from highly reactive or poorly soluble species retained by the lung for extended periods.

Exhaled breath has been applied for the identification of exposure biomarkers in different occupational/environmental settings, such as exposure to silica, toxic metals, and particulate matter. Targeted exposure in chamber studies is the subject matter of the next chapter.

32.2.1 Respirable crystalline silica

Traditionally, it has not been possible to use common biological monitoring matrices to quantify biomarkers of exposure to silica because of the relatively high concentrations of silicon (Si) in samples, such as urine, making it difficult to differentiate between unexposed and potentially exposed workers. Furthermore, measuring total silicon would not be a direct measurement of respirable crystalline silica, and if inhalation exposure had occurred, it is not known whether a urine sample would appropriately reflect this exposure.

Leese et al. performed a small pilot study that collected EBC from six quarry workers and six occupationally unexposed persons; the samples were analyzed using both single particle inductively coupled plasma mass spectrometry (spICP-MS) and transmission electron microscopy (TEM).[1] The results showed that EBC obtained from occupationally unexposed persons exhibited low background levels of dissolved silica, while silica particles of various sizes were present in samples from quarry workers.

32.2.2 Toxic metals

Respiratory diseases are strongly correlated with concentrations of atmospheric elements, and the bioaccumulation of heavy metals can be assessed noninvasively in sputum and breath. As EBC mainly consists of water that is practically free of potentially interfering solutes, it is an ideal biological fluid for elemental metal determination.

Goldoni et al. showed that EBC is a suitable matrix to assess lung dose and effects in workers exposed to cobalt (Co) and tungsten (W).[2] Specifically, data showed that concentrations of these elements in EBC were correlated with respective urinary levels, but moreover, Co in EBC positively correlated with biomarkers of lung damage, such as aldehydes derived from lipid peroxidation, which was not the case for urinary levels of Co.

The application of EBC in biological monitoring was further explored in chromium (Cr) plating workers;[3] Cr levels in EBC and urine were found to be higher in exposed subjects than in controls, and both were positively correlated with Cr levels in ambient air. Thus, both Cr in EBC (Cr-EBC) and in urine (Cr-U) may be considered valid biomarkers of exposure. Examining different correlations between variables, however, Cr-U and Cr-EBC levels evidently provide different information: only weak or no correlations were observed between Cr-U and biomarkers of effect, whereas Cr-EBC concentrations closely correlated with oxidative stress levels in EBC. Furthermore, Cr(VI) was directly measurable in EBC, thus giving an idea of the reducing power of the airways and information about the toxicokinetics of Cr(VI) in the respiratory system[4] (Fig. 32.2).

Interestingly, Hulo et al. applied EBC metal analysis in a group of welders and showed that the concentrations of manganese (Mn) and nickel (Ni) in EBC were significantly higher among welders than controls, whereas this difference was not significant for Mn in urine.[5] Felix et al. showed that following occupational exposure to lead (Pb), Pb-EBC levels appeared to reflect immediate exposure, providing a prompt signature of environmental Pb that may interact directly with lung burden.[6] Recently, Riccelli et al. investigated Cr and Ni content in EBC as biomarkers of exposure to stainless steel tungsten inert gas welding fumes.[7] Their results showed that Cr-EBC is a reliable biomarker of exposure, even at very low environmental concentrations, whereas Ni-EBC concentrations remained below the instrumental limit of detection.

32.2.3 Particulate matter

Particular matter (PM) enters the body predominantly through inhalation. After inhalation, PM is deposited in different parts of the respiratory system, depending

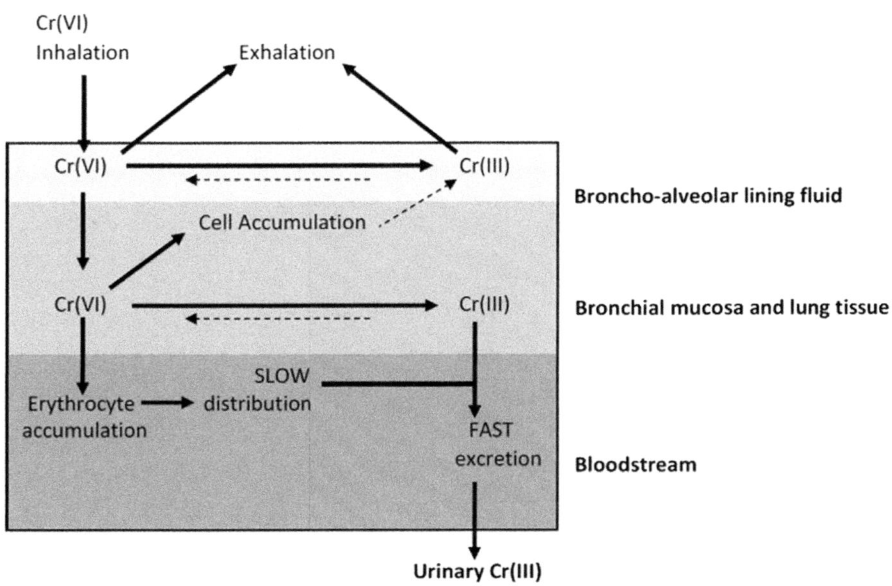

FIGURE 32.2

Kinetics of chromium (Cr(VI)) inhalation exposure.

on its chemical and physical properties: PM with aerodynamic diameter <10 μm (PM_{10}) can be inhaled and retained in the airways, whereas PM with aerodynamic diameter <2.5 μm ($PM_{2.5}$, fine particles) can penetrate deeper into the lower respiratory tract and reach the alveolar region (Fig. 32.3).

Measuring biomarkers relating to PM exposure is a difficult task because the exposure is usually chronic, with low-to-moderate exposure levels over a long period. The quantitative and qualitative analysis of particles recovered in airway samples could serve as a biological specimen for monitoring such exposure, mainly in those subjects acutely exposed to high PM levels. One device was developed to monitor the exhalation kinetics of particles in a single breath via online analysis. This application is particularly promising for studying individual susceptibility to airborne or model particles, as it allows repeated measures of different environmental pollutants in the same subjects.[9]

Recently, Fireman et al. directly measured PM in induced sputum and EBC samples from the airways of exposed individuals compared to nonexposed individuals.[10] The aim of the study was to find a suitable biomarker to monitor ultrafine particles (UFPs) accumulation in correlation with functional and inflammatory parameters. The study found that the percentage of small fraction UFP in EBC, but not induced sputum, was higher in exposed workers, and that EBC might therefore represent a sensitive medium to assess exposure to nanoparticles.

Sauvain et al. performed studies to validate a nanoparticle tracking analysis method for determining particle number concentrations and their hydrodynamic size distribution in EBC, and applied this method to EBC collected from workers exposed to soapstone or quartz dust, as compared to controls.[11] EBC from workers exposed to quartz, who experienced the largest exposure to silica particles, consistently exhibited a significantly higher concentration of particles in their EBC, with a

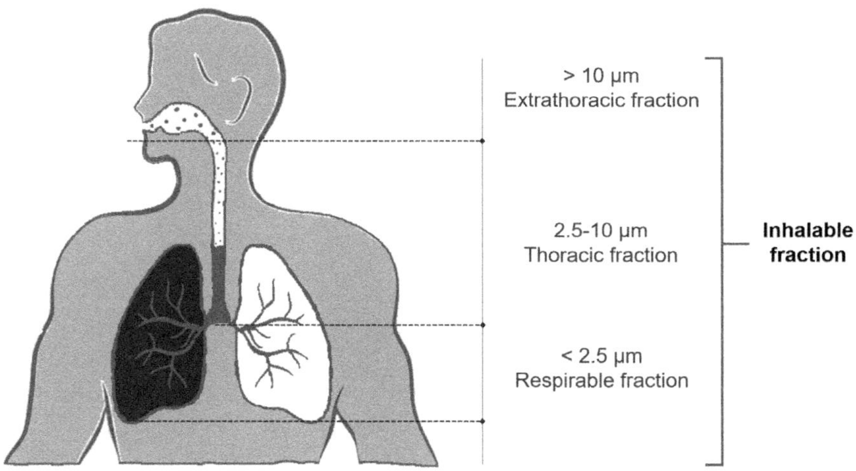

FIGURE 32.3

Particulate matter aerodynamic diameter and airways deposition.

size distribution shift toward larger values than in the other groups. Results showed that the nanoparticle tracking technique performed well for characterizing the size distribution and concentrations of particles in EBC.

A study on the exposure of airport workers to nanoparticles and metal evaluated EBC as a noninvasive representative biological matrix of the respiratory tract.[12] EBC was collected from 458 airport workers occupied on the apron (ramp workers) or in offices. Airport apron workers were exposed to higher particle number concentrations than office staff. The particulate content of EBC was ascertained by dynamic light scattering (DLS) and confirmed with scanning electron microscopy coupled to energy dispersive X-ray spectroscopy (SEM-EDS), but no difference was found between the two investigated groups.

32.3 Biomarkers of susceptibility

Biomarkers of susceptibility assess the variability of response by individuals to environmental stress, which mainly depend on the intrinsic characteristics of the individual. Biomarkers of susceptibility include genetic factors that may influence how the body interacts with a chemical, or other biological factors related to nutritional or health status, lifestyle, and life stage that may affect an individual's susceptibility to chemical exposure. A number of enzymes or other molecules exhibit polymorphisms, which may play a role in determining the extent and type of response to exposure.

Exhaled breath condensate is a suitable matrix to identify airway biomarkers of susceptibility, such as some transition elements that play a fundamental role in the respiratory chain (Fe), or are components of mitochondrial (e.g., Mn) or cytoplasmic (Cu or Zn) superoxide dismutases or glutathione peroxidase (Se). Given their key roles in the generation and detoxification of reactive oxygen species (ROS), and the scavenging of free radicals, transition elements modulate the response to toxic substances.

Mutti et al. showed that COPD patients had lower Fe-EBC and Cu-EBC levels than nonsmoker controls, but no between-group differences were observed in Mn-EBC or Se-EBC levels.[13] Similar results were reported by Ghio et al., who showed that total cigarette smoking in pack years showed a negative relationship with Fe-EBC concentration but not with Zn-EBC concentration. Fe-EBC from GOLD[a] stage II, III, and IV patients were all significantly decreased relative to those of nonsmoker and smoker controls.[14]

Metabolomics is the latest of the *omics* disciplines that employs state-of-the-art analytical instrumentation in conjunction with pattern recognition techniques to monitor and discover metabolic changes. For the reasons given above, this approach could help identify a specific profile in a patient that could act as a suitable surrogate biomarker of susceptibility.

[a] Global Initiative for Chronic Obstructive Lung Disease (GOLD).

Nuclear magnetic resonance (NMR) exploits the magnetic properties of atomic nuclei to identify molecules within a biofluid. It is an attractive technology because of its ability to provide qualitative and quantitative measurements and to study several compounds simultaneously in the same biologic fluid with minimal manipulation of the sample. NMR analysis has been applied to EBC in occupational medicine settings. Maniscalco et al. analyzed EBC samples of manual workers of a friction system plant for levels of carbon dust, phenol, formaldehyde, and VOCs; although these were found to be below regulatory limits, they were above the background levels.[15] Samples were collected from workers without wearing a mask (i.e., "exposed" to ambient air) and after wearing a mask (i.e., "not exposed"). Ten office worker controls were also included in the study for comparison. Using NMR-based metabolomics, the study found that the metabolic profile of EBC is modified after brief inhalation exposure to xenobiotics in the workplace. In particular, the effects of exposure to VOCs at ambient concentrations well below the current occupational exposure limit, or even below the detection limit, were detected. These data suggest that the metabolomics phenotype ("metabotype"), as derived from EBC, is highly sensitive as a group biomarker to this type of exposure. In addition, the different patterns of metabolites identified in different classes (controls, exposed without mask, and protected by mask) appear to stem from xenobiotics exposure only, since application of principal component analysis (PCA) within classes did not recognize outliers or distinct subsets.

Romano et al. investigated the metabolomic profiles of EBC from a group of Ni-exposed electroplater workers.[16] Ni in blood, urine, and EBC of workers and controls was measured. The ten most Ni-exposed workers were chosen for EBC metabolomic profile analysis, matched to controls by sex and smoking habits. All samples were analyzed using high-performance liquid chromatography tandem mass spectrometry (HPLC-MS/MS). Ni concentrations in EBC of the exposed workers were significantly higher compared to controls. The observation of the metabolomic profiles indicated a significantly different response pattern between the exposed and control groups.

Even if further studies are required to confirm such results, the analysis of the metabolomic profiles in the biological matrices should provide useful information, both in the clinical and prevention fields.

32.4 Biomarkers of effect

Biomarkers of effect either indicate early processes preceding disease, or predict the development and presence of disease (late effect). These biomarkers may be the products of different cellular responses following exposure to toxic substance that lead to the generation of a variety of biomolecules. In recent years, several biomarkers derived from the airways were tested as predictors for disease control, especially in asthma and COPD.

32.4.1 Nitric oxide in exhaled breath

Breath nitric oxide (NO) measurement (designated F_ENO, derived from fraction of exhaled NO) is strongly associated with eosinophilic airway inflammation. Increased levels of NO are detectable in exhaled breath of asthmatics and, to a lesser extent, in patients with allergic rhinitis, with a further increase during spontaneous or controlled asthma exacerbation and a decrease after antiinflammatory treatment (see Chapters 4 and 5).

F_ENO, though nonspecific, is increasingly used and considered more accurate than conventional tests (peak flow measurements, spirometry) for diagnostic orientation and respiratory monitoring of patients with occupational asthma.[17] In working populations, elevated F_ENO levels have been reported in sensitized subjects after specific challenge with inhaled isocyanates, latex, or in sensitized subjects exposed to laboratory animals, as well as in bakers, farmers, and health-care workers. Similarly, elevated F_ENO levels in lifeguards working in indoor swimming pools were associated with increased airway responsiveness.[18] These studies demonstrate that F_ENO measurement provides a means of assessing the effectiveness of preventative measures or treatment in the workplace.

Tafuro et al. validated biomarkers of airway inflammation in care workers of laboratory animals and found significantly elevated levels of F_ENO.[19] Comparison of F_ENO with specific parameters revealed that levels were not affected by sex, body mass index (BMI) or lung function, but they were influenced by exposure, sensitization, and age. Interestingly, F_ENO levels were >50 ppb (a cutoff value for eosinophilic airway inflammation) in many laboratory animal care workers, and concentrations were significantly correlated with hydrogen peroxide (H_2O_2) levels and surfactant protein D concentrations. It was concluded in the study that high bronchial F_ENO values should prompt occupational health practitioners to encourage workers to undergo a skin prick or methacholine test, especially when presenting symptoms.

The same group performed a cross-sectional study in employees of ham companies (exposed to storage mite allergens). F_ENO values did not correlate with age, anthropometric parameters, or lung function tests. Although there was no substantial difference in values observed in the exposed and control groups, however, those of the symptomatic sensitized participants were higher than those of healthy participants. A cutoff value distinguished the allergic and nonallergic participants in the exposed group and the control, with a sensitivity and specificity that were both close to 100%; in the population as a whole (i.e., regardless of exposure), a mean value of 22.5 ppb was found to be 100% sensitive and 99.4% specific.[20]

32.4.2 Oxidative stress biomarkers

A number of studies have shown that EBC contains significant indicators of oxidative stress. These are typically related to asthma pathogenesis, such as lower pH values, or high 8-isoprostane levels, which could be extrapolated and used for the diagnosis and monitoring of occupational asthma. One study investigated patients with suspected occupational asthma by monitoring EBC leukotriene and 8-isoprostane levels during bronchoprovocation tests with occupational allergens.[21] No changes in these parameters were found in the patients with negative bronchoprovocation tests, but control measurements (without provocation) showed significant differences in 8-isoprostane and leukotriene B4 levels within 24 h, as well as before and 5 h after the test, respectively.

Measuring EBC pH is considered a robust variable in determining the degree of EBC acidification in patients with inflammatory lung disorders, and there is evidence of airway acidification in acute asthma exacerbations from environmental exposure. Ferrazzoni et al. used EBC to detect pH in individuals with asthmatic reaction induced by isocyanates in order to test the possibility of monitoring airway responses and acidity changes after specific inhalation challenges.[22] However, no association between isocyanate-induced asthmatic reactions and EBC acidification was found. A study by Do et al. that examined airway acidity indications via EBC found an association between the duration of work and decreased pH and ammonia levels in the EBC of grain workers, but neither pH nor ammonia correlated with 8-isoprostane levels, suggesting that airway acidity and oxidative stress are distinct components of airway pathophysiology.[23]

A number of studies have investigated EBC oxidative stress measurements in asbestos-exposed (AEx) workers. This was demonstrated by Pelclova et al., for example, who reported significantly higher levels of EBC 8-isoprostane in AEx individuals compared to controls.[24] Recently, Syslova et al. tested a new and sensitive assay method consisting of pretreatment and solid-phase extraction (SPE) for the rapid and effective isolation of 8-iso-prostaglandin F2a, malondialdehyde (MDA), and 4-hydroxynonenal (HNE) from the body fluids (EBC, plasma, and urine) of patients with asbestosis, pleural hyalinosis, and silicosis.[25] Concentration differences between patients and controls were detected in all of these biomarkers in all three body fluids, although the difference in EBC was statistically most significant.

Exhaled breath condensate biomarkers of oxidative stress have also been assessed in relation to early lung damage in clay excavation workers exposed to free crystalline silica (FCS).[26] EBC 4-hydroxynonenale levels were not sufficiently sensitive to highlight a restrictive deficit, but did distinguish low and high rates of occupational exposure. Wu et al. used EBC to detect acute pulmonary responses to occupational wildfire smoke. EBC samples were collected from firefighters before, immediately after, and on the next morning following prescribed "burn" shift and regular work shifts on nonburning days.[27] There was limited evidence that smoke exposure caused an acute pulmonary response, since only the postshift 8-isoprostane concentrations were marginally higher than the preshift concentration on the burn days.

Two companion articles from the same research group examined inflammatory markers and markers of oxidative stress in EBC. Samples were collected from researchers exposed to nanomaterials generated during various tasks associated with nanocomposite synthesis, production, and postprocessing. All oxidation markers of nucleic acids and proteins were found to be elevated in the preshift

EBC samples compared to controls, and significant postshift elevations were documented in lipid oxidation markers. The most robust markers of lung inflammatory response were leukotriene B4 and tumor necrosis factor. Overall, the researchers were able to characterize the exposure to nanoparticles in term of mass concentrations and particle numbers. Since exposures to aerosols from mixed sources (originating from welding, smelting, and nanocomposite processing) were explored, however, the role of nanoparticles and other substances implicated in the lung oxidative stress and inflammatory response remains to be determined.

The current lack of uniformly agreed guidelines on malignant pleural mesothelioma (MPM) screening in AEx persons has led to the use of different imaging techniques, blood and other tests that are often executed inappropriately. Present research efforts seek a validated diagnostic blood biomarker, such as soluble mesothelin-related peptide (SMRP) or fibulin-3, but this has not been forthcoming, to date. A first-pass "rule-out" test would help to separate AEx persons at low risk (and not needing further screening) from those with early stage MPM requiring further follow-up and treatment. The analysis of VOCs in breath is a promising tool for this. A proof-of-concept study using gas chromatography—mass spectrometry GC-MS) identified cyclohexane as the discriminator in breath samples of patients with MPM compared to those with occupational AEx and nonexposed healthy controls, with 97.4% accuracy.[28] Other studies using pattern recognition with cross-reactive sensor technology discriminated patients with MPM from AEx subjects with acceptable accuracy. Using ion mobility spectrometry (IMS), patients with benign asbestos-related diseases (ARD) were discriminated from healthy controls with 99.9% accuracy, and patients with MPM were discriminated from healthy controls and AEx with 82% and 87% accuracy, respectively.[29] Finally, in a multicenter, cross-sectional, case—control study, it was shown that breath analysis using multicapillary column IMS (MCC-IMS) discriminated between participants with MPM and those with AEx, ARD, and benign non—asbestos-related lung diseases, with clinically relevant accuracy.[30]

32.5 Summary

The lung is the target organ for toxic substances that are introduced into the body by inhalation, leading to inflammation processes that can impair its functionality and induce lung diseases. For environmentally induced diseases, molecular biomarkers will play a key role in understanding the relationships between exposure to toxic environmental chemicals, the development of chronic human diseases, and identifying those individuals at increased risk for disease. Many direct approaches to evaluate lung inflammation in diseases have been proposed, including bronchial lung biopsies and lavages, but these methods are invasive. In recent years, less invasive techniques, such as induced sputum and exhaled breath analysis, have been successfully introduced for the evaluation of lung diseases. Biomarkers are key molecular or cellular events that link a specific environmental exposure to a health outcome.

A number of tools and techniques are available for measuring airway biomarkers of exposure, effect, and susceptibility for diagnostic, prognostic, and preventive strategies. New biomarkers of exposure in exhaled breath have been applied for targeting pneumotoxic substances at organ level, thereby improving the understanding of local consequences resulting from highly reactive or poorly soluble species retained by the lung for extended periods. The biological monitoring of Cr and Ni in occupationally exposed workers, for instance, is traditionally based on urine or blood measurements; but, although these reflect systemic metal exposure, they provide little information about pulmonary levels. Moreover, urinary Cr is completely reduced to a trivalent state and therefore provides no information on the real exposure to the carcinogenic fraction of Cr. EBC can be collected easily and noninvasively and is a suitable matrix for assessing lung tissue doses or detecting biomarkers of exposure and effect. EBC can also be used to measure the Cr(VI) fraction directly, despite its potential reduced to Cr(III) by airway lining fluid, since it may also persist in some workers for many hours after exposure and cause oxidative damage in the lungs. MDA and H_2O_2 are reliable and stable markers of airway oxidative stress, and their use in occupational medicine may help to assess any imbalance between the generation of ROS and antioxidative defences.

In the area of biomedical research, the technical advances in analytical equipment have spurned a huge boost to the field, with equipment that allows molecular features to be measured more precisely and in a more sensitive manner; the question of which technology or method is close to making it into clinical practice, however, remains unanswered. Although much progress has been made to identify potential biomarkers, the challenge remains to validate the accuracy, reproducibility, specificity, and sensitivity of biomarkers and to assess the feasibility and cost-effectiveness of applying biomarkers in large population-based studies. A wide spectrum of biomarkers is currently being incorporated as a clinically useful parameter; nevertheless, improvements in the field of biomarker research to gain a better understanding of the pathophysiology of diseases are still required. The application of proteomics and metabolomics approaches is already available and, in combination with genomic studies, will likely identify novel candidate biomarkers for noninvasive exploration of occupational and/or environmental exposures.

References

1. Leese E, Staff JF, Carolan VA, Morton J. Exhaled breath condensate: a novel matrix for biological monitoring to assess occupational exposure to respirable crystalline silica. *Ann Work Expo Heal* August 1, 2017;**61**(7):902–6.
2. Goldoni M, Catalani S, De Palma G, Manini P, Acampa O, Corradi M, et al. Exhaled breath condensate as a suitable matrix to assess lung dose and effects in workers exposed to cobalt and tungsten. *Environ Health Perspect* 2004;**112**(13):1293–8.

3. Caglieri A, Goldoni M, Acampa O, Andreoli R, Vettori MV, Corradi M, et al. The effect of inhaled chromium on different exhaled breath condensate biomarkers among chrome-plating workers. *Environ Health Perspect* 2006;**114**(4):542–6.
4. Goldoni M, Caglieri A, De Palma G, Acampa O, Gergelova P, Corradi M, et al. Chromium in exhaled breath condensate (EBC), erythrocytes, plasma and urine in the biomonitoring of chrome-plating workers exposed to soluble Cr(vi). *J Environ Monit* 2010;**12**(2):442–7.
5. Hulo S, Chérot-Kornobis N, Howsam M, Crucq S, de Broucker V, Sobaszek A, et al. Manganese in exhaled breath condensate: a new marker of exposure to welding fumes [Internet] *Toxicol Lett* 2014;**226**(1):63–9.
6. Félix PM, Almeida SM, Franco C, Almeida AB, Lopes C, Claro MI, et al. The suitability of EBC-Pb as a new biomarker to assess occupational exposure to lead. *Int J Environ Health Res* January 2, 2015;**25**(1):67–80.
7. Riccelli MG, Goldoni M, Andreoli R, Mozzoni P, Pinelli S, Alinovi R, et al. Biomarkers of exposure to stainless steel tungsten inert gas welding fumes and the effect of exposure on exhaled breath condensate. *Toxicol Lett* [Internet] 2018;**292** (November 2017): 108–114.
8. *Burning opportunity: clean household energy for health, sustainable development, and wellbeing of women and children*. World Health Organization; 2016. p. 1–113.
9. Goldoni M, Caglieri A, De Palma G, Longo S, Acampa O, Poli D, et al. Development and set-up of a portable device to monitor airway exhalation and deposition of particulate matter. *Biomarkers* August 2009;**14**(5):326–39.
10. Fireman E, Edelheit R, Stark M, Shai AB. Differential pattern of deposition of nanoparticles in the airways of exposed workers. *J Nanoparticle Res* 2017;**19**(2).
11. Sauvain JJ, Suarez G, Edmé JL, Bezerra OMPA, Silveira KG, Amaral LS, et al. Method validation of nanoparticle tracking analysis to measure pulmonary nanoparticle content: the size distribution in exhaled breath condensate depends on occupational exposure. *J Breath Res* March 1, 2017;**11**(1).
12. Marie-Desvergne C, Dubosson M, Touri L, Zimmermann E, Gaude-Môme M, Leclerc L, et al. Assessment of nanoparticles and metal exposure of airport workers using exhaled breath condensate. *J Breath Res* 2016;**10**(3).
13. Mutti A, Corradi M, Goldoni M, Vettori MV, Bernard A, Apostoli P. Exhaled metallic elements and serum pneumoproteins in asymptomatic smokers and patients with COPD or asthma. *Chest* 2006;**129**(5):1288–97.
14. Ghio AJ, Soukup JM, McGee J, Madden MC, Esther CR. Iron concentration in exhaled breath condensate decreases in ever-smokers and COPD patients. *J Breath Res* August 21, 2018;**12**(4).
15. Maniscalco M, Paris D, Melck D, Chiariello N, Di Napoli F, Manno M, et al. Biomonitoring of workers using nuclear magnetic resonance-based metabolomics of exhaled breath condensate: a pilot study [Internet] *Toxicol Lett* 2018;**298**(September): 4–12.
16. Romano A, Aiani MR, Brambilla M, Brambilla P, Carcano M, Galbiati E, et al. Metabolomic profiles of exhaled breath condensate of 39 nickel exposed workers [Internet] *G Ital Med Lav Ergon* 2012;**34**(3 Suppl. 1):682–6.
17. van Kampen V, Brüning T, Merget R. Serial fractional exhaled nitric oxide measurements off and at work in the diagnosis of occupational asthma. *Am J Ind Med* 2019; **62**(8):663–71.

18. Demange V, Bohadana A, Massin N, Wild P. Exhaled nitric oxide and airway hyperresponsiveness in workers: a preliminary study in lifeguards. *BMC Pulm Med* 2009;**9**:1–9.
19. Tafuro F, Selis L, Goldoni M, Stendardo M, Mozzoni P, Ridolo E, et al. Biomarkers of respiratory allergy in laboratory animal care workers: an observational study [Internet] *Int Arch Occup Environ Health* 2018;**91**(6):735–44.
20. Tafuro F, Ridolo E, Goldoni M, Montagni M, Mutti A, Corradi M. Work-related allergies to storage mites in Parma (Italy) ham workers. *BMJ Open* 2015;**5**(5):8–10.
21. Klusackova P, Lebedova J, Kacer P, Kuzma M, Brabec M, Pelclova D, et al. Leukotrienes and 8-isoprostane in exhaled breath condensate in bronchoprovocation tests with occupational allergens. *Prostaglandins Leukot Essent Fatty Acids* 2008;**78**(4–5):281–92.
22. Ferrazzoni S, Scarpa MC, Guarnieri G, Corradi M, Mutti A, Maestrelli P. Exhaled nitric oxide and breath condensate pH in asthmatic reactions induced by isocyanates [Internet] *Chest* 2009;**136**(1):155–62.
23. Do R, Bartlett KH, Dimich-Ward H, Chu W, Kennedy SM. Biomarkers of airway acidity and oxidative stress in exhaled breath condensate from grain workers. *Am J Respir Crit Care Med* 2008;**178**(10):1048–54.
24. Pelclová D, Fenclová Z, Kačer P, Kuzma M, Navrátil T, Lebedová J. Increased 8-isoprostane, a marker of oxidative stress in exhaled breath condensate in subjects with asbestos exposure. *Ind Health* 2008;**46**(5):484–9.
25. Syslová K, Kačer P, Kuzma M, Najmanová V, Fenclová Z, Vlčková Š, et al. Rapid and easy method for monitoring oxidative stress markers in body fluids of patients with asbestos or silica-induced lung diseases. *J Chromatogr B Anal Technol Biomed Life Sci* 2009;**877**(24):2477–86.
26. Corradi M, Folesani G, Robuschi B, Selis L, Riccelli MG, Andreoli R, et al. Non-invasive techniques to assess restrictive lung disease in workers exposed to free crystalline silica. *Med del Lav* 2019;**110**(2):83–92.
27. Wu CM, Adetona A, Song C (Chuck), Naeher L, Adetona O. *Measuring acute pulmonary responses to occupational wildland fire smoke exposure using exhaled breath condensate*. Archives of Environmental and Occupational Health; 2019.
28. Lamote K, Brinkman P, Vandermeersch L, Vynck M, Sterk PJ, Van Langenhove H, et al. Breath analysis by gas chromatography-mass spectrometry and electronic nose to screen for pleural mesothelioma: a cross-sectional case-control study. *Oncotarget* 2017;**8**(53): 91593–602.
29. Lamote K, Vynck M, Van Cleemput J, Thas O, Nackaerts K, Van Meerbeeck JP. Detection of malignant pleural mesothelioma in exhaled breath by multicapillary column/ion mobility spectrometry (MCC/IMS). *J Breath Res* September 26, 2016;**10**(4).
30. Lamote K, Vynck M, Thas O, Van Cleemput J, Nackaerts K, van Meerbeeck JP. Exhaled breath to screen for malignant pleural mesothelioma: a validation study [Internet] *Eur Respir J* 2017;**50**(6).

CHAPTER

Breath biomarkers in human xenobiotic exposure studies

33

Michael C. Madden

United States Environmental Protection Agency, Human Studies Facility, Chapel Hill, NC, United States

Chapter outline

33.1 Overview

Intentional exposures of humans to xenobiotic agents involve risks and benefits, ideally with the benefits outweighing the risks. A particular benefit of study designs that intentionally alter a participant's behavior is that they can offer essential insights on the effects of an agent on human biological outcomes and provide biological plausibility to epidemiological findings. Controlled exposure studies of healthy human volunteers to ozone (O_3), for example, were utilized extensively to set the initial O_3 regulatory standards. Equally, while epidemiological studies of ambient outdoor particulate matter (PM) pollution initially pointed to increased morbidity and mortality, controlled exposures and short-term field panel studies provided the underpinning of plausibility of the observed effects. Exposure studies can address diverse pursuits, for example, allowing exposure to an agent or pollutant, limiting medicine intake, or imposing a specific diet, and so forth. Different types of health effects are still being studied for inhaled indoor and outdoor agents, including extrapulmonary ones, such as biological effects induced in utero at different windows of development. The responsiveness of a population, and hence the size estimation of sensitive subgroup(s), can be discerned from

Breathborne Biomarkers and the Human Volatilome. https://doi.org/10.1016/B978-0-12-819967-1.00033-5
Copyright © 2020 Elsevier B.V. All rights reserved.

properly designed controlled and field studies. The mechanism or mechanisms behind the sensitivity can also be examined with these approaches. Clearly, despite some drawbacks, there is much to be gained from human exposure studies; for an extensive discussion of their advantages and disadvantages, the reader is directed to the literature,[1] since this is beyond the scope of this chapter.

Measurements made in the exhaled breath of exposed human subjects—the gaseous and nongaseous components—can be utilized to not only measure exposure biomarkers, but also determine how the lung and extrapulmonary organs and tissues respond by examining biomarkers of effect. Additionally, biomarkers of sensitivity can be factored into the study to determine potential predisposing factors that can make the dose or response exaggerated or even attenuated.

Besides air pollutant exposures, other xenobiotic exposures have been studied using controlled exposures. Swimming is common, with exposure to disinfection by-products (DBPs) such as trihalomethanes, whereby the effects on health are still uncertain, especially in relation to the development of asthma. Additionally, the current increasing trend in vaping and cannabis use has led to major concerns on exposure-induced health effects in users and bystanders, whether inhalation is worse than tobacco smoke alone or in combination with tobacco smoke, and the potential risk when nicotine levels are manipulated to greater doses than in tobacco smoke.

This chapter introduces the concept of human exposure studies, specifically in relation to utilizing exhaled breath to study exposure biomarkers and biomarkers of effect. Exhaled breath biomarkers from uncontrolled exposures, such as with retrospective epidemiological designed studies, are not covered in this chapter, nor are breath biomarkers from occupational exposures, which are the subject matter of the previous chapter. Table 33.1 lists the markers of exposure and of effects that have been utilized in exhaled breath and are discussed in this chapter.

33.2 Ambient air pollution

Ambient air pollution is generally ubiquitous, and the public health impacts are enormous. The World Health Organization (WHO) estimates that exposure to pollutants, such as PM, O_3, nitrogen dioxide (NO_2), and other pollutants, induces 4.2 million premature deaths per year. In addition, these exposures induce other primarily cardiopulmonary effects, such as hospitalizations for lung infections, asthma symptoms, and adverse cardiac events (arrhythmias, atrial fibrillation, etc.). Outside levels of some pollutants, e.g., O_3 concentrations, are increasing in some locations. Indoor air pollution is also as important a health issue as outdoors; the WHO estimates 3.8 million annual deaths worldwide from indoor household air pollutants mainly derived from cooking processes.

For centuries, various agencies charged with control of air pollution have implemented regulations, or at a minimum recommend concentrations, that sought to minimize public health impacts. As early as 1306, London issued a proclamation banning the use of sea coal in order to control smoke levels.[1] In attempting to establish regulations to protect human health, studies were performed where subjects

Table 33.1 Markers of exposure and markers of effect for different exposure scenarios in this chapter.

Marker	Abbreviation/ components	Biological process or emission source
Markers of exposure		
Volatile organic compounds	VOCs	Variety of sources: Anthropogenic and natural
Ozone	O_3	Photochemistry involving sunlight, heat, VOCs, NOx
Oxides of nitrogen	NOx: NO_2, NO	Man-made, fuel combustion
Particulate matter	PM	Generally coarse particles natural fine or smaller anthropogenic
Disinfection by-products	DSPs	Water additives, Br and/or Cl derived
Markers of effect		
Alkanes	Ethane, pentane, isoprene	Lipid peroxidation
Malondialdehyde	MDA	Lipid peroxidation, reactive oxygen species (ROS)
Isoprostane	8-Isoprostane, others	Lipid peroxidation, free radical species
Nitric oxide	F_ENO, NO	Inflammation (pro or anti); nitric oxide synthase
pH		Airway acidification
Carbon monoxide	$_ECO$, e%COHb	Exogenous combustion source emissions/ endogenous heme oxygenase

were exposed—intentionally and voluntarily—to air pollutants, either as a mixture or to a sole xenobiotic alone. These early intentional exposures generally attempted to establish concentration-response curves so that agencies could determine exposure levels that would not induce health effects and/or would impact an acceptably low number of exposed people. The general rationale for, and uses of, performing controlled exposures to air pollutants is further discussed in depth in a separate publication to which the reader is referred.[2]

Three approaches have been utilized in studying effects of exposures on human health and responses:

1. *Controlled exposures*, where the study design keeps pollutant levels relatively constant as the treatment,
2. *Field studies*, where pollutant levels can be variable but the study design provides some degree of control and treated exposures, and
3. *Retrospective epidemiological studies*, where controlling factors are very limited at best.

The first two approaches involve intentional alterations of participants' behavior including exposures, while the last approach does not alter subjects' behavior at the time of study. Hybrid designs of controlled and field approaches can also be performed.

One of the earliest controlled air pollutant exposure studies with human subjects dates back almost 70 years ago and utilized gaseous sulfur dioxide (SO_2) inhalation.[3] The findings of that study showed that in healthy subjects not regularly exposed to SO_2, the pulse rate and breathing frequency increased, while tidal volume (volume of exhaled breath) decreased when breathing 1 or 5 ppm SO_2. Interestingly, in boiler workers regularly exposed to this pollutant, the same physiological responses were attenuated, suggesting an adaptation derived from repeated exposures. Later, human volunteers were exposed to PM from whole diesel exhaust (gases plus PM) in a controlled manner[4] with no changes in lung physiology demonstrated. These examples show that intentional, controlled exposures of humans to air pollutants have been utilized for decades to study the types of responses, as well as the dose or concentration at which responses occur. Typical exposure systems are shown in Fig. 33.1.

Breath biomarkers have been utilized as surrogates of disease processes to examine the effects of the air pollutants on the volunteers' responses. As a surrogate marker, the biomarker reflects some stage between xenobiotic exposure and

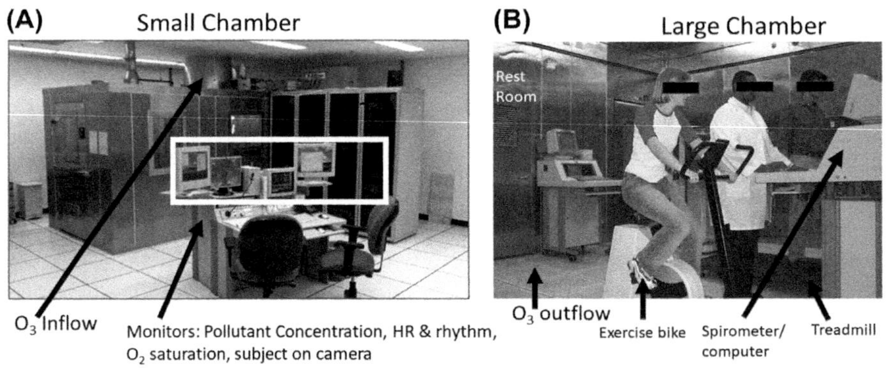

FIGURE 33.1

Typical airborne pollutant exposure systems. A small chamber, in this case for O_3 exposure, is shown on the left (A) and a larger one on the right (B). In a small chamber (A), one individual is exposed for a relatively brief duration with or without exercise on a stationary bike. Both chambers have an adjacent station outside where staff can monitor pollutant levels, physiological parameters, and the individual via a camera and/or window. Additionally, both chambers have antechambers to minimize pollutant level decreases when an individual enters the exposure chamber. Large chambers (B) can expose multiple individuals for longer periods (due to the presence of a rest room) with measurement equipment, such as a spirometer, inside the chamber. O_3 outflow is through the panels on the floor.

clinical disease. One of the earliest examples of this strategy was not to study effects of air pollutant toxicants on subjects per se, but to study "poisons" derived from human breath. Human exhaled breath condensate (EBC) was collected in a series of studies and injected into animal models to study anaphylaxis[5]; the conclusions by the investigators was that the poison(s) was/were not only likely protein in nature, either volatile or carried in solution in the "watery vapor," but also highly variable, appearing in only five of eight subjects' EBC. In the latter half of the 20th century, when controlled human pollutant exposures had been underway for more than two decades, most studies that analyzed exhaled breath used O_3 exposures in part due to the popularity of exposure studies based on the concerns of toxic effects of this main component of photochemical oxidant pollutants. Initially, the gas phase of exhaled breath was examined. The seminal reports of Dillard[6] in 1978 and Foster[7] in 1996 showed that breath pentane and isoprene were noninvasive markers of lung injury, such as peroxidation induced by the O_3 exposure. EBC was examined also from O_3-exposed subjects in a controlled exposure chamber setting for hydrogen peroxide (H_2O_2) and carbonyls as evidence of lung ozonolysis and/or reactive oxygen species (ROS) production.[8] Hydrogen peroxide was increased post-O_3 exposure by 72% if the subjects exercised (increasing the dose of O_3 to the lungs) compared to air exposure with exercise; no increase was noted without exercise. In intermittently exercising subjects, an analytical peak putatively identified as the dinitrophenylhydrazone derivative of acetaldehyde increased by 126% immediately post-O_3 exposure compared to 26% post air. With the successful use of exhaled breath to demonstrate changes induced by O_3 exposures with steady-state concentrations in inhalation chambers, the sensitivity of specific genotypes was examined to determine if O_3 induced greater changes than other genotypes. Individuals with a combination glutathione-S-transferase M1 null with AD(P)H:quinone oxidoreductase wild type had increases in EBC markers of oxidative damage (malondialdehyde, 8-isoprostane) and lipid mediator leukotriene B4.[9]

33.3 Controlled air pollution exposure studies

In another study with an individual pollutant exposure, normal, healthy volunteers were exposed to 0.24 ppm O_3 for 4 h with intermittent exercise, with a "sham" exposure of 0.04 ppm.[10] Values of fractional exhaled nitric oxide (F_ENO) and EBC pH were compared to baseline preexposure values. Both the O_3 and sham exposures induced an increase in pH immediately postexposure, which could be due to an exercise-induced effect. F_ENO decreased post-O_3 exposure (whereas the sham exposure had no change). Both markers returned to baseline levels by 16 h postexposure. The authors caution that the decreased F_ENO level may be possibly due to the reaction of O_3 with NO.

Exposures of healthy volunteers to complex pollutant mixtures in a controlled fashion have also utilized exhaled breath analyses for biological effects.

Using traffic particles ("soot"), an increase in F_ENO was the only endpoint observed; there were no changes in EBC nitrate, nitrite, or interleukin-6.[11] In another study, exposure to wood smoke did not induce an increase in 8-isoprostane but did increase malondialdehyde (MDA)[12]; both substances are indicators of lipid peroxidation.

33.4 Air pollution field studies

Field studies of the effects of air pollution exposures have incorporated both candidate and global biomarker approaches. As examples, volunteers were intentionally exposed to different levels and types of air pollutants at farms to well-trafficked urban sites in the Netherlands.[13] F_ENO was associated with the coarse and fine PM oxidative potential from these sites. In a small cohort ($n = 4$ volunteers), a 2 h rush hour Georgia highway drive was used to examine changes in the profile of thousands of proteins versus precommute samples.[14] Most EBC features (>60%) were present in either plasma or saliva, showing promise for use of EBC as a surrogate for other fluids; however, EBC also showed fewer unique components than the other two fluids. EBC components showed better sensitivity for a change in protein components after commuting, with a decrease after the day 1 commute.

Breath analyses have shown promise for noninvasive studies aimed at protective health interventions against air pollution exposures. Using traffic as the pollution source, individuals were driven in a car on a major New Jersey highway while wearing a face mask that filtered particles with a high-efficiency particle air (HEPA) filter at >99.9% efficiency or no filter as the control.[15] Use of the filter prevented immediate postexposure increases in EBC nitrite (nitrite + nitrate)[a] induced by PM exposure. No increase in MDA was observed from the traffic pollution exposure. Because no changes in cardiac parameters were observed between the filtered and unfiltered exposures, the NO production may itself represent a sensitive marker of traffic PM exposure. The lack of effects of the traffic pollutants on the heart may be due to relatively low PM concentrations and/or a low dose due to a lack of physical activity inside the car. In a similarly designed study, N95 face masks were worn by volunteers who walked along a heavily trafficked Beijing street, and their exhaled NO and EBC inflammation-associated cytokine levels were compared with no face mask present.[16] When wearing a face mask, F_ENO was attenuated ~38% immediately postwalk, and three EBC cytokines (out of six examined) were attenuated almost completely at 6 or 24 h postwalk compared to no face mask levels. Other biomarkers examined—urinary MDA and endothelial dysfunction (measured by pulse wave analysis of the radial artery)—were not significantly altered with pollutant exposure, possibly attesting to the sensitivity of the exhaled breath markers.

[a] Nitrate and nitrite were used as surrogate markers of lung NO production.

Investigators have incorporated physical exercise in their field study protocol designs in order to deliver more air pollutant dose to the subjects. Athletes in San Paulo ran five consecutive days for 45 min/day either through an urban forest or on urban streets, and EBC pH assessed post days one and five.[17] Runners in the more polluted street environment exhibited nasal mucociliary impairment and no change in EBC pH relative to their responses after running in the forest setting where EBC pH increased, which may play a role in preventing mucociliary clearance changes. Once again in a different study, $F_{E}NO$ was found to be one of the few sensitive biomarkers, decreasing 24 h after air pollution exposure (measured as black carbon) in three European cities (Antwerp, Barcelona, London).[18] $F_{E}NO$ also increased 2 h after exposure when physical activity was a factor combined with black carbon measurements but not with either factor alone.

Exhaled breath condensate was used for analyses of markers of exposure and effect in designed field study approaches. Wildland firefighters using no respiratory protection (in Georgia, USA) had EBC collected before and after (immediately and next day) planned burns. Only 8-isoprostane was increased slightly immediately post-exposure among the analytes examined, suggesting a mild lung inflammation.[19] In examining exposures to indoor cooking emissions based on carbon monoxide (CO) levels, in a cross-sectional study in Bangladesh, exhaled CO and indoor air levels were measured and then calculated as exhaled % carboxyhemoglobin (%eCOHb) and blood %COHb.[20] There was poor correlation between the two %COHb measurements, suggesting that improvements were needed to obtain noninvasive markers of CO exposure. Nevertheless, %eCOHb seemed to be somewhat dose responsive in that the %eCOHb increased with the number of meals cooked per day. EBC pH was not outside the normal range and did not correlate well with %eCOHb.

33.5 Breath biomarkers of vaping and cannabis smoking inhalation: primary and secondary exposures

Increased use of inhaling electronic cigarette (e-cigarettes) vapor, or vaping, and the increased use of inhaling cannabis due to legalization in several States in the USA, has elevated concerns about the safety of such practices. Long-term effects from the use of these two substances, as well as secondary exposures of bystanders, are unknown at present, and it is unclear if any effects would differ from tobacco smoke exposure. Nevertheless, even relatively "short-term" use of less than 2 years of vaping (when the vaping duration was known) can induce lung inflammation, as indicated by a neutrophilia in bronchoalveolar lavage fluid (BAL),[21] as can marijuana smoking,[22] giving credence to health concerns of these substances. Cannabis smoking is in some ways similar to inhaling ambient airborne PM and associated gases, but at an increased dose it could likely induce effects similar to air pollutant exposures by similar mechanisms.[23] Exhaled breath analyses can play a role in determining pharmacokinetics of components in the smoke and vapor, as well as providing overall exposure levels, especially during cessation periods.

The elimination kinetics of Δ(9)-tetrahydrocannabinol (THC) in exhaled breath was examined over 3 h in two studies. The first study showed elimination of THC to baseline in the 3 h span, although there was a wide interindividual variation in the kinetics, suggesting factors that can alter metabolism.[24] The metabolite 11-nor-9-carboxy-THC was not detected in breath. A second study also demonstrated a return to baseline after 3 h for THC in exhaled breath and blood after smoke inhalation of one cigarette.[25] Blood THC concentrations were highly correlated with exhaled breath measurements. The authors also observed high interindividual elimination kinetics. The findings from these studies suggest that breath analyses may be appropriate for examining uptake and elimination of THC from inhalation or ingestion. The detection of cannabis use via breath analysis is further discussed in Chapter 30.Cross-ref chapter 30

Exhaled CO ($_{E}CO$) in relation to cigarette smoking has been used mainly to therapeutically monitor tobacco smoking reductions. Individuals smoking e-cigarettes as a tobacco substitute had their $_{E}CO$ levels monitored for up to 6 weeks[26] or for as little as 5 days.[27,28] In the latter two studies, $_{E}CO$ decreased in the complete cessation and the e-cigarette—only groups and not the dual-use group (tobacco plus e-cigarette) compared to tobacco smoke baseline. The tobacco smokers who ceased or used solely vaping also had an increase in $F_{E}NO$,[28] which is usually suppressed in active cigarette smoking, possibly due to an inhibition of nitric oxide synthase (NOS). These observations indicate that exhaled breath measurements of NO and CO could be used to monitor decreased tobacco smoke exposure in users and secondary uptake in bystanders. The association of $F_{E}NO$ and/or $_{E}CO$ in relation to smoking activity is further discussed in Chapters 5, 6 and 19. Vaping in a controlled manner has also been observed to result in a lower exhaled breath ethylene concentration in active e-cigarette users compared to concentrations produced by active tobacco smokers.[29] Ethylene is therefore another potential exposure biomarker for tobacco smoking cessation compliance.

Aside from CO, NO, and ethylene, carbonyls and phenolics in exhaled breath from individuals vaping or smoking tobacco were examined as potential exposure markers.[30] Exhaled breath analyses of these classes of compounds showed essentially ambient indoor air concentration after vaping of two commercial e-cigarettes, in part due to the lower concentrations of these compounds produced by vaping. Conventional tobacco cigarette smoking resulted in detectable levels of total carbonyls at greater than background air concentrations. Phenolics were present in cigarette smoke but not detectable in exhaled breath, potentially due to near 100% deposition. The exhaled breath of nonvaping subjects in a room with no ventilation did not show increases in VOCs or nicotine that originated from vaping participants,[31] again supporting the concept of low doses of vaping emissions to bystanders.

33.6 Other exposures

Exhaled breath collection and analyses has been utilized with other intentional exposures for a variety of situations. Some examples are provided to demonstrate the utility of incorporating breath analyses into study designs for hazard identification, mechanistic insights (mode of action, adverse outcome pathway (AOC), etc.), risk assessment, and therapeutic monitoring.

Exhaled breath was collected from divers who reached a depth of 9 m for 60 min breathing either air or 100% oxygen (O_2). Sampling at 4 h post dive, the 100% O_2 exposure group produced increased gas-phase compounds (relative to a baseline pre-dive measurement); three of the compounds were methylated alkanes.[32] The production of these compounds induced by the 100% O_2 exposure may be due to a lipid peroxidation process and/or inflammation.

Several studies have examined uptake and elimination of DBPs and biological markers from swimming pool exposures using exhaled breath analyses, among other analyses. Other non-DBP compounds were also identified with the pool exposures via a global omics approach with biomarkers of exposure. These studies were aimed at validating known DBPs from the kinetics as well as finding unique ones. Additionally, potential biomarkers of sensitivity were incorporated into these studies to better understand the magnitude of biological responses and elimination kinetics.

Asthmatics were chosen for study with DBP exposures due to a concern over exacerbation of symptoms and also to examine potential links between the exposure and the induction of the disease. Using elite swimmers, production of mainly alkanes and carbonyls (putatively derived from lipid peroxidation) was captured from exhaled breath.[33] Both groups exhibited decreased production post swimming, though the nonasthmatic subjects tended to have a greater decrease. These findings show similarities of response in both groups, though it is unclear if the pathophysiology of the disease was a variable in the production and exhalation of the compounds examined. In a cross-sectional study that defined children as current, past, or nonswimmers, F_ENO was examined as a potential marker of eosinophilic inflammation.[34] Current swimmers had a greater F_ENO than nonswimmers, suggesting inflammation, along with a greater response to a beta-2 agonist (i.e., a bronchodilator) as measured as an improvement in forced expiratory volume in 1 s (FEV_1), which is typically seen in asthmatic subjects. The development of functional asthma in children was associated with early swimming (before the age of three). Use of breath markers such as F_ENO can potentially shed light on the mechanisms of asthma induction.

Using breath metabolomics, 293 features were found with at least one swimming exposure, whereby 13 of those features were identified as tryptophan-related metabolites.[35] The number of metabolic features in exhaled breath was close to that found in the water or urine, suggesting that breath collection and analyses provide adequate sensitivity compared to other compartments. There was no clear separation of these features between the physical activity and the actual chemical exposures.

In order to better understand whether swimming pool trihalomethane DBPs enter the human body primarily from inhalation or dermal routes with physical activity, "professional swimmers" were sampled after a 40 min period in pools that used either chlorine or bromine for disinfection (converted to mainly chloroform and bromoform, respectively, in the air).[36] Exposure while bathing served as a control. Trihalomethanes were measured in exhaled breath, air, and water. There was an exercise-related association with the trihalomethane levels in exhaled breath, based either on distance swam or energy expended. Using this study design, the authors were able to discern that air trihalomethane concentrations, but not water concentrations, were correlated with exhaled breath levels. These findings suggested that inhalation was the major route of intake. Greater bromination also had better retention in the lung, likely due to lower volatility. Using a similar study design but with nonprofessional swimmers and incorporating urine analyses, the same research group determined that sensitivity factors, such as sex, played a role in trihalomethanes dose.[37] Male subjects tended to incorporate more chloroform and trihalomethanes as seen in exhaled breath levels, while females incorporated more trichloroacetic acid. Uptake was not affected by age, body mass index (BMI), or body surface area. Some individuals with specific metabolizing genotypes also had different trihalomethanes levels in exhaled breath; for instance, glutathione-S-transferase T1 null genotypes had higher breath chloroform concentrations, while two SNPs mutations in the cytochrome P450 enzyme 2E1 had higher breath bromotrihalomethane levels. These controlled field studies demonstrated that some sensitivity factors can have an influence on DBP dose. The variations in the kinetics of elimination were examined in another study[38] using exhaled breath and demonstrated that some individuals (two out of 19) had a slow elimination for specifically dibromochloromethane and chloroform, but not bromodichloromethane, which was rapidly eliminated in all subjects. This finding again points to the existence of potentially sensitive subjects due to retention of the compounds. This study also reported the latent appearance of two potentially unique markers of the exposure at 500–600 min post swimming out of >600 volatile compounds. In another study, models identified over 1700 genes plus 23 microRNAs in blood associated with an exposure to at least one DBP while swimming.[39] Many of the microRNA are associated with bladder and colon cancer. Exhaled breath was utilized to determine exposure levels.

33.7 Summary

With several types of xenobiotic exposures ranging from air pollutants to swimming pool pollutants, to smoke or vapor inhalation, exhaled breath measurements have been used to indicate the level of the exposure and some effects over the last 6–7 years. Gas-phase markers appear to predominate the reports in the scientific literature, with F_ENO appearing frequently due to concerns related to the induction and development of asthma, as well as the exacerbation of symptoms in asthmatics. The use of F_ENO as one of the more sensitive effects biomarkers has or is moving past the validation stage in clinical and controlled exposure studies and into epidemiological studies. The incorporation of potential biomarkers of sensitivity in exhaled breath studies will help focus the efforts onto a more select population so thatstudy cohorts do not have to be as large.

Conversely, potentially sensitive populations could be unmasked by the use of breath analyses for pharmacokinetics of pollutants. Big data handling is now also evident among exhaled breath studies. The improvements in technology and software have allowed investigators to utilize *omics* approaches to examine markers of exposure and effects more comprehensively based on breath exposure markers.

Future research in the field of breath markers with pollutant exposures is likely to see an increasing move to use of clinical studies/controlled exposure studies of validated markers to more field research. Exposure to wildfire pollution, extensive in recent years in the United States, Australia, and Southeast Asia, as some examples, has led to many studies utilizing exhaled breath markers for putative health effects. The breath biomarkers that appear promising in air pollution–related studies can also be utilized in occupational settings, especially as evidence by many studies examining workers exposed to nanoparticles. These types of exposure—ambient airborne pollutants and occupational particle exposures—will provide ample opportunity for breath markers to make an important scientific impact as to measuring variable doses in real-world settings in addition to the biological responses from these exposures as dose-responsive and/or sensitivity-influenced responses.

Disclaimer

The views expressed in this chapter are those of the author and do not necessarily represent the views or policies of the U.S. Environmental Protection Agency. This is a work of the U.S. Government and is not subject to copyright protection in the United States. Foreign copyrights may apply.

References

1. Brimblecombe P. *The big smoke: a history of air pollution in London since medieval times*. New York: Methuen and Co., Ltd.; 1987. p. 9–10. 185 pp.
2. National Academies of Sciences, Engineering, and Medicine. *Controlled human inhalation-exposure studies at EPA*. Washington, DC: The National Academies Press; 2017. p. 158.
3. Amdur MO, Melvin Jr WW, Drinker P. Effects of inhalation of sulphur dioxide by man. *Lancet* October 10, 1953;**265**(6789):758–9.
4. Battigelli MC. Effects of diesel exhaust. *Arch Environ Health* February 1965;**10**:165–7.
5. Rosenau MJ, Amoss HL. Organic matter in the expired breath. *J Media Res* September 1911;**25**(1):35–84.
6. Dillard CJ, Litov RE, Savin WM, Dumelin EE, Tappel AL. Effects of exercise, vitamin E, and ozone on pulmonary function and lipid peroxidation. *J Appl Physiol Respir Environ Exerc Physiol* December 1978;**45**(6):927–32.

7. Foster WM, Jiang L, Stetkiewicz PT, Risby TH. Breath isoprene: temporal changes in respiratory output after exposure to ozone. *J Appl Physiol (1985)* February 1996;**80**(2): 706–10.
8. Madden MC, Hanley N, Harder S, Velez G, Raymer JH. Increased amounts of hydrogen peroxide in the exhaled breath of ozone-exposed human subjects. *Inhal Toxicol* 1997;**9**: 317–30.
9. Corradi M, Alinovi R, Goldoni M, Vettori M, Folesani G, Mozzoni P, et al. Biomarkers of oxidative stress after controlled human exposure to ozone. *Toxicol Lett* August 5, 2002; **134**(1–3):219–25.
10. Hoffmeyer F, Sucker K, Monsé C, Berresheim H, Jettkant B, Rosenkranz N, et al. Different patterns in changes of exhaled breath condensate pH and exhaled nitric oxide after ozone exposure. In: Pokorski M, editor. *Environment exposure to pollutants. Advances in experimental medicine and biology*, vol. 834. Cham: Springer; 2014.
11. Chen X, Sun Y, Zhao Q, Song X, Huang W, Han Y, et al. Design and characterization of human exposure to generated sulfate and soot particles in a pilot chamber study. *J Air Waste Manag Assoc* April 2016;**66**(4):366–76.
12. Murgia N, Barregard L, Sallsten G, Almstrand AC, Montuschi P, Ciabattoni G, et al. 8-isoprostane in exhaled breath condensate after experimental exposure to wood smoke in humans. *J Biol Regul Homeost Agents* 2016 ;**30**(1):263–70.
13. Janssen NA, Strak M, Yang A, Hellack B, Kelly FJ, Kuhlbusch TA, et al. Associations between three specific a-cellular measures of the oxidative potential of particulate matter and markers of acute airway and nasal inflammation in healthy volunteers. *Occup Environ Med* January 2015;**72**(1):49–56.
14. Ladva CN, Golan R, Greenwald R, Yu T, Sarnat SE, Flanders WD, et al. Metabolomic profiles of plasma, exhaled breath condensate, and saliva are correlated with potential for air toxics detection. *J Breath Res* December 6, 2017;**12**(1):016008.
15. Laumbach RJ, Kipen HM, Ko S, Kelly-McNeil K, Cepeda C, Pettit A. A controlled trial of acute effects of human exposure to traffic particles on pulmonary oxidative stress and heart rate variability. *Part Fibre Toxicol* November 1, 2014;**11**:45.
16. Guan T, Hu S, Han Y, Wang R, Zhu Q, Hu Y, et al. The effects of facemasks on airway inflammation and endothelial dysfunction in healthy young adults: a double-blind, randomized, controlled crossover study. *Part Fibre Toxicol* July 4, 2018;**15**(1):30.
17. Cavalcante de Sá M, Nakagawa NK, Saldiva de André CD, Carvalho-Oliveira R, de Santana Carvalho T, Nicola ML, et al. Aerobic exercise in polluted urban environments: effects on airway defense mechanisms in young healthy amateur runners. *J Breath Res* 2016;**10**(4):046018.
18. Laeremans M, Dons E, Avila-Palencia I, Carrasco-Turigas G, Orjuela JP, Anaya E, et al. Short-term effects of physical activity, air pollution and their interaction on the cardiovascular and respiratory system. *Environ Int* August 2018;**117**:82–90.
19. Wu CM, Adetona A, Song CC, Naeher L, Adetona O. Measuring acute pulmonary responses to occupational wildland fire smoke exposure using exhaled breath condensate. *Arch Environ Occup Health* 2020;**75**(2):65–9.
20. Lee A, Sanchez TR, Shahriar MH, Eunus M, Perzanowski M, Graziano J. A cross-sectional study of exhaled carbon monoxide as a biomarker of recent household air pollution exposure. *Environ Res* November 2015;**143**(Pt A):107–11.
21. Thakrar PD, Boyd KP, Swanson CP, Wideburg E, Kumbhar SS. E-cigarette, or vaping, product use-associated lung injury in adolescents: a review of imaging features. *Pediatr Radiol* January 2, 2020.

22. Barbers RG, Gong Jr H, Tashkin DP, Oishi J, Wallace JM. Differential examination of bronchoalveolar lavage cells in tobacco cigarette and marijuana smokers. *Am Rev Respir Dis* June 1987;**135**(6):1271–5.
23. Ghio AJ, Soukup JM, Madden MC. The toxicology of air pollution predicts its epidemiology. *Inhal Toxicol* 2018 Aug - Aug;**30**(9–10):327–34.
24. Coucke L, Massarini E, Ostijn Z, Beck O, Verstraete AG. Δ(9)-Tetrahydrocannabinol concentrations in exhaled breath and physiological effects following cannabis intake – a pilot study using illicit cannabis. *Clin Biochem* September 2016;**49**(13–14):1072–7.
25. Lynch KL, Luo YR, Hooshfar S, Yun C. Correlation of breath and blood Δ^9-tetrahydrocannabinol concentrations and release kinetics following controlled administration of smoked cannabis. *Clin Chem* September 2019;**65**(9):1171–9.
26. Rohsenow DJ, Tidey JW, Martin RA, Colby SM, Eissenberg T. Effects of six weeks of electronic cigarette use on smoking rate, CO, cigarette dependence, and motivation to quit smoking: a pilot study. *Addict Behav* May 2018;**80**:65–70.
27. D'Ruiz CD, Graff DW, Robinson E. Reductions in biomarkers of exposure, impacts on smoking urge and assessment of product use and tolerability in adult smokers following partial or complete substitution of cigarettes with electronic cigarettes. *BMC Publ Health* July 11, 2016;**16**:543.
28. O'Connell G, Graff DW, D'Ruiz CD. Reductions in biomarkers of exposure (BoE) to harmful or potentially harmful constituents (HPHCs) following partial or complete substitution of cigarettes with electronic cigarettes in adult smokers. *Toxicol Mech Methods* July 2016;**26**(6):443–54.
29. Popa C. Infrared spectroscopy study of the influence of inhaled vapors/smoke produced by cigarettes of active smokers. *J Biomed Optic* May 2015;**20**(5):051003.
30. Long GA. Comparison of select analytes in exhaled aerosol from e-cigarettes with exhaled smoke from a conventional cigarette and exhaled breaths. *Int J Environ Res Publ Health* October 27, 2014;**11**(11):11177–91.
31. van Drooge BL, Marco E, Perez N, Grimalt JO. Influence of electronic cigarette vaping on the composition of indoor organic pollutants, particles, and exhaled breath of bystanders. *Environ Sci Pollut Res Int* February 2019;**26**(5):4654–66.
32. Wingelaar TT, van Ooij PAM, Brinkman P, van Hulst RA. Pulmonary oxygen toxicity in navy divers: a crossover study using exhaled breath analysis after a one-hour air or oxygen dive at nine meters of sea water. *Front Physiol* January 25, 2019;**10**:10.
33. Couto M, Barbosa C, Silva D, Rudnitskaya A, Delgado L5, Moreira A, et al. Oxidative stress in asthmatic and non-asthmatic adolescent swimmers-A breathomics approach. *Pediatr Allergy Immunol* August 2017;**28**(5):452–7.
34. Cavaleiro Rufo J, Paciência I, Silva D, Martins C, Madureira J, Oliveira Fernandes E, et al. Swimming pool exposure is associated with autonomic changes and increased airway reactivity to a beta-2 agonist in school aged children: a cross-sectional survey. *PLoS One* March 12, 2018;**13**(3). e0193848.
35. van Veldhoven K, Keski-Rahkonen P, Barupal DK, Villanueva CM, Font-Ribera L, Scalbert A, et al. Effects of exposure to water disinfection by-products in a swimming pool: a metabolome-wide association study. *Environ Int* February 2018;**111**:60–70.
36. Marco E, Lourencetti C, Grimalt JO, Gari M, Fernández P, Font-Ribera L, et al. Influence of physical activity in the intake of trihalomethanes in indoor swimming pools. *Environ Res* July 2015;**140**:292–9.

37. Font-Ribera L, Kogevinas M, Schmalz C, Zwiener C, Marco E, Grimalt JO, et al. Environmental and personal determinants of the uptake of disinfection by-products during swimming. *Environ Res* August 2016;**149**:206–15.
38. Heaney LM, Kang S, Turner MA, Lindley MR, Thomas CLP. Evidence for alternative exhaled elimination profiles of disinfection by-products and potential markers of airway responses to swimming in a chlorinated pool environment. *Indoor Air* December 8, 2019.
39. Espín-Pérez A, Font-Ribera L, van Veldhoven K, Krauskopf J, Portengen L, Chadeau-Hyam M. Blood transcriptional and microRNA responses to short-term exposure to disinfection by-products in a swimming pool. *Environ Int* January 2018;**110**:42–50.

CHAPTER

Canine olfaction

34

Roger Giese[1], Klaus Hackner[2,3]

[1]*Department of Pharmaceutical Sciences and Barnett Institute, Northeastern University, Boston, MA, United States;* [2]*Department of Pneumology, University Hospital Krems, Karl Landsteiner University of Health Sciences, Krems, Austria;* [3]*Department of Internal Medicine II, Division of Cardiology, Medical University of Vienna, Vienna, Austria*

Chapter outline

Breathborne Biomarkers and the Human Volatilome. https://doi.org/10.1016/B978-0-12-819967-1.00034-7
Copyright © 2020 Elsevier B.V. All rights reserved.

34.1 Overview

The companionship and loyalty of domestic dogs towards people are well known. Interest has been building in the role for the dog as "clinical helper," in areas such as cancer sniffing and medical alerting. This is based on the unique olfactory capabilities of the dog. This chapter presents the concept of canine olfaction for biomarker detection, reports on progress in the field, and offers an outlook and recommendations for future developments. The reader is directed to a recent, thorough review by Jezierski et al. for further reading.[1]

Outside the scope of this chapter are uses of dogs for search and rescue, tracking criminals, detecting drugs or explosives, and providing disability assistance (e.g., guide dogs for blindness, hearing loss). These applications of dogs are widely practiced.

34.2 Nature of the dog

The highly developed olfactory abilities of dogs are evident. The canine olfactory system functions like a chemical sensor: it detects, identifies, and discriminates different types of odors. Further, dogs may demonstrate or develop *drive*, which is defined as the likelihood to perform a pattern of behavior based on certain stimuli. *Drive* may be increased or decreased externally relative to the dog, but not created or eliminated.[2] Humans take advantage of the combination of highly developed olfactory senses, eagerness to learn, and *drive* of dogs to train them for olfactory tasks, such as target location or scent discrimination. In several studies, dogs reportedly have been trained to discriminate odors from cancer and healthy specimens.[1] This skill has been sensationalized by the media and sometimes described as a "dog doctor giving a diagnosis." From a medical viewpoint, the phenomenon is not a diagnosis but a potential screening strategy.

A dog's sniffing response to a detected odor is of a binary character (*yes* or *no*). This makes it clear, aside from the issue of reliability, that using sniffer dogs for cancer screening and prophylaxis could have good prospects, if dogs are able to alert on very early (preclinical) cancer. This is a dream for dog-based diagnostics in the cancer arena, that a dog alert could trigger concern before detection by other means, thereby identifying people who should be monitored more carefully. Even a sensitivity (positivity in disease) of 80% or so might be valuable, as long as specificity (negativity in nondisease, i.e., few false positives) is very high (above 95%). Deciding these numbers is difficult, however, and no doubt dependent on the type of cancer and treatment available. It is hard to balance the trade-off of falsely scaring some people about cancer versus missing early detection in others. Sensitivity and specificity are discussed below in more detail.

There is a second general dream for dog diagnostics, discussed later in this chapter, and that is their help in discovering new biomarkers for cancer that can be measured instrumentally. In this scenario, the dog puts itself out of business. This is important since dogs are so complicated, like humans.

Another area of canine signaling concerns pathological events, such as seizures, glycemic aberrations, and migraines. These applications of dog sniffing can be very different from cancer detection because it is not clear how much the dogs are relying on odor versus changes in a patient's movement, behavior, or a combination of these.

When using canine scent discrimination for human diagnostics, the same parameters of detection accuracy need to be applied as for other analytical tests or instruments: *sensitivity* and *specificity* (Table 34.1). Dogs may give a correct positive indication for a sample having the disease of interest (True Positive, TP), or a correct negative for a sample lacking this disease (True Negative, TN). By the same token, a false positive indication may be given a nontarget sample (False Positive, FP), or a positive sample may be missed by the dog (False Negative, FN). Using these parameters, sensitivity and specificity then can be calculated (Table 34.1).

Sensitivity describes the ability of a diagnostic test to identify a disease of interest without missing anyone by leaving the disease undiagnosed. Thus, a highly sensitivity test has a high rate of true positives. Specificity describes the ability of a diagnostic test to be correctly negative in the absence of a disease without mislabeling anyone. Thus, a highly specific test has a high rate of true negatives. An ideal diagnostic test by canine olfaction would provide both high sensitivity and specificity, and thereby be rated high in detection accuracy.

These parameters enable an evaluation of the effectiveness of a given test and provide the background for comparative analysis. Especially for cancer detection studies, sensitivity and specificity must be evaluated and reported. Too many canine cancer studies have failed to do this.

Table 34.1 Calculation of sensitivity and specificity.

		Test sample	
		Disease	*No Disease*
Dogs indication	*Positive indication*	**True positive (TP)**	**False positive (FP)**
	Negative indication	**False negative (FN)**	**True negative (TN)**

Sensitivity = TP/(TP + FN), Specificity = TN/(TN + FP).

The situation is even worse for reports on canine response to other medical conditions (e.g., seizures, hypoglycemia), where little or no information is reported about sensitivity and specificity. These reports mainly have relied on observational parameters based on questionnaires or case reports, which lack depth or scope to report sensitivity and specificity.

The service dog industry, where trained dogs are sold for medical alerting, deserves some criticism, too. Individuals who buy such a dog expect that the dog will notify them in advance of, or during, a medical episode. The more responsible training organizations do not guarantee that the dogs will alert, since the present state of research indicates that not all dogs have the capability to detect such events in advance. Training methods for this are not standardized, and there is a lack of proof-of-concept trials. Some of the service dog industry may be a scam. The user should be cautious, and not pay for an overpriced companion dog.

The ensuing text presents an overview of the literature on canine scent detection for human diagnostics of cancer, infections, hypoglycemia, and seizures. The studies are evolving rather than mature.

34.2.1 First observations

Studies using canines for cancer detection were initially triggered in 1989 by a published letter to the editor of the *The Lancet* involving a case of spontaneous detection of a dermal cancer by an untrained dog. This report contained a description by a patient who first became aware of a lesion on her thigh because her dog constantly sniffed at it, frequently spending several minutes a day doing this and on one occasion even trying to bite it off, according to the report. When it finally got excised, the lesion turned out to be a malignant melanoma.[3] The authors, Pembroke and Williams, hypothesized that dogs are able to detect cancer in humans through an odor signature. Twelve years later, a similar report from Church and Williams appeared, adding the detail that the dog lost its interest in the area of the melanoma after the lesion was excised.[4] Church later organized a conference in 2003 in Saunderton, UK, to discuss the use of canines for cancer detection, and how to train such dogs. This conference also launched a period of experimental studies on cancer detection by trained dogs.

34.2.2 Cancer

During 2004 to 2018, at least 20 peer-reviewed studies were published in medical and animal behavioral scientific journals presenting results of cancer detection by trained canines.[5–27] Also, seven reviews[1,28–33] and one theoretical evaluation of the hypothesis that dogs can smell cancer[34] have been published. Different types of cancer have been investigated by canine scent detection: melanoma,[16] bladder,[6] breast,[27] lung,[25] prostate,[20] ovarian,[24] colorectal,[26] cervix,[26] and hepatocellular carcinoma.[23] Most frequently, patient breath samples, or combined breath and urine or breath and feces, were studied. Other odor samples included tissue scraps, urine, or blood plasma. It is difficult to summarize or compare the outcomes of these studies since they differed methodically in many ways, so they will be largely described individually.

The majority of cancer detection studies by trained canines was directed at lung cancer. Reported sensitivities varied from 56% to 99%, and specificities from 32% to 90%.[8,14,18,21,25] In 2006, McCulloch et al. reported an extraordinarily high detection rate of lung cancer in samples of exhaled breath, namely with a sensitivity and specificity of 99%.[8] The authors trained five dogs and used polypropylene (PP) tubes for breath collection. Montes et al. used tubes with hydrophobic and hydrophilic wool layers for sample collection, and reported a sensitivity of 95% and specificity of 98%.[25] The latter authors used an experimental setup similar to that of McCulloch et al., with a lineup of five containers with one known positive ("lineup of one in five"), and use of only one dog. Ehmann et al. reported a sensitivity of 71% and specificity of 93% for detection of lung cancer using breath samples collected with PP tubes. The results were reportedly independent of tobacco smoking, food odors, or chronic obstructive pulmonary disease (COPD). The authors concluded that canine scent detection could be a promising method for lung cancer screening, although they acknowledged that the number of subjects in their study was limited.[14]

Less promising results were obtained in two studies using a double-blind scenario with a randomized selected trial lineup, which differed from the prior studies. For non—small cell lung cancer, Hackner et al. reported a sensitivity of 79% and specificity of 35%.[21] Amundsen et al. showed similar results for sensitivity (70%) but an even lower specificity of 8.3%, and further reported that, after additional intense training, specificity increased to 33%, but sensitivity decreased to 60%.[18]

For detection of melanoma, Pickel et al. trained two dogs to identify used tissue scraps hidden in boxes. They reported a sensitivity of 100% (no specificity reported).[5]

For breast cancer, screening methods, such as computer tomography (CT) and mammography are available, at least in high-income countries. Canine scent detection might become a complementary method. In 2006, McCulloch reported a sensitivity of 88% and specificity of 98% using breath samples for breast cancer, using the same dogs trained to detect lung cancer. Again, a lineup of one positive out of five was used, and the breath samples were collected in PP tubes.[8] Gordon et al. used urine samples to train dogs for breast cancer detection, but achieved detection sensitivities worse than chance (sensitivity 22%). The authors stated that further studies should add more stringent training protocols, and a better management of urine samples.[9] In 2018, Thuleau et al. used compresses that had been worn overnight to discriminate breast cancer from healthy controls in a one out of four lineup.[27] It was reported that the two Belgian shepherds achieved a sensitivity of 90.3% (no specificity was reported), and the authors concluded that further studies with larger cohorts are needed.

More promising results for urine samples were reported for detection of prostate cancer. In 2011, Cornu et al. trained a Belgian Malinois and reported a value of 91% for both sensitivity and specificity.[13] According to the authors, when one of the subjects thought to be falsely classified as having prostate cancer by the dog was rebiopsied, prostate cancer was found. Three years before, in 2008, Gordon et al. was not able to achieve similar good results with four dogs and a reported sensitivity below chance (overall 18%).[9] The dogs in this latter study, however, were from breeds not commonly used as sniffer dogs (miniature goldendoodle, Pembroke Welsh corgi, Chihuahua mix, and border collie), and the specific training methods were not recorded systematically; they were left to the dog trainers. In 2014, Taverna et al. trained two dogs on urine samples for detection of prostate cancer and reported a very high detection accuracy (sensitivity 99.2%, specificity 97.1%), but no details on the training and testing were provided. In a follow-up trial in 2015, the same authors added further information and conducted an even larger study set (more than 300 prostate cancer urine samples). Training and testing were done with different samples, and a variable number of reference cancer samples during the test runs were used. Again, very high detection accuracy was reported with both dogs, ranging from 98.6% to 100% for sensitivity and 97.6%−98.7% for specificity.[20] In contrast, Elliker et al. reported training on urine samples for prostate cancer with ten dogs, but only two were able to complete the training successfully. However, the sensitivity of these dogs was very low (13%−25%), and the specificity was moderate (71%−75%).[19] As stated by the authors, a possible reason might be poor generalization of dogs on prostate cancer odors, i.e., the dogs did not build up much of an ability to detect the hypothetical odor components that "all" prostate urines share in common.

Urine samples were also used to discriminate bladder cancer from healthy controls. Willis et al. used a one out of seven lineup and reported an overall sensitivity of 41%.[6]

A research group from Sweden extensively studied canine scent detection of ovarian cancer. In 2008, Horvath et al. trained a dog to differentiate tissue samples (tumor, abdominal fat, and muscle) and reported a sensitivity of 100% and specificity of 97.5%.[10] In a follow-up trial in 2010, tissue samples and blood samples were used and the previously trained giant schnauzer, together with another detection dog, again were reported to achieve a very high detection accuracy (sensitivity 100%, specificity 98%).[11] Another study from the same group in 2013 used the same dogs and reported similar high detection rates for blood and tissue samples, along with a discrimination of samples during treatment as well as 3 and 6 months after treatment.[17] In a similar approach, Guerrero-Flores et al. trained a beagle to discriminate tissue samples, smears on glass slides, and surgical bandages of cervical cancer from healthy controls.[17,24] Overall detection accuracy was reported to be high (sensitivity and specificity >90%) for both the fresh smear samples and the bandages.

Although detection rates for early colorectal cancer have improved due to recommendations for regular screening examinations, such as colonoscopy, beginning at age 50, and other tests for this disease, Sonoda et al. investigated the use of canine scent detection in colorectal carcinoma. The authors trained a Labrador retriever and used breath samples and watery stool samples. A lineup of one in five was chosen, and the sensitivity was reported as 91%, and the specificity as 99%, for breath samples, along with a sensitivity of 97% and a specificity of 99% for stool samples. Detection accuracy was reported to be even higher for early-stage cancers.[12] In 2018, Seo et al. performed a cross detection study with cell cultures of colorectal cancer and breast cancer.[26] Two dogs were trained for either specimen and then tested on the other cancer, with the cell cultures and control samples placed in boxes with a hole. With a lineup of one out of five, the authors reported a sensitivity of 92% −94% and a specificity of 98% for both dogs. It was concluded that cancer cells share a metabolic waste that, in turn, makes them share an odor (yet unknown).

In 2018, Kitiyakara et al. trained a golden retriever to discriminate breath samples of hepatocellular carcinoma patients from healthy controls in a one out of four lineup.[23] Sensitivity was reported as 78% and the authors considered that they had achieved a proof of concept.

Finally, dogs were even evaluated to detect cancer within their own genus, for detection of canine urinary tract transitional cell carcinoma (TCC) in urine.[35] Although all four sniffer dogs were reported to master the training, during the testing with new samples only one dog was reported to correctly distinguish TCC from healthy controls. During double-blind tests, however, this dog was not able to indicate the cancer samples more frequently than expected by chance.

34.2.3 Infections

In 2012, Bomers et al. first described a dog trained to identify *Clostridium difficile* in stools with a reported sensitivity and specificity of 100%, and in patients (during detection rounds) with a sensitivity of 83% and a specificity of 98%.[36] The same authors said they strengthened their results 2 years later with another study on *C. difficile* detection by a trained dog during an outbreak in a Dutch hospital, and reported that the dogs correctly identified 12 out of 14 cases (sensitivity 86%, specificity 97%).[37] Later, in 2018, Taylor et al. evaluated the use of dogs as a point-of-care (PoC) tool for *C. difficile* in stool.[38] They compared the performance of two dogs: specificity was reported to be essentially the same for the dogs (84% and 85%), but sensitivity varied (78% and 93%). The authors stated that the practical value of dogs as a PoC tool for this purpose is limited, compared to molecular diagnostic tests.[38]

An ongoing study involves trained dogs for the diagnosis of malaria in asymptomatic school children in sub-Saharan Africa.[39] The training and testing samples are nylon socks worn overnight (15−24 h). A preliminary sensitivity around 70% and a specificity >90% for both trained dogs in the discrimination of malaria-infected children is reported, but details on the dog training and methodology have yet to be published.[39]

Koivusalo et al. published a study as a letter to the editor in 2017 about canine scent as a tool to distinguish methicillin-resistant *Staphylococcus aureus* in bacterial cultures.[40] The authors reported high specificity and sensitivity values for different bacterial streams. Nevertheless, they recommend that the gold-standard method (bacterial culture) should continue to be used.[40]

Detection of bacteriuria in urine by canines was studied by Maurer et al.[41] Indeed, sniffing urine is an innate behavior of dogs. Several lineups were carried out, including single- and double-blinded testing, and "zero trials" where only negative samples were presented. They reported sensitivities near 100% and specificities above 90% for different bacteria, such as *Escherichia coli*, *Enterococcus*, or *Klebsiella*.[41]

Overall, compared to cancer trials, canine olfaction studies for infections are in the minority. Again, the methodology used is not standardized, and confirmation studies with a larger number of samples are needed for each type of specimen. Nevertheless, in some cases, where the study methods were more fully defined, promising results have been reported.

34.2.4 Hypoglycemia

Canine alerting to hypoglycemic episodes has been the subject of articles for almost two decades. Chemical phenomena have been linked to this ability. It has been suggested that the dogs use their olfactory skills to react to olfactory changes in the chemical composition of their owners' sweat. However, reactions to behavioral alterations may play a role, and perhaps even fully explain the sensing by the dog.

In 2000, Chen et al. described three case reports of dogs that were able to detect hypoglycemic symptoms of their owners by behaving in a strange way. This occurred before the individuals were aware of any hypoglycemic symptoms. In two of the cases, the dogs even undertook actions to wake their owners during a nocturnal episode.[42] A further case report described a dog that repeatedly alarmed his owner of hypoglycemia during driving,[43] and detection of severe hypoglycemia in a nondiabetic patient was reported.[44]

Wells et al. later reported a study of 212 dog owners with primarily type 1 diabetes.[45] Thirty-six percent of the owners believed that their dogs reacted most of the times when they had low glucose levels. The dogs' behavioral responses varied and most dogs behaved by attracting their owners' attention, e.g., by vocalizing, licking them, or nuzzling. Some dogs showed responses suggestive of fear, including trembling, running away, or hyperventilation. The authors stated that increased sweating by people during hypoglycemic episodes might be a possible mechanism. However, it cannot be ruled out that dogs may respond to cues other than olfactory ones, such as subtle changes in their owner's mood, and the signals triggering a response may vary from one dog to another.

A study by Rooney et al. in 2013 involved patients with type 1 diabetes and dogs trained by Medical Detection Dogs, a UK organization.[46] The patients lived with trained dogs between 4 months and 7 years. Both the dogs' alerting behavior and blood sugar levels of the patients were recorded routinely and during diabetic episodes. It was reported that blood sugar levels from eight out of ten individuals taken after a dog's alert were significantly more likely to be out of range than during routine tests. Reports of unconsciousness and paramedic calls decreased with dog ownership. However, the study depended on self-reporting; there was a variable accuracy of times the dogs would alert; and the companionship provided by the dog could have played a role.

Another study, published in the same year, used an online survey to retrieve data from 36 patients with diabetic alert dogs, including parents of diabetic children. The authors reported significant decreases in the frequency of severe and moderate hypoglycemia and improvement of psychosocial outcomes, such as worry about hypoglycemia and improved quality of life due to diabetic alert dogs.[47] Another retrospective study, using an online survey that included 135 persons, was conducted by Petry et al. in 2015. Individuals who sought to train a dog to detect low glucose levels on their own, without professional involvement, were included. The authors reported that owners of self- and professionally trained dogs were very convinced about their dogs' abilities to alert them, while owners of dogs that learned entirely on their own reported more missed hypoglycemic episodes.[48]

A study by Dehlinger et al., which notably described the experimental setup and did not rely on self-reporting, showed less impressive results.[49] Three dogs were trained to push a bell when sniffing a container housing a swab taken from a subject during a hypoglycemic episode, but not for swabs taken during normoglycemic periods. The owners believed that all dogs were consistently able to detect hypoglycemia, but the study outcome revealed that correct indications ranged between 50% and 58%, with an overall sensitivity of 56% and an overall specificity of 53%. The authors concluded that future studies should include assessment of behavioral aspects of the subject and that the swabs should only be taken by the usual companion of the dogs.

The mechanisms (including olfactory changes and behavior alterations) involved in canine use for detection of hypoglycemia are unclear, so further research based on prospective and double-blinded studies is needed.

34.2.5 Seizures

Dogs have been reported to be sensitive to epileptic episodes, responding and alerting during or after a seizure. Rudy, a veterinarian, stated in a short note in 1995 that dogs might alert to slight changes in body language or very subtle changes in a person's odor prior to a seizure, and that this could happen even hours before an episode.[50]

In 2002, Strong et al. conducted a prospective study on tonic-clonic seizures and the use of seizure alert dogs for a period of 48 months, involving baseline, training, and follow-up periods.[51] The authors reported a decrease of seizure frequency to a 3-month mean of 9.7 (1.7–37) during the training phase, and a further drop at the follow-up period to 8.5 (2–30).[51] The authors stated that only dogs selected for their suitability, and with appropriate structured training, should be used, since there are reports of some dogs becoming highly agitated and aggressive during seizures.[52] In 2004, Kirton et al. performed a survey which indicated that 42% of the participants reported dogs with seizure-related behavior, such as licking (59%) or whimpering

(36%) in 20% of the families. The median anticipation time was 2.5 min (range 10 s–5 h) and there was no report of overt aggression.[53] In 2008, Kirton et al. also reported that spontaneous alerting behavior developed in 59% of their dogs, with 46% onset within 4 weeks and the remainder evolving over months.[54]

In 2005, videotaping was performed on two patients with seizure alert dogs at the epilepsy care unit of a Philadelphia hospital.[55] The authors found that the seizure alert of the first patient's dog was poor, and they believed that the dog-alerting of the second patient even contributed to her nonepileptic seizures and increased the frequency of the events. As a limitation, an epilepsy care unit might not be an ideal location to test a seizure dog for its capabilities, for example, other people in nearby rooms might distract the dogs. The authors suggested that visual cueing is the main mechanism for seizure anticipation, and that olfactory sensing seems less likely.[55]

Similar skepticism has been raised about studies on psychogenic nonepileptic seizures and dog alerting with the same concern that these patients might have seizures in response to certain canine behaviors. Self-reported and performance-based measurements in these patients often do not correlate with objective measurements.[56,57]

A very recent study, reported by Catala et al., performed a similar lineup trial as in the cancer detection studies, with five trained dogs finding the body odor (in opaque cans) from a seizure, with six other samples from the same patient, collected randomly during other occasions, including sports. The sensitivity reported ranged from 67% to 100% and the specificity from 95% to 100%. Three out of five dogs were claimed to give 100% sensitivity and specificity.[58]

Although there is no clear explanation for seizure alerting, the ability of dogs to perceive and interpret various human signals in this situation would not be surprising.[59] Furthermore, olfactory signals might result from neurotransmitters and hormones released during or prior to an epileptic seizure. Body smell might be modified by impaired nervous cell activity. A few studies have mentioned the possibility that dogs could alert on subjects other than their masters.

To date there have been no convincing, systematic clinical trials that establish the ability of dogs to alert ahead of seizures. Assuming that the dogs can do this, further investigations would be important to elucidate the mechanisms.[60]

34.3 Training

General concepts for dog training are provided below, with more specific recommendations made in Section 34.7.

34.3.1 Basic concepts

The underlying principle for canine olfaction training is a progressive sequence of classical conditioning, operant conditioning, reinforcement, and stimulus control.

The classical or Pavlovian conditioning uses a neutral stimulus paired with an unconditioned stimulus (one that gives a natural response, i.e., the response arises without any training). In Ivan Pavlov's own famous experiments in the 1920s, food was the unconditioned stimulus and a metronome was the neutral stimulus, which becomes associated with the presence of food for his experimental dogs.[61] The unconditioned stimulus (food) then led to an unconditioned response (salivation) in the presence of a neutral stimulus (metronome). Subsequently, the unconditioned response (salivation) could be elicited simply by the presence of the neutral stimulus (metronome) in the absence of the unconditioned stimulus (food). This basic reflex is a precondition for further odor detection learning.

During the initial odor training phase, the presence of an odor is followed by an unconditioned reinforcement, such as a toy or food. The odor then becomes a stimulus (like the metronome) and the food or toy becomes the unconditioned stimulus.[62] The conditioned response may vary, depending on the reinforcement, and may be salivation, scratching, or barking. Timing is of critical relevance. Maximum conditioned response may be available when the odor is presented and the food or toy is then presented while the odor is still present.[62] The timing aspects of the trials is also important (a trial is a period of time without interruption where a dog is sniffing one or more lineups of samples). There are three timing aspects to consider: trial duration, spacing (intermissions) between trials, and number of trials per day. Conducting a few training trials with several minutes break between each trial has been reported to be favorable for an imprinting effect compared to many *en bloc* trials.[63] Is the best strategy to subject the dog to a series of brief trials with intermittent breaks over an extended period (e.g., four 5 min trials with 40 min breaks over a period of 3 h, once per day), or to subject the dog to something like one 20 min trial per day? Perhaps either strategy, or one in between these examples, can be effective. Currently there is no convincing evidence to help make this decision.

While classical conditioning is central for the imprinting process during odor detection, to reach the desirable alerting behavior a different type of conditioning becomes important: operant conditioning. In classical conditioning, an association between two stimuli is created (e.g., odor and a food or toy). In operant conditioning, the dog's behavior is critical. Without the behavioral response, the dog will not receive the reward. In odor detection, the discriminative stimuli will be the target odor and the behavior will be a trained alerting response. The consequence is a food or toy, delivered by the trainer.[62]

The choice of reinforcement will be selected by the dog's trainer. Studies report that food is generally more effective than petting, praise or a toy.[64] The motivation of the food-motivated dog obviously will drop as it becomes satiated,[65] but this can be easily avoided. For example, one arranges for the dog to be hungry when it starts a trial, and then only rewards the dog up to half a meal as a series of snacks during the trial.

Another important aspect is blinding. The famous "Clever Hans effect" describes responses of the animal to unintentional cues provided by the trainer, or some other person knowing the target.[2] Dogs are competent in understanding human social and communicative behavior and are sensitive to human attention states. Dogs also seem to be sensitive to human nonolfactory signals, behaving differently depending on a person's emotional state.[65,66]

To avoid this, handlers must be blinded during testing, and ideally during the last part of the training phase, to ensure that the dogs are not alerting due to unintentional hints. Only the observer, by definition, should know the status of the samples during late training. The ideal condition for subsequent testing is a double-blind scenario, where the observer has been replaced by a monitor, who has set up the samples, and does not tell the handler when to reward a dog.

34.3.2 Training setup

Training the dogs to correctly identify target odors is a reward-based operant conditioning.[1] The fundamental part is a step-by-step learning and recognition of the target scent, supported by positive feedback, thus a reward. However, when using scent detection dogs on samples with unknown status (testing phase), controlled positive feedback is not possible. Therefore, a reinforcement strategy must be scheduled during training. The ultimate goal is to make the response unpredictable from reward delivery, but ensure that it averages out to a predetermined value. The question of "when to reward" is often discussed and of exceptional importance in this context.[67] From the dog's perspective, however, the odor detection is a sort of game to earn a reward and depends on motivation to get a reward and the ability to produce a firm association between the target odor and the reward. Therefore, the detector dogs may change their strategies during training or testing in order to get the reward at the lowest possible cost and in the simplest way. Keeping this in mind, the training and testing conditions, especially the lineup of target scents, becomes important. In a lineup of variable targets with full blinding, the handler will not necessarily have the opportunity to provide the dog with a reward on positive samples, to ensure that the dog will not be rewarded for a wrong response. Based on results from studies providing scenarios with variable numbers of target scents in the lineup and double-blinding (Fig. 34.1),[18,21] positive and negative controls should be included. Positive controls create an opportunity for the dog to earn a reward, and negative controls will confirm the correctness and eagerness of the dog. This can be accomplished by a "reward-enhanced single-blind test" where the observer signals the handler when to reward, since only the observer knows the positive samples. However, the pattern of the samples must be systemically exchanged to avoid memorization of the odors. Memorization is also an issue when using the same samples for training and actual testing phases.[68] It is no doubt important to have a large repository of training samples, both of the positive and negative types, for dog training, but evidence is lacking to suggest a precise number.

Let us consider, arbitrarily, that one has 50 positive and 50 negative samples at the outset of a training phase. There are three kinds of strategies that one can consider: The first one is to use all of the samples throughout this phase, so the

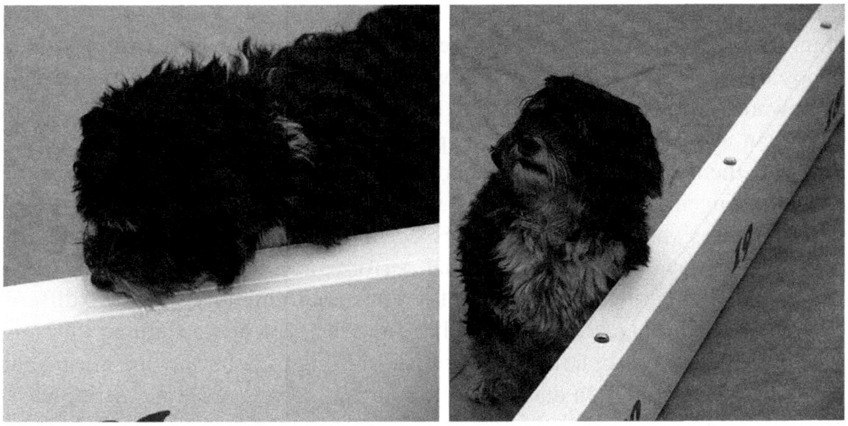

FIGURE 34.1

Odor detection training using tubes in a lineup setting.

dog is exposed repeatedly to the same samples. Does this risk that the dog is less able to generalize when confronted with new samples? In the second strategy, one starts the training phase with, say, 10 positive and 10 negative samples from the repository of 100, and progressively adds more of each during the training. In the third option, one follows either one of these strategies, but continues to acquire new samples that are used in the training phase. Since the dog will be encountering new samples in the testing phase, it seems to make sense to have this feature in the training phase, but evidence for this is lacking.

34.3.3 Dogs, time, teams

Individual characteristics of the dogs must be considered as important factors, since they play an intrinsic role in their trainability and performance. Several different breeds and crossbreds have been used for cancer detection on the basis of breath odor, with the most common breed being Labrador retrievers. While some breeds and crossbreeds are considered to be more effective for detection work than others, the selection of dogs mainly depends on their trainability, desire to work, eagerness to learn, performance, and the ability to ignore distractions. A main selection criterion might also be trainer's preference and availability. The animal's motivation for reinforcement or reward is of particular importance and the candidate dog should be strongly motivated by the reward the trainer prefers to use, e.g., food or toy. In addition to Labrador retrievers, the dogs used most frequently have been golden retrievers, German shepherds, and border collies. Further characterization of the dogs employed (e.g., by DNA sequencing) might be useful to guide future breeding toward optimal dogs. One seeks a dog with lots of drive, i.e., a high energy dog which works hard throughout each trial, and not all dogs can provide this.

Understanding a dog's personality and comprehension, and utilization of the way a dog interprets the world in which it lives, is critical. Allowing the dog to be a dog and enjoy the work in an unforced way will reduce its stress and add further motivation toward the training goals.[69] For the dog, olfactory diagnosis must be fun (a game), but under a high level of discipline provided by the trainer/handler.

Although there is some evidence that dogs may learn some tasks more quickly by reducing training intensity, training for odor detection in general can take much time, though dogs vary widely in this respect. It is essential to give dogs the time they need, and to evaluate their individual performance before advancing.[33] Opinions differ on the minimum time needed for detection training, as there are different training approaches. To get reliable results, however, typically a training time of at least 2 months has been conducted. Adequate sniffing time for the dog during each lineup is also important, both during the training and testing. One interesting idea is to shorten the training phase by "odor preexposure" (use of warm-up samples in the testing phase).[70]

Detection dogs and their trainers/handlers are usually a bonded team. Changing the handler is sometimes necessary (in certain organizations or due to change of ownerships), but may cause a team conflict and reduce overall performance. A handler may become so involved with the behavioral aspects of a dog that the animal's essential abilities of detection disappear when this person is removed.[71] Evidence suggests significant variation in detection performance when dogs are confronted with an unfamiliar handler. On the other hand, bonding of the handler and the dog positively correlates with higher team performance. It is unclear how long it takes for a dog to adjust to a new handler, and this needs more study.[72]

34.3.4 Training prerequisites

It is important to house the scent detector dogs in spacious temperature and humidity-controlled houses or kennels, feed them regularly, and provide enough water. Furthermore, the dogs need to receive enrichment in the form of outdoor walks and off-leash exercise, social interactions with people, and toys.[73] Since detection training is intense for the dog, mental and physical fitness of the dog is mandatory. The harder the dog works, the higher its body temperature, which may influence its performance. Therefore, ensuring the dogs' wellness prior to detection work is essential. Nutritional requirements are greater for detection dogs than normal pets. Adequate supplies of water are needed not only for the dog's thirst but also to help in maintaining a normal body temperature.[33]

Environmental conditions during training and testing must be selected and monitored carefully. In one approach, the conditions are as nondistracting as possible. For example, there is regular ventilation of the indoor experimental areas, along with a constant ambient temperature and a low level of constant airflow. Residual and lingering odors may influence the detection work, and variation in airflow and its odors might also have an effect.[74] On the other hand, one may view that the dog has the capacity to find positives samples against a variable background, and that the training and testing conditions in terms of background odors can be relaxed, such as not cleaning the exposure devices between each dog. This area of canine olfaction needs more study.

Thus, many important factors need to be considered when setting up a canine detection training and testing program for human diagnostics, including trial timing, schedules, commands, obedience, diversity in training and other motivational tools, types of trained alerts and indications, possible distractors, human involvement, boredom of the dog, end-of-session cues, and recordkeeping, as has been reviewed.[1,2] While the situation should feel casual for the dog, it needs to be planned and executed in a precise, detailed way that is recorded by the trainer and handler.

34.4 Recommendations for dog training and testing

The following proposes more specific recommendations with options for diagnostic dog training and testing. To begin, a disease, along with a corresponding type of training sample, needs to be selected. In most studies, a given kind of cancer is chosen, such as breast cancer. While there are subtypes for any cancer, it is usually not practical, nor sensible, to set up a subtype assay. Later in this chapter is a discussion of types of samples (e.g., breath, urine), along with recommendations for sample collection, containers (for sample presentation to the dog), and sample stability.

As a critical prerequisite, the clinical accuracy of the training samples (both disease-positive and disease-negative, where the latter can come from either healthy people or people with some other disease) must be known with complete or very high certainty. This is challenging, especially for negative healthy samples, so these should be selected carefully. The diversity and number of the disease-positive as well as the disease-negative samples should be maximized, so the dog does not develop an alert on samples simply because they come from a given setting, such as a hospital where a given background odor is present, or only learns to recognize a limited range of positive samples. While this section presents ideas and proposes suggestions for both the training and testing phases, the details must be customized to the characteristics of the dogs and their trainers/handlers. Complete sharing of techniques and results via detailed reports and publications among different trainers and handlers on an ongoing basis is important for advancing the field. This is a challenge. (When was the last time you were happy with the instructions on how to assemble something?)

The overall goal of the training phase, prior to the testing phase, is obvious, and the basic strategy throughout the training phase is to reward the dog when it alerts on a positive specimen, i.e., "rewarded operant conditioning." Three decisions need to be made to get underway. First is the reward; a bit of food (snack) is most popular, although a toy sometimes has been used. Second, a "reinforcer" is important, to intensify the reward so the dog truly gets the message that it has succeeded. (As described below, the reinforcer also can be used as a hint to help the dog learn the behavior response.) Examples of reinforcers are an audible signal as from a clicker or whistle, a verbal cue, such as "good dog," a tactile stimulus, such as chest rub or head pat, or a physical or visual signal, such as arm waving. (An analogy is to both tell someone and shake their hand when they have been successful, to make them feel really good.) It is proposed here that a unique reinforcer (e.g., clicker) should be used consistently, and that other options that the dog experiences in general are sometimes useful too, to make it more fun for the dog and the trainer. The preference of the dog is important too, and the trainer might vary the reinforcements in going from one stage of training to another (see below). Once the dog is trained, one might consider discontinuing with the clicker, and instead just praising the dog by saying "good dog." Third, a behavioral response needs to be selected for the dog, such as sitting down in front of the positive sample, barking, or pawing up.

In the following description of the six stages of the training phase, the terms "snack" and "clicker" are used instead of "reward" and "reinforcer." These stages are summarized in Table 34.2. The subsequent discussion will focus on the testing phase, when the dog, hopefully, provides highly accurate and early diagnostics.

During training stage I, the dog simply learns the selected behavior, e.g., sits when asked to do so, and immediately gets the snack and click when it complies, along with the audible "good dog." In training stage II, the dog is presented with a positive sample, and when the dog sniffs it, the trainer nudges the dog into the sit position, ideally without the "good dog" audible, and gives the snack and click. As needed, the trainer can initially just give the click, and then follow with the snack

Table 34.2 Overview of different stages of canine scent detection training.

Training stage	Training description
Stage I	Learning of the selected behavior (e.g., sit), by reward with an accompanying audible stimulus.
Stage II	Presentation of positive sample: when the dog sniffs it, nudge the dog to sit, and give the reward and audible stimulus.
Stage III	Same as stage II, but reward only if the dog sniffs by itself and give the audible stimulus.
Stage IV	Multiple empty containers, except for one positive; reward and audible stimulus only if the dog reacts to the positive sample. Reaction to empty samples are to be counted but not reinforced. Location of positive container is varied. Minimum error-free rounds are the criterion to pass this stage.
Stage V	As in stage IV, with "negative" (e.g., healthy or disease-free) samples replacing the empty containers. Carefully observe that the dog is not mistaking any sample. Minimum error-free rounds are the criterion to pass this stage.
Stage VI	Transition to single-blinded testing, where only a remote observer knows the location of the positive sample(s) in the lineup, and signals back to the trainer when a reward is appropriate.

once the dog sits, to help make this transition. In stage III, a positive sample is presented as before, but the dog has to sit by itself (maybe a click hint is given again) and then the dog gets the snack. In stage IV, there are multiple containers, and all are empty except one positive, and the dog gets the snack and click only when it alerts on the positive one. If the dog alerts on a negative sample, it gets a "no." The location of the positive container is varied randomly. A specific criterion needs to be selected for the dog to pass this stage. For example, one might require the dog to perform a minimum number of error-free rounds (e.g., at least 50 correct alerts in at least 100 rounds). In stage V, healthy and progressively "other disease" samples should gradually replace the empty containers, following the procedure for stage IV including the criterion for passing. The trainer must observe the dog carefully to make sure it has not missed any sample, nor given any sample inadequate attention. This, of course, also applies to the subsequent testing phase.

Up to this point, the snack and click have been provided by the dog trainer. Stage VI makes the transition to single-blind testing, where another person (the "observer") joins in remotely (i.e., not in the same room). Only the observer knows the location of the positive sample(s) in the lineup. When the dog alerts on a sample, the trainer signals this sample number with some sort of electronic device to the observer, who, in turn, signals back for a snack and click if the dog was correct. Dogs trained to select one positive sample in a lineup can be confused when more than one positive sample is present. This experience may need to be built into this stage VI. To pass this stage, a selected number of faultless rounds must be performed by the dog, as above. The dog should be returned to stage V if not successful.

Finally, for the testing phase, none of the samples in a lineup might be positive, and this is usually the case. However, if too many samples are negative, the dog will lose interest. Once the testing phase is underway, some positive samples should be included, but perhaps not in every lineup. (One can also consider "warming the dog up" with one or more lineups of training samples, each having one positive sample, before proceeding to lineups of test samples.) The handler needs to establish a schedule to make sure that the dog does not get bored. There is a danger that the dog will become frustrated and even alert on all of the samples to get a snack and click if the positives are too few and far between. One strategy is for the handler to add positive training samples to the testing samples, and decide on a schedule, e.g., include a positive sample in every third group of testing samples. The containers may need to be cleaned or changed between each round so the dog does not alert on a prior dog's excited breath, although it is not known if this can be a problem.

Periodically, a set of samples, including some known positive samples (from subjects known to have the clinical condition of interest), should be provided by a "monitor" who has no conflict of interest (unlike the typical observer) for the dog diagnostics study being conducted, where the clinical status of these samples is defined as much as possible, and this status is known only to the monitor until the testing is completed. Thus, the monitor replaces the observer, but the monitor does not signal to the handler when to reward the dog. This accomplishes double-blind testing. Ideally these samples are rich in "other disease" samples (as opposed to consisting of only healthy samples), and the positives come from early disease (a difficult task, but dog diagnostics has maximum value if it provides early detection). It is advised to conduct the testing phase with more than one dog (such as four), to reach a consensus decision. As pointed out above, proficiency samples also provide a way of accomplishing double-blind testing.

34.5 Sample collection

Sample collection for canine olfaction can be accomplished in two general ways: direct and indirect. In the former category, the dog directly sniffs the human. Potential detection of a seizure, hypoglycemia, malaria, or a tumor of an individual (e.g., potentially melanoma or maybe even a subcutaneous tumor) would fall into this category. In indirect detection, the dog sniffs a biosample, such as blood, plasma, urine, tissue, breath, sweat, or saliva. Of these kinds of samples, collection options and conditions seem to be most diverse for breath, as has been reviewed.[1] What have been mostly employed to date for breath collection in canine diagnostics are bags, tubes (possibly filled with a sorbent), masks, swipes, and cloths or garments.

Methods of breath collection for canine detection applications are varied, as they are for instrumental analysis (see Chapter 2). Devices leading to instrumental detection will likely increasingly be studied for canine detection, especially as instrumental analysis is directed toward learning what the dog is sniffing, a topic addressed later in this chapter, and elsewhere.[75]

Below are some examples, and potential examples, of devices for indirect diagnostic canine olfaction, with an emphasis on collection of breath. In general, the details provided in the original articles about the breath collection and dog sniffing are incomplete; this issue is stressed here because of the importance of increasing reproducibility in science in general, a topic returned to later.

34.5.1 Tubes

Breath has been collected in plastic tubes, such as those reported for breast cancer.[8] In that study, the tube was 6 inches (~15 cm) long and contained a 2 inch (~5 cm) long, removable insert of silicone oil–coated PP "wool" for breath accumulation. Subjects were asked to exhale long and deep breaths into the device three to five times, and the tubes then were capped and stored at room temperature prior to presentation to a dog. This device was also used in a study for detection of lung cancer,[76] where subjects were instructed to breathe into it two to three times. (In this latter study, breath samples were also collected in a Tedlar bag for comparative analysis by gas chromatography–mass spectrometry; GC-MS.) In both studies, the PP insert was placed in a hole-punched container for presentation to the dog, thereby preventing direct contact of the dog's nose with this insert.

Similarly, toward detection of lung cancer, glass tubes were assembled to have a filling of PP fleece impregnated with silicone oil with either hydrophilic or hydrophobic properties.[14] Each subject exhaled five times through the tube, which was then capped and stored at room temperature until presentation to a dog. For the dog sniffing, each tube was mounted vertically on the floor, with the end that had been breathed into at the bottom to minimize the potential complication of saliva. The dog made direct contact with the upper end of the tube during the testing.

34.5.2 Bag/tube

With a focus on detecting highly volatile components of breath via canine olfaction, subjects breathed normally to fill a 5 L polyethylene (PE) bag, and the collected breath was then pulled through an active charcoal adsorption tube at 300 mL/min using a vacuum pump.[21] The tube was sealed and stored at −18°C until subjected to analysis by dog sniffing. This strategy is similar to the use of Carbograph or Carboxen for breath collection, as has been reviewed in regard to instrumental detection (see also Chapter 2).[75] Such collection procedures provide a convenient way to focus on the more highly volatile chemicals in breath due to the high adsorption capacity of carbon-based materials.

34.5.3 Cloth

In principle, scents for canine diagnostics might be collected by placing a cloth or garment in contact with part of a patient's body. This has been studied for identification of people in regard to criminal investigations and is known for dogs trailing a suspect.[77] In the cited study, a 20 cm square of woven cotton fabric was placed in contact with part of a person's body or clothing for 30 min. In a study of canine olfaction for malaria, the sample sniffed by the dog was a sock that the child had worn.[39]

34.5.4 Containers

Routine containers have been used to house liquid or tissue samples for dog testing. Examples of some studies in this regard are as follows:

Plasma: For detection of ovarian cancer, plasma was obtained by centrifugation of blood in ethylenediaminetetraacetic acid (EDTA) tubes.[10,17] One drop of the plasma was placed in a small plastic dish, which in turn was placed in a box that was then presented to the dog. Responses were characterized as positive, negative, and uncertain.

Urine: For canine detection of prostate cancer, 2 mL of urine housed in a container was placed in a circularly perforated metal container, which in turn was housed in a thermally sealed plastic packet.[20]

Tissue: Ovarian carcinoma tissue samples (3 mm^3 cubes), as subfractions from an original sample stored at −80°C, each in a box, were sniffed by dogs.[10] Each cube was thawed at room temperature for 15–30 min prior to dog sniffing.

Masks. Routine hospital isolation face masks are being employed to collect breath (aerosol) for canine olfaction by cancer dogs. They were also used by Kitiyakara et al.[23] in a study directed at detection of hepatocellular carcinoma.

34.6 What is the dog sniffing?

34.6.1 Unknown basis of dog alerting

If only it was known what chemicals the trained dog is alerting on in a given test sample, after training on complex positive samples. How difficult will it be to determine this? The fact that a trained dog can alert on test samples that it has not encountered before might suggest that there is some pattern of chemicals that the trained dog has learned, as a "consensus pattern," and is applying this to the test samples. Assuming, further, that this hypothetical consensus pattern consists of a "relatively small" number of chemicals, then there is hope to discover what the dog is alerting on. This hypothetical dog will be referred to as a "consensus dog," keeping in mind that this no doubt is an oversimplification. One way in which it could be an oversimplification is that dogs might alert on similar chemicals to the ones they were trained on, a concept referred to as "generalization."[78] At the other extreme, perhaps the trained dog has learned a "large number" of chemical patterns in its training, each of which it remembers. This second hypothetical dog will be referred to as a "library dog." (After all, each of us has a large library of melodies in our head.) During testing, the library dog might be alerting on any one of many different patterns, each of which is close to one in its library. It will be very hard, if not impossible, to learn what the library dog is sniffing. However, cancers of a given type tend to share abnormal metabolism (arising from sharing a common genotype); this favors (in theory) the consensus dog. With this assumption in mind, there are four strategies to learning what the trained dog is alerting on during testing: nontargeted comparative analysis, hypothesis analysis, bioassay-directed chemical analysis, and high-throughput screening.

34.6.2 Nontargeted, comparative analysis

In nontargeted comparative analysis, biosamples are collected in one way or another (see above), from case and control subjects, for a given trait (such as lung cancer), and then tested both by trained dogs and by an advanced analytical technique, such as mass spectrometry (MS; obvious first choice) or nuclear magnetic resonance (NMR; less sensitive than MS). The data are subjected to statistical analysis (ideally by multiple statistical techniques, since each method can give different results) (see Chapter 38). It might help to employ mass tag MS, where the sample is derivatized to enhance and broaden analyte response,[79] potentially making it more likely to reach the sensitivity and scope of the dog. Because the dog sprays the sample it sniffs with its own aerosol, it may be important for test subjects to furnish two samples, one of which goes to the dog and the other to the MS. In this design, it would be interesting to then trade the samples for analysis by the other method. This approach is challenged by the huge number of peaks that can show up in MS analysis, many of which typically are unknowns in the field of metabolomics. Another challenge is to go beyond healthy control subjects, and include controls which have diseases other than the one being targeted, especially other diseases involving inflammation that so many diseases share in common. Many (perhaps most) canine olfaction studies have been compromised by testing only healthy subjects as controls.

34.6.3 Hypothesis analysis

In hypothesis analysis, it is assumed that a certain compound or class of compounds is what the dog is sniffing, and this is focused on in one way or another. Oxidized fatty acids might be targeted for analysis, since fatty acid oxidation is associated with tumor metastasis to lymph nodes.[80] Or the focus might be on alpha-dicarbonyl compounds, since these have been reported to be associated with the development of some of the complications of diabetes mellitus.[81]

34.6.4 Bioassay-directed chemical analysis

In bioassay-directed chemical analysis, a complex sample that is positive in some bioassay is fractionated, as by solvent extraction or chromatography, and the fractions are then similarly tested to see which one(s) have the bioassay activity. The structures of some important biomolecules, especially hormones (which are active at low concentrations), have been discovered in this way. Also, it has been applied to environmental samples to discover carcinogens (e.g., employ a bioassay for mutagenicity). Cumulative biosamples from one or more persons exhibiting the trait of interest are required to employ this kind of assay as each fractionation step only recovers some of the active ingredient(s) of interest, potentially leading to inadequate analyte quantities once purification has been achieved. Nevertheless, it would be exciting, for example, if an initial fractionation step of a biosample yielded a fraction(s) that the dog still alerted on, while the other fractions were negative. If reproducible, this would encourage a subsequent, scaled-up, continuation study.

34.6.5 High-throughput screening

Many drugs are discovered via high-throughput screening. In this approach, a biosample is collected (e.g., some cells, or a protein) that has or is a receptor of interest and plays a role (or is postulated to) in a given disease, and which can be made to yield an easily detected signal when this receptor is occupied (e.g., the fluorescent properties of the sample change). This biosample is repeated in many wells of many microtiter plates, and a different chemical (as a screening compound) from a chemical library is added to each well. There can be many hits if a large library is tested, and the ones with highest affinities (determined by the next round of testing) are rated as drug candidates for further study. For the question of what the dog is sniffing, many standard mixtures of chemicals are available commercially, and many studies of volatiles from humans have provided names of chemicals reported to correlate with the trait investigated. For validation, a multicenter study could be established in which defined chemical mixtures with de-identified compositions are sent to multiple laboratories with trained dogs investigating one trait or another. It would be exciting if even one sample tested positive at one laboratory, leading to further testing toward validation. However, there is a danger in this approach that the samples could be too artificial in one way or another for the dog to alert on.

34.6.6 Complementary or competing assays

The method of analyzing exosome liquid biopsies, once it matures, may compete with the dog for early detection of cancer, yet might equally present a complementary tool. Exosomes are very small vesicles that cells excrete for various purposes, and their analysis may reveal the presence of a tumor and its tissue of origin. Usually exosomes are isolated from blood. Liquid biopsy of exosomes represents an opportunity for dog testing to advance, and vice versa. Subjects who test positive for cancer by one technique can be considered for cotesting by the other; a nice collaboration.

34.7 Recommendations

There is a growing need to standardize nomenclature in the field of canine olfaction for disease detection, to avoid ambiguity and confusion when terms are used loosely and interchangeably in reporting scientific studies. For example, one encounters terms such as *handler*, *experimenter*, *helper*, *trainer*, *head trainer*, *test leader*, *assistant*, *observer*, etc., in the literature, sometimes mixed in a given article. Whatever terms an author selects, they should be defined precisely and used consistently. The following nomenclature is proposed here when reporting on canine olfaction studies (and summarized in Table 34.3). For the two successive phases of canine olfaction, the terms *training* and *testing* are recommended, where the dog is learning in the first phase, and potentially providing diagnostics in the second phase. In each phase, there are *trials*, where the dog sniffs samples without interruption. Each trial consists of one or more *lineups* of samples (typically five to seven samples), where the dog evaluates one lineup at a time. The training and testing phases are not always exclusive, since some samples known to be positive for the clinical condition of interest might be included in the testing phase to keep up the dog's interest, as discussed. Indeed, one can consider starting each trial in the testing phase with one or more lineup samples ordinarily used in the training phase to "warm the dog up." There are then two kinds of samples, *training samples*

Table 34.3 Recommended standardized nomenclature on canine olfaction studies.

	Term	Description
Phases	Training phase	Period when the dog is taught to react to the odor of interest with samples of known status (*training samples*); consists of several stages
	Testing phase	Period to test the dog for its ability to correctly respond to the trained odor of interest; ideally double-blinded
	Warm-up phase	Before or during testing phase: One or more *lineups* with samples with known status to reinforce the training and check for correct response
Structure	Lineup	Collection of known or unknown samples (or both) during each *trial*
	Trial	One or more *lineups* of *training* or *test samples*; sniffing without interruption
Samples	Training samples	Known positive or negative samples for the clinical condition of interest
	Test samples	Positive or negative samples for the clinical condition of interest; sample status only known to *observer* or *monitor*
Crew	Trainer	Teaches the dog by directly interacting with it during the training phase (*trainer* and *handler* may be the same person)
	Handler	Interacts with the dog during the *testing phase* (*trainer* and *handler* may be the same person)
	Observer	Observes during the *training phase* and *testing phases*; knows sample status (positive/negative), communicates to handler when to reward the dog
	Monitor	Same role as observer; must have no conflict of interest and must not communicate with the handler; essential for double-blind testing

and *test samples*, where either type may comprise positive or negative samples for the clinical condition of interest, but sample type in these terms is known beforehand, of course, only for the training samples. Of these latter samples, the ones that are negative may be so in two ways: healthy or diseased but with a different disease than the one being tested. The *trainer* teaches the dog by directly interacting with it during the training phase. The *handler* similarly interacts directly with the dog during the testing phase. Often the trainer and handler is the same person. The term *working* should be used for the dog in the testing phase, aside from whether some training samples are included. The *observer* enables single-blind testing, where only the observer knows which samples are positive and conveys this information instantly to the handler so that appropriate rewarding can be done, and inappropriate rewarding can be avoided. The observer can be involved during both the training and testing phases. Whenever the observer is part of the staff (as usually is the case), there is a potential for conflict of interest. This problem is solved by having a *monitor* who provides the same role as the observer, but has no conflict of interest, and thereby enables double-blind testing. In this case, the handler has no information at all about the samples, and simply rewards the dog whenever it alerts. Thus, the observer communicates to the handler when to reward a dog, but the monitor does not, by definition. Double-blind testing is necessary to fully define and validate a method. Just as with an observer, a monitor can be employed both during the training and testing phases, but is more likely to be used during the testing phase. Double-blind testing is also provided when proficiency samples are tested (known samples provided by an independent source acting, in effect, as a remote monitor).

One must be cautious about making recommendations when there is little or no evidence supporting them. The literature on dog olfaction contains many recommendations that fall into this category. Nevertheless, unsupported recommendations represent critical hypotheses to be tested. It is very important to be aware of the options for training and using dogs (a major purpose of this chapter), to help practitioners set up good experimental designs, including carefully planning, controlling, recording, and reporting the important experimental details. Taking these general considerations into account, here are some recommendations (largely placing further emphasis on recommendations covered throughout this chapter).

34.7.1 Sample/dog interaction

The dog subjects a sample that it sniffs to a "hurricane" of its own breath aerosol, no doubt to enhance its sniffing, including collection of "nonvolatiles." This must leave some dog biochemicals on the sample or sample site/container. Perhaps the dog deposits a special pattern of its own biochemicals on a sample or site when it alerts on it (pure speculation). Conceivably, a second dog might alert on a given training or test sample because the first dog did. Perhaps the sample containers should at least be washed, or fresh equipment used, between each dog for a given lineup. Repeated testing of a given lineup with a given dog with and without intermediate washing of the sample sites could be done to explore this. Since the dog is trained (more or less) to screen out signals other than positives, it is unclear how meticulous this approach should be. Certainly, search dogs are subject to all sorts of potential distractions when finding their targets, as are dogs that sniff luggage at airports, and have nevertheless proved to be useful. This area of canine olfaction needs further study.

34.7.2 Literature problem

Too many articles have been published in this field that inadequately describe the methods of the work involved. Some of these details may reveal defects that compromise the conclusions reached. As is true in the scientific literature in general, there is a great problem of irreproducibility, which has multiple causes. Experimenters, authors, reviewers, and journal editors all must work on minimizing this problem. It is hard to find an article on dog sniffing that discloses all of the details. A common shortcoming is failure to describe the sniffing container and station in sufficient detail. Information on how the handler deals with the various intermediate responses of a dog, such as when the dog hesitates on a sample, is often also vague. Perhaps videos of the dog training and testing should be made available as supplemental materials of articles and reports on dog sniffing endeavors.

34.7.3 Single- and double-blind studies

In a single-blind study involving dogs, there is both a handler and observer, and only the observer knows something about the samples. In a corresponding double-blind study, a monitor replaces the observer (as described at the beginning of this chapter). Blinded studies are intended to remove handler bias or influence. In either case, there is a challenge of not rewarding the dog when it alerts on a positive sample, since too much of this event counters training of the dog, and can make the dog lose interest. Four remedies for this are suggested, which can be combined: First, always or frequently include a positive sample in the testing set, to keep up the dog's interest. Second, limit the frequency and duration of the blind testing sessions when rewards are absent. Third, after a session in which the dog has not found any alerts, at least praise or reward it anyway once it is out of the testing room. Fourth, employ "reward-enhanced single-blind tests," where the observer is completely remote and signals the handler, in a discrete way relative to the dog, when the dog should be rewarded. In this way, the dog regularly enjoys rewards. It is important for the dog to enjoy many rewards in both the training and testing phases.

34.7.4 Prospective studies

In a prospective study, biosamples are collected periodically from a large population of people and stored under appropriate conditions (e.g., at low temperature) until a disease emerges (months to years). Once a disease of interest has reached a meaningful incidence, and a good hypothesis has emerged about a potential biomarker for the disease, then cases and controls can be compared with respect to that biomarker. This is a good strategy for evaluating the ability of dogs to reveal an early disease, which is the primary reason for dog-based testing. Many biorepositories are currently available which contain samples where some diseases in the donor population have reached a significant incidence, so this is an important opportunity for dog testing, especially because serial samples (samples collected through the years) are usually present in such collections.

34.7.5 Characterize the dog

Ideally the dog is fully characterized in every respect. It might be useful to sequence the DNA of the dogs that are employed. This might help to guide breeding toward the most effective dogs. It is well known that people perceive odors differently, and presumably this is similarly true for dogs.

34.7.6 Sample stability

It has been claimed that breath samples are sufficiently stable under ordinary conditions for testing by dogs, but there is no evidence to support this claim. Of particular concern is storage of breath samples under humid conditions at temperatures sufficiently high to support growth of microorganisms. In this case, the pattern of volatilizable chemicals will change with time. Perhaps a dog can overlook this to some degree, or over a particular duration, and still recognize when a cancer signature is present. Nevertheless, this remains to be proven, and more study is needed. In the meantime, breath (and other) samples should be tested fresh, or stabilized as much as possible, as by storage at a low temperature, such as −80°C (although recently it has been suggested that −70°C is adequate and beneficial by saving electricity[82]). A temperature of −20°C is inadequate for extended storage because of the presence of pockets of water due to high solute content where microorganisms can still thrive.

34.7.7 Proficiency testing

Proficiency testing is essential in the field of clinical diagnostics. Well-defined samples are sent for analysis to clinical laboratories, which then return their results for comparison with others. This might focus on accuracy of a given method for a given analyte when the samples are reference standards, or the range of values when the samples are less-well defined. When a laboratory reports a value that is far from the mean, the implication is that the laboratory has a problem. If the range is very broad, the instrument or method overall has a problem, assuming that there were no problems with the integrity of the proficiency sample. Proficiency testing is sorely needed in the field of canine olfaction as it represents the most robust way to conduct double-blind testing.

34.8 Summary

Training and using dogs for human diagnostics is challenging, fun, and important. The *challenge* is dealing with the complications of the dog, along with the need to make careful plans and keep detailed records to help the field advance. The *fun* is inherent because of the close bond between dog and handler. For the dog it is a game, although a much more serious and intense game than just playing with a pet dog. Nevertheless, much work is involved: typically it takes at least 2 months to train a dog (current common practice); there are many variables; and there are many points of view on exactly how it should be conducted, with scant evidence supporting some of the recommendations. The *importance* is the potential to (1) improve clinical diagnostics directly (the dog alerts early on a medical condition of interest, when prevention and treatment can be much easier) and (2) learn what the dog is sniffing, potentially leading to new biomarkers for instrumental analysis. Ideally canine olfaction diagnostics leads to the dog becoming obsolete in this way, since instrumental analysis, once automated, can be more cost-effective, stable, and reproducible. In reality, canine olfaction will likely be of interest for a long time. It will also continue to be challenging, since it is subjective in nature.

The history of this field is largely frustrating, with too many shortcomings in design, quality control, assessment of sensitivity and specificity in a meaningful way, or reporting of methods in sufficient detail for others to repeat the work. Nevertheless, the future is exciting, because some promise has been demonstrated in canine olfaction for human diagnostics.

Acknowledgments

Our knowledge of the subject has benefited greatly from discussions (including some disagreements) with others in the field through the years, mainly (alphabetical) Theodor Doll, Peter Errhalt, Glen Fergusson, Michael Müller, Andrea Schulheim, and Anton Schwarzhapl. Furthermore, we gratefully acknowledge the Team-Canine Vienna and Andrea Schulheim for the photos.

References

1. Jezierski T, Walczak M, Ligor T, Rudnicka J, Buszewski B. Study of the art: canine olfaction used for cancer detection on the basis of breath odour. Perspectives and limitations. *J Breath Res* 2015;**9**:027001.
2. Minhinnick S, Papet L, Stephenson CM, Stephenson MR. *Training fundamentals and the selection of dogs and personnel for detection work. Canine olfaction science and law: advances in forensic science, medicine, conservation, and environmental remediation.* 2016. p. 155–71.
3. Williams H, Pembroke A. Sniffer dogs in the melanoma clinic? *Lancet* 1989;**1**:734.
4. Church J, Williams H. Another sniffer dog for the clinic? *Lancet* 2001;**358**:930.
5. Pickel D, Manucy GP, Walker DB, Hall SB, Walker JC. Evidence for canine olfactory detection of melanoma. *Appl Anim Behav Sci* 2004;**89**:107–16.
6. Willis CM, Church SM, Guest CM, et al. Olfactory detection of human bladder cancer by dogs: proof of principle study. *Br Med J* 2004;**329**:712.
7. Leahy M. Olfactory detection of human bladder cancer by dogs: cause or association? *Br Med J* 2004;**329**:1286. Author reply 1286.
8. McCulloch M, Jezierski T, Broffman M, Hubbard A, Turner K, Janecki T. Diagnostic accuracy of canine scent detection in early- and late-stage lung and breast cancers. *Integr Cancer Ther* 2006;**5**:30–9.
9. Gordon RT, Schatz CB, Myers LJ, et al. The use of canines in the detection of human cancers. *J Altern Complement Med* 2008;**14**:61–7.
10. Horvath G, Jarverud GA, Jarverud S, Horvath I. Human ovarian carcinomas detected by specific odor. *Integr Cancer Ther* 2008;**7**:76–80.
11. Horvath G, Andersson H, Paulsson G. Characteristic odour in the blood reveals ovarian carcinoma. *BMC Cancer* 2010;**10**:643.

12. Sonoda H, Kohnoe S, Yamazato T, et al. Colorectal cancer screening with odour material by canine scent detection. *Gut* 2011;**60**:814–9.
13. Cornu JN, Cancel-Tassin G, Ondet V, Girardet C, Cussenot O. Olfactory detection of prostate cancer by dogs sniffing urine: a step forward in early diagnosis. *Eur Urol* 2011;**59**:197–201.
14. Ehmann R, Boedeker E, Friedrich U, et al. Canine scent detection in the diagnosis of lung cancer: revisiting a puzzling phenomenon. *Eur Respir J* 2012;**39**:669–76.
15. Walczak M, Jezierski T, Górecka-Bruzda A, Sobczyńska M, Ensminger J. Impact of individual training parameters and manner of taking breath samples on the reliability of canines as cancer screeners. *J Vet Behav Clin Appl Res* 2012;**7**:283–94.
16. Campbell LF, Farmery L, George SM, Farrant PB. Canine olfactory detection of malignant melanoma. *BMJ Case Rep* 2013;**2013**.
17. Horvath G, Andersson H, Nemes S. Cancer odor in the blood of ovarian cancer patients: a retrospective study of detection by dogs during treatment, 3 and 6 months afterward. *BMC Cancer* 2013;**13**:396.
18. Amundsen T, Sundstrom S, Buvik T, Gederaas OA, Haaverstad R. Can dogs smell lung cancer? First study using exhaled breath and urine screening in unselected patients with suspected lung cancer. *Acta Oncol* 2014;**53**:307–15.
19. Elliker KR, Sommerville BA, Broom DM, Neal DE, Armstrong S, Williams HC. Key considerations for the experimental training and evaluation of cancer odour detection dogs: lessons learnt from a double-blind, controlled trial of prostate cancer detection. *BMC Urol* 2014;**14**:22.
20. Taverna G, Tidu L, Grizzi F, et al. Olfactory system of highly trained dogs detects prostate cancer in urine samples. *J Urol* 2015;**193**:1382–7.
21. Hackner K, Errhalt P, Mueller MR, et al. Canine scent detection for the diagnosis of lung cancer in a screening-like situation. *J Breath Res* 2016;**10**:046003.
22. Amal H, Leja M, Funka K, et al. Breath testing as potential colorectal cancer screening tool. *Int J Cancer* 2016;**138**:229–36.
23. Kitiyakara T, Redmond S, Unwanatham N, et al. The detection of hepatocellular carcinoma (HCC) from patients' breath using canine scent detection: a proof-of-concept study. *J Breath Res* 2017;**11**:046002.
24. Guerrero-Flores H, Apresa-Garcia T, Garay-Villar O, et al. A non-invasive tool for detecting cervical cancer odor by trained scent dogs. *BMC Cancer* 2017;**17**:79.
25. Guirao Montes A, Molins Lopez-Rodo L, Ramon Rodriguez I, et al. Lung cancer diagnosis by trained dogs. *Eur J Cardio Thorac Surg* 2017;**52**:1206–10.
26. Seo IS, Lee HG, Koo B, et al. Cross detection for odor of metabolic waste between breast and colorectal cancer using canine olfaction. *PLoS One* 2018;**13**:e0192629.
27. Thuleau A, Gilbert C, Bauer P, et al. A new transcutaneous method for breast cancer detection with dogs. *Oncology* 2019;**96**:110–3.
28. Moser E, McCulloch M. Canine scent detection of human cancers: a review. *J Vet Behav Clin Appl Res* 2010;**5**:145–52.
29. Badhra A. Woof! Smells like cancer. *Curr Sci* 2011;**101**:480–3.
30. Lippi G, Cervellin G. Canine olfactory detection of cancer versus laboratory testing: myth or opportunity? *Clin Chem Lab Med* 2012;**50**:435–9.
31. Boedeker E, Friedel G, Walles T. Sniffer dogs as part of a bimodal bionic research approach to develop a lung cancer screening. *Interact Cardiovasc Thorac Surg* 2012;**14**:511–5.

32. McCulloch M, Turner K, Broffman M. Lung cancer detection by canine scent: will there be a lab in the lab? *Eur Respir J* 2012;**39**:511−2.
33. Hackner K, Pleil J. Canine olfaction as an alternative to analytical instruments for disease diagnosis: understanding 'dog personality' to achieve reproducible results. *J Breath Res* 2017;**11**:012001.
34. Balseiro SC, Correia HR. Is olfactory detection of human cancer by dogs based on major histocompatibility complex-dependent odour components?−A possible cure and a precocious diagnosis of cancer. *Med Hypotheses* 2006;**66**:270−2.
35. Dorman DC, Foster ML, Fernhoff KE, Hess PR. Canine scent detection of canine cancer: a feasibility study. *Vet Med (Auckl)* 2017;**8**:69−76.
36. Bomers MK, van Agtmael MA, Luik H, van Veen MC, Vandenbroucke-Grauls CM, Smulders YM. Using a dog's superior olfactory sensitivity to identify *Clostridium difficile* in stools and patients: proof of principle study. *Br Med J* 2012;**345**:e7396.
37. Bomers MK, van Agtmael MA, Luik H, Vandenbroucke-Grauls CM, Smulders YM. A detection dog to identify patients with *Clostridium difficile* infection during a hospital outbreak. *J Infect* 2014;**69**:456−61.
38. Taylor MT, McCready J, Broukhanski G, Kirpalaney S, Lutz H, Powis J. Using dog scent detection as a point-of-care tool to identify toxigenic *Clostridium difficile* in stool. *Open Forum Infect Dis* 2018;**5**:ofy179.
39. Guest C, Ponder M, Dogget M, et al. Trained dogs identify people with malaria parasites by their odour. *Lancet Infect Dis* 2019;**6**:578−80.
40. Koivusalo M, Vermeiren C, Yuen J, Reeve C, Gadbois S, Katz K. Canine scent detection as a tool to distinguish meticillin-resistant *Staphylococcus aureus*. *J Hosp Infect* 2017; **96**:93−5.
41. Maurer M, McCulloch M, Willey AM, Hirsch W, Dewey D. Detection of bacteriuria by canine olfaction. *Open Forum Infect Dis* 2016;**3**:ofw051.
42. Chen M, Daly M, Williams N, Williams S, Williams C, Williams G. Non-invasive detection of hypoglycaemia using a novel, fully biocompatible and patient friendly alarm system. *Br Med J* 2000;**321**:1565−6.
43. Tauveron I, Delcourt I, Desbiez F, Somda F, Thieblot P. Canine detection of hypoglycaemic episodes whilst driving. *Diabet Med* 2006;**23**:335.
44. O'Connor MB, O'Connor C, Walsh CH. A dog's detection of low blood sugar: a case report. *Ir J Med Sci* 2008;**177**:155−7.
45. Wells DL, Lawson SW, Siriwardena AN. Canine responses to hypoglycemia in patients with type 1 diabetes. *J Altern Complement Med* 2008;**14**:1235−41.
46. Rooney NJ, Morant S, Guest C. Investigation into the value of trained glycaemia alert dogs to clients with type I diabetes. *PLoS One* 2013;**8**:e69921.
47. Gonder-Frederick L, Rice P, Warren D, Vajda K, Shepard J. Diabetic alert dogs: a preliminary survey of current users. *Diabetes Care* 2013;**36**:e47.
48. Petry NM, Wagner JA, Rash CJ, Hood KK. Perceptions about professionally and nonprofessionally trained hypoglycemia detection dogs. *Diabetes Res Clin Pract* 2015; **109**:389−96.
49. Dehlinger K, Tarnowski K, House JL, et al. Can trained dogs detect a hypoglycemic scent in patients with type 1 diabetes? *Diabetes Care* 2013;**36**:e98−99.
50. Rudy L. *Service dogs for people with seizure disorders*, vol. 6. Alert: National Dog Center Newsletter; 1995.
51. Strong V, Brown S, Huyton M, Coyle H. Effect of trained Seizure Alert Dogs on frequency of tonic-clonic seizures. *Seizure* 2002;**11**:402−5.

52. Strong V, Brown SW. Should people with epilepsy have untrained dogs as pets? *Seizure* 2000;**9**:427–30.
53. Kirton A, Wirrell E, Zhang J, Hamiwka L. Seizure-alerting and -response behaviors in dogs living with epileptic children. *Neurology* 2004;**62**:2303–5.
54. Kirton A, Winter A, Wirrell E, Snead OC. Seizure response dogs: evaluation of a formal training program. *Epilepsy Behav* 2008;**13**:499–504.
55. Ortiz R, Liporace J. "Seizure-alert dogs": observations from an inpatient video/EEG unit. *Epilepsy Behav* 2005;**6**:620–2.
56. Krauss GL, Choi JS, Lesser RP. Pseudoseizure dogs. *Neurology* 2007;**68**:308–9.
57. Doherty MJ, Haltiner AM. Wag the dog: skepticism on seizure alert canines. *Neurology* 2007;**68**:309.
58. Catala A, Grandgeorge M, Schaff JL, Cousillas H, Hausberger M, Cattet J. Dogs demonstrate the existence of an epileptic seizure odour in humans. *Sci Rep* 2019;**9**:4103.
59. Catala A, Cousillas H, Hausberger M, Grandgeorge M. Dog alerting and/or responding to epileptic seizures: a scoping review. *PLoS One* 2018;**13**:e0208280.
60. Wells DL. Dogs as a diagnostic tool for ill health in humans. *Altern Ther Health Med* 2012;**18**:12–7.
61. Pavlov IP. *Conditioned reflexes: an investigation of the physiological activity of the cerebral cortex*. Oxford: Oxford University Press; 1927.
62. Hall NJ, Wynne C. *Canine olfactory learning and behavior. Canine olfaction science and law: advances in forensic science, medicine, conservation, and environmental remediation*. Taylor and Francis; 2016. p. 123–38.
63. Barela PB. Theoretical mechanisms underlying the trial-spacing effect in Pavlovian fear conditioning. *J Exp Psychol Anim Behav Process* 1999;**25**:177–93.
64. Feuerbacher EN, Wynne CD. Relative efficacy of human social interaction and food as reinforcers for domestic dogs and hand-reared wolves. *J Exp Anal Behav* 2012;**98**: 105–29.
65. D'Aniello B, Semin GR, Alterisio A, Aria M, Scandurra A. Interspecies transmission of emotional information via chemosignals: from humans to dogs (*Canis lupus* familiaris). *Anim Cognit* 2018;**21**:67–78.
66. Siniscalchi M, d'Ingeo S, Quaranta A. The dog nose "KNOWS" fear: asymmetric nostril use during sniffing at canine and human emotional stimuli. *Behav Brain Res* 2016;**304**: 34–41.
67. Sargisson RJ, McLean IG. The effect of reinforcement rate variations on hits and false alarms in remote explosive scent tracing with dogs. *J ERW Mine Action* 2010;**14**.
68. Williams M, Johnston JM. Training and maintaining the performance of dogs (*Canis familiaris*) on an increasing number of odor discriminations in a controlled setting. *Appl Anim Behav Sci* 2002;**78**:55–65.
69. Rooney N, Gaines S, Hiby E. A practitioner's guide to working dog welfare. *J Vet Behav Clin Appl Res* 2009;**4**:127–34.
70. Hall NJ, Smith DW, Wynne CD. Effect of odor preexposure on acquisition of an odor discrimination in dogs. *Learn Behav* 2014;**42**:144–52.
71. Kerepesi A, Jonsson GK, Miklosi A, Topal J, Csanyi V, Magnusson MS. Detection of temporal patterns in dog-human interaction. *Behav Process* 2005;**70**:69–79.
72. Jamieson TJ, Baxter GS, Murray PJ. You are not my handler! Impact of changing handlers on dogs' behaviours and detection performance. *Animals (Basel)* 2018;**8**.

73. Lazarowski L, Foster ML, Gruen ME, et al. Olfactory discrimination and generalization of ammonium nitrate and structurally related odorants in Labrador retrievers. *Anim Cognit* 2015;**18**:1255–65.
74. Rygg AD, Van Valkenburgh B, Craven BA. The influence of sniffing on airflow and odorant deposition in the canine nasal cavity. *Chem Senses* 2017;**42**:683–98.
75. Lawal O, Ahmed WM, Nijsen TME, Goodacre R, Fowler SJ. Exhaled breath analysis: a review of 'breath-taking' methods for off-line analysis. *Metabolomics* 2017;**13**:110.
76. Buszewski B, Ligor T, Jezierski T, Wenda-Piesik A, Walczak M, Rudnicka J. Identification of volatile lung cancer markers by gas chromatography-mass spectrometry: comparison with discrimination by canines. *Anal Bioanal Chem* 2012;**404**:141–6.
77. Settle RH, Sommerville BA, McCormick J, Broom DM. Human scent matching using specially trained dogs. *Anim Behav* 1994;**48**:1443–8.
78. Moser AY, Bizo L, Brown WY. Olfactory generalization in detector dogs. *Animals (Basel)* 2019;**9**:702–13.
79. Wang P, Gao J, Li G, Shimelis O, Giese RW. Nontargeted analysis of DNA adducts by mass-tag MS: reaction of p-benzoquinone with DNA. *Chem Res Toxicol* 2012;**25**: 2737–43.
80. Lee CK, Jeong SH, Jang C, et al. Tumor metastasis to lymph nodes requires YAP-dependent metabolic adaptation. *Science* 2019;**363**:644–9.
81. Wang XJ, Ma SB, Liu ZF, Li H, Gao WY. Elevated levels of alpha-dicarbonyl compounds in the plasma of type II diabetics and their relevance with diabetic nephropathy. *J Chromatogr B Analyt Technol Biomed Life Sci* 2019;**1106–1107**:19–25.
82. Matter of Degree DCA. *ACS Cent Sci* 2018;**4**:1294–7.

PART

Design and interpretation

G

CHAPTER

35 Clinical study design

Marc P. van der Schee[1], Silvano Dragonieri[2], Morad K. Nakhleh[1]

[1]*Owlstone Medical Ltd., Cambridge, United Kingdom;* [2]*Department of Respiratory Diseases, University of Bari, Bari, Italy*

Chapter outline

35.1 Overview: a practical guide to designing breath biomarker studies

Clinical trial design is primarily about strategy. Clinical studies are a critical tool to address the scientific questions relevant to progressing the field of breath research. The clinical study design is the strategic roadmap to how this will be achieved. As such, it is a nontrivial task that should not be underestimated. The literature—and breath is certainly no exception—is full of examples of (unintentional) poor trial design. Poor trial design results in a loss of time, money, and effort; it reduces the chances of successfully completing a trial, undermines the ability to publish any outcomes, and fundamentally does not progress science toward the clinic.

Breathborne Biomarkers and the Human Volatilome. https://doi.org/10.1016/B978-0-12-819967-1.00035-9
Copyright © 2020 Elsevier B.V. All rights reserved.

Bad trial designs are characterized by stating overly ambitious goals without providing detailed clarity on how these goals are to be achieved. Many papers have been published that cannot provide a definitive interpretation of results due to poor design choices. This chapter will outline a practical approach to designing a clinical study in relation to what is required to address the scientific question posed, what constraints cannot be violated, and what particular strengths can be leveraged to design a study that maximize the chances of extracting reliable findings that will progress the field. This chapter purposefully does not list potential designs, as each trial is unique and none of them would be exactly right for any particular study.

As a disclaimer, the content of this chapter draws from learnings from the many studies executed by the authors, including several poor approaches. The approach proposed herein is certainly not the only or best way to think about clinical study design. Nevertheless, this chapter hopefully will offer insights to help prevent common mistakes and avoid a future desire to travel back in time to improve flawed studies.

35.2 A three-step approach

Before undertaking a clinical trial, careful consideration must be made of several aspects. Three main questions can be posed to assist with this:

1. **What is the objective?** What question requires answering? What (clinical) problem is being addressed? What are the minimum requirements to address this question successfully?
2. **What is the context?** What is the context within which this objective is to be addressed? What are the constraints? Which particular strengths can be leveraged in the research environment?
3. **What is the best experiment?** What study design addresses these objectives best by respecting the defined constraints and leveraging the available strengths?

These three questions may seem obvious, yet there is a tendency among scientists to immediately delve into the third question and suggest trial designs without proper consideration of the objective. It is important to go through this three-step process by assembling a team that represents different disciplines relevant to the study: clinicians, analytical chemists, engineers, epidemiologists, statisticians, and so on. Breath research is highly multidisciplinary and requires the involvement of varied expertise in each of these steps.

Before addressing these questions, it is important to realize that in many cases the process is iterative (see Fig. 35.1); it is quite likely that after initially going through these steps it will become apparent that the objectives of a study are too ambitious or should be defined more exactly. This realization is a positive outcome, as it avoids future disappointment. The cycle of addressing the questions outlined

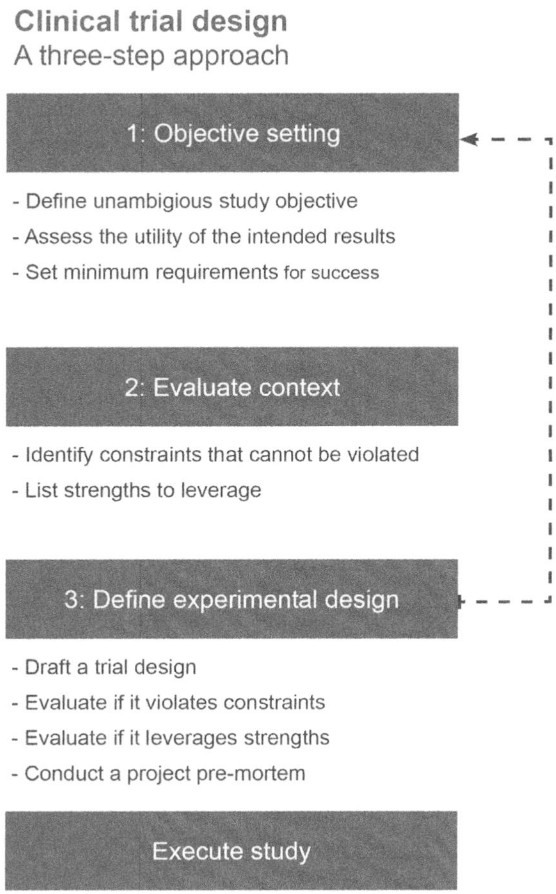

FIGURE 35.1

Flow chart depicting the suggested three-step approach to clinical trial design.

above can then be followed again (and again, and again) until the remaining risk has been minimized as best possible. It is only at this stage that a protocol should be fully developed and defined. In this context, a valuable lesson is to "*split the thinking from the writing*,"[1] in other words, the tasks of thinking through the study design and writing a protocol both deserve full attention, thus should not be combined into a single exercise.

35.3 The utility of breath tests: What are the objectives?

Designing a strong study should commence with asking the right question. This can be a difficult exercise; even though an objective is typically a single sentence in a protocol, it deserves a disproportionate amount of attention to get it right.

For the purposes of this chapter, the following assumption will be made: the primary objective of a clinical trial is to generate novel mechanistic insights and/or to progress a breath test closer to clinical practice, and not merely to publish a research paper. It is an unfortunate reality that the two frequently do not align, or even oppose each other when it comes to much needed validation studies.

A typical objective for a clinical breath study may be as follows:

Discover breath biomarkers for disease X.

Now, this question can be unpicked by simply going through the sentence and assessing the meaning of each part: what does finding breath biomarkers mean?; is the aim to diagnose a disease?; or phenotype it?; to predict therapy response?; or risk-stratify patients? Likewise, how is disease *X* defined for this study? It is essential to ensure that the meaning of this question is crystal clear.

After clarifying these aspects, the objective of a trial might look something like this:

The objective is to identify biomarkers that discriminate eosinophilic and neutrophilic airway inflammation in mild to moderate asthma by analyzing exhaled volatile organic compounds.

For any objective in the context of a clinical study, a key question to address is *assuming this works, would it be a useful test for patients and clinicians?* For understandable reasons, breath research tends to focus on whether the test is accurate at differentiating between two groups of interest. This is obviously very relevant, but an accurate test may not be useful to a patient at all, for instance, when there is no treatment available. Conversely, a test that is not very accurate may nevertheless be very useful if no better alternative available. Further, a noninvasive breath test might not be desirable in cases where an accurate and cheap test based on other clinical parameters already exists (as is discussed in the next chapter). Conversely, a breath test that increases access for patients or prevents more invasive procedures could be very useful (and desirable) despite modest accuracy.

Addressing the future utility of a (breath) test can be challenging and may seem over the top, especially for early-stage pilot studies. Lessons learned by thinking through how a test would be used, by whom, and what decisions it will impact in clinical practice, however, provide key information that help to improve the design of a study. For pilot studies, exploring this can be as straightforward as performing a literature search and talking to physicians and patients about the idea. For more advanced studies, a more rigorous approach is necessary, incorporating formal assessment of the health economics, desired performance characteristics, regulatory framework, barriers to adoption, clinical pathway positioning, pricing, reimbursement, and so on. At the very least, this process will deliver information on the point along the pathway at which patients and controls should be recruited, how accurate the test must be, and what the inclusion and exclusion criteria should be. In some cases, even highly accurate biomarkers would never be used in practice due to a lack of useful information generated for patients or clinicians. In those situations, it should be considered whether the objective should be altered to focus on the mechanistic understanding that a study might generate, whether a different position in the pathway should be chosen, or whether it should be abandoned altogether.

35.4 Setting minimum requirements

Assuming that addressing the objective would result in a clinically useful test, as confirmed by the aforementioned approach, the sensible next step is to assess the minimal requirements needed to address it. Simply put, if these requirements are not fulfilled, the answer to the question would be unreliable, and the probability of successful validation decreases significantly. These requirements will differ vastly in different studies, depending on its stage on the road to clinical practice. The requirements for a proof of concept study can be quite lenient, but a clinical implementation study will involve strict requirements that must be met, often enforced by regulatory bodies. The following questions are useful to identify study requirements.

35.4.1 Question 1: what breath compounds should be detected?

There are two main types of breath studies; evaluation of targeted panels of compounds that are known a priori (*targeted*) and biomarker discovery studies without predefined target compounds (*discovery*) (see Chapter 1). The requirements for these two types of study are very different.

For untargeted biomarker discovery studies, typically the aim will be to detect the largest number of breathborne chemicals in the most reliable way. Considering the very wide range and characteristics of exhaled compounds, biomarker discovery studies are the most demanding studies in terms of breath collection and analysis (see Chapter 2). Without an adequately developed and validated biomarker discovery platform, the risk of both false-negative and false-positive findings is considerable. An inconclusive outcome will lead researchers to ponder: are there no biomarkers for the target application?; or, was the analytical approach not sufficiently robust and sensitive to detect them?; or, can the candidate biomarkers found be trusted to relate to biology rather than instrument variability?

Based on experience, the authors would advocate that discovery studies must always have a semitargeted aspect. Provided the data are available, a discovery method could be based around compounds from literature and a fundamental understanding of the pathophysiology of the disease, by targeting specific groups of chemicals that are of interest. If relevant data are lacking, at the very least, it should be ensured that the end-to-end workflow reliably detects omnipresent breath compounds that are well established to be representative of the general population, such as acetone and isoprene, alkanes, hydrocarbons, and alcohols. Failing that, it could be questioned whether breath has been sampled and/or analyzed at all. If this can be demonstrated, then at the very least it will be possible to make definitive statements about this subset of compounds.

Such an approach is unfortunately not feasible for sensor-based approaches based on pattern recognition (see Chapter 13). This drastically increases the risk that a study will not provide novel insights. The tools to identify false negatives and false positives are very limited in these situations, and parallel analyses with analytical chemistry techniques to identify candidate compounds are highly recommended. It is without question that any discovered biomarker (profile) requires appropriate prospective blinded validation.

A requirement statement for an untargeted biomarker discovery study is challenging but worthwhile, and could look something like:

Reliably detect at least 50 compounds originating from the breath of each study subject, which must include acetone and isoprene for each subject.

Or, for a sensor-based study, a minimum requirement statement might be:

Detect at least five sensor features that reliably differentiate equipment blanks from breath samples.

Setting (and evaluating) requirements for studies by evaluating a list of targeted compounds is often more straightforward. This permits the creation of a list of compounds that should be detected, their expected concentrations, and the desired relative standard deviation. A requirement statement—relating to relative standard deviation, RSD—could look something like:

Identify and quantify propanol within a range of 20–40 ppb with an end-to-end process RSD<20%.

Such requirement statements will be very informative when thinking about appropriate breath collection, sample handling, and analysis techniques.

35.4.2 Question 2: what is the required diagnostic accuracy?

By thinking through the utility of an intended breath test, it should become relatively clear what the minimal required performance is in order for the test to be useful. Depending on the maturity of the test, this requirement can be treated more leniently. For clinical implementation studies, the required diagnostic accuracy should be predetermined.

For a pilot study, a requirement statement may simply be:

To estimate the sensitivity and specificity of the breath test with a maximum 5% error.

For a study designed for regulatory approval, it will be a more specific statement, such as

The breath test must have a sensitivity of 95% and specificity of 80% in the target population.

35.4.3 Question 3: what is the target population?

Again, consider that the utility of the breath test should lead to clarity in relation to the target case and control populations. In principle, a study population should reflect the intended use population as closely as possible. Unfortunately, this is not always possible (especially in early-stage studies), but should nevertheless always be considered carefully. The literature contains abundant examples of tests that deliver exciting results in very clearly defined subgroups but that never transition into clinical practice.

For early-stage and mechanistic studies, studying a specific subgroup can be a very useful strategy. A specific group might have more advanced disease, thereby potentially eliciting larger analytical signals. Other preferred groups may be more homogenous or have fewer comorbidities, thereby reducing noise that is unrelated to the target disease. Likewise, a study using an intervention, e.g., to induce or suppress a disease process such as inflammation, can be profoundly valuable to identify biomarkers in a controlled experiment. Sometimes these approaches are even inevitable when the cost of accessing the intended use population would be prohibitive at an early stage (e.g., cancer screening studies). Once candidate biomarkers have been identified, a targeted assay can be developed that can be evaluated in a subsequent validation study in the intended use population. Combining this with a detailed understanding of the origins of the target VOC enables the strategy to be derisked. Ultimately, this approach could aid in developing a bespoke assay that is specifically sensitive to target biomarkers, allowing detection of more subtle disease processes.

Nevertheless, it is important to be aware that identifying a biomarker in a specific subgroup assumes that the biomarkers will be the same in the intended use population. This assumption is certainly not without risk and is best avoided whenever possible by incorporating the intended use population early on during biomarker discovery. The latter is especially true for off-the-shelf available sensor-based systems, in which the detection system (i.e., the sensors) cannot be optimized for a set of target compounds that are not known a priori. As such, an assumption is made that the biomarker panel detected with that specific technology in an initial discovery phase directly translates, without alteration, to the intended use population and all future devices of the same type. Understandably, to date no one has been able to prove this without understanding the underlying compounds that elicit the signals on the sensory array. This challenging aspect makes studying the intended use population (both cases and controls) crucial for sensor-based biomarker discovery studies. Failing that, there is a high likelihood that any derived response profile will never be impactful in clinical practice. This will be less of a challenge for tailor-made sensors that are specifically sensitized to be selective for individual (subsets of) compounds.

An example requirement statement for a treatment response study is:

Mild to moderate asthma patients eligible for treatment with drug X, prior to initiating treatment must be recruited.

35.4.4 Question 4: how long will the study take?

Virtually all measurement tools drift and vary in their performance over time. This is true for a simple kitchen scale and definitely for highly sensitive and complex analytical equipment used for breath measurements. A particular requirement for breath studies is therefore to make careful consideration of the minimum time required to complete the study. Study duration obviously relates to the desired study sample size (see below), but could also be linked directly to the trial objective. Any longitudinal/interventional study with repeat sampling will have a minimum time for which the study must run. As the storage of breath samples is limited (see Chapter 2), the sampling platform and analytical equipment must meet predefined technical specifications that ensure stability and suitability for the duration of the study. The longer the study, the more challenging this requirement becomes to fulfill.

An example requirement statement is:

The end-to-end process RSD must be <30% for at least 6 months for 10 representative volatiles.

This list of questions pertaining to minimum requirements is by no means comprehensive, rather it is intended to touch upon some key aspects that will impact the trial design. Each study is likely to have additional requirements specific to that trial that will help guide the study design. Often, merely going through the steps outlined above will lead to a refinement and sometimes reconsideration of the objectives. Once the objective and related requirements are clear, it is very useful to think about the specific context within which this question will be addressed.

35.5 Constraints and context of the study

Consideration of the context in which the research objective will be addressed will enable explicit evaluation of the circumstances in which the results will be delivered. Competence in this task will lead to an increased likelihood of achieving impactful research. The procedure for considering context is quite straightforward, but calls for honesty and a willingness to address difficult questions relating to strengths and limitations.

35.5.1 Common research constraints

Every research project must operate within certain constraints that cannot be violated. Many of these will be specific to a research group and/or project, while others are specific to breath research. A conducive approach is to create a list of constraints that can be used as limitations when designing a study. The following list of questions are useful prompts to get started, but there will likely be specific issues relating to particular circumstances.

- Budget: How far will the money for this project stretch? It is wise to calculate some contingency for unexpected costs; 25% is realistic.
- Human resources: Is there staff available to execute the study? Is this a reasonable amount of work? Do they have the relevant skills?

- Timeframe: When must results be delivered to satisfy the funding body/be included in an academic qualification (e.g., Ph.D.)/support the next grant?
- Patient access: What type of subjects can be recruited and easily accessed?
- Sample size: What is the minimum number of subjects that should be recruited in relation to the expected effect size and target accuracy?
- Reference diagnosis: How accurate is the current diagnostic test? For diagnostic accuracy studies, the reference standard cannot be outperformed, but rather sets a hard limit on the maximum test performance obtainable. This also holds relevance for related power calculations.

35.5.2 Constraints specific to breath-based tests

If the maturity of breath research is compared to that of, say, genomics, it is still very much in its infancy. As such, breath research is an exciting field to work in, with the promise of major discoveries yet to be made. At the same time, however, this means that there is not yet a full understanding of the performance (limits) of the end-to-end analytical tool chains employed. This is compounded by the fact that many of these technologies are continuously evolving or are experimental prototypes, such as most sensor-based systems. Notwithstanding these constraints, the highly variable and volatile matrix of breath creates yet bigger issues. Other chapters in this book deal with the intricacies of specific analytical tools, thus this chapter will only cover general concepts when they are crucial in relation to study design.

A prototypical breath research study tries to detect minute changes in trace amounts of chemicals within a very complex matrix. This enforces strict requirements on the collection, storage, and analysis of breath (see Chapter 2). It is imperative to understand what a particular set of tools can and cannot achieve. A useful way to think about this is to "follow the path of the biomarker" and consider aspects such as:

- Sampling: How clean is the analytical setup? What volume should be sampled in the preconcentration step? How reliable and reproducible is the sampling? Does the setup remove VOCs or introduce artifacts?
- Storage: What is the impact of sample storage (and transportation)? How will the choice of sorbent affect sample stability? How well can the sample be preconcentrated?
- Analysis: How stable is the analytical technique? What is the accuracy and precision of the technique for target compounds? How many compounds can the technique discover in a typical breath sample? Can I use multiple analytical systems in a single study?

In its essence, it is a matter of finding the technology that best matches the requirements of the experiment, as has been outlined in the framework above. Some researchers will be fortunate enough to work in a laboratory with strong expertise in using a set of techniques for breath analysis, offering access to a wealth of knowledge required to understand instrument performance characteristics.

On the other hand, if this expertise is not directly available, it will be essential to find a collaborative academic or commercial partner that offers the relevant complementary expertise.

In relation to sensor-based systems, purchasing an off-the-shelf system can be an attractive proposition, yet it is essential to ensure that the supplier understands the performance characteristics and/or that adequate support is available, as well as allocating sufficient budget and time to perform the required performance evaluations. There is a great danger to assume that a device will be eternally stable, completely comparable to others, and will never break down; this is never the case.

Regardless of whether breath analyses are performed on or off site, it is a very valuable exercise to try to fill in the blanks in the following example statements, in order to understand the limits of a particular technology.

For an untargeted biomarker discovery study:

The chosen end-to-end workflow can detect approximately X exhaled breath compounds in more than Y % of a homogenous population, with an RSD < Z, for a period of T months.

For a targeted study:

The chosen end-to-end workflow can quantify target compound X with an RSD of Y at a concentration range of Z.

Obtaining exact numbers for the parameters above might be especially challenging for untargeted studies, yet order of magnitude estimates are usually sufficient to understand how they might affect a trial design. If even ballpark estimates are elusive because the method or technology is new, then it is highly advisable to run some standards, generate calibration curves, and analyze test samples to learn about the instrument performance. In any study, the performance should be quantified during the course of the complete study using quality control (QC) measures and calibration standards. In an ideal situation, this allows to correct for instrumental drift or variability. At the very least, it is necessary to quantify change over time and identify overt instrument failures. The requirement to run calibrations is also true for sensor-based studies, albeit with the additional challenge of not knowing which compounds are being detected, thus making reliable calibrations challenging if not impossible. This poses a significant challenge to overcome when conducting discovery studies using pattern recognition–based approaches.

As a general rule of thumb, the higher the variability that cannot be reliably corrected for, the smaller the power for detecting differences between groups. This type of variability increases over time and in multicenter studies, which has an important consequence: running a shorter study increases the power to find relevant differences. The variability of the toolchain employed may grow so much over time that running a shorter study with fewer subjects could result in having a higher power than a longer study with more subjects. Likewise, if the difference between the (collection) devices is large, using fewer sites for an initial discovery study might be better suited. These are crucial design considerations for a clinical study, thus understanding such constraints deserves close attention.

In the future, reliable storage solutions will be hopefully developed to allow batchwise analysis, thereby minimizing the challenge of temporal variation in discovery studies. For targeted studies, this is less of an issue, provided frequent calibrations with reliable standards for the target compounds are carried out.

35.5.3 Strengths

Strengths are as important as constraints. These will mostly be specific to a research group. Think through what specific skills and expertise are available for the study compared to other groups or previous studies. Thinking explicitly about those strengths allows them to be leveraged to benefit a study design. Strengths could include access to (rare) well-characterized patient populations, efficient recruitment, high accuracy and/or sensitivity of analytical tools, and so on. Importantly, collaboration networks and the breath community as a whole can be a great source of strength and support. Science is a team sport, so it is sensible to involve suitable experts early on in any study. External experts can aid in quantifying and alleviating constraints, by underpinning them with hard data. A failure to do this could cause issues when publishing the outcomes of a study, when scrutiny by the community cannot be countered by unsubstantiated assumptions.

35.6 Define experimental design

Working through the two previous steps may feel like a lot of effort prior to actually thinking about study design, but thorough treatment of the previous issues, with the objective and ground rules established, will make it much more straightforward to design a robust study.

It is strongly advised to approach thinking about study designs as a creative process. Gathering a variety of experts together to brainstorm on various aspects of the work is a great way to do this. Each scientific objective deserves a bespoke design, so recommending specific study designs in this chapter would be misleading. Published guidance documentation for evaluation of omics techniques are a useful reference here.[2] Once ideas have been collected, ask the following questions:

- Would this experiment address the study objective?
- Would these data be meaningful for clinical practice?
- Are all of the minimal requirements met?
- Are any of the constraints violated?
- Is existing expertise leveraged to its maximum extent?
- What assumptions are made and can these be substantiated?

It is likely that there will still be some outstanding challenges that need resolving at this stage. The following list is by no means comprehensive, but provides some common considerations that are useful when working through a study design:

- *Keep it simple*: A simple study design usually carries less risk. If the prospective biomarkers are not well established and the constraints not well understood, it is recommended to maintain a simple study design in order to avoid taking unnecessary risks. An adaptive trial design that includes prespecified decision points about continuation of the study can be a sensible approach.
- *Early-stage discovery versus validation study*: Early-stage biomarker discovery or mechanistic studies can benefit from more artificial design constraints and inclusion and exclusion criteria. It is important to note, however, that the less the criteria reflect the intended use population, the higher the risk that the biomarkers will not translate to clinical practice. Validation studies should always be conducted in the intended use population and introduce the variability of multiple sites, multiple users, multiple equipment, comorbidities, and so on.
- *Suitability of the analytical tools*: Different research questions require different tools to address them. For mechanistic studies, it is necessary to understand the chemical identities of compounds using sophisticated analytical chemistry instrumentation. Proof-of-concept studies can be undertaken using sensor-based devices, but if the objective is to validate biomarker signatures and move these to clinical practice, unequivocal identification of the target volatiles is unavoidable. Tailor-made sensor systems are highly suited for translation to the clinic, especially in acute settings when an immediate answer is required. Calibration of these devices, however, is an absolute requirement for translation to clinical practice to the point where it is a requirement enforced by regulatory authorities.
- *Technology readiness level (TRL)*: Be realistic about what a chosen technology can and cannot do. Understand that if a validation study is to be used for regulatory approval of a medical test, it will be necessary to go through assay development and validation.[2] This is a discipline in its own right, thus additional expertise is needed at this stage.
- *Recruitment rates*: A good rule of thumb is that recruitment rates will be half of the best estimates, unless the same population from the same site has been recruited before.
- *Study power*: Considerations on study power depend on many factors, such as the target population, or the intended outcome of a study. Specific considerations are as follows:
  - For a targeted study it is advisable to establish the intra- and interindividual variability of the target compounds to assess the index of individuality in support of a power calculation.
  - Collecting repeat samples in the same subjects is a useful procedure to deal with limited study power. It increases the power of a study without requiring recruitment of further patients.

- If the study population is small, the analytical signal is small, or there is considerable biological variability, it can be worth considering an exogenous VOC (EVOC) probe strategy.[3] This approach requires a detailed understanding of the biology and careful consideration of administration and dosing, but it offers the benefit of derisking a study, it reduces the required sample size, and it provides clear mechanistic insights into the origins of prospective biomarkers.
- When studying the intention to diagnose a population, power can be increased by evaluating possibilities to enrich the population for the target event, e.g., an exacerbation or cancer.
- Increase the volume of breath collected in order to detect lower concentrations of compounds, but bear in mind that high volumes could result in breakthrough when sampling on sorbent tubes.
- Be mindful of the "*winner's curse*": Any positive pilot study is more likely to have an overestimated effect size compared to the real between-group differences. A larger sample size than the initial calculation would suggest might be needed to validate a study.[4]

• *Time constraints on analytical stability*: Consider ways to boost recruitment or alternatively run stability validations prior to (1–2 months) collecting the clinical samples, thereby providing an indication of when continuation of recruitment becomes futile.

Besides these design considerations, the following guidelines can help to increase the chances of extracting meaningful knowledge from study data:

• *Biological validation*: Whenever possible, integrate other more established disease metrics: Can compounds be linked to some of these metrics? Does this reveal something about the underlying biology? Even if no clinically useful difference can be found, meaningful conclusions can nevertheless be made.
• *Statistical analysis plan* (*SAP*): Write an SAP after the study design has been completed. An SAP is tremendously useful to discover whether the data being collected will answer the research questions being posed. The involvement of a statistician in this process is essential.
• *Quality control strategy*: An integral part of any study design is a quality control and calibration strategy. In many cases, it will be beneficial to include extra methodological samples and/or blanks paired to patient samples.
• *Methodological aspects*: Adding a methodological evaluation step can improve a study in three ways: (1) it can allay concerns of unquantifiable constraints that might exist; (2) it contributes to general progress and understanding in the breath field; and (3) it can provide insights that are useful in designing the next study.
• *Leverage strengths*: Consider whether personal know-how and strengths are utilized to the fullest extent. Will the execution of the study offer a personal (unique) knowledge gain or the development of a new skill set, regardless of its outcome?

- *Reporting templates*: Use guidelines for reporting on diagnostic studies, such as the STARD initiative,[5] as a checklist when designing a clinical study. Submit the study design for evaluation by a regulatory body prior to initiation if it is to be used for regulatory approval.
- *Collaborate*: If certain skills, tools, or resources are missing from the pool of available expertise, collaborating with external experts is an excellent way to increase the probability of success.
- *Externally validate*: External validation is a crucial step to assess the validity of a breath compound as a biomarker. Incorporating a validation step into any clinical trial is typically a good idea. This can be achieved by incorporating specific aspects, such as collecting samples at multiple sites, running on multiple instruments, and collecting across multiple populations. Ultimately, true external validation to assess performance requires prospective recruitment of a novel population not utilized for the biomarker discovery; without this a reliable assessment of performance cannot be made.

Finally, after successfully establishing a satisfactory study design, running a study premortem can be of great benefit. Assuming a completed project did not provide any interpretable outcomes, go through the following exercise with all of the experts involved: Why did it happen? What aspect of the design might have led to this outcome? Is this outcome inevitable? Does it undermine the experiment? Or is it acceptable as a calculated risk? Can specific strengths of the experts mitigate this issue? At this stage it is useful to think this through, as it is the best way to clarify remaining risk and assumptions.

Following this process, hopefully there comes a point when the remaining risk can be considered acceptable. It is only at this stage that a full protocol for the planned study should be written. Since most aspects have now been considered in full, this task should be quite straightforward. If any doubt remains, consider adjusting the objective accordingly. A well-addressed, less ambitious objective is far more valuable than a poorly executed study with unrealistic objectives. Alternatively, consider running a small mechanistic pilot study to quantify some of the risks before commencing with the full study.

35.7 Summary

This chapter has outlined a three-step approach to thinking about clinical trial design. Firstly, think carefully through the study objective, as well as its relevance to patients, clinicians, and investors. Set minimum requirements for addressing this objective and try to specify these as best possible. Subsequently, think through what constraints must be conformed to and what strengths can be leveraged. Finally, design the study and carefully evaluate if it addresses the objective without violating the constraints while taking appropriate measures to assure validity of results. Taking this approach ensures that opportunities are more readily identified and risks can be more easily quantified. Although this approach may initially seem onerous, it is worth the investment. Ultimately, a systematic and thoughtful design process, as outlined here, will lead to improved studies that deliver tangible results more quickly, and equally reduces the risk of pursuing useless endeavors.

It is impossible to mitigate all risks and sometimes delaying the start of a trial is not sensible (despite potential risks), as waiting longer might worsen constraints (time, budget, etc.). Regardless, it is crucial to be aware of any remaining risks and their potential impact. If risks prior to study initiation are not considered, it will be impossible to mitigate any shortcomings and the likelihood of unexpected failure increases. With the concerted efforts of the breath, engineering, and analytical chemistry communities, the current constraints on breath research are likely to lessen over time. The only way to achieve that, however, is to respect current limitations. The steps outlined in this chapter are intended as a useful aid to help raise the bar on study design in breath research. Ultimately, optimum study design will help to expedite the translation of breath research into the hands of patients and doctors, where it can improve and save lives.

References

1. Sterk P, Rabe K. The joy of writing a paper. *Breathe − Eur Respir Soc* 2008;**4**:224−32.
2. Micheel CM, Nass SJ, Omenn GS, Policy HS. Evolution of translational omics lessons learned and the path forward. *Evolution* 2012:1−274.
3. Gaude E, Patassini S, Boschmans J, Nakhleh MK, van der Schee MP. Targeted breath analysis: exogenous volatile organic compounds (EVOC) as targeted metabolic probes in Breath Biopsy. *J Breath Res* 2018;**13**(3):032001.
4. Tugwell P, Knottnerus JA. A statistic to avoid being misled by the "winners curse". *J Clin Epidemiol* 2018;**103**. vi−viii.
5. Bossuyt PM, Reitsma JB, Bruns DE, Gatsonis CA, Glasziou PP, Irwig LM, et al. The STARD statement for reporting studies of diagnostic accuracy: explanation and elaboration. *Ann Intern Med* 2003;**138**(1):1−12.

CHAPTER

Challenges in clinical breath research development

36

Steven F. Solga[1], Lisa A. Spacek[2], Terence H. Risby[3]

[1]*Department of Medicine, Perelman School of Medicine, University of Pennsylvania, Philadelphia, PA, United States;* [2]*Department of Medicine, Sidney Kimmel Medical College, Thomas Jefferson University, Philadelphia, PA, United States;* [3]*Department of Environmental Health and Engineering, Bloomberg School of Public Health, Johns Hopkins University, Baltimore, MD, United States*

Chapter outline

36.1 Overview

Breath analysis has many advantages over other diagnostic tests. Moreover, the academic literature describes many possible breath metabolites that may be important for human health. Most, however, are not developed beyond the pilot stage, and few make it to commercial success. This reality leads to a "fatigue of unmet expectations," which further diminishes enthusiasm for research. Medicine is an imprecise art and science that relies on imperfect measurement standards, including the physical examination, blood and urine assays, pathology, and radiology tests. Though flawed, these standards, in many instances, have been in place for decades and their uses and limitations are generally understood. Clinicians are therefore comfortable that they can place familiar data in context and make sound, if not ideal, decisions.

Medical care is the practice of medicine to prevent or treat a disease and therefore involves high-stakes decision-making. Practitioners are duty bound to provide standard of care. Therefore, it is neither easy nor obvious for novel diagnostics to

Breathborne Biomarkers and the Human Volatilome. https://doi.org/10.1016/B978-0-12-819967-1.00036-0
Copyright © 2020 Elsevier B.V. All rights reserved.

Table 36.1 Clinical breath research challenges.

Technical/Scientific Factors	
Monitor/interface/biology	Too many interrelated unknowns
Unique data = uncertain utility	Relevance difficult to establish Relevance may not exist
Nontechnical Factors	
Inadequate teams	Engineers, chemists, physicians, statisticians
Inadequate synergy	Single-center efforts
Lack of focus	Too many diseases, too little strategy
Lack of common languages	Device development is not drug development
Few models of commercial success	Difficult to envision endgame

break in. This is equally true for a novel blood test, histology stain, or magnetic resonance imaging (MRI) acquisition sequence as it is for breath biomarkers: new tests must meaningfully address an unmet need to improve patients' lives.

While there are numerous challenges in clinical breath research development (Table 36.1), the most important are metabolite selection and development vision.

36.2 Metabolite selection

Wherever possible, novel biomarkers should be selected and classified by their (1) intended *purpose*: i.e., susceptibility/risk, predictive, prognostic, diagnostic, monitoring, or response, (2) intended *rationale*: i.e., the new measurement must offer some advantage over the existing standard, it can be faster, cheaper, more accurate, or easier to use, or it can describe some new putatively important biology, and (3) intended *user*: i.e., physician, hospital, medical laboratory, or consumer. Moreover, at the earliest possible development stages, plans should be in place to test biomarkers against gold standard tests, or validated surrogate and/or clinical endpoints, and always have an appropriate clinical context in mind. No one wants a new test unless there is a compelling reason.

One important practical obstacle to consider is an appraisal of the known biology.[1] It will be especially difficult, for example, to develop an empirically identified biomarker with an entirely unknown biochemistry. Clinicians must have some physiological basis to consider in using a new test, and the onus will be on the development team to put this forward. Indeed, even those that have been studied for decades require considerable effort to adapt to breath measurement. Ammonia, for example, has often been considered a good target for development because it is important to health and disease. Standard measurement in blood requires phlebotomy and is subject to significant error.[2] Therefore, a breath measurement option could provide opportunity for improved clinical management. This is especially true since ammonia physiology is changeable, and a safe, easy, point-of-care (PoC) test can enable repeated measures studies that previously have not been possible.

However, breath and blood ammonia cannot be easily compared: they are derived from different body compartments and have different kinetics and physiological influences, and their measurement has different sources of error. Blood ammonia measurement, for example, can be influenced by a variety of preanalytical factors, including tourniquet time and skin sweat during phlebotomy, concurrent levels of various enzymes within the blood, including alanine aminotransferase (ALT) and gamma glutamyl transferase, pH, temperature, hemolysis, and time to specimen processing.[3] Samples will vary according to arterial, venous, or capillary sources. In contrast, breath ammonia measurement can be influenced by other factors, including oral bacteria, dentition, salivary urea, rate of breathing, time since last oral ingestion or food intake, and tobacco use.[4] Hence, while blood and breath measurement monitors may be equally good at precisely measuring the same sample in the laboratory, there may be wide and meaningful variance when collecting samples from a person.

Furthermore, there are multiple physiological sources and sinks of ammonia, including the liver, kidneys, muscle tissue, gut mucosa, and gut bacteria.[5] Moreover, although increased systemic ammonia is pathologic in various disease states with important clinical consequences, there are no firm clinical endpoints that are easily measurable either (i.e., elevated ammonia can cause brain confusion, which is also difficult to accurately measure and has numerous influences, too).[6] Finally, despite decades of ammonia research resulting in many thousands of publications, there is almost no useable information about the ammonia response to routine physiological events and influences (e.g., the effect of food intake). Hence, even if a team hypothetically developed a "perfect breath ammonia monitor," there is no straightforward pathway to verify that this claim is true or establish its utility.

Similarly, the measurement of other metabolites will incur different specific challenges, but with the same global theme. Trace, endogenously produced gut-derived ethanol, for instance, might be pathogenically relevant,[7] but this has been only a hypothesis for many years, precisely because gut-derived metabolites are exceptionally challenging to measure by any method: gut ethanol is produced only at low levels and its production is influenced by factors including diet, gut transit time, gut bacterial composition, and the individual aldehyde dehydrogenase activity of the host.[8] There is no validated blood test for endogenous ethanol, and the usual "partition coefficient" of 2100:1 for breath/blood comparisons for exogenous ethanol cannot be assumed to apply[9] (see Chapter 29 for breath ethanol measurement in law enforcement).

Thus, the trials and opportunities of each metabolite must be considered individually and thoroughly. Nonmedical scientists and engineers commonly make the serious mistake of ignoring biological confounders. There has been a misguided but longstanding belief that breath analysis can identify and measure unique markers of disease. Unique markers in breath can only originate from the ingestion, inhalation, or dermal absorption of foreign substances or from bacterial, viral, or fungal metabolism. There are few yes/no tests in medicine, e.g., the presence or absence of a virus. In contrast, nearly all biomarker and metabolite tests measure a change in concentration that must be placed in a biological and clinical context.

Metabolite selection is, therefore, an exercise to determine if the unmet need is compelling enough to overcome development obstacles. And it is possible that the best answer could vary by indication. Taking breath ammonia as an example again, its measurement could be attractive to a research team as a diagnostic biomarker in patients with liver cirrhosis.[10] Because cirrhosis affects tens of millions of patients, the market potential is correspondingly quite large, and this may help raise funds for research. But cirrhotic patients always have a variety of comorbid conditions and confounders that affect ammonia physiology, such as kidney or lung disease, multiple medications, and so on. Therefore, development costs of a specific test could be high. In contrast, breath ammonia measurement could also be attractive for routine monitoring of patients with pediatric urea cycle disorders.[11] These patients are younger, have fewer medical conditions, and are thus more biologically straightforward. Further, the unmet need is compelling. On the other hand, pediatric urea cycle disorders affect only a few thousand patients worldwide, and therefore study recruitment may be difficult and return on investment may be low.

In summary, when considering candidate metabolites, some clinicians and investors may decide there is 'too much white space." Worse, others may not appreciate the white space at all. Ideally, successful teams will identify targets with manageable unknowns for an important unmet need.

36.3 Development vision

Breath monitors are noninvasive diagnostic medical devices. Unlike drugs or invasive medical devices, they are lightly regulated. This may be considered an upside because it results in a low barrier to entry. As a practical consequence, however, teams that are excused from the requirement to plan and fund a complete development pathway, from pilot studies to convincing pivotal clinical trials, tend not to. And, this mistake, i.e., lack of a development strategy, is both encouraged and compounded by a lack of a well-known development sequence.

Drug development, for example, progresses in a stepwise manner through a series of mandatory preclinical tests prior to first exposure in a human subject. Equally, clinical testing develops stepwise from phase I safety studies to phase II dose finding, safety, and initial efficacy studies to phase III pivotal randomized controlled trials. Pharmaceutical companies employ teams for each of these stages: laboratory development, clinical experimentation, regulatory submission, synthesis and mass production, packaging and distribution, and marketing and advertising. Exceptional individuals may serve as a liaison between teams or oversee the entire project but most stay within their domain.[12]

Because of this regulatory requirement and structure, companies must continually make "go or no-go" decisions and ensure funding and cohesion with the endgame in mind. Multiple disciplinary teams are essential. Companies that make the wrong decisions fail. Here, it should be appreciated that it costs approximately 600 million to over 2 billion US dollars to bring an average new drug to market, and this process takes between 15 and 20 years (split between preclinical and clinical development).[13] Further, more than 90 percent of new drugs fail in development.

Unfortunately, because breath diagnostics lack a regulatory structure and their attendant requirements, the first challenge is the recognition that these development decisions still exist. Here, we recall a cliché variously attributed to Benjamin Franklin or Winston Churchill, "failing to plan is planning to fail." History books are filled with anecdotes about small biotech companies with laboratory brilliance and promising products that failed due to poor development strategy.[12] While the costs and timelines for noninvasive diagnostics is considerably less than for new drugs, significant research and development are still required.

The second challenge is agreement on the development sequence and next steps. Here, there is risk for miscommunication among team members with diverse knowledge and skills. Because noninvasive device diagnostic development lacks a correlate to the familiar drug pathway (i.e., preclinical, phase I, II, or III clinical study), individuals may disagree on a matter as basic as the current status of a project and next steps. Monitor engineers, for example, may naively believe that laboratory precision testing (i.e., analytical validation) should suffice for clinical precision testing (i.e., clinical validation). Similarly, clinicians may naively believe that the existence of a single prototype indicates that engineers can easily scale up to multiple devices for multicenter studies. And both groups may underestimate the challenges, including the costs, timelines, and uncertainties, of the other.

A third challenge is the assembly and maintenance of a coherent team with the requisite expertise. For breath analysis, the core scientific team should include engineers or physicists, breath chemists, physicians or application specialists, and statisticians. The business team should include those with expertise, specific to medical devices, in raising capital, intellectual property (IP), and promotion. Honest and recurring appraisal is needed to address shortcomings, including critical feedback by third parties. University innovation centers and venture capital may present opportunities.

A final challenge may be the identification, protection, and sharing of IP and the potential rewards from commercialization. Again, breath biomarker development may be more challenging to organize and manage than new drug development. A pharmaceutical company will typically own all the IP related to a novel chemical compound and employ the entire development team (from preclinical to marketing). Therefore, IP and overall goals are easily harmonized. In contrast, breath biomarker measurement is less obviously protectable. Though there are many elements required for success (biomarker selection, clinical research, breath collection process, and monitor design), only the last of these may be protected and therefore have value. Unfortunately, even this value may be transient: because there are often multiple reasonable engineering approaches to monitor design, a successful demonstration by one outfit may result in multiple competitors or "me-too" reproductions.

A larger jeopardy, however, is team fracture: if the only meaningful value in the enterprise is monitor design and construction, then the engineers must be willing to share this value with the nonengineering members of the development team. This can be difficult because, as noted above, research costs, timelines, and output for both the engineering and clinical components are speculative. And, unlike the pharmaceutical company model, breath research team members are not usually employed by a single entity. Therefore, commitment and interest may wane if these issues are not addressed early and on a recurrent basis. Nonengineer team members, including clinicians, statisticians, advisors, and financial supporters, will not commit meaningful resources without a stake in the outcome.

36.4 Practical examples and costs in biomarker development

Metabolite selection and development vision are interrelated. An attractive idea may not be practical to develop or prove useful in the clinic. Further, while many researchers may have laudable academic or altruistic motives, serious clinical development requires sustained investment and therefore the product must have a realistic commercial goal.

A few examples are offered here in order to further illustrate this. Examples of poor choices for breath biomarker development are those designed to duplicate routine blood tests or serve as cancer detection biomarkers; the better choices are those designed for metabolism monitoring.

36.4.1 Poor choices

Many blood tests are imperfect, but this does not mean that even a good breath test will replace them. For example, there has been a longstanding interest in tests that evaluate liver inflammation or function. One such blood marker is ALT, where an increased ALT is used as a marker of liver injury. Though individual laboratories still define a normal value differently, the American Association for the Study of Liver Disease (AASLD) defines normal as 19 or 30 international units/liter for women or men, respectively.[14] Mild and moderate increases are variously defined as up to c. 200 or 800, respectively, while severe increases can reach >15,000. ALT is an imperfect biomarker: it does not always correlate with "gold standard" (liver biopsy) and offers little information about the etiology of inflammation, liver function, reserve, or prognosis.[15]

Hypothetically, imagine that through genius or luck, your team has developed a "perfect" breath replication of ALT (b-ALT). Will clinicians and thought leaders at the AASLD endorse it and incorporate it in their own research and practice? Probably not. Despite all of its known problems, blood ALT would continue to be used: its biology is reasonably well understood and has been in use in liver research and clinical care for decades. Moreover, the blood assay works well through the entire range, from normal to severely increased, and is proven valid across all demographic groups. Finally, and perhaps most importantly, it is part of a panel of blood tests that clinicians routinely order anyway (e.g., the comprehensive metabolic panel [CMP], which also includes multiple other tests of interest including sodium, potassium, and others). Since clinicians frequently desire these other results as well, they will continue to order the CMP regardless. Therefore, even a "perfect" b-ALT test for a common clinical need would be used rarely.

Cancer detection is another poor choice. It is reiterated here that there are no unique biomarkers of disease. All currently used cancer biomarkers (e.g., carcinoembryonic antigen; CEA) are normally present in healthy persons. Clinicians may only infer that cancer might be present based on increased absolute values or their change in relative concentration over time. For all tests, sensitivity and specificity vary according to cut points, and their ultimate utility must be placed in a clinical context.[16] As a practical matter, this requires a lot of research effort. In order to work well, a candidate test must be evaluated not only in a few putatively "healthy" subjects versus those with a target disease, but also in thousands of subjects in all demographic groups with various comorbid conditions. For example, the median CEA of five patients with metastatic colon cancer is much higher than the median value in five young healthy volunteers; the "p value" is impressive, and the result is reproducible. In order to be valid across a population, however, candidate tests must be evaluated in all demographic groups with a variety of comorbid conditions. A thorough evaluation of confounders should be done for any indication, but is especially important for high-stakes testing such as cancer detection.[17] Once this was done (there are several thousand academic publications on CEA and cancer), CEA was found to be useless as a screening test and is currently used only occasionally to monitor colon cancer progression once it is found by other methods (i.e., proven screening tests, such as colonoscopy or stool-based testing).[18] And this is despite the fact that the precision of the blood test is good and so is our understanding of the underlying biology.

Still, the development of CEA may be considered a rare success because it has any clinical use. Today, major universities have biorepositories with large numbers of banked samples, usually blood, to screen for new putative markers. In spite of this substantial research effort, new markers have not yet matured into use. Given the requirements of successful development, the reality that breath tests are often imprecise, and this competition, it seems unlikely that any breath biomarker will mature to clinical use in cancer detection.

36.4.2 Better choices

Breath biomarkers to monitor metabolism are likely a better and more promising choice for development. Breath testing is safe and noninvasive, and with engineering advances can become easy to use and offer immediate results. Therefore, its main advantage is that it is well suited for situations which require multiple measurements. An additional advantage is that breath measurement can be more sensitive than blood measurement (e.g., acetone) and/or a validated blood test correlate does not exist (e.g., endogenously produced ethanol).

Metabolic monitoring is one such scenario, and the metabolic response to food ingestion is a specific example. Little is known, for example, about the clinical impact of various quantities of macronutrients (e.g., simple vs. complex carbohydrates, fats, and proteins), or time of day influences on clinical endpoints, such as obesity, diabetes, and fatty liver. And this impact is personal, i.e., whether an afternoon or bedtime protein or carbohydrate snack is beneficial or harmful depends on a subject's individual metabolism, which is influenced by their genetics, comorbid conditions, and medications.[19]

Here, it is easy to imagine how to leverage breath's advantages: breath measurements can be obtained, as needed, repeatedly day or night and through fed or fasting conditions; this is not practical with blood-based testing. Moreover, protocols with repeated measurements can achieve high statistical power since each subject can serve as his or her own control. This is true even when the measurement result has a high standard deviation.

Thus, breath testing for metabolic monitoring can offer data that are unique, and research that is both pioneering and clinically relevant can be conducted. Moreover, metabolic monitoring can be applied to diverse applications, e.g., acetone monitoring is being used to evaluate dietary strategies for seizure prevention in patients with seizure disorders, weight loss regimens in obese patients, and exercise effectiveness in fitness enthusiasts (see Chapter 23).[20] Note that for these applications, breath acetone testing is not intended to "me-too" the blood test. Rather, with its capability for easy repeated measurements, it offers real, novel, and useful advantages over blood testing. Additionally, for the intended application of metabolism monitoring, these advantages are essential, whereas for liver inflammation evaluation, ALT is measured infrequently (usually monthly or less) and acetone would be measured multiple times per day. Moreover, breath acetone is more sensitive than its blood correlate.[20] Hence, this breath test as well as its application avoids the pitfalls of "b-ALT" versus (blood) ALT. Furthermore, breath acetone research and the development of these clinical applications builds incrementally upon an existing foundation of medical knowledge;[21] hence, it is not discovery. Further, breath metabolic testing can address common unmet needs without known alternatives. As such, breath test developers would be free of the jeopardy and burden of, for example, attempting to compete standard of care in a high-stakes endeavor such as cancer screening.

36.4.3 Clinical protocol costs

While simple proof-of-concept demonstrations can be done quickly and inexpensively, convincing clinical research is expensive.

A first step toward clinical validation is a thorough evaluation of the elements of precision, including repeatability, intermediate precision, and reproducibility. At a minimum, two identical monitors should be evaluated in tandem to determine agreement, and multiple testing sites should be employed. These steps are important because trace breath analysis is subject to contamination from ambient air, which may vary from site to site. Moreover, most metabolites are reactive and are being measured at low concentrations (often the low parts per billion range). Therefore, even if the monitor works well in laboratory testing using reference gases, clinical measurement incurs a battery of additional new variables (e.g., in breathing, pressure, humidity, and so on) that must be considered.

A second step is the exploration of the monitor's dynamic range. While the published literature may offer some guidance on the expected ranges of trace breath metabolites, each novel monitor and breath collection method must define its own limits and performance characteristics. This is because variability in sensors, monitor design, interface apparatus, etc., will impact results. Hence, breath ammonia studies that use monitors based on electrochemical sensors may significantly differ from those based on radiation. Here, successful execution of this step ideally requires some knowledge about the biology or host influences, which may lower or raise the metabolite of interest. For example, in breath ammonia, multiple oral factors, including pH and mouth rinses, influence breath measurement but are not reflective of change in systemic (host) ammonia levels (see Chapter 9). In contrast, vigorous exercise or high protein oral food ingestions are thought to increase systemic ammonia levels and thereby breath ammonia concentrations.

A third step is a characterization of normal ranges across demographic groups and various health and disease states. Because biology varies according to age, sex, race, weight, etc., diagnostic tests should be validated in multiple populations. Normal ALT ranges, for example, vary according to age, sex, or weight and are influenced by pregnancy. And baseline, fasting breath ammonia levels would be expected to vary by the absence or presence (and severity) of liver and kidney disease and nutritional status.

A final, and perhaps the most rewarding, step is the evaluation of the breath test for an intended application, e.g., breath acetone for weight loss management or breath ammonia for hepatic encephalopathy diagnosis. While this step can initially be a single-center exploratory study, clinicians and regulators will ultimately expect at least one, but preferably multiple, pivotal high-quality multicenter trials (i.e., the equivalent of a randomized phase III drug trial).

These steps do not always need to proceed exactly in sequence. After the first step (initial precision testing), it is possible to design clinical protocols for multiple purposes. For example, a protocol might evaluate time of day and demographic variables by having subjects provide multiple fasting breath collections from 8 a.m. to 11 a.m. at a single testing site. If this is successful, a follow-up protocol might have the same subjects return for a second testing day to receive a standardized meal and subsequently repeat the same timed collections. Even in this simple protocol, however, it would be important to recruit only healthy subjects within a narrow age range and to thoroughly prespecify eligibility requirements so that "healthy" is clearly defined. Moreover, all possible relevant demographic variables (e.g., weight, body mass index, medication usage, etc.) should be collected.

Furthermore, this should be done cautiously and only with the active engagement of dedicated clinicians and expert statisticians. A protocol with too many variables will fail and result in wasted time and resources, team demoralization, and missed opportunities. The most expensive clinical trial is the poorly designed protocol that does not answer any scientific question (see the previous chapter). And conversely, well-planned and executed research can be rewarded in unexpected ways. For example, while Pfizer was developing sildenafil for pulmonary hypertension, a serious but uncommon disorder, researchers were surprised in clinical trials that it could also be used to treat erectile dysfunction; it then became a blockbuster.

Quality clinical research requires money and time. For breath research, some costs will include

- Monitor maintenance, storage, and supplies (e.g., mouthpieces, etc.)
- Clinical research space, including waiting area and data collection area, utilities, access to computers, phones, etc.
- Institutional Review Board (IRB) fees for initial reviews, ongoing reviews, and protocol changes
- Research assistants trained in human subject research, data collection, and management
- Direct study costs (subject recruitment and reimbursement, supplies, phlebotomy, blood assays)
- Statistical support
- Insurance
- Publication costs and travel

Some costs are straightforward. Commercial IRBs, for example, will charge around \$5000–\$10,000 for an initial protocol review and around \$1000 for required annual reviews depending on a protocol's complexity. But even minor protocol changes and amendments require IRB approval and can cost an additional \$500–\$1000. Many universities will waive IRB fees for faculty pursuing investigator-initiated studies but will require that the faculty member both have and share IP with the institution. In general, IRBs additionally expect subject reimbursement, typically around \$20/h, with extra if phlebotomy is required.

Unfortunately, the most expensive costs are often less predictable, including study personnel and space. Professional research assistants and statisticians are paid around \$15–60/h and \$100–300/h, respectively, in the United States. Naturally, students and trainees can be paid less. But more experienced personnel have a greater likelihood of efficiently and correctly executing and analyzing a protocol with correspondingly less regulatory jeopardy than trainees. Here, note that noninvasive medical device trials are regulated as clinical trials. Therefore, despite minimal risks to human study subjects, protocols, case report forms, etc., must still adhere to Good Clinical Practice guidance as determined by international ethical and scientific quality standards.[22] Clinical research space costs can also be high but are inversely correlated to the time required to complete enrollment: if a research assistant has open access to space, subjects can be scheduled more quickly, but the cost to rent or purchase such access will be correspondingly higher.

Finally, insurance costs must also be considered. While breath research is generally safe, IRBs usually require insurance coverage for events and contingencies. Diabetic subjects, for example, have a small risk of dangerously low blood sugar for protocols that require an overnight fast, and therefore the principal investigator (PI) should have a prespecified plan to provide appropriate care. Similarly, the PI is responsible for the initial management of any significant health issues identified during the course of the study (e.g., if blood screen for liver disease is abnormal). Such responsibilities depend on regulations and can vary from country to country.

In conclusion, high-quality clinical research is inherently expensive. This is true despite the fact that breath research is generally safe and specimen processing is often low cost. Noninvasive device diagnostics are held to the same safety and evidence standards as other forms of clinical trials. And, poor quality research is often worthless. Naturally, there are logistic trade-offs between costs and timelines. An experienced team can maximize value and reduce uncertainties. At times, it is possible to attach a breath protocol as an ancillary to a larger, ongoing clinical trial that has data collection infrastructure already in place.

36.5 Summary

There are few models of successful clinical breath development. Therefore, this chapter necessarily focused instead on plentiful examples of failure. Appropriate metabolite selection is essential and, even once choices are made, should be reconsidered and challenged at every opportunity. A major upside of a focus on a clinical area, e.g., nutrition metabolism, is that it can be adapted to a variety of clinical scenarios and leverage the best properties of breath research; namely, easy repeated measures. It is not limiting.

Assembling and maintaining a multidisciplinary team is the next most important step. It is difficult to assemble an appropriately diverse group where each member maintains passion over time. Usually, team members contribute effort either part time or full time, but only for short periods. And when individual commitment vacillates, fractures and miscommunication can occur.

Finally, a coherent and realistic development plan is needed to maintain focus and interest. This plan must include the development, protection, and sharing of IP and, perhaps most importantly, aspirations for commercialization. Serious plans require equally serious funding and timelines. Otherwise, personal and financial investment will wane, and the project will fail. There is often a common but unfortunate and naïve notion that a clinician can make a meaningful demonstration with a single prototype monitor on a limited budget in a short time frame. In fact, this is akin to expecting the clinician to quickly traverse a lake in a leaky rowboat with a single broken paddle. Instead, unmet expectations are inevitable and lead to discouragement and disinterest, and the entire experience becomes counterproductive to progress. It is better when the entire team remains incentivized to create meaningful IP, with the potential for success and profit. Such a goal keeps the team together and focused. Finally, it is also perhaps best for various roles to be determined early, e.g., a "scientific champion" (often an engineer or chemist), a "medical champion" (often a clinician), and a "business champion" (often an experienced investor).

In conclusion, monitoring engineering and thorough analytical validation are only earliest and easiest steps in a successful clinical breath test development program. Rather, the most important and challenging steps are careful metabolite selection and the creation of a thorough but flexible development strategy for a compelling unmet need. This will necessarily involve multidisciplinary teams, significant and sustaining funding, and flexibility. Convincing clinical research is expensive.

References

1. Colburn WA, Lee JW. Biomarkers, validation and pharmacokinetic-pharmacodynamic modelling. *Clin Pharmacokinet* 2003;**42**(12):997–1022.
2. Yurdaydin C. Blood ammonia determination in cirrhosis: still confusing after all these years? *Hepatology* 2003;**38**(5):1307–10.
3. Nikolac N, Omazic J, Simundic A-M. The evidence based practice for optimal sample quality for ammonia measurement. *Clin Biochem* 2014;**47**(12): 991–5.
4. Solga SF, Mudalel M, Spacek LA, Lewicki R, Tittel F, Loccioni C, et al. Factors influencing breath ammonia determination. *J Breath Res* 2013;**7**(3):037101.
5. Adeva MM, Souto G, Blanco N, Donapetry C. Ammonium metabolism in humans. *Metabolism* 2012;**61**(11):1495–511.
6. Sturgeon JP, Shawcross DL. Recent insights into the pathogenesis of hepatic encephalopathy and treatments. *Expert Rev Gastroenterol Hepatol* 2014;**8**(1):83–100.
7. Cope K, Risby T, Diehl AM. Increased gastrointestinal ethanol production in obese mice: implications for fatty liver disease pathogenesis. *Gastroenterology* 2000;**119**(5):1340–7.
8. Cederbaum AI. Alcohol metabolism. *Clin Liver Dis* 2012;**16**(4):667–85.
9. Alobaidi TA, Hill DW, Payne JP. Significance of variations in blood: breath partition coefficient of alcohol. *Br Med J* 1976;**2**(6050):1479–81.
10. Adrover R, Cocozzella D, Ridruejo E, García A, Rome J, Podestá JJ. Breath-ammonia testing of healthy subjects and patients with cirrhosis. *Dig Dis Sci* 2012;**57**:189–95.

11. Spacek LA, Mudalel M, Tittel F, Risby TH, Solga SF. Clinical utility of breath ammonia for evaluation of ammonia physiology in healthy and cirrhotic adults. *J Breath Res* 2015;**9**(4).
12. Carpenter D. *Repuation and power. Organizational image and pharmaceutical regulation at the FDA*. Princeton University Press; 2010. p. 228–98.
13. Sertkaya A, Wong H-H, Jessup A, Beleche T. Key cost drivers of pharmaceutical clinical trials in the United States. *Clin Trials* 2016;**13**(2):117–26.
14. Kasarala G, Tillmann HL. Standard liver tests. *Clin Liver Dis* 2016;**8**(1):13–8 [cited 9 August 2019].
15. Sanyal AJ, Boyer TD, Terrault N, Lindor KD. Zakim and Boyer's hepatology: a textbook of liver disease. p. 1052.
16. Simon R. Sensitivity, specificity, PPV, and NPV for predictive biomarkers. *J Natl Cancer Inst* 2015;**107**(8). djv153.
17. Goossens N, Nakagawa S, Sun X, Hoshida Y. Cancer biomarker discovery and validation. *Transl Cancer Res* 2015;**4**(3):256–69.
18. Steele SR, Chang GJ, Hendren S, Weiser M, Irani J, Buie WD, et al. Practice guideline for the surveillance of patients after curative treatment of colon and rectal cancer. *Dis Colon Rectum* 2015;**58**(8):713–25.
19. Nielsen J. Systems biology of metabolism: a driver for developing personalized and precision medicine. *Cell Metab* 2017;**25**(3):572–9.
20. Anderson JC. Measuring breath acetone for monitoring fat loss: review. *Obesity (Silver Spring)* 2015;**23**(12):2327–34.
21. Kalapos MP. On the mammalian acetone metabolism: from chemistry to clinical implications. *Biochim Biophys Acta* 2003;**1621**(2):122–39.
22. Food and Drug Administration HHS. International Conference on Harmonisation; guidance on Q8(R1) Pharmaceutical Development; addition of annex; availability. Notice. *Fed Regist* 2009;**74**(109):27325–6.

CHAPTER

Mathematical interpretation of targeted volatilome data

37

Joachim D. Pleil[1], Peter P. Egeghy[2]

[1]*Department of Environmental Sciences and Engineering, Gillings School of Global Public Health, University of North Carolina, Chapel Hill, NC, United States;* [2]*Office of Research and Development, United States Environmental Protection Agency, Research Triangle Park, NC, United States*

Chapter outline

37.1 Overview

Breath research has developed into a broad and credible scientific arena complementing all types of environmental and clinical applications.[1,2] Of these, many breath research studies incorporate a targeted approach, wherein the analytes have been chosen a priori. Typically, the analytical methodology has been already established and previous studies have determined a list of compounds that are probative. If the study explores exposure to disinfection by-products from municipal water, for example, then certain trihalomethanes would be targeted; if, on the other hand, the study is concerned with metabolites indicating oxidative stress, then aldehydes, ketones, and alcohols would be targeted. The focus of this chapter is the targeted breath study, but the mathematics presented are broadly applicable to all manner of biomonitoring endeavors.

Breathborne Biomarkers and the Human Volatilome. https://doi.org/10.1016/B978-0-12-819967-1.00037-2
Copyright © 2020 Elsevier B.V. All rights reserved.

This chapter deals with the mathematical and statistical questions that are most important for understanding the outcomes of targeted breath studies. A set of tools for developing a statistically defensible interpretation of human biomonitoring data is provided with a focus on the nature and interpretation of the collected data. The organization of the chapter is designed to help answer a series of questions that may come up in any biomarker study, namely

1. Which mathematical methods are used for different study designs?
2. How do measurement distributions affect the mathematics?
3. How can different lognormal datasets be compared?
4. What is the value of repeated measures?
5. How could a single spot measurement be used to estimate risk level?
6. How can a new subject be assigned to the correct group?
7. Are outlier measurements useful or are they artifacts?

Regardless of the compounds of interest, the data from breath measurements must be interpreted appropriately to tell a story. This begins with the design of the study: the investigators should articulate a hypothesis, or at least a question to be answered (see Chapter 35). Then, the mathematical strategy should be laid out to help plan how the experiments should be structured to in order to provide the required answers. At this point, the number of subjects, samples per subject, number of groups, etc., are decided to assure that sufficient information will be acquired. In the following, each of the questions above will be discussed in detail.

37.2 Which mathematical methods are used for different study designs?

Study design depends on the nature of the research question or issue(s) being addressed. Targeted observational studies typically have cross-sectional, case–control, or longitudinal designs. Statistically, they are similar in that, ultimately, groups of subjects are compared to each other or to some external parameters.

Cross-sectional: In a cross-sectional design, randomly selected subjects are tested from a population of interest, and the results are compared internally among different subsets of that population. The subsets are purposely selected based upon existing differences in the sample. Suppose, for example, that exhaled isoprene—a compound primarily of endogenous origin and highly abundant in normal human breath—concentration is measured across 1000 individuals. Isoprene is known to be related to cholesterol synthesis; in the cross-sectional distribution, females could be compared to males to assess if they have differences in metabolism, or high body mass index (BMI) could be compared to those with low BMI, or older subjects to younger subjects, and so on. Comparisons are usually performed using tests of the central tendency and spread of data (e.g., t-test of the difference between the means of two groups, analysis of variance [ANOVA] for assessing how independent nominal variables like sex or genotype affect a continuous dependent variable), or with least squares regressions of biomarker measurements against independent continuous variables, such as age or BMI.

Case–control: In a case–control design, two different groups of individuals are assigned, which differ only with respect to the parameter to be tested. Case–control studies often seek to identify possible predictors of outcome and can be useful for studying rare outcomes because the controls are intentionally selected to be representative of the cases exclusive of the parameter of interest. If the question of whether smokers have higher exposures to benzene and toluene is to be addressed, for example, then the groups would be comprised of smokers (case) and nonsmokers (control) who are otherwise similar. Herein, group statistics (parametric t-test, nonparametric Wilcoxon–Mann–Whitney) are generally only the first step; ultimately, the possibility to assign smoking status to individuals based only on their benzene and toluene measurement data might be of interest, which may require further detailed confidence interval calculations and assessments of distribution.

Longitudinal: A longitudinal design tracks changes in time, collecting two or more observations of the same subjects over a specific period. Here, the individuals are the same throughout the study, but the data groups are based on when they were collected. If the influence of a particular intervention (exercise, flying on an airplane, drug regimen) on exhaled breath is to be determined, for example, the same individuals should be studied before and after the activity or event. These are considered paired data, and group differences are generally evaluated using paired t-test (two time points) or repeated measure ANOVA (three or more time points). If longitudinal studies are conducted over longer periods, then trends can be assessed using, for example, nonlinear regressions.

The main statistical difference among these study designs is that the longitudinal and (matched) case–control study allows paired tests, whereas the cross-sectional study does not; paired tests have fewer degrees of freedom and thus more statistical power for decision-making.

37.3 How do measurement distributions affect the mathematics (quantile-quantile plot)?

A major problem in breath data analysis is the failure by the researcher to recognize the lognormal character of biomonitoring data.[3] Statistical tests and their conclusions are only valid if underlying assumptions, such as normality, independence, and uniformity of variance, are met. Lack of normality is the highest risk for the assessment tools, such as linear regression, central tendency, confidence intervals, t-test, ANOVA, etc., as they lose accuracy if the data are erroneously treated as normally distributed.

The first indication that a lognormal transformation may be required is to look at the median, mean (μ), and standard deviation (σ) of the dataset. With an underlying normal (symmetrical) distribution, the mean and median are roughly equal. While the median is not very sensitive to extreme observations, the mean is extremely sensitive. If the mean is noticeably greater than the median, and/or if the coefficient of variation (cv = σ/μ) is greater than 0.2, then the underlying distribution is not normal but is positively skewed, suggesting a lognormal distribution; no further tests are required. As a complement to these simple observations, a subsequent normal quantile-quantile plot (Q–Q plot) provides a more comprehensive assessment of the character of the distribution; in fact, it is recommended to apply this as a primary step of any data exploration.

The Q–Q plot is a direct comparison between a measurement dataset on the *y*-axis plotted against the *z*-score of a normal distribution of the same size on the *x*-axis, where $z=(x-\mu)/\sigma$. It is relatively easy to construct using standard statistical functions, as found in business or scientific software products[4]; in this chapter, MS Excel (Microsoft, Inc, Redmond, WA, USA) is used to construct these plots.

Suppose the distribution of measurements of trace-level benzene in breath from 140 subjects should be characterized. The left panel of Fig. 37.1 shows the Q–Q plot of the measurement data in units of parts per billion, by volume (ppbv), with real-world mean, μ = 1.11, median = 0.54, and coefficient of variation, cv = 1.32. The right panel of Fig. 37.1 shows the same data after log transformation. The regression lines indicate the expected normal distribution. Obviously, the original data (left panel of Fig. 37.1) are not normally distributed as the points form a curve instead of a straight line (this also could have been surmised by the mean vs. median and cv rules of thumb). Upon log transformation, however, the data show much better agreement with the expected normal behavior (right panel of Fig. 37.1). Despite common practice in the scientific literature of characterizing measurement data by the arithmetic mean and standard deviation,[5] measurement data are never perfectly distributed, and generally the lognormal assumption is a better model for subsequent statistical evaluations. Furthermore, the Q–Q plot of the log-space

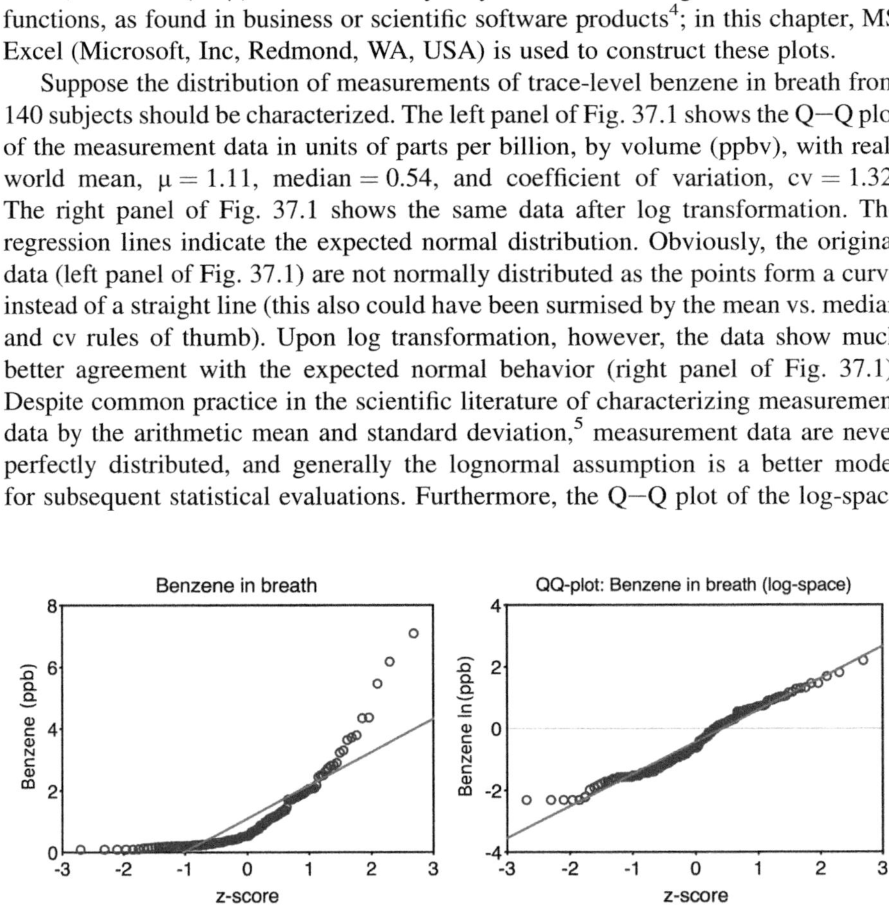

FIGURE 37.1

Q–Q plots of breath benzene concentrations drawn from previous studies; these are realistic values, but modified for these examples. The left panel indicates that original measurements are not normally distributed, whereas the log-transformed data in the right panel conform much better to normality.

data more clearly shows left-censored data (below detection limit imputations) that need to be considered in any future comparisons.

37.4 How can different lognormal datasets be compared (geometric mean and geometric standard deviation)?

Because biological and environmental data are almost always lognormally distributed,[6,7] the best parameters to describe the distribution are the geometric mean (GM) and geometric standard deviation (GSD).[8] GM and GSD are found by first natural log transforming the real-world measurements and then calculating mean μ' and standard deviation σ' of the transformed data; then $GM = \exp(\mu')$ and $GSD = \exp(\sigma')$ of the original data. Once GM and GSD are known, they describe the lognormal distribution much like the mean (μ) and standard deviation (σ) describe the normal distribution, except that unlike the usual arithmetic standard deviation, the GSD is a multiplicative factor. For normally distributed data, the interval $\mu \pm \sigma$ covers 68.3%, and for lognormal data, the corresponding multiplicative expression (GM/GSD, GM*GSD) similarly covers 68.3% of the distribution.[5]

The most basic comparison of two groups of data is made on their central tendencies (average or GM) and spread (standard deviation or GSD); in the case of lognormally distributed data, the mean μ' and standard deviation σ' of the transformed data are compared using a t-test, either paired or unpaired, depending on the structure of the original data. If there are three or more groups, standard one-way ANOVA tests are used. In either case, the test determines if the groups are statistically different but does not deliver information about individual measures or on how the "next" measurement should be interpreted.

For beyond-group statistics, the GM and GSD parameters of a lognormally distributed dataset can be used to determine levels of exceedance and confidence intervals (CI), analogously to how μ and σ are used for normally distributed data: for example, the 95% confidence limits of a normal distribution are determined as follows:

$$CI(2.5\%, 97.5\%) = \mu \pm 1.96\sigma \tag{37.1}$$

where 1.96 is the z-score for the two-sided 95% limit. For a lognormal distribution, the relationship

$$CI(2.5\%) = GM/GSD^{1.96} \text{ and } CI(97.5\%) = GM \times GSD^{1.96} \tag{37.2}$$

is used to calculate the two-sided 95% limit by raising GSD to the power of the z-score. By inserting other values of z-score, different upper and lower limits can be calculated for normally or lognormally distributed data using Eqs. (37.1) and (37.2), respectively. For example, if the cutoff for the 95th percentile of all values is to be determined, then $z = 1.645$; the 90th percentile corresponds to $z = 1.282$; and so on. Table 37.1 provides some useful critical values of z.

Table 37.1 Common critical values of z.

Tail area (α)	Central area ($1-2\alpha$)	z_α
0.35	0.30	$z_{.35} = 0.385$
0.25	0.50	$z_{.35} = 0.674$
0.10	0.80	$z_{.10} = 1.282$
0.05	0.90	$z_{.05} = 1.645$
0.025	0.95	$z_{.025} = 1.96$
0.01	0.98	$z_{.01} = 2.33$
0.005	0.99	$z_{.005} = 2.58$

37.5 What is the value of repeated measures (interclass correlation coefficient)?

Incorporating repeated measurements into study designs has become an important factor for assessing risk. The primary issue is that environmental risks or health states are predicted by long-term averages, not by single spot measures that may vary greatly from day to day. Suppose a measured breath biomarker (for example, the aldehyde hexanal) is associated with oxidative stress and inflammation in a group of patients. A distribution of values (most likely lognormal) that represents the total variability across the group is expected. However, it is not known if the differences are stable between patients (that is, if substantial variation exists between patients in the group) or if any given patient could have a measurement from anywhere within that distribution if measured at a different time.

Calculations of the intraclass correlation coefficient (ICC): The solution to this dilemma is to make repeated, but independent, measurements from the same group of subjects and then calculate the variances separately. There are a number of numerical methods to accomplish this,[9] including using the contents of an ANOVA table. Once the individual variances are available, total variance can be partitioned into intra- and interindividual components. The ICC can then be assessed. The ICC is a nondimensional parameter that indicates the relative magnitude of the within-subject variance, $\text{Var(w)} = \sigma_w^2$, to the total variance, $(\text{Var(w)} + \text{Var(b)})$, where the between-subject variance is $\text{Var(b)} = \sigma_b^2$.

$$\text{ICC} = \text{Var(b)}/[\text{Var(W)} + \text{Var(b)}] \tag{37.3}$$

From Eq. (37.3), it is evident that $0<\text{ICC}<1$ because the variances are constrained to positive numbers (or zero). When ICC is closer to 0, the between-subject variance is low, and so almost all of the variability is observed within individuals, and the individuals are indistinguishable from one another. Conversely, if the ICC is closer to 1, then most of the variability is between subjects, that is, an individual with a low concentration will tend to always have low concentrations, and an individual with a high concentration will tend to always have high concentrations. As such, the ICC is a measure of reliability over time. In the context of human parameters, suppose there is a group of subjects for which cholesterol and heart rate are measured at five random times each over a year. Any individual would be expected to have approximately the same cholesterol numbers at each measurement, and the differences between subjects would be expected to be preserved. However, their heart rate measurements could vary greatly depending on their recent activity level. As such, cholesterol would exhibit a high ICC and heart rate a low ICC.

37.6 How could a single spot measurement be used to estimate risk level?

Preclinical and long-term risks are generally thought to be associated with long-term averages—not instantaneous values—of biomarkers of exposure or biomarkers of effect. A single spot measure cannot assess an individual's average concentration without additional information. So, can the average value over time be predicted based on a lonely spot measurement?

ICCs and risk calculations: Under the assumption that individual biomarker compounds have internally consistent and generalizable ICCs, information from small studies with repeated measures could be used to apply the ICC to large studies with only a single measurement per subject. The ICC could then allow the total variance to be partitioned into Var(w) and Var(b) components.

For a lognormal dataset, GM and GSD can be easily calculated, as described above. Suppose that a particular concentration x_c is known to be of concern for health state, and it should be explored how this may fit into the study population. By rearranging Eq. (37.2) above (replacing 1.96 with a generic z), the z-score can be calculated, which is represented by x_c as follows:

$$z_c = \ln(x_c / \mathrm{GM}) / \ln(\mathrm{GSD}) \tag{37.4}$$

So, if a dataset with GM = 5 ppb and GSD = 2, and a concentration $x_c = 25$ ppb is of potential concern, then $z_c = \ln(x_c/\mathrm{GM})/\ln(\mathrm{GSD}) = \ln(25/5)/\ln(2) = 2.322$, which corresponds to the 99th percentile. If the level of concern was set at 16 ppb, then $z_c = 1.678$, which corresponds to the 95th percentile.

These calculations indicate only how many spot measures are above or below some cutoff value. They do not indicate how many *individuals* have an average value above the level of concern. If the compound has been reported to have high within-subject variance, then everyone would be clustered around GM = 5 and likely to have an average value close to that. If, on the other hand, the compound has high between-subject variance, someone with one high spot measurement would be expected to always have a high measurement, then over time that individual could likely have an average that exceeds the level of concern.[10]

Under the assumption that many measures of the ith individual are also lognormally distributed with $\mathrm{GM_i}$ and $\mathrm{GSD_i}$, and that an estimate for the ICC of that compound is available from earlier studies, the previously derived relationships for estimating individual $\mathrm{GM_i}$ and $\mathrm{GSD_i}$ are as follows:[10]

$$\mathrm{GM_i} = \mathrm{GM} \times (x_i / \mathrm{GM})\sqrt{(\mathrm{ICC})} \tag{37.5}$$

and

$$\mathrm{GSD_i} = \mathrm{GSD}\sqrt{(1 - \mathrm{ICC})} \tag{37.6}$$

This is an elegant solution for estimating average exceedances from spot measure datasets. Using the previous example, suppose an external estimate for ICC = 0.90, and further suppose that an individual has a spot measure $x_i = x_c$; what is the probability that this average value is greater than the spot measure, in this case 25? This can be calculated from Eqs. (37.5) and (37.6) as follows, first by finding the individual GM_i and GSD_i:

$$GM_i = GM \text{ x } (x_i / GM)\sqrt{(ICC)} = 5 \times (25/5)\hat{}sqrt(0.9) = 23.018$$

$$GSD_i = GSD\sqrt{(1\text{-}ICC)} = 2\hat{}(sqrt(1 - 0.9)) = 1.245$$

and then inserting these values into Eq. (37.4), which results in

$$Z_C = \ln(x_C / GM_i)/\ln(GSD_i) = \ln(25/23.018)/\ln(1.245) = 0.377$$

which roughly corresponds to the 65th percentile, meaning there is approximately a 35% probability that this individual is above the risk level defined by an average level of 25. If the ICC were estimated at 0.10 instead, the same calculations would result in $GM_i = 8.318$, $GSD_i = 1.930$, and $z_c = 1.674$, corresponding to the 95th percentile, and so the same individual would have only 5% chance of an average level above 25. These calculations demonstrate that the within-subject variance (defined by ICC) is important in defining risk estimates from single measures.

ICC risk summary: As shown above for one example, the risk is proportional to the average biomarker level over time, and a spot measure alone is not sufficient to assess whether any given individual is above some level of concern; this requires ICC as an additional piece of information.

Using the above example of a spot measure distribution of exhaled breath measurements with GM = 5 and GSD = 2, this approach can be expanded to estimate the percentiles of exceedances for different combinations of ICC and levels of concern. Table 37.2 presents the percentage of the population whose estimated average concentration would be above selected levels of concern for different ICC values using Eqs. (37.4)–(37.6). This reinforces the concept that knowing ICC is crucial in making population risk estimates based on spot measures and is particularly important when the level of concern is high with respect to the spot measure median.

Table 37.2 Estimated exceedances expressed as percent of population above level of concern for a spot measure distribution with GM = 5 and GSD = 2 dependent upon ICC.

	Percent exceedance of population (%)		
Level of concern (ppb)	**ICC = 0.1**	**ICC = 0.5**	**ICC = 0.9**
9	27	36	45
16	12	24	39
25	5	17	35
50	1	8	29

The dependence on both level of concern and ICC is not an obvious result! Often researchers stop with the original distribution and underestimate risk by using the population median as the average, or overestimate the risk by assigning the spot measurement to the average; neither is correct.

Practical examples of risk exceedance: As a complement to the results displayed in Table 37.2, suppose there is a concern about five different individuals who have spot measures of 2.5, 5, 10, 20, and 40 ppb, respectively. The question as follows: how much risk do they each carry for exceeding a level of concern set at 25 ppb?

The first step is to calculate the internal GM_i and GSD_i for each individual, as well as each hypothetical ICC value, using Eqs. (37.5) and (37.6) for each of the x_i spot measures. Table 37.3 shows these results.

Using Excel, a standard function can be applied to calculate a curve of x and f(x) values that correspond to the distributions determined by different GM_i and GSD_i combinations:

$$f(x) = \text{LOGNORM.DIST}(x, \ln(GM_i), \ln(GSD_i)\ \text{FALSE}) \qquad (37.7)$$

where x represents values from 0 to 50, in small increments to achieve a smooth curve. Fig. 37.2 presents the original spot measure distribution with GM = 5, GSD = 2 (top-left panel), and the calculated average distributions with GM_i and GSD_i for three example ICCs: 0.1, 0.5, and 0.9, (top-right, bottom-left, and bottom-right panels, respectively), as listed in Table 37.3.

Note, the function LOGNORM.DIST is a shortcut for the calculation of the standardized lognormal probability density function, as a function f(x):

$$f(x) = \frac{1}{x\sigma\sqrt{2\pi}} \exp\left(-\frac{(\ln x - \mu)^2}{2\sigma^2}\right)$$

where μ and σ are the mean and standard deviation of the log-transformed data, respectively, and the integral f(x) from 0 to infinity = 1. The Excel function is relied on here for simplicity.

Table 37.3 Predictions of average levels based on specific spot measures and knowledge of ICCs.

Spot measure	ICC = 0.1		ICC = 0.5		ICC = 0.9	
x_i (ppb)	GM_i	GSD_i	GM_i	GSD_i	GM_i	GSD_i
2.5	4.02	1.93	3.06	1.63	2.59	1.25
5	5.00	1.93	5.00	1.63	5.00	1.25
10	6.23	1.93	8.16	1.63	9.65	1.25
20	7.75	1.93	13.33	1.63	18.63	1.25
40	9.65	1.93	21.75	1.63	35.95	1.25

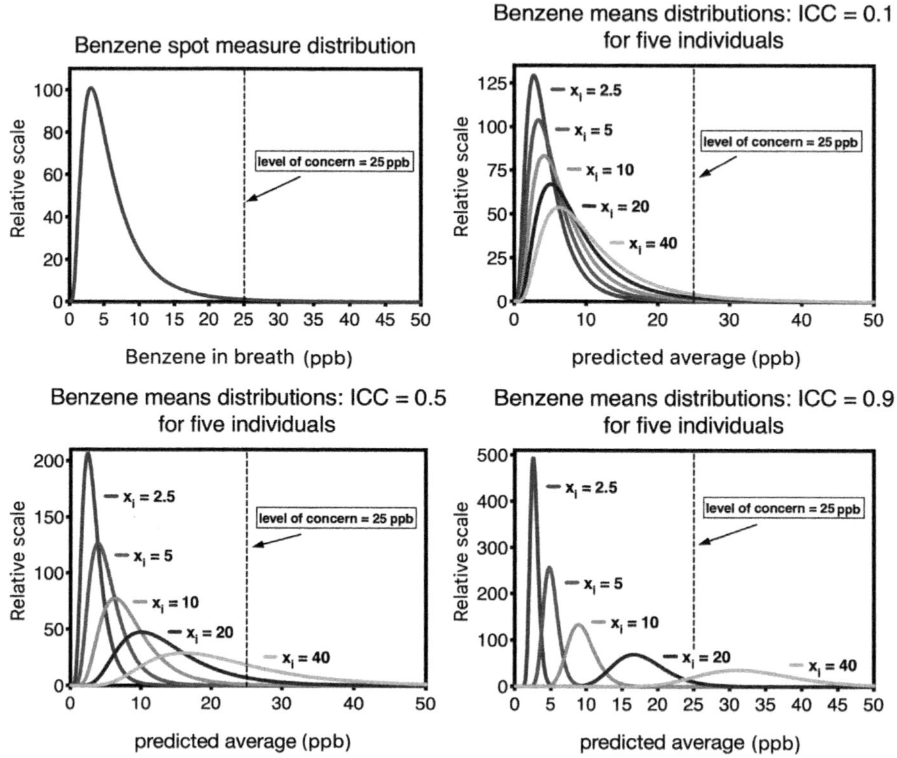

FIGURE 37.2

The figure panels compare how the mean distributions are affected by the underlying ICC for a particular compound. The top-left panel shows the distribution of the spot measures and the hypothetical level of concern at 25 ppb; very few spot measures are expected beyond the level of concern. Then, the five different measurements (2.5, 5, 10, 20, 40) are translated into means distributions for ICC $=0.1$, 0.5, and 0.9 (top-right, bottom-left, and bottom-right panels, respectively). In these panels, the proportion of area under the curve past the level of concern indicates the chance of exceeding that value. When ICC $=0.1$, there is very low chance that any of the five individuals would have an average value above 25 ppb; when ICC $=0.5$, it can be seen that individuals with spot measures of 40 and 20 ppb have about 25% and 10% chance of exceedance, respectively, and when ICC $=0.9$, then a spot measurement of 40 ppb essentially guarantees that this subject has a true average value above 25 ppb.

37.7 How can a new subject be assigned to the correct group?

One of the most important tasks in biomarker research is comparing datasets of individual measures to each other. This comes up in all types of studies: in cross-sectional assessments, comparison to a different previous study is often made;

in case-control studies, assessment of differences in groups that were designed to probe a specific characteristic are made, and in longitudinal studies, measurements before and after some intervention are typically compared. Regardless of the study type, the purpose is always twofold: first, determining whether the groups are different from each other, and second, the ability to place any new individuals into the correct group. The latter is the underlying concept for exposure assessment and preclinical disease diagnosis.

Consider that extensive case–control group measurements of patients with diagnosed lung cancer and patients deemed free of cancer have been made. Then, assume that a difference in a particular biomarker (compound C) is discerned in breath that is statistically higher in cases than controls, with $GM_{case} = 10$, $GSD_{case} = 2$ and $GM_{control} = 4$, $GSD_{control} = 2$, respectively. Upon calculating the unpaired, one-sided t-test statistic, it is found that the groups are significantly different at $P < .0001$; an excellent result!

In this hypothetical scenario, a large group of subjects who have suspicious coughs are encountered, but they cannot be confidently diagnosed with cancer without very expensive imaging or biopsy tests. The question is how well a particular noninvasive breath biomarker, compound C, can prioritize these individuals for further invasive testing.

Suppose compound C has been measured in this group of subjects and values for C for 11 example individuals are found to be $C_n = 0.5, 1, 2, 4, 6, 8, 10, 15, 20, 30$, and 50. The relative height (H) of the frequency distribution for each value in both case and control distributions can be calculated using Eq. (37.7), which is defined by the Excel function LOGNORM.DIST, thus

$$H_{control}(C_n) = \text{LOGNORM.DIST}(C_n, \ln(GM_{control}), \ln(GSD_{control}), \text{FALSE})$$

and

$$H_{case}(C_n) = \text{LOGNORM.DIST}(C_n, \ln(GM_{case}), \ln(GSD_{case}), \text{FALSE})$$

and the probability that any C_n falls into the control or case group can be calculated as follows:

$$\Pr(C_n): \text{case} = H_{case}(C_n)/[H_{case}(C_n) + H_{control}(C_n)]$$

and

$$\Pr(C_n): \text{control} = H_{control}(C_n)/[H_{case}(C_n) + H_{control}(C_n)]$$

Using the example numbers above, the probability that any particular new measurement falls into the case or control category can be calculated, as shown for a range of theoretical measurements in Table 37.4.

Note that there is always some small probability that a very low measurement may actually be a case and that a very high number may really be a control. The question remains for the investigator to decide how much error is acceptable. In this example, if a probability of at least 0.80 is accepted, then numbers above 15 or 20 are likely to be cases, and numbers below 2 or 3 are likely to be controls.

Table 37.4 Results from probability calculations for two distributions with $GM_{case} = 10$, $GSD_{case} = 2$ and $GM_{control} = 4$, $GSD_{control} = 2$.

	Values of "next" sample										
	0.5	**1**	**2**	**4**	**6**	**8**	**10**	**15**	**20**	**30**	**50**
Probability case	0.01	0.03	0.10	0.29	0.47	0.61	0.71	0.84	0.90	0.95	0.98
Probability control	0.99	0.97	0.90	0.71	0.53	0.39	0.29	0.16	0.10	0.05	0.02

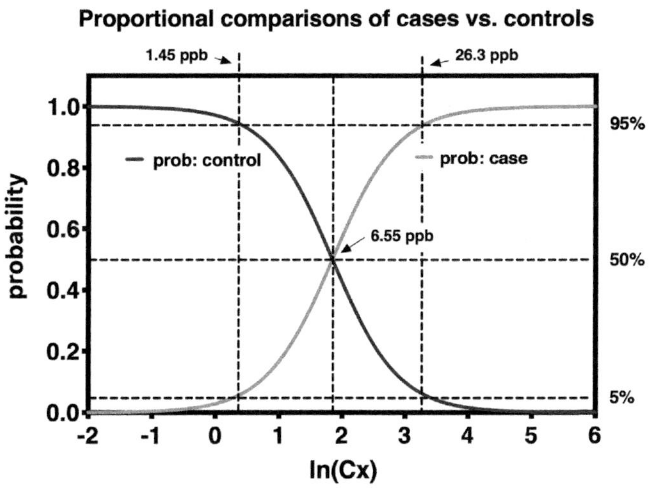

FIGURE 37.3

Visualization of probabilities that the "next" measurement belongs to either the case or control group. Note, for symmetry, the concentrations on the x-axis are in log space, where 0 = ln(Cx = 1 ppb), 1 = ln(Cx = 2.72 ppb), 2 = ln(Cx = 7.4 ppb), 3 = ln(Cx = 20.1 ppb), 4 = ln(Cx = 54 ppb), etc.

This analysis also shows that despite good distinction between the central tendencies (4 vs. 10), new values in that range, or even equal to the central tendencies, are statistically indistinguishable for group assignment.

Fig. 37.3 visualizes the concept developed for Table 37.4, with smooth functions for the respective case and control probabilities. This graphic is created using log-transformed space to make it symmetric, with the x-axis representing the natural log of the concentration. From this figure, it is easily discernible where any new measurement could be placed (into case or control) and how likely either choice is for being correct. The 95th and 5th percentiles of probability are annotated in Fig. 37.3, as are the real-world concentrations. The plot shows that an estimated new value below 1.45 ppb belongs in the control group, and one above 26.5 ppb belongs in the case group, with 95% confidence. Other confidence levels can be observed quickly from the graph, without resorting to the calculations shown for Table 37.2.

37.8 Are outlier measurements useful or are they artifacts?

All measurement groups will have occasional outlier data points, especially when dealing with biological media. In fact, many such datasets have unexpectedly large variances even under controlled study conditions and in nominally healthy adults.[11–13] This is one of the most difficult mathematical issues in biomonitoring; often, the temptation is to simply throw out such values because "*something is obviously wrong with that number*." That is not necessarily true, and it may well be that outliers provide important information for resolving a question in a hypothesis or raise issues about the overall experiment.

There is a simple process for making decisions about outlier values. First, revisit the raw data, instrument records, laboratory notebook, and other notes to see if there was a flag on the sample or mistake in label assignment, an instrument problem, low volume, wrong calculation, misidentification of compound, and so on. The doubt cast by such an error would be sufficient cause to discard a measurement. Barring an error, the measurement cannot just be ignored, regardless of how much it jars with the data. The next step is to do the statistical calculations with and without the outlier data. Often, it is found that a few outliers do not have sufficient influence to change broader conclusions and so it does not matter if they are kept.

If, however, the outliers do indeed modify the interpretation of the results, then further investigation for an unknown perturbation in the experimental design or some other real-world reason should be made. One particularly valuable approach is to implement the Q–Q plot strategy described above to highlight the outliers. If they lie near the regression line, they are likely part of the real distribution and should be retained in the dataset.

In some cases, even the oddest outliers may actually become part of the story. For example, US EPA and the US National Institute of Occupational Safety and Health (NIOSH) collaborated on a longitudinal exhaled breath study of firefighters before and after firefighting activities to determine the effectiveness of their protective gear. A small group of subjects had anomalously high measurements an hour after the fire suppression activity. Upon detailed investigation, it was found that it is possible to get reexposed from off-gassing from the protective gear after its removal; this ultimately led to a modification of the protocols for subsequent investigations and to recommendations for used gear decontamination.[14]

37.9 Summary

Breath measurements are a valuable tool for both environmental and medical/clinical assessments, but they are only as useful as the mathematical interpretations applied. This chapter has developed the rationale and tools for a series of numeric approaches for planning biomonitoring studies and interpreting measurement results. Specifically, it has focused on the following topics.

Planning: An important issue in developing an effective study is to plan ahead to assure that the hypotheses to be tested have sufficient and accurate data with the realization that distributions will likely be lognormal, and as such asymmetric, which influences sampling and analysis. Another essential aspect is to plan ahead to make appropriate repeat measurements if evidence (e.g., ICCs calculated in previous studies) suggests a substantial intraindividual component of variability. If that is the case, then a spot measure will not be an appropriate surrogate of the average concentration over a longer period, as it can lead to inaccurate assessment of risk.

Distributions: Decades of research have shown that essentially all biomonitoring measurement interpretations are better suited for calculations with log-transformed data. Using Q–Q plots is an excellent visualization tool for looking at the nature of the measurement data and deciding if log transformation is appropriate. In contrast to the standard tests for distribution (e.g., Shapiro–Wilk or Kolmogorov–Smirnov) based on a global assessment with a simple yes/no answer, the Q–Q plot allows the investigator more flexibility for interpreting anomalies (especially at the lower end of the measurement range) that might seem irrelevant but could skew summary statistics or calculations if ignored.

Comparisons: Often different distributions must be compared to each other, especially with respect to central tendency and confidence levels. As biomonitoring data are generally lognormally distributed, it is important to use GM and GSD calculated from transformed data rather than the standard arithmetic μ and σ calculated from the original measurements.

Repeat measures and risk: An oft-neglected feature of biomonitoring studies is the use of repeat measures; these are of immense value in discerning sources of variability that can modify the interpretation of the study outcome. The variance components (within- and between-subject) can be used to calculate the ICC, which has a further role in assessing long-term averages (related to health risk), and can be generalized to studies of the same compounds where only spot measures are available.

Group assignments: Case–control studies are a major tool in biomonitoring assessments; they are generally used to calibrate specific biomarkers for subsequent evaluation of cross-sectional measurements for exposure assessment or for preclinical health screening. Data from such studies are often misinterpreted or their value is not fully realized because investigators only consider the relative group statistic of central tendency, rather than the probability that the "next" individual measurement belongs to one group or the other. Accurate separation into groups is much more difficult than generally believed; detailed data interpretation, beyond-group statistics, is required but can be easily performed using GM and GSD together with cumulative distribution functions commonly available in spreadsheet programs.

Outliers: Biomonitoring studies tend to produce data with skewed distributions and large variances, which present as outlier values, and may be erroneously ignored; human exposures, activities, metabolism, and health state may all multiplicatively interact and result in a few samples with values distant from the expected central tendency or clustered at a particular value. These cannot be naively disregarded as they are likely to be genuine values and may be pertinent to elucidating aspects of the study.

As a final thought, these relatively simple mathematical methods are anticipated to be useful for properly interpreting biomonitoring data, and incorporating them into studies may help to avoid many of the issues that often arise in the peer-review process.

Disclaimer

This publication was reviewed and approved for publication by US Environmental Protection Agency (EPA). The views in this chapter are those of the authors and do not reflect policy of US EPA. References to products and trade names do not constitute endorsement for use.

References

1. Amann A, Smith D, editors. *Volatile biomarkers: non-invasive diagnosis in physiology and medicine*. Amsterdam: Elsevier B.V.; 2013.
2. Wallace MAG, Pleil JD. Evolution of clinical and environmental health applications of exhaled breath research: review of methods and instrumentation for gas-phase, condensate, and aerosols. *Anal Chim Acta* 2018;**1024**:18−38.
3. Pleil JD, Sobus JR, Stiegel MA, Hu D, Oliver KD, Olenick C, Strynar MJ, Clark M, Madden MC, Funk WE. Estimating common parameters of log-normally distributed environmental and biomonitoring data: harmonizing disparate statistics from publications. *J Toxicol Environ Health B Crit Rev* 2014;**17**:341−68.
4. Pleil JD. QQ-plots for assessing distributions of biomarker measurements and generating defensible summary statistics. *J Breath Res* 2016;**10**:035001.
5. Limpert E, Stahel WA, Abbt M. Log-normal distributions across the sciences: keys and clues. *BioScience* 2001;**51**(5):341−52.
6. Ott WR. A physical explanation of the lognormality of pollutant concentrations. *J Air Waste Manag Assoc* 1990;**40**(10):1378−83.
7. Zhang CL, Popp FA. Log-normal distribution of physiological parameters and the coherence of biological systems. *Med Hypotheses* 1994;**43**(1):11−6.
8. Rappaport SM. Assessment of long-term exposures to toxic substances in air. *Ann Occup Hyg* 1991;**35**(1):61−122.
9. Pleil JD, Stiegel MA, Wallace MAG, Funk WE. Human biomarker interpretation: the importance of intra-class correlation coefficients (ICC) and their calculations based on mixed models, ANOVA, and variance estimates. *J Toxicol Environ Health B Crit Rev* 2018;**21**(3):161−80.
10. Pleil JD, Sobus JR. Estimating lifetime risk from spot biomarker data and intra-class correlation coefficients (ICC). *J Toxicol Environ Health* 2013;**76**(12):747−66.
11. Pleil JD. Influence of systems biology response and environmental exposure level on between-subject variability in breath and blood biomarkers. *Biomarkers* 2009;**14**(8):560−71.

12. Sobus JR, Pleil JD, McClean MD, Herrick RF, Rappaport SM. Biomarker variance component estimation for exposure surrogate selection and toxicokinetic inference. *Toxicol Lett* 2010;**199**(3):247–53.
13. Stiegel MA, Pleil JD, Sobus JR, Stevens T, Madden MC. Linking physiological parameters to perturbations in the human exposome: environmental exposures modify blood pressure and lung function via inflammatory cytokine pathway. *J Toxicol Environ Health A* 2017;**80**(9):485–501.
14. Fent KW, Evans DE, Booher D, Pleil JD, Stiegel MA, Horn GP, Dalton J. Volatile organic compounds off-gassing from firefighters' personal protective equipment ensembles after use. *J Occup Environ Hyg* 2015;**12**(6):404–14.

CHAPTER

Preprocessing and analysis of volatilome data

38

Georgios Stavropoulos[1], Dahlia Salman[2], Yaser Alkhalifah[2], Frederik-Jan van Schooten[1], Agnieszka Smolinska[1]

[1]*Department of Pharmacology and Toxicology, NUTRIM School of Nutrition and Translational Research, Maastricht University, Maastricht, The Netherlands;* [2]*Centre for Analytical Science, Department of Chemistry, Loughborough University, Loughborough, United Kingdom*

Chapter outline

38.1 Overview

Volatilome research is a strongly emerging field that represents a new frontier in metabolomics. Research explores qualitative and/or quantitative changes in volatile organic compounds (VOCs) present within various biofluids, such as breath, blood, urine, saliva, feces, or excreted by cells or bacteria, and attempts to link these changes to health status. Consequently, the main objective in volatilome research is to discover patterns of VOCs that relate to deviant metabolic processes (for instance, inflammation) occurring in the human body. The emergence and continued development of high-throughput, high-resolution analytical platforms for VOC analysis has resulted in numerically large and biologically complex datasets for which sophisticated tools are required for their comprehensive processing and successful exploitation.

Volatilome data typically contain a mixture of endogenous compounds related to physiological, biochemical, and metabolic processes, and exogenous compounds derived from environmental exposure, bacteria, and viruses, amongst others (see Chapter 1). This leads to data with different sources of variance, i.e., variance of

Breathborne Biomarkers and the Human Volatilome. https://doi.org/10.1016/B978-0-12-819967-1.00038-4
Copyright © 2020 Elsevier B.V. All rights reserved.

interest as well as biologically irrelevant information or noise. The challenge in volatilome analysis is to separate those two sources of variation, by focusing on the compounds related to the studied problem and neglecting the redundant and irrelevant information in volatilome datasets. An array of sophisticated machine learning methods can be utilized to find the relevant set of VOCs. VOCs in volatilome data frequently correlate with one another, often because various metabolic pathways are involved that connect the measured VOCs. Since machine learning techniques are based solely on finding sets of important compounds, rather than single compounds, their usage should be at the core of statistical analysis of volatilome data.

This chapter presents and discusses the most common and promising machine learning techniques applicable for volatilome datasets, with the aim of providing a broad picture of the different methods, as well as an overview of their pros and cons. Many different data processing techniques exist, but a comprehensive overview of these is beyond the scope of this chapter. Moreover, there are no standard rules or clear instructions in machine learning as to which techniques should be applied to which datasets. As such, only the most important aspects of data preprocessing and analysis are described herein. The chapter commences with an illustration of different data preprocessing approaches that are suitable for a broad range of volatilome analytical instrumentation, followed by the relevant (occasionally abandoned) aspects of scaling, normalization, and transformation. Unsupervised (also known as explorative or descriptive)[1] and supervised approaches are presented, together with the importance of model validation.[2] The chapter concludes with an introduction and discussion on data fusion strategies, which are a relatively new but unmet area in volatilomics.

38.2 Data preprocessing

Data preprocessing is a crucial step in every data mining approach since it helps eliminate possible instrumental artifacts that may occur during the analysis, and most importantly, it improves and simplifies the data analysis. Note that diverse preprocessing procedures exist as either commercially or freely available packages, yet they all share very similar structure. The nature of preprocessing steps depends on the source of data, i.e., the instrumentation used in sample analyses, although many computations are common to all datasets. Preprocessing of analytical data typically comprises noise and baseline removal, correcting peak shift due to column aging (relevant to gas chromatography—mass spectrometry; GC-MS), temperature drift or biochemical interaction, and peak picking. The first step of noise removal can be done by wavelet transformations[3] followed by baseline correction via P-splines.[4] Noise removal and baseline correction are essential because, nowadays, analytical techniques, such as GC-MS or high-resolution MS (see Chapters 15 and 16, respectively) have become highly sensitive; consequently, they capture nonbiological information that ideally should be removed from datasets. The correction of peak shifts can be achieved via various techniques, as is reviewed in depth in the literature.[5] Peak picking, or binning, is the next step in preprocessing, which is a means of preliminary dimensionality reduction (i.e., reduction of the number of data points within each sample) by combining data points that relate to the same compound. The simplest method of peak picking is to sum up or average consecutive data points (e.g., 5–10) over the whole range of points in the samples. More advanced approaches, such as peak picking via peak correlation,[5] are implemented when coupled techniques are used (e.g., GC-MS). In such cases, local minima, maxima, and peak areas are calculated for every peak in each sample, and these are assigned as originating from the same compound based on their signal-to-noise ratio (S/N) and retention time. Ultimately, these peaks are represented as a single value.

Recently, Alkhalifah et al. suggested a procedure, named VOCCluster,[6] for peak picking solely for GC-MS data. This procedure groups the peaks, i.e., creates clusters, in GC-MS data arising from the same VOC using similarity measures based on cosine angular separation. The important aspect of the VOCCluster technique is that each VOC could change cluster membership as the algorithm progresses and could be reclustered into a different cluster, depending on the cosine similarity measurement of each VOC to the other VOCs. This reclustering enhances the accuracy of the clusters and does not depend on the order of samples. In addition, the VOCCluster approach calculates the similarity threshold (epsilon) that is used as an input parameter for the clustering process. This is helpful for untargeted volatilome studies as it allows such processes to be data driven rather than operator dependent.

Another important aspect of the VOCCluster procedure is its ability to take into account both retention index (RI) and mass spectrometric variations. For RI variation, the technique divides the data into different segments and calculates the RI variation for each segment. This leads to an RI range for each VOC within a sample, which is used to cluster VOCs into similar groups within that RI range. In the case of mass spectra variations, the technique takes into account the chemical nature of the VOC, signal intensity, and the number and/or order of extracted ions. The mass spectra variations are especially relevant for coeluting VOCs.

A widely applied software in metabolomics, but not yet extensively used in volatilomics, is XCMS, which is a freely available software for data preprocessing as a means of untargeted metabolite profiling between two sample groups.[7] The software was originally developed to preprocess liquid chromatography–mass spectrometry (LC-MS) data, but the developers state that it is equally applicable to GC-MS data. XCMS is an acronym for various (X) forms of chromatography–mass spectrometry techniques. The software has gained much attention in metabolomics analyses due to its simplicity and flexibility, as it allows for peak detection, peak matching (i.e., peak picking), misalignment correction, and metabolite identification in a semiautomated way and with minimal user interference. Baseline correction, or background subtraction, is not performed in XCMS due to the danger of this potentially adding more noise to the data and since overall the background typically remains constant from one analytical run to the next, thus its impact is assumed minimal. Peak detection is carried out by employing second-order derivatives and S/N. Importantly, XCMS deals with a common problem for most of the peak detection algorithms (i.e., to miss out peaks even though they are present) due to the inherent uncertainty close to the S/N cutoff value. In particular, XCMS fills in intensity values for every peak that is not detected by the peak detection algorithm by using information gathered from the raw data. Peak matching follows once all the peaks have been identified, and it is performed based on mass drifts (i.e., m/z values) rather than retention time. Peaks that are not matched in at least half of the samples are discarded, although different tie-breaking criteria can be chosen depending on the application. If it is known, for example, that the samples were obtained from two different groups (i.e., healthy and disease), then peaks that are not present in at least half of the samples of one of the groups are removed from that particular group. Finally, retention time alignment is performed. The algorithm does not use a reference signal to align the rest of the data but rather works purely with peak data. In fact, "well-behaved" peak regions are identified; these regions are areas in the chromatograms where very few samples have no peaks, and very few samples have more than one peak present. These well-behaved regions are more likely to be adequately matched and are therefore used as temporary standards (i.e., alignment references). XCMS permits for the identification of important metabolites by employing a univariate t-test and ranks the metabolites based on their *P*-values once all the preprocessing steps are finished. At the same time, it also allows for metabolite profiling based on exact masses, since it is linked to a metabolite database.

Regardless of the data preprocessing approach, the volatilome data can be described as a data matrix consisting of each measured observation in the row and detected VOCs in columns. In the volatilome, the majority of compounds (i.e., VOCs) are not present in all samples; therefore, only compounds that are present at a certain percentage (e.g., 10%–30%) of the samples are retained for further analysis.

38.3 Data normalization, transformation, and scaling

Data normalization, transformation, and scaling are often considered as data preprocessing steps; nonetheless, they are treated here as an additional step due to their significant influence on the outcome of the analysis. The issue of lognormality in biological datasets is treated in the previous chapter. Normalization is typically performed to remove effect size, i.e., unwanted variations between measured samples, and generates volatilome data with samples presented in an adequate and consistent way. Note that incorrect normalization might jeopardize putative differences between investigated cases. Data normalization can be implemented in a variety of ways. It is often performed before or after peak picking, and it is usually done by calculating a normalization factor per sample. Probabilistic quotient normalization and total area normalization (TAN) have been used in volatilome data.[8,9] TAN uses the assumption that the total profile that is measured is directly proportional to the total concentration of the sample and that total area is constant between samples. This is rarely the case, however, therefore TAN might lead to spurious correlations between VOCs.[10]

Data transformation follows the exclusion of zeros, and it accounts for correcting for data heteroscedasticity and skewness; both logarithmic and power transformations are the most broadly used.[11] As far as data scaling is concerned, many techniques exist, each of which exploits different aspects of the data at hand.[11] Scaling accounts for giving all the variables the same importance in the model; autoscaling and pareto scaling are the most common approaches. Moreover, it is essential to note that most of the multivariate approaches require the data to be scaled before running the algorithm. A few examples of powerful multivariate techniques that require prior data scaling are principal component analysis (PCA),[12] robust principal component analysis (R-PCA),[13] clustering, partial least squares (PLS) analysis,[14,15] and support vector machines (SVM) analysis.[16] Some multivariate approaches do not require prior scaling, including random forest (RF),[17] unsupervised random forest (URF),[18] and adaptive boosting (AdaBoost) analyses.[19]

38.4 Machine learning approaches

38.4.1 Unsupervised approaches

The human volatilome is characterized as a rather complicated, high-dimensionality data matrix. Often, conventional univariate statistical approaches, such as t-test, do not suffice when it comes to biomarker identification; therefore, multivariate statistical approaches are preferred. Multivariate approaches can be divided into two categories: *unsupervised* and *supervised* approaches. Unsupervised approaches are used for exploratory purposes, whereas supervised approaches are used for classification and regression purposes. The most common unsupervised approach is principal component analysis.[12] PCA captures most of the variation in the data by creating new variables, the *principal components* (PCs), which represent the original variables of the data linearly. This means that the PCs are linear combinations of the original variables, and by definition, they are orthogonal to each other. The first PC (PC1) captures the largest possible variation in the data, and each subsequent PC (PC2, PC3, etc.) captures most of the remaining orthogonal variation. PCA returns two matrices, the *score matrix* and the *loading matrix*. The scores are simply the coordinates of the samples in the PCA space, whereas the loadings show the importance (i.e., contribution) of every single original variable in generating each PC. Consequently, both score and loading figures can be drawn to visually identify clusters in the data and to explore which of the original variables play an essential role in creating these clusters, and to what extent. Furthermore, these two figures can be combined to generate what is called a *biplot* (Fig. 38.1), which offers a more comprehensive picture—and thereby a better understanding—of the data. For the volatilome data in Fig. 38.1, the biplot depicts each sample as a dot and each VOC as an arrow. In the PCA biplot, it is possible to visually appraise the relationships between individual VOCs by looking at the angle between them.

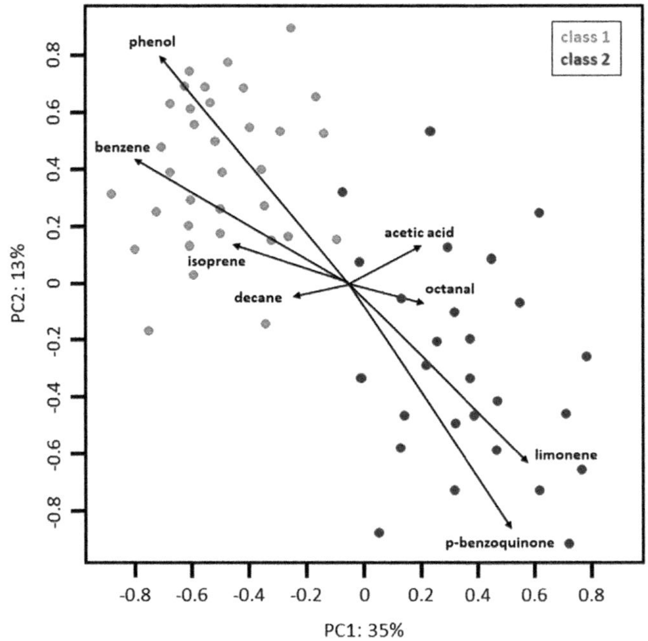

FIGURE 38.1

Example of a principal component analysis biplot illustrating two random classes and compounds plotted on their principal components (PCs). The *dots* are the scores, which represent the samples, and the *arrows* are the loadings, which represent the compounds present in the samples. The compounds that point toward a particular group are more important for that particular group. PC1 explains 35% of the original information, whereas PC2 explains 13% of the remaining original information.

A small, acute angle between two VOCs indicates a high positive correlation (e.g., isoprene and benzene in Fig. 38.1), whereas a wide, obtuse angle indicates an anticorrelation (e.g., phenol and p-benzoquinone in Fig. 38.1); two compounds are uncorrelated if the angle is close to a right angle (e.g., acetic acid and p-benzoquinone in Fig. 38.1). In addition, a biplot can provide useful information on the quantitative change in a VOC in relation to class membership. The relative concentration of a VOC in a sample is elevated if the signs of the PCs for the sample and the compound are the same, i.e., either both are negative or both are positive. Conversely, if the signs of the PCs for the sample and the VOC are opposite, i.e., one is positive and the other is negative, then the concentration in these samples is reduced.

Principal component analysis is also useful in helping detect outliers that are observable in the PCA space but are far away from the cloud of the sample observations. This type of outliers is called *good leverage outliers*. Frequently, however, the data also have what is called *orthogonal outliers*, which cannot be detected by PCA, and can therefore profoundly influence the analysis results and lead to false conclusions. These orthogonal outliers have orthogonal distances from the PCA space and it is mostly impossible to detect them by projecting them onto the PCA space. To overcome this issue, R-PCA was developed.[13] PCA maximizes variance by decomposing covariance in the data. R-PCA does the same but replaces the covariance matrix with a robust covariance estimator. PCA creates one PC at a time to capture most of the variation in the data, whereas R-PCA creates consecutive possible "interesting" PCs and then selects the one that best describes the data.

Unsupervized random forest[18] was introduced as a powerful unsupervised approach. URF is robust to any outliers as it examines variables, rather than the samples, as is done in PCA, R-PCA, and clustering. URF hypothesizes that if there is any hidden structure in the data, it should be possible to separate the data from a randomly generated version of themselves. Superiority of URF over other techniques is that it does not require any scaling, it provides variable importance, and most importantly, it returns a proximity matrix of the original data. Proximity matrices are excellent tools for visualization purposes and for directly comparing the samples at hand.

Another unsupervised approach that is widely used is clustering,[20] which consists of determining similarity measures, such as correlation or distance of the whole volatilome dataset. Several different approaches exist, including hierarchical cluster analysis (HCA), k-means, and c-means. HCA is based on iterative calculation of distance and combining the closest samples into one cluster. The procedure continues until all samples belong to the same cluster. The most important aspect of HCA is the selection of similarity measure (e.g., Euclidean, Manhattan, Mahalanobis, Minkowski distance, or correlation) and the way the clusters are created, i.e., average linkage (average distance), single (minimum distance), complete linkage (maximum distance), and Ward's method.[21] The similarity/dissimilarity of samples is then represented as a *dendrogram*. An example of a dendogram is shown in Fig. 38.2. As can be seen for a selected specific threshold on the distance measure

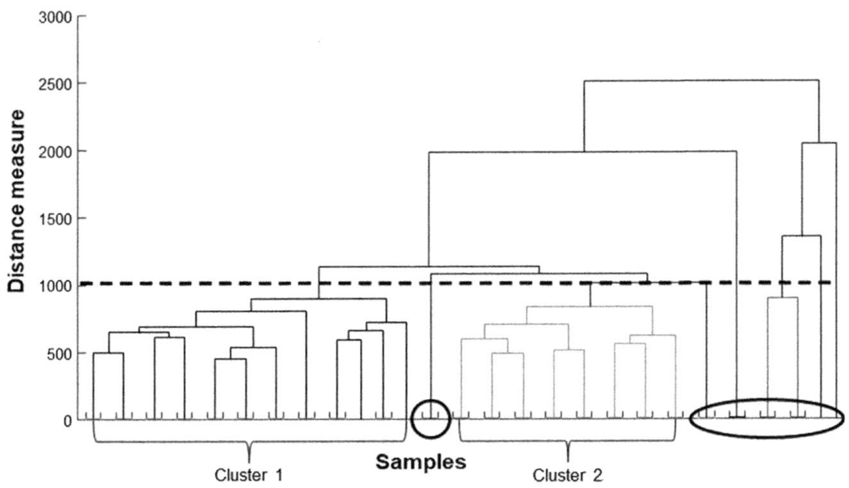

FIGURE 38.2

An example of dendrogram illustrating two main clusters of samples. Circled samples are allocated as outliers.

(a value of 1000, indicated as a horizontal dashed line), two main groups (clusters) are present. In addition, a few samples are included as separate clusters, suggesting that they are outliers (circled on the samples axis). HCA is the most suitable if a clear hierarchical structure is present in the data. An entirely different approach is taken by k-means,[22] which selects a number of centroids (i.e., clusters), defined by a user, in such a way that the overall distance of all samples to the centroids is minimized. Each sample is assigned to the closest centroid. The centroids are updated each time a new sample is assigned. The assignment of the samples to the centroids stops when convergence (i.e., no more changes in cluster content) is reached. An adaptation of k-means is c-means, which does not give a definitive assignment of each sample to a cluster but provides a probability of belonging to each cluster.

A drawback of clustering techniques is that they do not deliver information about compounds that are responsible for the resulting clusters. Moreover, they are very sensitive to outliers, and processing becomes very time-consuming for large datasets.

38.4.2 Supervised approaches

The most common supervised approach is partial least squares analysis.[14,15] PLS was initially developed to deal with regression problems, but later it was extended to additionally deal with classification problems. Similar to PCA, PLS creates new variables, the *latent variables* (LVs), that try to capture most of the information in the data with respect to a response/class vector Y in a linear way. Consequently, these LVs are linear combinations of the original variables that incorporate the response/class information of every sample. As with every supervised approach, PLS models must first be optimized (i.e., the optimal number of LVs to be used) and then validated before any final predictions/classifications are made. An overview of validation methods that can be used is given in the next section. PLS can be applied not only on binary but also on multiple class problems, although the latter show inferior prediction accuracy when compared to results from two-class or even multiple binary PLS models (i.e., one class against all the rest).[5] PLS exhibits high accuracy results when applied to collinear data but it has the tendency to overfit. As human volatilome data often demonstrate rather nonlinear relationships among variables, it may be preferable to implement supervised approaches that consider both linear and nonlinear relationships in the data; ensemble techniques are such approaches.[23] The most renowned and broadly used ensemble techniques are RF,[17] AdaBoost,[19] and gradient boosting, all of which are tree-based methods.

Random forest is a so-called *bagging* or *aggregative bootstrapping* algorithm that builds fully grown independent trees. Furthermore, each tree is built on a different subset of observations of the training dataset and a different subset of variables of these selected observations; therefore, different information is seen by different trees, thereby forcing the correlation among different outcomes to decrease. The observations that are not used to build a tree (i.e., out-of-bag observations) are used to assess this tree performance. The overall performance of the forest is assessed by the out-of-bag error (i.e., wrongly predicted/classified out-of-bag observations) of all the trees present in the forest, and in the end, the forest is evaluated by using a validation set. Finally, it is worth mentioning that the more trees that are added to the forest, the merrier and more robust the outcome is.

AdaBoost is representative of a boosting algorithm that creates stumps (i.e., a tree with two leaves) instead of fully grown trees. Most importantly, as the stumps are built sequentially, the mistakes a stump makes influence the way the next stump is built; thus, all stumps are dependent on each other. Each stump is built on a different subset of observations of the training dataset and a different subset of variables of these selected observations. Although the different subsets of observations are sampled without replacement, in RF they are sampled with replacement. This means that the same sample may be used in building more than one stump. The overall performance of the forest of stumps is evaluated with a validation set, and it should be mentioned that the more stumps that are added to the forest, the higher the chances to overfit. A method between random RF and AdaBoost is gradient boosting, which makes use of decision trees as weak classifiers that are constructed in a greedy manner.[24] In comparison to AdaBoost and RF that use stumps and fully grown trees, respectively, gradient boosting builds trees of four to eight levels (i.e., splits).

As a final note on the supervised approaches, they can all provide importance/contribution of the variables in the models, allowing for potential biomarker discovery.[5] An overview of various aspects of the multivariate methods, unsupervised and supervised, described here is presented in Table 38.1. The ensemble of techniques presented here are capable of finding linear and nonlinear relationships between compounds in volatilome data.

Another group of techniques capable of capturing nonlinear information in volatilome data are kernel-based techniques. The most common methods are support vector machines and kernel-PLS-and kernel-PLS-discriminant analysis (DA). The crucial step of kernel-based methods is transformation of the data via specific functions called kernel. This step allows the nonlinear problem in the original data to be mapped into a higher-dimensional feature space in a way that the problem becomes linear and thereby easily solvable. SVM was originally developed for two-class classification problems. The performance of the technique depends highly on kernel function used to transform the data. Kernel-PLS-DA and SVM have comparable prediction ability,[26] and both techniques have their disadvantage of losing compound information, since the data obtained after kernel transformation have dimension of samples. Despite this, both techniques have been applied in various human volatilome data treatment endeavors.[27,28] Note that every supervised technique presented so far can be used for regression as well as classification problems.

The growing importance of artificial neural networks (ANN) and consequently deep learning in medical diagnostics has become evident from various applications and publications. Both techniques are very powerful to model complex, nonlinear problems but require large sample size (in the case of deep learning, tens of thousands), and the interpretation of the results is difficult. ANN has been applied to human volatilome data,[29] but deep learning has not, to date.

Table 38.1 The main characteristics of the most common multivariate methods.

Method	Characteristics					
	Linear; nonlinear	Sensitive to			Compounds importance	
		Outliers	Scaling	Normal distribution		
Unsupervised						
PCA	Linear	Yes	Yes	No	Yes	
R-PCA	Linear	No	Yes	No	Yes	
URF	Nonlinear	No	No	No	Yes[a]	
HCA	Linear	No	No	No	No	
k-means; c-means	Linear	Yes	No	No	No	
Supervised						**Overfitting**
PLS/PLS-DA	Linear	Yes	Yes	No	Yes	Yes
AdaBoost	Nonlinear	No	No	No	Yes	Yes
RF	Nonlinear	No	No	No	Yes	No
Gradient boosting	Nonlinear	No	No	No	Yes	No
SVM	Nonlinear	Yes	No	No	Yes[a]	No
K-PLS/K-PLS-DA	Nonlinear	Yes	Yes	No	Yes[a]	Yes
ANN/deep learning	Nonlinear	No	No	No	No	No

AdaBoost, *adaptive boosting;* ANN, *artificial neural networks;* HCA, *hierarchical clustering;* K-PLS(-DA), *kernel-partial least square(-discriminant analysis);* PCA, *principal component analysis;* RF, *random forest;* R-PCA, *robust principal component analysis;* SVM, *support vector machines;* URF, *unsupervised random forest.*

[a] *Possible after applying pseudosample principle.*[25]

38.5 Data fusion

Data fusion, or data concatenation, refers to the process of combining data coming from different data platforms (e.g., GC-MS, nuclear magnetic resonance, 16S ribosomal RNA sequencing, etc.). The principle behind data fusion is that different types of samples (e.g., breath, blood, feces) are complementary to each other when a particular cohort is examined. The same holds when different techniques measure the same samples (e.g., breath samples measured by GC-MS and multicapillary column–ion mobility spectrometry [MCC-IMS]) because the different strengths and weaknesses of each technique with respect to their detection of compounds are exploited. Consequently, a better profile or increased prediction accuracy can be achieved for the cohort at hand by combining the data of these complementary samples and/or techniques. Data fusion is widely used in many research fields and holds promise for human volatilome research.

Data fusion can be performed at three different levels: *low-level*, *mid-level*, and *high-level* fusion.[30] Low-level fusion is the simplest of all since it concatenates the data by placing them next to each other as they are from different platforms without any prior analysis. This means that the fused matrix to be used for further analysis will consist of as many rows as the number of samples measured, and as many columns as the number of all the compounds measured by all different data platforms. Low-level fusion is not usually employed because concatenating all these hundreds or even thousands of variables in a single matrix increases the dimensionality of the data excessively, thereby making it difficult to analyze them.

Mid-level fusion fuses either variables or features; therefore, it requires prior analysis of each data platform separately. On the one hand, important variables per platform can be found, for example, via RF[17] or significance multivariate correlation,[31] and then, all the important variables from the platforms are placed next to each other to create the fused matrix to be used for further analysis. On the other hand, important features per platform can be found, for example, by implementing PCA[12] (and making use of the PCs), or PLS[14,15] (and making use of the LVs). Then, all the features from all the platforms are placed side by side to create the fused matrix. Mid-level fusion is the level of fusion that is most widely applied.

High-level fusion requires prior separate analysis on each data platform, and it is quite different from both low-level and mid-level fusion because it combines outcomes rather than actual data (i.e., variables or features) of the platforms. The most common approach for combining outcomes is majority voting.[32] For instance, if a sample of interest is classified as class 1 from the majority of the models (i.e., those built on the individual platforms), and as class 2 from the minority of the models, then the high-level fusion outcome will be class 1 for this sample of interest. High-level fusion can demonstrate excellent prediction results, which is to be expected because of the nature of this kind of fusion. Nonetheless, the possibility to discover potential biomarkers and to relate compound outputs from different platforms to each other is lost when high-level fusion is employed. This result is to be expected because high-level fusion does not work with actual data but with outcomes. Recently, a new and more advanced way of data fusion that uses kernels was proposed.[25] The authors named their approach multiple kernel learning data fusion because each platform is mapped onto kernel space once variable selection is performed individually on every platform. Next, all different kernels are combined in a linear way by using a weighted sum to obtain the fused matrix for use in further analysis. Using kernels is synonymous with using samples; thus, theoretically, investigation of the original variables (e.g., how they behave in the samples or how important they are) might be challenging. In practice, this problem can be overcome by using the pseudosample principle.[25] To conclude, data fusion can positively affect the outcome of a study, yet there is not a direct answer to identify which data fusion approach is the best, since this highly depends on the aim of each study.

38.6 Validation of supervised techniques

A crucial part of supervised methods is validation, which is a step that ensures the certainty of the findings. There are various ways of assessing the predictive ability of the constructed supervised model,[33–35] as outlined below.

If a studied problem consists of a sufficient number of samples, the data are divided into a training set to train the supervised model, a validation set to optimize the supervised model, and a test set to assess the prediction power of the model. The most straightforward approach consists of randomly assigning samples into those three sets. Another approach allows selecting representative training and/or test sets.[5] The most commonly applied strategy is cross-validation (CV). This approach is based on retaining a predefined number of samples and a building supervised model using the remaining samples. The simplest CV procedure is leave-one-out (LOO), where one sample is excluded and used as a test set, and n-1 samples are used to create a training set (where "n" is the total number of samples). In LOO CV, each sample becomes a test sample once, and therefore there are "n" rounds of testing. LOO CV is suitable when the number of available samples is relatively low, e.g., 20—30 samples. If more samples are available, k-fold CV can be used, where k is defined as a number, e.g., 5 or 10, or percentage of total samples, e.g., 10%. In the case of k-fold CV, k number of samples is removed from the data and used as the test set. The important aspect of CV is its overestimation of the predictive power of the supervised model. Hence, for data with a rather small number of samples, it is recommended to apply double CV (called nested CV) or *bootstrapping*.[33] The final prediction ability of the supervised model can be assessed by a permutation test, which checks whether division into defined classes is significantly better than any random division.[34] The permutation test consist of random rearrangement of the class labels in the training set and the construction of a classification model using these randomly permuted classes. In the subsequent step, the prediction for real test set samples is obtained and the number of misclassifications is obtained. The entire procedure is repeated many times, for instance, 1000 iterations. The assumption of the permutation test is that the test set should be wrongly predicted for randomly permuted classes in the training set.

A very important aspect of supervised techniques is scaling of the data after division into the training, validation, and test sets. It is of great importance that samples used for validating the performance of the machine learning model should be always scaled (e.g., by autoscaling or pareto scaling) using parameters obtained from the training samples. Although various approaches exist to test the quality of the predictive model, the ultimate manner of assessing prediction ability of a statistical model is to use an independent test set consisting of completely new samples from an independently sampled population.

38.7 Summary

The main steps involved in data preprocessing and multivariate analysis of volatilome data is reviewed in this chapter. Data preprocessing is a crucial step when dealing with numerically complex volatilome data. This procedure can be carried out via various approaches, such as VOCCluster or XCMS (used for data preprocessing of GC-MS–related data), and additionally involves the use of techniques for noise and baseline reduction that can be applied to different MS-based volatilome datasets. The importance of normalization and scaling of volatilome data has been emphasized in this chapter. The machine learning techniques covered in this chapter are limited to the most commonly applied approaches that have the highest potential for future applications. Moreover, the techniques presented here are reasonably easy to apply and potentially insightful. Strategies for data fusion are currently underused but should become more common in volatolomics in the future. Because of the increased complexity of the studied problems, it is favorable to combine multiple sources of information to gain a better understanding of the underlying problem. Therefore, combining multiple types of measurements, e.g., VOCs in breath as well as feces, in a single statistical analysis might increase not only the prediction accuracy, but also interpretability and comprehensiveness of the results. Appropriate utilization of machine learning techniques in the field of human volatilome research offers the potential for more robust discoveries of relevant VOC biomarkers.

References

1. Krzanowski WJ. *Principles of multivariate analysis*. Revised ed. 2000. New York.
2. Vandeginste BGM, Massart DL, Buydens LMC, Jong SD, Lewi PJ, Smeyers-Verbeke J. *Handbook of chemometrics and qualimerics: Part A*. Amsterdam: Elsevier; 1998.
3. Walczak B. *Wavelets in chemistry. Data handling in science and technology*. Elsevier Science & Technology; 2000.
4. Eilers PH. A perfect smoother. *Anal Chem* 2003;**75**(14):3631–6.
5. Smolinska A, Hauschild A-C, Fijten R, Dallinga J, Baumbach J, Van Schooten F. Current breathomics—a review on data pre-processing techniques and machine learning in metabolomics breath analysis. *J Breath Res* 2014;**8**(2):027105.
6. Alkhalifah Y, Phillips I, Soltoggio A, Darnley K, Nailon WH, McLaren D, et al. VOCCluster: untargeted metabolomics feature clustering approach for clinical breath gas chromatography/mass spectrometry data. *Anal Chem* 2020;**92**(4):2937–45.
7. Smith CA, Want EJ, O'Maille G, Abagyan R, Siuzdak G. XCMS: processing mass spectrometry data for metabolite profiling using nonlinear peak alignment, matching, and identification. *Anal Chem* 2006;**78**(3):779–87.
8. Smolinska A, Klaassen EM, Dallinga JW, van de Kant KD, Jobsis Q, Moonen EJ, et al. Profiling of volatile organic compounds in exhaled breath as a strategy to find early predictive signatures of asthma in children. *PLoS One* 2014;**9**(4):e95668.
9. Pijls KE, Smolinska A, Jonkers DM, Dallinga JW, Masclee AA, Koek GH, et al. A profile of volatile organic compounds in exhaled air as a potential non-invasive biomarker for liver cirrhosis. *Sci Rep* 2016;**6**:19903.
10. Filzmoser P, Walczak B. What can go wrong at the data normalization step for identification of biomarkers? *J Chromatogr A* 2014;**1362**:194–205.

11. van den Berg RA, Hoefsloot HC, Westerhuis JA, Smilde AK, van der Werf MJ. Centering, scaling, and transformations: improving the biological information content of metabolomics data. *BMC Genomics* 2006;**7**:142.
12. Bro R, Smilde AK. Principal component analysis. *Anal Methods* 2014;**6**(9):2812–31.
13. Hubert M, Rousseeuw PJ, Vanden Branden K. ROBPCA: a new approach to robust principal component analysis. *Technometrics* 2005;**47**(1):64–79.
14. Wold S, Sjöström M, Eriksson L. PLS-regression: a basic tool of chemometrics. *Chemometr Intell Lab Syst* 2001;**58**(2):109–30.
15. Gromski PS, Muhamadali H, Ellis DI, Xu Y, Correa E, Turner ML, et al. A tutorial review: metabolomics and partial least squares-discriminant analysis–a marriage of convenience or a shotgun wedding. *Anal Chim Acta* 2015;**879**:10–23.
16. Drucker H, Wu D, Vapnik VN. Support vector machines for spam categorization. *IEEE Trans Neural Network* 1999;**10**(5):1048–54.
17. Breiman L. Random forest. *Mach Learn* 2001;**45**:5–32.
18. Afanador NL, Smolinska A, Tran TN, Blanchet L. Unsupervised random forest: a tutorial with case studies. *J Chemometr* 2016;**30**(5):232–41.
19. Freund Y, Schapire RE, editors. *Experiments with a new boosting algorithm*. icml; 1996 [Citeseer].
20. Xu R, Wunsch D. *Clustering*. John Wiley & Sons; 2008.
21. Hastie T, Tibshirani R, Friedman JH, editors. *The elements of statistical learning*. 2nd ed. Springer.
22. Sinha A, Desiraju K, Aggarwal K, Kutum R, Roy S, Lodha R, et al. Exhaled breath condensate metabolome clusters for endotype discovery in asthma. *J Transl Med* 2017;**15**(1):262.
23. Ensemble methods in machine learning. In: Dietterich TG, editor. *International workshop on multiple classifier systems*. Springer; 2000.
24. Friedman JH. Greedy function approximation: a gradient boosting machine. *Ann Stat* 2001:1189–232.
25. Smolinska A, Blanchet L, Coulier L, Ampt KA, Luider T, Hintzen RQ, et al. Interpretation and visualization of non-linear data fusion in kernel space: study on metabolomic characterization of progression of multiple sclerosis. *PLoS One* 2012;**7**(6).
26. Smolinska A, Blanchet L, Coulier L, Ampt KAM, Luider T, Hintzen RQ, et al. Interpretation and visualization of non-linear data fusion in kernel space: study on metabolomic characterization of progression of multiple sclerosis. *PLoS One* 2012;**7**(6).
27. Purcaro G, Rees CA, Wieland-Alter WF, Schneider MJ, Wang X, Stefanuto PH, et al. Volatile fingerprinting of human respiratory viruses from cell culture. *J Breath Res* 2018;**12**(2):026015.
28. Smolinska A, Bodelier AG, Dallinga JW, Masclee AA, Jonkers DM, van Schooten FJ, et al. The potential of volatile organic compounds for the detection of active disease in patients with ulcerative colitis. *Aliment Pharmacol Ther* 2017;**45**(9):1244–54.
29. Kort S, Brusse-Keizer M, Gerritsen JW, van der Palen J. Data analysis of electronic nose technology in lung cancer: generating prediction models by means of Aethena. *J Breath Res* 2017;**11**(2):026006.
30. Borràs E, Ferré J, Boqué R, Mestres M, Aceña L, Busto O. Data fusion methodologies for food and beverage authentication and quality assessment—A review. *Anal Chim Acta* 2015;**891**:1–14.

31. Tran TN, Afanador NL, Buydens LM, Blanchet L. Interpretation of variable importance in partial least squares with significance multivariate correlation (sMC). *Chemometr Intell Lab Syst* 2014;**138**:153–60.
32. Penrose LS. The elementary statistics of majority voting. *J Roy Stat Soc* 1946;**109**(1): 53–7.
33. Anderssen E, Dyrstad K, Westad F, Martens H. Reducing over-optimism in variable selection by cross-model validation. *Chemometr Intell Lab Syst* 2006;**84**(1–2):69–74.
34. Westerhuis JA, Hoefsloot HCJ, Smit S, Vis DJ, Smilde AK, van Velzen EJJ, et al. Assessment of PLSDA cross validation. *Metabolomics* 2008;**4**(1):81–9.
35. Szymanska E, Saccenti E, Smilde AK, Westerhuis JA. Double-check: validation of diagnostic statistics for PLS-DA models in metabolomics studies. *Metabolomics* 2012; **8**(1):S3–16.

CHAPTER

39 Applications of the US EPA CompTox Chemicals Dashboard to support mass spectrometry and breath research

Antony J. Williams[1], Jon R. Sobus[2]

[1]*National Center for Computational Toxicology, United States Environmental Protection Agency, Research Triangle Park, NC, United States;* [2]*National Exposure Research Laboratory, United States Environmental Protection Agency, Research Triangle Park, NC, United States*

Chapter outline

39.1 Overview

Mass spectrometry (MS)—based monitoring techniques have long been used in public health research to assess chemical exposures and the responses of biological systems to those exposures. Various biological media have been the focus of MS monitoring studies. Breath-derived samples, in particular, provide a unique means of assessing exposure and health, as they contain both exogenous chemical stressors and endogenous response markers (see Chapter 1).[1] They further provide a near inexhaustive supply of sample media, which can be collected using a variety of noninvasive techniques (see Chapter 2). For many years, "targeted" MS methods have been used to characterize compounds of interest in breath samples. Here, chemicals of interest are identified prior to experimentation. Standards of these chemicals are then purchased or synthesized and used to develop methods with adequate sensitivity and specificity. The goal of targeted analysis is to generate quantitative measures for compounds of interest in support of exposure assessment, epidemiological

Copyright © 2020 Published by Elsevier B.V. All rights reserved.

evaluation, medical diagnosis, or a number of other applications. The primary limitation of targeted MS methods is the sole focus on a few (generally <100) predetermined compounds of interest. In other words, targeted methods do not focus on the discovery of "new" compounds of interest. To meet this need, "suspect screening" and "nontargeted analysis" (also referred to as "untargeted analysis," "nontarget screening," or "NTA") methods have been introduced in recent years[2] and are rapidly growing in popularity in the exposure and health science communities.

Suspect screening is a technique wherein experimental data (i.e., mass spectra) are acquired and compared against reference spectra for known compounds to identify possible "hits." Importantly, suspect screening applications make use of spectra that have been previously acquired for large numbers of target compounds and stored in accessible libraries.[3] Commercial libraries, some containing spectra for hundreds of thousands of known compounds, are available for purchase through the National Institute of Standards and Technology (NIST)[a] and Wiley[b]. Instrument vendors also have commercial data collections including Agilent PCDL (Personal Compounds, Databases and Libraries) collections[c] and Sciex spectral libraries[d]. Freely available spectral libraries also exist through sources such as MassBank[e], MassBank of North America (MoNA)[f], and METLIN[g].

In a suspect screening experiment, empirical MS or tandem MS (MS/MS) spectra are first acquired using the mass spectrometer then, through a series of data processing steps, these spectra are compared against library spectra to identify plausible matches. Mathematical calculations, in combination with manual review by the analyst, are ultimately performed to select the best compound match from a listing of plausible candidate matches. The presence or absence of specific compounds, or relative intensity of specific compounds, is then used to determine whether an exposure has occurred or whether a biological system has been perturbed as a function of some stressor.

The output of a suspect screening investigation is a listing of compounds believed to be present in a sample. While the number of "tentatively identified" candidates[4] generally far exceeds the number of compounds examined in a targeted analysis, the findings are inherently restricted to the chemical space of the suspect library, that is, the specific compounds for which reference spectra have been acquired and stored. "NTA" differs from suspect screening in that it is the process through which compounds can be identified in samples of interest *without* the use

[a] https://www.nist.gov/srd/nist-standard-reference-database-1a-v17.

[b] https://www.wiley.com/en-us/Wiley+Registry+of+Mass+Spectral+Data%2C+11th+Edition-p-9781119171010.

[c] https://www.agilent.com/search/?Ntt=PCDL.

[d] https://sciex.com/products/spectral-library.

[e] https://massbank.eu/MassBank/.

[f] http://mona.fiehnlab.ucdavis.edu/.

[g] https://metlin.scripps.edu/.

of reference spectra.[2] NTA is made possible using high-resolution mass spectrometers (HRMS)—most often with time-of-flight (TOF) or Orbitrap mass analyzers—paired with upstream gas chromatography (GC) or liquid chromatography (LC), "soft" ionization sources, such as electrospray ionization (ESI) and atmospheric pressure chemical ionization (APCI), and fragmentation capabilities.[5] In NTA experiments, "molecular features" (i.e., collections of related *m/z* peaks) are first identified in the sample matrix using peak picking and alignment procedures. A neutral monoisotopic mass is then estimated for each feature. This accurate mass estimate, along with the observed isotope profile, is used to propose a molecular formula for the unknown feature (see Chapter 16). These accurate mass and formula estimates are ultimately searched against public databases of known compounds. Certain databases, such as PubChem[h] and ChemSpider[i], contain chemistry information for tens of millions of compounds. Other resources, such as US EPA's Distributed Structure-Searchable Toxicity Database (DSSTox),[6] provide carefully curated chemistry content for hundreds of thousands of substances that are of specific interest to the US EPA and broader environmental health science communities. The EPA Comptox Chemicals Dashboard[7] is the web-based public interface to this database, with specific functions designed and built to service MS NTA research applications both within and outside EPA. The purpose of this chapter is to introduce the Dashboard and its potential applications to support breath researchers utilizing MS to identify chemicals of interest.

39.2 US EPA CompTox Chemicals Dashboard

The CompTox Chemicals Dashboard (hereafter the Dashboard) is a freely accessible web-based application[j] providing access to data associated with ~875,000 chemical substances. The substances include individual chemical compounds that can be represented as not only chemical structures but also substances of Unknown or Variable composition, Complex reaction products and Biological materials (UVCB substances). Such substances include, for example, hydrogenated palm oil[k] and complex surfactants, such as linear alkylbenzene sulfonates[l]. Associated with each chemical substance can be a myriad of different data types, including experimental hazard data, experimental and predicted physicochemical, fate and transport data, and in vitro bioassay data measured as part of the ToxCast effort conducted at EPA[m]. In the case of substances with associated chemical structures, quantitative structure–activity relationship (QSAR) predictions can often provide estimates of

[h] https://pubchem.ncbi.nlm.nih.gov/.
[i] http://www.chemspider.com/.
[j] https://comptox.epa.gov/dashboard.
[k] https://comptox.epa.gov/dashboard/DTXSID1028655.
[l] https://comptox.epa.gov/dashboard/DTXSID3020041.
[m] https://www.epa.gov/chemical-research/toxicity-forecasting.

a number of physicochemical and fate and transport endpoints, which are especially useful when experimental data are not available.[8] A full description of the Dashboard functionality and capabilities has been published elsewhere,[7] but this chapter will focus on the application of the Dashboard to support chemical identification in breath research.

As described earlier, chemical features observed using HRMS can be tentatively identified using online chemical reference databases by searching molecular formulae and monoisotopic masses and then rank-ordering the hits using appropriate relevance criteria. Search functionality within the Dashboard enables users to perform queries based on a single monoisotopic mass or molecular formula (via the Advanced Search screen[n]) or batches of many molecular formulae or *m/z* (via the Batch Search Screen[o]). The Advanced Search (see Fig. 39.1) includes three different search capabilities: (1) search for mass $\pm$ error (in either Daltons or parts per million error, as well as including the ability to allow for adducts from a specific list); (2) molecular formula search based on either "MS-Ready" formula (vide infra) or exact formula; and (3) the ability to generate potential formulae associated with an input mass $\pm$ error and allowing for the user to define potential elements and range of elements that can be included in the generated formulae. The resulting generated formulae are searched against the database automatically, and a frequency count for a particular formula is displayed.

Following searches based on either mass or formula(e), the search results can be rank-ordered (see Fig. 39.2) using various types of metadata, including the number of data sources associated with a particular chemical, the number of associated Pubmed articles, and the prevalence of the chemical in commerce, as identified through the number of consumer products containing the chemical(s).[9]

The performance of the Dashboard for identifying "known unknowns" has previously been evaluated[10] against that of the online ChemSpider database, one of the primary resources used by mass spectrometrists, using multiple previously studied datasets reported in the peer-reviewed literature and totaling 162[11]. The potential to identify these chemicals using searches conducted against both applications with either molecular formula or monoisotopic mass searches was investigated. The search results provided a list of candidate compounds that were then rank-ordered by associated references or data sources to investigate the potential of these limited metadata to appropriately rank the candidates. A greater percentage of chemicals ranked in the top position when using the Dashboard, indicating an advantage of this application over ChemSpider for identifying known unknowns using data source ranking. Comparing the Dashboard versus ChemSpider, respectively, the results were 85% versus 70% based on formula searching, and 88% versus 80% based on mass searching.

[n] https://comptox.epa.gov/dashboard/advanced_search/index.
[o] https://comptox.epa.gov/dashboard/dsstoxdb/batch_search.

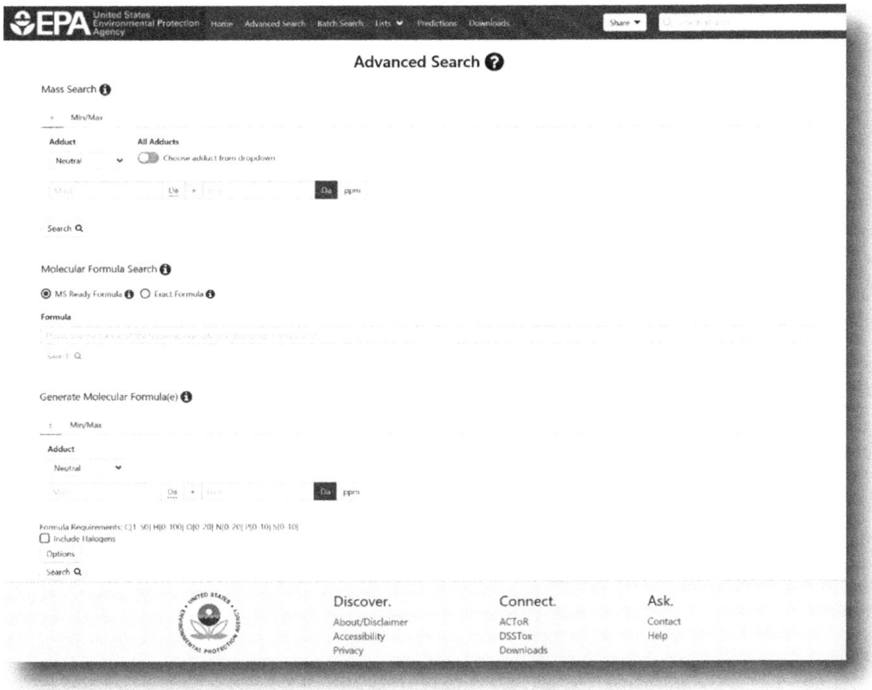

FIGURE 39.1

The Advanced Search includes three different search capabilities: (1) search for mass ± error (in either Daltons or parts per million error, as well as including the ability to allow for adducts from a specific list); (2) molecular formula search based on either 'MS-Ready' formula (vide infra) or exact formula; and (3) the ability to generate potential formulae associated with an input mass ± error and allowing for the user to define potential elements and range of elements that can be included in the generated formulae.

A Dashboard feature important for MS-based searches is the mapping of chemicals to their "MS-Ready" structures—structural forms that are desalted, desolvated, mixture-separated, and absent of stereochemistry, to match to the neutral form of a chemical observed by an analyst during MS data processing.[12] Searching the formulae for a list of unknowns against MS-Ready structures links instrument observations to all forms of a structure contained within DSSTox (e.g., the neutral form and the hydrochloride salt or solvate of a structure). The MS-ready mapping procedure is based on a series of transformation rules, and it is planned to make available to the community a service that can bulk process datafiles into an MS-Ready form. All chemical structures underpinning the Dashboard are available for download in MS-Ready format[p] for integration into third-party databases.

[p] https://comptox.epa.gov/dashboard/downloads.

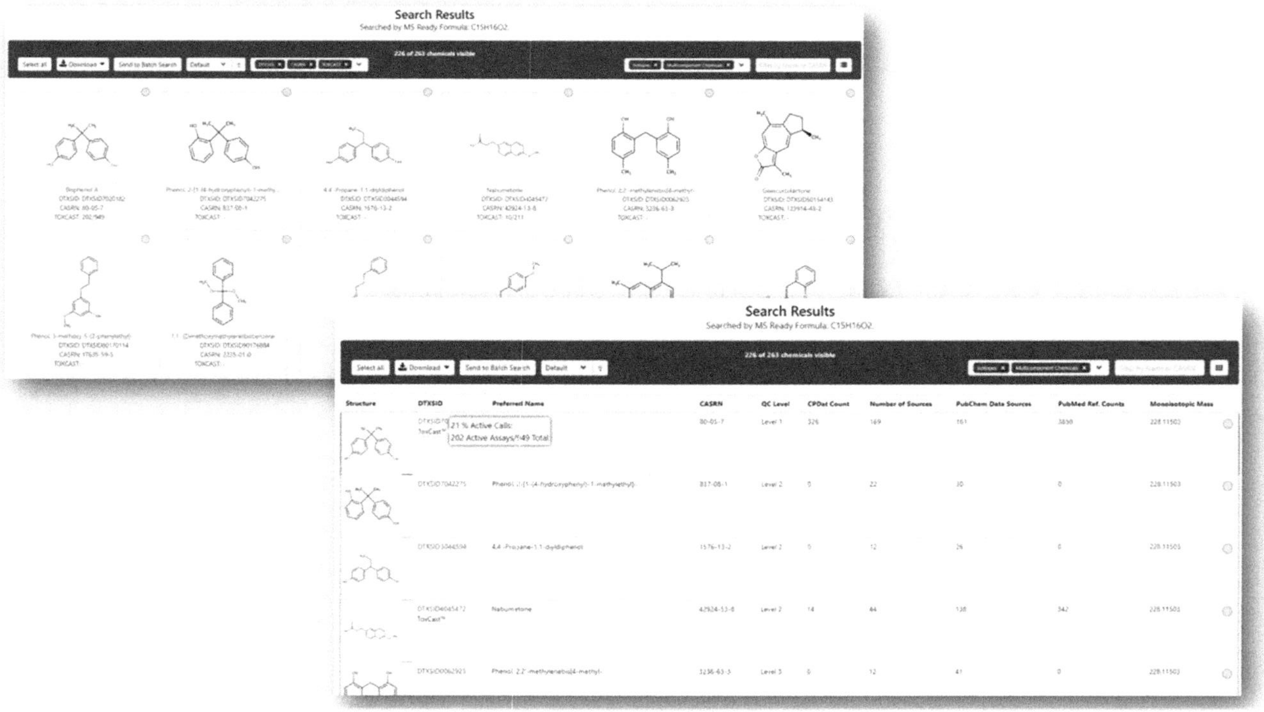

FIGURE 39.2

Search results for an MS-Ready formula search of $C_{15}H_{16}O_2$. The default view of the search results is a "tile view" (top left) of all hits that can be easily viewed and downloaded in multiple formats for further analysis (comma-separated value, Excel, or SDF file formats). The file can be sorted by multiple criteria including mass, number of associated data sources in the DSSTox database, number of PubMed articles, and other criteria. An alternative display mode is a "table view" (bottom right) that shows multiple criteria for sorting and ranking, including the number of separate data sources associated with a chemical and registered into the database, the number of articles associated with a chemical and referenced in PubMed (vide infra), the number of consumer products referenced in the Chemicals and Products Database (CPDat) that contains a particular chemical, and other associated metadata. These metadata are critical for candidate ranking and the process of identifying "known unknowns."

The chemical substances contained in DSSTox represent the assembly of data from multiple sources and include consumer products, manufacturing chemicals, drugs, and so on. For the purposes of research undertaken at the US EPA, there is interest in categorization of chemicals into lists to represent these research activities (e.g., ToxCast screening), data source providers, and specific categories and classes of chemicals. This assembly and categorization is approaching 200 individual public lists at the time of writing, with almost that many again available in-house to support multiple other projects. An index page listing a set of public DSSTox-registered chemical lists associated with a project, publication, source database, or other collections is accessed via the top banner menu "Lists" link on the Dashboard[q]. Each registered list is accompanied by the list title, the number of associated chemicals in the list, and a short summary. Some examples of available lists are a small algal toxin list[r], a list of chemical additives in cigarettes[s], and a list of chemicals in human blood[t]. All lists remain under constant curation and expansion, except those associated with point-in-time research articles. For each list, a detailed description, as well as a tabular view, of the chemical structures included in the list is accessed by clicking on the list name. Lists can be assembled in two ways: through a defined list curation process that registers the source list in the underlying DSSTox database[6], or at the application level where a list is generated by mapping to existing content, but the data are not yet fully curated or internally registered as a DSSTox list.

39.3 Developing a list of chemicals detected in breath

As described above, chemical lists offer an easy way to deliver a subset of data relevant to a particular area of study. The development of a list of chemicals detected in breath was initiated by extracting data from Table 1 of the article by Amann et al.[13] This was performed by saving the PDF file into Word docx format and filtering to identify the subset of breath-only chemicals. Of the 870 extracted chemicals in the list, 759 were available in the public version of the Dashboard (containing ~875 k chemicals, as of January 2020), as identified by performing a batch search[u] using the list of associated CAS Registry Numbers (CASRNs) as input. A batch search based on chemical names as inputs recovered only 549 chemicals from the database. Such a difference in retrieval is not surprising as CASRNs follow a fixed format, including a check digit, and therefore have only limited variation, for example, with or without dashes as separators. Chemical names, however, are highly variable, even when considering only systematic names. For example, even simple

[q] https://comptox.epa.gov/dashboard/chemical_lists.
[r] https://comptox.epa.gov/dashboard/chemical_lists/ALGALTOX.
[s] https://comptox.epa.gov/dashboard/chemical_lists/CIGARETTES.
[t] https://comptox.epa.gov/dashboard/chemical_lists/HUMANBLOOD.
[u] https://comptox.epa.gov/dashboard/dsstoxdb/batch_search.

alkenic bond orientations can be represented in multiple ways: cis/trans, E/Z, (cis-)/(trans-), and (E-)/(Z-), including placement at the beginning or end of the name. Name searches can be useful for gap-filling when CASRN searches fail. In this case, name searches identified seven additional chemicals only where the CASRN had failed. These differed in stereochemistry in general. It is difficult to confirm exactly what chemicals the authors were identifying at the time of publication, but it is possible that they were intending to identify nonstereoforms but retrieved stereospecific CASRN, a confusing situation that is common in public domain databases. The combined set of 766 chemicals were released as a chemical list[v], while a registration and curation process was initiated on the remaining ~100 chemicals. Additional chemicals discovered in the breath research conducted at EPA laboratories have since been added to the list, as shown in Fig. 39.3.

The dataset was registered into the chemical registration system, and automated checking for collisions between chemical names and the associated CASRNs was performed. The missing chemicals were registered and an updated list will be made available in a future release, at present scheduled for March 2020, in keeping with the biannual release schedule. While the initial "volatilome" article provided an initial list of chemicals of interest to breath researchers, this collection continues to expand with the addition of new chemicals on an ongoing basis and has already included in that list a number of chemicals in breath identified in laboratory studies at the EPA. In order to encourage community participation and contribution, a call to participation to contribute data to the growing collection of chemicals of interest to breath researchers has been issued.[14] A second more expansive list[w] of chemicals with the scope of having been detected in other human biological media, including blood, dried blood spots (DBS), urine, exhaled breath condensate (EBC), and exhaled breath aerosols (EBA), is also under development.

39.4 Work in progress

This article has reviewed the existing capabilities of the Dashboard, as of January 2020, in terms of supporting approaches to chemical identification using MS data in combination with a rich database of chemicals, associated metadata, and integrated cheminformatics approaches. Other research efforts and developments presently underway are likely to further facilitate the utility of the Dashboard to support breath research. Pleil et al. have reviewed additional developments that could be beneficial in terms of the application of MS to breath research,[15] including predicted fragmentation and utilizing spectral databases to assist in metadata ranking, as reported below.

[v] https://comptox.epa.gov/dashboard/chemical_lists/VOLATILOME.

[w] https://comptox.epa.gov/dashboard/chemical_lists/VOLATILOME2.

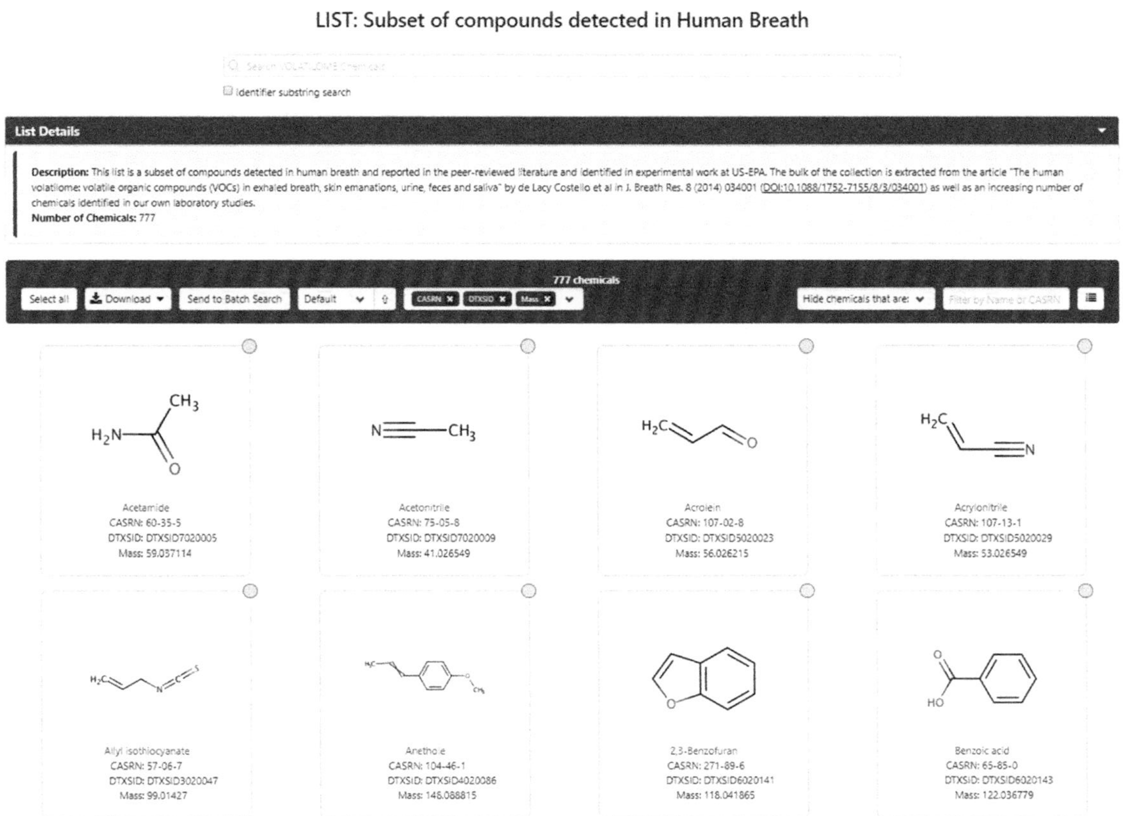

FIGURE 39.3

The volatilome list of chemicals viewed in the Dashboard. The list can be viewed in Tile or Table mode, filtered using various criteria, and exported in multiple formats. The list is under expansion based on published peer-reviewed articles and research conducted in EPA laboratories.

The application of computational fragmentation approaches to produce libraries of predicted spectra under different conditions is increasingly used as supporting evidence for structure identification when reference spectra are not available. There are a number of algorithmic approaches available, including commercial tools and an increasing number available as open source software. A comparison of a number of available tools and algorithms has been an output from the Critical Assessment of Small Molecule Identification (CASMI) Contests[x], and a useful summary of approaches is available in the CASMI 2016 publication.[16] The CFM-ID software (Competitive Fragmentation Modeling[17]) has been used to compute spectra for all MS-Ready structures in the DSSTox database to produce ESI-positive and ESI-negative ion spectra at three different collision energies (low/10 eV, medium/20 eV, and high/40 eV), as well as electron ionization (EI) spectra at a single energy of 70 eV. The performance of the GC-MS EI prediction algorithm has previously been compared against multiple other fragmentation prediction algorithms. Including a comparison of experimental versus predicted fragmentation will be used as an additional form of data to utilize in the candidate ranking.[18] The generation of fragmentation spectra for all available modes (EI, ESI+/−) for an MS-Ready form of a DSSTox dataset utilized a high-performance computer cluster to generate over 5 million predicted spectra.[19] A prototype user interface allowing for the searching of experimental spectra against predicted spectra is presently under testing, with the intention of providing one additional piece of data to assist in candidate ranking.

The External Links section of the Dashboard provides outlinks to public databases of spectra. For example, bisphenol A under the external links[y] provides links out to multiple public databases with available experimental spectra including mzCloud[z], MassBank[aa], and others. Some of these associated datasets have been registered as lists including, for example, MassBank[bb] and mzCloud[cc]. Having these lists of chemicals registered helps the user navigate directly online to reference spectra of interest and thereby assists the user in locating reference spectra generated using various instruments and methods. At present, work is underway to add metadata regarding which chemicals are represented in public and commercial databases, with the intention of including flags regarding details, such as whether the chemical has been characterized by GC-MS or LC-MS, and specific ionization modes, as these additional metadata details can be informative when ranking candidate hits acquired using specific measurement approaches.

The ongoing expansion of the list of chemicals detected in human breath is presently underway. The PubMed Abstract Sifter search capability surfaced in the

---

[x] http://casmi-contest.org/2017/index.shtml.

[y] https://comptox.epa.gov/dashboard/dsstoxdb/results?search=bisphenol%20A#links.

[z] https://www.mzcloud.org/.

[aa] https://massbank.eu/MassBank/.

[bb] https://comptox.epa.gov/dashboard/chemical_lists/MASSBANKREF.

[cc] https://comptox.epa.gov/dashboard/chemical_lists/MZCLOUD.

Dashboard is a limited implementation of work reported by Baker et al.[20] It utilizes the Medical Subject Heading (MeSH) controlled vocabulary thesaurus used for indexing articles to query against PubMed[dd] services. The Abstract Sifter employs a search query interface based on the highly curated CASRN and preferred name for a chemical substance, in conjunction with the more broadly inclusive MeSH name for the chemical. The sifter queries are focused on preloaded terms of interest to toxicology and exposure. Selection of a query term, for example, "Hazard," extends the chemical identifier list with a pregenerated MeSH query associated with the term of interest. For a substance, such as atrazine, a Hazard-based query would produce "1912-24-9" or "Atrazine" and NOAEL or NOEL or LOEL or RfD or "reference dose" or "reference concentration" or "adverse effect level" or "cancer slope factor" as input. The Sifter accesses external PubMed web services and returns an abstract count directly to the Dashboard interface. If a large number of results are retrieved, the user can refine the query by adding additional filter terms or downloading the set to the offline sifter application. Efforts are underway to expand the list of chemicals identified in breath using abstract sifting and specific queries against PubMed with the intention of including this information as additional metadata for candidate ranking. A similar approach has already been applied to investigate potential neurotoxicants and their investigation using HRMS[21] and by development of a specific list[ee].

In tandem with the expansion of candidate-ranking approaches, the inclusion of advanced search capabilities to featuring substructural search filters and selection of specific lists for searching will be added as Dashboard enhancements. Mass spectrometrists can often identify specific molecular fragments (substructures) from spectral signatures and will be able to use one or more of these substructural fragments as filters. A candidate hit list based on a formula or mass-based search can be filtered down by inclusion of chemicals containing, for example, a chlorobenzene fragment or a linear or branched alkyl chain of a specific length. Such filters can result in significant reduction in candidate lists. The ideal solution, of course, would be the intersection of candidate ranking using the expanding abundance of metadata and searching of instrument data against spectral databases of both experimental data and predicted fragmentation spectra.

39.5 Summary

This chapter has described the potential benefits of using the CompTox Chemicals Dashboard to support breath research, specifically those utilizing MS as an analytical technique. The Dashboard, in the form of a freely accessible web-based application, is a rich resource of chemical-related data and, with the search capabilities

[dd] https://www.ncbi.nlm.nih.gov/pubmed/.
[ee] https://comptox.epa.gov/dashboard/chemical_lists/LITMINEDNEURO.

already available, can be applied to structure identification using mass- and formula-based searching. This application to breath research is supported by the availability of appropriate chemical datasets associated with this domain of research, especially the expansion of a "volatilome" dataset. This is underway at present, with the community of researchers being encouraged to contribute to its growth. The addition of new data and functionality, especially the ability to search experimental spectra against in silico fragmentation spectra, offers an additional approach for ranking of candidates. By combining advanced search functionality and rich data streams to increase certainty of identification in NTA, the Dashboard provides a valuable resource for the MS NTA community and for breath research.

Disclaimer

The views expressed in this chapter are those of the authors and do not necessarily reflect the views or policies of the US Environmental Protection Agency. Mention of trade names or commercial products does not constitute endorsement or recommendation for use.

Supplementary information file

The volatilome file assembled and publicly available at https://comptox.epa.gov/dashboard/chemical_lists/VOLATILOME, as of January 2020, is also available as an Excel Spreadsheet and includes DTXSID identifiers, CASRNs, chemical names, formulae, masses, and other related data and is available at https://doi.org/10.23645/epacomptox.9762410.

References

1. Wallace MAG, Pleil JD. Evolution of clinical and environmental health applications of exhaled breath research: review of methods and instrumentation for gas-phase, condensate, and aerosols. *Anal Chim Acta* 2018;**1024**:18–38.
2. Krauss M, Singer H, Hollender J. LC-high resolution MS in environmental analysis: from target screening to the identification of unknowns. *Anal Bioanal Chem* 2010;**397**(3): 943–51.
3. Peisl BYL, Schymanski EL, Wilmes P. Dark matter in host-microbiome metabolomics: tackling the unknowns-A review. *Anal Chim Acta* 2018;**1037**:13–27.
4. Schymanski EL, Jeon J, Gulde R, Fenner K, Ruff M, Singer HP, et al. Identifying small molecules via high resolution mass spectrometry: communicating confidence. *Environ Sci Technol* 2014;**48**(4):2097–8.
5. Andra SS, Austin C, Patel D, Dolios G, Awawda M, Arora M. Trends in the application of high-resolution mass spectrometry for human biomonitoring: an analytical primer to

studying the environmental chemical space of the human exposome. *Environ Int* 2017; **100**:32–61.

6. Grulke CM, Williams AJ, Thillanadarajah I, Richard AM. EPA's DSSTox database: history of development of a curated chemistry resource supporting computational toxicology research. *Comp Toxicol* 2019;**12**:100096.
7. Williams AJ, Grulke CM, Edwards J, McEachran AD, Mansouri K, Baker NC, et al. The CompTox Chemistry Dashboard: a community data resource for environmental chemistry. *J Cheminform* 2017;**9**(1):61.
8. Mansouri K, Grulke CM, Judson RS, Williams AJ. OPERA models for predicting physicochemical properties and environmental fate endpoints. *J Cheminform* 2018;**10**(1):10.
9. Dionisio KL, Phillips K, Price PS, Grulke CM, Williams AJ, Biryol D, et al. The Chemical and Products Database, a resource for exposure-relevant data on chemicals in consumer products. *Sci Data* 2018;**5**:180125.
10. McEachran AD, Sobus JR, Williams AJ. Identifying known unknowns using the US EPA's CompTox Chemistry Dashboard. *Anal Bioanal Chem* 2017;**409**:1729–35.
11. Little JL, Williams AJ, Pshenichnov A, Tkachenko V. Identification of "known unknowns" utilizing accurate mass data and ChemSpider. *J Am Soc Mass Spectrom* 2012;**23**(1):179–85.
12. McEachran AD, Mansouri K, Grulke CM, Schymanski EM, Ruttkies C, Williams AJ. "MS-Ready" structures for non-targeted high-resolution mass spectrometry screening studies. *J Cheminf* 2018;**10**:45.
13. Amann A, de Lacy Costello B, Miekisch W, Schubert J, Buszewski B, Pleil J, et al. The human volatilome: volatile organic compounds (VOCs) in exhaled breath, skin emanations, urine, feces and saliva. *J Breath Res* 2014;**8**(3).
14. Pleil JD, Williams AJ. Centralized resource for chemicals from the human volatilome in an interactive open-sourced database. *J Breath Res* 2019;**13**(4). 040201.
15. Pleil JD, Wallace MAG, McCord J. Beyond monoisotopic accurate mass spectrometry: ancillary techniques for identifying unknown features in non-targeted discovery analysis. *J Breath Res* 2019;**13**:012001.
16. Schymanski EL, Ruttkies C, Krauss M, Brouard C, Kind T, Dührkop K, et al. Critical assessment of small molecule identification 2016: automated methods. *J Cheminform* 2017;**9**(1):22.
17. Allen F, Pon A, Wilson M, Greiner R, Wishart D. CFM-ID: a web server for annotation, spectrum prediction and metabolite identification from tandem mass spectra. *Nucleic Acids Res* 2014;**42**:W94–9 (Web Server issue).
18. Allen F, Pon A, Greiner R, Wishart D. Computational prediction of electron ionization mass spectra to assist in GC/MS compound identification. *Anal Chem* 2016;**88**(15): 7689–97.
19. McEachran AD, Balabin I, Cathey T, Transue TR, Al-Ghoul H, Grulke C, et al. Linking in silico MS/MS spectra with chemistry data to improve identification of unknowns. *Nat Sci Data* 2019;**6**:141.
20. Baker N, Knudsen T, Antony Williams AJ. Abstract Sifter: a comprehensive front-end system to pubMed. *F1000Research* 2017;**6**:2164 (Chem Inf Sci).
21. Schymanski EL, Baker NC, Williams AJ, Singh RR, Trezzi JP, Wilmes P, et al. Connecting environmental exposure and neurodegeneration using cheminformatics and high-resolution mass spectrometry: potential and challenges. *Environ. Sci.: Processes Impacts.* 2019;**21**: 1426–45.

Index

Note: 'Page numbers followed by "f" indicate figures and "t" indicates tables'.

C

F

Q

R

9780128199671